Chemistry

AN INTEGRATED APPROACH

Chemistry

An Integrated Approach

R. S. LOWRIE, M.A., D.PHIL., F.R.I.C.
Head of Science Department, Oxford School

and

H. J. CAMPBELL-FERGUSON, M.A., PH.D.
Head of Chemistry Department, Malvern College

A Division of Pergamon Press

A. Wheaton & Company Limited
A Division of Pergamon Press
Hennock Road, Exeter EX2 8RP
Pergamon Press Ltd
Headington Hill Hall, Oxford OX3 0BW
Pergamon Press Inc.
Maxwell House, Fairview Park, Elmsford, New York 10523
Pergamon of Canada Ltd
75 The East Mall, Toronto, Ontario M8Z 2L9
Pergamon Press (Australia) Pty Ltd
19a Boundary Street, Rushcutters Bay, N.S.W. 2011
Pergamon Press GmbH
6242 Kronberg/Taunus, Pferdstrasse 1, Frankfurt-am-Main,
West Germany

Library of Congress Catalog Card No. 76–142793
First edition 1975
Reprinted 1978 *(with revisions)*

Printed in Great Britain by A. Wheaton & Co. Ltd, Exeter

ISBN 0 08 017924 X flexi non net
0 08 018049 3 hard non net
0 08 017975 4 hard net

Contents

Preface

THIS textbook is a development from *Inorganic and Physical Chemistry—an Integrated Approach* by the same authors. We have retained much of the original text, and added nine new chapters devoted to organic chemistry. In planning this book we adopted the following general principles:

(1) Detailed factual information about individual substances has been excluded on the grounds that there are plenty of reference books and books of data available to the reader.
(2) At the end of each chapter, study questions have been provided. The purpose of these is not merely to test the reader's knowledge, but also to promote understanding and to stimulate further thought.
(3) Answers and notes on the study questions, as well as further notes on the text, lists of references, visual aids etc., are provided in a separate book entitled *Handbook with Answers*. This publication is intended for teachers, and students working on their own.
(4) Numerous diagrams and charts have been included in the text, and summaries are provided at the end of some chapters to reinforce the main points.

We would like to express our thanks to all those who have helped us in the preparation of this manuscript: Mr. C. Nicholls, Dr. T. E. Rogers, Dr. T. A. G. Silk, Mr. J. C. Simmons, Mr. R. C. B. Smith, Mr. A. Spiers and Dr. C. S. G. Phillips who read the manuscript of the first edition of *Inorganic and Physical Chemistry*, and especially to Miss H. H. James and Mr. M. R. Berzins who wrote to us with many valuable comments about that book and subsequently undertook a detailed reading of our new manuscript. We hope that as a result we have produced a text that will be of use with all the "modern" syllabuses which are being developed, including that of the Nuffield A-level chemistry project. We hope that there are no errors, but will be glad to hear from any reader with comments or criticisms.

Note In this reprinted edition a few minor corrections have been made. Some of the questions on kinetics have been revised and some more added. Answers to these questions will be found in Appendix VI on pages 464–5.

R. S. L.
H. J. C. F.

To the Reader

THE chapters are arranged in such an order that they can be taken in sequence. Chapters 12 and 13 may be omitted at a first reading and Chapter 17 may, if desired, be taken earlier. Chapters 1 to 9 are basic to the book as a whole.

The Inorganic chapters, 20 to 24, 34 and 35, may be taken in any order but the printed sequence will be found to be the most logical. A great deal of the "physical" part of Organic chemistry is dealt with in early chapters, and the main sequence of Organic chapters, 25 to 33, follows the chapter on carbon.

You will find that the Study Questions vary in difficulty. They are designed to help you to understand concepts rather than to test your knowledge of the book. Some of them you will find very easy; others will require rather more thought and sometimes more background reading.

Modern systematic names (IUPAC nomenclature) have been used throughout the book. Since you will undoubtedly be referring to older texts also, we have printed the older names alongside the new ones in the index. Use the index therefore as a reference for converting nomenclature when in difficulty. Appendix V summarizes the main rules for naming substances.

Acknowledgements

FIGURE 9.2 is reproduced with permission from a paper in the *Transactions of the Faraday Society* by Pedley, Skinner and Chernick, **53,** 1612 (1957), and Fig. 9.3 from a paper in the *Journal of Research of the National Bureau of Standards* by F. D. Rossini, **4,** 313 (1930). Figure 9.7 is redrawn with permission from *Senior Science for High School Students, Part 2, Chemistry*, ed. H. Messel, Pergamon, 1966, and Fig. 4.3 is redrawn with permission from the Unilever Educational Booklet *The Physics of Chemical Structure*. The periodic table is reproduced with permission from International Nickel Limited. Figure 33.4 is redrawn with kind permission from *The Nature of Biochemistry*, Baldwin E., Cambridge University Press, 1967. Figure 33.7 is redrawn with permission from *The Proteins*, Prof. R. E. Dickerson, ed. Neurath H., Academic Press, 1964.

Figures 25.2, 25.3, 26.1, 26.2, 26.5, 27.1, 32.1, 32.2, 32.3, 32.4, 32.6, 32.7, 33.1, 33.3, 33.5 and 33.6 depict skeletal molecular models made using the "Orbit" molecular building kit, manufactured by Cochranes of Oxford Ltd., Fairspear House, Leafield, Oxford.

The data used have been quoted from a variety of sources which it would be difficult to acknowledge individually. We would particularly acknowledge our debt to the *Chemical Data Book* published by the University of New South Wales, the *Book of Data* published for the Nuffield Advanced Science Project (Longmans), and to W. M. Latimer's *Oxidation Potentials*, 2nd ed., Prentice-Hall, 1952.

The following G.C.E. Examining Boards have kindly given us leave to quote A-level questions: London University (Nuffield), Southern Universities, and Cambridge Local. To them our thanks are also due.

CHAPTER 1

Periodicity

1.1 Introduction

This chapter deals with the important concept of **periodicity** in chemistry. There are now, in 1974, one hundred and seven known chemical elements, and with the development of new ways of synthesizing elements this number is bound to continue increasing. Their study would present quite a formidable problem but for the fortunate fact that they can be grouped together in "families" with similar or related properties. Ever since about 1869, when it first became clearly understood by Lothar Meyer and Mendeléeff, the concept of periodicity has been an important unifying factor in the study of inorganic chemistry. Much confusion existed at that time, however, due to the existence of undiscovered elements and incorrect atomic weights, and we shall therefore take up the study of periodicity and the periodic table from the twentieth-century standpoint.

1.2 The atomic nucleus

The year 1909 was a landmark in the development of ideas on the inner structure of the atom. Up to this time it was known that atoms were exceedingly small, of the order 0·1 or 0·2 nanometres (nm),* but little was known of how the atoms themselves were built up. In 1909 Geiger and Marsden did some experiments to investigate the effect which a very thin sheet of metal foil had on a beam of fast moving, positively charged particles, called **alpha-particles** (α-particles). These particles were known to be extremely small even compared to the sizes of atoms themselves, and it was hoped that if the metal was thin enough, a few particles might pass through the foil and be scattered in the process.

The result of this experiment was very surprising: despite the fact that the metal foil was several hundred atoms thick, *very nearly all the particles passed straight through the foil* without appreciable deflection; the metal foil behaved as if there were almost no obstructing matter there at all. This in itself was very remarkable, but even more interesting was the fact that a very small fraction of the α-particles were deflected considerably or even bounced back. In most of their experiments about one particle in twenty thousand was affected in this way. The metal foil itself never suffered any mechanical damage, and it thus appeared that the atoms themselves were unaffected by the passage of α-particles through their midst.

The experiments of Geiger and Marsden were examined further by Rutherford, who proposed a new theory of the structure of the atom to

* One nanometre (nm) equals 10^{-9} m. For a note on units, see Appendix IV.

account for these observations. Rutherford argued that the atom must contain a very small, positively charged **nucleus** where most of the atomic mass was concentrated. Surrounding this nucleus was a number of negatively charged particles of negligible mass, called **electrons** (Fig. 1.1).

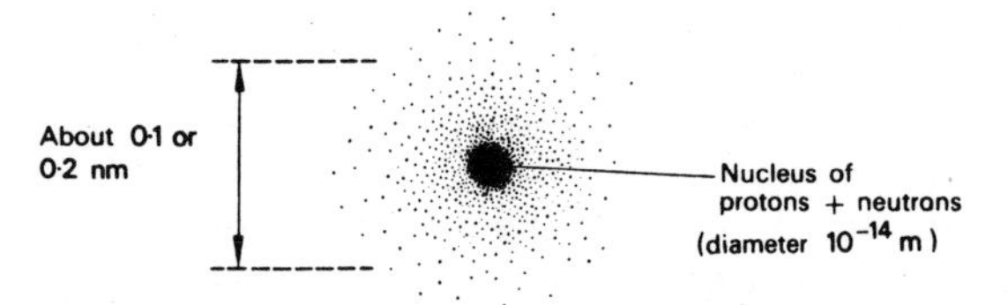

FIG. 1.1. The atom.

Nobody, even today, has ever succeeded in *seeing* inside an atom. It would be fairer to describe Rutherford's picture of it as a *model.* The nuclear model of the atom is something which scientists have found extremely useful in understanding chemistry, and we can safely say today that the existence of the atomic nucleus is a proven fact, even though we shall probably never succeed in "seeing" one. It is quite instructive to see how our model of the atom has changed over the years, and to bear in mind that scientists in the future will very likely be using a different model from us (Fig. 1.2).

1.3 Atomic number

Later experiments have confirmed Rutherford's view about the atomic nucleus. We now know that all atomic nuclei are positively charged, and it has proved possible (largely as a result of work on X-rays by Moseley) to measure the charges on nuclei. It has been found that all atoms have nuclear charges which are an exact whole number multiple of the charge on a hydrogen nucleus.

This observation enables us to extend our model of the atom a stage further, for if nuclear charges are related in this way it is extremely likely that atomic nuclei contain positively charged particles, and that each element is characterized by having a given number of these in its nucleus. We call these particles **protons**, and we refer to the total charge of the nucleus as the **atomic number**, which is defined as the number of protons in the nucleus.

Since the atom as a whole is uncharged, the charge of the protons in the nucleus must be exactly balanced by the negative charge of the electrons which surround it. We may say that:

(a) the positive charge on a proton is equal and opposite to that of the electron, and has been found to equal $1{\cdot}6 \times 10^{-19}$ coulombs;

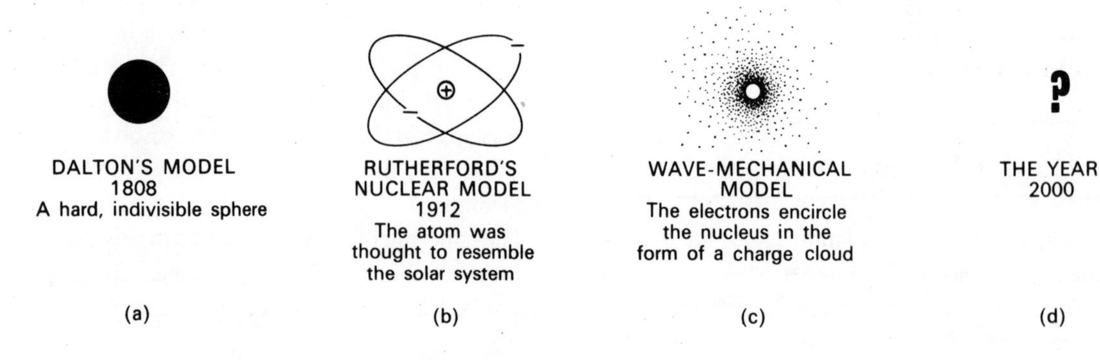

FIG. 1.2.

(b) the atomic number = the number of protons in the nucleus,
= the number of electrons in the neutral atom.

1.4 Mass number

Our model of the atom is still incomplete, for we have not yet taken account of the fact that atomic *masses* do not follow the exact pattern of atomic numbers. To take the simplest case, a helium atom has twice the nuclear charge of a hydrogen atom, and yet its atoms are *four* times greater in mass.

The masses of atoms can be measured in a device called a **mass spectrometer**, which was invented by Aston in 1919 (Fig. 1.3). An element can be made to vaporize into an evacuated tube where it forms separate atoms which are bombarded with electrons. This bombardment knocks one or more electrons off the atoms, so that they have an overall *positive* charge. The atoms are said to have formed **ions**, and the process is called **ionization**. For example, a helium atom in the gaseous state, denoted by the symbol He(g), can lose one or two electrons to form $He^+(g)$ or $He^{2+}(g)$. Chemical equations can be used to represent these processes:

$$He(g) \rightarrow He^+(g) + e^-$$
$$He(g) \rightarrow He^{2+}(g) + 2e^-$$

These ions can be accelerated by applying an electric potential of about 200 V, and focussed in the form of a beam of charged atoms. If now an electric or magnetic field (or usually a combination of both) is applied at *right angles* to the direction of travel, the particles will be deflected, and will follow a curved path. The more highly charged a particle is, the stronger will be the force acting on it and the greater the deflection. The more massive the ion, the smaller will be the deflection for a given charge, as a result of the relationship *force* = *mass* × *acceleration*.

Experiments with the mass spectrometer confirmed a fact that had already been observed for radioactive elements like uranium, namely that, while all atoms of the same element have the same atomic number, they do not all have the same atomic mass. Neon, for instance, is found to contain atoms which are, almost exactly, 20, 21 and 22 times the mass of the proton. The atomic number of neon is 10.

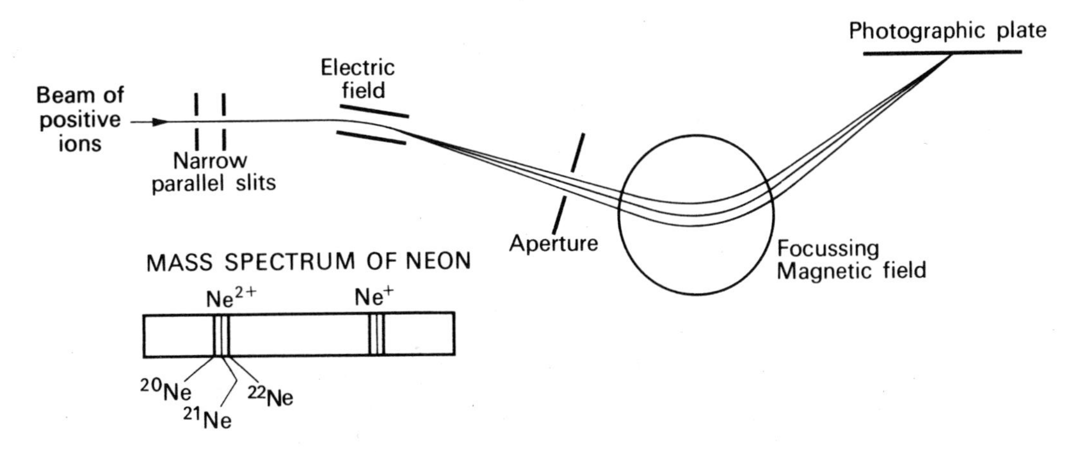

FIG. 1.3. The mass spectrometer.

This observation has been explained by postulating the existence of a third type of particle, called the **neutron**, whose mass is almost exactly the same as that of a proton, but which does not have any electrical charge. Thus atoms of neon always have 10 protons in their nucleus, but they may have either 10, 11 or 12 neutrons to make up the total mass.

Atoms of the same element which differ only in their number of neutrons are called **isotopes** (from the Greek *isos topos*, same place, meaning that isotopes occupy the same place in the periodic table). The existence of isotopes had been inferred from the study of radioactive materials, but it was not until the invention of the mass spectrometer that they were discovered for stable elements. In fact most elements have isotopes. Hydrogen for instance consists of two isotopes: the majority of hydrogen nuclei have a mass of 1 unit and a charge of +1, but a small fraction are "heavy" nuclei with a charge of +1 and a mass of 2 units. Heavy hydrogen atoms are known as **deuterium** atoms, and a deuterium nucleus can be regarded as made up of a proton plus a neutron (Figs. 1.4 and 1.5).

The existence of the **fundamental particles**, the electron, proton and neutron, can nowadays be taken as an established fact. Modern physics has revealed a bewildering array of other so-called "fundamental" particles such as mesons and positrons, but their existence does not directly concern the chemist. The fundamental particles which concern us, and the relationships between them, can be summarized thus:

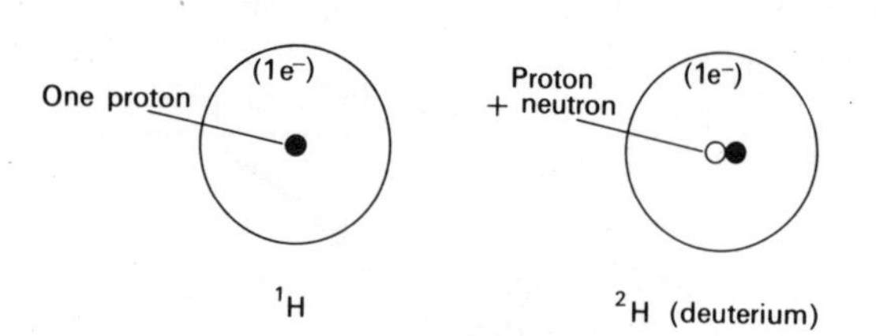

FIG. 1.4. Isotopes of hydrogen.

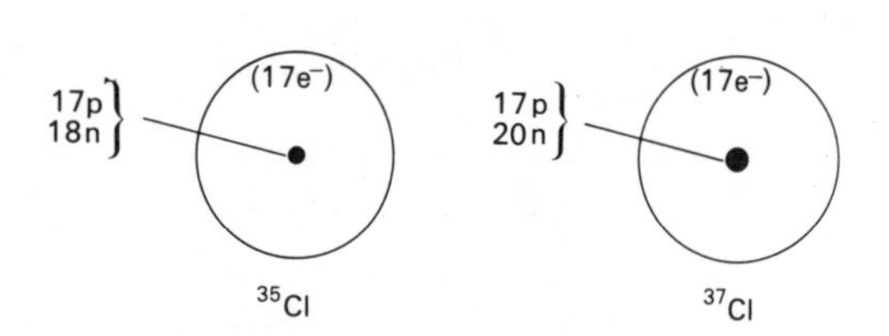

FIG. 1.5. Isotopes of chlorine.

Proton:	Charge = +1;	Mass = 1
Neutron:	Charge = 0;	Mass = 1
Electron:	Charge = −1;	Mass negligible (about 1/1840 unit).

The total mass of a nucleus is called its **mass number**, and it is related to other quantities as follows:

Mass number = number of protons plus number of neutrons in nucleus.

Number of neutrons = mass number minus atomic number.

Single isotopes of an element can be shown by writing the mass number as a superscript, for example ^{20}Ne, ^{21}Ne, ^{22}Ne.

1.5 Atomic weight

The so-called **atomic weight** (relative atomic mass) of an element is the average mass of the isotopes present.* For instance neon has an atomic weight of 20·2, which is the *weighted* average of the mass numbers of its isotopes:

Mass number	20	21	22
Abundance	90·4%	0·6%	9·0%

Weighted average =

$$= \frac{(20 \times 90{\cdot}4) + (21 \times 0{\cdot}6) + (22 \times 9{\cdot}0)}{100}$$

$$= 20{\cdot}2 = \text{atomic weight.}$$

* Some authorities recommend the term "relative atomic mass" as preferable to "atomic weight". The latter term seems likely to remain in use and it is used throughout this book.

Fortunately for chemists, the relative proportions of different isotopes in an element do not vary much in nature, so that the atomic weight of an element can be taken as a constant. In most of the light elements, a single isotope predominates, making the atomic weight an approximate whole number, but with heavier elements many different isotopes are present and the atomic weight is rarely a whole number. Tin for instance has eleven stable isotopes.

By international agreement, all atomic weights are expressed in terms of the isotope of carbon which has a mass number of 12, ^{12}C. Carbon itself contains traces of other isotopes making its atomic weight on this scale 12·01. Practically all atomic weights have now been checked, together with the exact isotopic composition of elements, using the mass spectrometer, for which the "carbon-12" standard is very convenient. A complete table of atomic weights is listed in Appendix II.

If the mass numbers and atomic numbers of elements are examined carefully, some interesting patterns can be traced. It is rather rare for instance for a nucleus to have an odd number of protons and an odd number of neutrons. The number of neutrons becomes greater than the atomic number as the element gets heavier, making it appear that the very highly charged nuclei need a large number of neutrons to hold them together. Even then the heaviest nuclei tend to disintegrate, and eject particles. This is the phenomenon of radioactivity (Chapter 2).

The atomic number of an element is a more fundamental property than its atomic weight, but nevertheless the atomic weight is of great practical value, since the chemist frequently determines the relative numbers of atoms which take part in a reaction by measuring weight changes. The existence of isotopes was a great source of confusion in nineteenth-century attempts to understand chemistry, since an arrangement of the elements in ascending order of their atomic weights is only approximately significant: chemical properties of elements depend upon their atomic number, not upon their atomic weight.

1.6 The periodic law

The properties of the elements are a periodic function of their atomic numbers. This is the periodic law, first stated by Lothar Meyer in 1869, though in fact he stated it for atomic *weights* where it is only approximately true. *Periodic* means repeating regularly after an interval (cf. the days of the week, which repeat after a *period* of seven days).

The law may be illustrated by plotting a graph of melting points (Fig. 1.6) or boiling points (Fig. 1.7) of the elements against their atomic numbers: despite the discrepancies, a pattern exists. Similar elements occupy corresponding positions on the curve—for instance the **noble gases** (inert gases) have the lowest melting points and boiling points. Immediately following each noble gas in the sequence there is a soft, highly reactive metal known as an **alkali metal**. Immediately preceding each noble gas (except helium) is a highly reactive non-metal known as a **halogen**—fluorine, chlorine, bromine, iodine or astatine. The elements can be arranged in chemical "families" known as **Groups**. There are places in Figs. 1.6 and 1.7 where the periodicity is slightly obscure. Thus the melting and boiling points of the **first row** of elements, lithium to neon, do not correspond all that closely with the **second row**, sodium to argon. Also, the periods are not all of equal length. The first and second rows of elements are termed **short periods**.* The

* Not all textbooks agree in their use of these terms; for instance, some refer to "lithium to neon" as the *first short period*, whereas others call the same row of elements the *second short period*. In this book the terms *first row* and *second row* are used in the way that they are defined above.

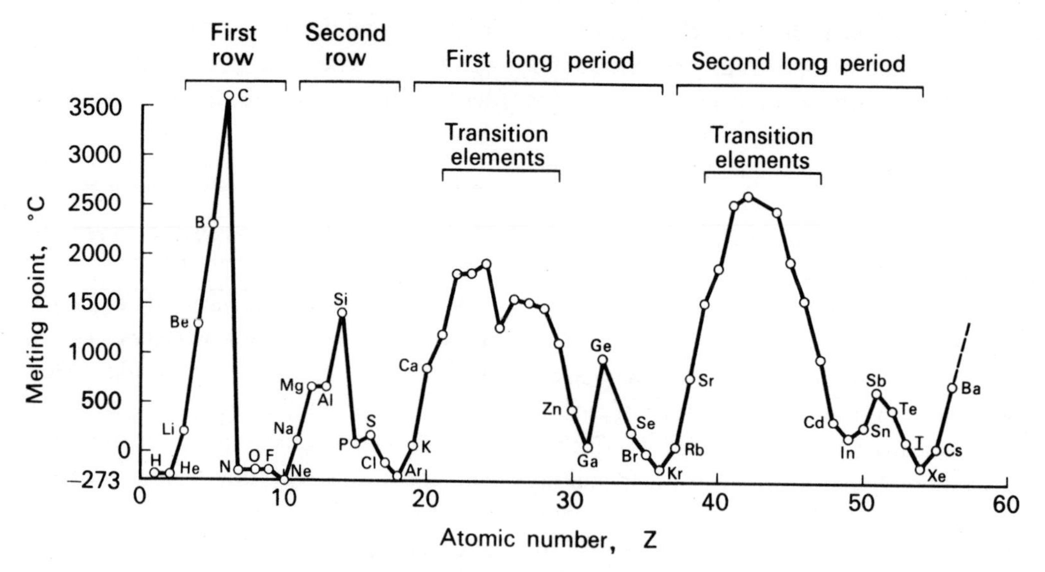

FIG. 1.6. Periodicity of the melting points of elements.

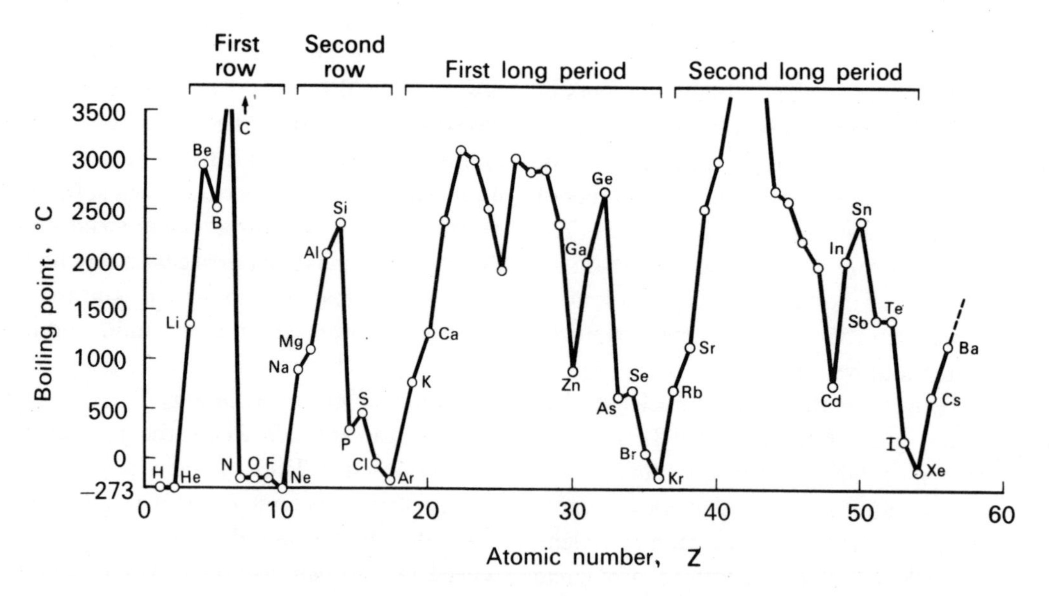

FIG. 1.7. Periodicity of the boiling points of elements.

third and fourth rows are termed **long periods**, since they contain eighteen elements each instead of eight. Hydrogen and helium do not fit in with this periodic pattern, and in a sense they comprise a complete period on their own.

The periodic law may be demonstrated more convincingly by the volumes occupied by atoms of the elements. Since atoms are inconveniently small, we will compare the volume occupied by *equal numbers* of atoms of all the elements. A

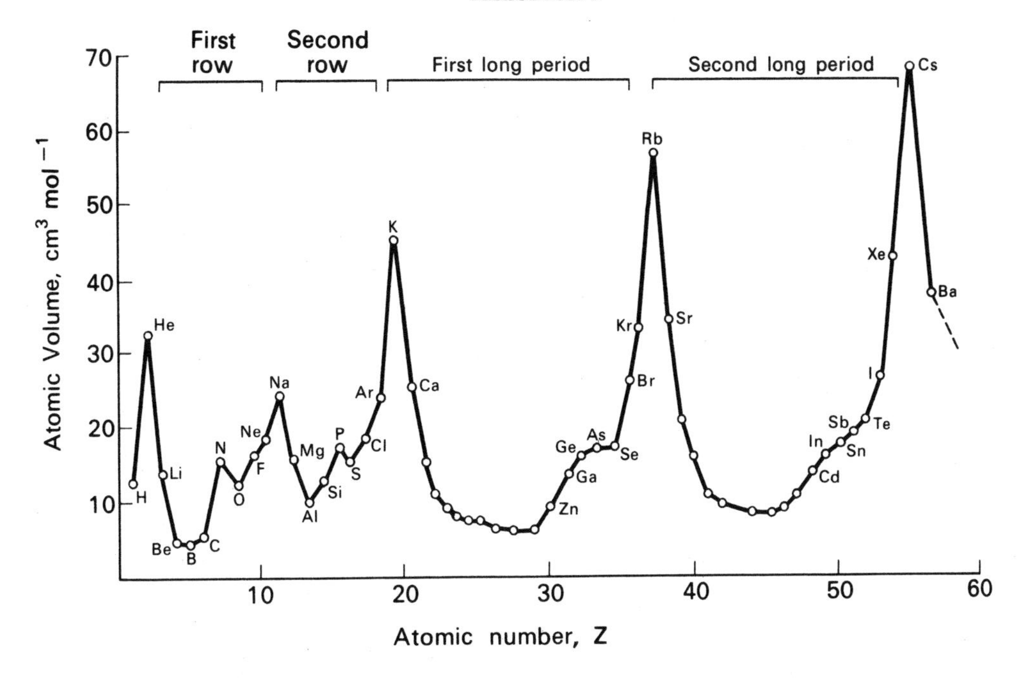

FIG. 1.8. The atomic volumes of the elements.

convenient number to take is 1 **mole*** of atoms, abbreviated as "mol". The mole is a large number, $6{\cdot}022 \times 10^{23}$. 1 mole of atoms of any element will always weigh a number of grams equal to its atomic weight, for instance:

1 mole of hydrogen atoms
= 1·008 g of hydrogen.

1 mole of carbon atoms
= 12·01 g of carbon.

1 mole of chlorine atoms
= 35·45 g of chlorine, etc.

In each case, one mole of atoms of the element is obtained by taking the atomic weight of the element in grams, and it always contains the *same number of atoms*. The number $6{\cdot}022 \times 10^{23}$ is a constant of great importance to the chemist, called the **Avogadro constant**, L.

The volume of one mole of atoms of any element in the solid state is called the **atomic volume**. The atomic volume of an element is a rough measure of the relative size of the atoms. Fig. 1.8 shows a plot of atomic volume of the elements against their atomic numbers. Again a clear periodicity is observed. The alkali metals turn out to have exceptionally large atomic volumes, suggesting that their atoms may be larger than those of neighbouring elements.

Atomic volume fails to take into account the fact that different elements can have their atoms packed together in different ways. In theory a contraction in atomic volume could represent either a decrease in atomic radius, or a closer mode of packing. The atomic radius, which can be taken as half the distance between adjacent nuclei in the element, is plotted against atomic number in Fig. 1.9.

Further examples of periodicity of physical properties will be met in later chapters. Chapter 3 examines the periodic variations in ionization energy, and Chapter 5 investigates the periodicity of binding energy.

* For a more precise definition of the mole, see Section 7.3.

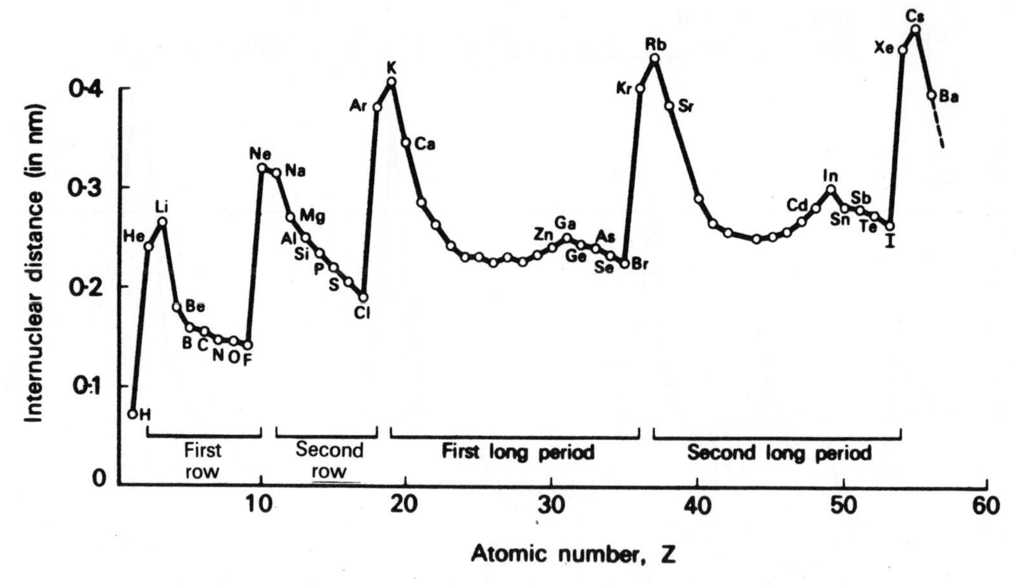

FIG. 1.9. The atomic radii of the elements.

1.7 Chemical periodicity

The periodic law is by no means confined to physical patterns—by far the greatest nineteenth-century contribution to inorganic chemistry was made by the Russian chemist, Mendeléeff, who was primarily concerned with the periodic variation in chemical properties. If an early period in the series of elements is examined, beginning with an alkali metal and ending with a noble gas, the same trend will be seen (Table 1.1).

At a later stage in the book we shall have to be more precise in the use of the word *reactive*. For the present, we may regard a reactive metal as one which enters vigorously into chemical combination with a non-metal such as oxygen, and a reactive non-metal as one which vigorously attacks metals.

Patterns of periodic behaviour exist not only in the reactivity of elements, but also in the composition of the compounds they form. Figure 1.10 plots the atomic number against the number of moles of hydrogen atoms which can combine chemically with the element to form a hydride. The noble gases do not combine at all. This is a clear example of periodicity, but it breaks down somewhat for the long periods. Similar plots can be constructed for oxides of elements (Study Question 13).

TABLE 1.1

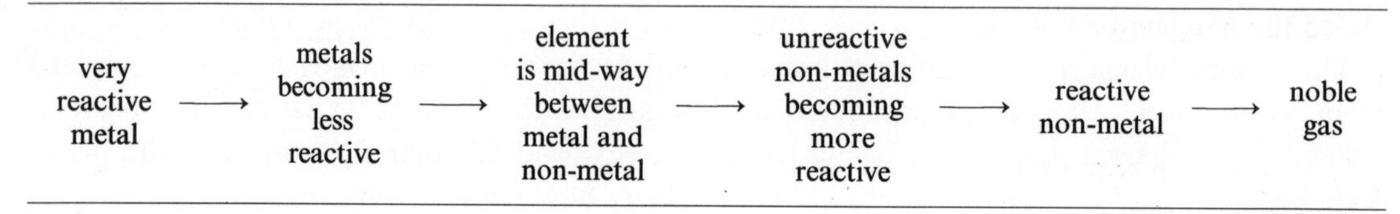

very reactive metal	⟶	metals becoming less reactive	⟶	element is mid-way between metal and non-metal	⟶	unreactive non-metals becoming more reactive	⟶	reactive non-metal	⟶	noble gas

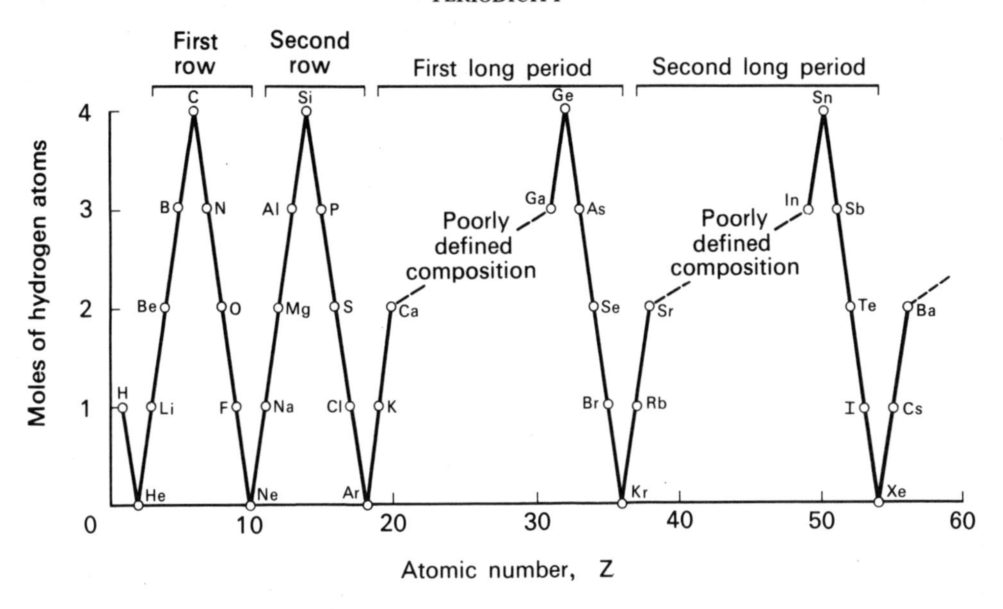

FIG. 1.10. Periodicity shown by the empirical formulae of hydrides.

1.8 The periodic table

The **periodic table** is a convenient way of arranging the elements so that similar elements occur as **Groups** in vertical columns.* Various ways of laying out the periodic table have been used but the so-called "long" form is used throughout this book (see inside cover, also Fig. 1.11). The final chapters of this book each deal with specific areas of the periodic table, and it is useful to be able to refer to the elements concerned by the following terminology:

(a) *Typical and transitional elements.* The elements in the first and second rows (as far as argon) are called **typical elements**. The term **transition element**† refers to the block of elements in the centre of the long periods, sometimes known as the *d*-block. The elements of Group IIB, zinc, cadmium and mercury, are *d*-block elements but are not strictly transition metals, since *d*-electrons are not used in bonding. The elements from lanthanum to lutecium and actinium to lawrencium are known as inner transition metals, or *f*-block elements.

(b) "A" *and* "B" *elements.* The elements on the left-hand side of Fig. 1.11 are known as "A" elements, and those on the right as "B" elements.

(c) *s-block, p-block and d-block elements.* The prefixes *s*, *p*, and *d* refer to the energy levels of the electrons within the atom, and it is often convenient to designate blocks of the periodic table by referring to their electronic structures.

A vertical column of elements in the periodic table is known as a Group. A column of transition metals, such as titanium, zirconium and hafnium, is sometimes called a **sub-Group**. The numbering of Groups refers mainly to the typical elements, though it is occasionally useful to refer to the sub-Groups by number as well.

* In this book the word "Group" with a capital G refers to a Group of the periodic table, to distinguish from other, non-specialized, uses of the same word.

† For a more detailed definition of "transition element", see Chapter 35.

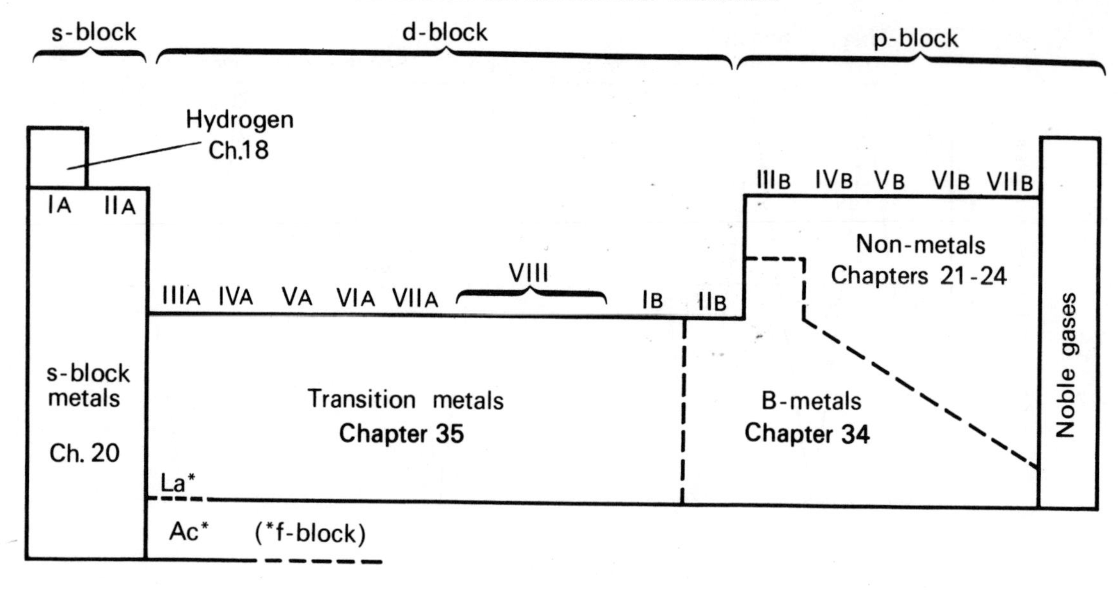

FIG. 1.11. The periodic table (long form) showing the plan of this book.

For instance, the Group IIB metals zinc, cadmium and mercury sometimes resemble the Group IIA metals.

1.9 Mendeléeff

The modern form of the periodic table is quite similar to the table introduced by Mendeléeff in 1869. The noble gases (Group 0) had not then been discovered, though this did not obscure the periodicity of the remaining elements. Both Lothar Meyer and Mendeléeff expressed the periodic law in terms of atomic weights.

Mendeléeff's contribution to chemical understanding was remarkable, in that he had the foresight to realize that where discrepancies occurred this was probably due to undiscovered elements. For instance he left a gap in Group IV which was filled only when germanium was discovered in 1885. The properties of germanium were found to be astonishingly close to those which Mendeléeff predicted (Study Question 15).

In one or two cases a pair of elements had to be exchanged in order that their atomic weights might "fit" the periodic law. Tellurium and iodine were such a pair, and these were termed an **inversion**. Inversions are caused by the variation in number of neutrons.

Study Questions

1. What would the results of the Geiger-Marsden experiment have been if matter had been composed of solid spherical atoms rather than particles with nuclei?

2. How is the atomic number related to (a) the number of protons, (b) the number of electrons, (c) the number of neutrons, in a neutral atom?

3. Bromine contains 50·53% ^{79}Br and 49·47% ^{81}Br.

(a) Calculate the atomic weight of bromine from these figures.
(b) Look up the atomic weight of bromine in a table of atomic weights.
(c) Why do the two values differ? (See also Chapter 2.)

4. Plot the following quantities against atomic number: (a) specific heat capacity, (b) the molar latent heat of fusion, (c) the molar latent heat of evaporation, (d) the difference between the melting and boiling point of an element.

(i) Show how the plots illustrate the principle of periodicity.
(ii) Comment on any interesting features in the plots.

5. Argon, atomic number 18, has an atomic weight of 39·95, while potassium, atomic number 19, has an atomic weight of only 39·10. How can you account for this? Use a periodic table and a table of atomic weights to discover two more pairs of elements which behave in the same way.

6. The densities of calcium and copper are 1·55 and 8·96 g cm^{-3} respectively. Calculate the atomic volume of each element.

7. Use a periodic table to represent the following in symbols:

(a) A nucleus with 9 protons and 10 neutrons.
(b) A nucleus with atomic number 74 and mass number 184.
(c) Strontium-90.
(d) A nucleus with 82 protons and 124 neutrons.
(e) A nucleus with 20 protons and 20 neutrons.

8. Do you think any elements with atomic weights of less than 250 remain to be discovered? Give reasons for your answer.

9. The similarity between calcium, strontium and barium was noted by Döbereiner in 1829, and he called the three elements a triad.

(a) Find a relationship between the atomic weights of the three elements.
(b) Make a list of any other properties of the elements that have the same numerical relationship.
(c) Find some examples of such triads in Groups I and VII that possess the same atomic weight relationship as calcium, strontium and barium.

10. What are the masses of the following?
(a) 1 mole of oxygen atoms.
(b) 0·2 mole of calcium atoms.
(c) 3 moles of lead atoms.
(d) 1 atom of hydrogen.
(e) $6{\cdot}022 \times 10^{23}$ atoms of carbon.
(f) $1{\cdot}2044 \times 10^{23}$ atoms of uranium.

11. The following elements are each composed of only one stable isotope: F, Na, Al, P, Sc, Mn, Co, As, Y, Nb, Rh, I, Cs, Pr, Tb, Ho, Tm, Ta, Au and Bi.

(a) How many (i) protons, (ii) neutrons does each isotope contain?
(b) What have these isotopes in common?
(c) The element Be is the only other element composed of a single stable isotope. How does it (i) resemble (ii) differ from, the elements above?

12. The mass spectrum of an element X consists of four lines with mass/charge ratios of 34·5, 35·5, 69 and 71 with relative intensities of 3:2:42:28 respectively.

(a) How many isotopes are present?
(b) How can you account for four lines?
(c) What is the approximate atomic weight of X?

13. Use the information in the later chapters in this book to plot the atomic number of the elements against the maximum number of moles of atoms of (i) chlorine, (ii) oxygen with which each element will combine.

(a) How do the plots differ from that obtained for hydrogen? (Fig. 1.10)
(b) Are the differences sufficient to make you suspect the periodic law?

14. List the ways in which boron and silicon resemble each other. In view of these similarities, justify the fact that they are included in different Groups in the periodic table.

15. The element germanium, which occurs between silicon and tin in Group IV was unknown to Mendeléeff, who attempted to predict its properties. Use the following data to do the same.

Property	Si	Sn
M.p.	1410°C	232°C
B.p.	2680°C	2270°C
Density, g cm^{-3}	2·33	7·30
Oxide m.p.	1700°C	1127°C
Hydride	b.p. −114°C	Decomposes easily
Chlorides	b.p. 57°C	$SnCl_4$ b.p. 113°C
		$SnCl_2$ b.p. 623°C
Fluoride	b.p. −96°C	Sublimes at 700°C

16. Gallium, the element below aluminium in the periodic table, was unknown to Mendeléeff. Aluminium melts at 660° and boils at 2450°C; it has a density of 2·70 g cm^{-3}. It forms an oxide Al_2O_3 (m.p. 2045°C), a chloride Al_2Cl_6 (b.p. 180°C) and a fluoride AlF_3 (m.p. 1040°C). From the properties of aluminium and the trends observed with silicon, germanium and tin (Q.15), predict the properties of gallium.

17. Atomic dimensions are so different from those to which we are used that it is hard for us even to begin to cope with them.

(a) If the nucleus of an atom is the size of a tennis ball, estimate how large the atom itself would be.

(b) Estimate how long you would take to count the atoms present in 1 g of hydrogen.

18. (a) Find out why the symbols for the following elements are not derived from their English names: sodium, potassium, gold, silver, mercury, iron, tin, lead and antimony.

(b) Find out the origins of the names rubidium, caesium, chlorine and iodine.

CHAPTER 2

Radioactivity

2.1 Forms of radioactivity

In 1896, Becquerel discovered that some elements have the property of emitting penetrating radiation rather similar to X-rays, which would fog photographic plates even when they were wrapped in black paper. These rays are called **gamma-rays** (γ-rays), and the property is called **radioactivity**. It was later found that high energy particles, either positively or negatively charged, were emitted at the same time, and that the element was itself converted to a new element.

Gamma-rays have extremely short wavelength and very high energy; when they are absorbed by chemical substances, decomposition occurs. The radiation can penetrate solid matter and can have a very damaging effect on living tissues.

Radioactivity is the disintegration of atomic nuclei, with the emission of some of the particles they contain. The factors which determine when a given nucleus is unstable are not yet fully understood, but it may be noted that *all* nuclei with atomic numbers greater than 83 are radioactive.

Two main types of particle are emitted from radioactive nuclei:

(i) **Alpha-particles** (α-particles). These may be regarded as being made up of two protons and two neutrons, and they are in effect helium nuclei. They have a charge of +2 and a mass number of 4, and can be written in symbolic form ${}^{4}_{2}He$.

(ii) **Beta-particles** (β-particles). These are simply very high energy electrons, having a charge of −1 unit, and negligible mass.*

Since a charged particle is always emitted from a radioactive nucleus, the atomic number must change. For instance radium-226, ${}^{226}_{88}Ra$ is alpha-active, emitting γ-rays and α-particles. The atomic number drops by two units, forming the radioactive gas, radon-222, ${}^{222}_{86}Rn$. The disintegration may be shown in the form of an equation:

$$ {}^{226}_{88}Ra \rightarrow {}^{4}_{2}He + {}^{222}_{86}Rn $$

Radium-226 α-particle   Radon-222

Similarly beta-active nuclei eject an electron (β-particle) from the nucleus, and the atomic number increases by one unit. One of the isotopes of lead, lead-214, is beta-active:

$$ {}^{214}_{82}Pb \rightarrow {}_{-1}^{0}e + {}^{214}_{83}Bi $$

In β-disintegrations the mass number does not

* Certain radioactive nuclei emit particles identical with electrons, but with positive charge, called *positrons* (see Study Question 1) but this is relatively unusual.

change. The process may be imagined as a nuclear rearrangement in which a neutron changes into a proton and ejects an electron. Both β- and α-emitters usually emit γ-rays as well.

In naturally occurring isotopes of heavy elements like radium, a **radioactive disintegration series** is observed. For instance radium-226 forms radon-222, which forms polonium-218, which forms lead-214. Lead-214, being beta-active, forms bismuth-214, which forms another isotope of polonium, of mass number 214. The process continues in a series of steps of this type until finally a stable species is reached, lead-206. Figure 2.1 illustrates this series, in slightly simplified form:

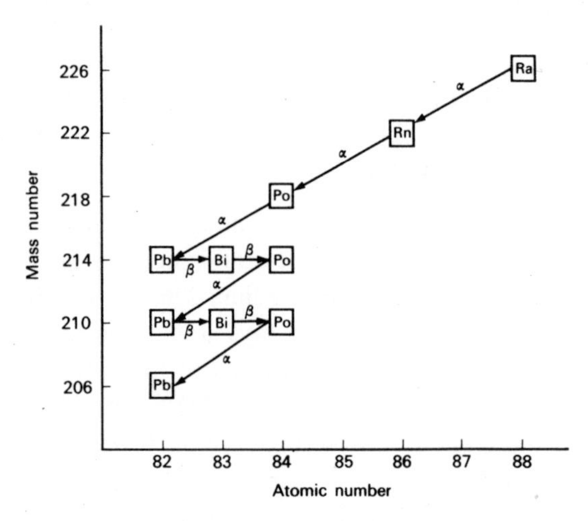

FIG. 2.1. A radioactive decay series.

A peculiar feature of radioactivity is that it is quite independent of the state of chemical combination of the element. The activity of radium for instance is the same whether it is the metal, or its chloride or hydroxide or any other compound. The rate of disintegration is found to be such that after a given time period, known as the **half-life period**, one-half of the nuclei will have disintegrated, no matter how many there were at the beginning of this period. After another half-life period the fraction left will be one-quarter, and so on. This form of disintegration is known as **exponential decay**. The mathematical treatment will be deferred until Chapter 17, though a simple graph showing the exponential decay of a radioactive element is given in Fig. 2.2.

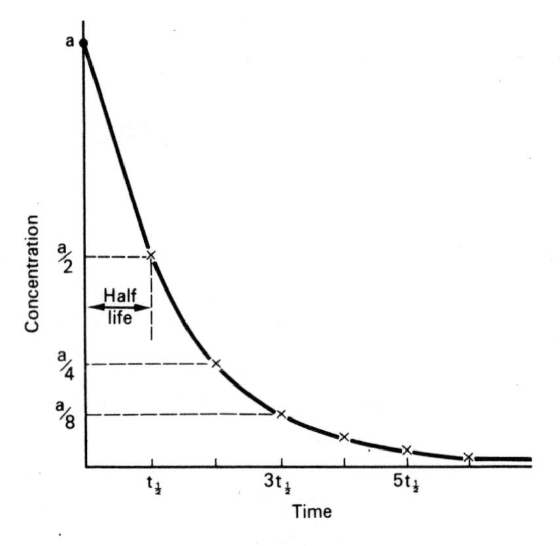

FIG. 2.2. Exponential decay.

The masses and charges of alpha- and beta-particles were determined originally in a device analogous to the mass spectrometer. Gamma-rays are not deflected by an electric or magnetic field since they do not have any charge (Fig. 2.3). The **Geiger counter**, and various other electronic devices which make use of the ionizing power of alpha- and beta-particles, can be used for counting the particles and hence estimating the amount of a radioactive element present.

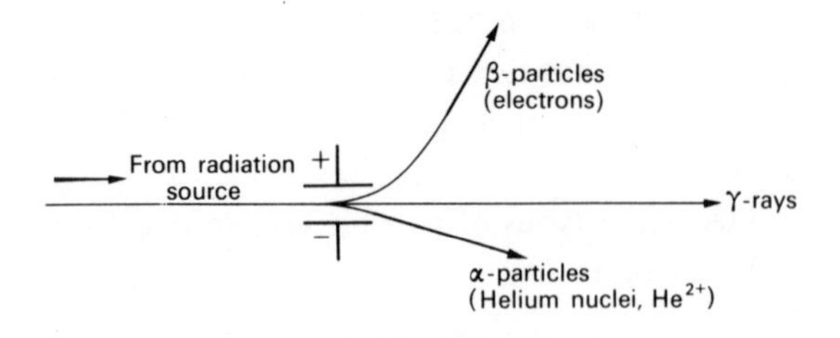

FIG. 2.3.

2.2 Nuclear reactions

The radioactive disintegrations considered above are the simplest form of nuclear reaction, in which a single nucleus decomposes spontaneously. More complex disintegrations can be effected by allowing particles such as α-particles to collide with another nucleus. Rutherford, in 1919, noticed that the normally stable nuclei of nitrogen-14 were disrupted by bombarding them with high energy α-particles. Protons were produced, and the nitrogen nuclei became oxygen-17, a stable isotope of oxygen:

$$^{4}_{2}He + ^{14}_{7}N \rightarrow ^{17}_{8}O + ^{1}_{1}H$$

α-particle + nitrogen-14 → oxygen-17 + proton

Many new isotopes have been synthesized by nuclear reactions of this type. Radioactive isotopes of the lighter elements can be made, by causing an abnormal number of neutrons to be present. Irradiating a stable element with a beam of neutrons may cause it to be converted to an unstable isotope, as in this example with aluminium:

$$^{27}_{13}Al + ^{1}_{0}n \rightarrow ^{28}_{13}Al + \gamma;$$

stable — unstable

$$^{28}_{13}Al \rightarrow ^{28}_{14}Si + ^{0}_{-1}e$$

stable — β-particle

An even more drastic form of nuclear disintegration is **nuclear fission**, which occurs when certain heavy isotopes are bombarded with neutrons. Uranium-235 is the most notorious example of this, since the energy released in its neutron-induced fission was utilized in the atomic bomb. The nucleus splits into two parts, and gives rise to several neutrons in addition. A typical reaction would be:

$$^{235}_{92}U + ^{1}_{0}n \rightarrow \underbrace{^{141}_{56}Ba + ^{92}_{36}Kr}_{\text{Radioactive fission products}} + \underbrace{3^{1}_{0}n}_{\text{three neutrons}}$$

Radioactive fission products (cause the "fall-out" from a nuclear bomb)

The energy released in this reaction is 16 000 000 000 J per mole of atoms of uranium-235.

Since each reacting neutron produces about three new neutrons which can react further, the process becomes a **chain reaction** where each successive stage gives rise to more and more neutrons (Fig. 2.4). In this way a single stray neutron can detonate a whole lump of uranium-235, once the lump exceeds a certain critical mass. To make the reaction yield energy in sufficient quantity for a bomb, it was necessary to

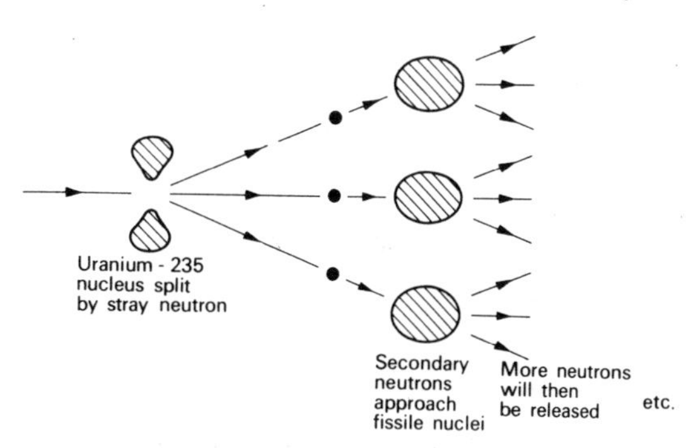

FIG. 2.4. A chain reaction.

separate the fissile isotope, ^{235}U, from the remainder of the uranium, mainly ^{238}U. This was a formidable problem since isotopes are chemically almost identical, and physically very similar. Pure uranium-235 was made during the Second World War by repeated diffusion of the volatile compound uranium hexafluoride, UF_6 (Chapter 8).

A chain reaction of this type can be made to produce energy at a controllable rate, instead of explosively. This is done in the **nuclear reactor** used in a modern nuclear power station. It is necessary to modify the design in such a way that some of the neutrons produced in the reaction are absorbed. In an atomic pile, a chain reaction is set up, and graphite rods can be lowered into the material if necessary to slow down the production of neutrons. The heat produced is used to raise steam to drive electrical generators.

A nuclear reaction which produces even more energy than uranium fission is the **fusion** of two light nuclei, for example:

$$^{3}_{1}H + ^{1}_{1}H \rightarrow ^{4}_{2}He; \quad \Delta E = -18{\cdot}9\times10^{11}\ J,$$

where ΔE = energy change per mole of helium atoms.

Reactions of this type are used in the H-bomb: they need an extremely high temperature to make them start—over 1 000 000°C, and a fusion reaction has to be "triggered" by means of an ordinary fission reaction. The energy released from the Sun is produced in this way: hydrogen and helium are the two main elements in the sun (the name helium comes from the Greek word *helios*, meaning sun).

2.3 Energy and mass

There is a small change in mass associated with any energy change, and in the case of nuclear reactions this mass change may be large enough to be detected. The relationship between mass and energy was stated by Einstein:

$$\Delta E = mc^2$$

where ΔE = change of energy, in J*
m = change of mass, in kg
c = velocity of light, in m s^{-1} (3×10^8 m s^{-1})

Different nuclei have different **nuclear binding energies**. Two effects result from this: (i) there is an energy change associated with any transformation from one nucleus to another (ii) the masses of atoms are not *exact* whole numbers, and there is a change in total mass, for a nuclear reaction. For instance in the foregoing reaction,

$$\text{mass change} = \frac{\Delta E}{c^2} = \left(\frac{18{\cdot}9\times10^{11}}{9\times10^{16}}\right)\text{kg} = 2{\cdot}1\times10^{-5}\ \text{kg (decrease per mole of helium atoms).}$$

* This symbol Δ means "change of".

2.4 Synthetic elements

Until the discovery of radioactivity it was not thought possible to change one element into another. The dream of the alchemists was to "transmute" base metals into gold. Transmutation of the elements is now commonplace, though it would be far more expensive to make gold by a nuclear reaction than to mine it! Until quite recently there were gaps in the atomic number series, but these have now been filled by synthesizing the elements in a nuclear reactor. Plutonium, ^{94}Pu, is nowadays made in quite large quantities—it does not occur naturally on Earth.

The periodic table is continually being extended by synthesizing "transuranic" elements, i.e. elements heavier than uranium. A series of inner transition elements has been completed with the discovery of element number 103, lawrencium, this series comprising the second set of *f*-block elements.

2.5 Isotope separation

The separation of a mixture of isotopes of an element presents problems because the properties of isotopes are practically identical in most cases. The extraction of deuterium from ordinary hydrogen is relatively easy because the mass number ratio is 2:1. The properties of deuterium oxide, for instance, are noticeably different from those of ordinary water.

The heavier isotope of an element will give rise to compounds of greater density and molecular weight, and separation by diffusion is a possibility. In the case of deuterium, however, use is made of the fact that deuterium gas is evolved more slowly than light hydrogen during electrolysis. In countries like Norway, where hydro-electric power is cheap, high purity deuterium oxide is made by the continual electrolysis of dilute sodium hydroxide solution—the heavier isotope remains behind as part of the solution.

TABLE 2.1

	Naturally occurring water (mainly ^1H_2O)	Heavy water (D_2O or ^2H_2O)
Melting point	0·0°C	3·8°C
Boiling point	100·0°C	101·4°C
Density at 25°C	0·997 g cm^{-3}	1·105 g cm^{-3}

The separation of uranium-235 from uranium-238 is relatively more difficult, since the mass ratio is close. The diffusion of UF_6 has already been mentioned. A recent development which is now superseding the diffusion process is the *gas centrifuge*. This is an ultra-fast centrifuge capable of producing appreciable enrichment of isotopes quite rapidly.

Enriched isotopes of most elements are available commercially and are used for various purposes. Such isotopes form conveniently "labelled" atoms. The progress of an element through a living system can be followed by making it radioactive—the working of the thyroid gland, for instance, can be observed by using radioactively labelled iodine. The mechanisms of chemical reactions can often be observed by labelling certain atoms.

Study Questions

1. Write balanced equations for the following nuclear reactions:

(a) The emission of a β-particle from $^{210}_{82}Pb$.

(b) The emission of a β-particle from ^{14}C.

(c) The emission of an α-particle from $^{233}_{92}U$.

(d) The emission of an α-particle from ^{232}Th.

(e) The capture of a neutron by ^{238}U followed by the emission of two β-particles consecutively.

(f) ^{14}N is bombarded with neutrons. The eventual product is ^{14}C.

(g) ^{65}Zn decays to give ^{65}Cu.

2. Identify **A**, **B**, **C**, and **D** from the following information.

(a) $^{239}_{94}Pu+^{1}_{0}n \rightarrow \mathbf{A}$.　(b) $\mathbf{A}+^{1}_{0}n \rightarrow \mathbf{B}$.

(c) $\mathbf{B} \rightarrow {}^{0}_{-1}\beta+\mathbf{C}$.　(d) $\mathbf{C} \rightarrow {}^{4}_{2}\alpha+\mathbf{D}$.

3. Tritium is the radioactive isotope of hydrogen with a mass number of 3. Would you expect it to be an α- or a β-emitter? Suggest an equation for the nuclear reaction that occurs.

4. (a) The radioactive decay of ^{63}Ni to give ^{63}Cu has a half-life of 120 years. How long will it take for (i) $\frac{3}{4}$, (ii) $\frac{15}{16}$, (iii) $\frac{9}{10}$, of the nickel to change into copper?

(b) How would this copper differ from the naturally occurring element?

5. (a) Why is helium found in certain radioactive minerals?

(b) Why is the atomic weight of lead often dependent on the source from which it is obtained?

6. (a) Why is the fallout from a uranium nuclear fission bomb itself radioactive?

(b) Find out why strontium-90 is a particularly dangerous product of nuclear bombs.

7. The isotopic masses of ^4_2He, ^7_3Li and ^1_1H are respectively 4·00390, 7·01818 and 1·00812. Calculate ΔE for the following fusion reaction:

$$^7_3Li+^1_1H \rightarrow 2^4_2He.$$

(Velocity of light $= 3\times10^8$ m s^{-1}.)

CHAPTER 3

Energy levels and bond formation

3.1 Energy units

Chemistry is largely concerned with the interplay between matter and *energy*, and in this chapter we shall be examining the energy states, or **energy levels**, of electrons in atoms.

In the mass spectrometer, atomic masses are found by ionizing an atom, that is, by knocking one or more electrons off it. A certain minimum quantity of energy is required to do this, known as the **ionization energy** of the atom. To remove the first electron from a gaseous helium atom, it is necessary to strike it with an electron which has been accelerated through a potential of at least 24·58 volts. Thus we say that the first ionization energy of helium is 24·58 **electron-volts**. To remove the second electron from helium will require more energy, since it must be removed from an ion which is already positively charged. The second ionization energy of helium is 54·40 electron-volts.

Many of the energy changes with which chemists are concerned are heat changes, and heat is generally measured in *joules* (J). One joule is defined as the energy converted as work when a charge of 1 coulomb is raised through a potential difference of 1V. In terms of heat, it is found that 4180 J, or 4·18 kJ, are required to raise the temperature of 1 kg of water through one kelvin degree (1 K).

In this book energies are quoted in kilojoules (kJ). Appendix IV gives a conversion table which can be used if data are required in other units.

One electron-volt is equivalent to an energy change of approximately 96 500 J, or 96·5 kJ per mole of atoms. This follows because 1 mole of electrons is equal to a charge of 96 500 coulombs. Thus an element with an ionization energy of 10 electron-volts will require 965 kJ to convert 1 mole of its atoms into singly charged positive ions. Similarly:

$$\mathrm{He(g)} \rightarrow \mathrm{He^{+}(g)} + \mathrm{e^{-}};$$
$$\Delta E = +2370 \text{ kJ mol}^{-1}$$
$$\text{or} \quad +24{\cdot}6 \text{ electron-volts per atom}$$

The symbol Δ is used to denote "change of" a quantity, so in this case ΔE denotes change of energy. The plus sign indicates that the change represents an increase in energy. The helium ion is in a state of higher energy than the uncharged atom (Fig. 3.1).

It might be expected that the first ionization energies of atoms would increase steadily with

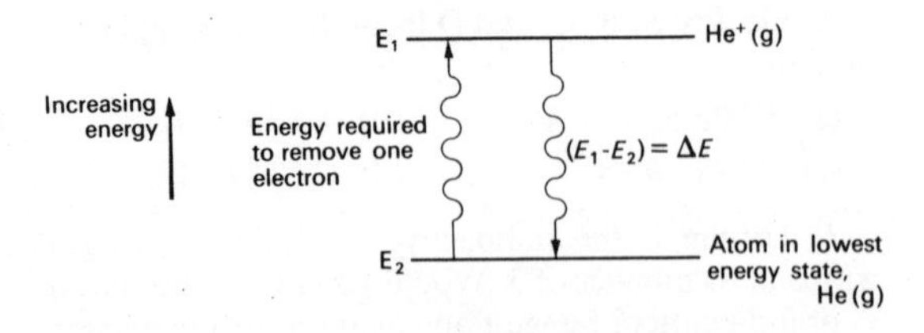

FIG. 3.1. An energy diagram.

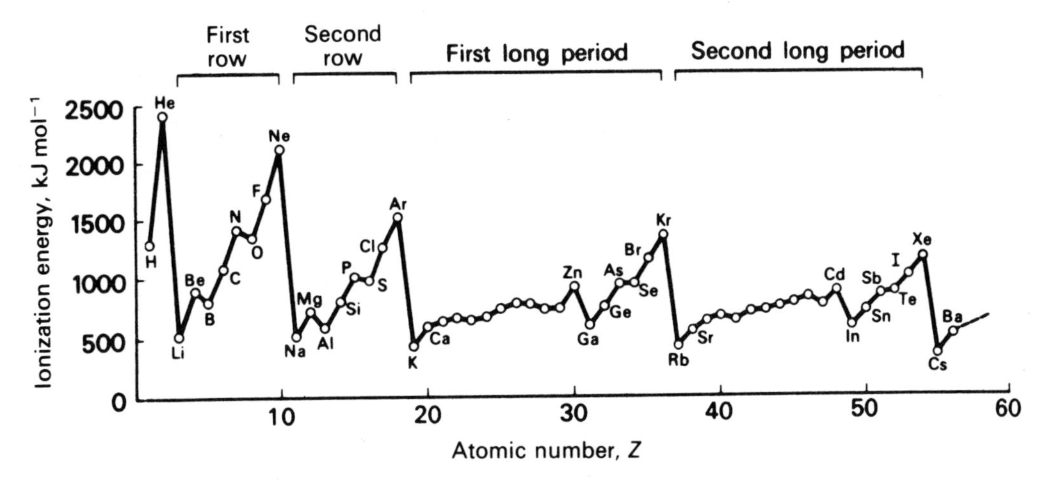

FIG. 3.2. The first ionization energies of the elements; $X(g) \rightarrow X^+(g) + e^-$.

increasing nuclear charge, but this is not found to be the case. True, the first ionization energy of helium is greater than that of hydrogen, yet that of lithium is *less* than that of helium. In fact ionization energies show a periodicity just like the properties mentioned in Chapter 1 (Fig. 3.2).

3.2 Energy levels

In order to explain the behaviour of electrons in atoms, it is postulated that they can only exist in certain energy states or **energy levels**. Evidence for this comes from two main sources:

(a) *Spectra*. If the gaseous atoms of an element are raised to a high state of energy, they will lose this energy by emitting it as radiation. However, the emission spectrum thus obtained does not consist of a continuous band of frequencies, but rather a series of single frequencies, each corresponding to a fixed energy value. The energy change, ΔE, is related to frequency ν by the expression

$$\Delta = E_1 - E_2 = h\nu$$

where h is a constant called Planck's constant.

Thus it would appear that atoms can only lose their energy in fixed quantities or "packets". The loss of energy is attributed to one or more electrons in the atom dropping from a state of high energy E_1 to a lower energy level E_2 (Fig. 3.3).

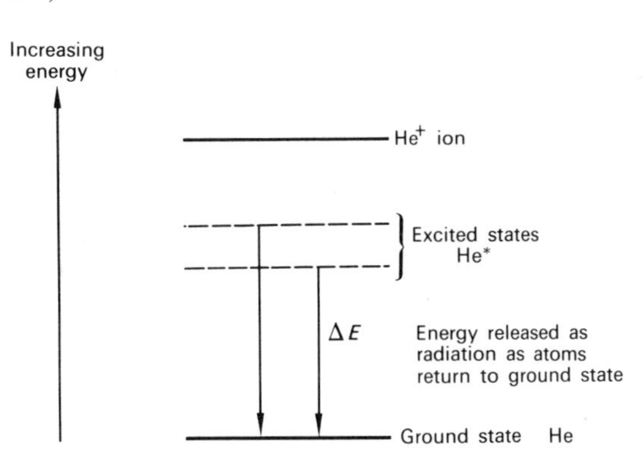

FIG. 3.3. Energy levels in helium.

The values of the energy levels in an atom can be determined by measuring the frequencies of light emitted when the "excited" atom returns to a lower energy state, for instance, the minimum energy state known as the **ground state**. The measurement of frequencies is done by splitting up the emitted light into a spectrum. This can be done either with a prism or a diffraction grating, and the result is known as an **emission spectrum**.

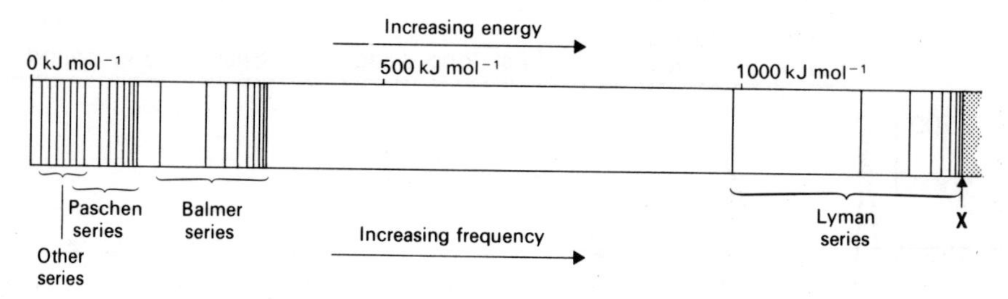

FIG. 3.4. The spectrum of the hydrogen atom.

Due to the fact that it has only one electron, the hydrogen atom possesses a relatively simple emission spectrum (Fig. 3.4).

The spectrum can be interpreted in terms of the various energy levels available to the electron. The **Lyman series** of lines corresponds to the electron returning to the lowest possible energy level from higher energy levels; the **Balmer series** to the electron returning from higher energy levels to the second lowest energy level; and so on (Fig. 3.5).

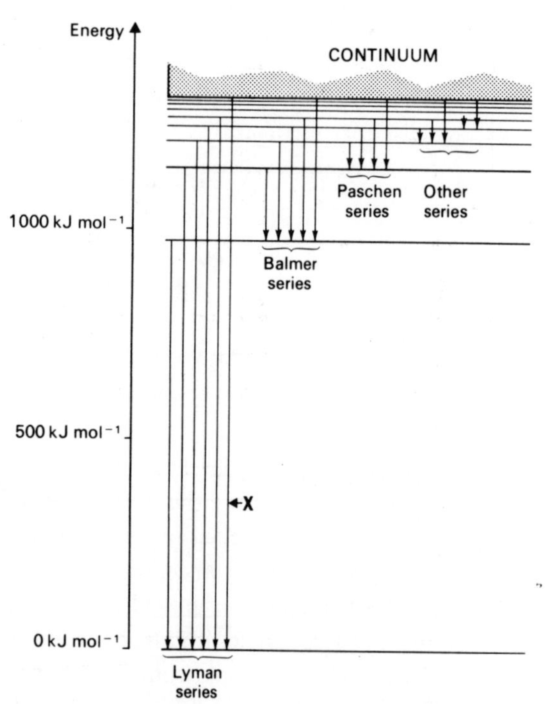

FIG. 3.5. Energy levels in the hydrogen atom.

The spectra of the heavier elements are far more complex for two main reasons:

(i) the atom contains more electrons;

(ii) the energy levels split (Fig. 3.9), thus increasing the number of lines in the spectrum.

The emission spectrum of an element is a form of fingerprint which can be used for identification purposes, in the procedure known as **spectroscopic analysis**. It is interesting to note that this is not a new technique— it was invented over a hundred years ago by Bunsen—and rubidium (meaning red) and caesium (meaning blue) were discovered in this way.

(b) *Excitation potentials*. If an electron of low energy collides with an atom it will undergo an elastic collision. That is to say it will simply bounce off, and due to its small mass it will not transmit any energy to the atom. If the energy of the electron is increased by applying a greater potential, there comes a critical point, known as **the excitation potential** where the collision with atoms becomes inelastic: the atom absorbs some of the electron's energy and is said to be in an **excited state**. The excitation potential is characteristic of the atom concerned, indicating that the atom can only absorb energy in this fixed amount.

If the applied potential is gradually increased, several excitation potentials may be observed, resulting finally in another electron being knocked off the atom altogether (which is the process of ionization described in Chapter 1 for the mass spectrometer) (Fig. 3.6).

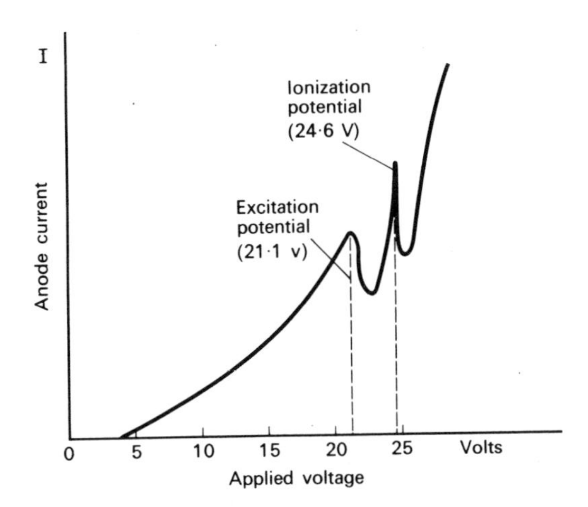

FIG. 3.6. Ionization of helium.

The excitation potentials of helium or argon can be investigated in the laboratory with a suitable circuit.

To summarize:

(i) Electrons in atoms and molecules can only exist with certain fixed energy values: they are said to occupy energy levels.

(ii) Electrons can only jump from one level to another by emission or absorption of the necessary amount of energy. This energy, ΔE, is fixed by the value of the energy levels, i.e. energy is emitted in discrete packets or **quanta**.

(iii) $\Delta E = hv$.

3.3 Measurement of ionization energy

Where the element can be converted directly into monatomic gas and introduced into a vacuum tube, its ionization energy can be determined in two main ways:

(a) The gas is irradiated with ultra-violet radiation of gradually increasing frequency until the energy, hv is sufficient to ionize it and make it conduct electricity.

(b) The gas is bombarded with electrons accelerated through a potential V which is gradually increased until ionization occurs and the gas itself conducts. The ionization energy is thus $96{\cdot}5V$ kJ mol^{-1}.

These methods are only applicable when the gas is *monatomic* (consists of single atoms). It is applicable for instance to the noble gases, and to alkali metal vapours, but not to a gas like oxygen which consists of *diatomic* molecules, $O_2(g)$.

In all other cases the ionization energies have to be calculated from the emission spectra of the elements. Not only is this method applicable to all the elements, it is also the most accurate. As an example consider the emission spectrum of hydrogen (Figs. 3.4 and 3.5). Ionization corresponds to the last line in the Lyman series (labelled X). Because the lines are so close together at this point the exact position of the line has to be extrapolated graphically and then the value is converted from spectroscopic units to conventional energy units (Study Question 15). Alternatively the ionization energy of hydrogen can be determined by taking the energy of the final line of the Balmer series ($\infty \rightarrow 2$) and adding to it the energy of the first line in the Lyman series ($2 \rightarrow 1$).

3.4 Magnetic properties of atoms

It is found that all substances can be broadly classified into two types on the basis of their magnetic properties:

(a) **diamagnetic**, i.e.: weakly repelled by a magnet; and

(b) **paramagnetic**, i.e.: weakly attracted by a magnet.

Substances such as iron and cobalt, which are strongly attracted, are called *ferromagnetic* but this property is limited to relatively few substances.

It is found that all diamagnetic substances contain an even number of electrons in their molecules, and this is the most common class. When a substance is paramagnetic this indicates at least one *unpaired* electron. Paramagnetic substances are much less common, and the property generally results from the presence of an odd number of electrons in the molecule.

Non-ferromagnetic substances require a very powerful magnetic field to produce a measurable force. The number of unpaired electrons in a substance can be measured by placing a sample in a **magnetic balance** (Gouy balance, Fig. 3.7) and measuring the force exerted by a given field.

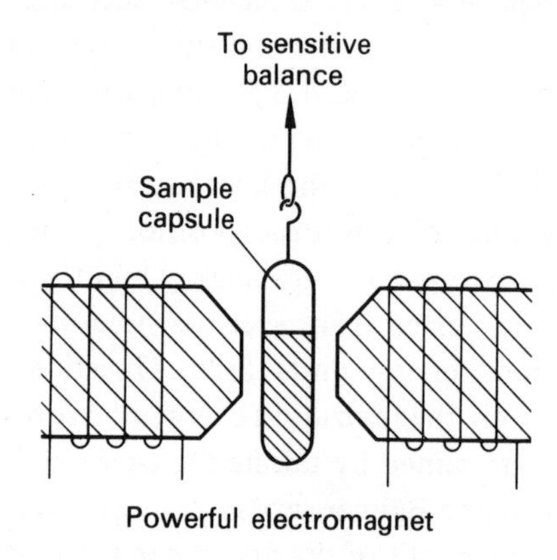

FIG. 3.7. The Gouy balance.

The association between number of electrons and magnetic behaviour is explained by postulating that electrons possess the property of **spin**. A single electron may be imagined as a spinning charge which behaves as a tiny magnet.*

When two electrons come together they sometimes pair off in such a way that their spins cancel each other out. A molecule made up entirely of paired electrons will have no overall spin, and will therefore be diamagnetic.

* We have previously likened an electron to a charge-cloud (Chapter 1). This cloud must not be thought of as spinning in the normal sense. The term "spin" is a convenient one to apply to a charge which also has magnetic properties, but it should not be taken too literally.

3.5 The first row elements

The fact that helium has a higher first ionization energy than hydrogen is easily explained by the higher nuclear charge which helium possesses. The low ionization energy of lithium is explained by postulating that the removed electron is in a higher energy level (Fig. 3.8).

If this postulate is correct, we should expect the second ionization energy of lithium to be rather high. The ion Li^+(g) is **isoelectronic** with (has the same number of electrons as) the helium atom He(g). We should expect the ionization energy of Li^+ to be greater than that of He on account of the positive charge.

$$Li^+ \rightarrow Li^{2+} + e^-;$$

$$\Delta E = 7300 \text{ kJ mol}^{-1}$$

or 75·6 electron-volts per atom.

The periodic variation of first ionization energy with increasing atomic number is explained in terms of two basic postulates:

(a) All the electrons in an atom occupy energy levels: in ionization the electrons which are highest in energy are removed from the atom first.

(b) There is a limit to the number of electrons a given energy level can contain. For instance, the lowest energy level in Fig. 3.8 can only hold two electrons, with their spins paired as indicated by the arrows ↑↓.

In the elements as far as neon there are three energy levels:

(a) The lowest energy level, labelled 1*s*, which can hold two electrons.

(b) The next energy level, labelled 2*s*, which can also hold two electrons only.

(c) Above the 2*s*, but not greatly exceeding it in energy, we have the 2*p*, which can hold six electrons, that is, three electron pairs. It is convenient to imagine that the electrons can be

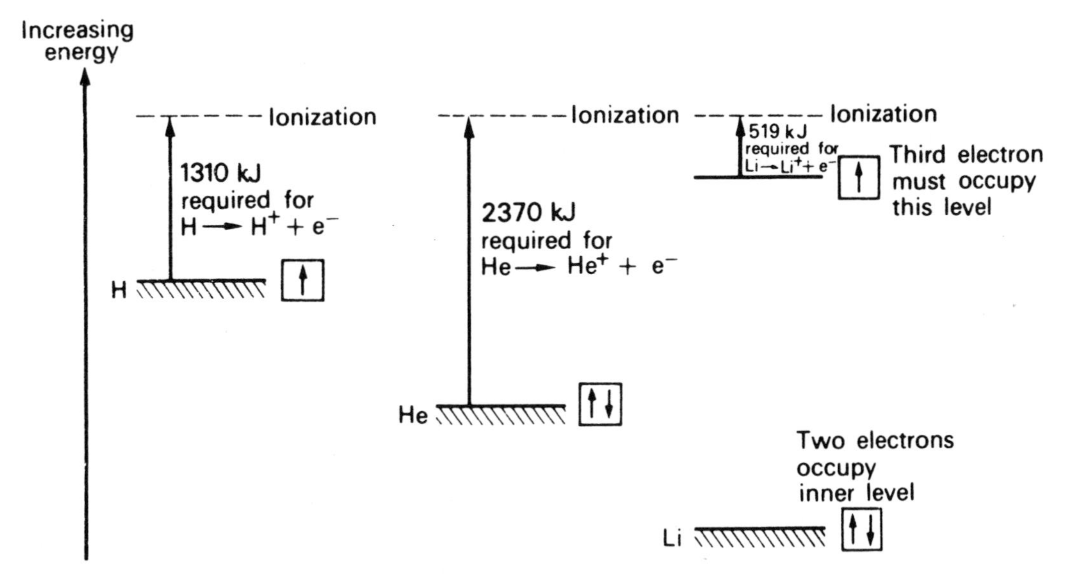

FIG. 3.8.

placed into "boxes" known as **orbitals** each able to hold *two* electrons with paired spins.*

An energy level within an atom is designated by:

(i) **Principal quantum number.** 2*s* and 2*p* both have a principal quantum number of 2. An atom can never have more than eight electrons with a principal quantum number of two.

(ii) **The letters s, p,** etc. This letter shows how many orbitals are available. For "*s*" electrons there is only one orbital of a given principal quantum number. For "*p*" electrons there are three orbitals. We shall later meet "*d*" electrons, where there are five orbitals, and "*f*" electrons where there are seven.

The energy levels in the hydrogen atom itself are very simple because *all* the states of a given quantum number have the same energy, no matter whether they are *s*, *p*, or *d*. This is only true for hydrogen however, and the levels become more complicated in larger atoms (Fig. 3.9).

The filling up of orbitals with electrons is governed primarily by the fact that electrons repel one another. This is well illustrated by the way in which the electrons are built up in the atoms as far as neon, which is summarized in Fig. 3.10.

We are now in a position to examine in detail the variations in first ionization energy which take place among the first row elements. Fig. 3.2 should be compared with Fig. 3.10.

(a) There is a general rise in ionization energy from lithium to neon, attributable to the change in nuclear charge.

(b) The dip observed in the plot at boron is due to the fact that the ionized electron in boron is lost from a higher energy level.

(c) The dip in the plot at oxygen can be explained by assuming that the electron lost in ionization is one of the paired ones. Since electrons repel one another, one of the paired electrons is lost more readily.

* The rule which states that not more than two electrons can be placed in the same orbital is known as the Pauli exclusion principle.

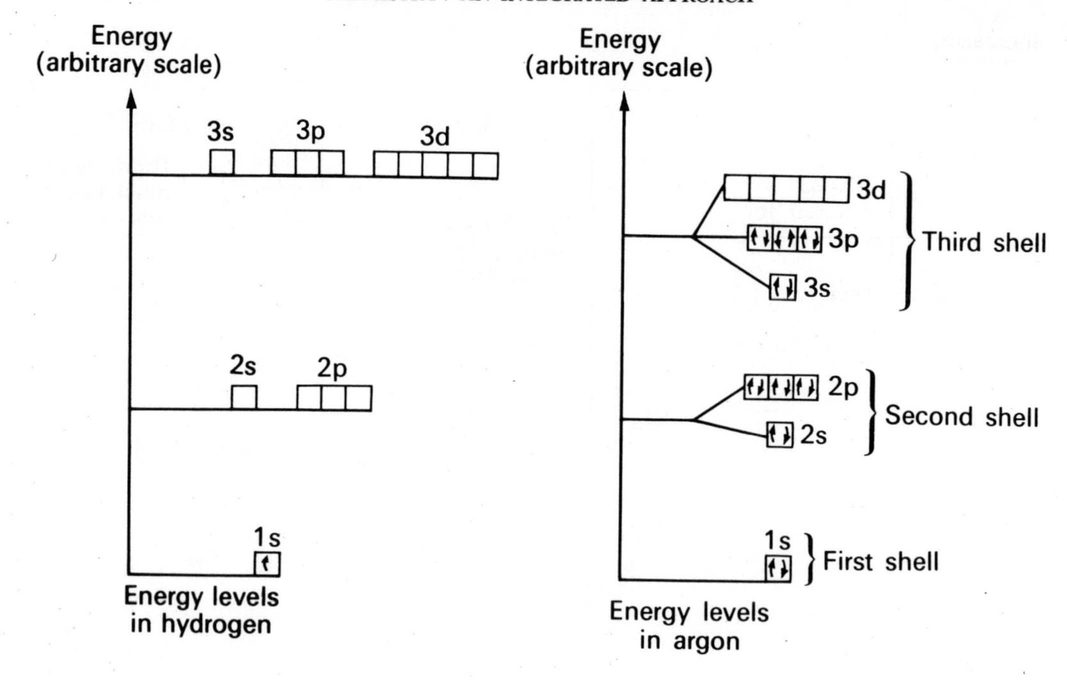

FIG. 3.9.

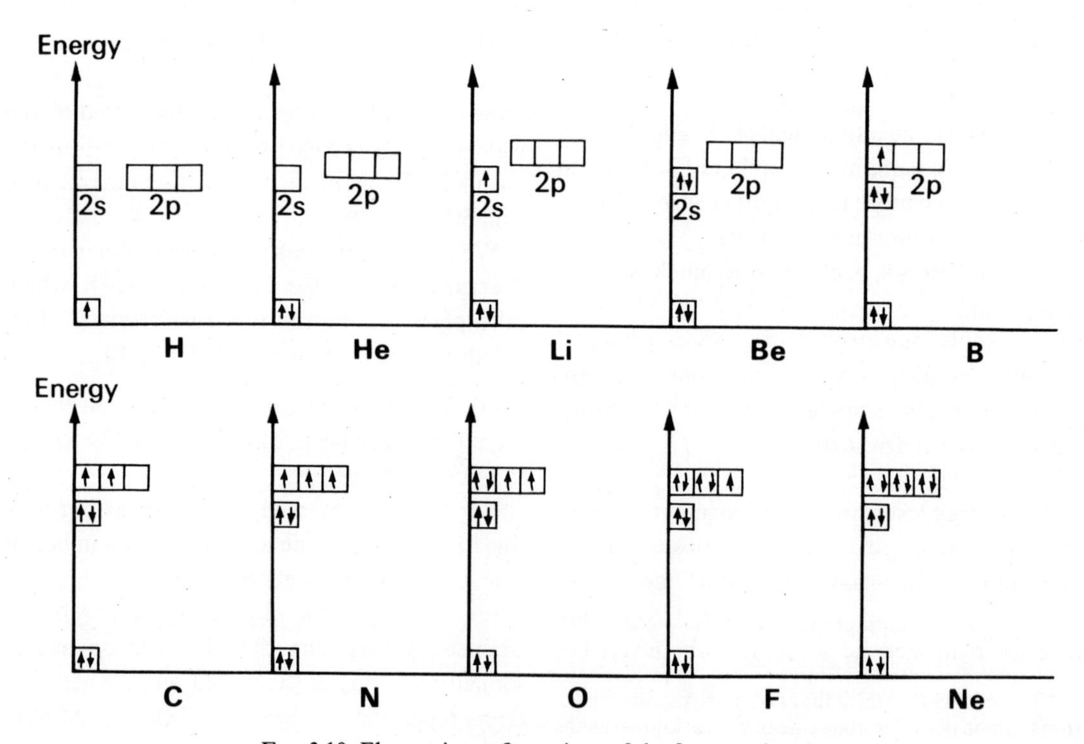

FIG. 3.10. Electronic configurations of the first ten elements.

3.6 The second row elements

The existence of three energy shells in a second row element can be clearly seen in Fig. 3.11,

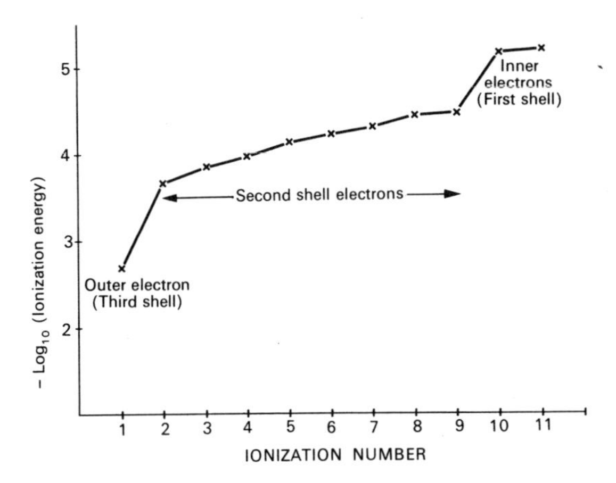

FIG. 3.11. The successive ionization energies of sodium.

where the successive ionization energies of sodium (Appendix III) are shown.

In Fig. 3.2 we can see that the second row elements show a very similar trend in ionization energies to the first row. This can be attributed to the existence of further energy levels, 3*s* and 3*p*, which are exactly analogous to the 2*s* and 2*p* levels which govern the behaviour of the first row. One of the consequences of this is that many of the chemical properties observed in the first row period are repeated in the second row. But there are also many chemical differences, some of which can be attributed to the existence of a group of energy levels, 3*d* (Fig. 3.9). Other factors such as size of atoms are also important.

3.7 Ways of representing electronic configurations of atoms

The **electronic configuration** (or electronic structure) of an atom or ion is the arrangement of its electrons within the energy levels. A shorthand notation is used in order to avoid drawing an energy level diagram for every atom. The number of electrons in a given energy level is shown by a small superscript numeral. For instance, five 2*p* electrons are denoted by the symbol $2p^5$. The ground state electronic configurations of the elements as far as argon are represented in Table 3.1.

A group of energy levels with the same principal quantum number is frequently referred to as a **shell**. For instance the electrons in an aluminium atom occupy three shells. The innermost shell, of principal quantum number 1, holds two electrons. The next shell, of principal quantum number 2, holds eight electrons. These two shells are **completed shells**. The shell of principal quantum number 3 is incomplete since it only contains three electrons. If we include the 3*d* orbitals, shell number 3 can hold $(2+6+10) = 18$ electrons.

3.8 Arrangement of electrons in space

We have so far avoided referring to the shapes of the "orbits" followed by the electrons surrounding atomic nuclei. Consider first the simplest case—that of the hydrogen atom in its ground state. There is a tiny nucleus with a charge of +1 "surrounded" by an electron of charge −1 and negligible mass. Bohr, in 1913, postulated that the electron followed a circular orbit around the nucleus, but this model has been abandoned since it did not give a consistent interpretation of the observed properties of hydrogen. We now think of the hydrogen electron as a **charge cloud**. Instead of visualizing it as a point charge we imagine it as a "smeared-out" charge. Its distribution around the nucleus is spherically symmetrical but the greatest concentration of charge is nearest the nucleus. Figure

TABLE 3.1

Element	Symbol	Complete electronic configuration	Abbreviated form
Hydrogen	H	$1s^1$	1
Helium	He	$1s^2$	2
Lithium	Li	$1s^2;\ 2s^1$	2, 1
Beryllium	Be	$1s^2;\ 2s^2$	2, 2
Boron	B	$1s^2;\ 2s^2,\ 2p^1$	2, 3
Carbon	C	$1s^2;\ 2s^2,\ 2p^2$	2, 4
Nitrogen	N	$1s^2;\ 2s^2,\ 2p^3$	2, 5
Oxygen	O	$1s^2;\ 2s^2,\ 2p^4$	2, 6
Fluorine	F	$1s^2;\ 2s^2,\ 2p^5$	2, 7
Neon	Ne	$1s^2;\ 2s^2,\ 2p^6$	2, 8
Sodium	Na	$1s^2;\ 2s^2,\ 2p^6\ 3s^1$	2, 8, 1
Magnesium	Mg	(neon "core") $3s^2$	2, 8, 2
Aluminium	Al	(neon "core") $3s^2,\ 3p^1$	2, 8, 3
Silicon	Si	(neon "core") $3s^2,\ 3p^2$	2, 8, 4
Phosphorus	P	(neon "core") $3s^2,\ 3p^3$	2, 8, 5
Sulphur	S	(neon "core") $3s^2,\ 3p^4$	2, 8, 6
Chlorine	Cl	(neon "core") $3s^2,\ 3p^5$	2, 8, 7
Argon	Ar	(neon "core") $3s^2,\ 3p^6$	2, 8, 8

1.2(c) is an attempt to represent the charge cloud picture of such an electron. Such a charge cloud shows the shape of an orbital. Figure 3.12 shows charge cloud distributions for helium, neon and argon.

It is misleading to think of an electron purely as a "particle" of charge—a charge cloud is a much better description. In fact the rules which appear to govern the shapes of electron charge clouds are found to be very similar to those which govern wave motions. A beam of electrons is found to be remarkably similar to a beam of light—it has a wavelength and can be made to undergo interference and diffraction just like a light beam.

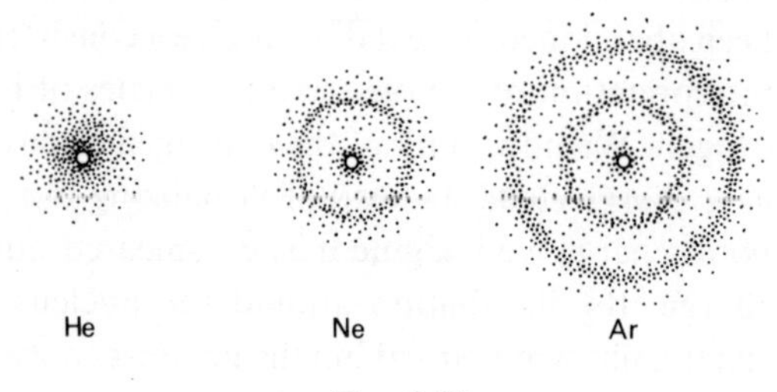

FIG. 3.12.

Despite the fact that electrons seem to have a split personality, they are still governed by the ordinary laws of electrostatics. Electrons repel one another and are attracted by positive charges such as nuclei.

3.9 Bonds between atoms

Although it is important to understand the behaviour of single atoms, chemistry is mainly concerned with the forces which hold atoms together. Single atoms are rather unusual—the only common examples at room temperature are the noble gases (Group 0).

Atoms generally have a tendency to join together as a consequence of the electrostatic forces between electrons and protons, but the noble gases have completely filled energy levels and their atoms have no tendency to join to one another.

Many common gaseous elements form diatomic molecules, for instance H_2(g), O_2(g), N_2(g), Cl_2(g) and F_2(g). Considerable energy has to be put into these substances to disrupt their molecules into atoms, and we therefore say that a **bond** exists between the atoms. This bond is often called a **covalent** bond. The properties of a bond are well illustrated by the simplest molecule of all, that which exists in hydrogen gas.

In hydrogen gas the distance between adjacent nuclei is always found to be the same, 0·074 nm. This distance is called the **bond length**. It is found that 435 kilojoules of energy are required to split up, or **dissociate**, one mole* of gaseous hydrogen molecules into atoms. In equation form:

$$H_2(g) \rightarrow 2H(g)$$

one mole of hydrogen molecules; that is, 2·016 g — two moles of hydrogen atoms; that is, 2·016 g

$$\Delta E = +435 \text{ kJ}$$

(positive sign indicates energy put into system)

Two hydrogen atoms will combine to form a molecule by taking up an arrangement in space where their electrons form a pair with opposed spins between the nuclei. A simple electrostatic

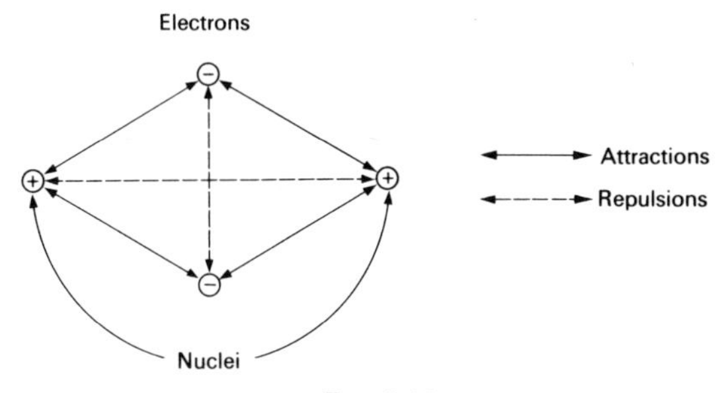

FIG. 3.13.

argument shows that the attractions will outweigh the repulsions, assuming the electrons to be point charges (Fig. 3.13).

Taking into account the charge-cloud nature of electrons, they are imagined to form a cloud

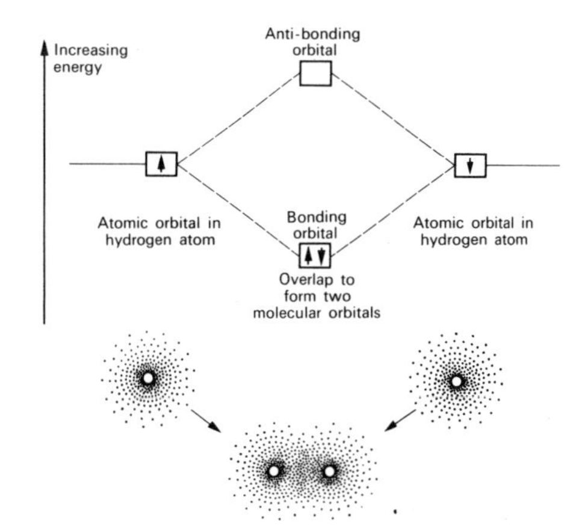

FIG. 3.14.

which has its greatest density on the axis between the nuclei (Fig. 3.14). In hydrogen atoms, the electrons occupied 1*s* **atomic orbitals**. In the hydrogen molecule they have combined together by pairing their spins, to occupy an energy level in the molecule, known as a **molecular orbital**. Hydrogen molecules are diamagnetic (their electron spins cancel out) indicating that the molecular orbital contains a pair of electrons.

The fact that molecules possess energy levels just like atoms explains why helium cannot form a molecule He_2. There are now four electrons to be placed in energy boxes, and this would mean placing two of them in a box of higher energy and this would involve a net loss in energy (Fig. 3.15). Since helium atoms cannot attain lower energy by combining they remain separate.

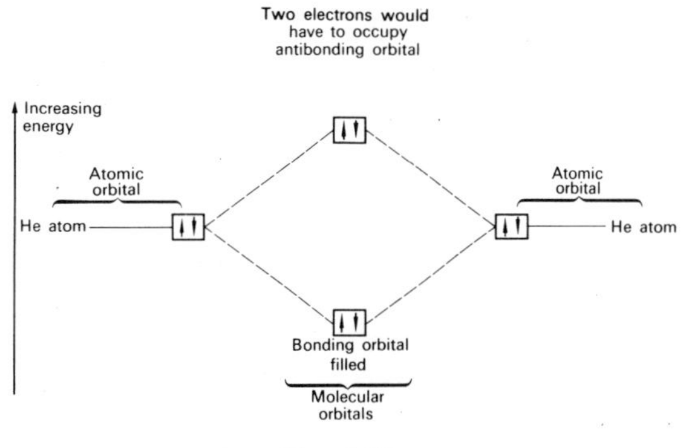

FIG. 3.15.

* The term **mole** may refer to $6{\cdot}022 \times 10^{23}$ *molecules*. Hence 1 mole of hydrogen molecules, H_2, weighs twice as much as 1 mole of hydrogen atoms.

A very similar argument explains why the elements of Group VIIB, the halogens, form diatomic molecules. The single atoms all have one "vacancy" in their uppermost *p* energy level. Two atoms combine by electron-sharing and a bond is formed in a similar way to hydrogen (Fig. 3.16).

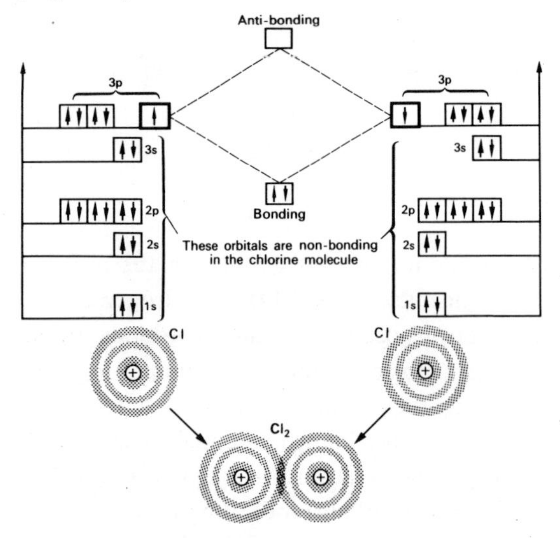

FIG. 3.16. The formation of a chlorine molecule.

3.10 Shorthand ways of representing bonds

It is important always to think of a bond as resulting from the electrostatic forces between electrons and nuclei, and the energy gained by placing electrons in energy levels. Nevertheless, it is rather cumbersome to use energy level diagrams for pictorial representations. A simple notation in which electrons are shown by dots is often helpful. For example:

$$H^{\cdot} + .H \rightarrow H\!:\!H \quad \text{or} \quad H\text{—}H$$

This is just a convenient way of counting electrons and energy levels— it is *not* a picture of what a molecule is supposed to be like. It ignores the fact that electrons are clouds of charge, not point charges, and also the fact that molecules are three-dimensional, and not flat like the paper they have to be drawn on.

In these diagrams, only the electrons in the uppermost, partially-filled energy shell are shown. This is known as the **valence shell**. For instance, in chlorine the inner electrons are hardly affected by the formation of a bond as they are deeply buried in the atom (Fig. 3.16).

Where we are interested in the overall structure and shape of a molecule, rather than its energy levels, an even simpler notation is used. Here the electron-pairs which form bonds are each represented by a line, and non-bonding electron pairs are omitted (Fig. 3.17).

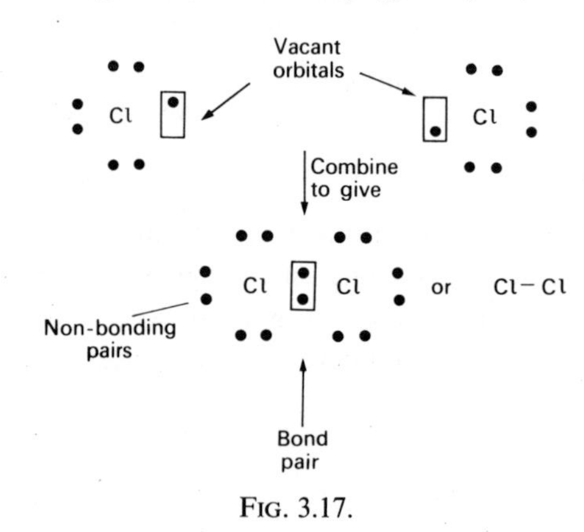

FIG. 3.17.

The bond between two nitrogen atoms in $N_2(g)$ is interesting; nitrogen is an extremely unreactive substance because considerable energy is required to separate the atoms. The dissociation energy of nitrogen into atoms is very high, 945 kJ per mole of N_2. Nitrogen atoms have three unpaired electrons, and it is thought that these all pair up when two atoms come together.

$$\underbrace{\dot{:}N\dot{:} + \dot{:}N\dot{:}}_{\text{unpaired electrons}} \rightarrow \underbrace{:N\vdots\vdots N:}_{\text{three bond pairs}}\ (N{\equiv}N)$$

We can imagine nitrogen atoms to be "welded" together by three electron-pair bonds. The bond length is short, only 0·1097 nm, which suggests a strong attraction between the electrons and the nuclei.

The oxygen molecule, $O_2(g)$, raises some fascinating problems. Its dissociation energy, 498 kJ per mole of O_2, and its bond length, 0·121 nm, are not exceptional, but oxygen is found to be paramagnetic despite the fact that it contains an even number of electrons! A test-tube containing liquid oxygen is readily attracted to a powerful magnet, indicating unpaired electrons. We have seen that this is what happens in atoms, so it ought not to be surprising that it happens in molecules as well. Paramagnetic molecules are, however, rather unusual.

3.11 Ionic substances

The removal of one or more electrons from an atom results in the formation of a positive ion. Negative ions can be formed in a similar fashion by the addition of electrons to an atom. A fluorine atom will quite readily form a negative ion by filling its $2p$ "vacancy" with a paired electron.

$$[:\ddot{\underset{\cdot\cdot}{F}}.] + e^- \rightarrow [:\ddot{\underset{\cdot\cdot}{F}}:]^-;$$

gaseous atom → gaseous ion

$$\Delta E = -350 \text{ kJ}$$

The energy released is called the **electron affinity** of fluorine. In this example the E.A. of fluorine is −350 kJ. (The ionization energy of the fluoride ion, F^-, is +350 kJ, the opposite of the electron affinity of the fluorine atom). Compare this energy with the ionization energy of an ionizable metal like potassium:

$$K^{\cdot}(g) \rightarrow K^+(g) + e^-;$$

atom (gaseous) → negative ion

$$\Delta E = +418 \text{ kJ mol}^{-1}$$

Comparing these two equations it might be expected that if a potassium and a fluorine atom are placed together, electron transfer would occur.

$$K^{\cdot}(g) + :\ddot{\underset{\cdot\cdot}{F}}.(g) \rightarrow K^+(g) + F^-(g);$$

$$\Delta E = -350 + 418$$
$$= +68 \text{ kJ}$$

Energy is released when a pair of opposite charges are brought together, making the overall transfer of an electron from potassium to fluorine, to form an ion-pair, an energy-releasing process:

$$K^+(g) + F^-(g) \rightarrow K^+F^-;$$

ion-pair

$$\Delta E = -585 \text{ kJ}$$

$$\therefore\ K^{\cdot}(g) + :\ddot{\underset{\cdot\cdot}{F}}^{\cdot}(g) \rightarrow K^+F^- \text{ (ion-pair)};$$
$$\Delta E = -517 \text{ kJ}$$

The compound potassium fluoride has properties which suggest that it is made up of ions. If it is heated until molten it may be **electrolysed**: an electric current can be passed through the melt by the movement of ions. Positive potassium ions move to the negative electrode where they become metallic potassium, and negative fluoride ions move to the anode where they lose electrons and form atoms. The atoms immediately combine to form molecules of fluorine gas, F_2.

Study Questions

1. Explain why the ionization energies of the alkali metals (Group IA) decrease with increasing atomic number.

2. The first ionization energy of beryllium is greater than that of lithium, but the second ionization energy of beryllium is very much less than the second ionization energy of lithium. Explain these facts.

3. Plot the atomic number of the elements as far as argon against their second ionization energies. Compare your plot with Fig. 3.2.

4. The following are data for the electron affinity of oxygen:

$$O(g)+e^- \rightarrow O^-(g); \quad \Delta E = -142 \text{ kJ}$$
$$O^-(g)+e^- \rightarrow O^{2-}(g); \Delta E = +790 \text{ kJ}.$$

Account for the differences in these two values.

5. What are the ground state electronic configurations of the following?

(a) a neon atom; (b) a lithium atom;
(c) a lithium cation, $Li^+(g)$;
(d) a fluorine atom; (e) a fluoride anion, $F^-(g)$.

6. To which elements do the following electronic configurations apply?

(a) $1s^2; 2s^2$.
(b) $1s^2; 2s^2p^6; 3s^2p^5$.
(c) $1s^2; 2s^2p^6; 3s^2p^6; 4s^1$.
(d) $1s^2; 2s^2p^6; 3s^2p^6d^8; 4s^2$.

7. Which of the following would you expect to be paramagnetic? Ne, NO, NO_2, N_2O_4, ClO_2, ClO_3, and Cl_2O_6.

8. (a) What have O^{2-}, F^-, Na^+, Mg^{2+}, and Al^{3+} in common?

(b) How would the sizes of these ions compare?

9. In a transition series, the radii of ions with the same charge decrease as the atomic number increases. Suggest an explanation for this behaviour.

10. (a) The first excitation potential of sodium is 2·1 V. Use the expression $\Delta E = h\nu$ to calculate the wavelength of radiation to which this corresponds. In what part of the spectrum does this radiation occur?

(b) The first excitation potential of argon is 13·0 V. What is the frequency and wavelength of the radiation to which this corresponds? (1 electron-volt $= 1{\cdot}6 \times 10^{-19}$ J; $h = 6{\cdot}62 \times 10^{-34}$ J s; $c = 3 \times 10^8$ m s^{-1})

11. Will a neon light obey Ohm's law? Why do you suppose that a neon light is red?

12. (a) The following are the first five ionization energies (kJ mol^{-1}) of five main group elements. In each case assign the element to its Group in the periodic table.

(i)	495	4560	6950	9550	13 400
(ii)	740	1500	7700	10 500	13 600
(iii)	790	1600	3200	4400	16 100
(iv)	800	2430	3680	25 000	32 800
(v)	590	1100	4900	6500	8100

(b) Two of the elements are in the same Group. State, giving your reasons, which of them has the larger atomic number.

13. In each of the following sets, place the atoms (or ions) in order of increasing first ionization energy. In each case give reasons to justify your order.

(a) Be, Mg, Ca. (e) Rn, Fr, Ra.
(b) Li, Be, B. (f) Fr, Fr^+, Fr^{2+}.
(c) B, C, N. (g) F^-, F, F^+.
(d) C, N, O. (h) F^-, Ne, Na^+.

14. Suggest explanations for the following observations:

(a) The first ionization energy of Al is less than that of Mg, but the first ionization energy of Sc is greater than that of Ca.

(b) The first ionization energy of Sn is less than that of Ge, but the first ionization energy of Pb is greater than that of Sn.

15. The first seven lines in the Lyman series in the hydrogen atom spectrum are ($\times 10^{14}$ s^{-1}): 24·66, 29·23, 30·83, 31·57, 31·97, 32·21 and 32·37 respectively. The first seven Balmer lines are ($\times 10^{14}$ s^{-1}); 4·57, 6·17, 6·91, 7·31, 7·55, 7·71 and 7·82.

(a) Use the Lyman series to work out the ionization energy of the hydrogen atom (in kJ mol^{-1}) by plotting either (i) each frequency against the difference between successive frequencies, or (ii) $\frac{1}{n^2}$ against the frequency where n is the number of the energy level from which the electron is returning. In each case extrapolation will be needed.

(b) Use the Balmer series to find the energy needed to ionize a hydrogen atom whose electron starts in the second energy level.

(c) Use your answer to (b) and the relevant line in the Lyman series to obtain a further value for the ionization energy of hydrogen.

CHAPTER 4

Methods of determining structure

4.1 Physical properties and structure

In Chapter 1 some of the aspects of atomic structure were considered, and Chapter 3 dealt with some of the forces which hold atoms together. This chapter is concerned with the different types of structure which matter can have as a result of the bonds between atoms, and the influence which structure has on physical properties.

In a solid the atoms are packed together in a rigid form; generally some regularity of pattern is observable in the packing, and the atoms are said to form a **lattice**. When a solid melts, this rigidity disappears and a slight increase in volume generally takes place. The liquid state is difficult to study experimentally since its structure is so complex, but it is known that the regular lattice structure of the solid breaks up due to vibration of the atoms. The atoms are still close together, though **translation** (that is, movement of particles from one place to another), and rotation, can take place freely. With sufficient energy, some of the molecules of a liquid can escape altogether from the surface—this is **evaporation**. In a gas or vapour the molecules are considerably further apart than in a liquid or solid. Attraction between molecules is relatively slight, and can often be ignored altogether. Gases are compressible as a result of the large amount of space between the molecules, whereas liquids and solids are only very slightly compressible—they are referred to as **condensed phases**.

Figure 4.1 shows the structure that is thought to exist in ice, water and steam. Water is rather

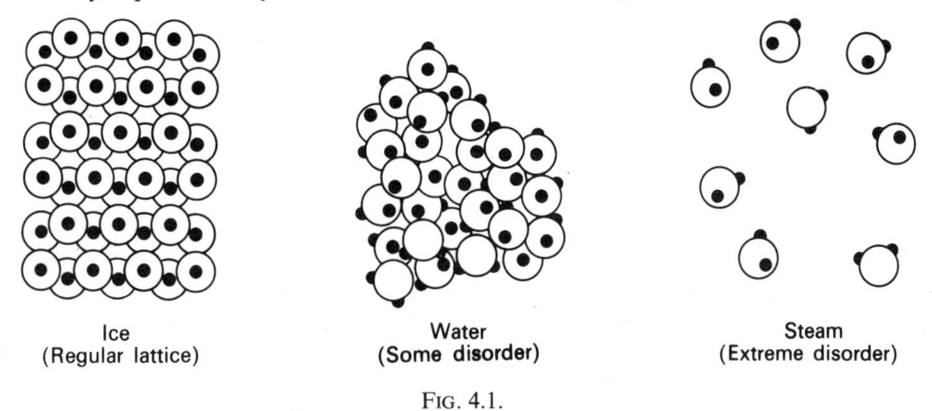

FIG. 4.1.

an unusual substance as it expands on solidification—a factor which causes pipes to burst and roads to be damaged in frosty weather. Figure 4.2 illustrates the fact that a gas is compressible: the pressure of a gas is due to the bombardment of the container walls by molecules, and the more highly compressed the gas is, the greater will be the number of molecules striking a given area in a given time and the greater the pressure.

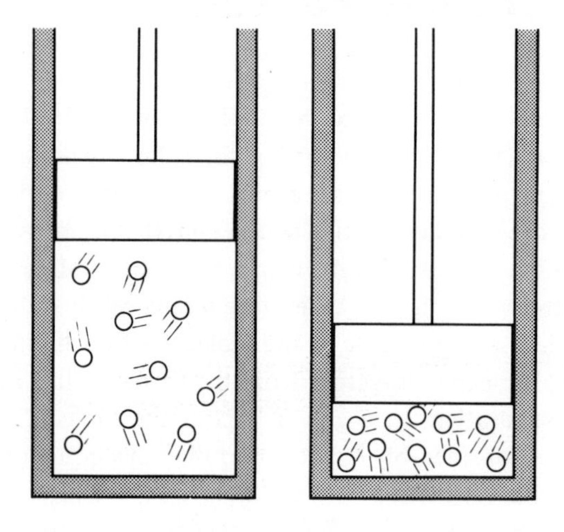

FIG. 4.2.

Two kinds of forces have to be considered in order to understand the structure of materials:

(a) Relatively large forces hold atoms together to form molecules: these we have already referred to as bonds. The electrostatic forces holding ions together in a lattice are of similar magnitude, and so are the forces which hold atoms together in a metal.

(b) Relatively weak forces act between separate molecules, especially where these molecules are small as in gases like carbon dioxide, CO_2, and methane CH_4. Weak forces are even observed in the noble gases where the "molecules" are in fact single atoms.

The precise nature of the forces in (a) and (b) will be examined in more detail later.

4.2 Methods of determining the structure of solids

The chief way of determining the positions of atoms in a solid lattice is by **X-ray-analysis**, first devised by von Laue in 1912. X-rays resemble light insofar as they are an electromagnetic radiation, but their wavelength is much shorter. A typical wavelength for visible light would be 589 nm (the wavelength of the yellow light emitted by sodium vapour in a flame), whereas X-rays have a wavelength in the region of 0·1 nm. Individual atoms are too small to diffract visible light, but if a beam of X-rays of single wavelength, that is a **monochromatic** beam,* is directed at a regular crystal, the crystal will act as a diffraction grating. For a given angle of incidence, X-rays will be reflected most strongly in certain preferred directions depending upon the distance between successive layers of the atoms in the crystal. If the angle of incidence is varied, an analysis of the *diffraction pattern* produced by

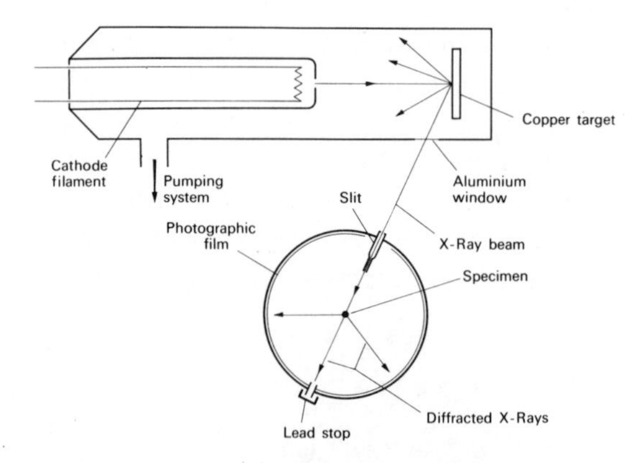

FIG. 4.3. X-ray analysis.

the crystal will enable all the interatomic distances to be computed. A simple arrangement for the examination of X-ray diffraction patterns is shown schematically in Fig. 4.3.

* Of one "colour" only, by analogy with visible light.

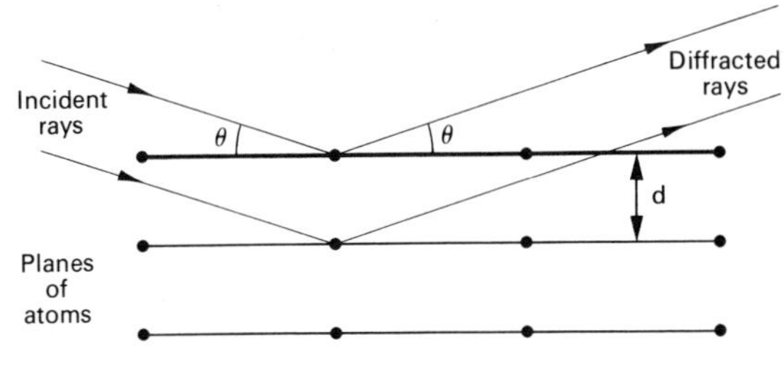

FIG. 4.4. Diffraction.

The distance between successive planes of atoms (Fig. 4.4) can be computed using the **Bragg equation**, $n\lambda = 2d \sin \theta$, in which:

n = an integer
λ = the wavelength of the X-rays
d = the interplanar distance
2θ = the angle between the incident and diffracted radiation.

Figure 4.5 shows the results obtained from a single crystal of the relatively simple compound, potassium chloride. By rotating the crystal three different patterns are obtained, each pattern corresponding to diffraction by a different set of planes. The interplanar distances in KCl can be calculated using the Bragg equation. For example, Fig. 4.5(b) shows two diffraction lines with $2\theta = 15°$ and $29°$

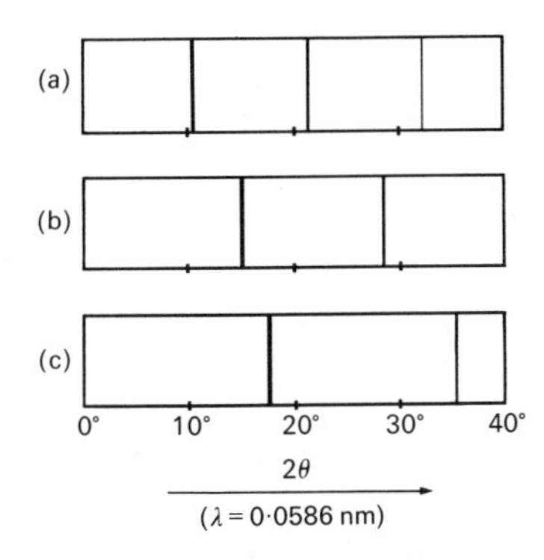

FIG. 4.5. X-ray diffraction data for KCl.

Hence $\theta = 7\frac{1}{2}°$ and $14\frac{1}{2}°$

Using $d = \dfrac{n\lambda}{2 \sin \theta}$, with $\lambda = 0{\cdot}0586$ nm, it is possible to show that $d = 0{\cdot}226n$ nm and $0{\cdot}117n$ nm in these two cases.

From these results it would seem probable that the distance between the planes is approximately 0·230 nm and that for the first line $n = 1$ and, for the second, $n = 2$. In the first case diffraction occurred between adjacent planes; in the second case, between planes twice as far apart. Study Question 4 gives practice in the use of the Bragg equation.

The diffraction of a finely ground powder produces a similar effect. Here the result is the same as would be obtained by superimposing all the pictures obtained by rotating the single crystal, since the powder crystals will be orientated in every direction.

With complicated substances, the analysis of X-ray patterns can be a very laborious mathematical chore, but now this can be overcome by the use of the computer. Some spectacular achievements have resulted from the use of computers to work through the calculations. For instance the Nobel Prize for Chemistry was awarded to Professor Dorothy Hodgkin in 1965, for the work she and her research team did in determining the complete structure of Vitamin B_{12}, a highly complex substance of molecular formula $C_{63}H_{90}O_{14}N_{14}PCo$.

In theory the method can tell us all we wish to know about the positions of *nuclei* in a lattice, but in practice it has its limitations. In particular hydrogen atoms are difficult to locate in the presence of other, larger nuclei on account of their small nuclear charge. Detailed calculations can lead to the plotting of an electron density map for a molecule or any other compound. From such a map, the positions of the different types of atom can be inferred; in addition, the electron density between the nuclei gives some information about the bonding.

A supplementary technique to X-ray diffraction is **neutron diffraction**. A beam of neutrons is found to behave as if it possessed a definite wavelength, and to give diffraction patterns. (Electrons also have properties which are wave-like, and this wave-particle duality is common to all the particles from which matter is built). Neutron diffraction can often enable hydrogen atoms to be located in cases where X-ray analysis fails.

4.3 Methods of determining the shapes of molecules

Before the shape of any molecule can be determined, it is first necessary to work out the molecular formula. This is done by carrying out experiments to discover the empirical formula and the molecular weight of the substance (see Chapter 7) and then combining these pieces of information. Once the molecular formula is known, further chemical and physical experiments are carried out to find out what bonds and which groups of atoms are present in the molecule. This information, combined with the molecular formula, leads to the structural formula (Fig. 4.6).

As an example, consider the compound phenylethene. Conventional carbon and hydrogen analysis leads to a carbon : hydrogen ratio of 1 : 1 and thus the empirical formula is CH. The molecular weight of 104 gives a molecular formula of C_8H_8. At this point several different methods can be used to deduce the structural formula. Chemical methods show the presence of a C=C double bond (Chapter 26) and the probable existence of a six-membered ring of carbon atoms (Chapter 27). In addition, this information can be deduced from a study of the infra-red spectrum of the compound, and the mass spectrum would certainly show the presence of a benzene ring. Nuclear magnetic resonance spectroscopy shows the relative positions of the hydrogen atoms. Combining the chemical and physical evidence, we arrive at a structural formula:

```
                 H
                 |
          H    C    H
           \ /   \ /
            C     C
            |  O  |
   H        C     C
    \     /  \   /  \
     C=C      C      H
    /    \    |
   H      H   H
```

In fact, we can only be certain of this structure and the shape of the molecule if we subject phenylethene to X-ray analysis. The disadvantage of X-ray analysis is that it is slow and laborious, whereas the other methods available are relatively quick and are easily interpreted. Let us consider these other methods in greater detail.

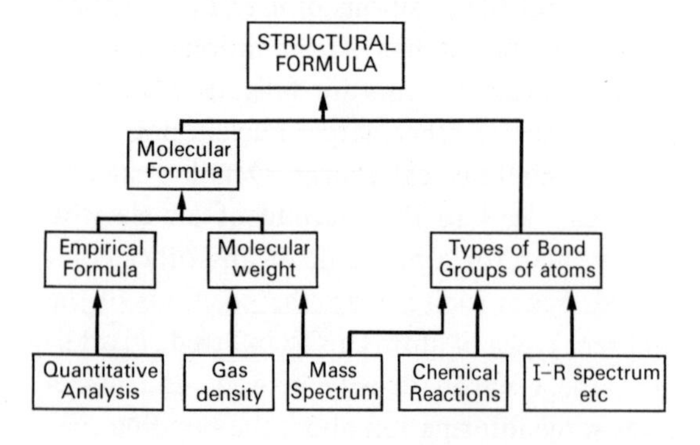

FIG. 4.6. The route to a structural formula.

(a) CLASSICAL METHODS

Classical methods of inferring the shape of a molecule depend on chemical reactions. We assume the structure of the molecule to be the one which explains its reactions and other properties most consistently. In many cases there is not enough information to decide between different alternatives by classical methods alone, but in other cases the classical arguments have proved to be triumphs of deductive reasoning. Figure 4.7 illustrates how chemists were able to deduce that the carbon atom in methane is tetra-

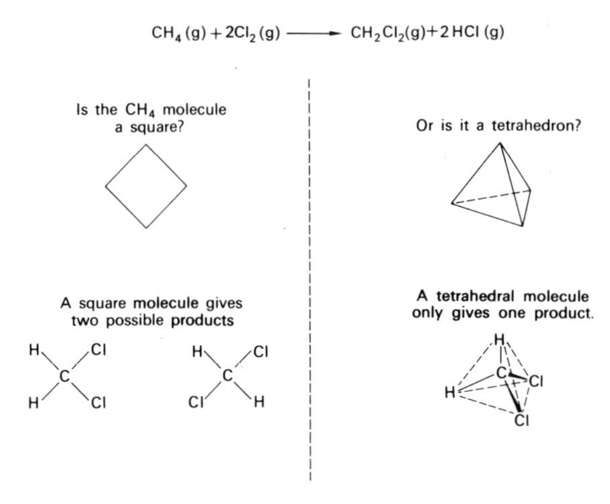

FIG. 4.7.

hedral rather than planar. A planar formula would have led to two *different* molecules both with the formula CH_2Cl_2 when methane was reacted with chlorine, but only one substance of this formula has ever been prepared. Similarly Kekulé was able to deduce the structure of benzene to be cyclic in 1865, long before the use of physical methods for structure determination.

(b) DIPOLE MOMENTS

In cases where the molecule is simple, or where other features of its structure are already known, determination of its **dipole moment** may enable its exact shape to be decided upon.

Any bond between two unequal atoms possesses an *electrical dipole*, due to the fact that the electron pair is attracted more strongly to one nucleus than to the other. The atom which attracts the electrons more strongly is said to be more **electronegative**, and it will acquire a small excess negative charge $\delta-$ and leave behind a small excess positive charge on the other atom (denoted by $\delta+$). For instance, the hydrogen chloride molecule, HCl, has a dipole moment because chlorine is more electronegative than hydrogen.

$$\underset{\delta+}{H}-\underset{\delta-}{Cl}$$

$$+ \rightarrow -$$

Direction of dipole moment

It is often possible to predict the direction of a dipole moment along a given bond. The more electronegative elements are those on the right-hand side and near the top of the periodic table, the most electronegative of all being fluorine (Fig. 4.8).

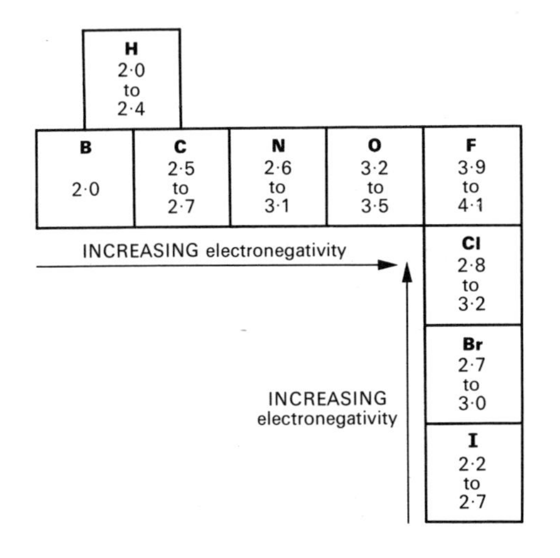

FIG. 4.8. Electronegativities.

Electronegativity can be defined as "the power of an atom in a molecule to attract electrons to itself". Electronegativity is only a qualitative concept and, although numerical values of electronegativities are sometimes quoted, they are not very meaningful since the electronegativity of an atom will vary from one molecule to another. In organic chemistry, the different electronegativities of atoms or groups of atoms give rise to electron movements, known as inductive effects. When compared with hydrogen atoms, halogen atoms exert a powerful electron-withdrawing inductive effect in organic molecules because of their large electronegativity values. On the other hand the methyl group, CH_3—, exerts an inductive effect in the opposite direction and it is electron-donating when compared with hydrogen.

A molecule composed of two identical atoms such as H_2, Cl_2 or F_2 will have no dipole moment. With molecules containing more than two atoms, a knowledge of the dipole moment combined with a knowledge of the electronegativities can often help us to decide between different possible structures. With triatomic (three atom) molecules, a knowledge of the dipole moment alone is often enough to distinguish between alternative shapes. For instance a consideration of the properties of hydrogen and oxygen atoms, and the reactions of water, does not tell us what *shape* the water molecule is, apart from saying that the oxygen atom is between two hydrogen atoms. There are two possibilities. The molecule might either be **linear** or **bent**.

If H_2O is a linear molecule we shall expect it to have no dipole moment, since the electrical dipoles will cancel one another out. If on the other hand the molecule is bent we shall expect a dipole moment in the direction illustrated:

$$\overset{2\delta-}{\mathrm{O}} \text{ bonded to } \underset{\delta+}{\mathrm{H}} \text{ and } \underset{\delta+}{\mathrm{H}} \qquad \overset{-}{\underset{+}{\uparrow}}$$

Direction of dipole moment

$$\overset{\delta-}{\mathrm{O}}=\overset{2\delta+}{\mathrm{C}}=\overset{\delta-}{\mathrm{O}}$$

No resultant dipole moment

In fact measurements on water show it to have quite a large dipole moment, therefore it must be bent. Unfortunately the measurement cannot tell us *how* bent.

Again consider carbon dioxide, CO_2. Measurements show this to have zero dipole moment. The molecule must therefore be *linear* with the carbon atom in the middle—any other arrangement of atoms would produce a dipole moment.

Provided the substance is a liquid, a very simple laboratory experiment will determine

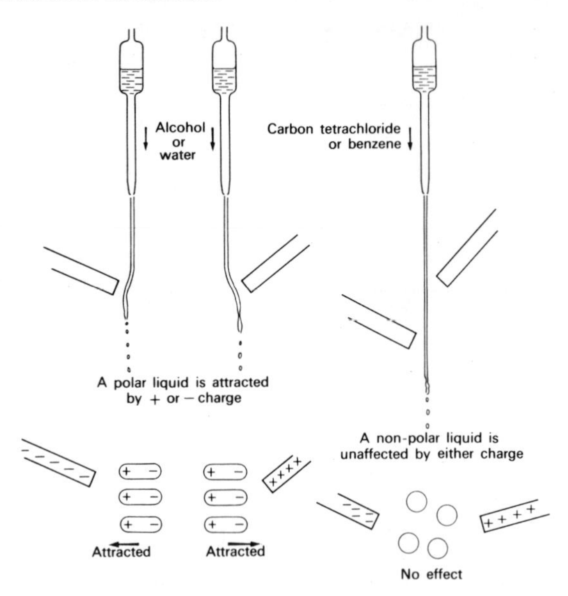

FIG. 4.9.

whether or not it has a dipole moment. Allow the liquid to flow as a fine jet from a pipette: **non-polar** substances (substances whose molecules have no dipole moment) are unaffected by bringing a charged rod near the jet, whereas **polar** substances (those with a dipole moment) will be attracted to the rod (Fig. 4.9). It does not matter whether the rod is positive or negative—a dipole is attracted to either charge. The substance under test should be absolutely dry. Since water itself has a high dipole moment, it will affect the properties of other substances it contaminates.

This simple laboratory test cannot be used for quantitative measurements. These are done by finding how the **relative permittivity** of the substance varies with temperature. Relative permittivity is the constant ε in the expression for the force of attraction or repulsion between two charges Q_1 and Q_2 a distance r apart, when placed in the given medium:

$$\text{Force} = \frac{Q_1 Q_2}{4\pi\varepsilon}$$

Direct electrostatic measurements are not practicable for measuring ε; an electronic method is used instead.*

(c) INFRA-RED SPECTROSCOPY

A molecule which can vibrate in such a way that its dipole moment alters while vibrating, can absorb electromagnetic radiation. The energy absorbed corresponds to radiation in the infra-red region of the electromagnetic spectrum ($\Delta E = h\nu$, Chapter 3). Simple molecules have infra-red absorption spectra which depend upon their shapes. The carbon dioxide molecule illustrates this. Although the molecule itself has no dipole moment, being linear, a dipole moment will appear if the molecule (a) bends as in Fig. 4.10a, or (b) stretches unsymmetrically as in Fig. 4.10b. Each of these modes of vibration will absorb infra-red energy of characteristic frequency. Note that the third possible mode of

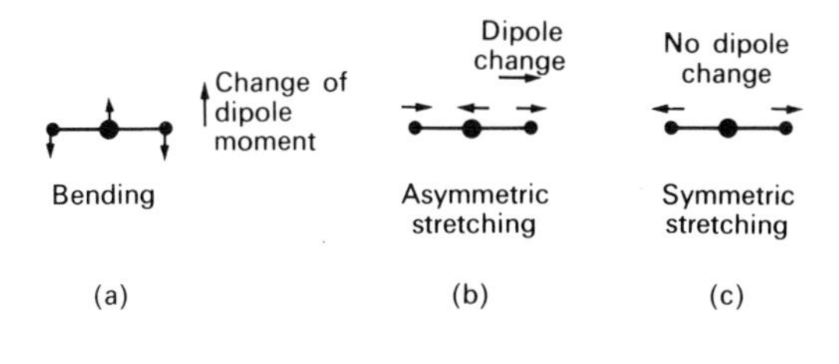

FIG. 4.10.

vibration (c), that of symmetric stretching of the bonds, cannot produce a dipole moment and will therefore not lead to absorption of infra-red radiation.

The water molecule is bent, and three possible modes of vibration can be expected to lead to the absorption of infra-red radiation (Fig. 4.11), since all modes lead to a variation in the dipole moment.

Both carbon dioxide and water are very simple

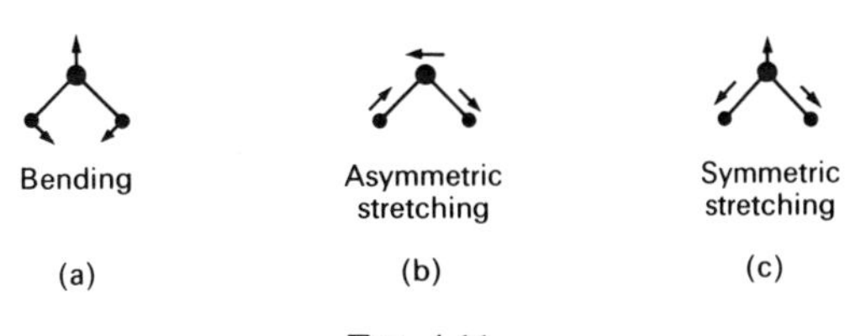

FIG. 4.11.

molecules, and the interpretation of their spectra is simple. A more complicated molecule with n atoms generally has $3n-6$ possible vibration frequencies with the consequence that the spectra of polyatomic molecules are very complex. Fortunately, however, individual bonds in a molecule tend to vibrate at frequencies that depend on the masses of the atoms forming the bond and also on the strength of the bond. Thus the stretching of the C—H single bond absorbs energy at about 3000 cm^{-1} (36 kJ mol^{-1}) while that of the Si—H bond absorbs at only 2200 cm^{-1} (26·5 kJ mol^{-1}). The C≡C stretching frequency is 2150 cm^{-1} (26 kJ mol^{-1}), while the weaker C=C stretching frequency is at about 1650 cm^{-1} (20 kJ mol^{-1}).

In organic chemistry, the infra-red absorption spectrum of a molecule can often be used to discover which functional groups are present. A functional group is a group of two or more atoms that behave characteristically in chemical reactions. Table 4.1 shows some functional groups together with their infra-red absorption frequencies.

TABLE 4.1

Functional group	Formula	Approximate infra-red absorption frequencies (cm^{-1})		($kJ\ mol^{-1}$)
Methyl	$—CH_3$	2950	(stretch)	35
		1450	(bend)	17
Hydroxyl	HO—	3600	(stretch)	43
Alkyne	—C≡C—	2150	(stretch)	26
Cyanide	—C≡N	2100	(stretch)	25
Carbonyl	C=O	1720	(stretch)	20·5
Alkene	C=C	1650	(stretch)	20

* For a detailed description of dipole moment measurements, see Mansel Davies, *Some Electrical and Optical Aspects of Molecular Behaviour*, Pergamon Press, 1965.

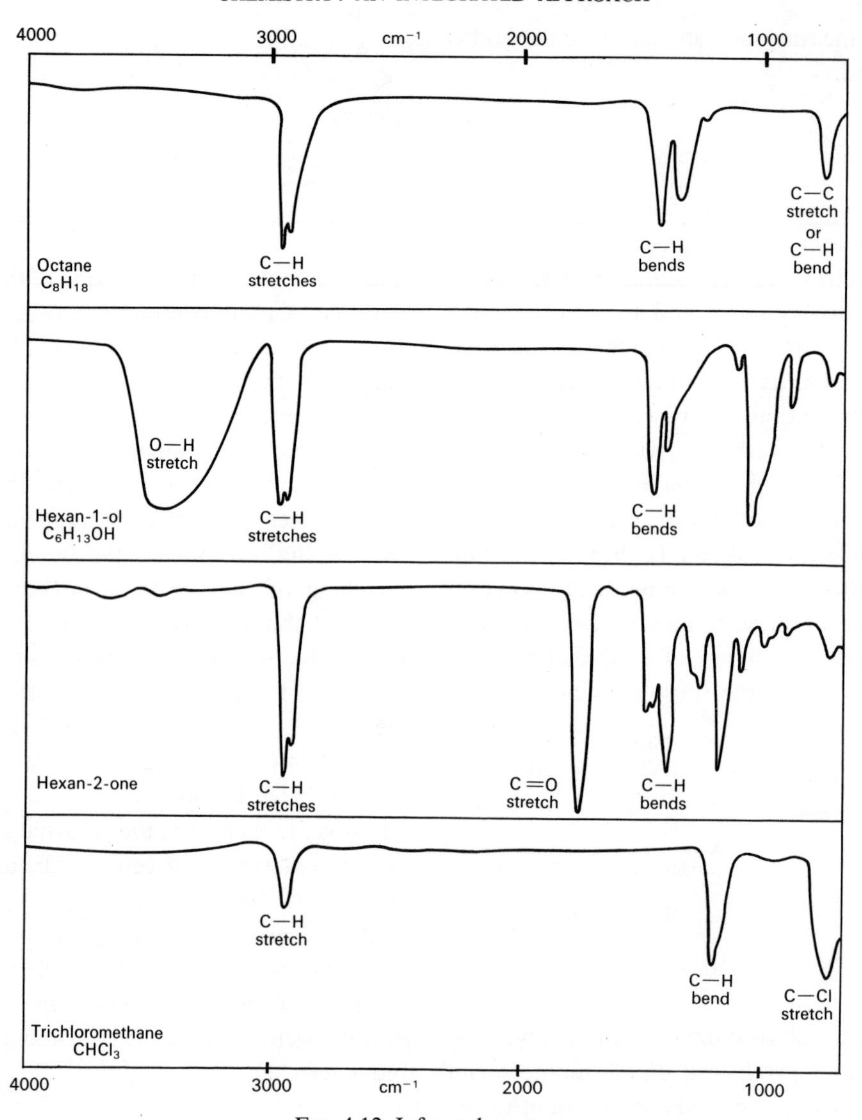

FIG. 4.12. Infra-red spectra.

The stretching vibrations between heavier atoms, and all bending frequencies, tend to occur below 1500 cm^{-1}. Figure 4.12 shows some typical organic infra-red spectra.

In addition to the identification of the functional groups likely to be present in a compound, infra-red spectra can also be used in other ways.

(i) If the structure of a molecule is known, then its infra-red spectrum becomes a "fingerprint" and the presence of a compound in a mixture can be deduced and its concentration estimated.

(ii) The detailed infra-red spectrum of a molecule in the gas phase can give us information about the geometry of that molecule, both with regard to the shape of the molecule and also the bond lengths within the molecule.

(d) Mass spectroscopy

The application of mass spectroscopy to the masses of the isotopes of the elements has already been discussed in Section 1.4. This technique can also be applied to molecules. The molecule to be investigated is converted into a positive ion in the usual way using electron bombardment. If the bombarding electrons are of low energy, then the molecule merely loses an electron of its own and its molecular weight can be discovered from the mass/charge ratio. However, if the bombarding electrons possess higher energy, then the molecule may fracture into several ionic fragments. By detecting these fragments in the usual way, some of the functional groups present in the original molecule can be deduced. Each molecule tends to have a typical **fragmentation pattern** that is produced in this way (Fig. 4.13).

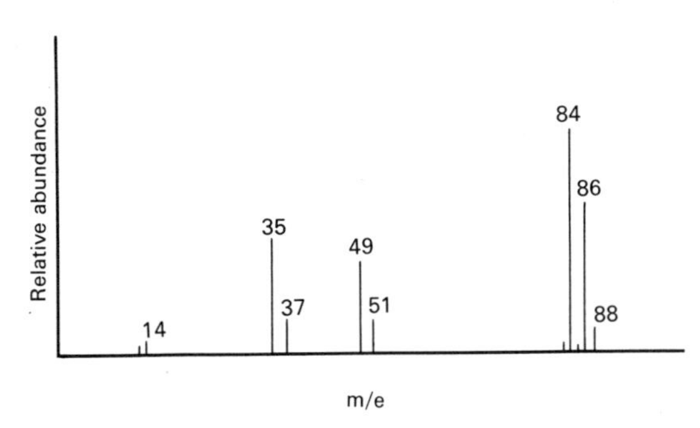

Fig. 4.13. The mass spectrum of dichloromethane CH_2Cl_2.

If the instrument is calibrated using standard substances, such as tetramethylsilane, $(CH_3)_4Si$, or perfluorobutane, C_4F_{10}, the masses of unknown fragments from a molecule can be determined very accurately. Where an element consists of more than one isotope, a pattern will be observed that depends on the relative abundance of the isotopes. Chlorine consists of 75% ^{35}Cl and 25% ^{37}Cl, so that fragments containing one chlorine atom should appear in pairs with a peak height ratio of 3:1 (Fig. 4.13).

(e) Diffraction techniques

X-ray analysis, mentioned above for the determination of macromolecular structures, is suitable also for determining the shape and mode of packing of molecules in molecular solids. The information can, however, be misleading if a substance undergoes a change of structure when it is melted or vaporized. For instance phosphorus(V) chloride (phosphorus pentachloride) has a vapour density corresponding to molecules of formula PCl_5, yet the solid is shown by X-ray analysis to be composed of separate structural units corresponding to the formulae $[PCl_4]^+$ and $[PCl_6]^-$. The structure of some molecules in the gaseous state can be determined by **electron diffraction**: a vapour will not diffract X-rays but it will diffract a beam of electrons, which behaves as if it has wave properties.

(f) Nuclear magnetic resonance

Information about the environment of atoms, particularly hydrogen atoms, can be derived from the **nuclear magnetic resonance** (NMR) spectrum of the compound. An atomic nucleus, like an electron, possesses the property of spin and can therefore behave as a tiny magnet, orienting itself in a magnetic field. The amount of energy absorbed when the orientation of a nucleus changes in a magnetic field is affected by the electrons surrounding the nucleus. For instance in a molecule of ethanol, $CH_3.CH_2.OH$ the pair of hydrogen nuclei on the $—CH_2—$ group will absorb at a different frequency from the three on the $CH_3—$ group, on account of their different environments; the hydrogen atom in the —OH group will absorb at yet another frequency (Fig. 4.14).

If the spectrum is recorded under conditions of high resolution, the peaks are split by neighbouring magnetic nuclei. In ethanol, the $CH_3—$ peak becomes a triplet due to the two hydrogens in the $—CH_2—$ group, and the $—CH_2—$ peak

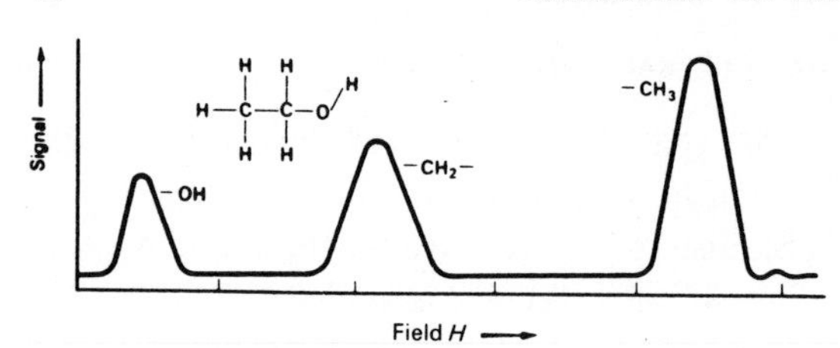

FIG. 4.14. Nuclear magnetic resonance.

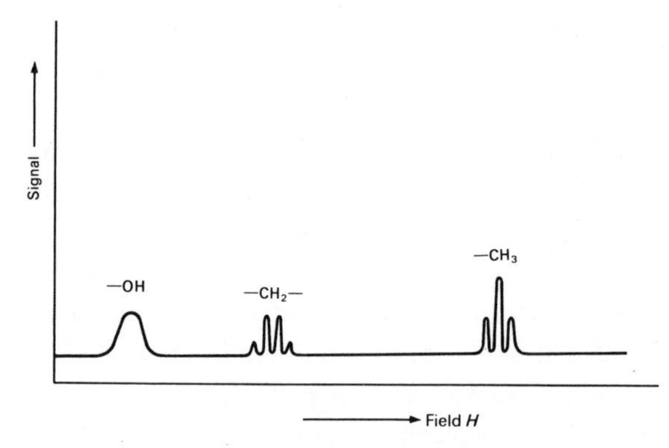

FIG. 4.15. High resolution nuclear magnetic resonance.

is split into a quartet by the three hydrogen atoms in CH_3— (Fig. 4.15).

Nuclear magnetic resonance is especially useful for the investigation of hydrogen atoms, since these cannot readily be detected using X-ray diffraction (Section 4.2).

(g) POLARIMETRY

Light which has passed through a piece of polaroid will be vibrating in one plane only, and is said to be **polarized**. If this light is then passed through a second piece of polaroid set at right angles to the first, then all the remaining light will be absorbed (Fig. 4.16). Two pieces of polaroid that cut out all the light from a source in this way are known as "crossed" polaroids.

Certain molecules possess the property of being able to rotate the plane of polarized light and are said to possess **optical activity**. If an optically active solution is placed between crossed polaroids, the polarized light is rotated through a certain angle which can be determined by rotating the analyser until no light passes through it (Fig. 4.16). Since different wavelengths are rotated to different extents, the light source must be monochromatic.

The **specific rotation**, $[\alpha]$, of a solute is defined by the expression:

$$\frac{\text{Observed rotation}}{\left(\begin{array}{c}\text{Length of solution (dm)}\\ \times\,\text{concentration (g cm}^{-3})\end{array}\right)}$$

For instance, the specific rotation of glutamic acid at 20°C using the light of the D line in the emission spectrum of sodium is given by the expression, $[\alpha]_D^{20} = +30{\cdot}4°$. The positive sign means that light in this case is rotated in the sense of a right-handed screw and glutamic acid is said to **dextrorotatory**. Substances that rotate in the left-handed sense are known as **lævorotatory**.

For a solute to show optical activity it must possess a structure which is **asymmetric**; that is, it must lack any plane, axis or centre of symmetry. This can arise in several ways but the commonest is when a carbon atom is attached to four different atoms or functional groups (Section 6.10).

4.4 Anisotropy

If polarized light passes through certain solids, for example NaCl or CaF_2 crystals, it is unaffected; on the other hand crystals such as calcite, $CaCO_3$, or mica rotate the plane of polarization. An observer looking through crossed polaroids will see light patterns, often with a rainbow effect if the light source is not monochromatic, when the crystal is placed between the polaroids. Crystals able to rotate the polarized light in this way are said to be optically **anisotropic**.

Salt is isotropic because it consists of spherical

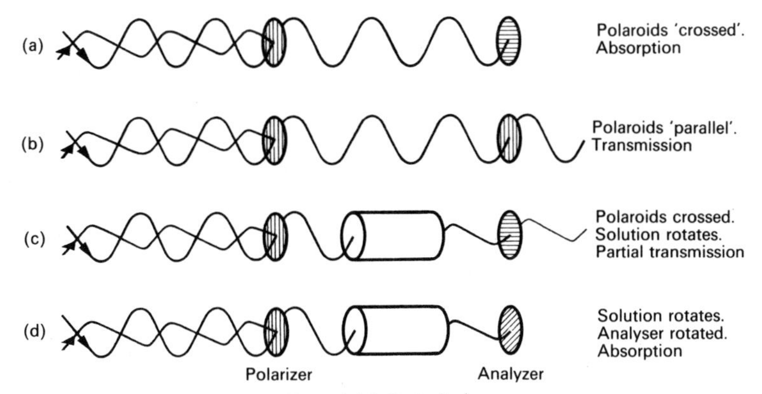

FIG. 4.16. Polarimetry.

ions in a regular cubic arrangement. A salt crystal is the same in all three directions, and for this reason it behaves isotropically. In a calcite crystal, the planar carbonate, CO_3^{2-}, ions are found in parallel layers and this makes the crystal anisotropic, since the physical properties along the layers will not necessarily be the same as the properties across the layers.

The rotation of polarized light by a crystal shows that the crystal is anisotropic. Other physical properties can be used to demonstrate anisotropy and some of these are dealt with in a discussion of the anisotropy of graphite in Section 5.6.

Study Questions

1. What can you deduce from the following?

(a) CH_4 has no dipole moment, but CH_2Cl_2 has.
(b) BF_3 has no dipole moment, but NF_3 has.
(c) CO_2 has no dipole moment, but SO_2 has.
(d) PF_5 has no dipole moment, but IF_5 has.

2. How would *you* distinguish between the two compounds:

```
Cl      Cl              Cl      H
  \    /                  \    /
   C=C         and         C=C        ?
  /    \                  /    \
H       H               H       Cl
```

3. A substance of empirical formula $C_4H_{10}O$ had a molecular weight of 74. I.R. spectroscopy showed that none of the hydrogen atoms was attached to an oxygen atom, and NMR spectroscopy showed two types of hydrogen atom, these being in the ratio of 3:2. What is the structural formula of the substance?

4. (a) Use the Bragg equation to determine the interplanar distances in KCl for the planes represented by Fig. 4.5(a) and 4.5(c).

(b) KCl has the simple cubic (NaCl) structure (Fig. 6.2). Work out which planes are represented by each set of lines in Fig. 4.5.

(c) Work out a value for the K—Cl internuclear distance.

5. Sketch the infra-red spectra you would expect for (a) CCl_4; (b) $H_2C{=}CH_2$; (c) CH_3CH_2OH

6. (a) Account for each of the labelled lines in Fig. 4.13.

(b) What would be the effect on the spectrum of increasing the energy of the bombarding electrons?

7. A compound of molecular formula, C_3H_6O, is thought to be one of the following: $(CH_3)_2C{=}O$, $CH_3.CH_2.CHO$ or $CH_2{=}CH.CH_2OH$.

(a) Which compound would have a very different infra-red spectrum from the other two? What would its spectrum look like?

(b) At low resolution how many peaks would there be in the NMR spectrum of each compound?

(c) Which of the compounds possess a dipole moment?

(d) Which of the techniques would be best for distinguishing between the three possible formulae?

8. A 0·1M solution of glutamic acid ($[\alpha]_D^{20} = +30{\cdot}4°$) was placed in a 10 cm long polarimeter tube. The molecular weight of glutamic acid is 147.

(a) Calculate the concentration of the solution in g cm^{-3}.

(b) What rotation would actually be observed in this case?

9. Suggest why the following crystals are anisotropic:

(a) KCN; (b) $NaNO_3$; (c) CaC_2.

CHAPTER 5

Structure of elements

5.1 The structure of non-metals

The elements in the top right-hand region of the periodic table are non-metals. With the exception of graphite, they are poor conductors of electricity. We can conclude from this that the electrons are not mobile, but must be firmly held in place, and that there are no ions present in the structure. Graphite is interesting in that it does conduct electricity better than any other non-metal, albeit rather badly. Graphite shows some of the physical properties of a metal, and its structure provides some clues about the electronic structure present in metals.

Amongst the non-metals some very striking differences in physical properties are found, pointing to profound differences in structure. Carbon for instance, is almost impossible to melt, yet its neighbour in the periodic table, nitrogen, is extremely difficult to liquefy! Often a single element can show more than one structural form. Phosphorus can be obtained as a volatile "waxy" translucent solid, known as white phosphorus or yellow phosphorus, or it can exist as a hard, dark red, rather involatile solid called red phosphorus. There is also a form called black phosphorus. Yellow phosphorus is a violently reactive substance which inflames spontaneously in the air, yet red phosphorus is fairly unreactive. Two or more distinct structural forms of the same element are called **allotropes**, and the property is called **allotropy**.

5.2 The noble gases

The noble gases represent a simple structural system, and their melting and boiling points are exceptionally low (Fig. 5.1). Helium cannot be solidified at normal atmospheric pressure no matter how much it is cooled.

The solid noble gases consist of atoms which are packed together as closely as possible. Each atom has twelve near neighbours, which is the maximum possible. It is easy to see why atoms form close-packed lattices by doing a simple experiment with a two-dimensional analogy, the *bubble raft*. If a uniformly sized stream of bubbles is allowed to collect on the surface of soap solution in a petri dish, the bubbles will be seen to adhere to one another, most of them being surrounded by six neighbours. This is close packing in two dimensions.

The number of near neighbours which an atom possesses is known as the **co-ordination number**. There are in fact *two* distinct ways in which atoms can arrange themselves in a close-packed lattice of co-ordination number 12, **cubic close-packing** (C.C.P.) and **hexagonal close-packing** (H.C.P.). It is not easy to see the difference in these two forms of packing without constructing a three-dimen-

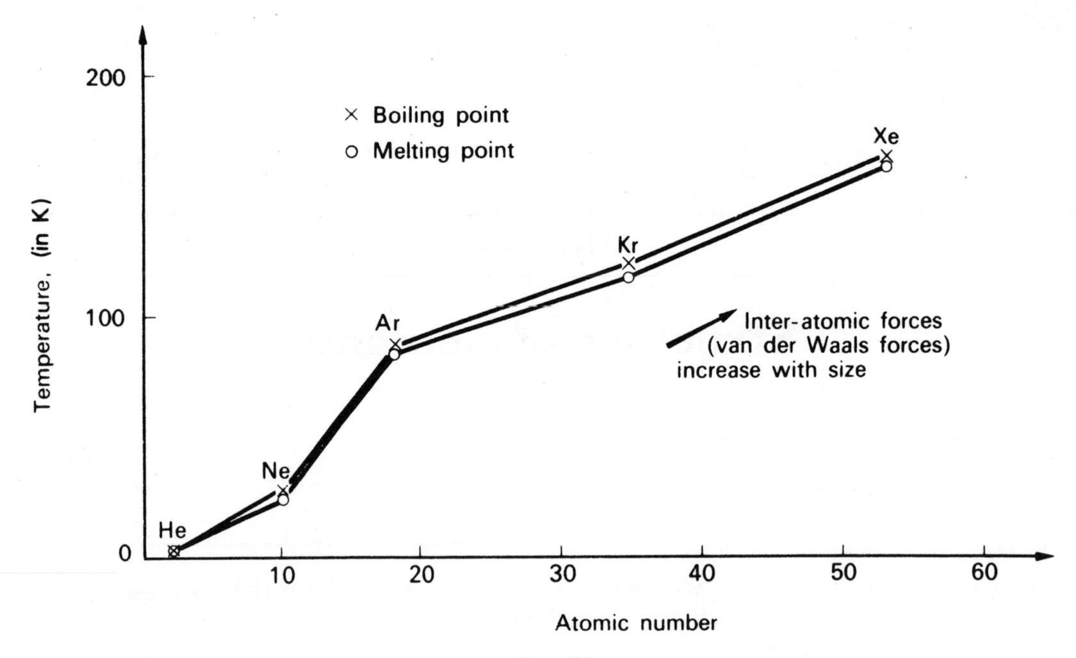

FIG. 5.1.

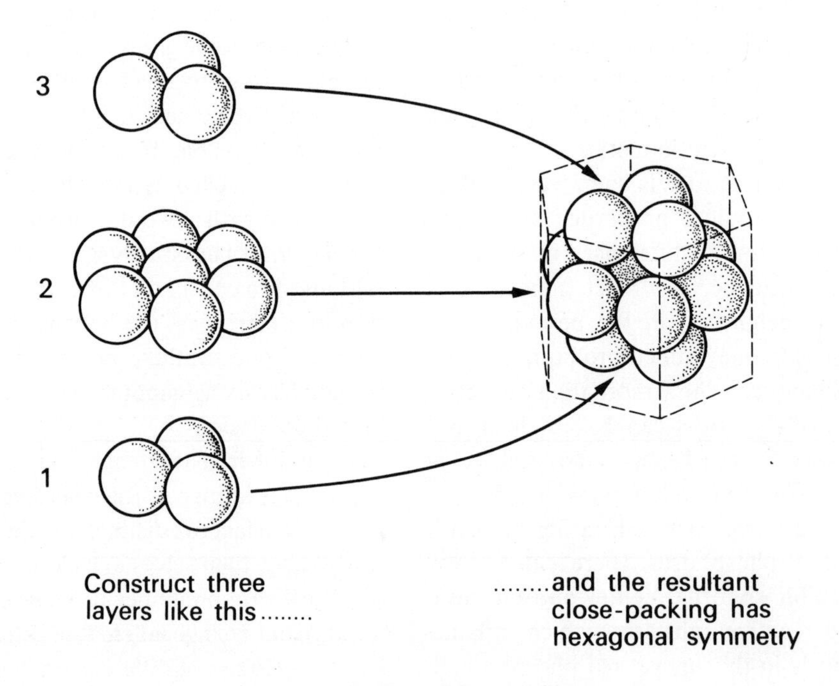

FIG. 5.2. Hexagonal close-packing.

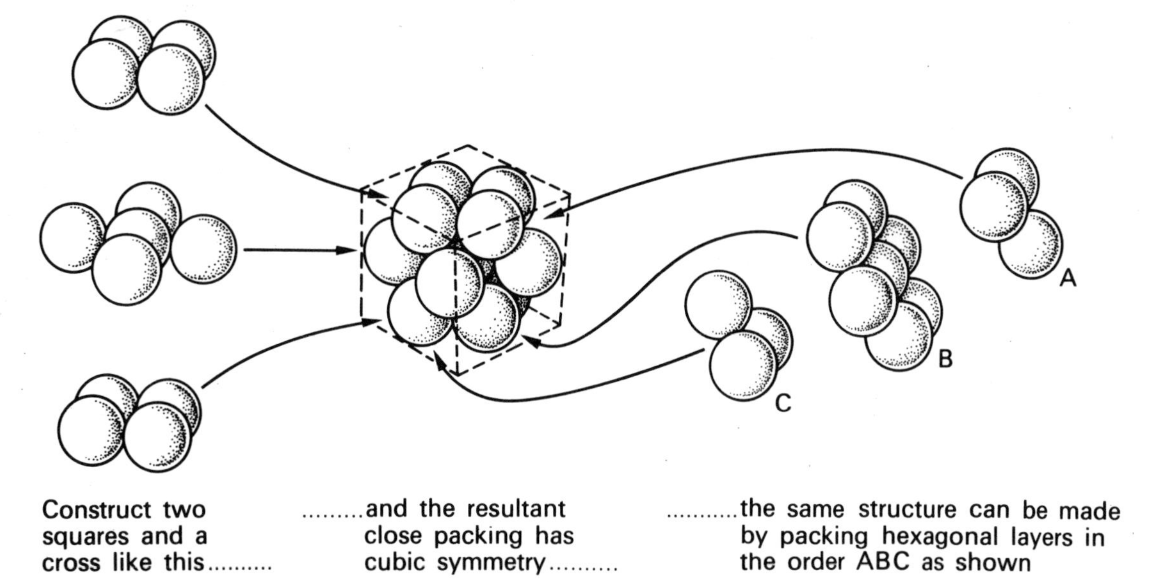

FIG. 5.3. Cubic close-packing.

sional model, for instance with polystyrene spheres. Figure 5.2 shows how this may be done. Join spheres together with pipe-cleaners or cement, to make the layers 1, 2, and 3 in the diagram. Now place layer 2 on top of 1. It will then be found that layer 3 can be added in two different ways, while still contacting the central atom in 2. If 3 is added so that it exactly corresponds to 1, the result will be a H.C.P. arrangement. It is easy to see that twelve spheres now touch the central sphere of layer 2. If layer 3 is rotated through 60 degrees and replaced, it will still be close-packed, but the resultant structure will have a cubic symmetry. The cubic symmetry may not be obvious at first sight but it may be brought out by constructing the C.C.P. form in a different way and comparing it. Construct three layers as in Fig. 5.3 and arrange these one above the other in order. If the resultant structure is tilted, it will be seen to correspond exactly with the form shown in Fig. 5.3.

Noble gas atoms do not join together to form molecules, on account of their already complete energy levels (Chapter 3). Relatively little energy is required to separate the solids into atoms, a typical value being that for argon:

$$\mathrm{Ar(s)} \rightarrow \mathrm{Ar(g)}; \quad \Delta H = 7{\cdot}4\ \mathrm{kJ\ mol^{-1}}.$$

The forces holding the atoms together are known as **van der Waals forces**. These are electrical in nature, even though the atoms themselves have no overall charge. Suppose that a helium atom underwent a momentary distortion of its electron cloud, making it behave as an electrical *dipole* with positive and negative charge slightly separated. This dipole would then *induce* a separation of charge in an adjacent atom, and the two resultant dipoles would then be attracting one another (Fig. 5.4). The induced dipole can induce further dipoles in other atoms and

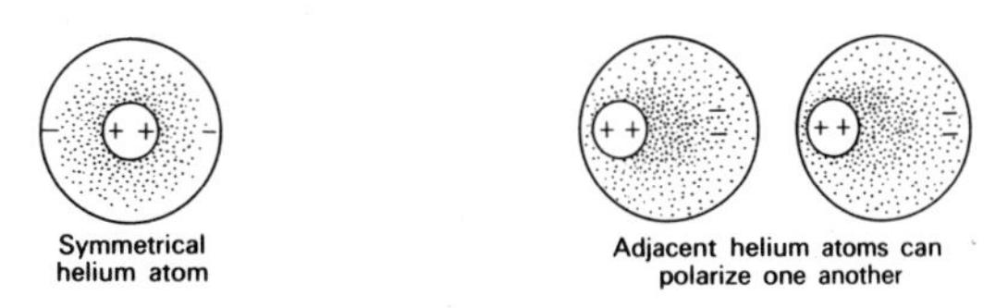

FIG. 5.4. Van der Waals force in a noble gas.

the overall effect will be attraction between all the atoms in solid helium. These dipoles are not permanent: it would be more correct to regard them as continuously oscillating in direction.

5.3 The structure of metals

Metals resemble the solid noble gases in structure in that they have high co-ordination numbers. Most metals have co-ordination numbers of twelve or eight. There the resemblance ends, for the solid noble gases are electrical insulators whereas metals conduct electricity freely. An electric current is a flow of charge, and it would appear that a metal structure contains a system of mobile electrons which can pass from atom to atom without disturbing the regular arrangement of the lattice. In addition, metals generally have the following properties:

(a) They are good conductors of heat and electricity. Electrical conductivity decreases as the temperature is raised.

(b) They are opaque, and are good reflectors of light.

(c) They are **malleable**, i.e. they can be flattened by rolling or hammering.

(d) They are **ductile**, i.e. they can be drawn into wire.

An element which has only some of these properties is not a metal; for instance iodine has a shiny "metallic" appearance but does not conduct electricity and does not have a closely packed structure; it is therefore a non-metal. To be classed as a metal, an element must have an electrical conductivity which decreases as the temperature is raised.

Aluminium is a typical example of a metal with a H.C.P. structure, while copper, silver and gold have C.C.P. structures.

The alkali metals (Li, Na, K, Rb and Cs) and some transition metals (such as Fe, Cr, Mo and W) have a more open structure with a co-ordination number of eight. The arrangement is known as **body-centred cubic** (Fig. 5.5). The more open mode of packing accounts in part for the lightness of the alkali metals, though a more important factor is their high atomic radius.

Observations of a bubble raft show many other features which simulate the arrangement of atoms in metals. There are places where bubbles are out of position, which may be called dislocations. There are groups of regularly arranged bubbles, separated from other groups by lines

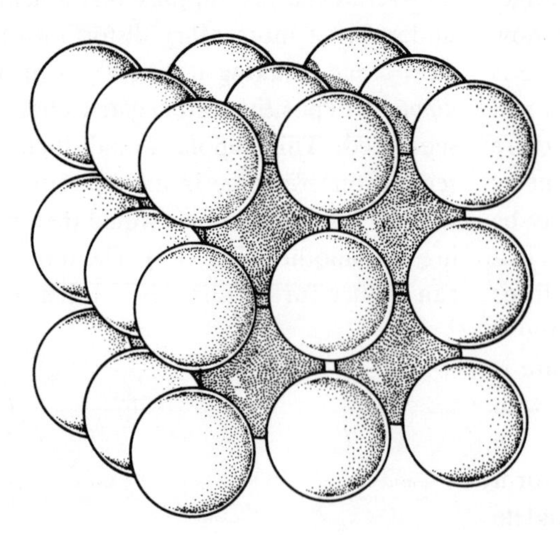

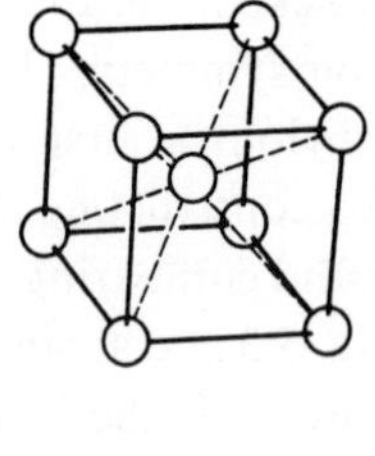

FIG. 5.5. Body-centred cubic structure.

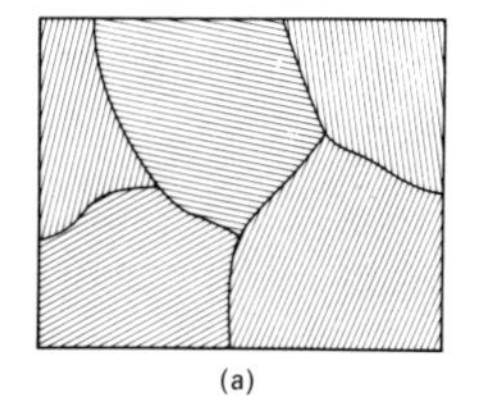
(a)

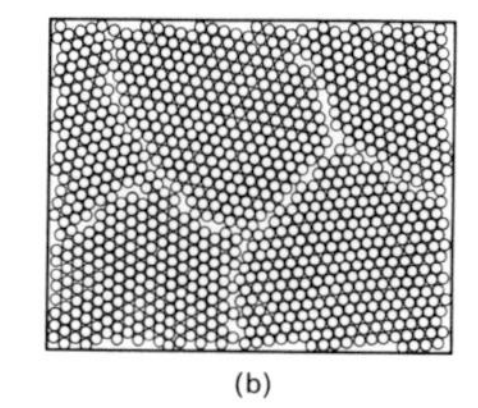
(b)

FIG. 5.6. (a) The appearance of a metal surface with grain boundaries. (b) "Grain boundaries" on a bubble raft.

along which the pattern fails to correspond. This situation occurs in metals, the boundaries being called **grain boundaries** (Fig. 5.6).

In view of the fact that metals contain atoms so regularly arranged, it is surprising at first sight that they do not appear to be crystalline, but all solid metals are in fact crystalline when viewed under magnification. The crystalline appearance may have been obscured by polishing of the surface. You have probably seen zinc crystals without realizing it—the surface of a "galvanized" bucket is patterned due to molten zinc having crystallized out on cooling. Metals such as lead and silver can be made to form beautiful crystals by displacing their ions from solution, in reactions such as:

$$Zn(s)+Pb^{2+}(aq) \rightarrow Pb(s)+Zn^{2+}(aq).$$

Even sodium and potassium form crystals when the molten metal is cooled.

A close-packed metal lattice is readily rearranged, and this accounts for the malleability and ductility of metals.

A well-known characteristic of metals is the way in which their physical properties can be modified by mechanical stress or heat treatment. Metals seem to vary in the way in which they respond: some types of steel become softer on bending, while copper becomes hard and brittle when repeatedly deformed—a phenomenon known as **work hardening**. Mixtures of metal atoms—alloys—often have special properties which the pure elements themselves lack.

Work hardening occurs when mechanical deformation shifts the grain boundaries in such a way as to produce interlocking grains of great mechanical rigidity. Heat treatment also modifies the grain structure by removing many of the defects present in the lattice. A metal is often found to be less deformable when it is impure—it is thought that the impurity atoms hinder the rearrangement of the grain boundaries and dislocations.

5.4 Bonding in metals

Figure 5.7 shows the heat of atomization of some metals plotted against the number of electrons in their valence shell, and it will be seen that for a given period there is an approximately linear relationship. The heat of atomization is the energy required to convert 1 mole of atoms of the solid into free atoms in the vapour state. This suggests that it is the valence electrons in a metal which somehow hold the atoms together—the more electrons the stronger the bond between atoms.

A metal lattice is thought of as containing positive ions held together by a mobile "electron gas". As an approximation sodium can be thought of as an array of Na^+ ions welded together by an equal number of electrons. Magnesium has a higher heat of atomization because it contains Mg^{2+} ions held together by *twice* as many electrons. Aluminium has Al^{3+} ions held together by three times the number of electrons.

The observed densities of metals are in accordance with this model. Table 5.1 gives figures for what might be called *metallic radii*, that is half the internuclear distance between near neighbours in a metal lattice, or the radius which the atoms would have if we assume them to be hard spheres in contact.

Where ions are isoelectronic, for example Na^+, Mg^{2+} and Al^{3+}, the ion with the highest

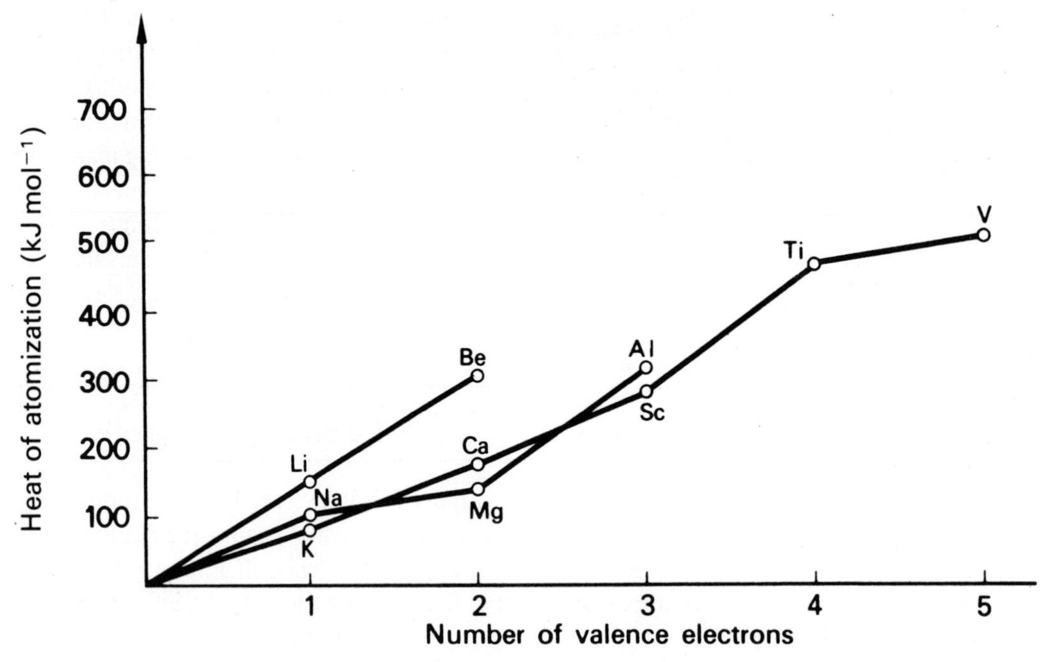

FIG. 5.7. The relation between number of valence electrons and binding energy.

TABLE 5.1
METALLIC RADII (nm)

Li 0·157	Be 0·112										
Na 0·191	Mg 0·160	Al 0·143									
K 0·235	Ca 0·197	Sc 0·164	Ti 0·147	V 0·135	Cr 0·129	Mn 0·137	Fe 0·126	Co 0·125	Ni 0·125	Cu 0·128	Zn 0·137

nuclear charge will be the smallest (Fig. 5.8). The weakly-bonded alkali metals, Group IA, are therefore the least dense metals in their respective periods. Lithium will even float on the oil in which it is customarily stored, and a large lump of lithium is so light that it feels as if it is hollow.

The conductivities of the alkali metals, of both heat and electricity, are extremely high. Conductivity is the result of the electrons being mobile and we can imagine that the electrons of Group IA metals are more mobile because the energy levels are less "crowded". This is an oversimplified picture, but it helps us to see why sodium, with only one mobile electron per atom is a better conductor of heat and electricity than calcium. Sodium is in fact used in nuclear reactors: being both easily melted and a good conductor of heat it can be used for heat transfer. It

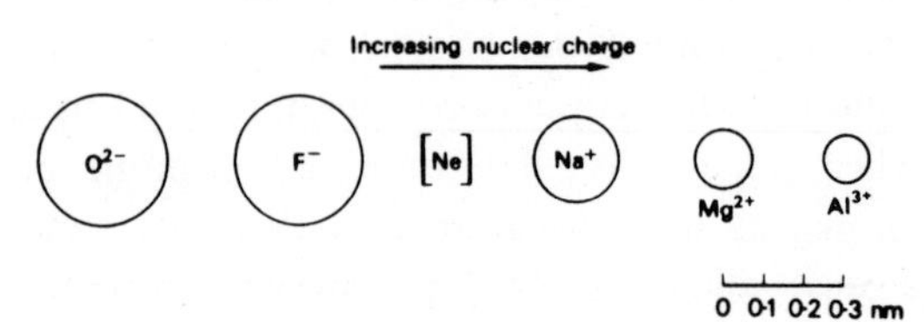

FIG. 5.8.

has also recently been proposed as a conductor to replace copper in special electricity cables.

5.5 Structure of non-metals

Excluding the noble gases which do not form bonds, the structure of non-metals is the result of electron-pair bonds forming between atoms. Two factors determine the number of bonds formed:

(i) The number of electrons in the valence shell (section 3.10).

(ii) The number of electrons which the valence shell can accommodate altogether.

A useful empirical rule is that if there are N electrons in the valence shell, the atoms in non-metallic elements form $8-N$ bonds with their neighbours. In the sections which follow, some non-metallic structures will be dealt with in more detail, providing illustrations of the rule.

5.6 Diamond and graphite

Carbon shows two distinct allotropes, diamond and graphite. It is readily shown by chemical tests, or by using the mass spectrometer, that they are the same element, and differ only in structure. Diamond is denser than graphite. Their physical properties are compared in Table 5.2.

TABLE 5.2

	Diamond	Graphite
M.p. (°C)	—	3730
B.p. (°C)	—	3830
Density g cm^{-3}	3·51	2·26
Internuclear distance	0·154 nm	0·142 nm (0·335 nm between layers)

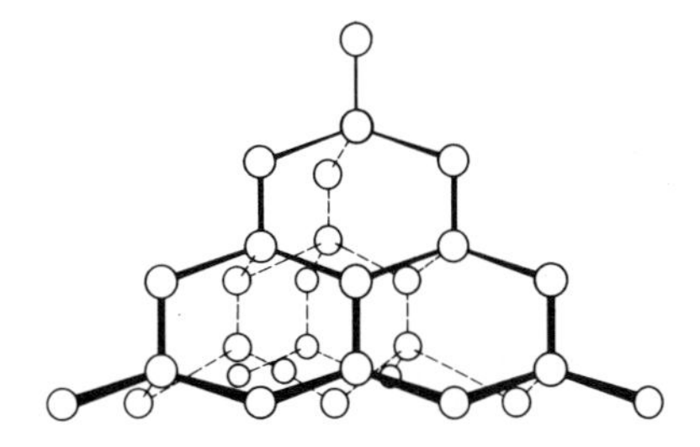

FIG. 5.9. The structure of diamond.

Figure 5.9 shows the structure of diamond—each carbon atom has *four* near neighbours, i.e. carbon has a co-ordination number of four in diamond.

Diamond is an electrical insulator, its electrons being rigidly held between atoms, with four electron-pair bonds per atom. The whole solid is a macromolecule of enormous rigidity and hardness. It is quite unlike a metal in physical characteristics. Apart from its insulating properties, it is transparent, and if attempts are made to bend a crystal or alter its shape it fractures in specific directions called **cleavage planes**. A cleavage plane is found to lie in the direction which cuts across the least number of bonds. It requires energy to fracture the crystal, and the crystal will come apart in the direction which requires least energy (Fig. 5.10).

Graphite is quite unlike diamond. It looks "metallic" although this can be deceptive, and many non-metallic substances can be mistaken for metals just because they happen to be shiny

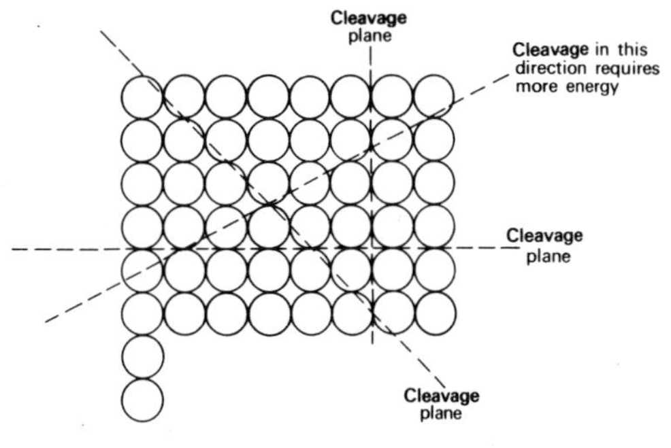

FIG. 5.10.

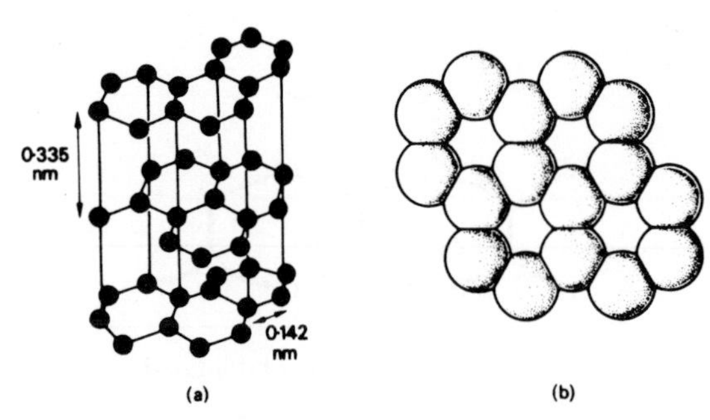

FIG. 5.11. The structure of graphite (a) showing that the distance between the atoms in different layers is greater than between adjacent atoms in a layer; (b) showing the hexagonal arrangement of one layer.

in appearance. A more convincing metallic property is that it conducts electricity. In other respects graphite is quite unlike a metal. Structurally it has a co-ordination number of three, and the atoms are arranged in hexagonal sheets, corresponding to a lattice rather like wire netting (Fig. 5.11).

Graphite is a much softer substance than diamond—indeed it is used as a solid lubricant, whereas diamond is used as an abrasive. Clearly there are weaknesses in the structure. The distance between adjacent sheets of carbon atoms is much greater than the distance between bonded carbon atoms, and the forces holding adjacent sheets together are much weaker than those holding the atoms together within a sheet.

If we assume that three electron-pair bonds are formed per atom in graphite, then each atom has a spare electron left over. Two properties of graphite must be explained: first, that it *conducts electricity* and second, that it is a diamagnetic substance. The spare electrons must be paired off in some way, otherwise graphite would be attracted by a magnet. However, the spare electrons must be free to wander about the lattice, rather than being rigidly held between pairs of atoms. These spare electrons are **delocalized**, just as they are in a metal. Electron pairs which remain in the vicinity of one atom, or in the space between a pair of atoms, are said to be localized. Graphite is a relatively poor conductor of electricity because the electrons can only flow freely within a given sheet of atoms—they cannot readily hop from one sheet to another. Metals are better conductors than graphite because their electrons are mobile in three dimensions.

The electrical conductivity of graphite is poor because of its anisotropy. Table 5.3 summarizes some of the anisotropic properties of graphite.

TABLE 5.3

Property	Across the sheets	Parallel to the sheets
Electrical conductivity	$4\ \Omega^{-1}\ cm^{-1}$	$5\times10^{3}\ \Omega^{-1}\ cm^{-1}$
Thermal expansion	$2\times10^{-5}\ K^{-1}$	$7\times10^{-7}\ K^{-1}$
Thermal conductivity	$2\times10^{2}\ W\ m^{-1}\ K^{-1}$	$2\ W\ m^{-1}\ K^{-1}$

So-called *amorphous carbon* is not really a third allotrope. A detailed examination of its structure shows it to be microcrystalline graphite. However, since its surface area is so very large for its overall mass it has very different physical properties from ordinary graphite, and it is often treated as a third allotrope. Special forms of carbon, such as carbon fibre and vitreous carbon, have also been prepared.

5.7 Nitrogen and phosphorus

The very great contrast between these two elements is in a sense a violation of the periodic law. Phosphorus is much more reactive, and its two common forms have structures which are different from that of nitrogen. Both elements can form three covalent bonds per atom, but they form them in quite different ways. Nitrogen, as we saw in Chapter 3, has a co-ordination number of *one*, that is, it forms diatomic molecules containing triple bonds. Phosphorus solves its energy-level problem in quite a different way. It appears that double and triple bonds of the type met with in nitrogen do not form at all easily, for phosphorus allotropes generally have a co-ordination number of three. The simplest way of explaining this is to postulate three electron-pair bonds and one lone pair per atom. Figure 5.12 illustrates the energy levels in an isolated phosphorus atom.

In white and black phosphorus a given phosphorus atom is connected to its three neighbours in a pyramidal manner, leaving the lone pair of electrons projecting away from the pyramid (Fig. 5.12).

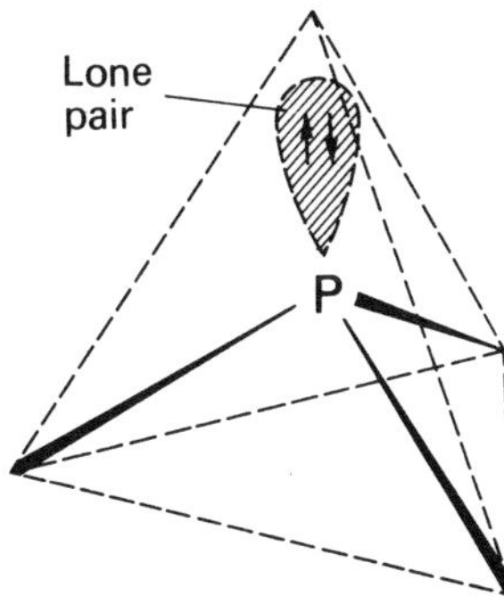

FIG. 5.12. The environment of a phosphorus atom.

White phosphorus melts very easily (m.p. 44°C), and vaporizes as tetrahedral clusters of four atoms, that is, P_4 molecules (Fig. 5.13(a)). These molecules are present in the solid state. It is not surprising that white phosphorus is "waxy" since the forces between its molecules are about the same as those in a hydrocarbon wax. Red phosphorus on the other hand exists as some sort of giant structure, which is less volatile, and similar to the structure of black phosphorus shown in Fig. 5.13(b). Notice that in each case the environment of a given phosphorus atom corresponds to that shown in Fig. 5.12.

When describing the energy levels in phosphorus molecules the 3*d* levels must be taken into account. At this stage it will suffice for us to treat the bonds between phosphorus atoms as if they were simple electron-pairs, but this is a slight over-simplification. In fact it is the very existence of available *d*-orbitals in phosphorus, arsenic, antimony and bismuth which probably accounts for the fact that these elements are so different from nitrogen. There are many differences between first row (lithium to neon) and second row (sodium to argon) elements which can be attributed to the non-existence of a 2*d* sub-shell in the first row elements.

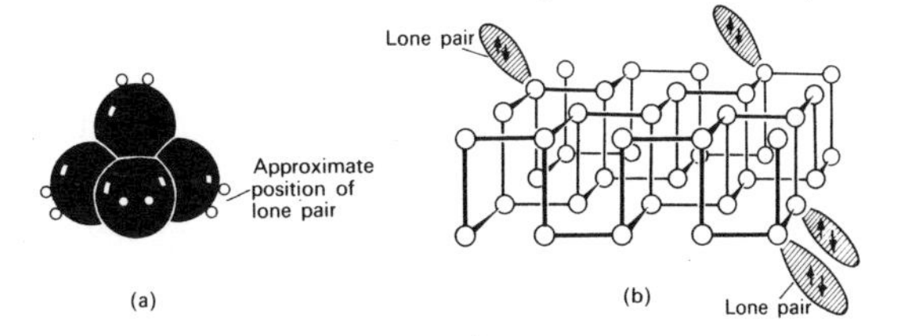

FIG. 5.13. (a) White phosphorus. (b) Black phosphorus: for clarity, only a few lone pairs are shown.

5.8 Oxygen and sulphur

The structural relationship between nitrogen and phosphorus is repeated here between oxygen and sulphur. Oxygen has a co-ordination number of *one* (but see ozone, below) while sulphur has a co-ordination number of *two*. Sulphur can

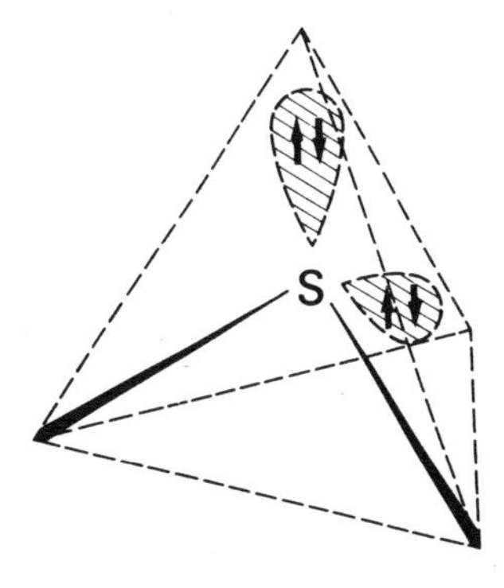

FIG. 5.14. The environment of a sulphur atom.

be expected to have two electron-pair bonds and two lone pairs (Fig. 5.14).

Given this environment for individual sulphur atoms, all sorts of structures are theoretically possible, and so it is not surprising that many allotropes of sulphur exist. The commonest basic units are:

(i) a puckered ring of eight atoms, S_8 (Fig. 5.15);

(ii) a long zig-zag chain of many atoms, made by S_8 rings opening out and the ends joining up.

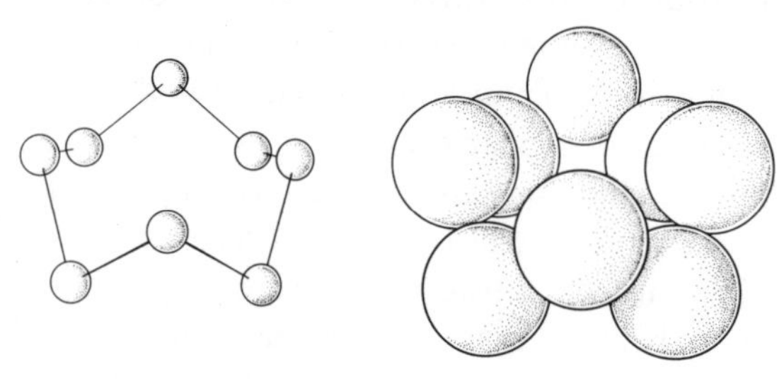

FIG. 5.15. S_8 rings which occur in rhombic and monoclinic sulphur.

The S_8 rings can pack together in various crystalline forms, two common ones being *rhombic sulphur*, stable below 96°C, and *monoclinic sulphur*, stable above this temperature. The temperature of change is called the **transition temperature**—it is analogous to a melting point or boiling point, but the changes from one phase to another, being between two solids, takes place more slowly.

Changes between two solid phases are bound to be slow, because translational motion of molecules is severely restricted. Once a "nucleus" is established on which the new structure can build up, transition from one crystalline form to another can take place quite rapidly.

If rhombic or monoclinic crystals are heated to just above their melting point, S_8 molecules persist in the liquid state. Further heating disrupts the rings, and the molecules *polymerize*, that is, many molecules join together to form a long chain. In the same way that small ethylene molecules C_2H_4 can be polymerized to form the plastic "polythene", *plastic sulphur* is produced when the strongly heated liquid is rapidly cooled by pouring into cold water. Unfortunately plastic sulphur has little practical use—articles manufactured from it would rapidly disintegrate at room temperature as the rings closed up again forming micro-crystalline rhombic sulphur!

If sulphur is slowly heated all sorts of changes take place in the viscosity and colour of the liquid. On boiling, a dark red vapour is produced containing various molecules, S_8, S_4 and S_2, the proportion of each depending on the temperature.

Oxygen, as we normally know it, exists as diatomic molecules, O_2. The absorption of ultra-violet radiation in the upper atmosphere dissociates many of these molecules and *atomic* oxygen is produced.

$$O_2(g) \xrightarrow{h\nu} 2O(g); \quad \Delta E = +498 \text{ kJ}.$$

The energy for this reaction can alternatively be provided by passing an electric discharge. At room temperature and normal atmospheric pressure, atomic oxygen is too reactive to persist, and some O_3 molecules are produced:

$$O(g)+O_2(g) \rightarrow O_3(g).$$

O_3 is called *ozone* or *trioxygen*. It has a characteristic smell, which can often be detected in the region of electrical equipment. Contrary to popular belief it is far from health-giving, and, being highly reactive, is harmful to living tissues in high concentrations. The "ozone" smell at the seaside can generally be attributed to rotting seaweed!

5.9 Borderline elements

Among the elements two distinct structural types can be distinguished: on the one hand metals of high co-ordination number, and on the other non-metals of low co-ordination number. Near the centre of the *p*-block of the periodic table there exist many elements with intermediate properties and structures. For instance, arsenic, antimony and bismuth have layer structures which are fairly close-packed, but with only three near neighbours, and the structure becomes more "metallic" down Group V. This is in accordance with the 8–*N* rule, and would suggest that covalent bonds form between atoms. However, the elements conduct electricity and their electrons are therefore mobile. The truth is that the bonding is intermediate between metallic and covalent.

A similar state of affairs exists with the elements of Group IVB, which exhibits a change-over in structure and physical properties with germanium on the borderline. Borderline elements like germanium are **semiconductors**.

TABLE 5.4

SOME PROPERTIES OF THE GROUP IVB ELEMENTS

Element	Melting point	Bond length (nm)	Electrical conductivity	Remarks
Carbon	3730°C (graphite)	0·154	non-conductor	These data refer to diamond-like forms of the elements.
Silicon	1420°C	0·234	semi-conductor	
Germanium	937°C	0·244	semi-conductor	
Tin	232°C	0·280 (grey) 0·302 (white)	semi-conductor (grey) conductor (white)	Grey form with diamond structure. Changes to white metallic form at 13°C.
Lead	327°C	0·350 (metallic)	conductor	Diamond-like form does not exist.

Study Questions

1. How are the properties of the following elements related to their structure: (a) diamond, (b) graphite, (c) plastic sulphur, (d) copper, (e) tin?

2. In each of the following sets, arrange the elements in order of increasing melting and boiling points:

(a) Ar, Br, Cr.
(b) Na, Ne, Ni.
(c) Cl_2, Br_2, I_2.
(d) Li, Na, K.
(e) K, Ca, Sc, Ti.

Explain any trends you observe.

3. Give examples of elements which have the following co-ordination numbers: 1, 2, 3, 4, 8 and 12.

4. Place the elements W, X, Y and Z as accurately as you can in the periodic table:

(a) W is a red solid containing puckered rings of atoms.

(b) Solid X has a close-packed structure and does not conduct electricity.

(c) Y has a body-centred cubic structure and a density of 0·53 g cm^{-3}.

(d) Z consists of puckered sheets in which each atom is bonded to three neighbours, the sheets being held together by metallic bonding.

5. Solid aluminium and solid argon crystallize with similar structures.

Explain why:

(a) Al is a much better conductor than Ar(s).
(b) Al has a much higher boiling point than Ar.

6. (a) What three factors affect the density of a solid element?

(b) The alkali metals all have body-centred cubic structures. Explain why Na is denser than Li.

(c) K is less dense than Na. Suggest a reason why this should be so.

(d) How would the density of Na be affected if it were face-centred, rather than body-centred, cubic?

CHAPTER 6

Structure of compounds

6.1 Ionic lattices

It was shown in Chapter 3 that if a potassium atom and a fluorine atom are brought together, complete transfer of an electron is energetically favoured. The presence of ions can actually be demonstrated in the molten state by electrolysis, but their presence in the solid can only be inferred from physical properties.

The halides of the alkali metals provide the simplest possible ionic systems. A few facts about them are summarized below:

(i) They all have high melting and boiling points. This suggests that they are giant lattices, not separate molecules in the solid phase. The melting and boiling points show regular trends related to the position of the elements in the periodic table. Figure 6.1 shows how the boiling points vary.

(ii) They are all soluble in water, the solubility being least for the atoms of lowest atomic weight. Lithium fluoride, which represents the lowest atomic weight combination of all, is only sparingly soluble (0·13 g per 100 g of water at 20°C).

(iii) Their crystals are cubic in shape. X-ray analysis shows that in nearly all cases the co-ordination number is six. Each alkali metal atom has six halogen atoms as near neighbours, and vice versa—the 6:6 structure. Caesium chloride, bromide and iodide are cubic but with a co-ordination number of eight—the 8:8 structure. These structures are not close-packed; a coordination number of 12 would be impossible for ions of opposite charge.

One reason for the different structures of NaCl and CsCl is that their **radius ratios** are different. It is possible to pack more chloride ions (radius about 0·181 nm) around the larger caesium ion (radius about 0·167 nm) than around the smaller

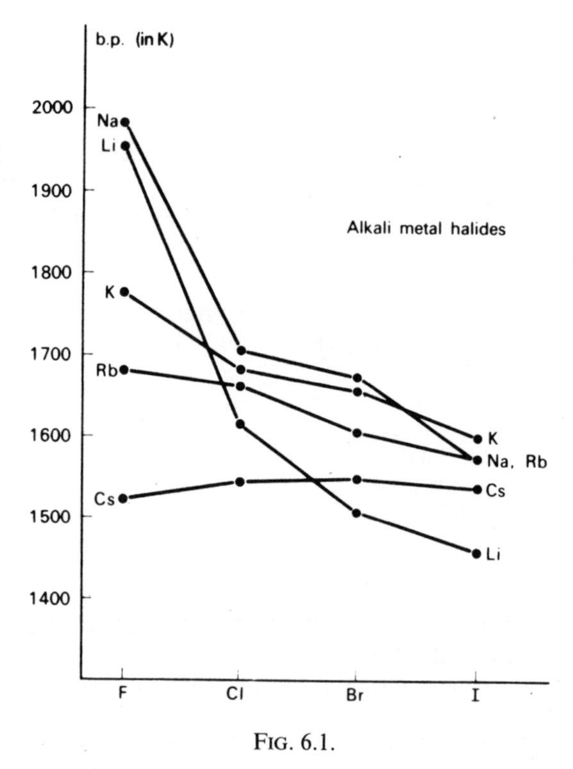

FIG. 6.1.

sodium ion (radius about 0·098 nm). When the cation gets very small, the coordination becomes 4:4 as in ZnS (Zn^{2+} = 0·074 nm; S^{2-} = 0·185 nm). Small coordination numbers are rarely observed with truly ionic compounds, and the bonding in zinc sulphide is partly covalent.

(iv) The solids conduct electricity very badly. Therefore there can be no mobile electrons in the lattice. In fact there is very slight electrical conductance arising from the presence of mobile M^+ ions.

(v) When molten or when dissolved in water the alkali metal halides will all conduct electricity, with electrolysis, i.e. they are electrolytes. This again is not *proof* that ions exist in the solid, and we shall later meet cases where it is more reasonable to suppose that ions exist in the melt but *not* in the solid. Nevertheless the ionic model gives a good interpretation of the behaviour of the alkali metal halides.

(vi) Measurement of internuclear distances between pairs of ions, by X-ray analysis, shows that the ions *behave roughly as if each had its own separate radius*. For instance the internuclear distances have been measured in sodium chloride, sodium bromide, potassium chloride and potassium bromide with the following results:

Salt	Distance between nuclei (nm)	Salt	Distance between nuclei (nm)
NaCl	0·2814	NaBr	0·2981
KCl	0·3139	KBr	0·3293

$$0{\cdot}2981 - 0{\cdot}2814 = 0{\cdot}0167 \text{ nm},$$
$$0{\cdot}3293 - 0{\cdot}3139 = 0{\cdot}0154 \text{ nm}.$$

The differences, approximately 0·016 nm, can be attributed to the difference in radius between Cl^- and Br^-.

$$0{\cdot}3139 - 0{\cdot}2814 = 0{\cdot}0325 \text{ nm}$$
$$0{\cdot}3293 - 0{\cdot}2981 = 0{\cdot}0312 \text{ nm}$$

The differences, approximately 0·032 nm, can be attributed to the difference in radius between Na^+ and K^+.

Not even X-ray analysis enables us to measure the radii of *separate* ions, and the discrepancy of the figures shows that the "radius" of an ion appears to vary with its environment. Various methods of calculating single ionic radii are available, and the following figures are generally accepted as the radii of the ions assuming 6:6 co-ordination:

Na^+ 0·098 nm	Cl^- 0·181 nm
K^+ 0·133 nm	Br^- 0·196 nm

The 6:6 structure is shown in Fig. 6.2 and the 8:8 structure in Fig. 6.3.

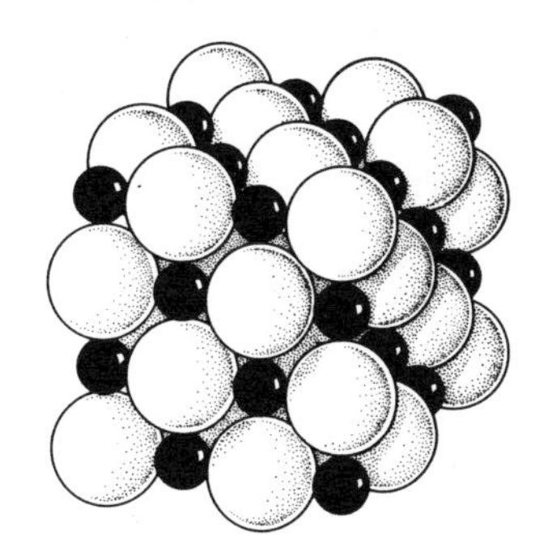

FIG. 6.2. The 6:6 (sodium chloride) structure.

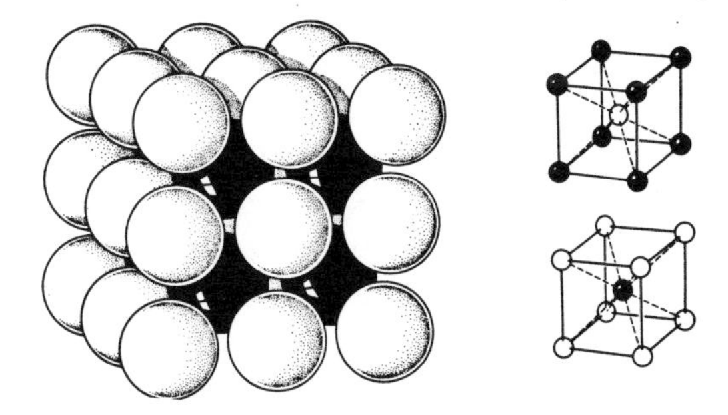

FIG. 6.3. The 8:8 (caesium chloride) structure.

(vii) The crystals are brittle, and readily fractured along planes of cleavage. This property does not in itself point specifically to the presence of ions, but it does suggest that a close-packed

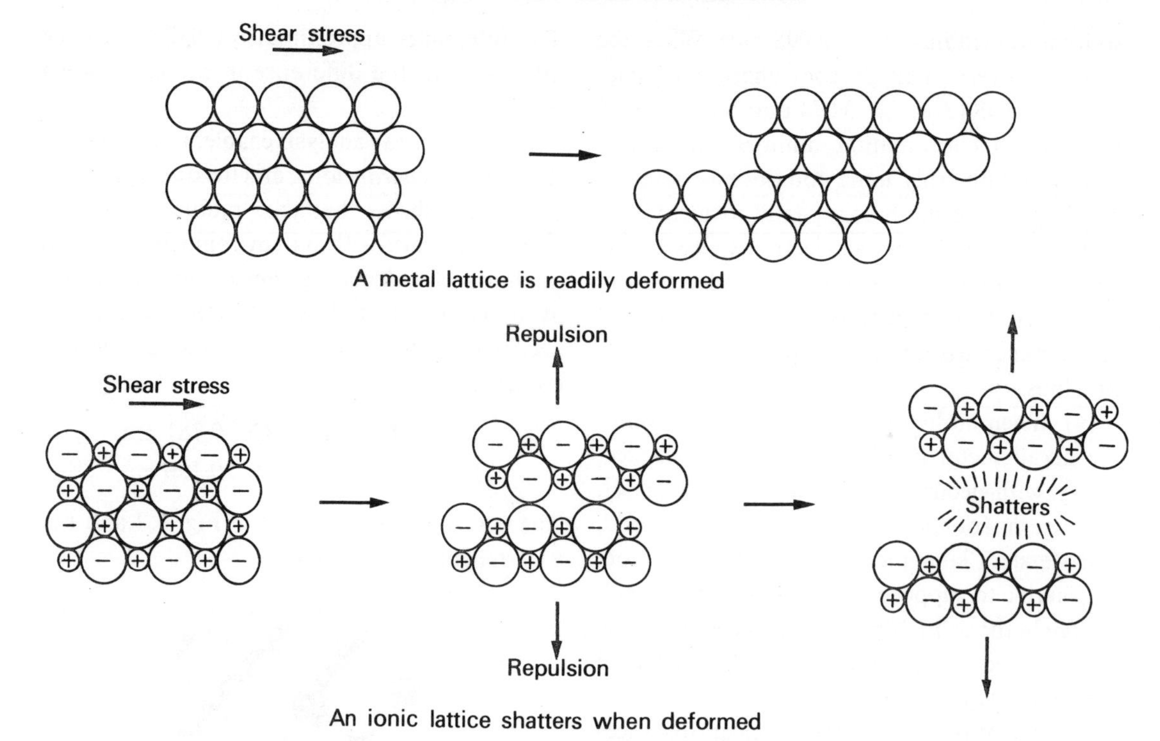

FIG. 6.4.

metallic arrangement is absent. The fracture of a crystal is readily interpreted on the ionic model. If adjacent planes of ions are dislocated, repulsion between ions of similar charge will force the layers apart (Fig. 6.4).

The above list gives *some* of the reasons why we believe that a salt like sodium chloride consists of ions. In Chapter 9 we shall test this hypothesis quantitatively, and it will be shown that the ionic model fits the energy data extremely well, and gives a good interpretation of observations such as solubility in water and heat of solution. Other salts, for instance the halides of zinc, do not give such good agreement, suggesting that the ionic model is less applicable. Their structures are less simple, and cannot be interpreted as being due to the packing together of charged spheres. Some ions are thought to be more readily polarized (distorted) than others, causing the structures to be modified (Fig. 6.5). Polarization occurs most readily in compounds that contain small, highly charged cations and large anions.

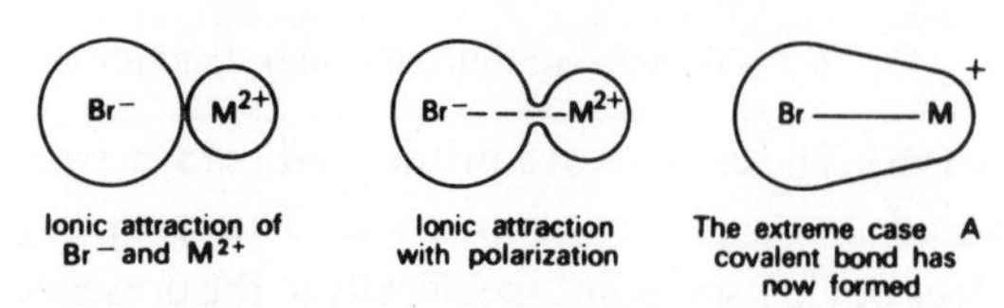

FIG. 6.5.

6.2 Electrolysis

The properties of alkali metal halides are best interpreted by assuming that each ion carries a single charge. It is found by experiment that when a molten alkali metal halide is electrolysed, the same number of moles of atoms of element are always liberated provided the total *charge* passed through the cell is the same.

The liberation of one mole of atoms of an alkali metal, or of a halogen, always requires 96 490 coulombs of charge. One coulomb is the charge which flows when 1 ampère of current flows for one second. To liberate one mole of atoms of an alkali metal requires one ampère for 96 490 seconds, or about $26\frac{1}{2}$ ampère-hours. (One 2-volt accumulator of the type commonly used for laboratory d.c. supplies will yield approximately this amount of charge.) The quantity 96 490 coulombs is called one **faraday**.

If a salt like lead(II) bromide* is electrolysed, *two* faradays of charge are required to liberate one mole of lead atoms. This is interpreted by postulating that the lead ions carry two positive charges whereas the alkali metal ions only carry one:

Na^+	+	e^-	$\rightarrow$	Na(l)
ions in melt		one faraday		one mole of sodium atoms

Pb^{2+}	+	$2e^-$	$\rightarrow$	Pb(l)
ions in melt		two faradays		one mole of lead atoms

Quantitative observations on the amounts of substances liberated when an ionic melt or solution is electrolysed, were made by Michael Faraday (1832). Faraday coined the term "electrochemical equivalent" for the mass of an element liberated by 1 coulomb in electrolysis, and framed three laws of electrolysis to describe his observations. The term electrochemical equivalent has now largely fallen into disuse, and we can re-state Faraday's second law in more modern terms.

Faraday's laws of electrolysis (modern form):

(1) The mass of a substance liberated at an electrode during electrolysis is directly proportional to the charge which flows through the cell.

(2) The quantity of electricity needed to liberate one mole of atoms of an element at an electrode is 96 490 coulombs, or a small whole-number multiple of this amount.

(3) Chemical changes during electrolysis occur at the electrodes, and not within the body of the cell.

Electrolytes, unlike metals, do not obey Ohm's law: a certain minimum potential difference has to be applied to an electrolyte in order for current to flow at all. Below this minimum, loosely termed **discharge potential**, no electrolysis occurs. This suggests that a certain minimum energy has to be attained before the electrolyte will split up into its simpler electrolysis products. We can make a rough estimate of this energy by measuring the discharge potential, after making due allowance for side effects such as the extra voltage needed to overcome the internal resistance of the electrolytic cell.

If one coulomb of charge is raised through a potential of one volt, one *joule* of work is done. To decompose one mole of sodium chloride, NaCl, completely into its elements, 96 490 coulombs will be required, at a minimum discharge potential of 4·0 V.

$$\begin{aligned}\text{Hence energy required} &= 96\,490 \times 4{\cdot}0 \\ &= 3{\cdot}86 \times 10^5 \text{ J mol}^{-1} \\ &= 386 \text{ kJ mol}^{-1}\end{aligned}$$

This figure is only approximate, since various inaccuracies can occur when using a discharge potential as a measure of energy, but it illustrates the magnitude of the quantities involved.

6.3 Conductivity of electrolytes

The electrical conductivity of an electrolyte is generally less than that of a metal because the

* The Roman numeral (II) is included in the name because lead does not *always* exist as ions with two positive charges. The numeral indicates that in this case the ion has a charge of +2.

current is carried by bodily movement of ions rather than by electrons. Measurements of conductivity can give useful information about the nature of the ions in an electrolyte and the number of ions present in a given solution.

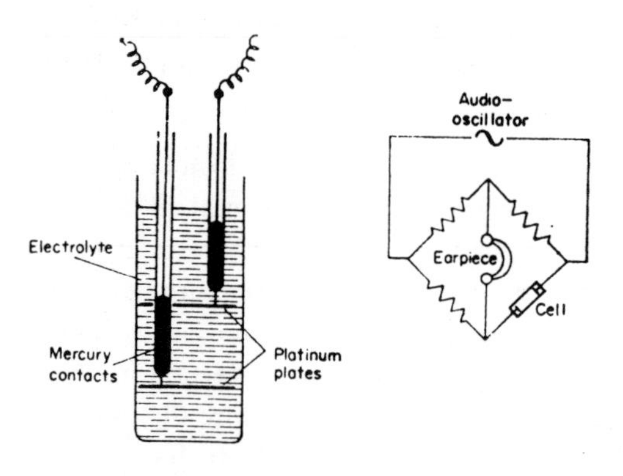

FIG. 6.6. Conductivity cell and Kohlrausch bridge.

The conductivity of a solution is measured in a cell of known dimensions, called a conductivity cell, in a Kohlrausch bridge circuit (Fig. 6.6). This is similar to a Wheatstone bridge but with the following modifications:

(a) a.c. has to be used instead of d.c. to prevent electrolysis taking place. The a.c. source is usually an oscillator.

(b) An ordinary moving coil galvanometer cannot be used for a.c. For rough work a telephone earpiece is used, and the operator balances the bridge for minimum sound. For more accurate work a cathode ray oscilloscope is used.

It is convenient to define a quantity known as the **resistivity** of a substance. This is the electrical resistance, in Ω, of a cube of material of unit side, the potential difference being applied to two opposite faces of the cube. The S.I. unit of resistivity is the ohm metre (Ω m), this being the resistance of a metre cube; however, in the section which follows a more convenient unit is the ohm cm, or the resistance of a centimetre cube.

In a conductivity cell,

$$\text{resistivity} = k \times \text{measured resistance}$$

where k is a constant of proportionality that depends upon the dimensions of the cell and is called the **cell constant**. The cell constant is best determined by measuring the resistance in the cell of an electrolyte of known resistivity, such as potassium chloride solution of known concentration.

The reciprocal of resistivity is called the **conductivity** of the cell.

6.4 Molar conductance

The **molar conductivity** (molar conductance) of a solution is defined as the conductivity, in $\Omega^{-1}\,\text{cm}^{-1}$, multiplied by the volume in cm^3 containing one mole of dissolved electrolyte. It is denoted by the symbol Λ. The significance of molar conductivity can best be understood by imagining a cell with plates 1 cm apart, and of such area that they enclose 1 mole of electrolyte (Fig. 6.7).

If an ionic substance dissolves in water completely giving ions, the molar conductivity ought not to alter if the solution is diluted. If the area of the plates is increased as the solution is diluted, so that 1 mole of electrolyte is enclosed throughout, then the same ions are responsible for the conductivity throughout, and they have to conduct across a path of 1 cm.

If this experiment is done, it is found that many electrolytes do in fact give constant values for Λ at a given temperature provided the solution is dilute. At high concentrations Λ decreases somewhat. This can be interpreted by assuming that in the concentrated solutions the ions are attracting one another and hence impeding each other's progress. In dilute solutions interionic attractions become negligible as the ions are further apart.

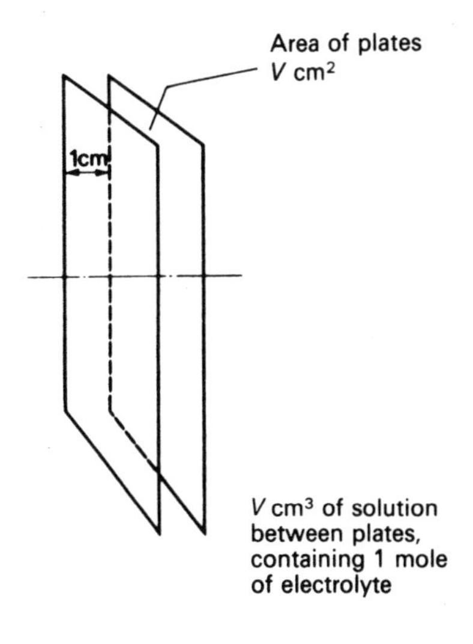

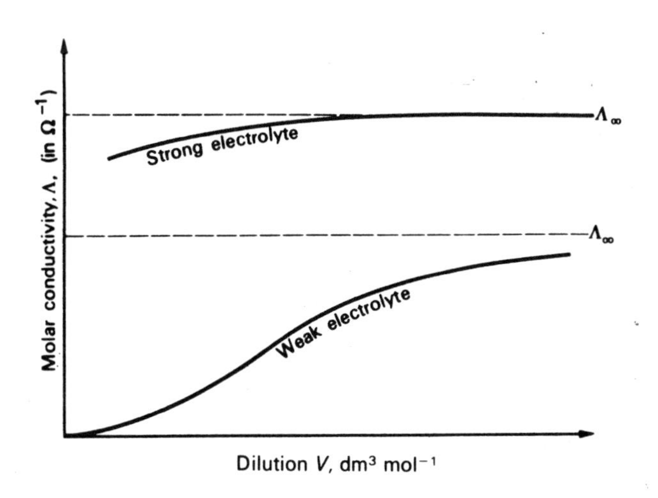

FIG. 6.7.

The conductivity of an electrolyte increases with temperature, and this is interpreted by imagining that the ions move with greater velocity at the higher temperature.

Some electrolytes show a very marked change in molar conductivity as their concentration is changed, and these are known as *weak electrolytes*. It is supposed that in these cases ionization is incomplete and the degree of ionization increases as the solution is diluted.

6.5 Ionic conductance

It is found experimentally that, provided solutions are very dilute, molar conductances are additive in nature, and it appears that the molar conductance can be represented as the sum of the separate conductances of the positive and negative ions.

$$\Lambda_\infty = \Lambda_+ + \Lambda_-$$

where Λ_∞ = the molar conductance at "infinite dilution";

Λ_+ = the ionic conductance of the positive ion, assumed to be singly charged;

Λ_- = the ionic conductance of the negative ion, assumed to be singly charged.*

The above relation is called *Kohlrausch's law of ionic conductances*. The data in Table 6.1 illustrates the law.

TABLE 6.1

Electrolyte	Λ_∞	Electrolyte	Λ_∞	Difference
KCl	130·0	NaCl	108·9	21·1
KNO_3	126·3	$NaNO_3$	105·2	21·1

A given ion has the same ionic conductance, no matter what other ions are present, and the Λ_∞ of a given salt is the sum of the ionic conductances of the ions contained in it. Table 6.2 gives some typical values for ionic conductances at 25°C.

The values for the hydrogen and hydroxide ions are rather higher than for other ions, and this is because these ions can move more rapidly

* The argument will have to be modified if multiple-charged ions are present, since it is assumed that there are equal numbers of each ion present.

TABLE 6.2

Cation	Conductance Ω^{-1} cm^2 mol^{-1}	Anion	Conductance Ω^{-1} cm^2 mol^{-1}
H^+	349·8	OH^-	198·5
K^+	73·5	Br^-	78·4
NH_4^+	73·4	I^-	76·8
Ag^+	61·9	Cl^-	76·3
Na^+	50·1	NO_3^-	71·4
Li^+	38·7	$CH_3CO_2^-$	40·9

through water. Hydrogen ions are thought to attach to water molecules forming species such as H_3O^+, and to move through the electrolyte by a kind of "relay race" mechanism:

```
H+ H   H          H  H    H
 \/   /            \/    /
 O   O     ──→     O    O      ──→ etc.
  \   \                /\
   H   H             +H  H
```

Similarly with hydroxide ions:

```
H     H  H         H  H    H
 \     \/           \/    /
  O-   O    ──→     O   -O     ──→ etc.
```

Most other ions have rather similar values for ionic conductance, suggesting that they move with roughly the same velocity down a given potential gradient. Differences are observed however, and these are not always of the type predicted by considering the simple ionic sizes. It might be imagined for instance that the ionic conductance of Li^+ would be greater than that of Na^+ because the former ion is smaller, but in fact the reverse is found to be true. This observation is explained by assuming that ions in solution attract water molecules electrostatically to themselves, forming a *hydration sheath*. Since the lithium ion produces a more intense electrostatic field at its surface it is hydrated to a greater degree and is therefore less mobile in water. This is one of the reasons why it is important to write $Li^+(aq)$ and $Na^+(aq)$ for the ions in solution: the state symbol (aq) refers to the presence of the hydration sheath. The average number of water molecules associated with an ion in aqueous solution is termed the **hydration number**.

6.6 Principles underlying shapes of molecules

Although the shapes of different molecules are very varied, they are determined by quite simple principles, which may be understood by considering the electrostatic repulsions which exist between electrons. As an example consider methane, CH_4.

Figure 6.8 shows that an isolated carbon atom has vacant energy levels. Four electrons are quite readily accommodated. One way of achieving this is by forming four electron-pair bonds

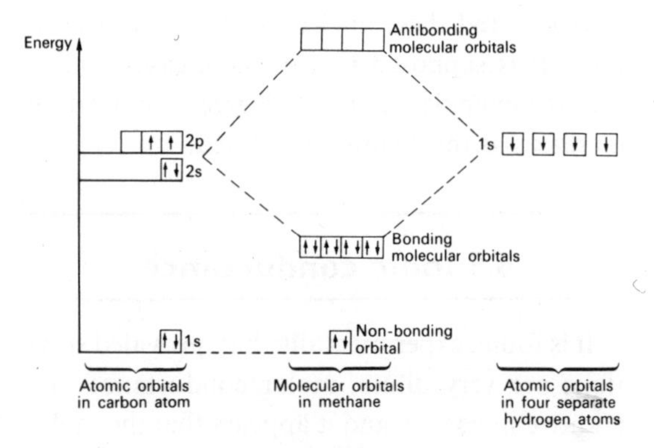

FIG. 6.8.

with hydrogen atoms. The 2*s* and 2*p* atomic orbitals in carbon, together with the four 1*s* atomic orbitals in the four hydrogen atoms, have now become eight molecular orbitals in the CH_4 molecule, four of which are low energy (bonding) and four high energy (antibonding). Figure 6.9 shows the same process in electron-dot notation.

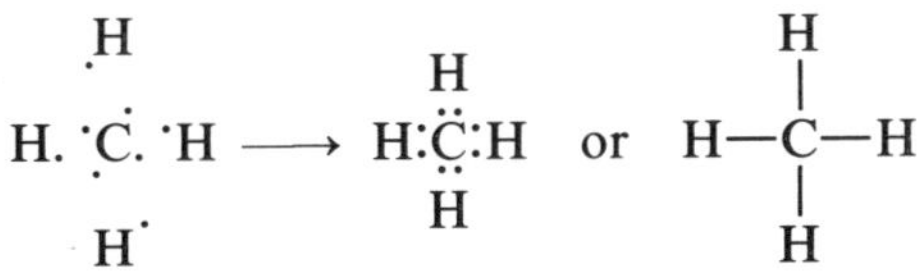

$$\begin{array}{c}\mathrm{H}\\ \mathrm{H}\cdot\dot{\mathrm{C}}\cdot\mathrm{H}\\ \mathrm{H}\end{array} \longrightarrow \begin{array}{c}\mathrm{H}\\ \mathrm{H}:\ddot{\mathrm{C}}:\mathrm{H}\\ \mathrm{H}\end{array} \text{ or } \begin{array}{c}\mathrm{H}\\ |\\ \mathrm{H}-\mathrm{C}-\mathrm{H}\\ |\\ \mathrm{H}\end{array}$$

FIG. 6.9.

The electron pairs in methane repel one another, and so the molecule, instead of being a flat square, has the shape of a tetrahedron (triangular pyramid). Figure 6.10 shows various ways in which the solid three-dimensional shape of a methane molecule can be represented on a flat piece of paper, but the best way to visualize it is by making a model.

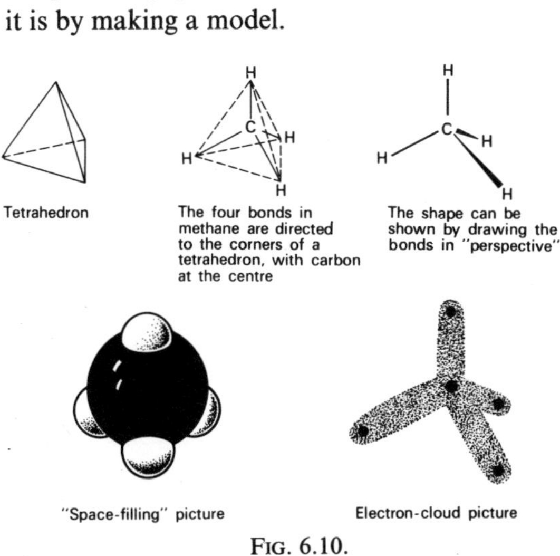

FIG. 6.10.

A simple model, using toy balloons, can be used to demonstrate the effect of repulsion between bonds. If four long balloons are attached in the manner shown in Fig. 6.11 they will automatically assume a tetrahedral shape, making

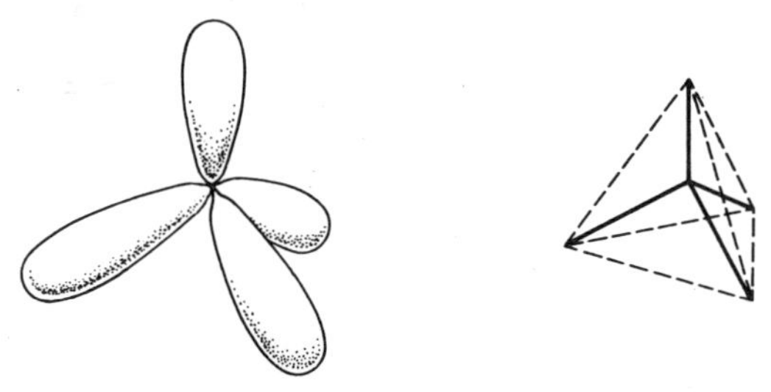

FIG. 6.11. Four balloons tied together assume a tetrahedral shape.

angles of approximately 109° with one another. Note that a square planar shape, although symmetrical, has angles of only 90° between adjacent bonds (balloons), and that this configuration is therefore not the most stable. A similar balloon experiment would suggest that three bonds will form a flat triangular molecule (Fig. 6.12). A number of molecules and ions, for instance boron trifluoride BF_3, and sulphur trioxide SO_3, do indeed have this shape.

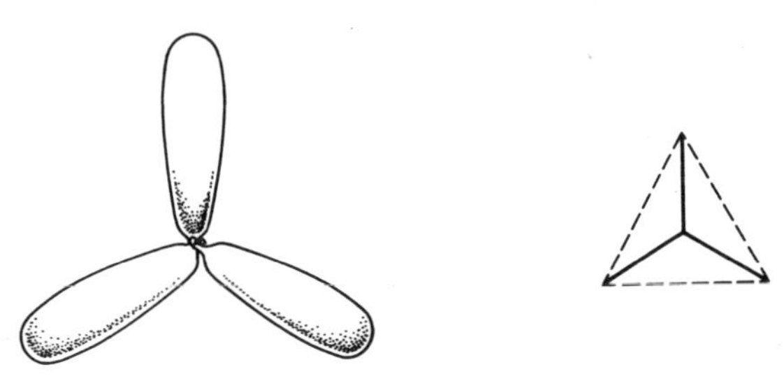

FIG. 6.12. Three balloons tied together assume a planar shape.

A very large number of molecules and ions are found to be not planar but pyramidal, such as ammonia NH_3, and the sulphite ion SO_3^{2-}. The reason for this difference becomes apparent upon examining the electronic configurations of the molecules concerned (Fig. 6.13). It is found that molecules of the type AX_3 are

(i) planar, if A does not have any **lone pairs**;

(ii) pyramidal, if A has one lone pair in addition to its **bond pairs**. As far as shape is concerned it does not matter whether the bond A—X is a single bond (that is, one using only one electron pair) or a double bond (using two electron pairs). Ethene, for instance, is a planar molecule, with bond angles of approximately 120° (Fig. 6.14).

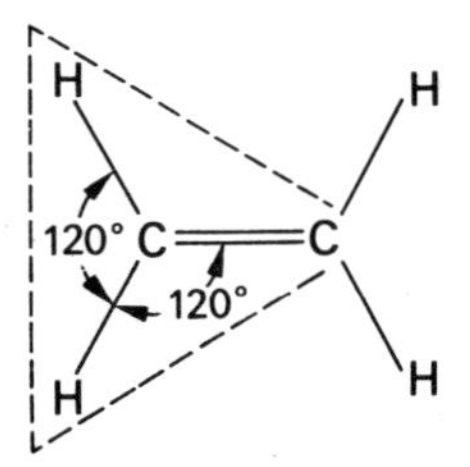

FIG. 6.14.

BF_3 No lone pairs: planar

NH_3 pyramidal

Lone pair

SO_3 No lone pairs: planar

SO_3^{2-} The two extra electrons now form a lone pair making a pyramidal shape

Lone pair

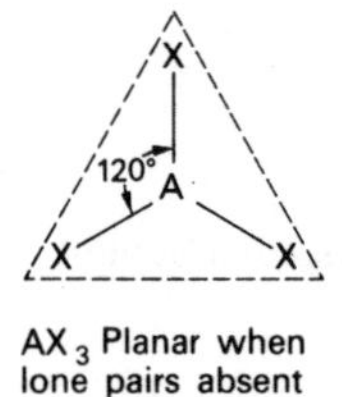

AX_3 Planar when lone pairs absent

AX_3 Pyramidal when one lone pair present

FIG. 6.13(a).

The shapes of molecules of the type AX_2 can be similarly explained. Our balloon model suggests that such a molecule will be linear, but this is only found to be true in fact for molecules where A has no lone pairs, such as mercury(II) chloride, $HgCl_2$, and carbon dioxide, CO_2. When A has one or more lone pairs the molecule is found to be *bent*, as in water, H_2O, hydrogen sulphide, H_2S, or sulphur dioxide, SO_2.

Linear

These molecules have no lone pairs on the central atom

Bent

Two lone pairs

One lone pair

FIG. 6.13(b).

There are surprisingly few instances where a consideration of the number of lone pairs and bond pairs on the central atom fails to give a correct prediction of the molecular shape. All that is necessary is to write down the electronic structure correctly, which can be done by observing the following rules:

(a) The valence shell in hydrogen can only accommodate 2 electrons.

(b) First row elements cannot accommodate more than eight electrons in their valence shell.

(c) Later elements can accommodate more than eight electrons in their valence shell.

The rules of electron pair repulsion can be applied to compounds of the type AX_5, AX_6, and even AX_7. Type AX_6 is quite simple to understand, and sulphur hexafluoride, SF_6, provides an example. A preliminary experiment with balloons predicts an octahedral shape, with exactly 90° between adjacent bonds, and this is in fact correct (Fig. 6.15).

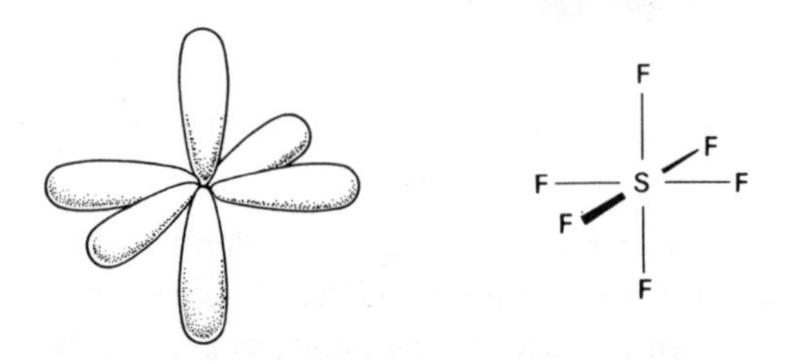

FIG. 6.15. The formation of an octahedral structure illustrated with balloons.

The balloon experiments are generally too rough and ready to give a reliable answer for AX_5 and AX_7. Phosphorus pentachloride, PCl_5, has molecules shaped like a trigonal bipyramid, while iodine heptafluoride, IF_7, has the shape of pentagonal bipyramid (Fig. 6.16).

The complex ions of transition metals sometimes have distorted shapes, due to the presence

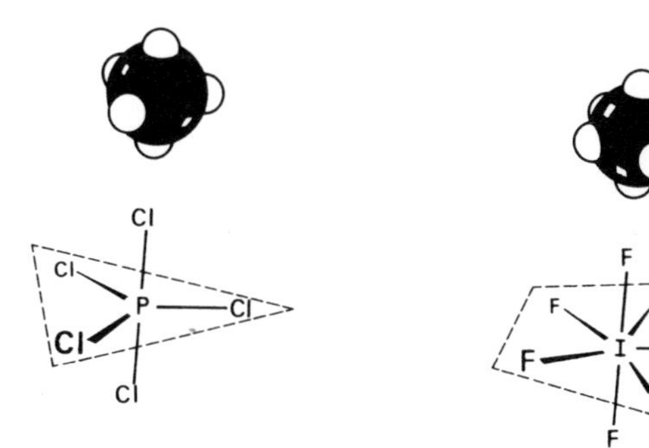

FIG. 6.16.

of lone pairs, or lone electrons. For instance the ion tetramminecopper(II), $Cu(NH_3)_4^{2+}$, has the nitrogen atoms arranged around the central copper atom in a plane.

6.7 The iso-electronic rule

A consequence of the above rules for molecular shape is that molecules which are **iso-electronic**, that is, have the same number of outer electrons and the same number of atoms, tend to have the same shape. It will be left as an exercise for the student to work out the reasons for the rules which follow, but they are a consequence of the principles laid down in section 6.6:

(a) Species of the type AX_2 containing 16 valence electrons are linear; more than 16 electrons lead to a bent structure.

Linear:
CO_2; number of valence electrons $= 4+6+6 = 16$
N_2O; number of valence electrons $= 5+5+6 = 16$
$BeCl_2$; number of valence electrons $= 2+7+7 = 16$
NO_2^+; number of valence electrons $= 5+6+6-1 = 16$

Bent:
NO_2; number of valence electrons $= 5+6+6 = 17$
NO_2^-; number of valence electrons $= 5+6+6+1 = 18$

(b) Species of the type AX_3 containing 24 valence electrons are planar triangular; more than 24 electrons lead to a pyramidal structure.

Triangular plane: BF_3; $3+(3\times 7) = 24$
BO_3^{3-}; $3+(3\times 6)+3 = 24$
CO_3^{2-}; $4+(3\times 6)+2 = 24$
NO_3^-; $5+(3\times 6)+1 = 24$
SO_3; $6+(3\times 6) = 24$

Pyramid: SO_3^{2-}; $6+(3\times 6)+2 = 26$
PCl_3; $5+(3\times 7) = 26$

(c) Species of the type AX_4 containing 32 valence electrons are tetrahedral:

CCl_4; $4+(4\times 7) = 32$
PO_4^{3-}; $5+(4\times 6)+3 = 32$
SO_4^{2-}; $6+(4\times 6)+2 = 32$
ClO_4^-; $7+(4\times 6)+1 = 32$

6.8 Structure of hydrocarbons

Hydrocarbons are compounds of carbon and hydrogen only and all contain C—H bonds (section 6.6). Methane, CH_4, is the simplest and is a gas at room temperature (b.p. $-160°C$). Very little energy is required to separate methane molecules from one another—we say that intermolecular attraction is small:

$$CH_4(l) \rightarrow CH_4(g)$$

$CH_4(l)$	$CH_4(g)$
one mole (16 g) of liquid methane molecules	one mole of gaseous methane

$$\Delta H = +9{\cdot}22 \text{ kJ at } -160°C.$$

ΔH denotes change in heat energy and the positive sign indicates that heat is added to the system. In contrast to this, an enormous amount of energy would be needed to split up the same number of molecules into separate atoms of carbon and hydrogen. So much energy is needed that it cannot be done in practice, though the energy is calculated to be:

$$CH_4(g) \rightarrow C(g) + 4H(g);$$

$CH_4(g)$	$C(g)$	$4H(g)$
one gram-molecule of methane gas	carbon vapour (atoms)	hydrogen in the form of separate atoms

$$\Delta H = +1660 \text{ kJ}$$

Carbon atoms can also form electron-pair bonds with each other and a whole range of hydrocarbons can be built up from carbon and hydrogen atoms. The simplest hydrocarbon containing a C—C bond is ethane, C_2H_6; the molecule containing a chain of three carbon atoms is the gas propane, C_3H_8.

A whole series of substances is known, the molecules of which consist of chains of carbon atoms joined by electron-pair bonds. If n is the number of carbon atoms, there will be $2n+2$ hydrogen atoms. Such compounds are called **alkanes**. The name of an unbranched alkane consists of two parts; a *stem* which indicates the number of carbon atoms and the *suffix* **-ane** which shows that the compound is a hydrocarbon consisting entirely of single bonds. Some unbranched hydrocarbons may contain C=C and C≡C bonds. In these cases the suffixes are **-ene** and **-yne** respectively. Figure 6.17 shows the electron distribution in the three types of carbon-carbon bond.

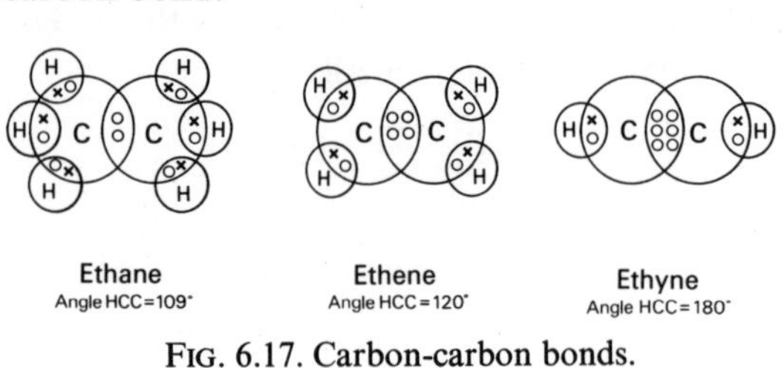

FIG. 6.17. Carbon-carbon bonds.

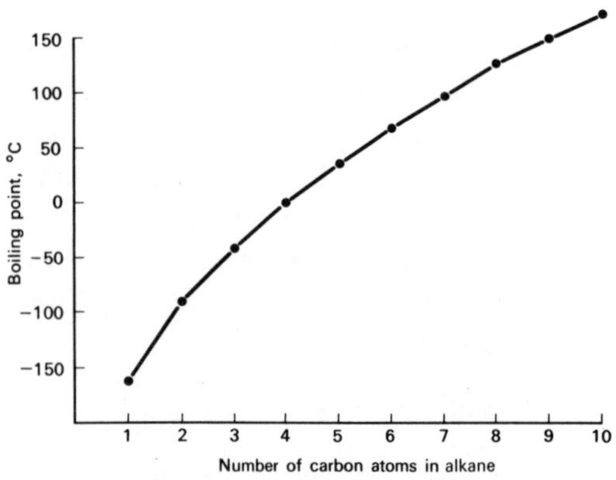

FIG. 6.18. Force of attraction between hydrocarbon molecules increases with chain length.

The size and shape of molecules have a considerable effect on their physical properties. If the boiling point of an alkane is plotted against number of carbon atoms, a gradual rise is noted (Fig. 6.18) as the force of attraction between molecules become larger. This is illustrated by comparing the heat required to vaporize 1 gram-molecule of each of the hydrocarbons (Table 6.3).

TABLE 6.3

Formula	Name	Heat required to vaporize 1 mole (kJ)	Boiling point (°C)
CH_4	Methane	9·2	−160
C_2H_6	Ethane	13·8	−88
C_3H_8	Propane	18·4	−41·4
C_4H_{10}	Butane	22·2	+0·8

The effect of structure on physical properties is well illustrated by comparing the characteristics of the main products obtained by distilling crude petroleum at an oil refinery (Table 6.4).

TABLE 6.4

Product	Number of carbon atoms	Properties
"Calor" gas	3 or 4	Gas, liquefiable under pressure
Motor fuel	about 8	Liquid, easily evaporates
Diesel oil	about 14	Liquid, does not evaporate easily
Lubricating oil	about 18	Thick liquid, does not evaporate
Pitch (road tar)	20 or more	Semi-solid black substance

Petroleum consists of a wide range of hydrocarbons, most of them alkanes. The larger molecules from substances which are thick and treacly in the liquid state, due to the molecules becoming tangled up. Translational motion is difficult, and we say the liquid is *viscous*, or has a high viscosity.

If a hydrocarbon has very long chains the result is a more rigid solid structure. *Polyethene* (polythene) $(C_2H_4)_x$ is an example of this.

$$H_2C{=}CH_2 \longrightarrow \left(-\overset{H}{\underset{H}{C}}-\overset{H}{\underset{H}{C}}- \right)_x$$

ethene gas → solid polythene

Rubber affords another illustration of the connection between structure and physical properties. Natural rubber is a hydrocarbon which consists of long zig-zag chains of carbon atoms tangled together. When rubber is stretched some

(a) Single rubber molecule (diagrammatic)

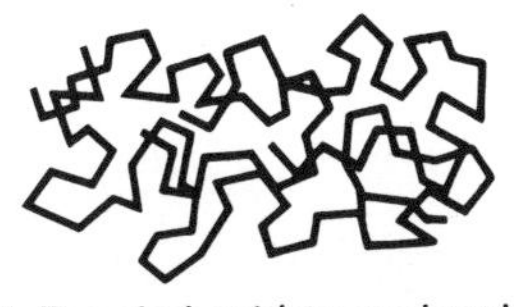

(b) Tangled rubber molecules before stretching

(c) Rubber molecules when stretched

FIG. 6.19.

straightening out of molecules takes place (Fig. 6.19). By studying the structure of the molecules of natural rubber, chemists have been able to develop the far superior synthetic rubbers which are now commonly used.

6.9 Functional groups

A wide variety of compounds can be formed from hydrocarbons by substituting for the hydrogen atoms other non-metallic atoms or groups of atoms. It is found that the same group of atoms tends to confer similar properties on all the molecules in which it occurs. Such groups are known as **functional groups**; they can often be identified from their infra-red spectra (section 4.3(c)). Table 6.5 gives the structural formulae of some common functional groups. Figure 6.20 shows how different classes of organic compounds are related to one another.

TABLE 6.5

Functional group	Formula	Class of compound
Carbonyl	$>C{=}O$	Aldehyde or ketone
Hydroxyl	$-O-H$	Alcohol or phenol
Carboxyl	$-C({=}O)-O-H$	Carboxylic acid
Amino	$-NH_2$	Amine
Cyano	$-C{\equiv}N$	Nitrile

As well as by observation of its infra-red spectrum, a particular functional group can usually be identified by performing suitable chemical tests. A series of compounds containing the same functional group attached to alkane chains of different lengths is known as a **homologous series**. Each member of the series differs from the next by the addition of a CH_2-group. Thus the alcohols CH_3OH, $CH_3.CH_2OH$, $CH_3.CH_2.CH_2OH$ etc. form a homologous series. The physical properties of the members of such a series vary in a regular manner and the

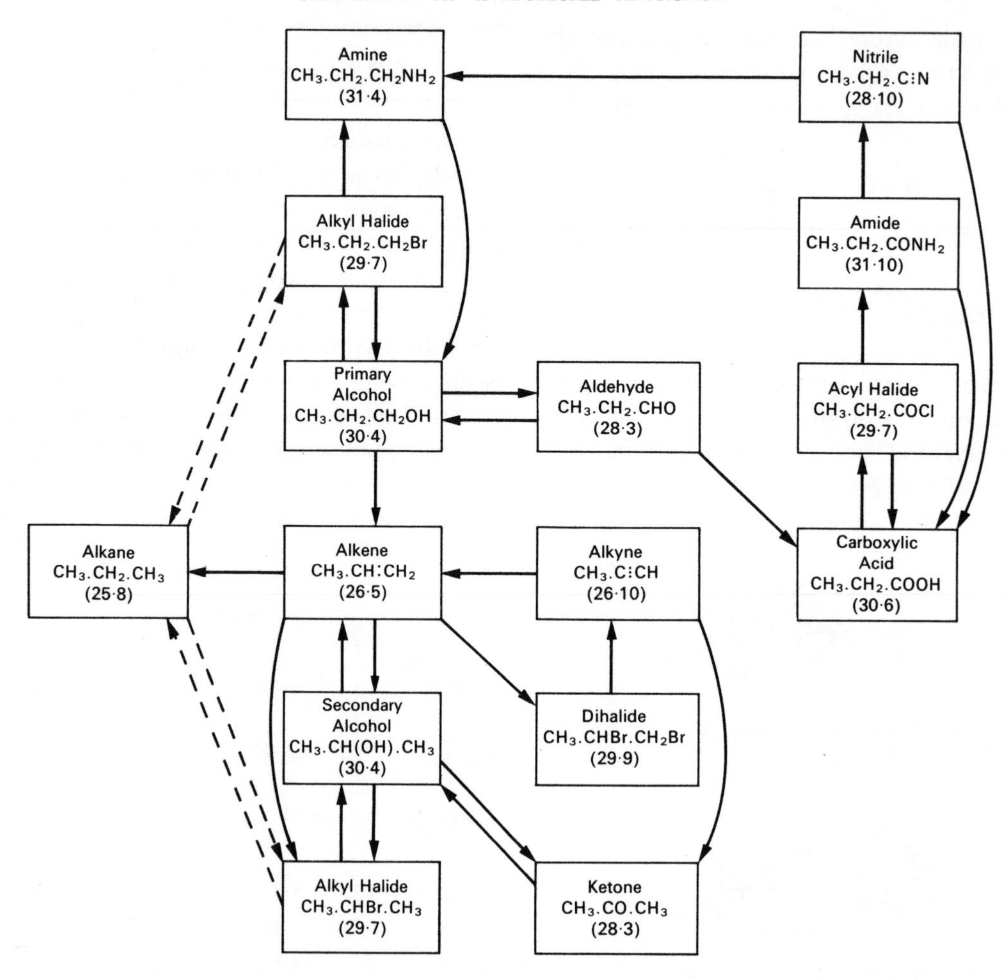

FIG. 6.20. The main classes of organic compounds. The numbers in brackets refer to sections in the book where the compounds are discussed.

boiling points of members of four series are plotted on Fig. 6.21.

In every homologous series an increase in the boiling point occurs as the molecular weight increases. Since the boiling points are a function of the intermolecular forces, it can be seen that the strongest attraction occurs between the molecules of the carboxylic acids (—COOH) and the alcohols (—OH). The forces between alkane molecules are relatively weaker than the forces in other homologous series. The reasons for these differences are discussed in Chapter 15.

6.10 Isomerism

Isomers are compounds having the same molecular formula, but different structural formulae. This section is concerned with three of the more important types of isomerism, structural, geometric and optical.

(a) **Structural isomerism.** Structural isomers occur when the arrangement of the atoms and the bonds holding them together are not the same. For example the molecular formula C_4H_{10} can

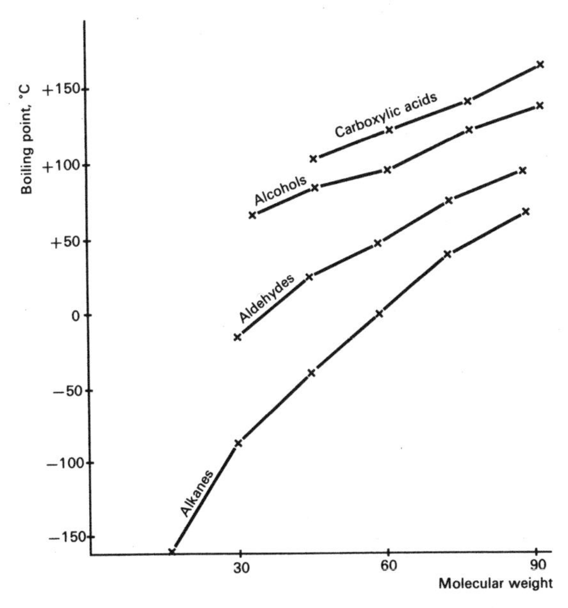

FIG. 6.21. The boiling-points of members of four homologous series.

represent either butane, $CH_3.CH_2.CH_2.CH_3$, or methylpropane, $(CH_3)_3CH$. The formula C_2H_6O can represent ethanol, $CH_3.CH_2.OH$, or methoxymethane, CH_3OCH_3, two very different compounds. The formula C_4H_8 represents five isomers, three containing double bonds and two containing rings of carbon atoms.

(b) **Geometric isomerism.** One of the isomers of C_4H_8, but-2-ene, $CH_3.CH{:}CH.CH_3$, itself exists as two compounds. The reason for this is that, whereas free rotation about single bonds is possible, no such rotation can occur about double bonds unless they are broken by the process. The angles about the two central carbon atoms in but-2-ene are 120° (Fig. 6.17), and the two isomers are known as **cis** in which the two methyl groups are on the same side of the double bond and **trans**, in which the two methyl groups are on opposite sides of the double bond.

H_3C CH_3	H_3C H
$C{=}C$	$C{=}C$
H H	H CH_3
cis but-2-ene	*trans* but-2-ene

This type of isomerism, which is also known as **cis-trans isomerism** occurs when there is a double bond, provided that two different atoms or groups are attached to each of the carbon atoms forming the double bond. Cis-trans isomerism is not confined to C=C bonds, and the compound $CH_3.CH{:}NOH$ shows geometric isomerism because of the lone pair of electrons on the nitrogen atoms.

The physical and chemical properties of geometric isomers often differ considerably. Thus *trans*-but-2-ene has no dipole moment in contrast to the large dipole moment of *cis*-but-2-ene. The *cis* form of the acid $HO_2C.CH{:}CH.CO_2H$ (maleic acid) readily loses water when it is heated, while the *trans* form (fumaric acid) is unaffected (Chapter 32).

Geometric isomerism is not confined to organic chemistry and it is often encountered in the chemistry of octahedral and square planar transition-metal complexes.

(c) **Optical isomerism.** In section 4.3(g) the idea of an asymmetric molecule, one which possesses no plane or centre of symmetry, was introduced. A common form of asymmetry involves an asymmetric carbon atom, an atom which is attached to four different atoms or functional groups. Such a molecule is 2-hydroxypropanoic acid (lactic acid) (Fig. 6.22) which exists as two different optical isomers or **enantiomers**; these isomers rotate polarized light in opposite directions.

The two optical isomers of lactic acid, being asymmetric, cannot be made to coincide with

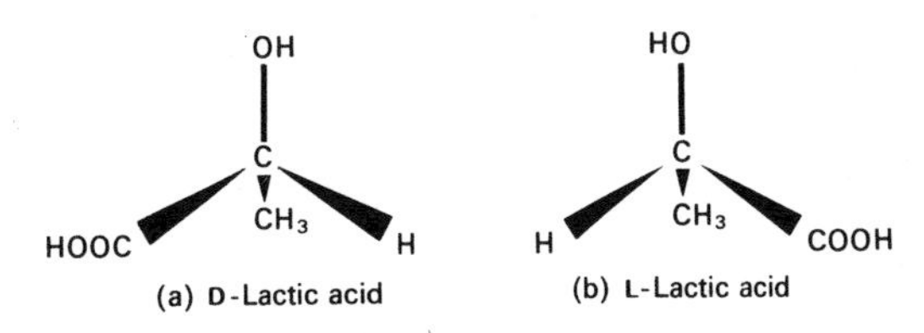

FIG 6.22. The isomers of lactic acid. (a) D-lactic acid (b) L-lactic acid. (By convention the thicker part of a bond is taken to be nearer to the observer).

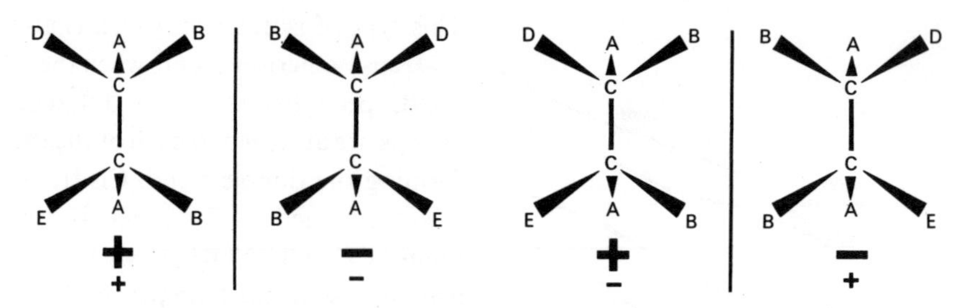

FIG. 6.23. The four isomers of a molecule containing two asymmetric carbon atoms.

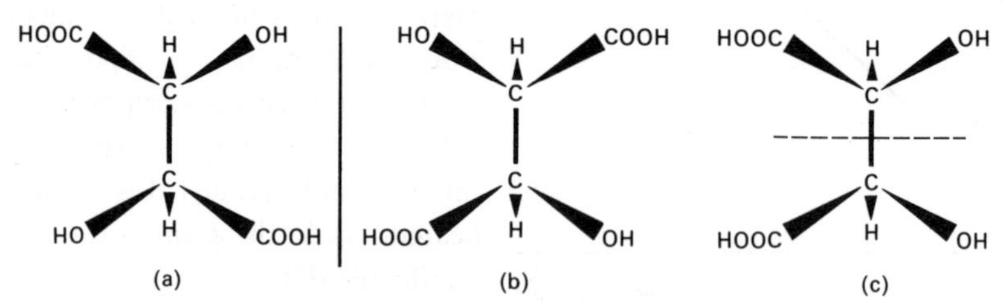

FIG. 6.24. The isomers of tartaric acid. (a) L-tartaric acid (b) D-tartaric acid (c) *Meso*-tartaric acid (note the plane of symmetry).

each other, just as your asymmetric left hand cannot be made to coincide with your right hand. But, just as your right hand is a mirror image of your left hand, so is D-lactic acid a mirror image of L-lactic acid; this can be seen by imagining a mirror placed between the two isomers in Fig. 6.22.

When an optically active compound is synthesized in a laboratory, the product is often optically inactive because D- and L-forms are produced in equal amounts. Such a mixture is known as a **racemic mixture** and its components usually have to be separated, by reacting them with a compound that is itself optically active, separating the products and then reversing the initial reaction. Optically active compounds that occur naturally are not racemic mixtures but usually consist of one of the isomers only.

A molecule containing two asymmetric carbon atoms will tend to form four optical isomers (+ +, + −, − +, − −) and these are shown in Fig. 6.23.

In general, a compound containing n asymmetric carbon atoms can form 2^n optical isomers. But 2,3-dihydroxybutanedioic acid (tartaric acid) and similar compounds do not form four isomers because the two asymmetric carbon atoms are identical. In this case the + − form is identical with the − + form and, because it contains a plane of symmetry, it is optically inactive and is known as *meso*-tartaric acid (Fig. 6.24).

6.11 Giant structures (macromolecules)

Previous sections have dealt with the shapes of small molecules and the same principles of shape govern larger molecules, as it is necessary simply to consider the number of bonds and lone-pairs associated with each atom in turn.

The element carbon, in the form of diamond, has each atom joined to its neighbours by four covalent bonds arranged tetrahedrally. In other words the same principle of electron-pair repulsion applies. The result is an absolutely rigid macromolecule of enormous strength—diamond is one of the hardest substances known to man. Since the whole solid is in effect a single molecule a very large amount of energy is required to vaporize it. Bonds have to be broken to produce carbon vapour, and a temperature of over 3000°C has to be attained.

$$\underset{\text{1 mole of atoms of solid carbon}}{C(s)} \rightarrow \underset{\text{1 mole of atoms of gaseous carbon}}{C(g)}; \quad \Delta H = +715 \text{ kJ}.$$

If a solid is very difficult to melt, then it must be some sort of giant structure in which the atoms are joined throughout by relatively large forces, and not an assembly of molecules held together by relatively weak forces (van der Waals forces). Rather more evidence about the properties of the solid is needed before concluding that the bonds between atoms are covalent. In many cases the forces are of a different nature, as we shall see in the next section.

6.12 Bonding in solids

This chapter has so far been largely concerned with *structure*, that is the stereochemical arrangement, of molecules. It is also important to consider the question of **bonding**, which is concerned with the *nature* of the forces between atoms. The principal types of bonding which exist can be illustrated with examples:

(a) Bonds between non-metallic atoms are **essentially covalent**. The physical properties of such substances depend upon whether they consist of finite molecules or of macromolecules, but in general they are electrical insulators both as solids and when molten.

(b) Compounds between elements situated on the extreme left and extreme right of the periodic table respectively, such as sodium chloride, are **essentially ionic**. Such substances have structures which appear to be related to the relative sizes of ions, rather than to electron-pair repulsions. They do not conduct electricity in the solid state, but they can be electrolysed when molten.

(c) Compounds between metals are **metallic** in character. By this we mean that they conduct electricity, without decomposition, in the solid state. Metals generally have a lustrous appearance, and are malleable and ductile. The stoichiometry of intermetallic compounds is complicated, and effects such as solid solutions are common. The general term **alloy** can be taken as applying to all systems of two or more metals, whether true compounds exist or not.

Electron-density maps give some information about bonding. In metallic and truly ionic compounds the electron density falls to zero between the nuclei. In covalent compounds, the "bond" means that a considerable electron density will be observed between the nuclei.

Very many compounds cannot be classified into these three categories, and intermediate types of bonding are very common.

Study Questions

1. Predict the shapes and write down electron dot structures for the following molecules and ions:

(a) BCl_3, CCl_4, NCl_3, PCl_5, SCl_4, SF_6, IF_5.

(b) H_3O^+, AlH_4^-, NH_4^+, PCl_4^+ and PCl_6^-.

2. Suggest three ions that would be isoelectronic with SF_6. What will the shape of the ions be?

3. Predict the shapes of the following three-atom systems: O_3, NOCl, N_3^-, SO_2, ICN and SCN^-.

4. Two shapes are theoretically possible for the ion ICl_4^-. What are these? Which shape is the ion most likely to adopt?

5. (a) The internuclear distance in RbCl is 0·3285 nm. Use this information and the data given in the chapter to predict the internuclear distance in RbBr.

(b) The internuclear distances in NaF and KF are 0·2307 and 0·2664 nm. Predict the internuclear distance in RbF.

6. The melting and boiling points of the chlorides of five consecutive elements are:

Empirical formula of compound	m.p. (°C)	b.p. (°C)
NaCl	800	1465
$MgCl_2$	712	1418
$AlCl_3$	190	180
$SiCl_4$	−68	57
PCl_5	148	164

What can you say about the probable structure and bonding of these compounds? What further experiments would you carry out to confirm your suspicions?

7. The melting point of the Group IVB dioxides are: CO_2, −57°C under pressure; SiO_2, 1720°C; GeO_2, 1120°C; SnO_2, 1130°C. Comment on the structures of these compounds in the light of these data.

8. Magnesium oxide, MgO, is harder than sodium fluoride, NaF. It also melts at a higher temperature (2640°C as compared with 990°C). How can you explain these facts?

9. Make an estimate of the radii of the ions Si^{4+} and C^{4-}. The co-ordination number in silicon carbide, SiC, is found to be 4. Do you think it likely that silicon carbide contains these ions?

10. Diamond, by virtue of its structure, is a very hard substance. Suggest another substance which might be about as hard as diamond.

11. Copper, silver, and gold are in Group IB of the periodic table. Separate solutions containing ions of these metals were electrolysed in series, and the following results were obtained:

	Copper cathode	Silver cathode	Gold cathode
Weight before electrolysis (g)	1·243	1·163	1·435
Weight after electrolysis (g)	1·370	1·594	1·698

What conclusions can you draw about the magnitude of the charges on the three ions?

12. A current of 0·1 A is passed through a solution of nickel ions for 16 minutes 5 seconds, and 0·0294 g of nickel was deposited on the cathode.

(a) How many moles of nickel atoms were deposited?
(b) How many faradays were passed?
(c) What is the charge on the nickel ion in this solution?

13. A minimum discharge potential of 3·34 V is required to decompose potassium iodide completely into its elements. Assuming that the ions are singly charged, calculate the energy required in kJ mol^{-1}.

14. Polymers called silicones are derived from the reactions of methyl silicon chlorides with water. Hydrogen chloride is eliminated and Si—O—Si links are formed, for instance:

$$2(CH_3)_3SiCl - H_2O \rightarrow (CH_3)_3Si{-}O{-}Si(CH_3)_3.$$

With $(CH_3)_2SiCl_2$ as the reactant, two types of polymer consisting of rings or chains are formed.

(a) Draw the ring and chain structures.
(b) What would the physical properties of the two types of polymer be?
(c) At high temperatures the chains become rings. Why does this limit the use of silicones as high temperature plastics?
(d) What polymer would form from the reaction of CH_3SiCl_3 with water? How would its physical properties differ from those of the ring and chain polymers?

15. Write down the structural formulae of as many isomers as you can of each of the following molecules:

(a) C_3H_8;
(b) C_3H_6;
(c) C_3H_4;
(d) C_5H_{12};
(e) C_5H_8 (unbranched chains only).

16. How many isomers are there of each of the following? What sort of isomerism is observed in each case?

(a) $CH_3.CH(OH).COOH$
(b) ClBrCH.CHClI
(c) ClBrCH.CHClBr
(d) Hex-2-ene
(e) Hexa-1,3-diene
(f) Hexa-2,4-diene

17. The co-ordination of an ionic compound is thought to depend on the ratio of the radii of the ions present. The change from one type of co-ordination to another will depend on certain limiting cases where all the anions are in contact with a cation and are also just touching each other.

(a) Work out the limiting ratio for 6:6 co-ordination.
(b) Work out the limit for 8:8 co-ordination.
(c) Calculate the radius ratios for NaCl, CsCl and ZnS (for the radii, see section 6.1).
(d) Do the actual co-ordination numbers in NaCl, CsCl and ZnS agree with the predictions made in (c)?

CHAPTER 7

Formulae

7.1 Atoms, molecules and moles

When elements combine together to form a compound, the composition of the compound is denoted by its chemical formula. Two distinct meanings may be attached to this formula, according to the context in which it is used. Taking ammonia, NH_3, as an example:

(i) the formula denotes that a *molecule* of ammonia consists of a unit containing three hydrogen atoms and one nitrogen atom; or

(ii) the formula denotes that the compound ammonia is formed when one *mole* of nitrogen atoms is combined with three *moles* of hydrogen atoms.

The context makes it perfectly clear which meaning is intended: If we are discussing the structure of the ammonia molecule, then the symbol N denotes one atom of nitrogen in the molecule NH_3. If, on the other hand, we are interested in the heat of a chemical reaction, then the symbols in a chemical equation refer to mole quantities:

$$N_2 + 3H_2 \rightleftharpoons 2NH_3; \quad \Delta H = -92 \text{ kJ}$$

This equation means that 92 kJ of heat are evolved when 28 g of nitrogen combine with 6 g of hydrogen, 34 g of ammonia being formed. Here the symbol N does not denote a single atom of nitrogen—what it really denotes is a mole of atoms. Similarly the formula NH_3 may be called a mole of NH_3 molecules.

The concept of the mole was introduced in Chapter 1. One mole represents a given number of atoms, namely the Avogadro number, $6{\cdot}022 \times 10^{23}$. Thus a mole of H_2(g) contains $6{\cdot}022 \times 10^{23}$ molecules, and has a mass of approximately 2 g. A mole of NH_3(g) contains $6{\cdot}022 \times 10^{23}$ molecules of ammonia, with a mass of approximately 17 g. *The formula NH_3 can be used either to denote a single molecule of ammonia or a mole of ammonia molecules.*

In a giant lattice the formula should be taken as referring to the mole. Sodium chloride, NaCl, does not exist as separate molecules composed of ion pairs (except in the vapour phase), and when we write its formula as NaCl we are stating that one mole of sodium atoms and one mole of chlorine atoms combine together to produce one mole of sodium chloride, NaCl. The term **gram-formula weight** can also be used to denote a mole of a substance.

7.2 Stoichiometric and non-stoichiometric compounds

A compound whose composition can be represented by a simple whole number formula is said to be **stoichiometric**. Any pure compound which is made up of molecules is bound to be stoichiometric, since the molecules are all identical and atoms cannot be split up.

Giant structures are more often non-stoichiometric. A perfectly crystalline solid such as pure sodium chloride will be stoichiometric since the rules of geometry determine the relative numbers of each kind of particle in the structure. Such perfection is frequently not attained in giant structures however, due to lattice defects. It is meaningless for instance to talk of "pure" silver oxide since a whole range of solids, Ag_xO_y, of continuously variable composition can be made, depending upon the pressure of oxygen gas which is applied. We may at times write Ag_2O as the formula of silver oxide, but this exact ratio of moles is only attained at one particular oxygen pressure.

Many compounds are non-stoichiometric due to their lattices being imperfect: the oxides and sulphides of many metals, particularly transition metals, often deviate considerably from exact stoichiometry, and the metal hydroxides precipitated by the addition of alkali, $OH^-(aq)$, to a solution of a metal salt have an indefinite amount of water in their composition. Thus for convenience we may write aluminium hydroxide as $Al(OH)_3$, but in practice we should be writing $Al(OH)_3 . x H_2O$ where x is not a whole number.

Metal alloys are interesting in that "compounds" of definite composition and stoichiometry can often be identified. The number ratios which occur often seem to be quite arbitrary, but a closer study has revealed that there are rules underlying the formulae of alloys. Typical formulae of inter-metallic compounds are $MgZn_2$, Cu_3Sn, $CuZn$, Cu_5Zn_8 and $CuZn_3$.

In very many cases the deviation from exact stoichiometry of solids is so slight that only very accurate measurements of combining weights can detect it. In the nineteenth century atomic weights were calculated assuming the law of constant composition, first stated in 1802 by Proust. We now realize that the law is not universal and that non-stoichiometric compounds are widespread where giant lattices are involved.

7.3 Determination of the Avogadro constant

The **Avogadro constant** is the number of particles in a mole of a substance.* It is generally given the symbol L. The actual value of this number was not known during the nineteenth century, even though the existence of atoms was firmly established. Essentially some method of determining the actual size of atoms was required.

(a) *X-ray method.* In describing X-ray diffraction (Chapter 4) it was stated that in order to measure absolute values of internuclear distances, the actual wavelength of the X-rays used must be known. This can be done by diffracting X-rays by reflection at the surface of a ruled metal diffraction grating. Normally for diffraction to occur the spacing between the rulings on the gratings must be comparable to the wavelength undergoing diffraction. However, if the glancing angle (the complement of the angle of incidence) is made less than about 1°, then a very finely ruled grating will diffract X-rays.

Having established the wavelength of the X-rays on an absolute scale, the same X-rays are diffracted by the adjacent planes of atoms in a crystal. Once the distance between atoms has been established, it is easy to calculate the volume which the atoms occupy, and measurements of density establish the volume occupied

* The mole is defined as the amount of substance in a system which contains as many "elementary entities" (e.g. atoms, ions or molecules) as there are atoms in exactly 12 g (0·012 kg) of ^{12}C.

by a *mole* of the substance. The following calculation illustrates how this has been used to determine the value of the Avogadro constant.

Density of potassium chloride, KCl,
$= 2{\cdot}01\ \text{g cm}^{-3}$ approximately

Distance between K and Cl nuclei
$= 3{\cdot}14 \times 10^{-8}$ cm (from X-ray data)

Volume of 1 mole of potassium chloride
= volume of $(39{\cdot}0 + 35{\cdot}5)$ g of KCl

$$= \frac{74{\cdot}5}{2{\cdot}01} = 37{\cdot}1\ \text{cm}^3.$$

Since potassium chloride has a simple cubic structure (Chapter 6), each atom may be regarded as being at the centre of a cube, of side $3{\cdot}14 \times 10^{-8}$ cm.

$\therefore$ volume occupied by each atom
$= (3{\cdot}14 \times 10^{-8})^3 = 30{\cdot}9 \times 10^{-24}\ \text{cm}^3$.

$\therefore$ in potassium chloride $\dfrac{1}{30{\cdot}9 \times 10^{-24}}$ atoms occupy 1 cm^3;

$\therefore \dfrac{37{\cdot}1}{30{\cdot}9 \times 10^{-24}}$ atoms occupy 37·1 cm^3 (one mole) $= 1{\cdot}20 \times 10^{24}$ atoms.

However, the Avogadro constant is *half* this figure, for it is the number of ion pairs, KCl, which go to make up one mole of potassium chloride. This works out at approximately 6×10^{23}.

We do not need to know the true volume occupied by a single potassium ion or a single chloride ion separately. We know the internuclear K—Cl distance, but we do not need the separate radii of the ions.

(b) *Radioactivity method.* When a substance emits α-particles, helium gas is produced by the reaction:

He^{2+}	+	$2e^-$	→	He(g)
alpha particles		stray electrons		gaseous helium

In one year, a gram of radium emits $11{\cdot}6 \times 10^{17}$ α-particles. This figure is estimated by counting the particles with a Geiger counter over a suitable time interval, making allowances for other particles emitted by secondary disintegration products of radium. Careful measurement has shown that over the same period of time, one gram of radium gives rise to $7{\cdot}67 \times 10^{-6}$ g of helium.

$7{\cdot}67 \times 10^{-6}$ g of helium contain $11{\cdot}6 \times 10^{17}$ particles

4 g (one mole) of helium contains

$$\frac{11{\cdot}6 \times 10^{17} \times 4}{7{\cdot}67 \times 10^{-6}} = 6{\cdot}05 \times 10^{23}$$

molecules mol^{-1}.

The fact that the figure obtained by radioactivity measurements ties up well with the figure obtained by X-ray diffraction measurements is one of the best pieces of evidence we can obtain for the atomic nature of matter. The figure $6{\cdot}022 \times 10^{23}$ molecules mol^{-1} is the accepted value for L, the Avogadro constant.

(c) *From the charge on an electron.* Millikan, in 1913, succeeded in determining the charge of an electron by careful observation of oil drops produced by a finely divided spray in an electrostatic field. The apparatus which he employed is shown diagrammatically in Fig. 7.1: the vessel, containing air, was immersed in a thermostat. A finely divided spray of oil was introduced through the atomizer, and some passed through

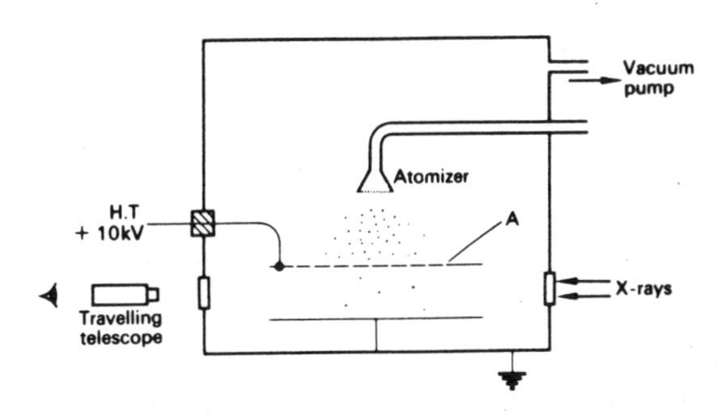

FIG. 7.1. Millikan's apparatus (diagrammatic) for determining the charge on an electron.

the holes in the upper plate A. Observations on a single oil drop could be made with a travelling telescope; when an oil drop was observed the air was ionized by passing X-rays in through the window on the right. The velocity with which the drop fell under gravity alone was measured, and then an electric field of about 10 000 V was applied so as to make the drop move upwards. From these measurements the magnitude of the charge carried by the oil drop could be calculated. The results showed that an alteration in velocity frequently occurred, due to the oil drop capturing different numbers of ions on successive occasions. The values for the charge were found always to be whole number multiples of about $1{\cdot}6 \times 10^{-19}$ coulombs. This figure was taken to be the charge on a single electron.

Since one faraday is a mole of electrons, the Avogadro constant is the ratio

$$L = \frac{\text{charge of one faraday}}{\text{charge of one electron}} = \frac{96\,490}{1{\cdot}6 \times 10^{-19}}$$
$$= 6{\cdot}03 \times 10^{23}$$

There are other methods of determining L, for instance from Brownian motion, or from observations on scattered light. The fact that all these methods give approximately the same value for the charge on an electron is our best direct evidence for believing in the existence of atoms.

7.4 Calculations involving changes in weight

Accurate measurements of changes in weight during a chemical reaction were important to the historical development of chemistry, since atomic weights were originally determined from weight changes in chemical reactions. Nowadays accurate measurements of weight, gravimetric measurements, are mainly of analytical importance, for instance in determining the chemical formula of an unknown pure substance, or for determining the percentage composition of a mixture. The methods used in calculations of this type are described in this section. Provided that relevant atomic weights are known, weighing may enable the *stoichiometry* (relative numbers of reacting particles) of a reaction to be determined.

Worked Example 1. 0·53 g of iron filings was placed in an excess of silver nitrate solution. Silver was precipitated which, after filtering and drying, was found to weigh 2·05 g. Write a chemical equation for the reaction.

We shall assume that silver nitrate solution consists of the ions $Ag^+(aq)$ and $NO_3^-(aq)$. All common nitrates are soluble in water, and therefore $NO_3^-(aq)$ ions are **spectator ions**, that is, they do not participate in the reaction.

$$\text{Atomic weight of silver} = 107{\cdot}9$$
$$\text{Atomic weight of iron} = 55{\cdot}8$$

∴ Number of moles of iron atoms added

$$= \frac{\text{weight}}{\text{atomic weight}} = \frac{0{\cdot}53}{55{\cdot}8} = 0{\cdot}0095$$

Number of moles of silver atoms precipitated

$$= \frac{2{\cdot}05}{107{\cdot}9} = 0{\cdot}0188$$

From this it is seen that the ratio

$$\frac{\text{number of moles of silver atoms}}{\text{number of moles of iron atoms}} = 2, \text{ approximately.}$$

We must write an equation which shows two moles of silver atoms, on the right-hand side, produced from one mole of iron atoms, on the left-hand side.

That is, $Fe \rightarrow 2Ag$ (incomplete).

The silver is produced from silver ions Ag^+, and in order to conserve positive charge in the reaction, iron must produce ions which are positively charged, which we may provisionally write Fe^{n+}.

$$Fe+2Ag^{+} \rightarrow 2Ag+Fe^{n+}.$$

The charges are balanced by making $n = 2$. The mole quantities balance also, and the completed equation, with state symbols, reads:

$$Fe(s)+2Ag^{+}(aq) \rightarrow 2Ag(s)+Fe^{2+}(aq).$$

It is important to make some sort of measurement in order to establish a chemical equation, and to establish what substances are actually formed. If it is *known* that the iron dissolves to form iron(II) ions, Fe^{2+}, the equation can be written down straight away. If iron had formed the species iron(III), Fe^{3+}, the stoichiometry would have been different.

Gravimetric measurements are not confined to solids and liquids. Gases are readily weighed if they can be absorbed into a solid or liquid phase. For instance, the carbon dioxide produced in the combustion of carbon compounds can be absorbed in previously weighed soda-lime and the increase in weight noted. Water vapour can similarly be absorbed in silica gel.

Worked Example 2. 5·601 g of a barium salt was heated. Oxygen gas only was evolved, and the residue was found to be barium chloride weighing 3·834 g. Write down the name and formula of the original salt. The barium ion carries two positive charges and the chloride ion one negative charge.

Since the charges must balance in barium chloride, the compound must contain two moles of Cl^- for every one of Ba^{2+}. Therefore the formula is $BaCl_2$. Since there was a weight loss, and only oxygen was evolved, the original barium salt must be written $BaCl_2O_x$.

Atomic weight of barium = 137·3
Atomic weight of chlorine = 35·5
Atomic weight of oxygen = 16·0

We need to know the number of *moles* of barium chloride formed. It is convenient to talk about the **formula weight**, or strictly the gram-formula weight, of barium chloride, namely the sum of the atomic weights Ba+2Cl.

g-formula weight of barium chloride

$$= 137{\cdot}3+(2\times 35{\cdot}5) = 208{\cdot}3$$

∴ Number of moles of barium chloride

$$= \frac{3{\cdot}834}{208{\cdot}3} = 0{\cdot}0184$$

Number of moles of oxygen atoms evolved

$$= \frac{\text{loss in weight}}{\text{atomic weight of oxygen}} = \frac{5{\cdot}601-3{\cdot}834}{16{\cdot}0} = 0{\cdot}110$$

∴ Number of moles of oxygen atoms which would be evolved per mole of barium chloride

$$= \frac{0{\cdot}110}{0{\cdot}0184} = 6.$$

The equation may be written:

$$BaCl_2O_x \rightarrow \underset{\text{one mole}}{BaCl_2} + \underset{\text{six moles of atoms}}{6[O]} \quad \text{(incomplete)}$$

Balancing up the oxygen atoms, and taking into account the fact that oxygen is evolved as diatomic molecules, the complete equation, with state symbols, becomes:

$$BaCl_2O_6(s) \rightarrow BaCl_2(s)+3O_2(g)$$

In fact the salt is barium chlorate(V), and its formula is more correctly written $Ba(ClO_3)_2$, to take into account that it gives chlorate(V) ions $ClO_3^-(aq)$, when dissolved in water.

The formula of an organic compound can be established by a procedure such as **combustion analysis**. In this process, which is nowadays completely automated, the substance is completely burned in excess oxygen forming carbon dioxide and water vapour. The water is absorbed in a suitable drying agent, such as magnesium chlorate(VII) and the carbon dioxide is absorbed in soda-lime (a mixture of calcium hydroxide and sodium hydroxide).

$$CO_2(g)+NaOH(s) \rightarrow NaHCO_3(s).$$

Measurements involving solely changes in weight can be used to deduce the so-called **empirical formula** of a substance, that is, the relative numbers of moles of elements which combine. It will not however give the **molecular formula** of a molecular substance. To give an example, measurements on the weights of carbon dioxide and water evolved when benzene is burned lead to the knowledge that benzene contains one mole of carbon for every mole of hydrogen atoms. That is to say,

$$\text{empirical formula of benzene} = CH.$$

If the molecular weight is determined (section 7.7), it is found to be 78. Therefore the molecular formula of benzene is C_6H_6. When we talk about one mole of benzene, we mean $6{\cdot}022 \times 10^{23}$ molecules of C_6H_6, that is, 78 g. We do not mean $6{\cdot}022 \times 10^{23}$ units of CH or 13 g.

Worked Example 3. A hydrocarbon X of molecular weight 128 was subjected to combustion analysis. 64 mg were completely burned in oxygen. The increase in weight of the soda-lime tube was 220 mg, and the absorbed water weighed 36 mg. Deduce the molecular formula of the hydrocarbon.

220 mg of carbon dioxide

$$= \frac{220}{44} = 5 \text{ millimoles (mmol)}$$

36 mg of water

$$= \frac{36}{18} = 2 \text{ mmol}$$

64 mg of X

$$= \frac{64}{128} = 0{\cdot}5 \text{ mmol}$$

The above ratios can now be expressed in terms of a chemical equation:

$$0{\cdot}5 \text{ mmol X} + \text{oxygen} \rightarrow 5 \text{ mmol } CO_2 + 2 \text{ mmol } H_2O$$

$$\therefore 1 \text{ mol of X} + \text{oxygen} \rightarrow 10 \text{ mol } CO_2 + 4 \text{ mol } H_2O$$

$$C_xH_y + \text{oxygen} \rightarrow 10CO_2 + 4H_2O$$

$$\therefore x = 10$$

$$y = 8.$$

The completed equation, balancing the oxygen, becomes:

$$C_{10}H_8 + 12O_2 \rightarrow 10CO_2 + 4H_2O.$$

The hydrocarbon in this problem is probably naphthalene.

7.5 Calculations involving solutions

Reactions which occur in solution are often investigated quantitatively using aqueous (or sometimes non-aqueous) solutions of known molar concentration in the procedure known as **volumetric analysis**. As the name implies, the volumes of solutions are measured, using the pipette and burette, in a **titration**.

Unless other units are specified, the **concentration** of a solution may be taken as meaning **molar** concentration or **molarity**, that is, the number of moles of solute dissolved in a litre (1000 cm^3) of solution. A solution containing *one* mole of sodium hydroxide, NaOH, per litre (that is $23+16+1 = 40$ g dm^{-3}*) is written "1 M NaOH". Similarly $4{\cdot}0$ g dm^{-3} of sodium hydroxide is written $0{\cdot}1$ M NaOH.

Worked Example 4. Calculate the mass of solute in the following solutions:

(a) *One dm^3 of 2 M nitric acid, HNO_3.*

* One litre is equal to 1000 cm^3 or 1 dm^3. Throughout this book dm^3 will be used as an abbreviation for litre.

$$\text{Formula weight} = \text{weight of one mole} = 1+14+48 = 63 \text{ g}$$

$\therefore$ 2 mol weigh $2 \times 63 = 126$ g

$\therefore$ One dm^3 of 2 M HNO_3 contains 126 g of nitric acid.

(b) *50 cm^3 of 0·5 M sodium carbonate, Na_2CO_3.*

$$\text{Formula weight} = (2 \times 23)+12+(3 \times 16) \quad 106 \text{ g}$$

$\therefore$ Mass of solute per $dm^3 = 0{\cdot}5 \times 106 = 53$ g

$\therefore$ Mass of solute in 50 cm^3

$$= 53 \times \frac{50}{1000} = 2{\cdot}65 \text{ g.}$$

Worked Example 5. Calculate the concentration of the solution produced by adding water to 50 cm^3 of 2 M sulphuric acid, H_2SO_4 to give 200 cm^3 of solution.

50 cm^3 of 2 M acid contain $2 \times \frac{50}{1000}$ moles.

New volume $= 200$ cm^3.

$$\therefore 1 \text{ dm}^3 \text{ contains } 2 \times \frac{50}{\cancel{1000}} \times \frac{\cancel{1000}}{200}$$

$$= 2 \times \frac{50}{200} = 0{\cdot}5 \text{ mol}$$

$\therefore$ The diluted solution is 0·5 M.

This result is easily remembered and understood in the form: volume before dilution $\times$ concentration before dilution $=$ volume after dilution $\times$ concentration after dilution.

Worked Example 6. To what volume would 500 cm^3 of 0·5 M potassium manganate(VII), $KMnO_4$, have to be diluted in order to prepare a 0·2 M solution?

Let total volume after dilution $= V$ cm^3
Using the above relationship,

$$V \times 0{\cdot}2 = 0{\cdot}5 \times 500$$

$$\therefore V = \frac{0{\cdot}5}{0{\cdot}2} \times 500 = 1250 \text{ cm}^3.$$

Worked Example 7.

(a) *25 cm^3 of 0·1 M sodium carbonate solution, from a pipette, were titrated with 0·2 M hydrochloric acid from a burette. The indicator was phenolphthalein. The burette readings were: initial reading 1·6 cm^3, final reading 14·1 cm^3. How many moles of hydrochloric acid have reacted with one mole of sodium carbonate?*

Volume of HCl added $= 14{\cdot}1 - 1{\cdot}6 = 12{\cdot}5$ cm^3.

$\therefore$ Number of moles of HCl added

$$= \frac{12{\cdot}5}{1000} \times 0{\cdot}2 = 0{\cdot}0025 \text{ mole.}$$

Number of moles of Na_2CO_3 used

$$= \frac{25}{1000} \times 0{\cdot}1 = 0{\cdot}0025 \text{ mole.}$$

$\therefore$ The answer is one mole of hydrochloric acid.

The equation for this reaction may be written:

$$Na_2CO_3(aq) + HCl(aq) \rightarrow NaCl(aq) + NaHCO_3(aq).$$

(b) *The titration was repeated using methyl orange indicator, and this time it was found that twice the volume of acid had to be added to produce a colour change. Explain!*

Since twice the volume of acid was required, it follows that 2 moles of acid were required to react with one mole of sodium carbonate. The equation is:

$$Na_2CO_3(aq) + 2HCl(aq) \rightarrow 2NaCl(aq) + CO_2(aq) + H_2O.$$

This subject is discussed further in section 19.14.

Worked Example 8. The concentration of chloride ion, $Cl^-(aq)$, in a solution can be estimated by titrating it with silver nitrate from a burette. Silver chloride, AgCl, is precipitated. The endpoint is detected by adding a few drops of potassium chromate which gives brick-red silver chromate immediately a slight excess of silver ions has been added. Silver nitrate solu-

tion, 34 g dm^{-3}, was used to titrate 20 cm^3 portions of a solution of chloride ion. Use the burette readings to estimate the concentration of chloride ion.

Burette readings (cm^3):

	(1)	(2)	(3)	(4)
Final	26·0	26·1	28·2	29·4
Initial	0·5	1·1	3·1	4·4
Volume	25·5	25·0	25·1	25·0

Titration (1) is taken as inaccurate, due to overshooting of the end-point. Within the limits of experimental error, the mean of titrations (2) to (4) is 25·0 cm^3.

Equation:

$$Ag^+(aq)+Cl^-(aq) \rightarrow AgCl(s)$$

i.e. 1 mole reacts with 1 mole

Concentration of silver nitrate

$$= 34 \text{ g dm}^{-3}$$

$$= \frac{34}{170} \text{ mol dm}^{-3} = 0{\cdot}2 \text{ M}.$$

Number of moles of Ag^+ added in titration

$$= 0{\cdot}2 \times \frac{25{\cdot}0}{1000}$$

∴ Number of moles of Cl^- present in 20 cm^3 is equal to this.

∴ Number of moles of Cl^- per dm^3 of solution

$$= 0{\cdot}2 \times \frac{25{\cdot}0}{1000} \times \frac{1000}{20{\cdot}0} = 0{\cdot}25$$

Worked Example 9. The solution of chloride ions in the previous question was made up by dissolving 3·02 g of rubidium chloride in water and making the solution up to 100 cm^3. What is the formula of rubidium chloride?

Since there are 0·25 moles of Cl^- per 1 dm^3 there must be 0·025 moles of Cl^- per 100 cm^3. In this volume, the mass of

$$Cl^- = 0{\cdot}025 \times 35{\cdot}5 = 0{\cdot}89 \text{ g}.$$

The mass of rubidium is therefore

$$3{\cdot}02-0{\cdot}89 = 2{\cdot}13 \text{ g}.$$

$$\text{This is } \frac{2{\cdot}13}{85{\cdot}5} = 0{\cdot}025 \text{ moles} = \text{moles of } Cl^-$$

Since the number of moles of Rb and Cl are the same, the formula of rubidium chloride = RbCl.

7.6 Calculations involving gas volumes

Quantities of gases are usually measured more conveniently by volume than by weight, owing to their low density. Avogadro, in 1811, noticed that all gases obeyed approximately the same laws relating volume to temperature and pressure, and this observation led to the hypothesis that *equal volumes of gases at the same temperature and pressure contain equal numbers of molecules*. Subsequent observations have proved that this is true to a reasonable approximation, so we now refer to it as **Avogadro's law**. It holds most accurately for gases of very low boiling point.

The volume of 1 mole of a gas, that is, L molecules, at 273 K and 760 mm of mercury pressure, is called the **molar gas volume**. It is approximately 22 400 cm^3. 273 K and 760 mm represent standard temperature and pressure (**s.t.p.**). At room temperature 1 mole of any permanent gas will occupy approximately 24 000 cm^3 (24 dm^3). This volume is easily visualized as it is somewhere near to one cubic foot.

Worked Example 10. Hydrogen peroxide decomposes when a catalyst of manganese dioxide powder is added, to give water and oxygen only. Calculate the concentration of hydrogen peroxide solution, in g dm^{-3}, which will give 10 times its own volume of oxygen when decomposed, at s.t.p.

$$H_2O_2(aq) \rightarrow H_2O+\tfrac{1}{2}O_2(g)$$

In this equation, $\frac{1}{2}O_2$ denotes half a *mole* of

oxygen molecules, not half a molecule! Half a mole of O_2 = 11·2 dm^3 at s.t.p. = 16 g. From the equation, this quantity of oxygen is given by 1 mole of hydrogen peroxide, = 34 g.

For the solution to give ten times its volume of oxygen,

$$\text{Volume of solution} = \frac{11\cdot 2}{10}\ \text{dm}^3$$

Concentration of solution

$$= 34\ \text{g dissolved in } \frac{11\cdot 2}{10}\ \text{dm}^3$$

$$= 34 \times \frac{10}{11\cdot 2}\ \text{g dm}^{-3} = 30\cdot 4\ \text{g dm}^{-3}.$$

It is generally not convenient in the laboratory to make measurements at s.t.p. In practice, the volume of a gas is measured at whatever temperature and pressure are convenient. The volume that the gas would occupy at s.t.p. is then calculated by applying:

(1) **Boyle's law.** At constant temperature, the volume of a gas is inversely proportional to its pressure.

pV = constant,
where p is the pressure and V the volume.

(2) **Charles' law.** At constant pressure, the volume of a gas is directly proportional to its absolute temperature.

$$V \propto T,$$

where T is measured in degrees Kelvin (K). 0°C is 273 K approximately.

These two laws are conveniently brought together in a single mathematical equation, using a new constant of proportionality, R, termed the **gas constant**.

$$pV = RT.$$

R has the same numerical value for one mole of any permanent gas. For n moles of a gas, it is often convenient to use the form:

$$pV = nRT$$

where V now equals the volume of n moles. If a gas, occupying a volume v_1 at pressure p_1 and temperature T_1 is subjected to altered conditions v_2, p_2 and T_2, it follows that

$$\frac{p_1 v_1}{T_1} = \frac{p_2 v_2}{T_2}.$$

The above relationship is useful for Boyle's law and Charles' law calculations.

Worked Example 11. Calculate the molecular weights of the following gases from the data given.

(a) *2·24 dm^3 of an oxide of nitrogen weigh 4·4 g at 273 K and 760 mm Hg.*

Weight of 22·4 dm^3 at s.t.p.

$$= \frac{4\cdot 4 \times 22\cdot 4}{2\cdot 24} = 44\ \text{g}$$

∴ Molecular weight = 44. (The gas must be N_2O.)

(b) *350 cm^3 of an oxide of carbon weigh 0·416 g at 298 K and 745 mm Hg.*

$$\underset{\substack{\uparrow \\ \text{actual conditions}}}{\frac{p_1 v_1}{T_1}} = \underset{\substack{\uparrow \\ \text{s.t.p. conditions}}}{\frac{p_2 v_2}{T_2}}$$

Substituting,

$$\frac{745 \times 350}{298} = \frac{760 \times v_2}{273}$$

$$\therefore\ v_2 = \frac{745 \times 350 \times 273}{298 \times 760} = 315\ \text{cm}^3$$

$$\therefore\ 22\,400\ \text{cm}^3\ \text{weigh}\ \frac{0\cdot 416 \times 22\,400}{315} = 28\ \text{g}$$

∴ Molecular weight = 28. (The gas must be CO.)

Measurements of gas volumes are sometimes used to determine molecular formulae of gases, and the composition of gaseous mixtures of known formula. In the case of hydrocarbon gases, the gas is sparked in a eudiometer tube

(Fig. 7.2) with excess oxygen. The products are water, whose volume at room temperature is negligible in comparison with the gas volumes involved, and carbon dioxide, which is estimated by introducing potassium hydroxide solution and measuring the decrease in volume.

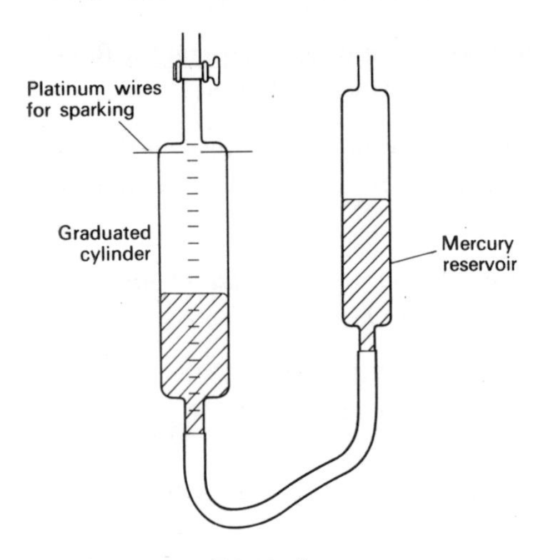

FIG. 7.2. Eudiometer.

The formula of the gas is then determined by applying Avogadro's law, together with **Gay-Lussac's law of combining volumes**, which states that when gases react they do so in volumes which bear a simple ratio to one another, and to the volume of any gaseous products formed, all measurements being made at the same temperature and pressure. Gay-Lussac's law is a consequence of Avogadro's law, as the following worked example shows:

Worked Example 12. 10 cm³ of a gaseous hydrocarbon were exploded with 75 cm³ of oxygen, and the gaseous products occupied 65 cm³ when cool. Potassium hydroxide solution caused the volume to be reduced to 45 cm³. The residual gas was shown to be oxygen by absorbing it completely in alkaline pyrogallol (an organic reducing agent). Determine the molecular formula of the hydrocarbon.

Let the hydrocarbon be C_xH_y. The equation for its combustion is

$$C_xH_y+\left(x+\frac{y}{4}\right)O_2 \rightarrow xCO_2+\left(\frac{y}{2}\right)H_2O$$

Applying Avogadro's law:

$$\underset{\text{mole}}{1} + \underset{\text{moles}}{\left(x+\frac{y}{4}\right)} \rightarrow \underset{\text{moles}}{x} + \underset{\text{moles}}{\left(\frac{y}{2}\right)}$$

$$\therefore \underset{\text{volume}}{1} + \underset{\text{volumes}}{\left(x+\frac{y}{4}\right)} \rightarrow \underset{\text{volumes}}{x} + \text{negligible volume of liquid } H_2O$$

Substituting:

$$10 \text{ cm}^3+(75-45) \text{ cm}^3 \rightarrow 20 \text{ cm}^3$$
$$\therefore 1 \text{ cm}^3+3 \text{ cm}^3 \rightarrow 2 \text{ cm}^3$$

$\therefore x = 2$

$\therefore y/4 = 3-2. \therefore y = 4.$

$\therefore$ The molecular formula is C_2H_4.

Worked Example 13. 25 cm³ of a mixture of hydrogen, methane and carbon dioxide were exploded with 30 cm³ of oxygen; the total volume decreased to 22·5 cm³. Potassium hydroxide solution caused the volume to be reduced further to 12·5 cm³. The residue was completely absorbed by alkaline pyrogallol. Deduce the volume composition of the mixture.

Let x = number of cm³ of hydrogen,
y = number of cm³ of methane,
z = number of cm³ of carbon dioxide originally present.

$$H_2+\tfrac{1}{2}O_2 \rightarrow H_2O$$

Volumes: x $x/2$ negligible

$$CH_4+2O_2 \rightarrow CO_2+2H_2O$$

Volumes: y $2y$ y negligible

Final volume of carbon dioxide after exploding $= y+z$.

$y+z$ = contraction in volume on adding KOH
$= 22·5-12·5$
$= 10$

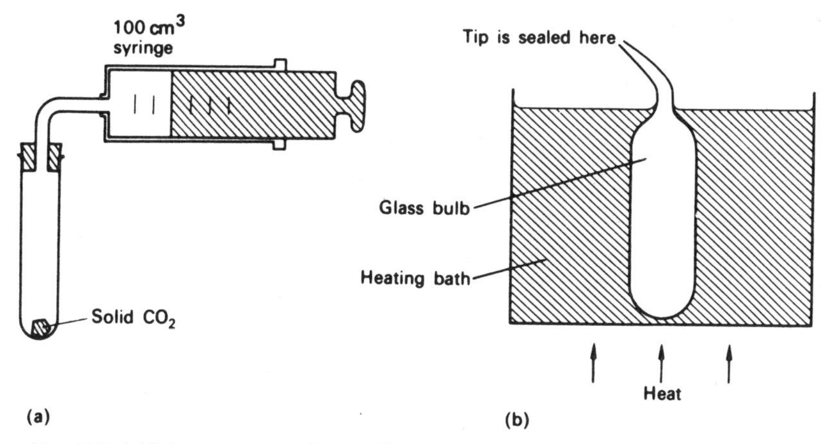

FIG. 7.3. (a) Measurement of a gas density with a syringe. (b) Dumas' method.

Original volume of mixture before adding oxygen

$$= x+y+z = 25$$

$$\therefore x = 15$$

Volume of oxygen used, from the equations

$$= x/2+2y = 7{\cdot}5+2y$$

By experiment, volume of oxygen used

$$= 30-12{\cdot}5 = 17{\cdot}5$$

$$2y = 17{\cdot}5-7{\cdot}5$$

$$\therefore y = 5$$

$$\therefore z = 10-y = 5$$

∴ The gas consisted of 15 cm^3 hydrogen, 5 cm^3 methane, and 5 cm^3 carbon dioxide.

7.7 Molecular weights of gases from density measurements

For an ideal gas, $pV = nRT$, so that at constant temperature and pressure $V = kn$, where k is a constant. If both sides of this equation are divided by the mass of gas, m, we get:

$$\frac{V}{m} = \frac{kn}{m}$$

Since $\frac{m}{V}$ is the density of the gas, and $\frac{m}{n}$ is the molecular weight, it can be seen that the molecular weight of a gas will be proportional to its density. Hence the molecular weight of a gas may be obtained simply by weighing a known volume.

A rough laboratory determination of the molecular weight of a gas may be obtained by weighing a flask (i) evacuated, and (ii) containing the gas at known temperature and pressure. The weight which will occupy 22 400 cm^3 at s.t.p. is then calculated—this is the molecular weight.

Measurement may be done in a flask of known volume, or alternatively in a graduated syringe such as that shown in Fig. 7.3. For instance, the molecular weight of carbon dioxide can be roughly checked by rapidly weighing a piece of solid carbon dioxide and allowing it to evaporate into a syringe.

More accurate determinations were done by Regnault in 1845. Two identical globes fitted with stopcocks were counterpoised on a balance, one having been previously evacuated. The evacuated globe was then filled with hydrogen at a known temperature and pressure, and reweighed. The globe was then filled with the given gas and reweighed. The relative density of the gas, and hence its molecular weight, was then calculated.

The vapour of a volatile liquid may be weighed

by **Dumas' method**. The sequence of operations is as follows:

(i) A glass bulb or boiling tube with a drawn off tip is first weighed full of air (W_1 g).

(ii) A few cm^3 of the volatile liquid are introduced by warming the bulb and cooling it with the tip held under the surface of the liquid.

(iii) The bulb is placed in a heating bath to volatilize the liquid completely, until no more vapour issues from the bulb.

(iv) Still keeping the bulb in the heating bath, the tip is sealed off with a small flame.

(v) The bulb is cooled, dried, and weighed (W_2 g).

(vi) The tip of the bulb is broken off under water, causing water to rush in and fill the bulb completely.

(vii) The bulb is dried externally, and reweighed together with the broken off tip (W_3 g).

$$\text{Volume of bulb} = (W_3 - W_1)\ \text{cm}^3.$$

The mass of air occupying this volume is calculated, at laboratory temperature and pressure (W_4 g). This is subtracted from W_1 giving the mass the bulb would have had if evacuated. Hence

$$\text{Mass of vapour in bulb} = W_2 - (W_1 - W_4)\ \text{g}.$$

The mass of vapour occupying 22 400 cm^3 at s.t.p. is then calculated: this is the molecular weight.

A heated syringe may also be used to determine the molecular weight of a volatile liquid.

7.8 Determination of atomic weights

An exact knowledge of atomic weights is of importance to the analyst when finding the formula of a compound or the composition of a mixture by means of gravimetric or volumetric measurements. Nowadays very precise values for atomic weights, expressed on the carbon-12 scale, have been determined by means of the mass spectrometer; the accurate atomic masses of the isotopes of an element, and their relative abundance, are determined (section 1.5).

Before the mass spectrometer was invented chemists had to rely on other, often very ingenious, methods of finding atomic weights. Most of these are now of academic interest only, but one method of outstanding interest is **Cannizzaro's method** (1858). Cannizzaro based his argument upon the hypothesis of Avogadro (which we now know as Avogadro's law, but which at that time was not susceptible to experimental verification). Cannizzaro determined the molecular weight of a series of compounds of an element by assuming that the density of a vapour relative to hydrogen under the same conditions is always one-half of the molecular weight. Cannizzaro then determined the number of grams of the element in one mole of each compound. The series of weights thus obtained were found to be whole number multiples of a weight which he took to be the weight of one mole of atoms.

The data in Table 7.1 illustrate the argument which enabled Cannizzaro to fix 35·5 as the atomic weight of chlorine:

TABLE 7.1

Substance	Molecular weight	Mass of chlorine in 1 mole
Chlorine	71	71 g = 2 × 35·5 g
Hydrogen chloride	36·5	35·5 g
Mercury(II) chloride	271	71 g = 2 × 35·5 g
Arsenic(III) chloride	181·5	106·5 g = 3 × 35·5 g
Phosphorus(III) chloride	138·5	106·5 g = 3 × 35·5 g
Iron(III) chloride	325	213 g = 6 × 35·5 g

The argument does not *prove* that the atomic-weight of chlorine is 35·5, but the more compounds of chlorine are taken the more reliable the method becomes. Since no one has ever discovered a compound of chlorine in which one mole contains *less* than 35·5 g of chlorine, the atomic weight of chlorine may safely be assumed to be the least weight present.

The reasoning is in a sense analogous to that applied to Millikan's oil-drop method for determining the charge on an electron: the measurements there do not *prove* that the unit charges gained or lost by an oil drop was equal to the charge on an electron, since there is always a possibility that electrons are gained or lost in pairs, or threes. We therefore take the unit amount of charge gained or lost as the most *probable* value for the charge on an electron, and there is plenty of corroborative evidence to support this assumption.

7.9 Summary

Formulae may be derived in the following ways:

(a) Using *masses*.

$$\text{Moles} = \frac{\text{Mass}}{\text{Molar weight}} = \frac{\text{Volume} \times \text{density}}{\text{Molar weight}} \text{ etc.}$$

The ratio of the numbers of moles of each type of atom gives the formula.

(b) Using *volumes of solutions*.

$$\text{Moles} = \frac{\text{Volume in cm}^3 \times \text{concentration}}{1\,000}$$

Or millimoles (mmol) = vol in $cm^3 \times$ concentration.

(c) Using *gas volumes*.

$$\text{Moles} = \frac{\text{Vol in cm}^3}{22\,400} \text{ (at s.t.p.)}$$
$$= \frac{\text{Vol in cm}^3 \times p \times 273}{22\,400 \times 760 \times T}, \text{ at } p \text{ (in mmHg) and } T \text{ (in K)}$$

(d) In *electrolysis*, a faraday = a mole of electrons.

$$\text{Moles of electrons} = \frac{\text{amps} \times \text{seconds}}{96\,500}$$

Study Questions

1. What are the masses of the following?

(a) 1 mole of carbon atoms.
(b) 5 moles of silicon atoms.
(c) 0·1 mole of germanium atoms.
(d) 0·2 mole of tin atoms.
(e) 0·5 mole of lead atoms.
(f) 1 mole of CO_2.
(g) 1 mole of COS.
(h) 3 moles of $C_6H_{12}O_6$.
(i) 1 mole of $MnCl_2$.
(j) 0·1 mole of $Al_2(SO_4)_3$.

2. How many atoms are there in the following?

(a) 1 molecule of $(CH_3)_2CO$. (c) 15 molecules of CH_4.
(b) 1 mole of $(CH_3)_2CO$. (d) 8 g of CH_4.

3. (a) What is meant by a mole of electrons?

(b) What is (i) the charge, (ii) the mass of a mole of α-particles?

4. The internuclear distance in NaBr is 0·298 nm. The salt crystallizes with the simple cubic (6:6) structure. Work out a value for the density of NaBr.

5. Write down the equations for the following reactions:

(a) 0·653 g of zinc dust were added to an excess of copper sulphate ($CuSO_4$) solution. The copper that formed was filtered, dried and found to weigh 0·635 g.
(b) When 0·52 g of chromium was added to excess silver nitrate ($AgNO_3$) solution, 3·234 g of silver were formed.
(c) 20 cm^3 of M KOH was titrated with 0·4 M H_2SO_4. The reaction was complete when 25 cm^3 of the latter had been added.
(d) 7·17 g of a lead oxide was reduced in a current of hydrogen to give 6·21 g of metallic lead.
(e) 5 cm^3 of M KI reacted with exactly 2·5 cm^3 of M $Pb(NO_3)_2$ to give a yellow precipitate of lead iodide.

6. 6·98 g of a colourless liquid was heated; decomposition into chlorine and 5·56 g of a white solid, $PbCl_2$, occurred readily.

(a) What is the formula of the liquid?
(b) What structure would you expect the liquid to have?

7. When 2·74 g of barium was burned in oxygen, 3·38 g of a white solid was formed.

(a) Deduce the formula of the white solid.
(b) Is this result consistent with your ideas of the periodic law?

8. 2·6 g of a white crystalline compound was heated to 400°C in a vacuum, when it decomposed into 0·92 g of sodium, and nitrogen. What is the formula of the compound?

9. 0·54 g of aluminium was heated in a stream of chlorine to give 2·67 g of a white product, which sublimed at about 200°C. The whole of the product occupied a volume of 224 cm^3 at 760 mmHg and 273 K.

(a) Calculate the empirical formula of the product.
(b) Calculate the molecular formula of the product.

10. 3·2 g of sulphur was burned in air to give 6·4 g of A. A reacted with oxygen in the presence of a catalyst to give 8·0 g of B. With water B gave first 8·9 g of C, and then 9·8 g of D. Identify A, B, C and D.

11. 11·90 g of salt containing only potassium, sulphur and oxygen was heated. Only an acidic reducing gas was evolved, and 8·70 g of a white solid remained. The solid was dissolved in an excess of dilute nitric acid and an excess of barium nitrate added: 11·65 g of a white precipitate formed. Only the potassium ions from the original salt now remained in solution.

(a) What is the empirical formula of the salt?
(b) Write down the equation for the thermal decomposition of the salt.

12. A hydrocarbon of molecular weight 92 was subjected to combustion analysis. When 184 mg was completely burnt in oxygen, 616 mg of CO_2 and 144 mg of water were formed. Calculate (a) the empirical formula, and (b) the molecular formula of the hydrocarbon.

13. A compound was found to contain carbon and hydrogen. When 120 mg of it was completely burned in oxygen, the evolved CO_2 was found to weigh 176 mg and the water 72 mg. Since all tests for other elements proved negative, the experimenter assumed that the compound also contained oxygen.

(a) Deduce the empirical formula of the compound.
(b) A further experiment showed that the molecular weight of the compound was 180. What is the molecular formula?

14. (a) Calculate the mass of solute in the following:

(i) 1 dm^3 of 2 M NaOH.
(ii) 25 cm^3 of 0·1 M H_2SO_4.
(iii) 40 cm^3 of M $KMnO_4$.
(iv) 20 cm^3 of 0·02 M KI.
(v) 4 dm^3 of M/2 HCl.

(b) To what volume would 250 cm^3 of 2 M HCl have to be diluted in order to obtain a 0·1 M solution?
(c) 20 cm^3 of 2 M KOH solution was diluted to 500 cm^3. What is the concentration after dilution?

15. The concentration of thiocyanate ions, CNS^-, in solution can be estimated using silver nitrate solution, $AgNO_3(aq)$. Silver thiocyanate, AgCNS, is precipitated and the end point can be detected because thiocyanate ions form a deep red complex with $Fe^{3+}(aq)$. A thiocyanate solution was titrated into 25 cm^3 of a solution of silver nitrate containing 17 g dm^{-3}. Successive burette readings were 23·9, 23·34, 23·36 and 23·32 cm^3.

(a) Estimate the thiocyanate ion concentration.
(b) Suggest why the first reading was higher than the others.

16. The molar volume of a gas at s.t.p. is 22 400 cm^3. What would be the volume of a mole of an ideal gas at 23°C and 735 mmHg?

17. The following analytical figures were obtained for certain compounds. (a) Write empirical formulae for each compound. (b) Which are stoichiometric and which non-stoichiometric?

(a) 50% S and 50% O.
(b) 38·7% Ti and 61·3% F.
(c) 77·3% Ti and 22·7% O.
(d) 75% Ti and 25% O.
(e) 72·9% Ti and 27·1% O.
(f) 60·9% Fe and 39·1% S.
(g) 76·5% Fe and 23·5% O.

(O = 16, S = 32, Ti = 48, Fe = 56, F = 19.)

18. 15 cm^3 of a gaseous hydrocarbon X was exploded with 100 cm^3 of oxygen. When cool, the resulting gas mixture occupied 70 cm^3; the carbon dioxide was absorbed by potassium hydroxide solution, and the gas volume decreased to 25 cm^3. The remaining gas was shown to be oxygen by absorbing it in alkaline pyrogallol. What is the molecular formula of the hydrocarbon?

19. A mixture of ethane, C_2H_6, ethene, C_2H_4, and hydrogen was obtained after an experiment to test a new catalyst. 24 cm^3 of this was exploded with 100 cm^3 of oxygen. The volume after reaction was 70 cm^3. Excess potassium hydroxide solution absorbed 40 cm^3 of this and alkaline pyrogallol the remainder. What was the volume composition of the mixture?

CHAPTER 8

Molecular motion

8.1 The kinetic theory of gases

According to Boyle's law (Chapter 7) the volume of a given mass of gas should be inversely proportional to its pressure, but this is only approximately true, even for the so-called "permanent" gases. The extent of the deviation is greatest for those gases which are most easily liquefied. Figure 8.1 shows the variation of pV with pressure for three common gases. For an **ideal gas** there would be no variation—Boyle's law would be exactly obeyed—but a gas like nitrogen is found to be only about half as compressible at 1000 atm as Boyle's law would predict. Carbon dioxide at 40°C is found to be *more* compressible than predicted; in fact, at room temperature carbon dioxide condenses to the liquid state on applying pressure alone.

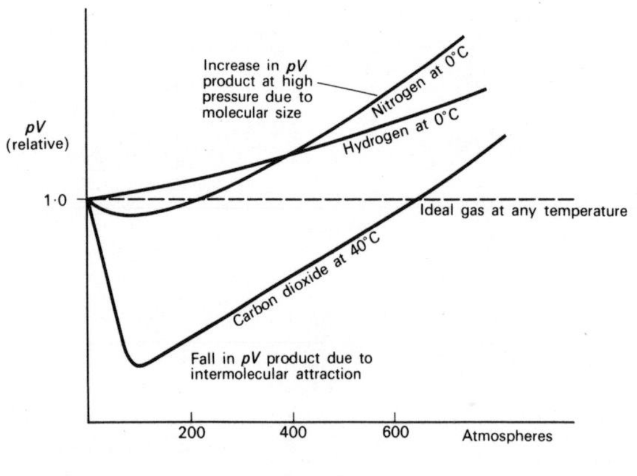

FIG. 8.1.

The **kinetic theory** was developed mainly in the nineteenth century, and is a good example of the scientist's *model*, devised here in order to rationalize the behaviour of gases. The kinetic theory of gases is based upon the following postulates:

(1) Pressure is due to the bombardment of the walls of the containing vessel by molecules of the gas. Evidence for this comes from **Brownian motion**, for instance in smoke particles or colloidal sulphur. Observation with a microscope shows the particles to be in continual erratic motion, due to molecular bombardment.

(2) The average kinetic energy of the molecules in a gas increases as the temperature is raised. Heat energy is kinetic energy of molecular motion.

(3) Collisions occur between molecules and the walls of the containing vessel. Such collisions are perfectly *elastic*, i.e. no energy is dissipated in any other form. If this were not so, the pressure and temperature of a given volume of gas would gradually drop as the molecules slowed down.

(4) Attraction between molecules is negligible. (It will later be shown that this is only true at low pressures where the distance between gas molecules is large.)

(5) The molecules are taken as being infinitely small.

This simple kinetic theory leads directly to Boyle's law and Charles' law. As far as the chemist is concerned, one of the most interesting things is that real gases do *not* obey the gas laws. The kinetic theory can help to explain the deviations.

8.2 Diffusion and effusion

theory came from the work of the Scottish chemist, Graham, in 1846. If a gas is allowed to escape from its container through a small hole into a vacuum, the rate at which molecules escape will depend upon the rate at which they reach the area represented by the hole. This process is called **effusion**. An analogous process is that of **diffusion**, which is the passage of a gas through a porous partition, such as a porous pot. Diffusion is similar to effusion, with a large number of tiny holes through which the molecules can escape instead of one single hole.

Graham's work on the effusion and diffusion of gases led him to the following law which applies equally well to either process:

The rates of effusion (diffusion) of gases under given conditions are inversely proportional to the square roots of their densities.

Effusion rates are most readily compared by taking two gases and allowing them in turn to pass through the same effusion hole, under identical pressure and temperature conditions, comparing the times for the same volume of each gas to pass through. It is found that:

$$\frac{\text{time}_1}{\text{time}_2} = \sqrt{\frac{d_1}{d_2}} \quad \text{or} \quad \frac{\text{rate}_2}{\text{rate}_1} = \sqrt{\frac{d_1}{d_2}}$$

It follows from Avogadro's law that the density, d, of a gas is proportional to its molecular weight, or to the mass m of its molecules; therefore

$$\frac{\text{rate}_2}{\text{rate}_1} = \sqrt{\frac{m_1}{m_2}} \qquad (1)$$

Does Graham's law support the elementary kinetic model of gases? If we were comparing equal volumes of two different gases under identical pressure and temperature conditions, they would contain equal numbers of molecules. Translated into molecular terms therefore, Graham's law suggests that the *number* of molecules of a gas which can effuse per unit time is inversely proportional to the square root of the molecular mass. But the number of molecules passing through the effusion hole per second is directly proportional to the *mean velocity* with which molecules impinge upon the area of the hole. Graham's law in fact suggests that

$$\frac{\text{mean velocity of molecules of gas}_2}{\text{mean velocity of molecules of gas}_1} = \sqrt{\frac{m_1}{m_2}}$$

Let c_1 = mean velocity of the molecules of gas_1
c_2 = mean velocity of the molecules of gas_2

Then,
$$\frac{c_2}{c_1} = \sqrt{\frac{m_1}{m_2}}$$
$$\therefore m_1 c_1^2 = m_2 c_2^2$$
$$\therefore \tfrac{1}{2} m_1 c_1^2 = \tfrac{1}{2} m_2 c_2^2 \qquad (2)$$

Graham's experiments therefore suggest that, if the kinetic theory is true, the mean kinetic energy of the molecules of different gases are the same, at a given temperature.

Diffusion is employed as a method of separating gases or vapours of different molecular weights. It is particularly valuable for separating isotopes, for example, uranium-235 (Chapter 2).

8.3 Boyle's law

Boyle's law is reauiiy explained by the simple kinetic theory. The pressure of a gas is assumed to be due to the bombardment of molecules on the containing vessel. Now,

pressure = force exerted per unit area,
force = rate of change of momentum in a given direction.

If we compress a gas from a volume V to a volume $V/2$ at constant temperature, it may be assumed that, since the temperature does not change, the mean kinetic energy of the molecules remains the same. However, since the gas is now occupying only half its original volume, a given volume of gas will now contain twice as many molecules. Consequently a given area of surface will now receive twice as many collisions per unit time as it did before. The number of collisions per unit time determines the total rate of change of momentum of the particles when they collide. Therefore

Rate of change of momentum per unit area $\propto 1/V$

that is, force per unit area $\propto 1/V$

that is, pressure $\propto 1/V$.

8.4 The pressure exerted by a gas

Consider a gas enclosed in a sphere of radius a (Fig. 8.2). The diagram represents any particle chosen at random, moving with velocity v.

Simple kinetic theory assumes collisions to be perfectly elastic: on this assumption the angle θ at which the particle bounces off the wall equals the angle at which it strikes the wall.

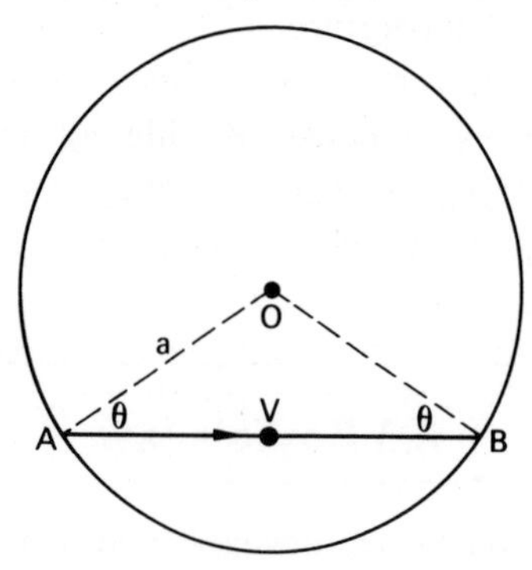

FIG. 8.2.

Triangle OAB is isosceles, therefore the particle always meets the surface at the same angle θ.

Time between collisions, for the particle shown

$$= \frac{AB}{v} = \frac{2a\cos\theta}{v}$$

The contribution of this one particle to the pressure

$$= \frac{\text{rate of change of momentum}}{\text{area of surface}}$$

$$= \frac{2mv\cos\theta}{2a\cos\theta} \cdot \frac{v}{4\pi a^2}$$

$$= \frac{mv^2}{4\pi a^3}$$

where m = mass of the particle.

The total pressure, p, = the sum of the pressure due to all the particles present

$$\therefore p = \Sigma\left(\frac{mv^2}{4\pi a^3}\right)$$

The mean square velocity, $\overline{c^2}$, is equal to the sum of the squares of the velocities divided by the number of molecules, n.

$$\overline{c^2} = \frac{1}{n}\Sigma v^2$$

$$\therefore p = \frac{nm\overline{c^2}}{4\pi a^3}$$

The density of the gas, ρ = mass ÷ volume

$$\rho = \frac{mn}{\left(\frac{4\pi a^3}{3}\right)}$$

$$\therefore p = \tfrac{1}{3}\rho\overline{c^2}$$

Also,

$$pV = \tfrac{1}{3}Lm\overline{c^2}$$

where V = volume of one mole,
L = the number of molecules in a mole (the Avogadro constant).

8.5 Kinetic energy and temperature

The experimental observations of the gas laws lead to the equation

$$pV = RT$$

where V = volume of 1 mole of the gas,
R = the gas constant, which is approximately the same for all gases.

In the previous section we derived the relationship

$$pV = \tfrac{1}{3}Lmc^2$$

where L is now the number of molecules in a mole, the Avogadro constant.

$$pV = \tfrac{2}{3}\times\tfrac{1}{2}Lm\overline{c^2}$$
$$= \tfrac{2}{3}\times \text{total kinetic energy of the molecules.}$$

Therefore the kinetic theory can be reconciled with the statement $pV = RT$ by making the simple assumption that

$$T \propto \text{total kinetic energy of the molecules.}$$

8.6 Avogadro's law and the Avogadro constant

The assumption that the mean kinetic energy of the molecules of gases at a given temperature will be the same, is also in accord with Avogadro's law. Let suffixes denote quantities relating to gas$_1$ and gas$_2$:

$$p_1V_1 = \tfrac{1}{3}n_1m_1\overline{c_1^2} \quad \text{and} \quad p_2V_2 = \tfrac{1}{3}n_2m_2\overline{c_2^2}$$

But,

$$\tfrac{1}{2}m_1\overline{c_1^2} = \tfrac{1}{2}m_2\overline{c_2^2} \quad (\text{since } T_1 = T_2)$$

Therefore, if we take equal volumes of gases ($V_1 = V_2$) at the same temperature and pressure ($p_1 = p_2$), they will contain equal numbers of molecules, i.e.:

$$n_1 = n_2.$$

8.7 Dalton's law of partial pressures

Dalton (1808) noticed that when two or more gases are mixed together, the total pressure exerted by the gas mixture is equal to the sum of the pressures which each separate gas would exert by occupying the space alone. Dalton coined the term **partial pressure** to describe this property. The partial pressure of a gas in a mixture is the pressure which that gas would exert if it alone occupied the same volume. Dalton's law of partial pressures states that the pressure exerted by a mixture of gases is equal to the sum of the partial pressures of the component gases.

Simple kinetic theory explains this behaviour. The mean kinetic energy of the molecules of a given gas depends solely on the temperature, and is not affected by the presence of different molecules. Hence each gas bombards the walls of the containing vessel independently of each other gas. Intermolecular collisions away from the vessel walls will not have any effect on the pressure.

8.8 The van der Waals equation

The fact that real gases do not obey the relationship $pV = RT$ exactly shows that some, at least, of the assumptions of the simple kinetic theory do not always apply. It is necessary to examine some of these assumptions more closely:

(a) *Collisions between molecules, and collisions with the vessel wall, are taken to be perfectly elastic*. This is reasonable for, if it were not so, the molecules would gradually slow down and finally settle out at the bottom of the vessel.

(b) *The molecules are taken to be infinitely small*. For real gases the validity of this assumption is bound to be questioned. The point is whether the volume occupied by actual molecules is negligible compared to the total volume occupied by the gas. It would be a fair assumption with small molecules at very low pressures. Van der Waals (1873) suggested a modified form of the gas equation in which the volume term V was replaced by a term $(V-b)$ where b represents a constant correction term which allows for the actual volume of the molecules. b becomes important when V becomes small, that is, when the gas is highly compressed.

(c) *Intermolecular attraction is taken to be zero*. This assumption becomes progressively less true the nearer the gas molecules are to one another. Again the attraction term becomes strongest when the gas is highly compressed. Van der Waals proposed modifying the pressure term p in the gas equation. If p is the actual pressure which a gas exerts on the containing vessel, this pressure will be *less* than the "ideal" pressure, due to intermolecular attraction. The "ideal" pressure (the pressure that the gas would exert if intermolecular attraction were absent) may be written $\left(p+\frac{a}{V^2}\right)$.

The van der Waals pressure term $\left(p+\frac{a}{V^2}\right)$ is not entirely satisfactory, and several other attempts have been made to improve on it. It now seems likely that no simple equation will ever adequately describe the real pressure exerted by a gas.

Inserting both correction terms, the van der Waals equation reads:

$$\left(p+\frac{a}{V^2}\right)(V-b) = RT$$

The term a will be relatively large for molecules with permanent dipole moments such as hydrogen chloride. What is surprising is that it is also quite large for some molecules which have no permanent dipole moment, such as chlorine. a increases with ease of liquefaction of the gas. Attractive forces between molecules are known as van der Waals forces.

Figure 8.3 shows a comparison between three sets of plots of pressure against volume. (a) is for an ideal gas: it will be seen that no effect such as liquefaction is predicted. (b) shows the actual plots for a real gas, carbon dioxide. It will be

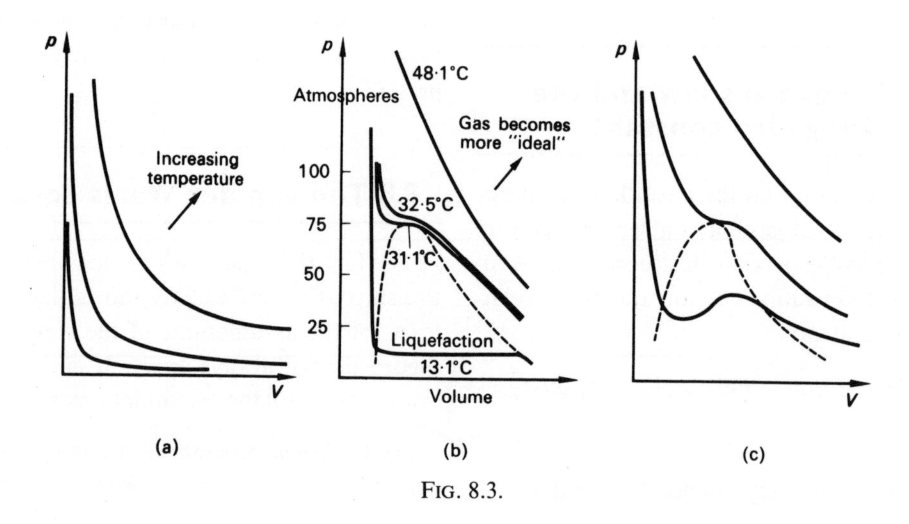

FIG. 8.3.

seen that below 31·1°C, the gas can be liquefied by applying pressure alone. 31·1°C is called the **critical temperature** of carbon dioxide. (c) shows a set of plots using the van der Waals equation, with values of a and b chosen to give the closest possible fit to real conditions. Although graph (c) bears only superficial resemblance to (b) it will be seen that something akin to liquefaction and a critical temperature are in fact predicted.

It is instructive to re-examine Fig. 8.1 at this stage, and attempt to see which of the factors a and b are responsible for the deviations in the plot of pV against p for the gases considered.

8.9 The behaviour of liquids

The liquid state is the least understood of the three states of matter. The kinetic theory of matter must explain how liquids flow, and why they are almost incompressible. Liquids are often described as possessing *short-range order*, and long-range disorder. In other words if a very small region of a liquid, say 1 or 2 nm in diameter, were examined closely, it would appear to be relatively ordered—almost as well ordered as a crystal. If, on the other hand, a larger volume were observed it would appear to be highly disordered.

The disorder of a liquid is thought to be due to the movement of "holes" through the system. The molecules in a liquid can move rapidly relative to one another, provided there are vacancies into which they can move. The "holes" cannot contribute more than a small amount to the total volume of the liquid, otherwise liquids would be compressible.

As the liquid state is so complex, it is hardly surprising that no simple generalizations can be made about their heat capacities. Translation, rotation, and vibration all contribute to the degrees of freedom available to liquid molecules, to varying extents.

8.10 The heat capacity of solids

Despite the wide diversity in structure which solid elements possess, the remarkable fact is that practically all of them have similar heat capacities per mole. This quantity is often referred to as the **atomic heat capacity**. For most solid elements it is in the region 25 to 27 J mol^{-1}, as shown by the data in Table 8.1.

This fact was first discovered as early as 1819 by Dulong and Petit, and is commonly known as **Dulong and Petit's law**. It was later shown that certain elements, notably beryllium, boron, carbon and silicon, do not obey the law at room temperature, though they do as the temperature is raised. Further work has shown that the law breaks down for all elements at very low temperatures. Nevertheless, the fact that elements

TABLE 8.1

Element	Atomic weight	Specific heat (J g^{-1} K^{-1})	Atomic heat J mol^{-1} K^{-1}
Li	7	3·8	27
Al	27	0·87	23
Ca	40	0·63	25
Fe	56	0·43	24
Ag	108	0·23	25
I	127	0·22	28
U	238	0·11	26

with such widely different atomic weights as lithium and uranium should have approximately the same atomic heat capacity cannot be mere coincidence, and it is due to these elements behaving as ideal solids (compare with ideal gases) at high temperatures. In an ideal solid, the atoms in the lattice vibrate *independently*, each in three possible directions; *ideal* solids will all have the same atomic heat capacity and it can be shown theoretically that this value should be about 25 J mol^{-1} K^{-1}, the average value in Table 8.1.

8.11 Kinetic theory and rate of chemical reactions

Most chemical reactions proceed faster as the temperature is raised; this is what would be expected from simple kinetic theory. Simple kinetic theory fails however to account for the very pronounced effect which a rise in temperature has on most chemical reactions. The rates of many chemical reactions are approximately doubled by raising the temperature about 10 K.

An understanding of simple kinetic theory—the various ways in which the particles of matter can move in different states—is essential to the broader understanding of chemical processes. **Chemical kinetics** (Chapter 17) is the study of rates of chemical reactions, and the deduction of the nature of collision processes and other effects from them.

Study Questions

1. (a) Which of the following would diffuse most quickly under comparable conditions?

(i) Ar, Kr or Xe.
(ii) H_2O, HDO or D_2O.
(iii) $^{235}UF_6$ or $^{238}UF_6$.

(b) In (iii), calculate the relative rates of diffusion under comparable conditions.

2. (a) Under comparable conditions, 100 cm^3 of nitrogen diffused in 21 s while 100 cm^3 of a coloured vapour diffused in 63 s.

(i) Calculate the approximate molecular weight of the vapour.
(ii) Suggest what the colour of the vapour might be.

(b) A sample of 50 cm^3 of neon diffused in 26 s, while a sample of 50 cm^3 of a compound of empirical formula BNH_2 diffused in 52 s under comparable conditions. Calculate the molecular formula of the compound.

3. Predict the specific heat capacities, in J g^{-1} K^{-1}, of cobalt (at. wt. 59), osmium (at. wt. 190), silver (at. wt. 108), and sodium (at. wt. 23).

4. The following densities (all in g cm^{-3}) were obtained for nitrogen and water:

	Solid	Liquid	Gas
Nitrogen	1·02	0·81	0·0012
Water	0·91	1·00	0·0008

(a) Estimate the approximate volume that is taken up by a molecule in each phase.
(b) Contrast the results obtained for each substance.
(c) Is it nitrogen or water that is behaving in an unexpected manner?

5. Although Avogadro first formulated his hypothesis in 1811, it remained unknown for many years. Use your library to find out (a) why this was, and (b) when Avogadro's hypothesis was first accepted by chemists.

6. Explain the deviations in the plots of pV against p in Fig. 8.1 in terms of the van der Waals parameters a and b.

7. The rate of decomposition of HI at 400°C is almost three hundred times as fast as it is at 300°C (Bodenstein, 1899). Can this be accounted for by the increased number of collisions that occur at the higher temperature due to the increased molecular speeds?

CHAPTER 9

Enthalpy

9.1 The importance of energy changes in chemistry

Two important things which a chemist must understand about the behaviour of matter concern chemical reactions. He must ask:

(i) How *fast* does a chemical reaction go?
(ii) How *far* does it go?

The heat energy change of a chemical reaction may give a *partial* explanation of its kinetics (how fast), and its position of equilibrium (how far), although heat is not the *only* factor involved. It is a matter of common experience that chemical processes tend to reach a position of equilibrium, after which no further spontaneous change occurs. Quite often a spontaneous change takes place with the absorption of heat, indicating that the system has attained a state of higher energy. Cooling by evaporation is a well known example; another is the dissolution of sodium nitrate in water to form an aqueous solution, where a very noticeable cooling can be observed. Figure 9.1 represents the change

$$NaNO_3(s) + aq \rightarrow Na^+(aq) + NO_3^-(aq); \quad \Delta H = +21 \text{ kJ mol}^{-1}$$

by means of an energy diagram.

Other spontaneous changes take place with the evolution of heat energy, for instance the burning of hydrogen in oxygen. Although heat energy is not the only factor influencing chemical change, it is very important.

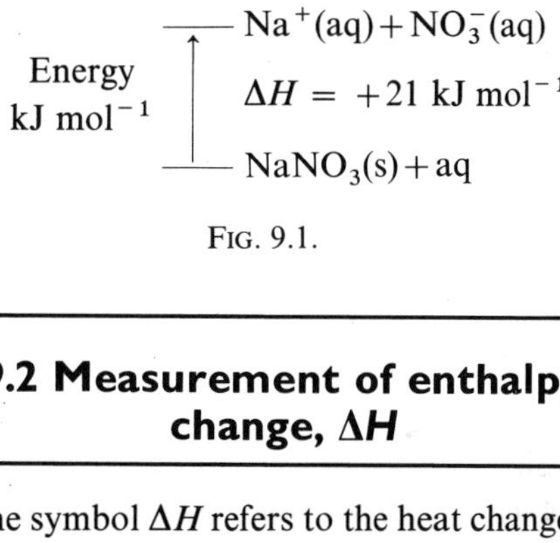

Energy kJ mol^{-1}	
	$Na^+(aq) + NO_3^-(aq)$
↑	$\Delta H = +21$ kJ mol^{-1}
	$NaNO_3(s) + aq$

FIG. 9.1.

9.2 Measurement of enthalpy change, ΔH

The symbol ΔH refers to the heat change of a process at constant pressure, often known as the **enthalpy** change.* Various ways are available for measuring ΔH, depending on the reaction conditions. Whatever quantities are actually used in the experiment, the enthalpy change is generally converted into molar units. For instance a calorimetric measurement may show that 33 J are required to convert 0·1 g of ice at 0°C into water at the same temperature. This is often expressed

* Many data books use composite symbols, in which a superscript $^\ominus$ denotes that standard conditions (1 atm) are observed, and a subscript numeral (usually 298) denotes the temperature to which the measurement refers. For instance $\Delta H^\ominus_{298}$ denotes the standard enthalpy change at 298K (25°C). In this book all ΔH data are quoted for 25°C unless otherwise stated.

by saying that the specific latent heat of fusion of ice is 330 J g^{-1}. However the chemist expresses quantities per mole. ΔH for the reaction

$$H_2O(s) \rightarrow H_2O(l) \text{ is}$$

$$\frac{(330 \times 18)}{1000} = +5{\cdot}94 \text{ kJ mol}^{-1}.$$

(a) Calorimetric measurements

Since many chemical reactions occur in dilute aqueous solutions, the specific heat capacity of which can be assumed approximately to equal 4·2 J g^{-1}, a simple measurement of temperature change when two aqueous solutions are mixed gives a rapid estimate of ΔH. For instance, suppose 1 mole (98 g) of concentrated sulphuric acid (100%) was added to water to make 1 litre of solution, and a temperature rise of 17 K was noted. If we assume the specific heat capacity of a molar solution of sulphuric acid to be 4·2 J g^{-1}, we may write

$$H_2SO_4(l) + aq \rightarrow H_2SO_4(aq, \text{M});$$

$$\Delta H = -(17 \times 4{\cdot}2) \simeq -71 \text{ kJ mol}^{-1}.$$

Note that the rise in temperature in degrees is numerically equal to the heat evolved in kJ $mol^{-1} \div 4{\cdot}2$, in this case.

Similarly the neutralization of an acid with an alkali can be observed in this way. Suppose 100 cm^3 of M NaOH(aq) were added to 100 cm^3 of 0·5 M H_2SO_4(aq), and a rise in temperature of 6·8 K was noted. Assuming the specific heat capacities of the solutions to be 4·2 J g^{-1},

$$\text{Heat evolved} = 6{\cdot}8 \times 200 \times 4{\cdot}2 \simeq 5600 \text{ J}$$

This is for mixing 0·1 mole of OH^-(aq) with 0·1 mole of H^+(aq)—note that every mole of H_2SO_4 gives rise to 2 moles of H^+(aq)—and so we may write:

$$H^+(aq) + OH^-(aq) \rightarrow H_2O(l);$$

$$\Delta H = -56 \text{ kJ mol}^{-1}.$$

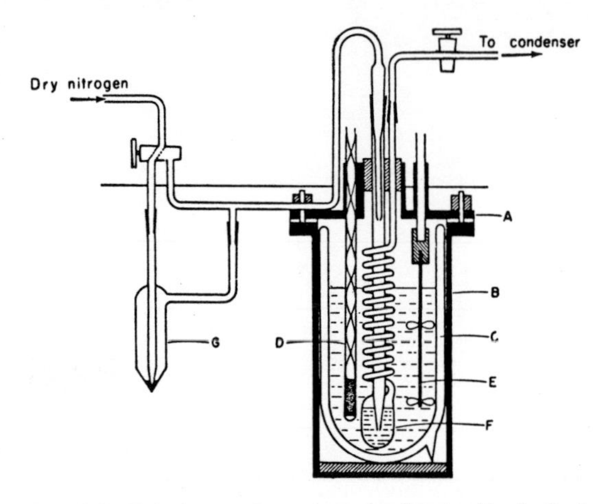

Fig. 9.2. Calorimeter for measuring heats of hydrolysis (with water vapour, at 25°C). A, lid; B, can; C Dewar-vessel; D, thermistor; E, stirrer; F, reaction chamber; G, water saturator.

For molar solutions the rise in temperature in degrees is numerically *one half* of the heat evolved in kJ $mol^{-1} \div 4{\cdot}2$, unlike the first case.

It is possible to find the end-point of an acid-alkali titration by taking various volumes of solutions and mixing them, noting the temperature rise in each case. This forms the basis of a **thermometric titration** (see Study Question 6).

The apparatus used for calorimetric measurements depends on the degree of accuracy required. For very rough measurements nothing more elaborate is needed than a thin plastic beaker or bottle, mounted in an insulating support, such as a block of foam polystyrene. For more accurate work a Dewar flask is essential. Figure 9.2 shows a typical experiment.

(b) Electrical measurements

An arrangement is sometimes employed, in which the heat produced by a chemical reaction is directly compared with the heat generated by an electric heating coil under identical conditions. In this way errors due to heat loss may be compensated. This type of measurement gives the result directly in joules:

Current (amperes) × time (seconds) × P.D. (volts)
= charge (coulombs) × P.D. (volts)
= energy (joules).

(c) The flame calorimeter

Simple measurements of heats of combustion require nothing more sophisticated than a burner with a wick, arranged to heat a known mass of water in a suitable heat exchanger. The weight of fuel burned is noted. This method is limited to inflammable liquids, and is in any case rather inaccurate, but is instructive for comparing such things as the molar heats of combustion of members of a homologous series, for instance aliphatic alcohols (see Study Questions 1, 2, 3). Figure 9.3 shows a more accurate arrangement, used for measuring the heat of combustion of a gas like hydrogen.

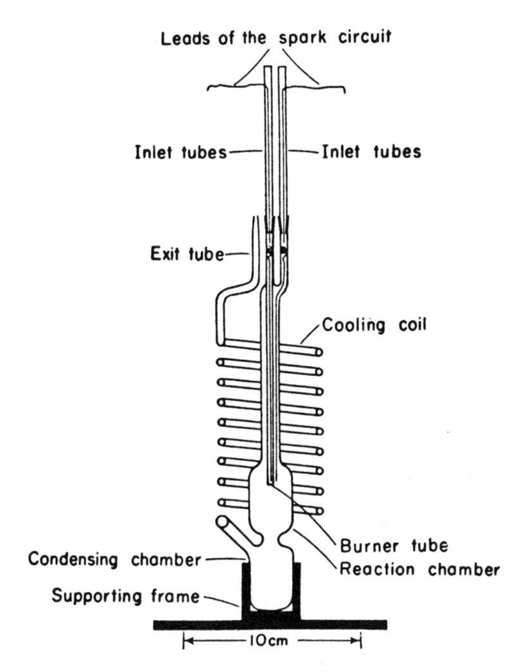

Fig. 9.3. A flame calorimeter.

(d) The bomb calorimeter

For more accurate work a bomb calorimeter is used, the original design being due to Berthelot (1867). The substance is placed in a platinum crucible in a cylindrical steel bomb lined with a resistant enamel, which is immersed in water. The bomb is filled with oxygen at a pressure greater than atmospheric, and the combustion is triggered off with a small platinum resistance wire. The heat evolved is measured by noting the rise in temperature of the water, and a small correction is made for the heating effect of the platinum resistance wire. Since the bomb calorimeter measures the temperature rise at constant volume rather than at constant pressure, a further correction has to be made to obtain a value of ΔH. The advantage of using pure oxygen is that the combustion is very rapid, and is certainly complete (Fig. 9.4).

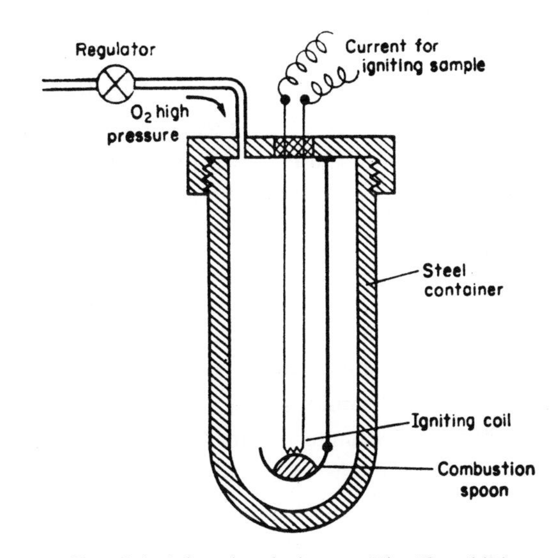

Fig. 9.4. A bomb calorimeter. The "bomb" is immersed in water (not shown).

(e) Spectroscopic measurements

The heats of reaction of many processes involving ionization in the gas phase can be de-

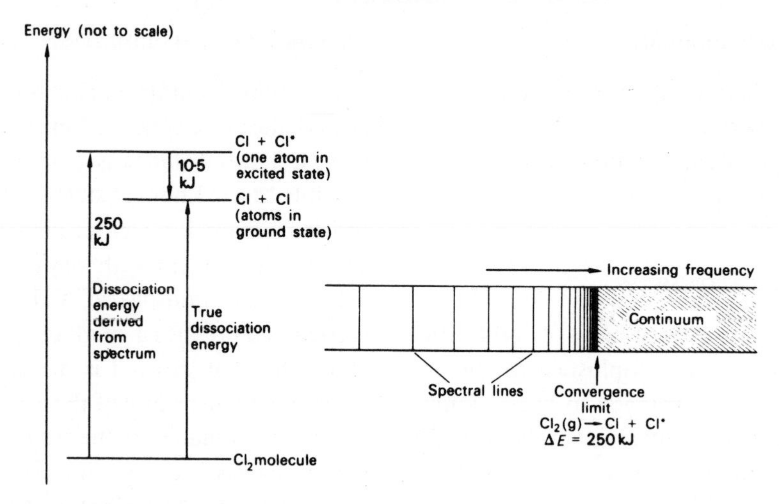

FIG. 9.5. Idealized vibrational spectrum of a gas; (real gases often show more than one series of lines).

termined by measuring the wavelength and hence the frequency of the radiation absorbed in the change. In Chapter 3, the principles of the method used to determine the ionization energies of the alkali metals (e.g. $Na(g) \rightarrow Na^+(g)$) were outlined.

Spectroscopic measurements can also be used to determine the energy needed to dissociate molecules into atoms in the gas phase. As an example consider chlorine, Cl_2. When the chlorine molecule is made to vibrate, it absorbs a given quantity of energy and this gives rise to the absorption of radiation of definite frequency. The vibrational spectrum of chlorine consists of a number of lines, each at a definite frequency, which converge to a limit, beyond which energy is continuously absorbed (Fig. 9.5). The energy of this convergence limit corresponds to the dissociation energy, the energy necessary to split the molecule into atoms.

If this were the whole story, it would be a simple matter to compute dissociation energies from spectroscopic measurements: unfortunately some of the energy is also used to form an electronically excited chlorine atom, Cl*, and this energy must be accounted for in the calculation of the dissociation energy of chlorine.

$$Cl_2(g) \rightarrow Cl + \underset{\text{excited atom}}{Cl^*}$$

The convergence limit for chlorine is 478·5 nm. Using the relationship $\Delta E = h\nu$, we have:

$$\text{Wavelength, } \lambda, = 478{\cdot}5 \times 10^{-9} \text{ m}$$

$$\text{Frequency, } \nu, = \frac{3 \times 10^8}{478{\cdot}5 \times 10^{-9}} = 6{\cdot}27 \times 10^{14} \text{ s}^{-1}.$$

$$\Delta E = h\nu$$
$$h = 6{\cdot}62 \times 10^{-34} \text{ J s}$$
$$\therefore \Delta E = 6{\cdot}62 \times 10^{-34} \times 6{\cdot}27 \times 10^{14}$$
$$= 4{\cdot}15 \times 10^{-19} \text{ J molecule}^{-1}$$
$$= 4{\cdot}15 \times 10^{-19} \times 6{\cdot}022 \times 10^{23} \text{ J mol}^{-1}$$
$$= 2{\cdot}50 \times 10^5 \text{ J mol}^{-1}$$
$$= 250 \text{ kJ mol}^{-1}$$

Atomic excitation energy is 10·5 kJ, so the dissociation energy of chlorine is 239·5 kJ mol^{-1}, in good agreement with the value of 238 obtained from thermal measurements.

The method, however, has severe limitations; it can only be applied to very simple molecules, and it is necessary to know the electronic states of the atoms after dissociation so that corrections can be applied. Within these limitations, the method is potentially extremely accurate.

(f) CALCULATION OF ΔH FROM EQUILIBRIUM CONSTANT MEASUREMENTS

In many cases an accurate value for the heat change of a reaction can be derived from measuring the way in which the equilibrium composition of a mixture in a reversible reaction varies with temperature (section 10.9).

(g) DERIVATION OF ΔH USING AN ENERGY CYCLE

It frequently happens that a measurement cannot be made directly on a chemical change, though separate measurements can be made on the values of ΔH for the process carried out in stages.

Hess's law of constant heat summation states that **the total heat evolved or absorbed in a given chemical reaction is the same whether the reaction proceeds directly or in a series of stages, and is dependent solely on the states of the initial reactants and final products**. Hess's law is really just another way of stating the law of conservation of energy: if it were not true it would be possible to reverse a given reaction using a different route with a different corresponding heat change. This is impossible since it would involve either the creation or the destruction of energy.

An important assumption which forms a part of Hess's law, and which is readily demonstrated experimentally in many cases, is that if a reaction is reversed then ΔH of the reverse reaction will be *minus* the value for the forward reaction. For instance, it is not at all easy to make direct measurements of the heat of reaction of $HCl(g) \rightarrow \frac{1}{2}H_2(g) + \frac{1}{2}Cl_2(g)$. The reverse process is, however, readily carried out: when one mole of hydrogen chloride is formed by burning hydrogen in chlorine, 92 kJ are evolved. Hence ΔH for the decomposition of hydrogen chloride is $+92$ kJ mol^{-1}.

Hess's law may be illustrated by means of an energy level diagram, Fig. 9.6. ΔH is the algebraic sum of ΔH_1, ΔH_2, ΔH_3 and ΔH_4.

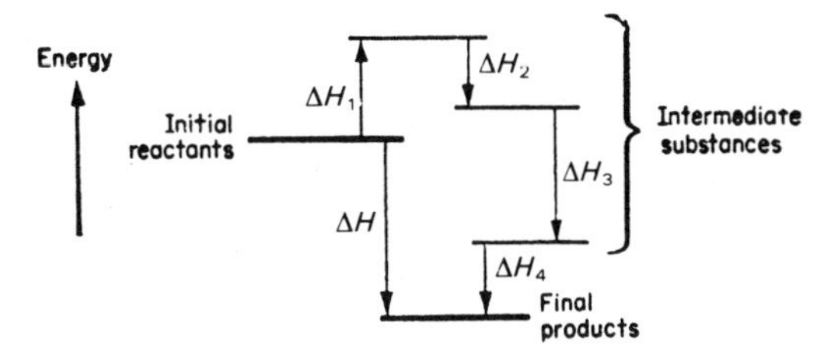

Hess's law: $\Delta H = \Delta H_1 + \Delta H_2 + \Delta H_3 + \Delta H_4$ (Algebraic sum)

FIG. 9.6.

9.3 Heat of formation of a substance

Since energy plays such an important part in our understanding of chemical substances, it is important that we should have some idea of the energy "stored up" in substances. For instance, if hydrogen gas is blown through an electric arc struck between carbon electrodes, some ethyne, C_2H_2, is formed. An extremely high temperature (above 2000°C) is needed in order to form appreciable quantities of ethyne and it is not possible to measure the heat of reaction directly:

$$2C(s) + H_2(g) \rightleftharpoons C_2H_2(g); \quad \Delta H = +240 \text{ kJ}$$

The heat of the above reaction, termed the **heat of formation** of ethyne, can be deduced from separate measurements of the **heats of combustion** of the three substances involved, carbon, hydrogen and ethyne. **Heat of combustion is defined as the heat evolved when one mole of a**

substance is completely burned in oxygen. (A minus sign denotes that heat is evolved.)

$$C(s)+O_2(g) \rightarrow CO_2(g); \quad (1)$$
$$\Delta H = -395 \text{ kJ}$$
$$H_2(g)+\tfrac{1}{2}O_2(g) \rightarrow H_2O(l); \quad (2)$$
$$\Delta H = -280 \text{ kJ}$$
$$C_2H_2(g)+\tfrac{5}{2}O_2(g) \rightarrow 2CO_2(g)+H_2O(l); \quad (3)$$
$$\Delta H = -1310 \text{ kJ}$$

Remembering that reactions can be theoretically reversed, and the sign of their ΔH changed, the heat of formation of ethyne can be derived by taking $2 \times (1)+(2)-(3)$.

In order that data may be listed for reference purposes, heats of formation are quoted at a standard temperature, usually 25°C (298 K) which is conveniently near the temperature of the average laboratory. The substances must be in standard states and if there is any ambiguity, the state is given. For elements the standard used is that state of the element at 1 atm pressure and in its stable allotropic form.

Heat of formation is defined as the quantity of heat evolved or absorbed when one mole of a compound is formed from its elements in their standard states. It is often denoted by the symbol $\Delta H_f^\ominus$.

A reaction which evolves heat is called an **exothermic reaction**. In an exothermic reaction the products will be shown on an energy level diagram *below* the reactants. A compound formed exothermically from its elements is called an *exothermic compound.* Carbon dioxide and water are examples of exothermic compounds.

A reaction which takes in heat is called an **endothermic reaction**. For instance the formation of ethyne from its elements is an endothermic reaction, and ethyne itself is called an *endothermic compound.* An endothermic compound can be regarded as having a larger amount of stored energy (it is higher up on the energy level diagram), and will give out a considerable amount of heat when burned in oxygen. The oxygen-ethyne (oxy-acetylene) flame produces an extremely intense heat (about 3000°C) for this reason.

A reaction which neither evolves nor absorbs heat is sometimes termed a **thermoneutral** reaction.

9.4 Photochemical reactions

The endothermic formation of ethyne from its elements was caused to occur spontaneously by the choice of a very high temperature, in the example chosen above. Many endothermic compounds can in fact be formed in this way, and it is a fact of common experience that endothermic species tend to be stable at high temperatures. For instance if a diatomic gas is heated strongly enough it will dissociate into atoms, and these atoms are formed endothermically. Iodine vapour above 1000°C is present largely as atoms —an instance of the endothermic species being stable at the higher temperature. However, it is often not possible to form an endothermic compound simply by heating the appropriate elements together to a high enough temperature.

Some endothermic reactions proceed as a result of the absorption of electromagnetic radiation. Such reactions are known as **photochemical reactions**. Ultra-violet energy is the most useful for this purpose—X-rays have so much energy that they tend to disrupt molecules completely, while visible and infra-red light frequently contains insufficient energy. In the leaves of plants carbohydrates are formed endothermically from carbon dioxide and water. Light is essential for this reaction, it being absorbed in plant cells by coloured substances of which the green compounds, the chlorophylls, are examples. The detailed mechanism of the formation of carbohydrates is not yet understood.

Biochemistry provides many other examples of endothermic reactions, such as the formation of the metabolic substance, adenosine triphosphate, ATP, a substance which supplies energy which can be used in our muscles to do mechanical work. This is illustrated with an energy level diagram in Chapter 12, Fig. 12.3.

A mixture of hydrogen and chlorine will explode in the presence of sunlight. Even though the overall process is exothermic, the *mechanism* involves an endothermic stage, namely the dissociation of chlorine into atoms by the absorption of ultra-violet light. Chlorine atoms are highly reactive and give rise to a chain reaction with a sudden release of large amounts of energy. Chain reactions are dealt with in Chapter 17.

9.5 Energy changes when substances dissolve

The most useful generalization which can be made about the tendency of substances to dissolve in one another is *like dissolves like*. Even this is only a very rough rule-of-thumb, however, and we need to examine the energy changes a little more closely in an attempt to understand what is really happening.

(a) *Molecular substances*, such as iodine, white phosphorus, paraffin wax and so on, tend to dissolve in liquids which are themselves molecular, and non-polar. Highly polar liquids such as water are very poor solvents for essentially non-polar molecules.

(b) *Giant lattices* may or may not dissolve in solvents: those with extremely strong forces holding the atoms together have little tendency to dissolve in anything at all, as too much energy is required to break up the lattice; those which can break up to form ions may dissolve in polar solvents such as water or liquid ammonia. Metallic lattices do not dissolve except in other metals, such as mercury: mercury will dissolve sodium to form an alloy known as an *amalgam*. When sodium is said to "dissolve" in water the word "dissolve" is being used rather loosely because a violent chemical reaction has occurred in which the sodium has formed $Na^+(aq)$ and the water has been changed into $OH^-(aq)$ and $H_2(g)$.

Lattices held together by covalent bonds, rather than by ionic forces, are generally insoluble in all solvents.

The dissolving of an ionic solid in a solvent can be visualized as taking place in two separate stages:

(i) The lattice must be broken into separate ions. The energy per mole required to effect this is known as the **lattice energy**.

(ii) The separate ions attach themselves electrostatically to the molecules of solvent. This is the attraction between a single charge and a dipole, known as an ion-dipole interaction. Energy will be released at this stage of the process. The amount of energy released when one mole of ions is combined with the solvent in this way is termed the **solvation energy**.

We can understand the process better with the aid of another energy level diagram, though it is important to bear in mind that stages (i) and (ii) do *not* take place separately. We can imagine, using Hess's law, that a mole of sodium chloride could first be converted *theoretically* into gaseous ions and that these ions could then be added to water. Hess's law states that the overall energy change is the same whatever the route chosen. Figure 9.7 is an attempt to show diagrammatically what really does happen, as far as we know, when sodium chloride dissolves in water. Figure 9.8 is an enthalpy diagram which analyses the process into stages (i) and (ii) above.

Experiments show that when sodium chloride is dissolved in water the rise or fall in temperature is negligible—we say that its heat of solution in water is practically zero ($\Delta H = +5{\cdot}3$ kJ mol^{-1}).

Heat of solution is defined as the heat change

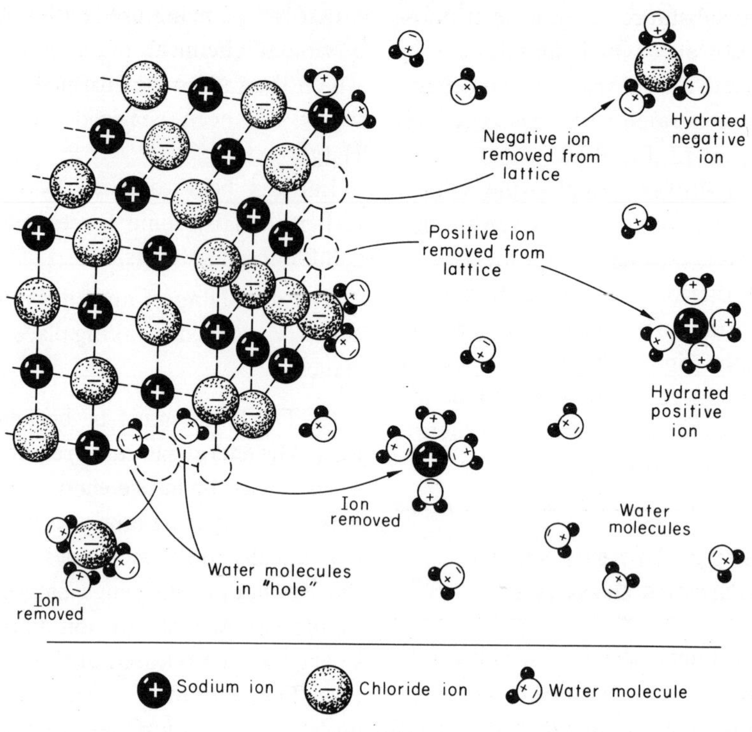

FIG. 9.7.

which takes place when one mole of a solute is added to a solvent to make a solution of stated concentration. If a given solution is diluted further there is usually a further small heat change, and this is referred to as heat of dilution. The more dilute a solution becomes the smaller will be the heat changes on further dilution, and for purposes of quoting data the heat of solution at infinite dilution may be referred to.

If the solvation energy of a substance is greater than its lattice energy, the substance will dissolve exothermically in the solvent. Anhydrous copper (II) sulphate and sodium hydroxide are examples. If the solvation energy is less than the lattice energy then the substance will dissolve endothermically. Sodium nitrate, ammonium chloride, and many other highly soluble substances provide examples.

An ionic salt like sodium chloride fails to dissolve in a non-polar solvent, because there is no solvation energy. Water has a very large dipole moment, so that ion-dipole interactions will be high. A solvent like benzene or hexane is not attracted to ions in this way, and there is no gain in solvation energy to overcome the lattice energy term. Uncharged, non-polar solvent molecules such as benzene do not have the ability to insert themselves in the ionic lattice in order to break it up.

9.6 Other factors involved when substances dissolve

A simple consideration of ΔH does not provide us with all the explanations of why changes occur. Why for instance does iodine dissolve in benzene? Clearly the energy factor here is very small, because iodine is a molecular lattice and the molecules I_2 and C_6H_6 are both non-polar.

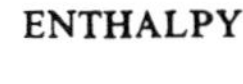

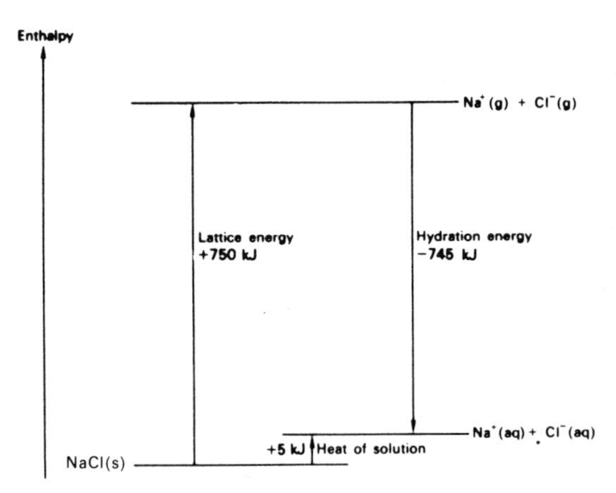

FIG. 9.8. Heat of solution of sodium chloride.

An even more extreme case is the question why gases dissolve in one another. Why does neon gas mix with argon gas completely in all proportions? There is no enthalpy loss or gain at all here, and the problem certainly cannot be treated in terms of energy levels. This aspect of the problem will be taken up in Chapter 12.

9.7 The Born–Haber cycle

Provided that large energy changes are involved, ΔH is generally the overriding factor determining why chemical reactions take place. One chemical reaction of general interest is the combination of a metal with a non-metal to form an ionic solid, which releases considerable energy. For instance, the alkali metals (Group IA) and the alkaline earth metals (Group IIA) combine vigorously with the more electronegative non-metals such as fluorine, oxygen and chlorine. The question why such elements combine exothermically, can be answered by constructing an enthalphy diagram, in which the process is broken up into separate theoretical stages. Hess's law is assumed to apply. The whole process is referred to as an energy cycle, this particular sequence being named the **Born–Haber cycle** (1919).

The reaction between sodium and fluorine forming solid sodium fluoride is taken as a typical example, though a similar argument could be applied to similar pairs of elements. Direct measurement of ΔH gives

$$Na(s)+\tfrac{1}{2}F_2(g) \rightarrow NaF(s); \quad \Delta H = -570 \text{ kJ mol}^{-1}.$$

The process can be imagined in various stages (Fig. 9.9).

(1) Sublimation: $Na(s) \rightarrow Na(g)$; $\quad \Delta H_1 = +109$ kJ

(2) Ionization: $Na(g) \rightarrow Na^+(g)+e^-$; $\quad \Delta H_2 = +495$ kJ

(3) Dissociation: $\tfrac{1}{2}F_2(g) \rightarrow F(g)$; $\quad \Delta H_3 = +75$ kJ

(4) Electron capture: $F(g)+e^- \rightarrow F^-(g)$; $\quad \Delta H_4 = -338$ kJ

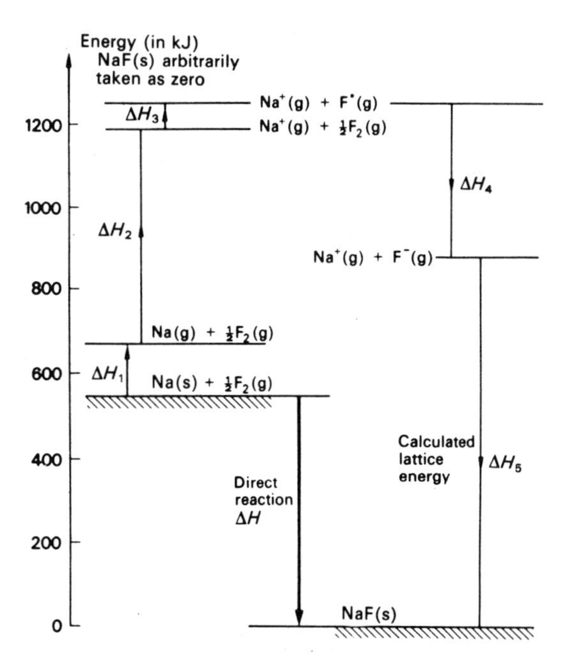

FIG. 9.9. The Born–Haber cycle for sodium fluoride, shown as an enthalpy diagram.

(5) Lattice formation: $Na^+(g)+F^-(g) \rightarrow NaF(s)$;

$$\Delta H_5 = -900 \text{ kJ}$$

ΔH_5 is the **lattice energy** of sodium fluoride. Figure 9.9 shows that $\Delta H = \Delta H_1+\Delta H_2+\Delta H_3+\Delta H_4+\Delta H_5$. The same result could of course be obtained without actually constructing an energy level diagram simply by treating equations (1) to (5) algebraically.

The lattice energy of sodium fluoride, ΔH_5, cannot be measured directly, but can be calculated using the expression:

$$\text{Lattice energy} = \frac{-Lz^+z^-e^2M}{4\pi\varepsilon_0 r}\left(1-\frac{1}{n}\right),$$

where L = the Avogadro constant
z^+ and z^- = the charges on the cation and anion
e = the electronic charge
ε_0 = the permittivity of a vacuum
r = the internuclear distance. (The sum of the anionic and cationic radii.)

$\left(1-\frac{1}{n}\right)$ is a term introduced to allow for the repulsive forces caused by compressing the crystal. M, the **Madelung constant**, is a dimensionless number that depends only on the type of co-ordination in the crystal. Lattice energies are greatest when the ionic charges are high and when the ions are small.

The Born–Haber cycle provides a very good test of whether the ionic model which we use for sodium fluoride is valid. Direct measurement of ΔH gives -570 kJ mol^{-1}. Application of the Born–Haber cycle gives $\Delta H = 109+75+495-900-338 = -559$ kJ mol^{-1}. The assumption made in the Born–Haber cycle is that solid sodium fluoride can be regarded as being made up of singly charged ions. The very good agreement with experiment shows the assumption to be a good one in this case, though in many of the cases met later in this book the agreement is less good, suggesting that the ionic model does not apply.

Born–Haber cycles for CaF, CaF_2 and CaF_3 can help to show why the only binary compound of calcium and fluorine so far prepared is CaF_2. The various terms in the cycle are summarized in Table 9.1.

The results are shown on Fig. 9.10. The valucs of the lattice energies of CaF and CaF_3 have been estimated.

The large endothermic heat of formation of CaF_3 means that it is most unlikely that it will ever be prepared. The main reason for this is apparent from Fig. 9.10: a great deal of energy is needed to ionize calcium's third electron from an inner shell.

CaF, on the other hand, would be formed exothermically, but its decomposition into Ca and CaF_2 would also be appreciably exothermic:

$$2CaF(s) \rightarrow Ca(s)+CaF_2(s) \qquad \Delta H = -611 \text{ kJ}$$

9.8 Bond energy

The **bond energy** of a diatomic molecule is easy to understand: it is simply the molar energy of dissociation of the molecule into atoms. Several values have already been quoted in this book.

The energy of a bond in a polyatomic molecule is not always something we can directly measure. For instance, if water is dissociated into atoms, the process can occur in two stages, each with a different intake of energy:

$$H_2O(g) \rightarrow H(g)+OH(g); \qquad \Delta H = +490 \text{ kJ} \quad (1)$$

$$OH(g) \rightarrow H(g)+O(g); \qquad \Delta H = +424 \text{ kJ} \quad (2)$$

This does not mean that the two H—O bonds

TABLE 9.1

Term	→Description	$n = 1$ (kJ)	$n = 2$ (kJ)	$n = 3$ (kJ)
ΔH_1	$Ca(s) + \frac{n}{2}F_2(g) \rightarrow Ca(g) + \frac{n}{2}F_2(g)$	+177	+177	+177
ΔH_2	$Ca(g) + \frac{n}{2}F_2(g) \rightarrow Ca^{n+}(g) + \frac{n}{2}F_2(g)$	+590	+1730	+6670
ΔH_3	$Ca^{n+}(g) + \frac{n}{2}F_2(g) \rightarrow Ca^{n+}(g) + nF(g)$	+79	+158	+237
ΔH_3	$Ca^{n+}(g) + nF(g) \rightarrow Ca^{n+}(g) + nF^-(g)$	−335	−670	−1005
ΔH_5	$Ca^{n+}(g) + .\quad {}^{-}(g) \rightarrow CaF_n(s)$	−800?	−2580	−5400?
ΔH	$Ca(s) + \frac{n}{2}F_2(g) \rightarrow CaF_n(s)$	−287	−1185	+679

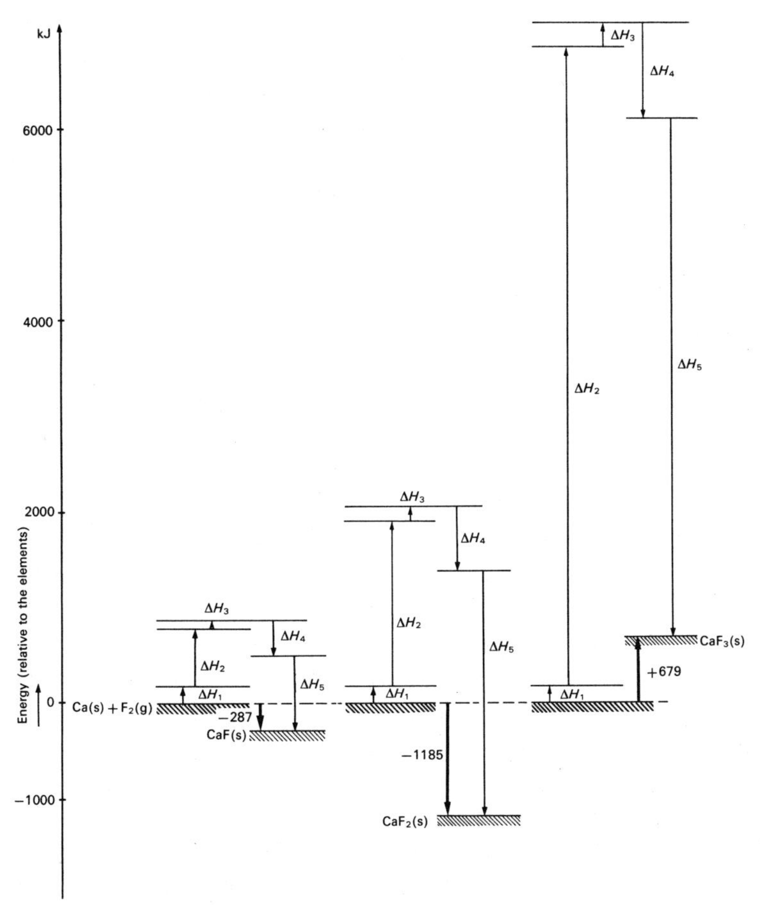

FIG. 9.10. The heats of formation of CaF, CaF_2 and CaF_3.

in a water molecule differ from one another; it simply means that we are considering two different reactions in (1) and (2) above.

The bond energy of O—H in water is one-half of the energy required to atomize the entire molecule in one step, breaking both bonds simultaneously. The energy for this process cannot be directly measured, but it can be calculated by applying Hess's law, and adding equations (1) and (2).

$$H_2O(g) \rightarrow 2H(g) + O(g); \quad \Delta H = 914 \text{ kJ}$$

$$\therefore \text{ Bond energy of O—H in water} = \frac{914}{2} = 457 \text{ kJ mol}^{-1}.$$

Similarly the bond energy of C—H in methane is one-quarter of the total energy needed to atomize a mole of methane:

$$CH_4(g) \rightarrow C(g) + 4H(g);$$
$$\Delta H = 1660 \text{ kJ (calculated using Hess's law).}$$

$$\therefore \text{ Bond energy of C—H in methane} = 1660/4 = 415 \text{ kJ mol}^{-1}.$$

The concept of bond energy is an important one to chemists even though, like the concept of atomic radius, it turns out to be only approximate in its applicability. The calculations done above for water and methane can be extended to other substances, and a complete set of bond energies derived. It is found that, to a first approximation, the bond energy of a bond between two given atoms is independent of other bonds in the molecule: For instance the same value of C—H bond energy applies approximately to all alkanes, though in ethyne the value is somewhat larger. The precise value of the bond energy term in a particular molecule depends on the other atoms and bonds that are present, but this does not detract from the usefulness of average bond energy terms in thermochemical calculations.

Where there is a pronounced discrepancy between the experimental heat of combustion and that predicted by bond energy, some structural peculiarity must be sought. For instance, if the bonds are under strain, some of the potential energy—**strain energy**—might be released as extra heat of combustion.

9.9 Le Chatelier's principle

The fact that chemical reactions can reach a position of equilibrium, instead of reaching completion in either direction, is a proof that ΔH cannot be the only factor which determines whether a reaction will go or not. An equilibrium can be approached from either direction, starting with the pure materials on either side of the equation, and this suggests that it represents a balance between opposing reactions. ΔH for the backward reaction must be numerically the same as, but opposite in sign to, the forward reaction.

If a substance is dissolved in water it does not matter whether the heat of solution is positive or negative—sooner or later a concentration will be reached at a given temperature, when no more solid will dissolve and equilibrium exists between the aqueous phase and the undissolved solid. The solution is said to be **saturated** with solute at that temperature. The mass of solute dissolved by a stated mass, or volume, of solvent at a given temperature, is defined as the solubility. ΔH cannot give any information about the solubility of a substance: many water-soluble substances dissolve exothermically in water, while many others of comparable solubility dissolve endothermically.

In other chemical processes where equilibrium is involved, it is also found that measurements of ΔH cannot give any direct information about the composition of the equilibrium mixture.

Despite all this, the magnitude of ΔH *does* have an effect on the way equilibrium composition varies with temperature. Take a very simple case, namely the dissociation of dinitrogen tetroxide into nitrogen dioxide:

$$\underset{\text{pale yellow}}{N_2O_4(g)} \rightleftharpoons \underset{\text{brown}}{2NO_2(g)}; \quad \Delta H = +61{\cdot}5 \text{ kJ}$$

This reaction is readily studied by taking the gas in a sealed tube and observing the darkening in colour when the tube is immersed in hot water. The result of the experiment can be summarized in an enthalpy diagram (Fig. 9.11). It shows that the species of higher energy, in this case $NO_2(g)$, is favoured at the higher temperature.

Many other examples can be taken, for instance:

(a) $NH_4Cl(s) \rightleftharpoons NH_3(g) + HCl(g); \quad \Delta H = +177 \text{ kJ}$

(b) $PCl_5(g) \rightleftharpoons PCl_3(g) + Cl_2(g); \quad \Delta H = +120 \text{ kJ}$

(c) $I_2(g) \rightleftharpoons 2I(g); \quad \Delta H = +214 \text{ kJ}$

These are all examples of **thermal dissociation.** The effect of temperature on such equilibria led Le Chatelier, in 1885, to propose the following general rule, known as **Le Chatelier's principle:**

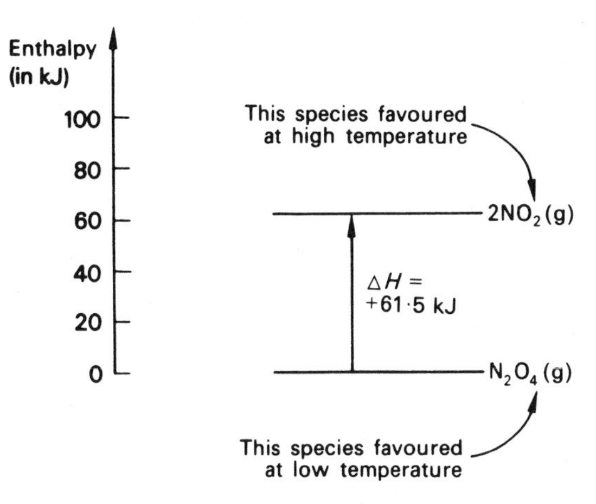

FIG. 9.11. Le Chatelier's principle.

when a constraint is applied to a system in equilibrium, the system will change in such a way as to try to remove the constraint. The "constraint" which concerns us here is that of increasing the temperature. If the temperature is increased, the system will behave as if it is trying to lower the temperature again by performing an endothermic reaction. Hence at higher temperatures the system will favour the species of higher energy. If ΔH for an equilibrium is very small, then changing the temperature will not have much effect on the equilibrium composition. For instance, the heat of solution of sodium chloride in water is extremely small, and as a consequence, the solubility of sodium chloride in water hardly alters with temperature.

Study Questions

1. The heats of combustion of C, H_2, CH_4, C_2H_6 and C_3H_8 are -395, -287, -880, -1545 and -2210 kJ mol^{-1} respectively.

(a) Calculate the heat of formation of each hydrocarbon.

(b) Predict the values of the heat of combustion and the heat of formation of butane, C_4H_{10}.

2. The heats of combustion of butylamine, propylamine and ethylamine are -2970, -2340 and -1710 kJ mol^{-1} respectively. What value would you expect for the heat of combustion of methylamine?

3. A number of alcohols were burned and the following results were obtained:

Alcohol	Weight loss	Heat given out
Methyl, CH_3OH	3·2 g	71 kJ
Ethyl, C_2H_5OH	9·2 g	268 kJ
Propyl, C_3H_7OH	7·5 g	252 kJ
Butyl, C_4H_9OH	7·4 g	263 kJ
2-Methylpropyl, C_4H_9OH	8·2 g	288 kJ

(a) How many moles of each alcohol were burned?

(b) How many kJ were given out per mole of each alcohol?

(c) What trend do you observe in your results?

(d) Comment on the values obtained for the two alcohols with formulae C_4H_9OH.

4. When 8 g of NH_4NO_3(s) was dissolved to give 100 cm^3 of solution, the temperature fell from 19° to 14·5°C. Calculate ΔH for the reaction

$$NH_4NO_3(s) \rightarrow NH_4NO_3\ (aq, M)$$

5. A series of neutralization experiments using molar solutions of sulphuric acid, nitric acid and sodium hydroxide, all initially at 20°C, gave the following results:

Acid	Base	Final temperature (°C)
100 cm^3 H_2SO_4	100 cm^3 NaOH	26·8
100 cm^3 H_2SO_4	200 cm^3 NaOH	29·1
100 cm^3 HNO_3	100 cm^3 NaOH	26·8
100 cm^3 HNO_3	200 cm^3 NaOH	24·6

(a) Account for these results.

(b) Calculate values for the heats of neutralization.

(c) What would the temperature rise have been in each case if the sodium hydroxide solution had been only M/2?

6. 25 cm^3 of HCl of unknown concentration was titrated against M NaOH, both being initially at 20°C. The base was added at regular half-minute intervals, one cm^3 at a time and the temperature was then taken. One mole of HCl exactly neutralizes one mole of NaOH.

cm^3 base	temp. (°C)	cm^3 base	temp. (°C)	cm^3 base	temp. (°C)
1	20·5	6	22·4	11	23·9
2	20·9	7	22·7	12	24·1
3	21·3	8	23·0	13	24·2
4	21·7	9	23·3	14	24·1
5	22·1	10	23·6	15	24·0

(a) What is the approximate concentration of the acid?

(b) Suggest two reasons why the temperature rise was less nearer the end-point than it was at the start of the reaction.

7. 1·16 g of acetone was burned in a bomb calorimeter. The same temperature rise was also caused when a current of 2 A at a P.D. of 10 V was passed through a resistance for 30 min.

(a) How many joules were needed to heat the water?

(b) How many moles of acetone were burned?

(c) What is the heat of combustion of acetone?

8. (a) The vibrational spectrum of hydrogen has a convergence limit at 84·9 nm due to dissociation. 980 kJ are evolved when the dissociated atoms return to their ground state. Calculate the dissociation energy of hydrogen.

(b) The convergence limit in the spectrum of iodine occurs at 499·1 nm. The dissociation energy of iodine can be determined thermochemically at 148 kJ.

(i) Calculate the spectroscopic dissociation energy of iodine.

(ii) Comment on the value that you obtain.

9. Construct energy level diagrams to solve the following problems:

(a) The heats of formation of CO_2 from graphite and diamond under identical physical conditions are -393 and -396 kJ respectively. Calculate the heat of transition from graphite to diamond.

(b) The heat of formation of SnO_2 from white tin is -580 kJ. The heat of transition from white to grey tin is $+2{\cdot}5$ kJ. Calculate the heat of formation of SnO_2 from grey tin.

(c) The heats of formation of Hg_2Cl_2 and $HgCl_2$ are -265 and -230 kJ mol^{-1} respectively. Calculate the heat of the reaction $Hg_2Cl_2 + Cl_2 \rightarrow 2HgCl_2$.

(d) The heats of formation of HF, HCl, HBr and HI are -268, -92, -36 and $+26$ kJ mol^{-1} respectively.

(i) What is the heat of the reaction

$$2HBr + Cl_2 \rightarrow 2HCl + Br_2?$$

(ii) Will chlorine displace bromine from HBr?

(iii) Arrange the halogens in a displacement order.

10. Sulphuric acid cannot be made directly from its elements. Using Chapter 22, state what data you would need to calculate the heat of formation of $H_2SO_4(l)$.

11. Suggest suitable solvents for (a) naphthalene, $C_{10}H_8$, (b) caesium iodide, CsI, (c) zinc, (d) carborundum, SiC.

12. (a) Which of the following will be more soluble in a given suitable solvent:

(i) White or red phosphorus?

(ii) Rhombic or plastic sulphur?

(b) White phosphorus is often stored under water while potassium is often stored under paraffin. Why are these particular liquids chosen?

(c) What would happen if the potassium were stored under water, and the phosphorus under paraffin?

13. Account for the following observations:

(a) Phenylamine, $C_6H_5NH_2$, is readily soluble in ether, but not in water.

(b) Phenylammonium chloride, $C_6H_5NH_3^+Cl^-$, is readily soluble in water, but not in ether.

(c) Silica, SiO_2, is insoluble in both water and ether.

(d) Sugar is a molecular substance, yet it dissolves in water.

(e) HCl(g), a molecular substance, dissolves in water to give a conducting solution.

(f) Water and benzene, C_6H_6, are immiscible.

14. The lattice energy of potassium bromide is -656 kJ mol^{-1}, and the sum of the hydration energies of the gaseous ions is -640 kJ mol^{-1}.

(a) Draw an energy level diagram for this process.

(b) Calculate the heat of solution of potassium bromide in water.

15. (a) Construct an energy level diagram for the heat of formation of HCl in terms of the bond dissociation energies of H_2, Cl_2 and HCl.

(b) The bond dissociation energies of H_2, Cl_2 and HCl are 435, 242 and 431 kJ mol^{-1} respectively. Calculate the heat of formation of HCl.

16. The heat of formation of AgCl $= -126$ kJ.
The first ionization energy of Ag $= +730$ kJ.
The bond dissociation energy of Cl_2 $= +242$ kJ.
The electron affinity of chlorine $= -364$ kJ.
The heat of sublimation of Ag $= +272$ kJ.

(a) Calculate the lattice energy of AgCl from these figures.

(b) Assuming that AgCl is ionic, the lattice energy can be calculated to be -835 kJ. Comment on this.

17. The second ionization energy of Ag $= 2100$ kJ mol^{-1}. The calculated lattice energy of $AgCl_2$ would be about 2500 kJ mol^{-1}.

(a) Use these values and those given in Question 16 to calculate the heat of formation of $AgCl_2$.

(b) Why do you think that AgCl is formed in preference to $AgCl_2$?

18. (a) $N_2 + O_2 \rightleftharpoons 2NO$; $\Delta H = +90$ kJ
(b) $N_2 + 3H_2 \rightleftharpoons 2NH_3$; $\Delta H = -92$ kJ.

What would be the effect of altering the temperature on each of these equilibria?

19. The heats of combustion of the simple cyclic hydrocarbons of formula $(CH_2)_n$ are as follows (kJ mol^{-1}).

C_2H_4	-1420	C_3H_6	-2120
C_4H_8	-2780	C_5H_{10}	-3330
C_6H_{12}	-3930	C_7H_{14}	-4600

(a) Work out the heat of combustion "per CH_2 group" for each substance.

(b) How do these compare with the values given in question 1?

(c) What is the angle between adjacent C—C bonds in each compound?

(d) Can you discover a relationship between the bond angle and the heat of combustion "per CH_2-group"?

CHAPTER 10

Equilibrium

10.1 Steady state and equilibrium

A system is said to be in **equilibrium** when it has reached a "steady state" at a given temperature, with no matter passing into or out of the system. Some systems can be described as "steady states" even though strictly they are not in equilibrium: an example is the flame of a Bunsen burner with a constant flow of gas. Here the "macroscopic" properties of the flame, such as its temperature, colour, size and density, do not alter, but the flame is not a *closed system*, because matter is passing continually into and out of it.

When the properties of a closed system, such as a liquid and its vapour in a closed flask, reach a steady state at a given temperature, equilibrium is said to exist. Although individual molecules are continually passing between the liquid and vapour phases, measurement of large-scale properties (macroscopic properties) shows no evidence of overall change. In other words, microscopic changes are continuing but with no change in macroscopic properties.

The connection between the composition of a system at equilibrium and its energy has already been mentioned in Chapter 9. Here it was clearly shown that an equilibrium can exist between two sets of substances even when there is a large energy difference between them. The species of higher energy are favoured by high temperature, nevertheless equilibrium does exist at all temperatures. Equilibria which are *thermoneutral* do not have their compositions affected much by altering the temperature.

It can be shown theoretically that *all* chemical reactions are in reality equilibria. In this book we often write a single arrow $\rightarrow$ in a chemical equation, which implies that the backward reaction does not occur. Strictly we should always write the equilibrium sign $\rightleftharpoons$ for any chemical process. Where the equilibrium sign is not written it may be assumed that the forward reaction is the one which predominates.

10.2 Representation of equilibrium by equations

The equilibrium between liquid water and its vapour is a balance between two opposing changes. On the one hand, molecules of water are leaving the liquid surface and changing into vapour, with consequent absorption of latent heat:

$$H_2O(l) \rightarrow H_2O(g); \quad \Delta H = +44 \text{ kJ mol}^{-1} \text{ at } 25°C.$$

On the other hand, molecules of water vapour

are returning to the liquid phase with consequent evolution of heat:

$$H_2O(g) \rightarrow H_2O(l); \quad \Delta H = -44 \text{ kJ mol}^{-1} \text{ at } 25°C.$$

At equilibrium the two rates balance, and there is no overall heat change. Where the change in heat content has to be shown, it is expressed for the *forward* reaction, i.e. the reaction as written from left to right:

$$H_2O(l) \rightleftharpoons H_2O(g); \quad \Delta H = +44 \text{ kJ mol}^{-1} \text{ at } 25°C.$$

The chemical equation gives information about the number of moles of substances which react in either direction, but it does *not* give any information about the *composition* of the reaction mixture at equilibrium. The above equation, for example, does not tell us whether there is a large amount of liquid water in equilibrium with a small amount of steam, or vice versa.

Examples of other phase equilibria which can be represented in the form of chemical equations are:

$$CO_2(s) \rightleftharpoons CO_2(g); \quad \Delta H = +32{\cdot}2 \text{ kJ mol}^{-1}$$

$$S(\text{rhombic}) \rightleftharpoons S(\text{monoclinic}); \quad \Delta H = +0{\cdot}4 \text{ kJ mol}^{-1} \text{ at } 25°C.$$

The first equation shows the equilibrium which can exist between solid carbon dioxide and the gas. At normal pressure carbon dioxide *sublimes* without intervention of the liquid phase. The second equation shows the equilibrium which can exist between two crystalline forms of sulphur.

All equilibria involve some form of chemical change, even though in the case of a simple phase change on a single chemical substance the chemical change may be simply the breaking and reforming of relatively weak hydrogen bonds or dipolar forces. Although changes are conventionally classified as "chemical" and "physical", the distinction is by no means a rigid one.

10.3 Rate of attainment of equilibrium

Many everyday systems are not true equilibria, even though they may not appear to change over a long time. Petrol in contact with air is one such system: equilibrium is not attained unless a spark is applied.

It is often necessary to examine a system very carefully in order to determine whether equilibrium truly exists, for in many cases the rate of reaction may be extremely slow. This is true for the transition between monoclinic and rhombic sulphur mentioned above. Monoclinic crystals of sulphur can be cooled to room temperature without any visible transformation to rhombic occurring, even though true equilibrium cannot exist at room temperature. The change from one solid phase to another is frequently slow since comparatively large forces have to be overcome in order to rearrange the crystal lattice. Similarly equilibrium cannot be attained at room temperature, in the reaction

$$N_2(g) + 3H_2(g) \rightleftharpoons 2NH_3(g); \quad \Delta H = -96 \text{ kJ}$$

because the rate of reaction between nitrogen and hydrogen at this temperature is very slow indeed.

10.4 Le Chatelier's principle

If a system in equilibrium is caused to change, for example by altering the conditions such as temperature or concentration of the reactants, then *processes occur which tend to nullify the change* (le Chatelier's principle, Chapter 9). This enables us to predict what will happen to the equilibrium composition of a system, if an external constraint is applied to it. The following examples illustrate the principle.

(a) *The effect of concentration.* It is found experimentally that an equilibrium exists in aqueous solution between chromate(VI) ions $CrO_4^{2-}(aq)$ and dichromate(VI) ions $Cr_2O_7^{2-}(aq)$ as follows:

$$\underset{\text{yellow}}{2H^+(aq)+2CrO_4^{2-}(aq)} \rightleftharpoons \underset{\text{orange}}{Cr_2O_7^{2-}(aq)}+H_2O.$$

If the solution in equilibrium is acidified, i.e. $H^+(aq)$ ions are added, the change which occurs tends to remove the added hydrogen ions. The solution is seen to become deep orange. If sufficient hydrogen ions are added it is possible to effect an almost complete conversion of chromate into dichromate ions. If we now add an alkali to remove the hydrogen ions, the equilibrium composition promptly shifts back in an attempt to replenish some of the hydrogen ions removed.

$$H^+(aq)+OH^-(aq) \rightarrow H_2O$$

Since at the same time this forms chromate ions at the expense of dichromate ions, the solution turns pale yellow. This process may be reversed any number of times by the addition of acid or alkali.

Another instance of reversible equilibrium is provided by the addition of acid or alkali to an indicator. A simple experiment shows that litmus, methyl orange, or any other acid-base indicator can undergo repeated colour changes to and fro, depending on whether the acid or alkali is present in excess. This is explained by the fact that an indicator is a weak acid, which can lose a proton to give an anion of different colour. Addition of $H^+(aq)$ will cause the acid form HX to predominate, while addition of $OH^-(aq)$ will shift the equilibrium back to the right, making the observed colour that of the anion $X^-(aq)$:

$$HX \rightleftharpoons X^-(aq)+H^+(aq).$$

(b) *The effect of pressure.* In the case of a gaseous system, which is compressible, the concentration of a reactant can be altered by varying the pressure. Consider again the equilibrium between nitrogen and hydrogen forming ammonia. The reaction is immeasurably slow at room temperature, but at about 500°C in the presence of a catalyst it proceeds quite rapidly.

$$N_2(g)+3H_2(g) \rightleftharpoons 2NH_3(g); \quad \Delta H = -96 \text{ kJ at } 25°C.$$

The equation means that one mole of $N_2(g)$ would react with three moles of $H_2(g)$ to form two moles of ammonia, if the reaction were able to proceed completely from left to right. The equation does *not* tell us about the composition at equilibrium.

Since the molar volumes of gases are approximately equal under given conditions of temperature and pressure, the total volume of gas must decrease if nitrogen and hydrogen are made to react at a given constant pressure, because the number of moles of ammonia formed is less than the total number of moles of nitrogen plus hydrogen at the start.

Le Chatelier's principle predicts that increasing the pressure of the above equilibrium mixture will be to some extent nullified by the formation of more ammonia at the expense of nitrogen and hydrogen.

In general, an equilibrium involving a change in volume of reactants will be altered by increasing the pressure in such a direction as to favour the side of the equation with the smaller volume.

The same argument applies if the system is **heterogeneous**, that is, if there is more than one phase present. Consider the heterogeneous equilibrium between calcium carbonate and its decomposition products, calcium oxide and carbon dioxide:

$$CaCO_3(s) \rightleftharpoons CaO(s) + CO_2(g); \quad \Delta H = +178 \text{ kJ at } 25°C.$$

If calcium carbonate is heated in a closed container, carbon dioxide will be evolved until equilibrium is reached: thereafter the pressure of carbon dioxide will not alter at that temperature. If the volume of the container is now reduced, the equilibrium composition will shift in order to try to restore the original pressure. In other words carbon dioxide will react with calcium oxide to form calcium carbonate. Conversely if the volume of the container is increased the reaction will proceed from left to right, forming more gas.

(c) *The effect of temperature.* Le Chatelier's principle predicts that an endothermic change will be favoured by high temperature, while the reverse exothermic change will be favoured by low temperature. This is borne out by experiment. It is a familiar experimental fact, for example, that increasing the temperature of calcium carbonate causes it to decompose more readily to the oxide. If this experiment is done in a closed container, and the equilibrium pressure of carbon dioxide is measured with a manometer, the pressure is seen to rise with temperature. If heat is added to the system, a change will take place in such a way as to try to lower the temperature again. The endothermic change is therefore favoured, and more carbon dioxide is present in the equilibrium mixture.

10.5 Equilibrium as a balance between opposing changes

A chemical system in equilibrium has been described as a closed system in a steady state. It should not, however, be thought of as static. There is a constant interchange of matter due to the existence of equal and opposite changes which balance out. For instance, when a liquid is in equilibrium with its vapour the vapour is said to be **saturated**. The vapour pressure of a given pure substance is determined by temperature, and when a liquid is in equilibrium with its saturated vapour we can say that the rate at which molecules leave the liquid surface = the rate at which other molecules return to the liquid from vapour.

A similar state of affairs occurs with a saturated solution in equilibrium with solute. For instance the dissolving of a simple ionic salt in water can be represented as a balance between two opposing reactions:

$$A^+B^-(s) \rightleftharpoons A^+(aq) + B^-(aq).$$

It is impossible to state with any precision, by a mere examination of the above equation, the exact *mechanism* by which a salt dissolves in water. Beware of trying to guess the mechanism of a reaction by looking at the equilibrium equation. The familiar Haber equilibrium is indeed the result of opposing reactions balancing, but the forward reaction is *not* the result of three hydrogen molecules colliding simultaneously with one nitrogen molecule, nor does the backward reaction result from the immediate conversion of $2NH_3(g)$ into $3H_2(g) + N_2(g)$. Many intermediate stages are involved, and the overall equation does not show these.

Simple experiments can be done to show that the composition of an equilibrium mixture is the same regardless of the direction from which the composition is approached. These experiments do not *prove* that equilibrium is dynamic however. To be certain of this we need some way of "labelling" atoms in order to follow their progress during the exchange process. Use of a radioactive isotope is one way of doing this. Thorium salts contain radioactive lead, produced by disintegration of the thorium nucleus. If we precipitate the sparingly soluble salt lead(II) chloride, $PbCl_2$, by mixing a solution of Cl^-(aq) with a solution of Pb^{2+}(aq) which also contains some thorium ions, and hence radioactive lead ions, which we will denote as $*Pb^{2+}$(aq), the solid lead chloride now contains some *labelled* lead atoms—it is a mixture of $PbCl_2$(s) with some $*PbCl_2$(s). Now prepare separately a saturated solution of unlabelled lead(II) chloride by allowing it to remain in contact with pure water for several days with occasional shaking. Test this solution in a radioactive counter and it will be found to be non-radioactive. Now decant some of the saturated solution and add some of the labelled solid lead(II) chloride to it. Since the solution was saturated the overall equilibrium composition cannot change. The question is whether an exchange of atoms can occur. To find out, test samples of the solution from time to time in the radioactive counter. It will be found to have become radioactive. Since the concentration cannot have altered (provided the temperature was kept constant) an exchange must have taken place. The radioactive lead(II) ions must have come from the labelled solid, by the forward reaction

$$*PbCl_2(s) \rightarrow *Pb^{2+}(aq) + 2Cl^-(aq)$$

The existence of dynamic equilibrium in more complex chemical processes can be similarly illustrated, provided that a suitable radioactive isotope is available. If all the available radioactive isotopes of a given element are too short-lived to be useful (as in the case of aluminium for instance) we can use a compound which has been enriched in one of its stable isotopes. Detection of the isotope is then more of a problem and the products have to be analysed in a mass spectrometer. The use of radioactive tracers and labelled compounds with enriched stable isotopes is very common in studies of reaction *mechanisms*. It can be shown for instance that the rate of exchange in an equilibrium process is more rapid at a higher temperature.

10.6 The equilibrium law

A typical gaseous equilibrium, which was first thoroughly investigated by Bodenstein in 1897, is the following:

$$H_2(g) + I_2(g) \rightleftharpoons 2HI(g).$$

Bodenstein found that, at a given temperature, the concentrations in mol dm^{-3} of the three substances present, denoted by square brackets $[H_2]$, $[I_2]$ and $[HI]$, are always related by the following approximate expression:

$$\frac{[HI]^2}{[H_2][I_2]} = K_c, \text{ a constant.}$$

The constant K_c is called the equilibrium constant. The subscript c is used to denote that molar concentration units are employed. Another equilibrium constant, K_p, can be written, in which the amounts of gases present are stated in terms of their partial pressures p_{H_2}, p_{I_2} and p_{HI}.

$$\frac{p_{HI}^2}{p_{H_2} \cdot p_{I_2}} = K_p.$$

In the case of the above equilibrium K_c and K_p are equal. This must be so, by the following argument: assuming the gases to be ideal, their molar concentrations will be directly proportional to their partial pressures. Partial pressure is therefore just another unit by which molar gas concentrations can be represented.

Worked Example 1. In an equilibrium mixture the following concentrations were observed at 0°C:

$$[H_2] = x \text{ mol dm}^{-3}$$
$$[I_2] = y \text{ mol dm}^{-3}$$
$$[HI] = z \text{ mol dm}^{-3}$$

Calculate (i) K_c, (ii) K_p.

$$\text{(i)} \quad K_c = \frac{[HI]^2}{[I_2][H_2]} = \frac{z^2}{xy}$$

(ii) $[H_2] = 22{\cdot}4x$ mol per 22·4 dm^3.

$\therefore p_{H_2} = 22{\cdot}4x$ atmospheres.

Similarly,

$$p_{I_2} = 22{\cdot}4y \text{ atmospheres}$$
$$p_{HI} = 22{\cdot}4z \text{ atmospheres.}$$
$$K_p = \frac{p_{HI}^2}{p_{H_2} \cdot p_{I_2}} = \frac{z^2}{xy} = K_c$$

Bodenstein was able to show that altering the pressure did not have any appreciable effect on the equilibrium composition, and he also showed that the equilibrium composition for a given set of conditions was the same whether he started with pure hydrogen iodide or with pure hydrogen and pure iodine.

Accurate measurements by later workers have shown that the equilibrium law, where concentrations are used, is only approximate. The following figures obtained by Taylor and Crist (1941) illustrate this point (Table 10.1).

TABLE 10.1
TEMPERATURE = 457·6°C

$[H_2]$ mol dm^{-3}	$[I_2]$ mol dm^{-3}	[HI] mol dm^{-3}	K_c
$5{\cdot}617 \times 10^{-3}$	$0{\cdot}5936 \times 10^{-3}$	$1{\cdot}270 \times 10^{-2}$	48·38
3·841	1·524	1·687	48·61
4·580	0·9733	1·486	49·54
1·696	1·696	1·181	48·48
1·433	1·433	1·000	48·71
4·213	4·213	2·943	48·81

The first three rows of figures were obtained starting with pure hydrogen and pure iodine, and the remainder by starting with pure hydrogen iodide.

It will be noticed that in the expression for the equilibrium constant, the concentration of hydrogen iodide is raised to a power equal to its coefficient in the chemical equation. This is found to be generally true; e.g. for the Haber synthesis of ammonia the equilibrium constant is found to be

$$K_c = \frac{[NH_3]^2}{[H_2]^3[N_2]}.$$

Note that in this case K_c and K_p are *not* equal, as there are four concentration terms in the denominator and only two in the numerator.

Worked Example 2. In an equilibrium mixture the following concentrations were observed at 0°C:

$$[H_2] = a \text{ mol dm}^{-3}$$
$$[N_2] = b \text{ mol dm}^{-3}$$
$$[NH_3] = c \text{ mol dm}^{-3}$$

Calculate (i) K_c, (ii) K_p.

$$\text{(i)}\quad K_c = \frac{[NH_3]^2}{[H_2]^3[N_2]} = \frac{c^2}{a^3 b} \text{ mol}^{-2}\text{ dm}^6$$

(ii) $p_{H_2} = 22{\cdot}4a$ atmospheres (see Worked Example 1).

Similarly,

$$p_{N_2} = 22{\cdot}4b$$
$$p_{NH_3} = 22{\cdot}4c$$
$$\therefore K_p = \frac{p_{NH_3}^2}{p_{H_2}^3 \cdot p_{N_2}} = \frac{22{\cdot}4^2 \times c^2}{22{\cdot}4^3 \times a^3 \times 22{\cdot}4 \times b}$$
$$= \frac{1}{22{\cdot}4^2} \times \frac{c^2}{a^3 b}$$
$$= \frac{K_c}{22{\cdot}4^2} \text{ atmospheres}^{-2}$$
$$\therefore K_c \neq K_p.$$

Units. For convenience, the units of equilibrium constants are generally omitted. In the case of ammonia above, the dimensions of K_c are

$$K_c = \frac{(\text{mol dm}^{-3})^2}{(\text{mol dm}^{-3})^4}$$
$$= \text{mol}^{-2}\text{ dm}^6.$$

The dimensions of K_p are $(\text{atmospheres})^{-2}$, or $N^{-2}\ m^4$.

In the case of the hydrogen iodide equilibrium both K_p and K_c are dimensionless quantities since the concentrations and partial pressures cancel out.

The equilibrium law can be summarized by stating that for an equilibrium

$$aX+bY+cZ+\ldots \rightleftharpoons pL+qM+rN+\ldots$$

the expression $\frac{[L]^p[M]^q[N]^r\ldots}{[X]^a[Y]^b[Z]^c\ldots}$ will be constant at a given temperature. By convention, the substances on the right-hand side of the equation are written in the numerator.

10.7 Activity

The equilibrium law, as stated above, is only approximately true. It can be shown theoretically that, quite apart from experimental error, deviations are to be expected in the case of real gases on account of their failure to obey the gas laws. If we were always dealing with ideal gases the law would always hold exactly, but real gases are far from ideal in their behaviour, especially at high pressures and low temperatures.

It is also found that similar deviations occur in equilibria involving solutions, though for very dilute solutions the agreement is sufficient for the law to be of practical value.

In the accurate statement of the equilibrium law, a quantity known as the **activity** of each reactant is used in place of concentration. The activity is a quantity obtained by making allowance for non-ideal behaviour of the gas or solution, and is thus an "effective concentration" of a component of a component. The measurement of activities is dealt with in section 11.6.

The activity of a pure liquid or solid is a constant at a given temperature and is generally taken as unity; (the concentration of a pure solid might be represented by its density).

10.8 The effect of a catalyst on equilibrium composition

A catalyst is a substance which can increase the rate of a chemical reaction without itself undergoing any overall chemical change. The effect of a catalyst on an equilibrium system will be to accelerate the attainment of equilibrium: it is thus acting as a kind of chemical lubricant. A catalyst cannot affect the equilibrium composition of a system since it does not alter its energy content.

When the equilibrium composition of a system is altered, there will be a change in energy: heat will either be evolved or absorbed depending on whether the forward reaction is exothermic or endothermic. Conversely if there is an energy change in an equilibrium system the equilibrium composition will alter. Since the catalyst is itself not undergoing any permanent chemical change it cannot contribute to the energy of the system, and hence the equilibrium composition must stay the same.

An everyday example of a catalyst is the platinum catalyst used in some automatic gas lighters. Living systems also use biological catalysts, called **enzymes**, for instance to enable "fuels" such as sugar to undergo "combustion" at body temperature. Catalysts are used extensively in manufacturing processes. The theory of catalysis is discussed in Chapter 17.

10.9 The effect of temperature on equilibrium

In section 10.4, le Chatelier's principle was applied to show that the endothermic change in an equilibrium reaction is favoured by increasing the temperature. The actual relationship between equilibrium constant, K_p, and temperature is given by the expression:

$$\frac{d(\log_e K_p)}{dT} = \frac{\Delta H^\ominus}{RT^2} \qquad (1)$$

where $\Delta H^\ominus$ is the heat change for the reaction and R the gas constant. The expression is more often used in its integrated form:

$$\log_e K_p = \text{constant} - \frac{\Delta H^\ominus}{R}\left(\frac{1}{T}\right) \qquad (2)$$

$$\text{or} \quad \log_{10} K_p = c - \frac{\Delta H^\ominus}{2{\cdot}303R}\left(\frac{1}{T}\right) \qquad (3)$$

The integration step assumes that $\Delta H^\ominus$ does not change with temperature, which is approximately true for most reactions, provided that none of the reactants or products changes its state within the temperature range under consideration. If $\log_{10} K_p$ is plotted against $\frac{1}{T}$ then the resulting graph will have a gradient of $\frac{-\Delta H^\ominus}{2{\cdot}303R}$; this is an excellent method of determining $\Delta H^\ominus$ in an equilibrium reaction.

On the other hand the equilibrium constant may only be known at one temperature and its value at a second temperature may be required. In this case, if $\Delta H^\ominus$ is known, K_p can be calculated by using equation (3) in a slightly different form:

$$\log_{10}\left(\frac{K_2}{K_1}\right) = \frac{\Delta H^\ominus}{2{\cdot}303R}\left(\frac{1}{T_1} - \frac{1}{T_2}\right) \qquad (4)$$

where K_1 is the equilibrium constant at temperature T_1, etc.

Worked Example 3. For the reaction $N_2(g) + 3H_2(g) \rightleftharpoons 2NH_3(g)$, K_p at 300 K is $6{\cdot}3 \times 10^5$ atm^{-2} and ΔH is about -100 kJ. Calculate K_p at 600 K. ($R = 8{\cdot}3 \times 10^{-3}$ kJ K^{-1} mol^{-1}).

Substituting the values into equation (4):

$$\log_{10} K_{p,600} - \log_{10}(6{\cdot}3 \times 10^5) = \frac{-100}{2{\cdot}3 \times 8{\cdot}3 \times 10^{-3}}\left(\frac{1}{300} - \frac{1}{600}\right)$$

$$\log_{10} K_{p,600} - 5{\cdot}8 = \frac{-10^5}{19{\cdot}1}\left(\frac{1}{600}\right)$$

$$\log K_{p,600} = -8{\cdot}73 + 5{\cdot}8 = -2{\cdot}93 = \bar{3}{\cdot}07$$

$$K_{p,600} = 1{\cdot}2 \times 10^{-3} \text{ atm}^{-2}.$$

The quantitative calculation is in agreement with the qualitative prediction that can be made using le Chatelier's principle.

10.10 Solution equilibria

The equilibrium law can also be applied to solutions. Consider a solid which dissolves as molecules. Examples of this are naphthalene in benzene, and sucrose, $C_{12}H_{22}O_{11}$, in water:

$$C_{12}H_{22}O_{11}(s) \rightleftharpoons C_{12}H_{22}O_{11}(aq)$$

The activity of solid sucrose is a constant, and the equilibrium law predicts that the concentration of solute in the saturated solution will be a constant at a given temperature. In other words the solubility of a substance in a solvent at a given temperature is independent of the mass of undissolved solute present, or upon its surface area. This is borne out by experiment. Increasing the surface area of undissolved solute will increase the rate at which molecules leave the solid, but it will also increase the rate at which they return.

10.11 Henry's law

The solubility of a gas in a liquid, in mol dm^{-3}, is found to be directly proportional to the partial pressure (section 8.7) of the gas at a given temperature. This relationship, known as **Henry's law**, is also a consequence of the equilibrium law. For example, with nitrogen dissolving in water we have

$$N_2(g) \rightleftharpoons N_2(aq)$$

and therefore $$\frac{[N_2(aq)]}{[N_2(g)]} = K$$

$$\therefore \frac{[N_2(aq)]}{p_{N_2}} = \text{constant. (Henry's law.)}$$

where p_{N_2} = partial pressure of nitrogen. Soda water (carbon dioxide dissolved in water) provides a good illustration of Henry's law.

10.12 The distribution law (partition law)

If a solute is added to two immiscible liquids, being soluble in both, it will distribute itself between them in such a way that its concentration in one solvent is directly proportional to the concentration in the other solvent at a given temperature. This relationship is known as the distribution law or partition law, and again it is a form of the equilibrium law. Iodine, for example, dissolves both in benzene (readily) and in water (sparingly).

$$I_2(aq) \rightleftharpoons I_2(benzene)$$

$$\therefore \frac{[I_2(benzene)]}{[I_2(aq)]} = K.$$

For such systems, the constant ratio K between the molar concentrations is known as the **distribution coefficient** (or **partition coefficient**).

The distribution law is only obeyed accurately at low concentrations. This is because strictly we should write activities rather than concentrations.

A complication arises when molecules dissociate into ions in one solvent but not the other. For example hydrogen chloride ionizes in water but not in benzene so that

$$K = \frac{[HCl(aq)]^2}{[HCl(benzene)]}.$$

Sometimes association rather than dissociation occurs as when ethanoic acid dimerizes in benzene (section 14.22) but not in water, so that

$$K = \frac{[CH_3COOH(aq)]}{[CH_3COOH(benzene)]^{\frac{1}{2}}}.$$

10.13 Solubility product

Many solids dissolve in water to form ions. In the case of substances which are sparingly soluble in water the equilibrium law can again be applied. With silver chloride, for example, the solubility at room temperature is only about 10^{-5} mol dm^{-3} and we can write:

$$AgCl(s) \rightleftharpoons Ag^+(aq) + Cl^-(aq)$$

$$\therefore \frac{[Ag^+][Cl^-]}{[AgCl]} = K_c.$$

Since AgCl is a solid, [AgCl] is a constant.

$$\therefore [Ag^+][Cl^-] = \text{another constant, } K_s,$$

known as the **solubility product**. (The activity of a pure solid is a constant, which can be put equal to unity.)

Similarly for lead(II) chloride, we have

$$PbCl_2(s) \rightleftharpoons Pb^{2+}(aq) + 2Cl^-(aq)$$

and therefore $K_s = [Pb^{2+}][Cl^-]^2$. Similarly, with aluminium hydroxide $Al(OH)_3$, the relationship is $K_s = [Al^{3+}][OH^-]^3$.

The solubility product enables the effect on solubility of adding a *common ion* to be predicted. Returning once more to silver chloride, we have seen that its solubility at room temperature is 10^{-5} mol dm^{-3}. From this we can deduce the solubility product:

$$AgCl(s) \rightleftharpoons Ag^+(aq) + Cl^-(aq)$$

$AgCl(s)$	$Ag^+(aq)$	$Cl^-(aq)$
10^{-5} mol dissolves per dm^3	10^{-5} mol dm^{-3}	10^{-5} mol dm^{-3}

$$\therefore K_s = 10^{-5} \times 10^{-5} = 10^{-10}.$$

Now let us calculate the solubility of silver

chloride in molar sodium chloride solution. Let the solubility of the silver chloride be x mol dm^{-3}. We therefore have x mol dm^{-3} of silver ions and x mol dm^{-3} of chloride ions in solution. In addition the fully ionized NaCl gives 1 mol dm^{-3} of chloride ions in solution so that the total chloride ion concentration is $(1+x)$ mol dm^{-3}.

Since $\quad K_s = [Ag^+][Cl^-]$,

then $\quad K_s = x(1+x)$.

Since x is very small compared with 1, we can take $(1+x)$ to be 1, so that $x = K_s = 10^{-10}$ mol dm^3. The presence of a common ion, the chloride ion, lowers the solubility of AgCl from 10^{-5} to 10^{-10} mol dm^{-3}, a factor of 100 000. This reduction in solubility is called the **common ion effect**, and similar results would be obtained if we considered the solubility of AgCl in silver nitrate solution.

The solubility of silver chloride in a dilute solution of other ions, e.g. a dilute solution of potassium nitrate will be the same as the solubility in pure water, since a non-common ion does not enter into the solubility product relationship.

If two solutions containing ions are mixed, and the solubility product is exceeded for one pair of oppositely charged ions in the mixed solutions, then that substance will have to *precipitate* in order to restore equilibrium. In the above case, if a 1·0 M solution of silver ions was mixed with a 1·0 M solution of chloride ions the solubility product would momentarily be exceeded by a factor of 10^{10}, and very rapid precipitation would occur.

10.14 Limitations of the solubility product principle

The solubility product principle should only be used as an approximate guide when predicting whether substances will precipitate from solution. At least three factors cause deviations in practice:

(i) The use of ionic *concentrations* in the expression for K_s is only strictly permissible when the concentrations are low. We would not expect a substance such as potassium nitrate to obey the principle since it is too highly soluble, and indeed the deviations are very considerable. Similarly the calculation given above for the solubility of silver chloride in molar sodium chloride is only approximate, because even at a concentration of $[Cl^-] = 1$ mol dm^{-3} the difference between concentration and activity is very noticeable. Nevertheless the law holds qualitatively.*

(ii) In many cases, even though the solubility product of a substance may be exceeded, a precipitate may not form due to the slow *rate* of precipitation. This applies particularly to substances with very low solubility products.

An interesting experiment to illustrate this point is to take solutions of $Ba^{2+}(aq)$ (barium chloride or nitrate) and SO_4^{2-} (sodium sulphate) at concentrations varying from 10^{-1} to 10^{-4} mol dm^{-3}. It will be found that as the solutions are made more dilute the precipitate appears more slowly, and that with very dilute solutions it will not appear at all even though the solubility product is exceeded (K_s for barium sulphate $\simeq 10^{-10}$ mol^2 dm^{-6}).

Remember again therefore that the equilibrium law tells us nothing about the rate of a chemical process: it tells us *how far* a given chemical reaction will go, but *not how fast*.

(iii) In other cases the formation of a *complex ion* may be a complicating factor. With lead(II)

* An interesting quantitative study of the "common ion" effect has been carried out as a pupils' research project at Marlborough College. The results showed convincingly that the deviations from ideal behaviour of solutions can be very considerable (*Educ. in Chem.*, vol. 3, No. 4, p. 164).

chloride, for example, a rather curious situation results: the substance is less soluble in dilute hydrochloric acid than in water, as one would expect, but on the addition of concentrated hydrochloric acid, the lead chloride becomes *more* soluble. We have in fact two equilibria to consider:

(a) The dissolving of the solid to form aqueous ions:

$$PbCl_2(s) \rightleftharpoons Pb^{2+}(aq) + 2Cl^-(aq)$$

(b) The formation of a complex ion:

$$Pb^{2+}(aq) + 4Cl^-(aq) \rightleftharpoons PbCl_4^{2-}(aq)$$

$$\frac{[PbCl_4^{2-}]}{[Pb^{2+}][Cl^-]^4} = K.$$

If $[Cl^-]$ is made very large (e.g. 10 mol dm^{-3}) then the effect will be most marked in the equilibrium involving the complex ion, for now $[Cl^-]^4$ will equal 10^4. A larger amount of the complex ion $PbCl_4^{2-}$ has to form, and this has the effect of lowering the concentration of $Pb^{2+}(aq)$. More solid lead chloride therefore dissolves to restore equilibrium between itself and the simple aqueous ions.

Silver chloride, AgCl(s), although only sparingly soluble in water, dissolves readily in aqueous ammonia (ammonium hydroxide). This is due to the formation of the complex ion diamminesilver(I), which is quite stable:

$$\underset{\text{simple hydrated silver(I) ion}}{Ag^+(aq)} + 2NH_3(aq) \rightleftharpoons \underset{\text{diammine silver(I) ion}}{Ag(NH_3)_2^+(aq)}$$

$$K = \frac{[Ag(NH_3)_2^+]}{[Ag^+][NH_3]^2}$$

Application of le Chatelier's principle to this reaction shows that the equilibrium composition will move to the right, favouring the complex ion, if the concentration of ammonia, $[NH_3(aq)]$, is increased. For the same reason, a high concentration of ammonia will reduce the concentration of $Ag^+(aq)$. Since the ammine complex of silver is quite stable ($K \simeq 10^7$), the solubility product of silver chloride, $[Ag^+][Cl^-]$, is not attained, and silver chloride therefore dissolves. The constant K is called the **stability constant** of the complex ion.

The dissolution of an insoluble precipitate in a reagent can often be attributed to the formation of a complex ion. Other examples will be found in the text, and by reading the exercises at the end of this chapter.

10.15 Auto-ionization and ionic product

Some liquids, although they consist almost entirely of molecules, are found to have a very small electrical conductivity due to the presence of traces of ions. Water is a very good example: no matter how carefully it is purified it will always show a small residual conductivity, and a high enough applied potential will cause electrolysis, giving hydrogen at the cathode and oxygen at the anode. This phenomenon is due to an equilibrium between water molecules and the ions $H^+(aq)$ and $OH^-(aq)$.

$$H_2O(l) \rightleftharpoons H^+(aq) + OH^-(aq); \quad \Delta H = +58 \text{ kJ mol}^{-1}$$

$$\therefore K = \frac{[H^+][OH^-]}{[H_2O]}$$

$[H_2O]$ is constant, since the extent of ionization is small.

$\therefore [H^+][OH^-]$ = a constant, K_w, termed the **ionic product** of water.

The reaction from left to right is an endothermic process with an enthalpy change of +58 kJ mol^{-1}. Application of le Chatelier's principle to this shows that the formation of ions will be favoured by high temperature, and this is borne out by experimental measurements on the variation of conductivity of pure water with temperature. Table 10.2 gives values for K_w at temperatures between 0°C and 100°C.

It will be noticed that at 25°C the value for K_w is very near to 10^{-14}. Since this is close to normal room temperature it is a useful figure to remember, and for approximate calculations involving K_w the figure 10^{-14} is often taken.

TABLE 10.2

Temperature (°C)	$K_w \times 10^{14}$ $mol^2\ dm^{-6}$
0	0·11
25	1·01
50	5·47
100	51·3

Some other liquids auto-ionize like water: in every case the concentration of ions is very low. Liquid ammonia (obtained by cooling ammonia to −33°C, not to be confused with an aqueous solution of ammonia) is a good example, and so is anhydrous ethanoic acid. Table 10.3 summarizes some of the liquids which auto-ionize.

The phenomenon of auto-ionization plays a very important part in the behaviour of acids and bases, and is discussed further in Chapter 19.

TABLE 10.3

Liquid	Equation	Ionic product ($mol^2\ dm^{-6}$)
Water	$H_2O \rightleftharpoons H^+(aq) + OH^-(aq)$	10^{-14} at 25°C
Ammonia	$NH_3 \rightleftharpoons NH_4^+ + NH_2^-$	10^{-22} at −33°C
Ethanoic acid	$CH_3COOH \rightleftharpoons CH_3COOH_2^+ + CH_3COO^-$	10^{-17} at 25°C

Study Questions

1. The following equilibrium exists between iron, steam, hydrogen and iron(II)diiron(III) oxide:

$$Fe_3O_4(s) + 4H_2(g) \rightleftharpoons 3Fe(s) + 4H_2O(g); \quad \Delta H = +138 \text{ kJ}.$$

(a) What would be the effect of (i) increasing the amount of steam present in a closed vessel, (ii) passing a continuous flow of hydrogen over the heated iron oxide, removing the steam as it formed, (iii) passing a continuous flow of steam over heated iron and removing the hydrogen as it formed?

(b) What information about the reaction does the chemical equation *not* contain?

2. Bismuth(III) chloride is soluble in water, but aqueous Bi^{3+} ions are hydrolysed according to the equation:

$$Bi^{3+}(aq) + Cl^-(aq) + H_2O \rightleftharpoons BiOCl(s) + 2H^+(aq)$$

(a) What would happen if a solution of bismuth(III) chloride were poured into an excess of water?

(b) What would happen if solid bismuth(III) chloride oxide (BiOCl) were shaken with a solution of nitric acid?

(c) What is meant by the statement that this equilibrium is "dynamic"?

(d) Suggest an experiment which would show that this equilibrium was dynamic.

3. Dinitrogen tetroxide, N_2O_4, is colourless when pure. In the vapour phase it dissociates into brown nitrogen dioxide, NO_2.

$$N_2O_4(g) \rightleftharpoons 2NO_2(g)$$

(a) The vapour darkened on warming to 100°C, but when cooled, it became pale again. What can you say about ΔH for this reaction?

(b) Would the vapour darken or grow paler if the pressure was decreased?

4. Magnesium carbonate, $MgCO_3$, dissociates as follows:

$$MgCO_3(s) \rightleftharpoons MgO(s) + CO_2(g) \quad \Delta H = +117 \text{ kJ}$$

In a closed vessel, how will the pressure of carbon dioxide alter:

(a) If the temperature is raised?

(b) If an inert gas, such as argon, is pumped into the container, the temperature remaining constant?

(c) If more magnesium carbonate is added?

5. What effect would the following changes have on the equilibrium concentration of ammonia in the reaction:

$$4NH_3(g)+5O_2(g) \rightleftharpoons 4NO(g)+6H_2O(g); \quad \Delta H = -900 \text{ kJ.}$$

(a) Increasing the temperature?
(b) Decreasing the total pressure?
(c) Increasing the volume of the reaction vessel?
(d) Decreasing the oxygen concentration?
(e) Adding a catalyst?

6. $\log_{10}K_p$ for the reaction $N_2+3H_2 \rightleftharpoons 2NH_3$ is 5·8 at 300 K, −1·0 at 500 K and −4·1 at 700 K. What will the effect of the equilibrium composition be if

(a) More nitrogen is pumped into the system?
(b) The pressure is increased?
(c) The temperature is increased?
(d) Calculate a value of ΔH for this reaction.

7. 531 g of ammonia will dissolve in 1000 cm^3 of water at 20°C, giving an alkaline solution.

(a) Suggest an experiment to investigate the conditions under which ammonia obeys Henry's law.

(b) How closely would you expect ammonia to obey the law?

8. Ethanoic acid, CH_3COOH, is soluble in both water and carbon tetrachloride, CCl_4, which are themselves immiscible. If solutions of different concentrations are shaken together, it is found that

$$\frac{(\text{concentration in } CCl_4)}{(\text{concentration in water})^2} \text{ is a constant.}$$

(a) What can be deduced from this result?

(b) Suggest how you could determine the value of the constant.

9. Write down expressions for the solubility products of

(a) Silver(I) bromide.
(b) Barium sulphate.
(c) Lead(II) iodide.
(d) Bismuth(III) sulphide.
(e) Aluminium hydroxide.
(f) Calcium carbonate.

10. The solubility products of CuS and MnS are 10^{-36} and 10^{-15} mol^2 dm^{-6} respectively.

(a) Calculate the solubilities of the two compounds in g dm^{-3}.

(b) What would happen if a 10^{-10} M solution of copper ions was mixed with a 10^{-10} M solution of sulphide ions?

(c) What would happen if a 10^{-10} M solution of manganese ions was mixed with a 10^{-10} M solution of sulphide ions?

11. The solubility products of barium and calcium sulphate are 10^{-10} and 2×10^{-5} mol^2 dm^{-6} respectively.

(a) What is the concentration of a saturated solution of each salt in g dm^{-3}?

(b) Both salts can be used to determine sulphate gravimetrically. Which is preferable?

12. In the manufacture of soaps, such as sodium stearate, common salt is often added to precipitate the solid. Why is this?

13. At 25°C the solubility product for AgCl is 2×10^{-10}, while at 100°C it is $1{\cdot}4\times10^{-4}$ mol^2 dm^{-6}.

(a) Calculate the solubility of silver chloride at 100°C, in g dm^{-3}.

(b) What can you say about the heat of solution of AgCl in water?

14. The solubility of iron(II) hydroxide is 0·1 g dm^{-3} at 20°C. When a solution of Fe^{2+} ions is boiled with potassium cyanide solution, the complex ion, $Fe(CN)_6^{4-}$, hexacyanoferrate(II), is formed. If NaOH (aq, M) is added to an aqueous solution of hexacyanoferrate(II) ions, no precipitate forms.

(a) Write down an expression for the solubility product of iron(II) hydroxide.

(b) Calculate the solubility of iron(II) hydroxide in mol dm^{-3}.

(c) Calculate the solubility product of iron(II) hydroxide, giving the correct units.

(d) Write down an expression for the stability constant of the hexacyanoferrate(II) ion.

(e) What can you say about the magnitude of this stability constant?

15. (a) What happens if gaseous hydrogen chloride is added to a saturated solution of sodium chloride?

(b) How would you try to determine the solubility product of sodium chloride in the presence of varying concentrations of hydrochloric acid?

(c) Why would the values obtained in (b) not be constant?

16. For the reaction $H_2O(g)+C(s) \rightleftharpoons H_2(g)+CO(g)$, K_p at 300 K = 10^{-16} atm. and ΔH is about +140 kJ mol^{-1}.

(a) What is the value of K_p at 900 K?
(b) What is the value of K_p at 1500 K?
(c) What conditions are necessary in order to produce hydrogen from steam and coke?

CHAPTER 11

Cells

11.1 The importance of electrical measurements in chemistry

This chapter is concerned with the measurement of the e.m.f.s of cells as a means of obtaining information about chemical processes. It is not always easy in chemistry to make accurate quantitative measurements, but voltage measurements can be made with a high degree of precision, leading to precise knowledge of energy changes.

Chapter 9 was concerned mainly with measurements of heat energy, and it was shown that exothermicity was not a criterion of whether a reaction would proceed spontaneously or not. It will be shown in this chapter that measurement of e.m.f. *does* provide a criterion: if a cell gives an e.m.f. then the cell reaction will proceed spontaneously. The larger the e.m.f. the greater the "driving force" of the chemical process.

The use and design of cells is of great technological importance today. Apart from the familiar examples such as the dry battery and the lead accumulator used in motor cars, recent years have seen the development of miniature cells (such as those for hearing aids) and fuel cells capable of delivering considerable electrical power.

11.2 The cell

You will probably have done experiments with a Daniell cell consisting of copper and zinc plates (electrodes) each dipping into an electrolyte (Fig. 11.1). An electromotive force (e.m.f.)

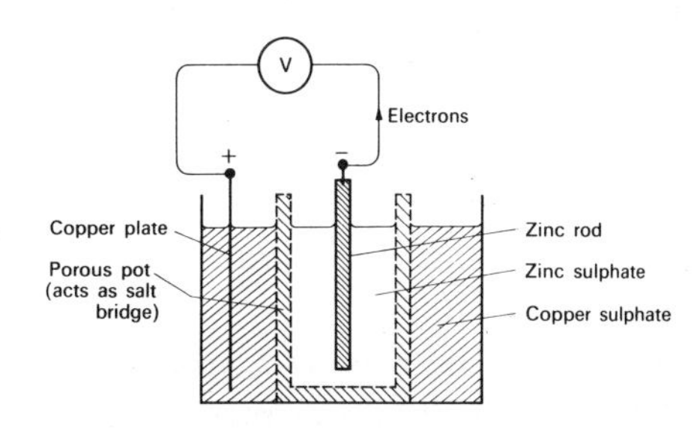

FIG. 11.1. The Daniell cell.

is developed and the current generated is sufficient to light a torch bulb. If a voltmeter is inserted in the circuit, a potential difference of about 1·1 V is observed, the zinc plate being negative with respect to the copper.

Varying the size of the plates and their distance apart does not affect the e.m.f. produced, though it does affect the capacity of the cell for giving current. Varying the concentration and composition of the electrolytes does have an effect, however, and it is necessary to devise some standard condition if meaningful measurements are to be made.

The processes which occur at the electrodes may be summarized by equations:

zinc electrode: $Zn(s) \rightarrow Zn^{2+}(aq) + 2e^-$;

copper electrode: $Cu^{2+}(aq) + 2e^- \rightarrow Cu(s)$;

The zinc electrode is acting as a source of electrons and these flow around the external circuit, and the electrode dissolves forming ions, $Zn^{2+}(aq)$. At the copper electrode the electrons combine with $Cu^{2+}(aq)$ ions in the solution and copper is deposited.

The above equations represent **half-cell reactions**. It is impossible for a half-cell reaction to occur on its own—there must always be two half-reactions together to make up a complete chemical process. In the zinc-copper case above the overall chemical process which occurs is that which results from adding the two half-reaction equations algebraically, in such a way that the electrons "cancel out":

$$Zn(s) + Cu^{2+}(aq) \rightleftharpoons Zn^{2+}(aq) + Cu(s)$$

In this reaction the zinc has lost electrons, and is said to have been **oxidized**. The copper ions have gained electrons, and are said to have been **reduced**.

The loss of electrons by a species is termed oxidation. The gain of electrons by a species is termed reduction.

The reaction between zinc metal and copper ions occurs when zinc is added to copper sulphate solution. It is a typical example of a **redox** (*red*uction-*ox*idation) reaction:

loses $2e^-$ ∴ oxidized

$$Zn(s) + Cu^{2+}(aq) \rightarrow Zn^{2+}(aq) + Cu(s)$$

gains $2e^-$ ∴ reduced

This reaction, when carried out in a test tube, gives out heat. When put to use in a simple cell it gives rise to electrical energy.

If two different metals are made to touch under the surface of an electrolyte they really constitute a "short-circuited" cell. A current flows between the two metals and a reaction occurs.

Everyday life contains many examples of cell action. Chromium plated articles corrode very rapidly once the chromium layer has been penetrated, and so do tin-plated metal cans, the corrosion taking place where the two metals are simultaneously in contact with air and moisture. Corrosion of car fittings is rapid in winter when salt has been put down on the roads—chromium and iron are the two metals, and sodium chloride is the electrolyte in a cell. Corrosion of iron is oxidation forming rust—hydrated iron oxide—assisted by the electrolytic process.

11.3 Displacement reactions of metals

Metals may be classified according to their ability to *displace* the ions of other metals from solution. Adding zinc powder to copper(II) sulphate solution, which contains $Cu^{2+}(aq)$, causes metallic copper to be deposited and the zinc to dissolve:

$$Zn(s) + Cu^{2+}(aq) \rightarrow Cu(s) + Zn^{2+}(aq); \quad \Delta H = -210 \text{ kJ}.$$

There will also be a certain amount of hydrogen evolved, due to the fact that a solution of $Cu^{2+}(aq)$ contains some $H^+(aq)$ as well, but this does not detract from the main argument.

If zinc powder is added to a series of aqueous solutions of metal ions, it will be found to displace certain ones with ease, such as silver(I) $Ag^+(aq)$, mercury(II) $Hg^{2+}(aq)$, lead(II) $Pb^{2+}(aq)$, and tin(II) $Sn^{2+}(aq)$. Others it will not displace at all, such as sodium $Na^+(aq)$, magnesium $Mg^{2+}(aq)$, and aluminium $Al^{3+}(aq)$. The metals can in fact be arranged in a sequence in such a way that a given metal will displace the ions of all metals below it in the series. This

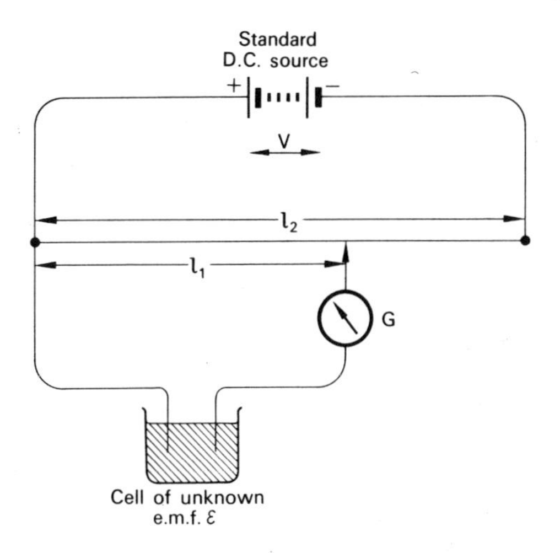

FIG. 11.2. A potentiometer: $\mathcal{E} = V(l_1/l_2)$.

series is known as the **electrochemical series**, which will be considered in section 11.9.

The "driving force" of a displacement reaction is obtained by constructing a cell in which the reaction can occur, and finding the magnitude of an external opposing e.m.f. which has to be applied in order *just* to prevent the reaction happening. A potentiometer circuit is ideal for this, provided that it has an accurate source of d.c. supply as a standard (Fig. 11.2). Alternatively a valve voltmeter may be employed. The chief problem lies not in measuring the e.m.f., but in constructing a cell which performs the desired reaction.

Each electrode is placed in a solution of its own ions at a standard concentration (molar will do, but strictly we should use solutions which are at "unit activity"—see section 11.6.) A piece of wire is not suitable for connecting these two half-cells together as it would introduce its own voltages at the points where it contacts the different electrolytes. A much better circuit is made by employing a **salt bridge**. This is generally an inverted U-tube containing a concentrated, non-reacting electrolyte such as potassium chloride solution KCl(aq). Bulk diffusion between the liquids is eliminated either by cotton wool plugs or, better, sintered glass discs at the two ends of the U-tube. If a valve voltmeter is used, strips of filter paper soaked in electrolyte suffice.

Figure 11.3 shows the cell which would be required in order to measure the e.m.f. of the displacement reaction between zinc and copper(II). The slide wire of the potentiometer is adjusted until the opposing e.m.f. just prevents current flowing through the galvanometer. The reactants are then being *artificially held at equilibrium* by the applied e.m.f. We can therefore say that the applied e.m.f. is a measure of the "driving force" of the displacement reaction.

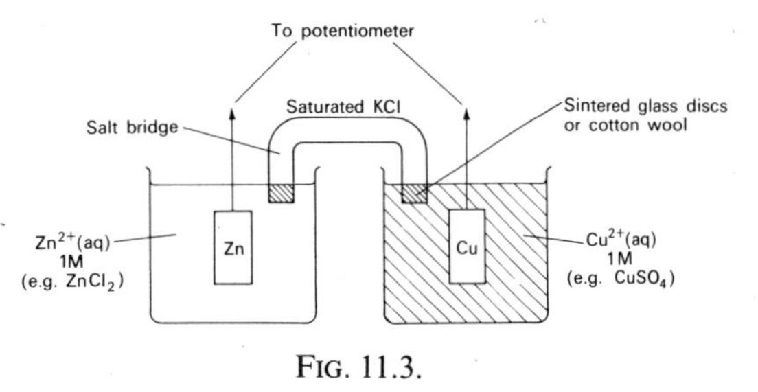

FIG. 11.3.

11.4 Conventions used in writing cell reactions

In order to save space, and the labour of drawing a separate diagram for each cell we set up, a shorthand notation is used in describing cells. The cell used in Fig. 11.3 is written down thus:

$$\mathrm{Zn}|\mathrm{Zn}^{2+}(\mathrm{aq}, 1\ \mathrm{M}) \vdots \mathrm{Cu}^{2+}(\mathrm{aq}, 1\ \mathrm{M})|\mathrm{Cu}; \quad \mathcal{E} = +1{\cdot}10\ \mathrm{V}.$$

For brevity, symbols showing the states of reactants are omitted when these are obvious, but concentrations or activities of the electrolytes are written in. The symbol ℰ is used to denote e.m.f. in this book, and should not be confused with E for energy.

Sign convention. The sign of the e.m.f. is the sign of the right-hand electrode. In the above case the copper terminal is positive with respect to the zinc.

Many displacement reactions can be investigated in a cell, as well as in the test tube. For instance the reaction

$$Cu(s)+2Ag^+(aq) \rightleftharpoons Cu^{2+}(aq)+2Ag(s),$$

which gives rise to very beautiful crystals of metallic silver when done in a test tube, can be investigated quantitatively in the cell:

$$Cu|Cu^{2+}(aq) \vdots Ag^+(aq)|Ag; \quad \mathcal{E} = +0{\cdot}46 \text{ V}.$$

A positive sign means that the reaction will proceed spontaneously from left to right if carried out in a test tube. A negative value of $\mathcal{E}$ means that the reaction would proceed from right to left. A value of $\mathcal{E} = 0$ represents an equilibrium constant of unity.

11.5 Concentration cells

We have seen that the e.m.f. of a cell is a measure of the driving force of the cell reaction. If this is reliable, then *any* spontaneous change ought to be capable of being harnessed to give an e.m.f. Take, for instance, the mixing of two solutions of an electrolyte at different concentrations. This will occur spontaneously, quite regardless of whether heat energy is evolved or absorbed in the process. It is possible to set up a cell in which the driving force of the mixing process is harnessed: such a cell is called a **concentration cell**.

If 0·01 M copper(II) sulphate solution is added to 0·1 M copper(II) sulphate, mixing will occur until the concentration is uniform; thereafter spontaneous change will cease. Such a mixing process is represented by the following cell (Fig. 11.4).

$$Cu|Cu^{2+}(0{\cdot}01 \text{ M}) \vdots Cu^{2+}(0{\cdot}1 \text{ M})|Cu; \quad \mathcal{E} = +0{\cdot}029 \text{ V}.$$

The sense of the e.m.f. is such as to cause some copper from the left-hand electrode to dissolve, thereby increasing the concentration of $Cu^{2+}(aq)$ in the electrolyte, and to cause some of the copper ions in the solution of greater concentration to deposit on the right-hand electrode. The driving force of this process can be measured by applying an external e.m.f. from a potentiometer in order *just* to prevent current flowing. A sensitive potentiometer is needed, showing that the driv-

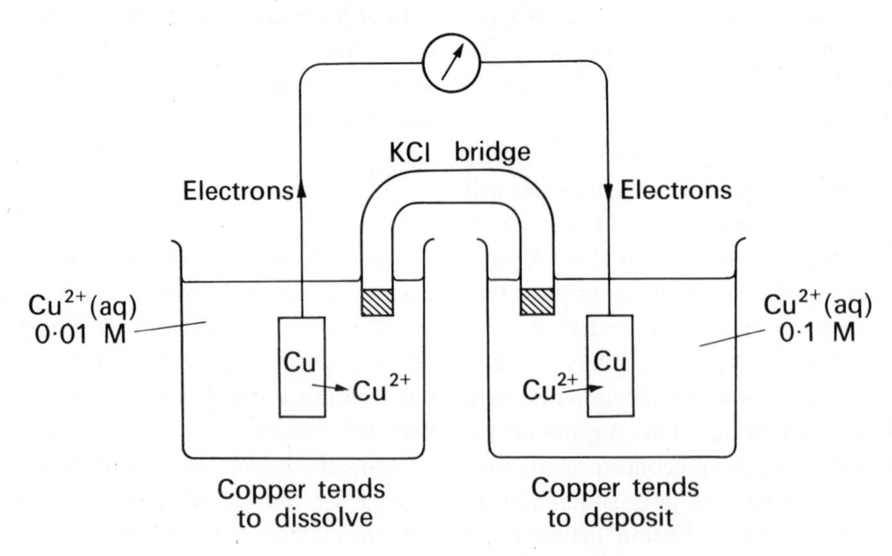

FIG. 11.4. A concentration cell.

ing force in a concentration cell is very much smaller than when two different metals are used in a displacement reaction.

Further experiments show that, provided the solutions are dilute, it is the *ratio* of ion concentrations which determines the e.m.f. and hence the driving force. A mathematical expression can be derived relating the concentrations to the e.m.f., and this is found to be logarithmic in form.

$$\mathcal{E} = \frac{RT}{nF}\log_e \frac{c_1}{c_2} \quad \text{(provided the solutions are } \textit{dilute}\text{)}$$

where $\mathcal{E}$ = e.m.f. of cell,
T = temperature in K,
F = the faraday constant, 96 490 coulombs,
n = the number of faradays transferred per mole,
R = the gas constant, expressed in $\text{J K}^{-1}\ \text{mol}^{-1}$,
c_1 = concentration, in mol dm^{-3}, of the solution in the left-hand cell,
c_2 = concentration, in mol dm^{-3}, of the solution in the right-hand cell.

Concentration cells are of little practical importance, as the e.m.f.s they give are very feeble, though they play an important role in living systems, for instance in muscles. There are two important things to be learned from studying them however:

(i) *Concentration of electrolyte does have an effect on the e.m.f. of a cell. If meaningful data are to be quoted, the ion concentration must be known.*

(ii) *Even in the simple case of "physical" mixing of two electrolytes, it is possible to measure the "driving force" of the process quantitatively. Some solutions mix endothermically, some exothermically; others do not give a detectable temperature change. Thus whereas heat change is not a criterion of spontaneous change, e.m.f. is.*

11.6 The concept of activity

In the argument above it was stressed that the solutions were dilute. Concentration cells can certainly be made to work at higher concentrations, but their e.m.f.s no longer fit the mathematical relationship given above. This is one of the reasons why chemists often use the term **activity**. At very low concentrations the terms activity and concentration may be taken as synonymous—a 0·001 M solution of Cu^{2+}(aq) can be taken as having an activity of $a = 0{\cdot}001$ —but at higher concentrations this may not be so.

One way of *defining* activity would be "that quantity which fits the equation

$$\mathcal{E} = \frac{RT}{nF}\log_e \frac{a_1}{a_2}$$

in a concentration cell". A concentration cell can be used to *measure* activities, once we have defined activity in this way.

11.7 Standard electrode potentials

We shall in future frequently refer to solutions of unit activity. To a very rough approximation, activity = molar concentration. For instance it is calculated that a molar solution of sodium chloride has an activity of 0·68. For practical purposes activity may be regarded as a sort of "effective concentration".

In order to quote meaningful data for electrodes in cells, the concentration of ions has to be stated. In theory any concentration could be taken, but in practice we choose that concentration which corresponds to unit activity.

It is not possible to quote the potential of a single electrode: the only way of measuring it would be to insert a second electrode, and we

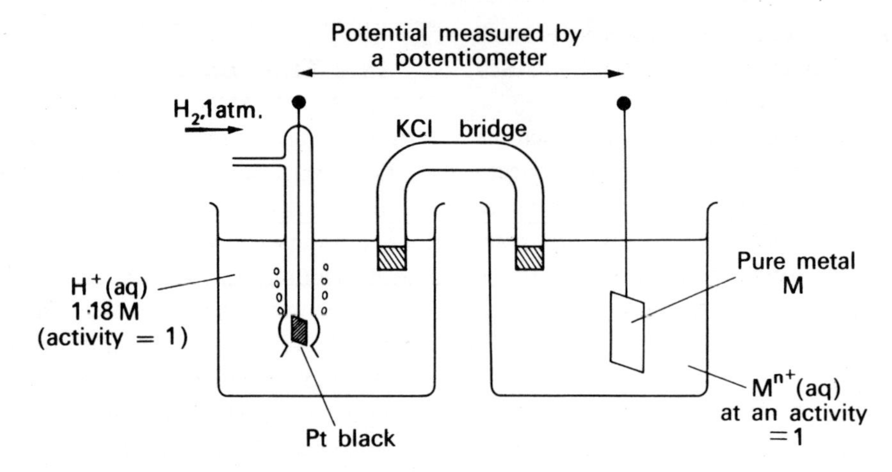

FIG. 11.5. Standard electrode potential on the hydrogen scale.

would then be measuring the difference between two electrode potentials! This being so, we must choose one electrode arbitrarily as a standard, and measure the potential difference between this standard electrode and any other standard electrode we require.

By general agreement the **standard hydrogen electrode** is taken as the reference point from which data on other electrodes are quoted. Since hydrogen is a gas at room temperature we cannot use it as an electrode in the ordinary way, and instead we use a piece of platinum coated with finely divided "platinum black" containing adsorbed hydrogen, which catalyses the half-cell reaction:

$$2H^+(aq) + 2e^- \rightleftharpoons H_2(g).$$

Pure hydrogen at 1 atm pressure must be used, and this is bubbled continuously over the platinum black electrode to maintain equilibrium (except momentarily when taking the actual e.m.f. reading). A solution of dilute hydrochloric acid is generally used, having a concentration of hydrogen ions of 1·18 mol dm^{-3}, which corresponds to unit activity.

Figure 11.5 shows the arrangement which would be used to measure the standard electrode potential of a metal M on the hydrogen scale. The data in Table 11.1, and in Appendix 1, are quoted for these conditions. The symbol $\mathcal{E}^\ominus$ is used when giving standard potentials.

The cell arrangement may be written down thus:

$$H_2(g), Pt|H^+(aq, a = 1) \vdots M^{n+}(aq, a = 1)|M, \quad (\text{e.m.f.} = \mathcal{E}^\ominus)$$

where n = number of positive charges on the metal ion.

TABLE 11.1
SOME STANDARD HALF-CELL POTENTIALS

Metal	Half-reaction	$\mathcal{E}^\ominus$ (in V)
Sodium	$Na^+ + e^- \rightleftharpoons Na$	−2·714
Magnesium	$Mg^{2+} + 2e^- \rightleftharpoons Mg$	−2·370
Zinc	$Zn^{2+} + 2e^- \rightleftharpoons Zn$	−0·763
Iron	$Fe^{2+} + 2e^- \rightleftharpoons Fe$	−0·440
Hydrogen	$2H^+ + 2e^- \rightleftharpoons H_2$	0·000
Copper	$Cu^{2+} + 2e^- \rightleftharpoons Cu$	+0·337
Silver	$Ag^+ + e^- \rightleftharpoons Ag$	+0·799

The half-cell potentials quoted in Table 11.1 are *standard* potentials, which we denote by using the superscript zero in the symbol $\mathcal{E}^\ominus$. Section 11.5 showed that e.m.f.s are affected by concentration according to a logarithmic relationship. The expression for a half-cell potential,

$\mathcal{E}$, of any couple in which *oxidized form + n electrons* $\rightleftharpoons$ *reduced form* is similar to that quoted for concentration cells:

$$\mathcal{E} = \mathcal{E}^{\ominus} + \frac{RT}{nF} \log_e \frac{[\text{oxidized form}]}{[\text{reduced form}]}$$

where square brackets denote activities, or for approximate purposes, concentrations. In the case of a metal in equilibrium with its ion, this expression reduces to

$$\mathcal{E} = \mathcal{E}^{\ominus} + \frac{RT}{nF} \log_e [\text{ion}]$$

For practical purposes, it is more convenient to use logarithms to base 10, and since data are usually quoted at 25°C, we may write

$$\mathcal{E} = \mathcal{E}^{\ominus} + \frac{2{\cdot}3RT}{nF} \log_{10} \frac{[\text{oxidized form}]}{[\text{reduced form}]}$$

$$= \mathcal{E}^{\ominus} + \frac{0{\cdot}059}{n} \log_{10} \frac{[\text{oxidized form}]}{[\text{reduced form}]}.$$

These equations are all forms of the **Nernst equation.***

11.8 Using half-cell potentials to predict reactions

Although $\mathcal{E}^{\ominus}$ values are quoted relative to the normal hydrogen electrode they can be also used to predict the direction of reactions other than those involving hydrogen as well. First, however, we will show how the $\mathcal{E}^{\ominus}$ values predict whether or not metals will reduce hydrogen ions.

(a) Displacement of hydrogen by metals

Consider the reaction which occurs when magnesium is added to dilute acid:

$$Mg(s) + 2H^+(aq) \rightleftharpoons Mg^{2+}(aq) + H_2(g).$$

This reaction may be investigated in various ways, for instance:

(i) In a test tube, rapid and vigorous evolution of hydrogen is observed. Heat is evolved, that is, ΔH is negative.

(ii) A standard cell might be set up, and an external e.m.f. applied, in order *just* to prevent reaction taking place. Table 11.1 tells us that the e.m.f. required will be 2·37 V. If the opposing e.m.f. is removed, spontaneous reaction will occur in the direction left to right in the equation above. We may term this e.m.f. the standard e.m.f. of reaction. In this book it is denoted by the symbol $\Delta\mathcal{E}^{\ominus}$.

Observation (ii) is more valuable, because it tells us accurately how great is the "driving force" of the reaction. The procedure for predicting a reaction from half-cell potentials is as follows:

(1) Write down the half-reactions, together with their $\mathcal{E}^{\ominus}$ values from a book of data:

$$2H^+ + 2e^- \rightleftharpoons H_2; \quad \mathcal{E}^{\ominus} = 0{\cdot}00 \text{ V, by definition}$$

$$Mg^{2+} + 2e^- \rightleftharpoons Mg; \quad \mathcal{E}^{\ominus} = -2{\cdot}37 \text{ V.}$$

(2) Obtain the overall equation by subtraction of the second half-reaction from the first, and obtain the reaction e.m.f. $\Delta\mathcal{E}^{\ominus}$, by subtracting the second $\mathcal{E}^{\ominus}$ from the first:

$$2H^+ + Mg \rightleftharpoons Mg^{2+} + H_2; \quad \Delta\mathcal{E}^{\ominus} = 0{\cdot}00-(-2{\cdot}37) = +2{\cdot}37 \text{ V.}$$

(3) If $\Delta\mathcal{E}^{\ominus}$ is positive, the reaction will proceed from left to right as written. If it is negative the equilibrium composition will lie to the left. If the e.m.f. is almost zero, then the reactants will come to equilibrium with all species present in appreciable amounts.

* By international agreement the sign given to $\mathcal{E}^{\ominus}$ values is the sign of the charge on the *electrode*, relative to the hydrogen electrode. Alternatively, it is the e.m.f. of the cell written down with the standard electrode on the left, and the electrode being measured on the right.

Consider another case: predict from $\mathcal{E}^\ominus$ data whether metallic silver will reduce the hydrogen ion in dilute acid. Following the procedure above:

(1) Write down the half-reactions:

$$2H^+ + 2e^- \rightleftharpoons H_2; \quad \mathcal{E}^\ominus = 0{\cdot}00 \text{ V}. \quad \text{(i)}$$

$$Ag^+ + e^- \rightleftharpoons Ag; \quad \mathcal{E}^\ominus = +0{\cdot}799 \text{ V}. \quad \text{(ii)}$$

(2) Subtract as before. In this case equation (ii) has to be doubled in order that the electrons "cancel out" in the overall equation. This has no effect on its $\mathcal{E}^\ominus$ value.

$$2H^+ + 2Ag \rightleftharpoons 2Ag^+ + H_2;$$
$$\Delta\mathcal{E}^\ominus = -0{\cdot}799 \text{ V}.$$

(3) Since the resultant e.m.f. is negative the reaction will *not* proceed from left to right. Instead the equilibrium composition will lie heavily to the left. The $\mathcal{E}^\ominus$ data predict that if hydrogen gas is bubbled into a solution of, say, silver nitrate, metallic silver should be precipitated.

This prediction is however not borne out by experiment: in the absence of a catalyst there is no observable change when hydrogen is bubbled into Ag^+(aq). This illustrates one of the limitations of using half-cell potentials, namely that they cannot predict reaction *rates*.

(b) Displacement of one metal by another

Although $\mathcal{E}^\ominus$ values are quoted relative to the hydrogen electrode the same data can be used to predict the outcome of reactions which do not involve hydrogen. Suppose for instance we wish to predict the outcome of adding zinc powder to a solution of lead(II) ions, Pb^{2+}(aq). A data book gives two half-reactions which are relevant, namely:

$$Pb^{2+} + 2e^- \rightleftharpoons Pb; \quad \mathcal{E}^\ominus = -0{\cdot}126 \text{ V}.$$

$$Zn^{2+} + 2e^- \rightleftharpoons Zn; \quad \mathcal{E}^\ominus = -0{\cdot}763 \text{ V}.$$

We require the e.m.f. of the reaction

$$Pb^{2+} + Zn \rightleftharpoons Zn^{2+} + Pb;$$
$$\Delta\mathcal{E}^\ominus = -0{\cdot}126 - (-0{\cdot}763) = +0{\cdot}537 \text{ V}.$$

This equation is obtained by subtracting the second equation from the first, and hence the reaction e.m.f. is obtained by subtracting $-0{\cdot}763$ from $-0{\cdot}126$ V. Since the e.m.f. is positive for the equation as written, the data predict that adding zinc powder to Pb^{2+}(aq) should cause metallic lead to be precipitated. In fact this reaction is a very effective way of growing crystals of lead—a so-called "lead tree' is produced.

11.9 The electrochemical series of metals

The metals in Table 11.1 are arranged in the order of their $\mathcal{E}^\ominus$ values. The order is seen to be the same as the electrochemical series produced by conducting simple displacement reactions. This is not surprising, for we have seen that a comparison of $\mathcal{E}^\ominus$ values enables us to predict the direction of displacement reactions. A given metal in the series will *reduce* the ions of all metals below it, and a given ion will *oxidize* all the metals above it. The advantage of using e.m.f. data is that it enables the electrochemical series to be placed on a quantitative basis. For instance, simple displacement reactions in a test tube are not accurate enough to distinguish between tin and lead. $\mathcal{E}^\ominus$ measurements, made to $\pm 0{\cdot}001$ V or better, show that tin comes just above lead in the series.

In a case like tin and lead, the reaction e.m.f. is extremely low, and we should expect to be able to observe an equilibrium rather than a practically complete displacement of one metal by the other:

$$Sn^{2+} + 2e^- \rightleftharpoons Sn; \quad \mathcal{E}^\ominus = -0{\cdot}136 \text{ V}$$

$$Pb^{2+} + 2e^- \rightleftharpoons Pb; \quad \mathcal{E}^\ominus = -0{\cdot}126 \text{ V}$$

Subtracting,

$$Sn^{2+} + Pb \rightleftharpoons Sn + Pb^{2+};$$

$$\Delta\mathcal{E}^{\ominus} = -0{\cdot}136 - (-0{\cdot}126) = -0{\cdot}01 \text{ V}.$$

Hence only 0·01 V would need to be applied to a cell to prevent reaction occurring. We therefore expect an equilibrium, slightly favouring Sn^{2+} and Pb. The equilibrium constant can be determined experimentally for such a system:

$$K = \frac{[Sn][Pb^{2+}]}{[Sn^{2+}][Pb]} = \frac{[Pb^{2+}]}{[Sn^{2+}]} = 0{\cdot}46.$$

11.10 Prediction of equilibrium constants from e.m.f. measurements

It can be shown theoretically, and verified by experiment in many cases, that the equilibrium constant, K, of a reaction is related to the reaction e.m.f., $\Delta\mathcal{E}^{\ominus}$, by an expression which is similar to that given earlier for concentration cells:

$$\Delta\mathcal{E}^{\ominus} = \frac{RT}{nF}\log_e K = \frac{2{\cdot}303\,RT}{nF}\log_{10} K$$

$$= \frac{0{\cdot}059}{n}\log_{10} K \text{ at } 25°\text{C}.$$

Figure 11.6 shows this relationship graphically and may be used for the problems on equilibrium constants at the end of this chapter, as well as elsewhere in the book.

Substituting in this equation, using the data for tin and lead,

$$\log_{10} K = \frac{n\Delta\mathcal{E}^{\ominus}}{0{\cdot}059} = \frac{2\times 0{\cdot}010}{0{\cdot}059} = 0{\cdot}339$$

$$\therefore K = 2{\cdot}18 = \frac{[Sn^{2+}]}{[Pb^{2+}]}.$$

Worked Example 1. Calculate the equilibrium constant of the reaction between cobalt and nickel(II).

Half-reactions: $Ni^{2+} + 2e^- \rightleftharpoons Ni;$

$$\mathcal{E}^{\ominus} = -0{\cdot}250 \text{ V}$$

$$Co^{2+} + 2e^- \rightleftharpoons Co;$$

$$\mathcal{E}^{\ominus} = -0{\cdot}277 \text{ V}$$

Subtract for complete reaction:

$$Ni^{2+} + Co \rightleftharpoons Ni + Co^{2+};$$

$$\Delta\mathcal{E}^{\ominus} = -0{\cdot}250 - (-0{\cdot}277) = +0{\cdot}027 \text{ V}.$$

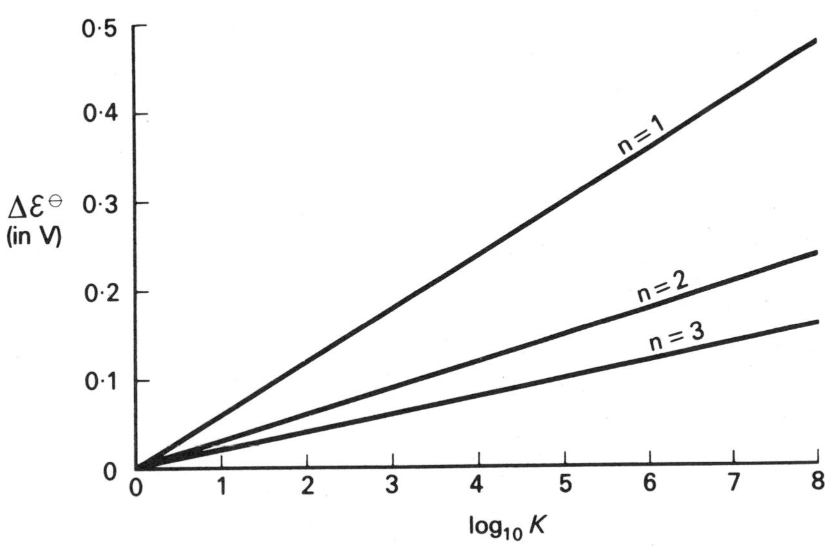

FIG. 11.6. Graph for obtaining equilibrium constant from reaction e.m.f.

$$K = \frac{[Co^{2+}][Ni]}{[Co][Ni^{2+}]} = \frac{[Co^{2+}]}{[Ni^{2+}]}$$

$$\frac{2{\cdot}303\ RT}{nF}\log_{10}K = \frac{0{\cdot}059}{n}\log_{10}K = 0{\cdot}027$$

$$\log_{10}K = \frac{0{\cdot}027\times 2}{0{\cdot}059} = 0{\cdot}915$$

$$K = 8{\cdot}23.$$

The positive value of the reaction e.m.f. $\Delta\mathcal{E}^\ominus$ indicates that cobalt(II) ions should reduce metallic nickel. The fact that it is a low e.m.f. suggests that a measurable equilibrium ought to exist. The calculation shows us that $K = 8{\cdot}23$, and hence the reaction does not proceed completely to the right. A reaction is virtually complete when $K > 10^6$. This corresponds to a $\Delta\mathcal{E}^\ominus$ value of 0·36 V where $n = 1$.

A similar calculation can be applied to any reaction for which $\Delta\mathcal{E}^\ominus$ is known. Remember that a favourable equilibrium does not necessarily *prove* that the reaction is feasible, for it might be too slow, or there might be side reactions which interfere. However, the calculations do tell us with complete certainty those reactions which are *impossible* to carry out; this in itself is very important. For instance, a chemical engineer may save his company many thousands of pounds by carrying out a five-minute calculation to show that a proposed reaction is impossible!

11.11 Redox reactions among non-metals

So far this chapter has been concerned with redox reactions among metals, apart from hydrogen (which is "metallic" insofar as it forms positive ions in aqueous solution). Displacement reactions among metals are easily carried out in a test tube and are readily studied quantitatively by means of cells.

It is quite possible to set up electrodes for the study of non-metals, as has already been done in the case of hydrogen, and redox reactions similar to those among metals occur here also. It is quite natural therefore to extend the redox series to include non-metallic half-reactions. A standard electrode can be constructed for say, chlorine, representing the following:

$$Cl_2(g) + 2e^- \rightleftharpoons 2Cl^-(aq); \quad \mathcal{E}^\ominus = +1{\cdot}36\ V.$$

Chlorine has a strong tendency to lose electrons and form $Cl^-(aq)$ in solution, as its highly positive half-cell potential indicates. It is therefore a *strong oxidizing agent*.

Chlorine is an element of Group VIIB of the periodic table, one of the halogens. The halogens show the similarity expected of members of the same group or chemical family, and also a gradation in properties. The most powerful oxidizing agent of all is fluorine, and the oxidizing power decreases down the series.

$$F_2(g) + 2e^- \rightleftharpoons 2F^-(aq); \quad \mathcal{E}^\ominus = +2{\cdot}87\ V$$

$$Cl_2(g) + 2e^- \rightleftharpoons 2Cl^-(aq); \quad \mathcal{E}^\ominus = +1{\cdot}36\ V$$

$$Br_2(l) + 2e^- \rightleftharpoons 2Br^-(aq); \quad \mathcal{E}^\ominus = +1{\cdot}07\ V$$

$$I_2(s) + 2e^- \rightleftharpoons 2I^-(aq); \quad \mathcal{E}^\ominus = +0{\cdot}54\ V$$

The halogens are an example of a redox series exactly analogous to the electrochemical series of metals. Examination of the above series indicates that fluorine gas ought to be capable of oxidizing all the other halogen ions $X^-(aq)$. Bromine ought to be capable of oxidizing $I^-(aq)$ to iodine, but will have no effect on $Cl^-(aq)$ or $F^-(aq)$.

An examination of the reaction e.m.f. $\Delta\mathcal{E}^\ominus$ gives us a closer insight into these reactions. Let us see for example what might happen when chlorine gas is bubbled into a solution of potassium bromide, which contains $Br^-(aq)$:

$$Cl_2+2Br^- \rightleftharpoons Br_2+2Cl^-$$

$$\Delta\mathcal{E}^\ominus = 1{\cdot}36-1{\cdot}07 = +0{\cdot}29\ \text{V}.$$

A positive value for $\Delta\mathcal{E}^\ominus$ indicates that the reaction will proceed from left to right as written. Although strictly an equilibrium it will be complete for practical purposes ($K \doteq 10^5$).

The displacement reactions of the halogens (except fluorine gas) are readily carried out on a test-tube scale in the laboratory.

11.12 The general redox series

Appendix 1 gives a selection of standard half-cell potentials, $\mathcal{E}^\ominus$, for a wide variety of elements under various conditions. It contains both metals and non-metals, and there is considerable overlap although metals like the alkali metals occur at the "negative" end of the scale, and the most active non-metals like fluorine and oxygen occur at the "positive" end.

These data give an enormous amount of information about redox reactions when correctly used. If we wish to determine whether a given substance will oxidize or reduce another substance, then it is simply necessary to imagine setting up a cell. If the $\mathcal{E}^\ominus$ values of the relevant half-reactions differ by a large amount, indicating that a large e.m.f. would have to be applied in opposition to the cell by means of a potentiometer, to prevent reaction occurring, then we may predict a strong tendency for the reaction to occur. If the reaction e.m.f. is very small then we predict incomplete reaction.

The data in Appendix 1 have been obtained by a variety of methods, but in most cases it is possible to set up some sort of cell. For instance an electrode might be set up to investigate the reducing power of iron(II) ions, which are readily oxidized to iron(III):

$$Fe^{3+}(aq)+e^- \rightleftharpoons Fe^{2+}(aq);$$

$$\mathcal{E}^\ominus = +0{\cdot}771\ \text{V}$$

The standard electrode for this would be a piece of platinum dipping into a solution containing iron(II) and iron(III) ions mixed together, each at a concentration corresponding to unit activity (Fig. 11.7).

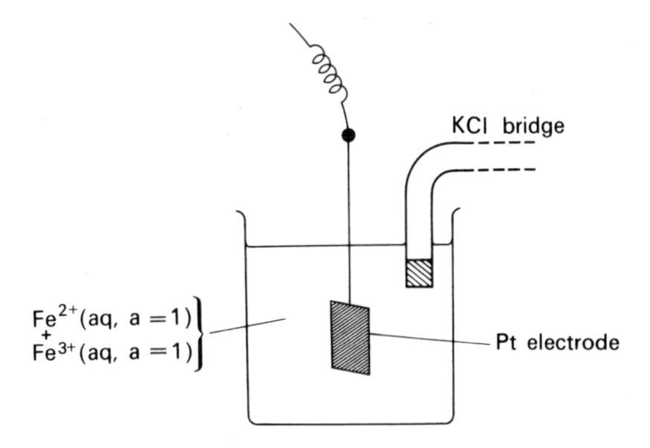

FIG. 11.7. Standard half-cell for the reaction $Fe^{3+}+e^- \rightleftharpoons Fe^{2+}$.

Similarly, an electrode can be set up to measure the oxidizing power of manganate(VII) ions, $MnO_4^-(aq)$, in acid solution. Potassium manganate(VII) is a powerful oxidizing agent, and is itself reduced to manganese(II) ions, $Mn^{2+}(aq)$. The half-reaction is as follows:

$$MnO_4^-(aq)+8H^+(aq)+5e^- \rightleftharpoons Mn^{2+}(aq)+4H_2O;$$

$$\mathcal{E}^\ominus = +1{\cdot}51\ \text{V}$$

The standard electrode for this would be a platinum plate dipping into a solution of $MnO_4^-(aq)$, $Mn^{2+}(aq)$ and $H^+(aq)$, all at unit activity.

Worked Example 2. Predict from redox data whether manganate(VII) ions should oxidize iron(II) ions in acid solution, and obtain a balanced equation.

To obtain the overall equation, the electrons must "cancel out", and we must therefore take five moles of $Fe^{2+}(aq)$ for every mole of $MnO_4^-(aq)$:

$$MnO_4^- + 8H^+ + 5e^- \rightleftharpoons Mn^{2+} + 4H_2O; \quad \mathcal{E}^\ominus = +1{\cdot}51\ V$$

$$5Fe^{3+} + 5e^- \rightleftharpoons 5Fe^{2+}; \quad \mathcal{E}^\ominus = +0{\cdot}77\ V$$

Subtract:

$$MnO_4^- + 8H^+ + 5Fe^{2+} \rightleftharpoons Mn^{2+} + 5Fe^{3+} + 4H_2O.$$

$$\Delta\mathcal{E}^\ominus = 1{\cdot}51 - 0{\cdot}77 = +0{\cdot}74\ V.$$

The equilibrium will therefore lie completely to the right for practical purposes, and iron(II) is quantitatively oxidized by manganate(VII) ions.

This reaction is frequently performed as a titration, for estimating the concentration of iron(II) ions in a solution.

Worked Example 3. 20 cm^3 of a solution of $Fe^{2+}(aq)$ were titrated with 0·01 M $KMnO_4(aq)$ until reaction was complete. 21·6 cm^3 of manganate(VII) solution were required. Calculate the molarity of the iron(II) solution.

21·6 cm^3 of 0·01 M $KMnO_4$ contain

$$\left(\frac{21{\cdot}6}{1000} \times 0{\cdot}01\right) \text{ moles } MnO_4^-.$$

∴ the 20 cm^3 of iron(II) solution contained

$$\left(5 \times \frac{21{\cdot}6}{1000} \times 0{\cdot}01\right) \text{ moles of } Fe^{2+}$$

∴ 1 dm^3 (1000 cm^3) of iron(II) solution contains

$$5 \times \frac{21{\cdot}6}{1000} \times 0{\cdot}01 \times \frac{1000}{20} \text{ moles} = 0{\cdot}054 \text{ moles}.$$

∴ the concentration of $Fe^{2+}(aq)$ is 0·054 M.

It is sometimes a little complicated to set up standard cells in the laboratory for determining the redox potentials given in Appendix 1, though it is relatively easy to set up a cell to demonstrate qualitatively that a redox reaction can generate an e.m.f. Figure 11.8 shows how this can be done for the reaction in which MnO_4^- oxidizes $Fe^{2+}(aq)$. Such a cell reaction is sometimes described as *oxidation at a distance*. If a trace of ammonium thiocyanate is added (NH_4CNS, which provides thiocyanate ions, CNS^-) a blood-red colour is seen to develop around the platinum plate in the iron(II) solution. Although iron(II) ions do not react with CNS^- ions, iron(III) react to form deep blood-red complexes, such as $Fe(CNS)_6^{3-}$:

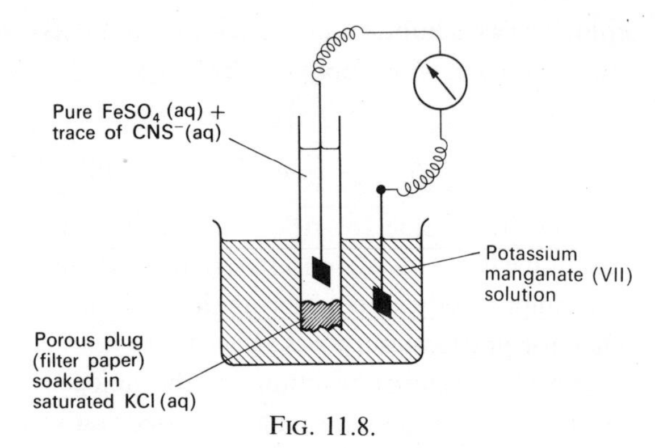

FIG. 11.8.

$$\underset{\text{iron(III)}}{Fe^{3+}(aq)} + 6CNS^-(aq) \rightleftharpoons \underset{\substack{\text{blood-red}\\\text{complex ion}}}{Fe(CNS)_6^{3-}}$$

Note that the blood-red colour forms at the platinum electrode, not at the salt bridge. This is a clear indication that the oxidation is taking place at a distance, due to electron transfer round the external wire.

11.13 Oxidation number

The concept of oxidation number is an important one, especially as a "book-keeping" device for balancing equations and for doing calculations involving oxidation and reduction. In the previous section we referred to the existence of two aqueous ions for iron, namely $Fe^{2+}(aq)$ and $Fe^{3+}(aq)$. These are named iron(II) and iron(III), and in these cases iron has an **oxidation number** of +2 or +3 respectively. Similarly the oxidation number of chlorine is −1 in $Cl^-(aq)$.

The concept of oxidation number is not necessarily limited to ionic compounds. Silver chloride, for instance, shows few, if any, of the properties of an ionic substance. Nevertheless, the oxidation number of silver is still +1 in silver chloride as it is in the aqueous ion; the oxidation number of chlorine is again −1.

$$\begin{array}{llll} & Ag^+(aq) + & Cl^-(aq) \rightarrow & AgCl(s) \\ \text{O.N.:} & +1 & -1 & +1\ -1 \end{array}$$

Magnesium oxide, MgO, has properties which suggest that it is made up of ions Mg^{2+} and O^{2-}, and they are assigned oxidation numbers of +2 and −2. By analogy, other oxides (with one or two exceptions, see below) are assigned oxidation numbers on the basis that oxygen always has an oxidation number of −2. For instance, FeO. O.N. of iron = +2, therefore the compound is named iron(II) oxide. The oxidation numbers must add up algebraically to zero. Fe_2O_3 is iron(III) oxide.

$$\begin{array}{lcc} & Fe_2 & O_3 \\ \text{O.N.:} & (2\times +3) + & (3\times -2) = 0. \end{array}$$

The compound Fe_3O_4, magnetic iron oxide, presents problems: the oxidation number of iron works out to $+2\frac{2}{3}$. Alternatively it can be regarded as a mixture of two oxidation states, as if it were $FeO + Fe_2O_3$. This is the method adopted in naming this compound systematically, and it may be called iron(II)diiron(III) oxide. Similarly lead forms two simple oxides, lead(II) oxide PbO, and lead(IV) oxide PbO_2. Red lead oxide, Pb_3O_4, can be regarded as a compound of these two simple oxides, $2PbO + PbO_2$, and the name becomes di-lead(II)lead(IV) oxide.

Compounds which are markedly non-stoichiometric have to be assigned fractional oxidation numbers, and here the system consequently finds only limited application. Silver oxide for instance is only approximately Ag_2O, and silver is only approximately +1.

The system of oxidation numbers also applies to elements, which are assigned the number zero. To some extent the assignment of oxidation numbers is arbitrary, but for convenience, where an electron-pair bond exists, the oxidation number is assigned by transferring the electrons completely to the more electronegative element. In hydrogen chloride:

less electronegative → H—Cl ← more electronegative

$$\delta+ \quad \delta-$$

Transferring the electron pair completely to the chlorine atom, we have:

$$H^+ \qquad :Cl^-$$

Therefore the oxidation numbers are: hydrogen = +1, chlorine = −1. Where two identical atoms are covalently linked, the electron-pair is assumed equally shared. This point is brought out by comparing water and hydrogen peroxide:

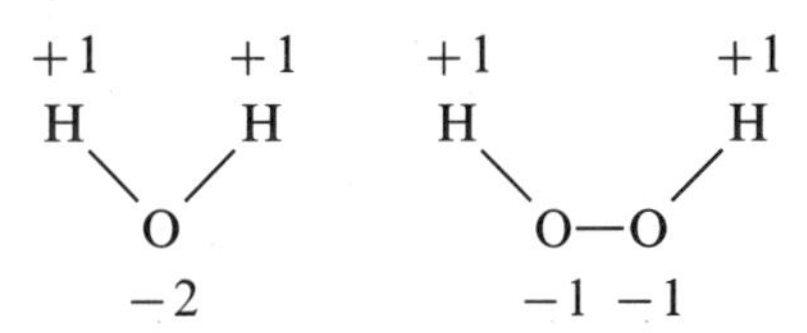

Compounds which contain the peroxo-link, —O—O—, are an exception to the general rule that oxygen atoms are assigned −2. Barium peroxide, BaO_2, is correctly regarded as barium = +2, oxygen = −1, since its properties suggest that it is structurally built up of ions, Ba^{2+} and $[O—O]^{2-}$.

The oxidation numbers in a neutral molecule must add up algebraically to zero; in a complex ion they must add up to the overall charge on

TABLE 11.2

Formula of acid	Formula of ion	O.N. of chlorine	Systematic name of ion	Common name of ion
$HClO$	ClO^-	+1	Chlorate(I)	Hypochlorite
$HClO_2$	ClO_2^-	+3	Chlorate(III)	Chlorite
$HClO_3$	ClO_3^-	+5	Chlorate(V)	Chlorate
$HClO_4$	ClO_4^-	+7	Chlorate(VII)	Perchlorate

the ion. The complex ion $CuCl_4^{2-}$ has copper in an oxidation state of +2:

$$[CuCl_4]^{2-} + 2 + (4 \times -1) = -2.$$

This ion is named tetrachlorocuprate(II), the ending *-ate* implying an anion. The ion $[Cu(NH_3)_4]^{2+}$ is named tetraammine copper(II). The four ammonia molecules are neutral with an overall oxidation number of zero, so the oxidation number of the copper atom is the overall charge on the ion.

The oxidation number of the central atom in an oxo-ion is calculated on the assumption that the oxygen atoms are −2. Many oxo-ions have common names which are rather unsystematic and are falling accordingly into disuse, but they can also be named systematically on the basis of their oxidation states. The oxo-ions of chlorine illustrate this point (Table 11.2). This system of nomenclature has been developed by the International Union of Pure and Applied Chemistry to avoid the confusion which common names often cause. It is often referred to as IUPAC nomenclature. The salts of vanadium and manganese, for instance, occur in a wide variety of oxidation states which often make the older names inadequate.

It is customary to define oxidation of a substance as the removal of electrons, and reduction as the addition of electrons. A more general definition is:

oxidation = increase in oxidation number
reduction = decrease in oxidation number

Thus the definition of oxidation and reduction is not restricted to ionic substances.

Worked Example 4. For each of the following changes, write "oxidized", "reduced" or "no change".

(a) $SnCl_2 \rightarrow SnCl_4$

Oxidation number of Sn increases from +2 to +4.

∴ oxidized.

(b) $MnO_4^-(aq) \rightarrow Mn^{2+}(aq)$

Oxidation number of Mn decreases from +7 to +2.

∴ reduced.

(c) $MnO_4^{2-}(aq) \rightarrow MnO_4^-(aq)$

Oxidation number of Mn increases from +6 to +7.

∴ oxidized.

(d) $2CrO_4^{2-}(aq) \rightarrow Cr_2O_7^{2-}(aq)$

Oxidation number of Cr is +6 on both sides.

∴ no change.

Study Questions

1. (a) Show by means of labelled sketches how you would set up the following cells. Assign the polarity of each electrode by reference to Appendix 1.

(i) $Zn|Zn^{2+}(M) \vdots Sn^{2+}(M)|Sn$

(ii) $Pb|Pb^{2+}(2\ M) \vdots Pb^{2+}(0{\cdot}1\ M)|Pb$

(iii) Pt, H_2(1 atm.)$|H^+(M) \vdots H^+(10^{-7}\ M)|H_2$(1 atm.), Pt

(b) Write a complete equation for the cell reaction that you would expect to occur in (i).

2. Account for the following observations using electrochemical theory.

(a) If a tin-can (iron dipped in molten tin) is scratched, rusting occurs where the tin surface has been penetrated: if a galvanized bucket (iron dipped in zinc) is scratched, no rusting occurs, but the zinc is slowly eaten away.

(b) Chromium plated articles, such as steel motor car fittings, only remain rustproof as long as the chromium surface is unbroken.

(c) Corrosion of motor cars is a greater problem in the winter, when salt is spread on the roads to melt ice and snow.

(d) Corrosion of the iron in the hull of a ship can be prevented by attaching pieces of zinc to the hull.

(e) Copper and silver will not react at all with dilute sulphuric acid. On the other hand, zinc and magnesium react rapidly.

(f) Aqueous solutions of potassium manganate(VII), $KMnO_4$, and of chlorine will decompose if kept for a long time.

(g) If chlorine is bubbled through a solution of Br^-(aq), the colour of bromine is immediately observed. The addition of a crystal of iodine to Br^-(aq) does not liberate bromine.

3. Suggest a simple way of preparing metallic lead from a solution of lead nitrate, $Pb(NO_3)_2$(aq).

4. Suppose that the half-cell $Cu^{2+}(aq)+2e^- \rightleftharpoons Cu(s)$ had been selected as the "standard electrode". Calculate the standard potentials, $\mathcal{E}^*$, for the following half cells, $\mathcal{E}^*$ for Cu^{2+}/Cu being zero.

(a) $Mn^{2+}(aq)+2e^- \rightleftharpoons Mn(s)$.

(b) $Cl_2(g)+2e^- \rightleftharpoons 2Cl^-(aq)$.

(c) $2H^+(aq)+2e^- \rightleftharpoons H_2(g)$.

5. Explain why it is not possible to measure a single electrode potential, such as the potential difference that is supposed to exist between a copper rod and a solution of Cu^{2+} ions.

6. (a) What could happen if a nickel spatula were used to stir a solution of copper(II) sulphate, $CuSO_4$(aq)?

(b) Could silver(I) nitrate, $AgNO_3$(aq), solution be stored in a copper container?

7. Use the data in Appendix 1 to calculate the equilibrium constants for the reactions:

(a) $Ni+Sn^{2+} \rightleftharpoons Ni^{2+}+Sn$.

(b) $Cr^{3+}+Al \rightleftharpoons Cr+Al^{3+}$.

8. Can you discover any relationship between a metal's position in the electrochemical series and its position in the periodic table?

9. How would $\mathcal{E}^\ominus$ for Na^+/Na be affected (qualitatively) if the sodium was present as a 1% amalgam in mercury?

10. 20 cm^3 of 0·1 M iron(II) ammonium sulphate (containing Fe^{2+}) was titrated with an acid solution of potassium dichromate(VI) (containing $Cr_2O_7^{2-}$) of unknown concentration. 25·36 cm^3 of the latter were needed.

(a) Six Fe^{2+} ions react with one $Cr_2O_7^{2-}$ ion. Write down an ionic equation for this reaction.

(b) What is the concentration of the dichromate solution?

11. Use your library to list the different kinds of cell that are in everyday use. Try to discover the chemical reactions that each depends on.

12. Calculate the oxidation numbers in the following:

(a) N in NH_3, N_2H_4, NH_2OH, N_2, N_2O, NO, NO_2^+, NO_2, NO_2^-, NF_3, NCl_3 and NOCl.

(b) Mn in $MnCl_2$, MnO_2, Mn_2O_7, $KMnO_4$, MnO_4^{2-}, MnF_6^{2-}, $Mn(CN)_6^{4-}$, $MnSO_4$, $Mn_2(CO)_{10}$ and Mn.

13. (a) Label the atoms in the following equations with oxidation numbers:

(i) $2CO \rightarrow C+CO_2$.

(ii) $2HI+H_2O_2 \rightarrow I_2+2H_2O$.

(iii) $2H_2O_2 \rightarrow 2H_2O+O_2$.

(iv) $SiH_4+HI \rightarrow SiH_3I+H_2$.

(v) $NH_3+HCl \rightarrow NH_4Cl$.

(vi) $Ca+H_2 \rightarrow CaH_2$.

(b) Deduce from your numbers which species are oxidized and which are reduced.

CHAPTER 12

Free energy

12.1 The limitations of ΔH measurements

A chemical engineer frequently needs to know the heat evolved or absorbed in a chemical process. He may require information on the efficiency of a new fuel, or the energy input required to maintain a reaction in a chemical plant. A simple measurement of ΔH will however only give a limited amount of *theoretical* information about a reaction. Calculation (by Hess's law) of the ΔH of a new, uninvestigated reaction will *not* tell him whether or not the reaction will go. Spontaneous chemical changes can be either exothermic, endothermic or thermoneutral, hence ΔH is no criterion.

ΔH measurements will not tell us how far a chemical change will proceed towards equilibrium. For instance, data on the heat of solution of a salt in water give no information on whether that salt is highly soluble or sparingly soluble.

If, then, ΔH measurements are apparently so limited in their application to theoretical problems, why are we so concerned with energy measurements in chemistry? It was shown in the last chapter that we *can* measure the "driving force" of a reaction, in terms of the e.m.f. which has to be applied to a cell in order to prevent the reaction taking place. The e.m.f. of a cell *is* a criterion of whether or not the reaction will proceed. It is now time to investigate the nature of this "driving force" in more detail.

12.2 The nature of driving force of a reaction—free energy

When we use the term *driving force* we are using the word force rather loosely. The term e.m.f. (electromotive *force*) is similarly confusing. It is necessary to think of the driving force in terms of *energy* units.

Imagine once again a cell in which standard electrodes of silver and zinc are coupled together. Let us calculate the energy *per mole* of reactants which will be liberated, if a small amount of reaction (that is, insufficient to run the cell down) is allowed to occur.

$$Zn(s)+2Ag^{+}(aq) \rightleftharpoons 2Ag(s)+Zn^{2+}(aq); \quad \Delta\mathcal{E}^{\ominus} = +1{\cdot}54\ V.$$

Charge which flows per mole $= 2\,F$ coulombs where $F =$ the faraday constant, 96 490 coulombs.

(Note that *two* faradays of charge are transferred per mole, because the half-reactions involve two electrons per ion.)

$\therefore$ Energy change in J per mole $= -2\,F\Delta\mathcal{E}^{\ominus}$

(Remember: energy change in joules = charge × potential through which that charge is transferred. The minus sign is inserted to indicate that there is a decrease in energy as the reaction proceeds, by analogy with the convention used for ΔH.)

$\therefore$ energy change under equilibrium conditions

$$= -297 \text{ kJ mol}^{-1}$$

This energy change at *equilibrium* is called the **free energy** of reaction. It is given the symbol ΔG. In the above case a cell containing *standard* electrodes was considered, and the free energy change is then termed the **standard free energy of reaction**, and is denoted by adding a superscript $\Delta G^{\ominus}$. In general,

$$\Delta G^{\ominus} = -nF\Delta \mathcal{E}^{\ominus}$$

where n = number of faradays of charge transferred per gram-equation.

ΔG normally has a different numerical value from ΔH. It may even be of opposite sign.

12.3 ΔG as a criterion of spontaneous change

It has already been shown that a reaction e.m.f. is a measure of the tendency of a reaction to proceed in a particular direction. It follows from the relationship $\Delta G^{\ominus} = -nF\Delta \mathcal{E}^{\ominus}$ that free energy change is an alternative criterion. A cell reaction proceeds spontaneously in the direction determined by its positive e.m.f., which is in the direction of *negative* ΔG.

Spontaneous changes occur with a decrease in free energy (ΔG negative).

It appears then that ΔG, not ΔH, is the chemical analogue of "potential energy" of mechanical systems. In mechanical systems, bodies tend to move downhill under the action of gravity—a decrease in potential energy. In chemistry, substances move "downhill" in terms of free energy.

12.4 Factors which determine the magnitude of ΔG

The enthalpy change, ΔH, of a reaction appears nevertheless to have some importance in deciding the outcome of chemical reactions. It is a matter of common experience that the *majority* of vigorous chemical changes at room temperature seem to give out heat, and it has already been observed when considering le Chatelier's principle that endothermic reactions seem to become more important as the temperature is increased. ΔH is one factor involved, but not the only one.

Another factor which influences spontaneous change is the *tendency of matter to reach a state of maximum disorder* (Fig. 12.1). A few examples may be quoted to illustrate this:

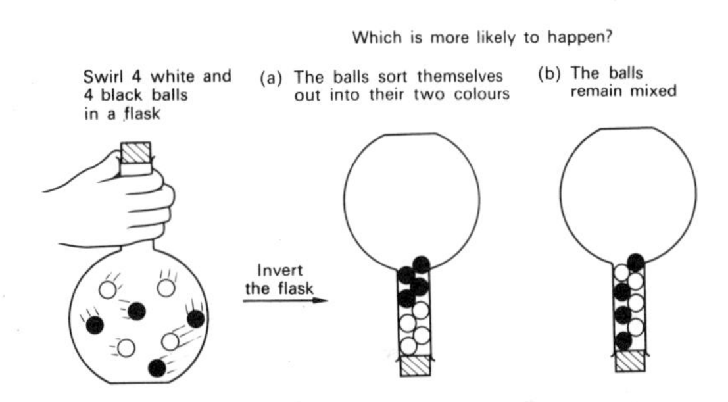

FIG. 12.1.

(i) If oxygen and nitrogen are mixed together in a vessel, the random motion of molecules determines the mixing process. There is no energy change. We cannot say for certain that the gases will never spontaneously "unmix". We can only say that it would be exceedingly improbable. Note that there is no energy released when oxygen and nitrogen are mixed. Note also that theoretically we should not require to use up any energy in separating oxygen and nitrogen. Yet in practice, the separation of oxygen and nitrogen from liquid air consumes quite a considerable expenditure of energy.

(ii) If ink is dissolved in water there is practically no change in temperature (ΔH is approximately zero). Ink dissolves because the system reaches a state of greater randomness or disorder when the molecules disperse in the water.

(iii) When a liquid boils there is an increase in enthalpy. For instance:

$$H_2O(l) \rightleftharpoons H_2O(g); \quad \Delta H = +41 \text{ kJ mol}^{-1} \text{ at } 100°C.$$

Nevertheless, the vapour state is more disordered, more random, than the liquid state since the molecules have freedom to move. This effect becomes more pronounced at a higher temperature.

(iv) If iodine vapour is heated it dissociates. This again is endothermic:

$$I_2(g) \rightleftharpoons 2I(g); \quad \Delta H = +150 \text{ kJ mol}^{-1} \text{ at } 25°C.$$

Note that iodine atoms represent a state of greater disorder than the diatomic molecule.

We may summarize these observations by saying

(tendency for spontaneous change)
= (tendency for system to give out heat)
+ (tendency for system to become disordered)

12.5 Entropy

The degree of disorder of a system, or its randomness, is measured by a quantity called the **entropy**, S. It is in fact easier to talk about *changes* in entropy, and these are shown by the symbol ΔS.

ΔS positive means *system becomes more disordered.*

ΔS negative means *system becomes less disordered.*

A few examples of physical and chemical change ought to make this clear:

(i) $H_2O(l) \rightarrow H_2O(g)$; ΔS positive.
(ii) $H_2O(g) \rightarrow H_2O(l)$; ΔS negative.
(iii) $Cl_2(g) \rightarrow 2Cl(\text{atoms})$; ΔS positive.
(iv) $O_2(\text{pure}) + N_2(\text{pure}) \rightarrow \text{mixture}$; ΔS positive.
(v) $C(s) + CO_2(g) \rightarrow 2CO(g)$; ΔS positive.

The relationship summarized at the end of section 12.4 may now be restated:

(tendency for spontaneous change)
= (tendency for system to give out heat)
+ (tendency for system to increase in entropy)

The entropy factor becomes more important at higher temperatures. Tendency to spontaneous change is free energy, and free energy consists of a "heat factor" and an "entropy factor". The mathematical relationship between them is as follows:

$$\mathbf{\Delta G = \Delta H - T\Delta S}$$

When a system is in equilibrium $\Delta G = 0$. This must be the case, since ΔG determines whether spontaneous change will take place. To summarize, not all reactions proceed spontaneously with the evolution of heat, but *all* spontaneous reactions involve *either* the evolution of heat *or* increase in disorder.

12.6 Entropy of vaporization

Consider again the example of boiling water in equilibrium with steam at 1 atm pressure:

$$H_2O(l) \rightleftharpoons H_2O(g); \quad \Delta G = 0.$$
$$\Delta H = +41 \text{ kJ mol}^{-1}$$
$$\therefore T\Delta S = \Delta H - \Delta G,$$
$$= +41 \text{ kJ mol}^{-1}.$$

∴ under equilibrium conditions,

$$\Delta S = \frac{\Delta H}{T} = \frac{41}{373} = 0{\cdot}11 \text{ kJ mol}^{-1} \text{ K}^{-1}.$$

This quantity is the entropy change which occurs when water is vaporized. It may seem strange that entropy change is measured in the same units as specific heat, but remember that ΔS is only a *measure* of the disorder of a system. We now have a way of expressing ΔS numerically, and calculating it in simple cases.

How does the entropy of vaporization of water at 1 atm pressure compare with other liquids? If measurements are made at the same pressure it is simply necessary to take the molar latent heat of vaporization at the boiling point and divide it by the boiling point in degrees Kelvin. Figure 12.2 shows a graph of ΔH against T for a number of common liquids. They mostly lie on the same straight line, indicating that the entropy of vaporization of all liquids is approximately constant. This constant is known as **Trouton's constant**. It is approximately 88 J mol^{-1} K^{-1}.

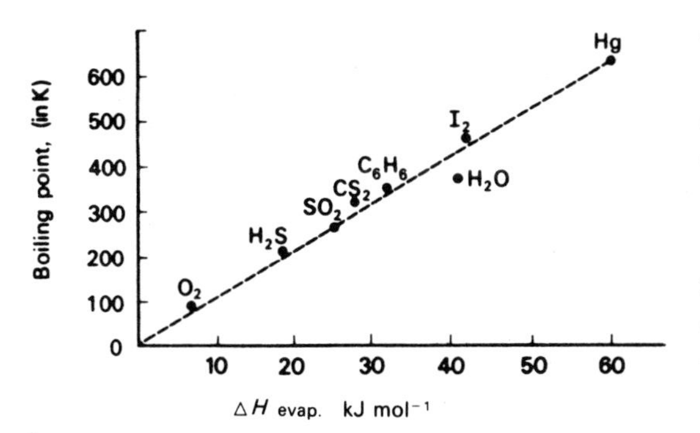

FIG. 12.2. The entropy of vaporization of all liquids is approximately constant (see text).

Water is an exception, but in general it is true to say that all liquids when they become vapour become disordered to approximately the same extent. Water is rather more highly ordered in the liquid state than most other liquids due to the formation of **hydrogen bonds** which have to be broken when water vaporizes, and its entropy of vaporization is abnormally high.

12.7 Free energy and equilibrium constant

In section 11.10 the relationship between K and the standard reaction e.m.f. $\Delta\mathcal{E}^\ominus$ was stated:

$$\Delta\mathcal{E}^\ominus = \frac{RT}{nF}\log_e K$$

Combining this with the relation $\Delta G^\ominus = nF\Delta\mathcal{E}^\ominus$, we have

$$\Delta G^\ominus = -RT\log_e K$$

This equation is known as the **van't Hoff isotherm**. To derive it rigorously would take us even further into the realm of chemical thermodynamics, but it is a simple relationship and very useful. If we can calculate the standard free energy change, $\Delta G^\ominus$, of a reaction, then it is easy to derive its equilibrium constant. This sort of calculation is frequently done by chemists and chemical engineers. For instance, a metallurgist may need to find out the optimum temperature at which to carry out a reaction. To do this he must first calculate $\Delta G^\ominus$ and then work out the equilibrium constant. Chapter 13 shows how this reasoning is put into effect in the extraction of metals from their ores.

Worked Example 1. In the reaction $Cu(s) + 2Ag^+(aq) \rightleftharpoons Cu^{2+}(aq) + 2Ag(s)$, the following values were obtained at 25°C:

(a) *$\Delta\mathcal{E}^\ominus$ for the cell = 0·46 V.*
(b) *$\Delta H = -121$ kJ mol^{-1}.*

Calculate (i) $\Delta G^\ominus$, (ii) ΔS.

(i) $\Delta G^{\ominus} = -nF\Delta\mathcal{E}^{\ominus}$

$= -2 \times 96\,490 \times 0{\cdot}46 \text{ J mol}^{-1}$

$\simeq -89 \text{ kJ mol}^{-1}$.

(ii) $\Delta S = \dfrac{\Delta H - \Delta G^{\ominus}}{T}$

$= \dfrac{-121+89}{298}$

$\simeq -0{\cdot}108 \text{ kJ K}^{-1} \text{ mol}^{-1}$

$= -108 \text{ J K}^{-1} \text{ mol}^{-1}$.

In this example there is a decrease in entropy, due to the decrease in randomness on the right-hand side of the equation. There are less copper ions than silver ions.

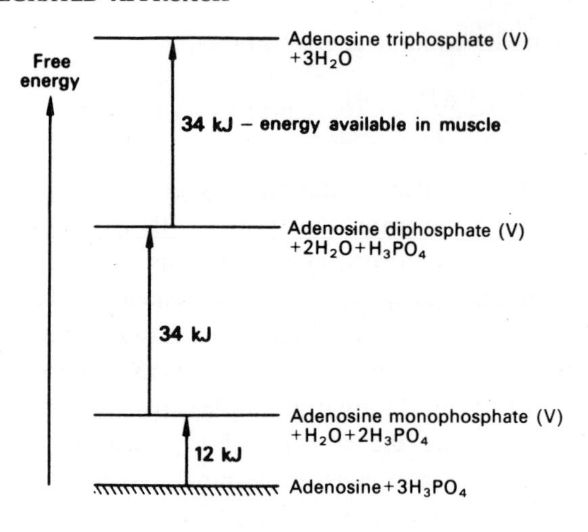

FIG. 12.3

12.8 Free energy and living systems

Free energy is stored chemically in muscles in the form of adenosine triphosphate (ATP). This can be illustrated by the use of a free energy diagram (Fig. 12.3) analogous to the enthalpy diagrams used in Chapter 9. Figure 12.3 shows that there is 34 kJ mol^{-1} of *available* energy in ATP.

Study Questions

1. Predict whether the following changes involve an increase or a decrease in entropy:

(a) $Br_2(g) \rightarrow Br_2(l)$.
(b) $Br_2(g) \rightarrow 2Br(g)$.
(c) $O_2(g)+2CO(g) \rightarrow 2CO_2(g)$.
(d) $NaCl(s) \rightarrow NaCl(aq)$.
(e) $NH_4Cl(s) \rightarrow NH_3(g)+HCl(g)$.
(f) $C(s)+ZnO(s) \rightarrow Zn(g)+CO(g)$.
(g) $2Al(s)+3Fe^{2+}(aq) \rightarrow 2Al^{3+}(aq)+3Fe(s)$.
(h) Naphthalene dissolving in benzene.
(i) The crystallization of salt from brine.

2. When ammonium nitrate dissolves in water, there is an appreciable drop in temperature. Explain why the salt dissolves, when the process is absorbing so much heat.

3. (a) At 25°C, ΔH for the reaction $I_2(g) \rightarrow 2I(g)$ is $+151$ kJ, and ΔS is $+101 \text{ J mol}^{-1} \text{ K}^{-1}$. Calculate ΔG, and hence the equilibrium constant at this temperature.

(b) At 2000K, ΔG for the same reaction is -50 kJ. What is the equilibrium constant at this temperature?

(c) Can iodine atoms be produced from iodine molecules by the action of heat?

4. (a) Use the data given in the Appendix 1 to calculate ΔG for the following reactions.

(i) $Zn(s)+Cu^{2+}(aq) \rightarrow Cu(s)+Zn^{2+}(aq)$.
(ii) $Cl_2(g)+2Br^-(aq) \rightarrow Br_2(g)+2Cl^-(aq)$.
(iii) $Au^{3+}(aq)+3Ag(s) \rightarrow 3Ag^+(aq)+Au(s)$.

(b) What can you say about ΔH in each case?

5. Methanol can be synthesized according to the equation: $CO(g) + 2H_2(g) \rightleftharpoons CH_3OH(g)$: $\Delta H = -90$ kJ as $\Delta S = -220$ J K^{-1} at 25°C.

(a) Calculate $\Delta G^{\ominus}$, and hence the equilibrium constant, at 25°C.

(b) Since reactions proceed faster at higher temperatures, a chemical firm is thinking of making methanol by this method at 700°C, at which temperature ΔG for the reaction is +125 kJ. What advice would you give to the firm?

6. $C(s) + CO_2(g) \rightleftharpoons 2CO(g)$: $\Delta H = +173$ kJ and $\Delta S = +176$ J K^{-1} at 25°C.

(a) How will ΔG and the equilibrium constant be affected by increasing the temperature?

(b) Are your conclusions consistent with le Chatelier's principle?

7. (a) Work out the entropies of vaporization of the following substances:

Substance	Boiling point (K)	Latent heat of evaporation (kJ mol^{-1})
Hydrogen	20	0·84
Methane	108	8·15
Ethoxyethane	307	27·0
Bromine	332	30·5
Ethanol	351	39·5
Zinc	1180	116

(b) Comment on the values that you obtain.

(c) Can you suggest a reason for any exceptional results that you have obtained?

CHAPTER 13

Extraction of metals from their ores

Sections 13.1 to 13.7 deal with the underlying principles of the extraction of metals, in terms of free energy. They may be omitted by readers whose syllabuses do not require this treatment.

13.1 The free energy of formation of metal oxides

It may be taken as a general guide that the most electropositive metals in nature occur in combination with the most electronegative anions such as chloride and nitrate, while the "weaker" metals (lead, zinc, mercury for instance) are found in combination with "weaker" anions such as sulphide, S^{2-}. The weakest metals of all, gold and the platinum metals, generally occur *native*, that is to say, as the free elements.

The Earth's crust is a very complex system, and the study of the occurrence and distribution of the elements is part of the subject of **geochemistry**. The factor which enables the whole problem to be approached rationally is, perhaps not surprisingly, that of free energy. In order fully to understand what is happening, we should have to consider the relative free energies of formation of a great many compounds of metals, such as chlorides, sulphides, silicates, sulphates and so on, as well as oxides. However, the problem of extracting a metal from its ore is very often essentially concerned with decomposing the *oxide* of the metal, since (apart from simple binary compounds such as metal sulphides and chlorides which occur in nature) most metal ores consist essentially of a metal oxide in association with one or more non-metal oxides.

Hence one of the important factors in determining whether a metal can be extracted from, say a carbonate or silicate, is the standard free energy of formation of the oxide. $\Delta G_f^\ominus$ is the standard free energy change of the reaction:

$$xM+\frac{y}{2}O_2 \rightarrow M_xO_y; \qquad (1)$$

$\Delta G_f^\ominus$ = standard free energy of formation *per mole of oxide*

or

$$\frac{2x}{y}M+O_2 \rightarrow \frac{2}{y}M_xO_y; \qquad (2)$$

$\Delta G_f^\ominus$ = free energy of formation *per mole of $O_2(g)$*

If the standard free energy of formation, $\Delta G_f^\ominus$, has a minus sign at a given temperature then the oxide can be expected to form spontaneously from the metal plus oxygen. If $\Delta G_f^\ominus$ has a positive sign, the oxide will be expected to decompose spontaneously into its elements.

13.2 Factors determining free energy of formation of oxides

Unfortunately we cannot devise a cell capable of obtaining an e.m.f. for equation (2) in the majority of cases. $\mathcal{E}^\ominus$ measurements were referred to in Chapter 12, for reactions occurring in aqueous solutions, but here we are dealing with a metal oxide which may or may not be molten, and with a metal which may be solid, liquid, or gaseous, depending on the temperature.

However, $\Delta H_f^\ominus$ is very easily measured by burning the metal in oxygen. If the overall heat change is obtained with the reactants starting at, say 25°C, and with the products finally cooled to 25°C, the heat of reaction will be $\Delta H_f^\ominus$ for 25°C.*

The heats of formation of a number of oxides, *expressed per mole of* $O_2(g)$, are listed in Table 13.1. Notice that their order roughly parallels the $\mathcal{E}^\ominus$ values for the appropriate aqueous ions, which are given for comparison.

TABLE 13.1
THE HEATS OF FORMATION OF SOME METAL OXIDES

Oxide	$\Delta H_f^\ominus$ (per mole of $O_2(g)$)	$\mathcal{E}^\ominus$ (M^{n+}/M) for comparison
MgO	−1200 kJ	−2·37 V
Al_2O_3	−1110 kJ	−1·66 V
ZnO	− 695 kJ	−0·76 V
NiO	− 490 kJ	−0·25 V
CuO	− 310 kJ	+0·34 V
Ag_2O	− 63 kJ	+0·78 V

The free energy of formation of an oxide can now be determined, provided we know its standard entropy of formation, $\Delta S_f^\ominus$.

*Since the specific heats of reactants and products are not equal, $\Delta H_f^\ominus$ will change slightly with temperature but we can disregard this relatively small effect.

$$\Delta G_f^\ominus = \Delta H_f^\ominus - T\Delta S_f^\ominus$$

($\Delta H_f^\ominus$: can be measured; $\Delta S_f^\ominus$: can be calculated)

The entropy change which takes place is essentially that which occurs when one mole of gas phase $O_2(g)$ is removed from the system, provided that neither the metal nor its oxide M_xO_y is vaporized:

$$\frac{2x}{y}M + O_2(g) \rightarrow \frac{2}{y}M_xO_y; \qquad (2)$$

(solid or molten metal; one mole of gas; solid or molten oxide)

$\Delta S_f^\ominus$ negative

There will be small changes in disorder, and hence entropy, associated with the change from the metal phase to the oxide phase, but the *main* factor influencing $\Delta S_f^\ominus$ is the using up of the highly disordered gas phase.

For this reason, the entropy change of reaction (2) is roughly the same for all metal oxide systems, provided that the boiling point of neither metal nor oxide is exceeded. It is approximately 200 J mol^{-1} K^{-1}.

This information enables us to plot the variation of $\Delta G_f^\ominus$ with temperature for metal oxides (Fig. 13.1). Notice that, below the boiling point of the metal, the slopes of all the graphs are *roughly* the same, since the $T\Delta S$ factor is the same whatever the metal. Where the boiling point of the metal is exceeded however, the slope increases since the reaction is now involving a bigger entropy change. For instance, above 1110°C, three moles of gas phase are converted into solid phase, in the reaction:

$$2Mg(g) + O_2(g) \rightarrow 2MgO(s).$$

(magnesium vapour; gaseous oxygen; solid oxide)

Above a certain temperature, $\Delta G_f^\ominus$ becomes positive for some of the oxides. This explains why

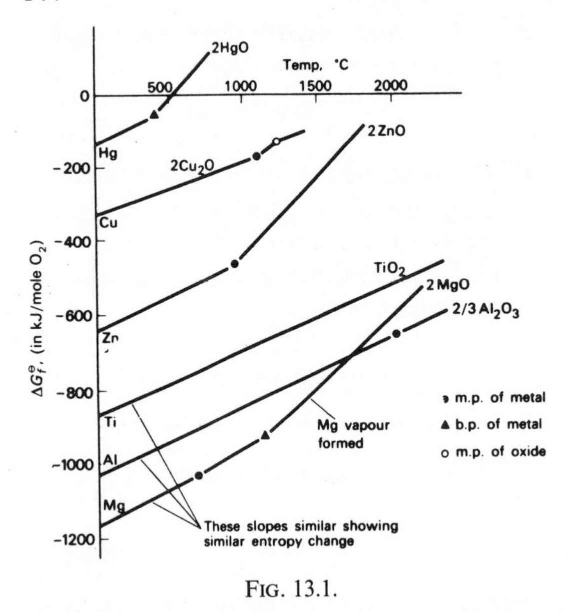

FIG. 13.1.

mercury(II) oxide, for instance, decomposes spontaneously into its elements when heated. The diagram predicts that zinc oxide ought to decompose if heated strongly enough, but it does not hold out much hope for obtaining, say, pure magnesium by straightforward heating of the oxide to a high temperature.

The main points to be learned from Fig. 13.1 are therefore as follows:

(a) The entropy of formation of a metal oxide, expressed per mole of $O_2(g)$, is approximately the same whatever the metal, as shown by the similar slopes of the plots of $\Delta G^\ominus_f$ against temperature.

(b) The thermal stability of metal oxides depends on the way in which $\Delta G^\ominus_f$ varies with temperature. Those with the least negative heat of formation have the lowest stability towards heat.

(c) Heat alone is insufficient to decompose the oxides of most metals.

(d) The stability order of the oxides of metals parallels roughly, but not exactly, their order in the redox series.

13.3 The "Thermit" process

Figure 13.1 also enables us to predict whether a given metal will reduce the oxide of another metal. Consider the reaction between chromium(III) oxide, and aluminium.

$$\tfrac{4}{3}Al + O_2 \rightarrow \tfrac{2}{3}Al_2O_3; \quad \Delta H^\ominus = -1110 \text{ kJ at } 25°C \qquad (3)$$

$$\tfrac{4}{3}Cr + O_2 \rightarrow \tfrac{2}{3}Cr_2O_3; \quad \Delta H^\ominus = -755 \text{ kJ at } 25°C \qquad (4)$$

$$\tfrac{4}{3}Al + \tfrac{2}{3}Cr_2O_3 \rightarrow \tfrac{2}{3}Al_2O_3 + \tfrac{4}{3}Cr;$$
$$\Delta H^\ominus = (-1110+755) = -355 \text{ kJ at } 25°C \qquad (5)$$

The standard free energy, $\Delta G^\ominus$, of this reaction can similarly be derived by subtracting the two relevant $\Delta G^\ominus_f$ values at a given temperature. Figure 13.2 shows the two curves plotted alone. At all accessible temperatures, ΔG is markedly negative, and we should expect the reaction to proceed.

In fact the entropy change of this reaction is quite small, for the simple reason that no gaseous

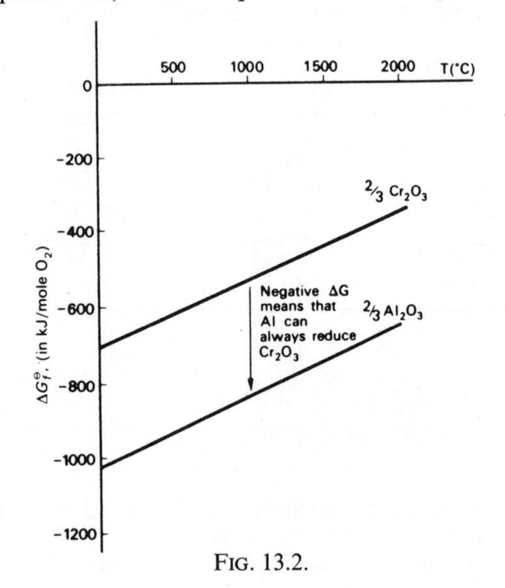

FIG. 13.2.

products or reactants are involved. The quite large entropy terms in the oxide-formation reactions cancel out, and we are left with comparatively small effects due to the different structures of the various phases.

$\Delta G^{\ominus}$ is approximately the same at room temperature as at higher temperatures, but the reaction needs to be raised to a high temperature to "trigger it off". This can be done in the laboratory by priming it with magnesium ribbon and barium peroxide. Once started the reaction is highly exothermic, and very intense temperatures are reached. Care must be taken with this reaction—it is essential either to do it out of doors or to provide a good area of asbestos board and a sand tray.

The "Thermit" reaction, as reduction with aluminium is called, finds relatively little application on an industrial scale, because cheaper reducing agents than aluminium are available (see next section). However, some manganese and chromium are produced in this way. Earlier in this century, before it became more economical to use portable welding equipment, the reaction between iron(III) oxide and aluminium was used for making welded joints, for instance in tramway track.

Another name for the "thermit" reaction is the Goldschmidt process. It could, in principle, be applied to all but the most electropositive metals, but in practice other methods are employed on a large scale.

The conversion of a metal oxide to the metal is reduction, and the reverse process, addition of oxygen, is oxidation. Oxidation is the removal of electrons, and reduction is the addition of electrons. This covers the reaction of metal oxides if we assume them to be ionic, or at any rate partially ionic.

electrons added

$$Fe^{3+} + Al \rightleftharpoons Fe + Al^{3+} \qquad (6)$$

electrons removed

The concept is made more general by using the term oxidation number. The rules for assigning oxidation numbers were given in section 11.13. The system can be expressed as a redox process as follows:

decrease in oxidation number ∴ reduction

$$Fe(III) + Al(0) \rightarrow Fe(0) + Al(III)$$

increase in oxidation number ∴ oxidation

13.4 Carbon as a reducing agent

Since ancient times, carbon has been used as a reducing agent in the extraction of metals like iron, lead and copper, for carbon is the only reducing element which occurs native in large enough quantities. Below about 700°C carbon burns in oxygen to form carbon dioxide, CO_2, with little entropy change because one mole of gaseous O_2 forms one mole of gaseous CO_2. The disappearance of the well-ordered solid phase lattice of carbon has little effect on the entropy change $\Delta S^{\ominus}$.

$$C(s) + O_2(g) \rightleftharpoons CO_2(g); \quad \Delta G^{\ominus} \simeq \Delta H^{\ominus} = -395 \text{ kJ at } 25°C \quad \Delta S^{\ominus} = +3 \text{ J K}^{-1} \qquad (7)$$

However, there is a possible reaction between carbon and carbon dioxide to form carbon monoxide, which is favoured at high temperatures because it involves an increase in the disorder. $\Delta S^{\ominus}$ is positive because *two* moles of CO(g) are formed from only one of CO_2(g):

$$C(s) + CO_2(g) \rightleftharpoons 2CO(g); \quad \Delta H^{\ominus} = +173 \text{ kJ at } 25°C \quad \Delta S^{\ominus} = +176 \text{ J K}^{-1} \qquad (8)$$

Figure 13.4 explains what happens in practice. Adding (7) and (8) we see that equation (9) also has a favourable entropy change which will make the entropy term $T\Delta S^\ominus$ more significant at higher temperatures.

$$2C(s)+O_2(g) \rightleftharpoons 2CO(g);$$
$$\Delta H^\ominus = (-395+173) = -222 \text{ kJ mol}^{-1}$$
$$\Delta S^\ominus = +179 \text{ J K}^{-1} \quad (9)$$

Figure 13.3 shows that, of the two possible reactions between carbon and oxygen, the one which actually occurs at a given temperature is the one which has the more negative $\Delta G^\ominus$. The bold line on the free energy graph indicates this. It is the positive $\Delta S^\ominus$ for equation (9) which causes the free energy plot to slope downwards, above about 700°C.

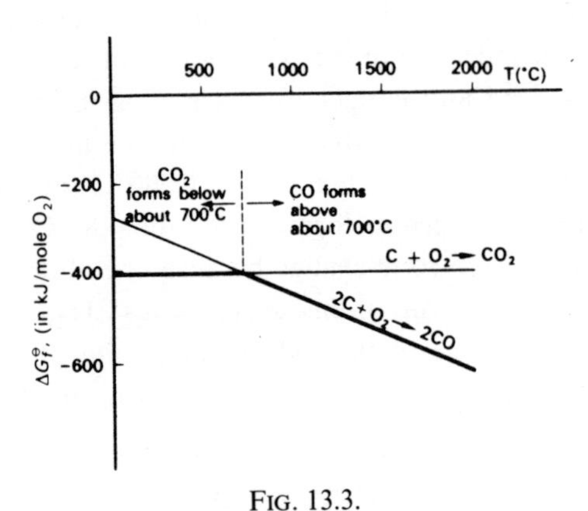

FIG. 13.3.

In the region around 700°C, where the two reactions (7) and (9) both have approximately equal $\Delta G^\ominus$ values (where the two lines cross), the products of combustion will be a mixture of $CO(g)+CO_2(g)$.

The downward slope of the carbon-oxygen graph shows carbon to reduce the oxides of most metals, provided the temperature is high enough. Consider the possible reaction.

$$MgO+C \rightleftharpoons Mg+CO; \quad (10)$$
$$\Delta H^\ominus = +492 \text{ kJ at } 25°C$$

The free energy of reaction can be derived by considering the separate $\Delta G^\ominus$ values for the reactions:

$$2C+O_2 \rightleftharpoons 2CO \quad (9, \text{above})$$
$$\text{subtract} \quad 2Mg+O_2 \rightleftharpoons 2MgO \quad (10)$$

Figure 13.4 shows that this reaction has $\Delta G = 0$ at the point where the two lines representing (9) and (10) cross over. At 1900°C, from the graphs, the equilibrium constant equals unity. Above this temperature ΔG for the reaction becomes negative and the reaction is favoured. Note that the formation of magnesium vapour steepens the "MgO" curve, and makes the temperature for

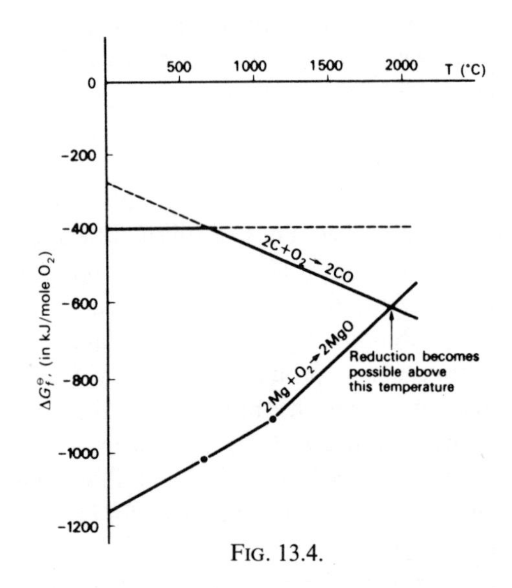

FIG. 13.4.

reduction lower than it would otherwise be.

A qualitative argument using entropy can be applied to this reaction: above the boiling point of magnesium metal, there is an increase of two moles of gas and therefore the ΔS of reaction is strongly positive. Provided that T is high enough, the $T\Delta S$ term will be sufficient to offset the adverse positive ΔH value. This is another example of an endothermic reaction being favoured by high temperature.

ΔG	=	ΔH	−	$T\Delta S$
reaction feasible only if this is negative		this is positive opposing the reaction		ΔS is positive, so $T\Delta S$ overcomes ΔH if T high enough

This qualitative argument leads to the same conclusion as does le Chatelier's principle. This would predict that the position of equilibrium would shift in such a direction as to tend to lower the temperature. The species of higher energy content [r.h.s. of equation (10)] will be favoured by the higher temperature.

Figure 13.5 shows a more complete diagram of the free energies of some metal oxide systems,

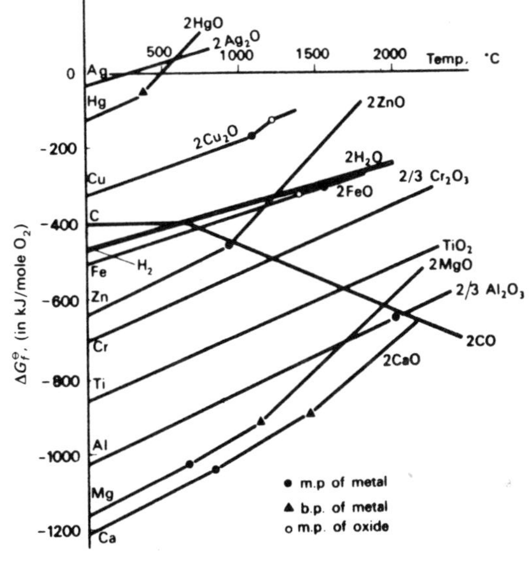

FIG. 13.5.

together with carbon (and also hydrogen). The temperature at which the "carbon" curve intersects the "metal" curve is the temperature at which the equilibrium constant of the reduction becomes unity. The diagram shows, for instance, that $TiO_2(s)$ requires about 1700°C for reduction to titanium, and $ZnO(s)$ only about 1000°C. In the case of the reduction of copper(I) oxide the diagram predicts that carbon dioxide, rather than the monoxide, will be the main product of the reaction:

$$2Cu_2O(s) + C(s) \rightleftharpoons 4Cu(s) + CO_2(g).$$

Table 13.2 gives a summary of the methods used in practice for the extraction of metals.

13.5 Hydrogen as a reducing agent

Hydrogen is not a very effective reducing agent for obtaining metals from their oxides, as Fig. 13.5 shows. The reason is that ΔS is negative for the reaction:

$$\underbrace{2H_2(g) + O_2(g)}_{\text{3 moles of gas}} \rightleftharpoons \underbrace{2H_2O(g)}_{\text{2 moles of gas}};$$

ΔS negative; r.h.s. less disordered.

The plot of ΔG against T therefore rises with temperature, meaning that not many metal oxide plots are intersected. Hydrogen will therefore reduce oxides such as copper(I) oxide, and mercury(II) oxide, but not the oxides of aluminium, magnesium and calcium. Oxides of iron are reduced only with difficulty. In the case of magnetic iron oxide, Fe_3O_4, an equilibrium composition is readily established.

13.6 Reduction of sulphides and chlorides of metals

Although carbon reduces oxides quite effectively, it is not so effective when used directly on sulphides or chlorides of metals. In the case of chlorides, reaction is prevented by the fact that the entropy of reaction is not sufficiently favourable. To take a hypothetical case, there is not sufficient increase in entropy to make the reaction

$$2MgCl_2 + C \rightleftharpoons CCl_4(g) + 2Mg$$

go, even at high temperatures, despite the fact that ΔS is positive.

A similar case occurs with sulphides. Carbon forms only one stable sulphide, CS_2 (b.p. 46°C). The situation can be likened to the analogous case which would occur if carbon was incapable of forming CO. The downward-sloping part of the curve in Fig. 13.3 would not exist. Hence carbon will reduce sulphides directly but only with difficulty. The situation is overcome by **roasting** the sulphide ore by heating it in a stream of air. This reaction converts it to the oxide, with the formation of sulphur dioxide, $SO_2(g)$. Lead(II) sulphide, found in the ore *galena*, is converted into the oxide in this way:

$$PbS + \tfrac{3}{2}O_2 \rightarrow PbO + SO_2.$$

Metallic lead may then be obtained by incomplete roasting, to produce a mixture of PbS and PbO, followed by heating in the absence of air:

$$PbS(s) + 2PbO(s) \rightarrow 3Pb(l) + SO_2(g); \quad \Delta S \text{ positive.}$$

Such a method is possible for lead since the formation of PbO is not very exothermic. It would not be possible for metals where ΔH_f of the oxide is strongly negative—in this case a reducing agent such as carbon must be used.

Mercury(II) oxide, which occurs as the ore *cinnabar*, converts directly to the metal when roasted:

$$HgS + O_2 \rightarrow Hg + SO_2.$$

This happens in preference to the reaction observed for lead above, because at the roasting temperature the free energy of formation of HgO is positive. HgO therefore decomposes into $Hg + \frac{1}{2}O_2$.

13.7 Slag formation

In many metal extraction processes, an oxide is added deliberately to combine with other impurities and form a stable molten phase immiscible with the molten metal, called a **slag**. The principle of slag formation is essentially

$$\underset{\text{(acidic oxide)}}{\text{non-metal oxide}} + \underset{\text{(basic oxide)}}{\text{metal oxide}} \rightarrow \text{fusible (easily melted) slag.}$$

Two instances will be given: in the first case an acidic oxide is added to remove basic oxide impurities, and in the second case a basic oxide is added to remove acidic oxide impurities.

(1) *Removal of unwanted basic oxides*. If metal oxides are present in an ore that would otherwise interfere with the main extraction process, an acidic oxide can sometimes be used to remove them as slag. Sand, which is silicon(IV) oxide, SiO_2, (silicon dioxide, silica) is chosen because it is cheap, involatile, and leads to silicates which are themselves stable though fusible.

For example, a common source of copper is chalcopyrite, $CuFeS_2$. On roasting, this mixed sulphide produces a mixture of oxides of copper and iron. The iron oxides are not wanted, and since they are more basic than those of copper they combine preferentially with added sand:

$$3SiO_2(s) + Fe_2O_3(s) \rightarrow \underset{\text{iron(III) silicate slag}}{Fe_2(SiO_3)_3(l)}$$

Alternatively the iron(III) oxide is allowed to combine with the surface of the converter in which the ore is roasted.

Once an oxide is "tied up" as silicate it is exceedingly difficult to recover it since the free energy of formation of silicates is very negative. Aluminium occurs very widely in nature as clays and other minerals which are silicates, but these cannot be used economically for the extraction of aluminium.

(2) *Removal of unwanted acidic oxides*. Interfering acidic oxide impurities which can occur in metal ores, such as the oxides of phosphorus and silica (in combination) can be removed by the addition of a strongly basic oxide. For this purpose calcium oxide is used, obtained from limestone, $CaCO_3$.

In the blast furnace for extracting iron there are many complex reactions occurring simultaneously, but essentially iron(III) oxide is reduced to iron by carbon monoxide:

$$Fe_2O_3(s) + 3CO(g) \rightleftharpoons 2Fe(l) + 3CO_2(g) \quad (11)$$

This reaction is achieved by charging the blast furnace with a mixture of coke, iron(III) oxide and limestone, and heating it with a blast of air going through. The heat liberated is sufficient to maintain the temperature once the reactions have started (Fig. 13.6). At the bottom of the

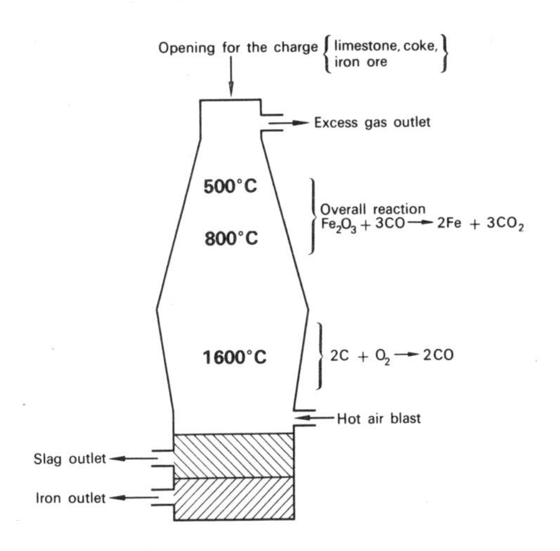

FIG. 13.6. The blast furnace.

furnace the combustion of coke generates carbon monoxide, with intense heat. This intense heat decomposes the limestone into $CaO + CO_2$. The carbon dioxide passes from the top of the furnace, and the calcium oxide combines with unwanted acidic impurities, for instance:

$3CaO$	+	P_2O_5	→	$Ca_3(PO_4)_2(l)$
from lime stone		phosphorus(V) oxide in combination with metal oxides		fusible slag ("basic slag")

Note that carbon monoxide, rather than carbon itself, is the reducing agent at the top of the furnace. Nevertheless the *overall* reaction is the reduction of oxides of iron by carbon. The free energy diagrams such as Fig. 13.5 can be applied to the overall process, even if the reaction is thought to proceed in stages. Considerations of free energy have no bearing on the *mechanism* of the reaction. It is indeed highly improbable that two solid phases will react directly together as implied by the overall equation,

$$Fe_2O_3(s) + 3C(s) \rightleftharpoons 2Fe(l) + 3CO(g).$$

The same consideration applies to the reduction of other oxides with carbon considered in section 13.4.

13.8 Factors affecting choice of extraction method

(1) *Economic considerations*. The method of reduction must be as cheap as possible consistent with producing a product of the required degree of purity. In some cases the purity required is the main factor determining the choice of method. Carbon (in the form of coke) is undoubtedly the cheapest reducing agent, but it does not give a very pure product in all cases owing to the formation of carbides. Iron from a blast furnace contains a few per cent of carbon, and this has to be removed in a later process because it makes the metal brittle. An extreme case is calcium, which fails to form the metal at all, no matter how high the temperature. Instead,

$$CaO(s) + 3C(s) \rightarrow CaC_2(s) + CO(g).$$

In many cases electrolytic extraction is used in preference to chemical reduction, particularly when a pure product is needed. In other cases, an impure product is obtained by carbon reduction, and is refined where it is required pure. Impure copper, although good enough for many purposes, is not good enough for electrical conduction where high conductivity is required; electrolytic refining is therefore used (section 13.9).

Aluminium is only chosen as reducing agent in one or two special cases, where the quantity of metal is not critical, and where the cost of raw materials is high anyway (Thermit process). One or two interesting cases may be instanced:

(i) an Fe–Ti alloy, *ferrotitanium*, is made by the Thermit reaction between aluminium and the ore *ilmenite*, $FeTiO_3$; this alloy is more useful for adding to steel than pure titanium itself, since the overall cost is less and iron is present anyway.

(ii) a Cr–Fe alloy, *ferrochrome*, is made similarly by the reduction of the ore *chromite*, $FeCr_2O_4$, with aluminium.

(2) *Practical considerations*. The choice of a reducing method for extracting a metal depends upon the availability of suitable ores. Many metals occur in unsuitable ores—the case of aluminium has already been mentioned, and silicates in general do not afford promising material despite their widespread abundance.

The metals at the top of the electrochemical series (which also have large negative values for $\Delta H_f^\ominus$ of oxide, see Table 13.1) tend to occur mainly with electronegative anions. The alkali metal cations associate with singly charged anions on the whole, and are found largely as chlorides and nitrates. The alkaline earth metals (Group IIA of the periodic table) occur commonly with doubly charged anions, such as SO_4^{2-} and CO_3^{2-}. Further down the electrochemical series the preference appears to be for silicates, sulphides, S^{2-} and oxides O^{2-}, while the metals at the bottom of the series may occur native.

(3) *Solvent extraction*. In recent years a process known as **solvent extraction** has begun to assume greater importance, particularly in the extraction of copper. Essentially, it is a type of ion-exchange process (section 16.2) used to increase the concentration of the wanted metal from low-grade ores. Originally the methods were developed for the extraction of radioactive metals such as uranium for World War II, and they have since been used also for the extraction of the lanthanide elements, and metals such as zirconium and tantalum.

The solvent for the extraction process is generally kerosene or some similar non-polar solvent immiscible with water. This contains a complexing agent which combines selectively with the metal required, thereby removing its ions from the aqueous phase and concentrating them in complexed form in the organic phase. Very often the precise chemical constitution of the organic complexing agent is a commercial secret.

(4) *Recycling of resources*. Assuming present-day consumption rates, our reserves of nearly all the major metals will be exhausted by the end of this century. (The exceptions are Al, Cr and Fe.) Hence the recovery of metals from low-grade ores and scrap will become increasingly necessary in future.

13.9 Specific methods for extracting metals

Potassium. Highly reactive, occurring in the ore *carnallite*, $KCl.MgCl_2.6H_2O$, and in living matter. Chemical reduction is insufficiently powerful. The chloride can be obtained from carnallite but it is unsuitable for electrolysis when fused, largely because potassium is volatile at the temperature involved. Potassium hydroxide is first made, and this is made anhydrous, fused (m.p. 360°C), and electrolysed. Demand for metallic potassium is not great—it is used in certain specialized reductions—and so the cost of the process is not an overriding factor.

Sodium. Large quantities of sodium are required for a diversity of purposes—as a heat exchanger in nuclear reactors, as a chemical reducing agent, and even as an electrical conductor—and it is essential that the process be as cheap as possible. Originally the electrolysis of fused sodium chloride (Downs process) failed in competition with the electrolysis of fused sodium hydroxide. The higher cost of producing sodium hydroxide was offset by the technical problems concerned in producing sodium from the fused chloride. Sodium is appreciably volatile unless the melt temperature is kept low, and it tends to disperse in the melt in the form of tiny droplets instead of collecting as a liquid at the cathode. Nowadays these technical difficulties have been overcome, and a modern Downs cell uses added calcium chloride to lower the melting point of the electrolyte. The discharge potential of calcium is very close to that of sodium in such a melt, and both metals are discharged simultaneously. They are however immiscible, the less dense sodium rising to the top, and they can be readily separated (Fig. 13.7). Reduction with an element like carbon might be just feasible at a high enough temperature, but could not compete economically with the electrolysis of the fused chloride.

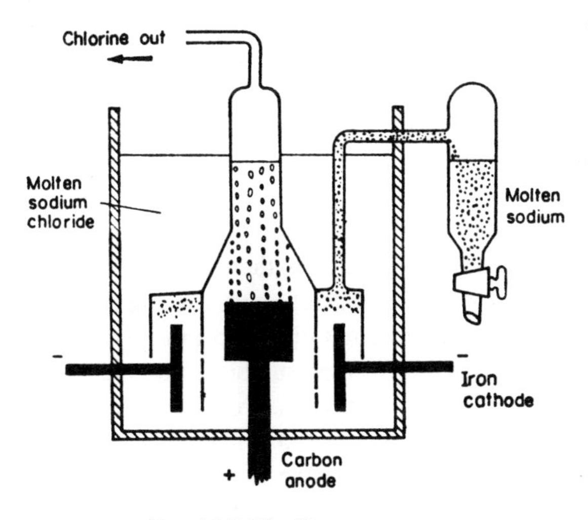

FIG. 13.7. The Downs process.

Magnesium. A variety of processes have been used for making magnesium, which is an important metal today for making light alloys for use as aircraft materials. Magnesium is found widely as carbonates such as *magnesite*, $MgCO_3$, and *dolomite*, $MgCO_3.CaCO_3$. The ore is first heated to give the oxide, then water is added to give magnesium hydroxide, which is filtered off and dissolved in dilute hydrochloric acid. The hydrated chloride is dehydrated by evaporation in a stream of hydrogen chloride gas, the function of which is to repress the hydrolysis reaction:

$$MgCl_2.6H_2O \rightleftharpoons MgO+2HCl+5H_2O$$

In the electrolytic process, steel cathodes and carbon anodes are immersed in an electrolyte bath of about 25% magnesium chloride together with calcium and sodium chlorides to lower the melting point. As with sodium, a voltage of about 6 V and a current of about 30 000 A are employed.

Study of the Ellingham diagram, Figure 13.5, shows that thermal reduction of magnesium oxide using carbon becomes feasible at about 2000°C. In the USA and Canada this reaction is carried out in an electric-arc furnace. The main technical problem is preventing the reverse reaction from setting in as soon as the products start to cool down. Rapid quenching has to be employed, using large quantities of a reducing gas such as hydrogen as a coolant.

Aluminium. Aluminium oxide, Al_2O_3, has a very high negative (exothermic) heat of formation. Furthermore the binding energy of aluminium metal, with three bonding electrons per atom, is rather higher than that of magnesium metal with only two bonding electrons per atom. This means that aluminium is not easily vaporized, unlike magnesium, and the reduction of the oxide with carbon is therefore not feasible (Fig. 13.5).

In default of a chemical reducing agent, we are forced to use electrolytic reduction, where there is no limit to the reducing cathode potential which can be applied. However, there are problems here too: aluminium chloride is structurally unsuitable, being a molecular solid Al_2Cl_6 (b.p. 180°C, sublimes), while aluminium oxide has an extremely high lattice energy and hence cannot be easily melted. Other salts of aluminium present similar problems—the most abundant of all, the silicates in clay, are far too stable and cannot be electrolysed—and we are forced to electrolyse the oxide under rather special conditions.

It is found that the oxide dissolves to form ions when added to molten cryolite, Na_3AlF_6. Ions Al^{3+} and O^{2-} are produced. (The ions Na^+ and the complex $[AlF_6^{3-}]$ are also present but they are not discharged in electrolysis.) The source of oxide is the ore *bauxite*, $Al_2O_3.xH_2O$, which is refined by:

(i) Dissolving in concentrated NaOH(aq) to separate from Fe_2O_3 impurity (see section 20.15 for an explanation of why this occurs).

(ii) Reprecipitating the aluminium as $Al(OH)_3.xH_2O(s)$, by passing carbon dioxide in to remove the excess OH^- ions.

(iii) Heating the hydroxide to give pure oxide. The pure oxide, dissolved in cryolite, is electrolysed at 800–900°C at high current density. Molten aluminium *sinks* to the bottom (contrast magnesium which floats). Oxygen is liberated at the carbon anodes, and slowly burns these away.

Cathode half-reaction: $Al^{3+} + 3e^- \rightarrow Al(l)$

Anode half-reaction: $O^{2-} \rightarrow \frac{1}{2}O_2(g) + 2e^-$;

followed by $\frac{1}{2}O_2(g) + C \rightarrow CO(g)$.

Titanium. Very large amounts of titanium are consumed as the dioxide which is a white pigment of exceptional covering power. The metal is remarkable, being the fourth most plentiful in the earth's crust and having outstanding strength and corrosion resistance. It is expensive to extract though, and its uses remain specialized. It is used in space technology and in modern supersonic aircraft such as Concorde.

The main sources of titanium are *ilmenite*, iron(II)titanate(IV), $FeTiO_3$, and *rutile*, titanium(IV) oxide, TiO_2. The Kroll process for making titanium involves first the heating of the ore with chlorine and coke at 800°C:

$$TiO_2 + 2C + 2Cl_2 \rightarrow TiCl_4 + 2CO.$$

The chloride is molecular and therefore volatile, and is purified by distillation. It is reduced by heating with a powerfully reducing metal such as magnesium, or sodium, in an inert atmosphere of argon.

The high cost of titanium is due to the technical difficulties of handling the molten metal—it attacks most refractory furnace linings and reacts directly with gases such as oxygen and nitrogen.

Copper. The commonest ore of copper is *chalcopyrite*, $CuFeS_2$. Other sulphide ores of copper are important also, such as the combined nickel-copper sulphides found in the Sudbury mines of Ontario. Another source of copper is *malachite*, $CuCO_3.Cu(OH)_2$, though this is not found in such large quantities. Most copper ores contain a high proportion of useless rocky material, and the extraction process is therefore largely concerned with methods of concentrating the ore, generally by froth flotation. Incomplete roasting is carried out, forming sulphides and oxides of copper, mainly copper(I). Iron sulphide, forming iron oxide, is removed as iron silicate slag by adding sand. On heating these together in the absence of air, copper is produced (the reaction is feasible because $\Delta G_f^\ominus$ of Cu_2O is only slightly negative, Fig. 13.5):

$$2Cu_2O + Cu_2S \rightarrow 6Cu + SO_2(g).$$

Alternatively, carbon may be used to assist the reduction of oxide.

Copper of high purity is made by electrolytic refining. Crude copper from the furnace is moulded to form anodes which are immersed in an electrolyte bath of copper(II) sulphate and dilute sulphuric acid. The cathodes are thin sheets of pure copper. The process is an application of the principle of preferential discharge: the ions with half-cell potentials more positive than Cu^{2+}/Cu do not dissolve at the anode, since they require too much free energy. Impurities such as silver and gold sink to the bottom of the cell as a sludge (anode slime). Of the ions which do dissolve, only $Cu^{2+}(aq)$ deposits on the cathode, because again the discharge potential for the remainder is too negative. Electrolytic refining is nowadays becoming much more common for metals which are relatively low in the electrochemical series (see also zinc, and chromium below). Solvent extraction is also becoming an increasingly important technique in the extraction of copper (Section 13.8).

Zinc. The common ore of zinc is the sulphide, *zinc blende*, ZnS. Zinc oxide is obtained by roasting the ore, which is then reduced with anthracite (impure carbon) at 1100°C. At this temperature the zinc vapour is distilled.

Chromium. Chromium occurs as *chromite*, $FeCr_2O_4$. Direct reduction with carbon or aluminium gives the alloy *ferrochrome*. To obtain the pure metal, the ore is converted to Cr_2O_3, which is then reduced with aluminium.

Nickel. Nickel is a useful metal on account of its remarkable resistance to attack by corrosive substances. Its alloy with copper, *cupro-nickel*, is used for making "silver" coinage. The most important source of nickel is at Sudbury in Ontario, Canada, where it occurs as a mixed sulphide with iron. Extraction processes are somewhat complicated, and the initial processes for concentrating the low-grade ore are especially important, involving magnetic separation and froth flotation. The result of this is reasonably high-grade nickel(II) sulphide. This is roasted to give the oxide which is dissolved in sulphuric acid and electrolysed using cathodes of pure nickel. The principles of electrolytic refining of nickel are the same as for copper, and precious metals are recovered from the anode sludge in a similar way.

Another important process for refining nickel is the Mond process. Crude nickel is first produced by reducing nickel oxide with water gas (hydrogen plus carbon monoxide, section 18.3). The process is rather unusual, for nickel forms a compound with carbon monoxide, tetracarbonylnickel(0) $Ni(CO)_4$. This substance is molecular in structure, and readily volatilized (b.p. 43°C). It is made by heating nickel powder to 50°C in a stream of carbon monoxide, and is then decomposed at 200°C. The sequence of reactions is

$$\underset{\text{steam}}{H_2O(g)} + \underset{\text{coke}}{C(s)} \longrightarrow \underbrace{CO(g) + H_2(g)}_{\text{"water gas"}};$$

$$\Delta H^\ominus = +131 \text{ kJ}$$
$$\Delta S^\ominus = -420 \text{ J K}^{-1}$$
$$\Delta G^\ominus = -38 \text{ kJ}$$

$$Ni(s) + 4CO(g) \xrightarrow{50^\circ} Ni(CO)_4(g); \quad \Delta H^\ominus = -164 \text{ kJ}$$

$$Ni(CO)_4(g) \xrightarrow{200^\circ} Ni(s) + 4CO(g)$$

Many other transition metals form carbonyls, but none so readily and rapidly as nickel. Carbonyls are analogous to complex ions, but are uncharged and have the metal in zero oxidation state. The decomposition of nickel carbonyl is accompanied by a large entropy increase (four moles of gas formed from one mole) and hence it takes place spontaneously at higher temperatures, when the $T\Delta S$ term becomes significant.

Silver. Silver occurs native in Norway, Chile and Peru. In combination it is found chiefly as the sulphide, Ag_2S, and the chloride, AgCl. It is extracted by adding aqueous sodium cyanide, NaCN(aq): in whatever form silver occurs, cyanide ions will dissolve it, due to the formation of the very stable ion $Ag(CN)_2^-(aq)$, ($K = 10^{19}$):

$$AgCl(s) + 2CN^-(aq) \rightarrow Ag(CN)_2^-(aq) + Cl^-(aq)$$

Silver is then precipitated by reducing this ion with aluminium or zinc, in alkaline solution:

$$2Ag(CN)_2^-(aq) + Zn(s) + 4OH^-(aq) \rightarrow$$

reduced oxidized

$$2Ag(s) + 4CN^-(aq) + Zn(OH)_4^{2-}(aq)$$

13.10 Summary of methods used for extraction of metals

Table 13.2 summarizes the methods used for extracting metals from their ores. It is only to be used as a general guide, and the details will be found either in this chapter, or in the relevant chapter dealing with the metal in question.

TABLE 13.2

Metal	Common ores	Source used for extraction	Method of extraction
K	$KCl.MgCl_2.6H_6O$	KOH	Electrolysis of the fused salt using a carbon anode and usually an iron cathode. Another salt is added to lower the melting point.
Na	$NaCl$, $NaNO_3$	$NaCl$	
Ca	$CaCO_3$, $CaSO_4$	$CaCl_2$	
Mg	$MgCO_3$, $MgCl_2$	$MgCl_2$	
Al	$Al_2O_3.xH_2O$	Al_2O_3	
Ti	TiO_2, $FeTiO_3$	TiO_2	Kroll process.
Mn	MnO_2	MnO_2	Thermit process or reduction with carbon.
Cr	$FeCr_2O_4$	Cr_2O_3 or $FeCr_2O_4$	
Zn	$ZnCO_3$, ZnS	ZnO	Sulphide roasted to oxide, and the oxide is then heated with coke. Slag formation is important.
Fe	Fe_3O_4, Fe_2O_3	Fe_2O_3	
Ni	$NiFeS_2$	NiO	
Sn	SnO_2	SnO_2	
Pb	PbS	PbO	
Cu	$CuFeS_2$	$CuFeS_2$	
Hg	HgS	HgS	Roasting, and thermal decomposition of HgO.
Ag	Ag, $AgCl$, Ag_2S	$AgCl$, Ag_2S	Cyanide process.

Study Questions

1. Use Fig. 13.5 to determine approximately what temperatures are needed for (i) carbon (ii) hydrogen to reduce the following oxides to the metals:

(a) CaO.
(b) FeO.
(c) TiO_2.

2. (a) In what form would you expect the following metals to occur in nature?

(i) Bismuth, (ii) cadmium, (iii) cobalt, (iv) germanium, (v) platinum, (vi) rubidium, (vii) strontium, (viii) tungsten.

(b) How is each metal likely to be extracted?

3. What factors must be taken into consideration when deciding on a suitable reducing agent for (a) metal oxides, (b) metal chlorides?

4. What relationships can you discover between the method used to extract a metal and

(a) its position in the periodic table?
(b) its position in the electrochemical series?

5. (a) Use a data book to arrange the metals in order of decreasing heats of formation of (i) chlorides, (ii) oxides.

(b) What other series do these orders resemble?

6. (a) Use the following information to sketch the Ellingham diagram for sulphides.

Compound	$\Delta G_f^\ominus$ at 0°C	$\Delta G_f^\ominus$ at 1000°C	$\Delta G_f^\ominus$ at 2000°C
CS_2	− 42	− 42	− 42
HgS	− 188	−63 (500°C)	
Cu_2S	− 230	−188	−167
FeS	− 272	−176	− 84
ZnS	− 418	−230	
MnS	− 480	−335	−188
SO_2	− 690	−542	−398
H_2S	− 146	− 42	+ 63
CaS	−1020	−815	−522

(All values in kJ per mole of S_2(g).)

(b) Which of these sulphides can be reduced directly to the element using carbon?

(c) Why is the line for calcium sulphide not straight?

(d) Ignoring changes in $\Delta H^\ominus$, what is $\Delta S_f^\ominus$ for (i) CS_2, (ii) FeS?

(e) Why are these figures not valid below the boiling point of sulphur?

(f) Use Fig. 13.7 and your sketch to work out which sulphides can be converted to the oxides by roasting in air.

(g) Comment on the methods used for the extraction of (i) copper (ii) mercury (iii) zinc, in the light of these results.

7. Library project. Discover what you can about the extraction of (a) Be (b) W and (c) U. How do these elements fit into Table 13.2?

CHAPTER 14

Equilibria between phases

14.1 Definitions

This chapter is concerned with systems in equilibrium in which there is more than one **phase** present.

A phase is a homogeneous region of a system, separated from other phases by a boundary surface. For instance, a system containing pure water in equilibrium with its vapour constitutes two phases. A solution of ethanol in water is one phase, while an immiscible "mixture" of carbon tetrachloride and water represents two phases. A saturated solution of sodium chloride in equilibrium with undissolved solid constitutes two phases: the liquid phase (solution) is homogeneous even though it contains two substances, and the solid phase is regarded as one phase even though it may consist of many separate crystals.

The term **homogeneous** requires comment. Homogeneous means *uniform* in properties, but since matter is composed of atoms and molecules which are definitely non-uniform when viewed under sufficiently high magnification, we must be clear what this means. A solution is regarded as homogeneous because the particles are fully dispersed as separate molecules or ions. It is possible to buy milk which is said to be "homogenized", but it is not truly homogeneous since it consists of tiny fat globules, whose dimensions greatly exceed those of the molecules which they contain, suspended in water. Milk is therefore an **emulsion**, and consists of two phases, fat and aqueous. Both these phases are homogeneous, but milk itself is not.

A system contains one or more **components. The number of components in a system is defined as the least number of substances whose quantity must be determined in order to fix the composition of all the phases present at equilibrium, at a given temperature.** Some examples are given to make this clear:

(a) *Water* $\rightleftharpoons$ *steam*. Two phases but only one component, namely H_2O.

(b) *A solution of ethanol in water, in equilibrium with vapour*. Two phases (liquid and vapour) and two components (C_2H_5OH and H_2O). If the amounts of ethanol and water in the liquid phase are defined, the composition of the vapour phase is fixed also. Conversely if the composition of the vapour is fixed, then only one composition of liquid can be in equilibrium with it.

(c) *Liquid nitrogen(IV) oxide in equilibrium with its vapour*. In both phases there is an equilibrium present:

$$N_2O_4 \rightleftharpoons 2NO_2.$$

However, there is only one component. The concentration of NO_2 molecules in a given phase is fixed once the concentration of N_2O_4 has been fixed, and once the composition of one phase is fixed, the composition of the other phase is fixed too. It does not matter whether we regard the component of the system as being NO_2 or N_2O_4. The important thing is that the *number* of components is one.

14.2 Vapour pressure

A pure liquid in equilibrium with its vapour constitutes a two-phase, one-component system. Water can remain in equilibrium with steam, and if the temperature is specified then only one vapour pressure is possible. Similarly ice can be equilibrated with vapour, even when there is no liquid water present—snow will evaporate slowly in a dry wind even when the temperature is well below freezing—and ice too has a vapour pressure curve.

At the melting point of ice, its vapour pressure must *equal* that of water. The two vapour pressure curves must meet at that point (Fig. 14.1). This can be proved by the following argument (Fig. 14.2). Suppose we have some ice in equilibrium with its vapour, at a vapour pressure p. Let the ice be in equilibrium with water at the same time. Let the vapour pressure of the water

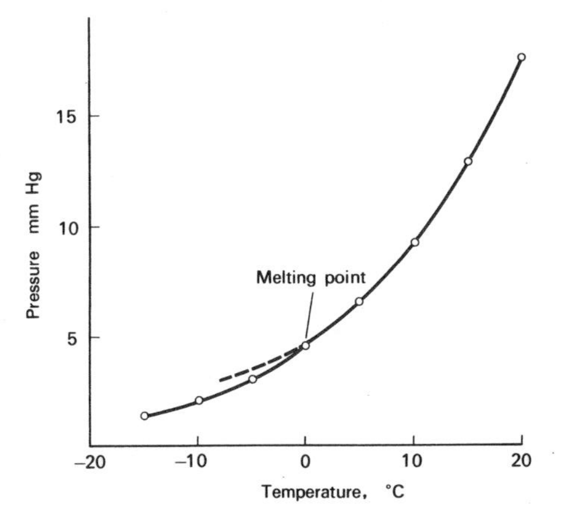

FIG. 14.1. Vapour pressure of water and ice.

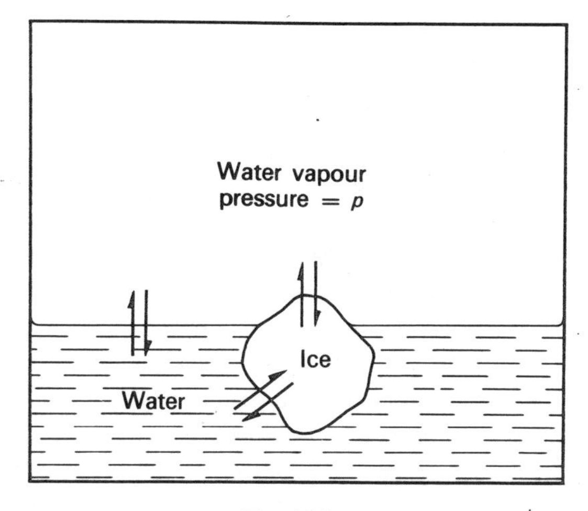

FIG. 14.2.

be p', and suppose that $p' > p$, H_2O molecules will evaporate from the liquid phase to establish a pressure of p' in the vapour phase. The vapour pressure of ice is now exceeded (because $p' > p$) and so H_2O molecules will condense on to the ice phase. However, this cannot happen, because we stated that the water and ice phases were in equilibrium. Therefore p' cannot be greater than p. Conversely, if $p' > p$, H_2O molecules would have to evaporate from the ice phase and condense in the water phase, and again this would contradict the fact that the water and ice are in equilibrium. It follows therefore that, for equilibrium $p = p'$.

When a solid and liquid phase are in equilibrium with one another, they both exert the same vapour pressure.

14.3 Phase diagrams

Figure 14.1 was a plot of temperature against pressure, showing the vapour pressure curves of ice and water. It is possible to add a third line to Fig. 14.1, namely the set of conditions where ice and water are in equilibrium with one another. The resultant plot is now a complete **phase diagram** for the H_2O system (Fig. 14.3).

The temperature at which ice melts to form water depends upon the pressure. If the pressure is increased, melting point is lowered. The creeping movement of glaciers is attributed to this effect. Application of le Chatelier's principle suggests that the phase of higher density will be favoured by increasing the pressure, and this is exactly what happens, as the slope of the line *TY* in Fig. 14.3 indicates.

The point *T* is termed the **triple point** of the H_2O system. It is the *unique* set of physical conditions (4 mm Hg, +0·0075°C) where all three phases can co-exist in equilibrium.

A phase diagram is more meaningful than a simple vapour pressure curve. In Fig. 14.3, every point on the diagram has a significance. Each *area* represents the conditions of temperature and pressure under which a given phase is stable. Each *line* represents the conditions under which two given phases may be in equilibrium. The *point* at which three areas are in contact is the condition for all three phases being in equilibrium.

The H_2O system has been chosen because it illustrates a simple phase diagram, but the same idea will be applied in this chapter to other systems, including two-component systems.

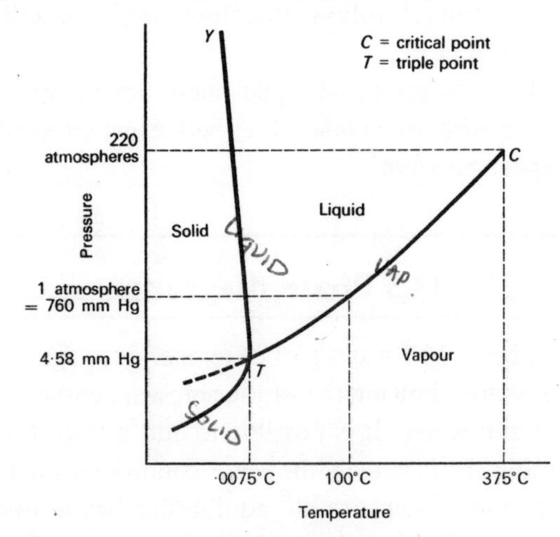

FIG. 14.3. A phase diagram for water (not to scale) (cf. Fig. 14.1).

14.4 Allotropy of elements

If the density of sulphur is investigated at different temperatures, by heating it in a diatometer such as that shown schematically in Fig. 14.4 then, provided the heating is carried out *slowly*, an abrupt decrease in density is observed around 96°C. Sulphur above 96° is less dense than that below, once equilibrium is established.

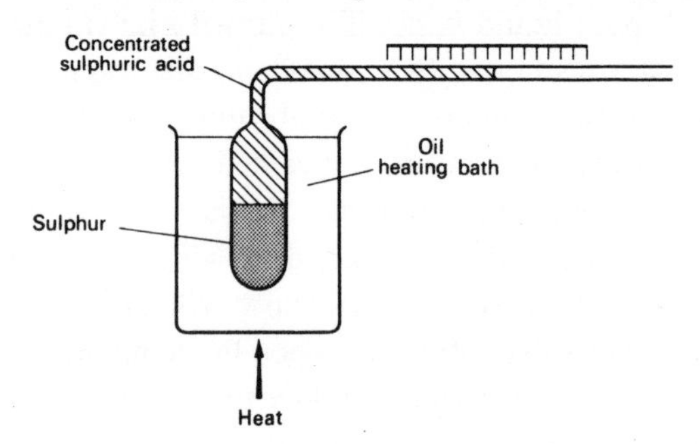

FIG. 14.4. Dilatometer.

Furthermore the crystals which form above 96°C when a solution of sulphur is cooled are monoclinic (needle-shaped) while those which form below 96°C are rhombic. Careful measurements of vapour pressure of sulphur indicate a discontinuity in the curve at 96°C, as shown at *B* in Fig. 14.5.

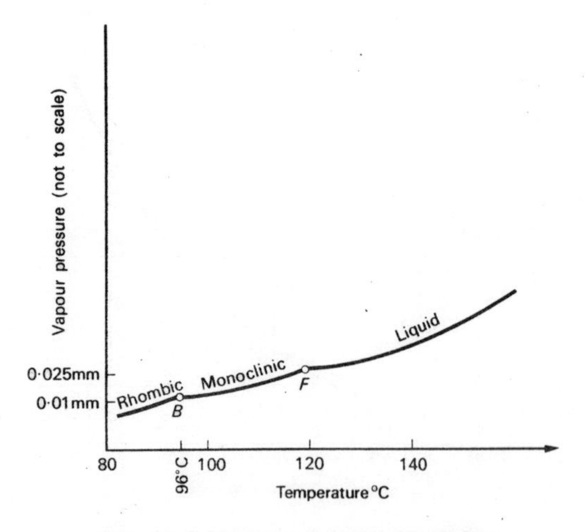

FIG. 14.5. Vapour pressure of sulphur.

Here is an example of an element capable of existing in two different structural forms, **allotropes**, each of which forms a different phase. 96°C is the **transition temperature** between the two phases: it is the temperature at which both phases can be in equilibrium because they exert the same vapour pressure.

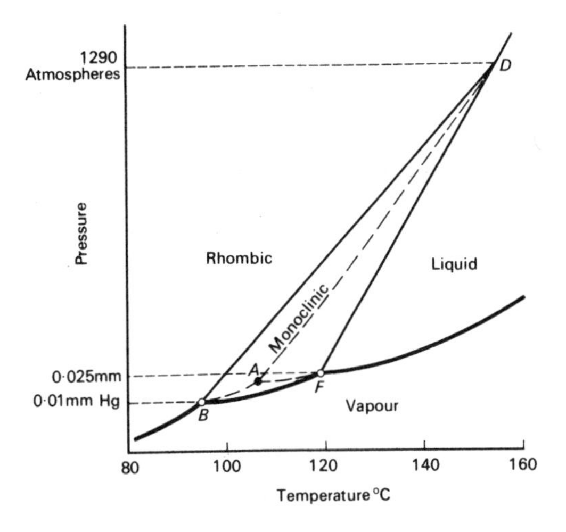

FIG. 14.6. Complete phase diagram for sulphur.

Figure 14.6 shows the same data as Fig. 14.5, but with additional information to make a complete phase diagram for the sulphur system. It is possible to heat rhombic sulphur above 96°C without it changing, because the rate of transition is slow. When it is outside its stable range of conditions, but is slow to change, it is said to be **metastable**. The dotted line *AB* shows the vapour pressure curve for metastable rhombic sulphur. Similarly the dotted line *AF* shows the vapour pressure of metastable monoclinic sulphur. *Note that the metastable allotrope has a higher vapour pressure than the stable form.* The densities of the phases are in the order

rhombic > monoclinic > molten.

This explains why the lines *BD* and *FD* slope as they do (the slope is exaggerated in Fig. 14.6 for the sake of clarity). For instance, if molten sulphur at 140°C is subjected to about 1000 atm pressure, monoclinic sulphur will form because the volume is then less.

14.5 Enantiotropy

The property of elements whose allotropes possess a transition temperature is known as enantiotropy. Other examples of enantiotropic behaviour are:

(i) Grey tin $\rightleftharpoons$ metallic tin, transition temperature 13·2°C. Grey tin is non-metallic in properties and objects made of tin disintegrate to powder if kept at prolonged low temperatures. Transition is more rapid once a nucleus of the new phase has become established to assist crystallization, giving the change the appearance of a "disease" afflicting the tin. Organ pipes, which in affluent times were often made of fairly pure tin, are said to have been subject to "tin plague" in cold churches. In more recent times, the leakage of fuel cans for Scott's antarctic expedition has been attributed to the failure of the soldered seams (Pb/Sn alloy) due to allotropic change.

(ii) Iron loses its ferromagnetism at about 730°C. Above this temperature iron cannot be magnetized, nor is it attracted to a magnet.

14.6 Monotropy

Allotropes of many elements are metastable with respect to the stable form under all conditions. Such allotropes are said to be *monotropes*, or to show monotropy. Thus white phosphorus has a vapour pressure greatly exceeding that of red, because it is composed of molecules (section 5.7). There can be no triple point at which red and white phosphorus are mutually in equilibrium with vapour (Fig. 14.7).

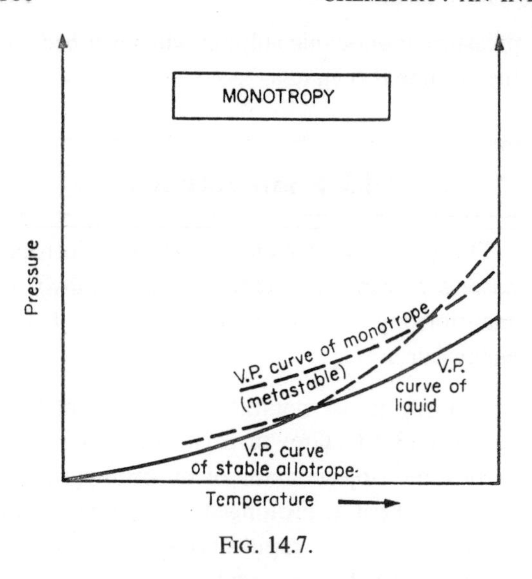

FIG. 14.7.

14.7 ΔG and ΔH for allotropic changes

When a metastable allotrope changes into its stable form, heat is usually evolved. It is often not practicable to measure this directly because of the slow rate of change, but Hess's law may be applied to the separate heats of combustion or reaction of the two forms. For instance:

$$\text{C(diamond)} + \text{O}_2\text{(g)} \rightarrow \text{CO}_2\text{(g)}; \quad \Delta H_d^\ominus = -395 \text{ kJ}$$

$$\text{C(graphite)} + \text{O}_2\text{(g)} \rightarrow \text{CO}_2\text{(g)}; \quad \Delta H_g^\ominus = -393 \text{ kJ}$$

Hence, C(diamond) → C(graphite);

$$\Delta H^\ominus = \Delta H_d^\ominus - \Delta H_g^\ominus = -2 \text{ kJ (Fig. 14.8)}$$

The change from a metastable to a stable allotrope must be accompanied by a decrease in free energy (ΔG negative). The fact that ΔH is often observed to be negative as well is an indication that the entropy changes are generally fairly small. The transition from one ordered crystal lattice to another of different order, does not involve very great change in the total disorder of the system.

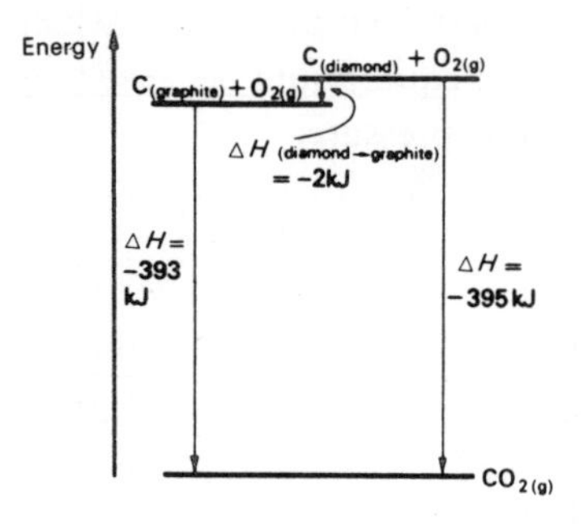

FIG. 14.8. Heat of transition: diamond → graphite (not to scale).

14.8 Polymorphism of compounds

The arguments given above for elements apply equally well to compounds. Indeed the term allotropy can equally well be applied to compounds, though the term is not so common here. The term **polymorphism** (of many shapes) is a general term which can be taken to describe the property of any substance, element or compound, existing in more than one crystalline form.* It should not be thought that elements have some special property which enables them to exhibit phase transitions.

Chalk and limestone are crystalline forms of $CaCO_3$ which occur naturally as *sedimentary* rocks. Crystalline modifications have occurred due to the intense pressures and temperatures which prevailed, giving rise to *metamorphic* (changed shape) forms of calcium carbonate such as marble.

* Some books use the word allotropy even when no crystalline structures are involved, for instance to describe the various molecular states (S_2, S_4, S_8, etc.) which occur in molten sulphur.

14.9 Vapour pressure of salt hydrates

The solubility curves of many salts in water show discontinuities analogous in some respects to the discontinuities in vapour pressure curves, such as those in Fig. 14.5. Closer investigation reveals that each part of the curve represents a different phase, generally due to the incorporation of varying numbers of moles of water in the crystalline phase. Many salts possess this **water of crystallization** and are known as **salt hydrates**. Where only one salt hydrate is formed, there will be no discontinuity in the solubility curve.

Figure 14.9 shows the solubility curve of sodium sulphate in water. There is a marked discontinuity at 32·4°C (point *Q*). Above this temperature sodium sulphate separates out as a solid crystalline phase of composition Na_2SO_4. Below 32·4°C the composition of the solid phase in equilibrium is $Na_2SO_4.10H_2O$.

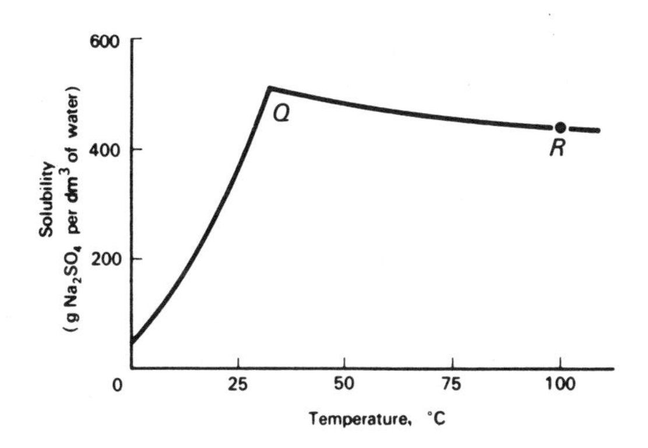

FIG. 14.9. Solubility of sodium sulphate in water.

Figure 14.9 can be converted into a phase diagram, with the addition of a little more information. This is done in Fig. 14.10, in which the original solubility curve is shown in bolder lines. Note how this phase diagram differs from the single-component diagrams shown in Fig. 14.3, 14.5 and 14.7: here there is an additional component, but pressure is no longer a significant variable so it is disregarded. Figure 14.10 plots temperature against composition instead of temperature against pressure. By convention composition is generally given the *x*-axis on phase diagrams, and so Fig. 14.9 has been turned through 90 degrees in making Fig. 14.10.

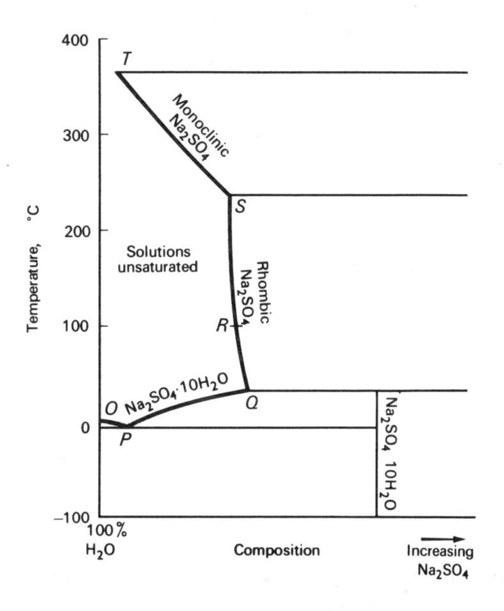

FIG. 14.10. Complete phase diagram for sodium sulphate.

In a phase diagram, areas represent different compositions and these are labelled. The line *PQRST* represents the composition of aqueous phase (solution) which is in equilibrium with the solid phase at various temperatures. The line *OP* is added because, if the solution is cooled sufficiently ice will separate out as the solid phase.

At the point *Q*, it is possible for the two different solid phases to be in equilibrium with the same solution. Assuming constant pressure, there is only one temperature at which this can occur. This is again called the *transition temperature*.

14.10 Efflorescence and deliquescence

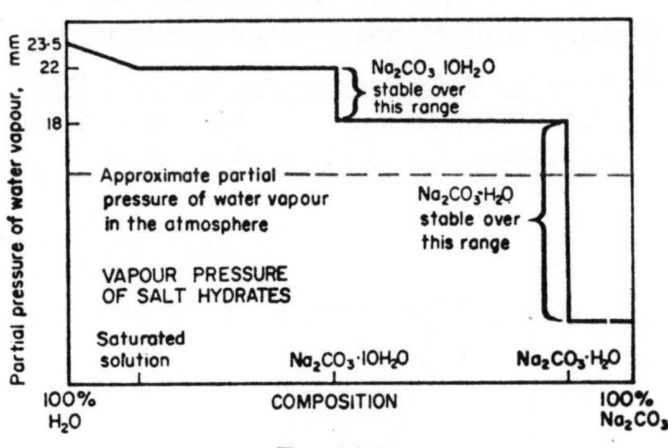

FIG. 14.11.

Figure 14.11 illustrates another type of phase diagram for a two-component system. This time temperature is assumed constant and omitted from the graph, and the *y*-axis represents pressure. The graph illustrates how the partial pressure of water vapour in equilibrium with sodium carbonate varies with the degree of hydration of the salt.

In normal atmospheric conditions washing soda, $Na_2CO_3.10H_2O$, produced by crystallizing sodium carbonate from aqueous solution, loses water. The decahydrate forms the monohydrate as follows:

$$Na_2CO_3.10H_2O(s) \underset{\text{above 18 mm}}{\overset{\text{below 18 mm}}{\rightleftharpoons}} Na_2CO_3.H_2O(s) + 9H_2O(g)$$

The crystals take on a powdery appearance and the phenomenon is known as **efflorescence**.

If sodium carbonate decahydrate is subjected to a very high concentration of water vapour, it will undergo a phase change in the opposite direction, taking up water to form a solution. Figure 14.11 shows that this would only happen if the partial pressure of water vapour exceeded 22 mmHg; such conditions are not normally met in the atmosphere though they could be reproduced artificially. Many other salts, particularly highly soluble ones, take up water vapour to form a solution at a much lower partial pressure, and will do this if left to stand in the laboratory. The phenomenon is known as **deliquescence**. Sodium carbonate would never deliquesce in the laboratory, but many salts, for example $CaCl_2(s)$ have to be kept in a desiccator or well-stoppered bottle. To summarize:

Efflorescence is the loss of water of crystallization to form a lower hydrate or anhydrous salt.

Deliquescence is the absorption of water vapour by a solid to form a solution.

Both properties depend upon prevailing conditions of humidity.

The word *hygroscopic* is a general term applied to substances which absorb water vapour. For instance:

(i) Anhydrous copper sulphate (white) absorbs water to become $CuSO_4.5H_2O$ (blue) on standing, but remains solid.

(ii) Many liquids, for example sulphuric acid and ethanol, absorb water while remaining liquid.

(iii) A common drying agent is *silica gel*; it can absorb considerable quantities of water without becoming wet.

14.11 Vapour pressure of miscible liquids and Raoult's law

When two liquids are shaken together the resultant system may consist of a single phase (liquids miscible) or two phases (liquids immiscible). Generally it is observed that "like dissolves like". If two liquids whose chemical properties are very similar are mixed it is found that each exerts a vapour pressure proportional to its **mole fraction**. Systems which obey this rule are said to obey **Raoult's law**. The mole fraction of a component is defined as

$$\frac{\text{number of moles of that component}}{\text{total number of moles of all components present}}$$

Figure 14.12 shows a plot of vapour pressure against composition for a mixture which obeys

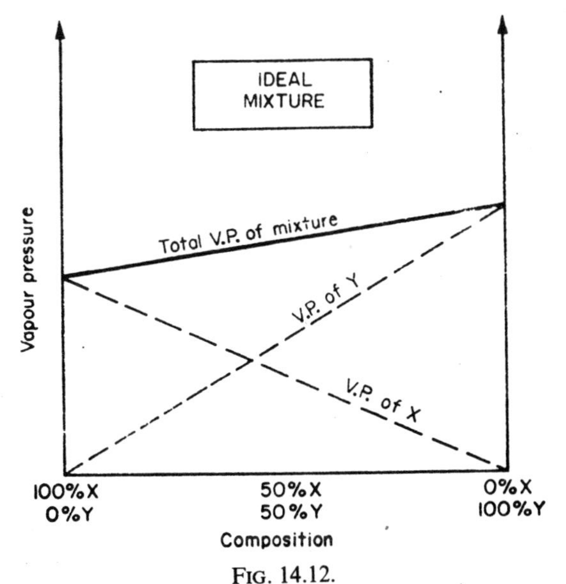

FIG. 14.12.

Raoult's law, temperature being assumed constant. The overall pressure is given by

$$P = p_x n_x + p_y n_y$$

where p_x, p_y = partial pressures of pure X and Y
and n_x, n_y = their respective mole fractions.

Typical examples of such a mixture, known as an **ideal mixture**, include the following:

(i) A mixture of $O_2(1)$ and $N_2(1)$ (liquid air) is nearly ideal.

(ii) A mixture of adjacent organic homologues, such as hexane and heptane, is almost ideal in behaviour.

14.12 Boiling points and molecular weights

The boiling point of a liquid can be defined as that temperature at which the saturated vapour pressure of the liquid is equal to the atmospheric pressure; the **standard boiling point** is that temperature at which the substance exerts a saturated vapour pressure of 1 atm (760 mmHg). Because of the relationship between the vapour pressure and the composition of an ideal mixture (Fig. 14.12), the boiling point of an ideal mixture will also be related to its composition. A special case arises when one of the constituents of the mixture (say Y) is involatile (exerts no vapour pressure), since the total pressure will then depend solely on the mole fraction of the volatile constituent, X:

$$\text{total pressure} = p_x n_x.$$

The lowering of the vapour pressure and the consequent elevation of the boiling point (Fig. 14.13) will depend on the mole fraction of Y.

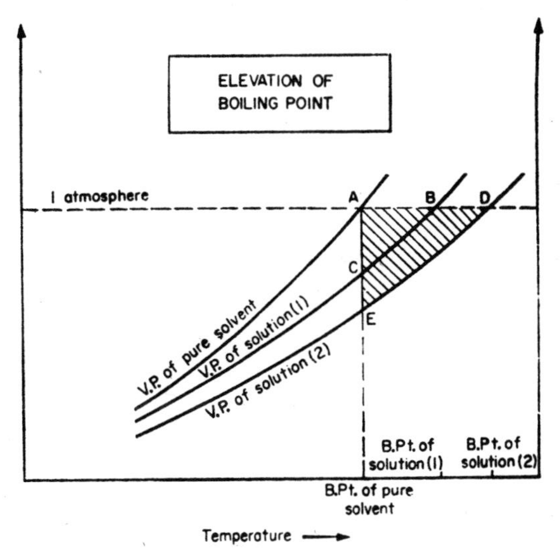

FIG. 14.13.

Most solutions only behave "ideally" when they are dilute, and we state the relationship between the boiling point of a solution and the amount of solute added as follows:

the elevation of boiling point of a solution is proportional to the molar concentration of the solute, provided the solution is dilute.

Measurement of the boiling point of a solution therefore affords a convenient way of counting the number of moles of solute. Elevation is determined solely by the number of moles present and not upon their nature. In order to measure a small temperature change, a thermometer is required which can register precisely very small *differences* in temperature. Such a thermometer need not necessarily give actual values in °C, but it must be able to measure the difference in boiling point between a pure solvent and a dilute solution. A good thermometer for this purpose is the Beckmann thermometer, which has a large bulb but a very fine column, giving perhaps 1 degree per inch, or better. The scale is divided into hundredths of a degree, and at the top of the thermometer is a small mercury reservoir which enables the total quantity of mercury in the operating part of the thermometer to be varied. In this way, a Beckmann thermometer can be set for, say, boiling point elevations using water in the range 99°–104°C, or freezing point measurements in the region 0°C and below.

When taking the boiling point of a solution, the thermometer must be placed in the solution, *not* in the vapour above. If the bulb is in the vapour, it will register the boiling point of pure solvent which condenses on it. A device is needed which bathes the bulb in solution, and Cottrell's apparatus, shown in Fig. 14.14, is one way of doing this. The force of boiling solution sends some of it up the tubes by the side of the thermometer bulb.

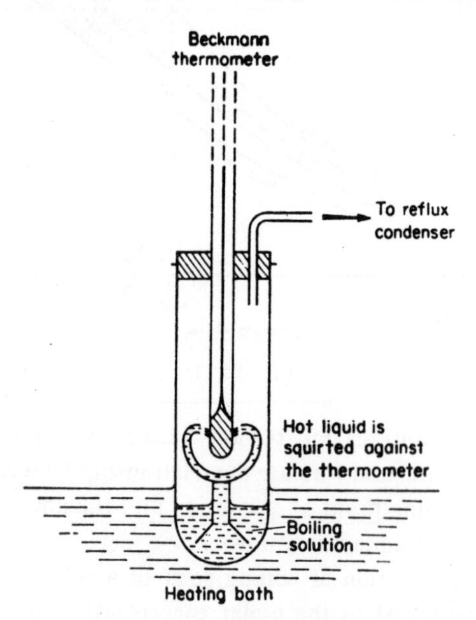

FIG. 14.14. Cottrell's apparatus.

The elevation of boiling point is measured:

(i) with a solution of known molarity, solution (A), and

(ii) with a solution whose concentration in g dm^{-3} is known, but whose molar concentration is unknown, solution (B).

Hence,

$$\frac{\text{molar concentration of (A)}}{\text{molar concentration of (B)}} = \frac{\text{elevation of b.p. of (A)}}{\text{elevation of b.p. of (B)}}$$

In Cottrell's apparatus the weight of solvent is best found by observing the *volume* of solution actually present when the measurement is taken, knowing its density. A graduated tube is useful for this purpose. The weight of solute is subtracted from the weight of solution thus determined.

Solvents other than water are also used for measurements of boiling point elevation. It is not necessary to take a second measurement on a liquid of known molar concentration, provided the ebullioscopic constant of the solvent is known (section 14.21).

14.13 Distillation of an ideal mixture of two liquids

Figure 14.15 shows the phase diagram corresponding to Fig. 14.12, but this time with temperature as a variable, pressure being assumed

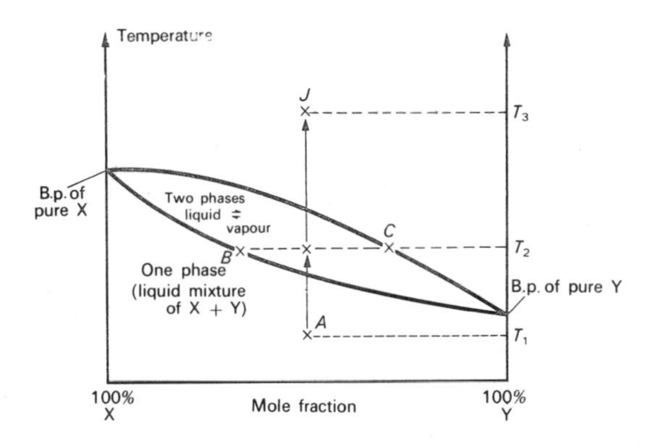

FIG. 14.15. Phase diagram for an ideal mixture of liquids (see text).

constant. The two components X and Y will be in equilibrium with vapour at a given temperature. In the diagram, Y is more volatile, and therefore the vapour composition will be richer in Y than the liquid from which it is derived. There are therefore two boundary curves on the phase diagram. Points between the two curves represent two phases of composition shown by the ends of the "tie-line". For instance, if liquid of composition represented by the point A at temperature T_1 is heated to T_2 it will form two phases, of composition B (liquid phase) and C (vapour phase). On further heating to T_3 (point J) only one phase will be present, the whole of the mixture having been vaporized.

Figure 14.16 shows what will happen if the same mixture is distilled from a fractionating column. The labels on the diagram correspond to the stages listed below:

(1) The mixture of X and Y is heated until it just boils, at T_4°C (point D).

(2) The mixture comes into equilibrium with its vapour at T_4°C, on the bottom plate of the column as shown. The vapour will be richer in component Y (point E).

(3) The vapour cools as it reaches the next plate of the column, reaching a temperature of T_5°C (point F).

(4) Hotter vapour from below causes this condensed liquid to re-evaporate: the vapour is now richer still in Y (point G).

(5) This vapour again condenses on the next plate in the column at a temperature T_6.

And so the process continues. Equilibrium is eventually established with a temperature gradient existing up the column. If the column is a good one, practically pure Y will emerge from the top of it, and the mixture which drips back into the flask will become progressively richer in X.

The efficiency of a column is often expressed in terms of its **theoretical plate number**. This is the number of successive theoretical equilibrations which would produce a product of the observed purity. In Fig. 14.16 it is seen that only three theoretical plates produce a product which is better than 90% pure. A good laboratory fractionating column may have perhaps twenty theoretical plates, while a fractionation tower for the refinement of crude oil might have many more.

The process of fractionation has been studied in detail because of its great practical importance. *All* laboratory distillations should, if properly carried out, involve some degree of fractionation. Even in a simple distillation flask, the

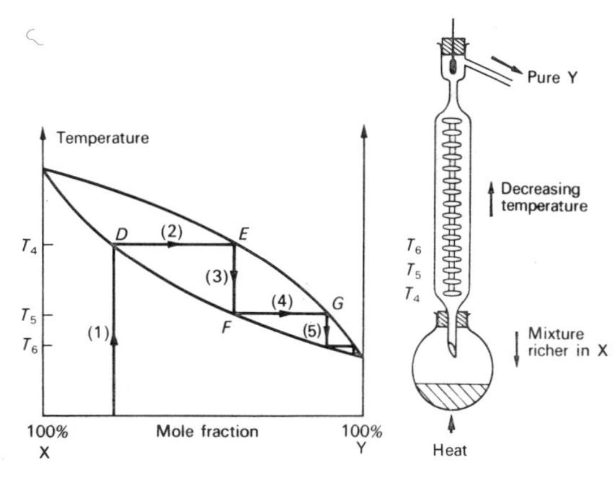

FIG. 14.16. Fractionation of an ideal mixture of liquids (see text).

neck allows some refluxing to occur, and special care should be taken when using small-scale apparatus to ensure that heat is not too strongly applied, or an impure product will distil over.

Provided the mixture does not depart too far from Raoult's law, it will distil in the manner just indicated, enabling a fairly complete separation to be obtained.

If the two liquids are too close together in boiling point, distillation will not separate them because the number of theoretical plates which a fractionating column can achieve are insufficient. Liquids with boiling points less than 20 degrees apart require very efficient fractionation. For such cases a much more powerful method of separation is *gas chromatography* (section 14.26).

14.14 Positive deviations from Raoult's law

Suppose two liquids are miscible in all proportions, but do not have great chemical affinity for one another. Evaporation of molecules from such a mixture might be expected to take place more readily than if the mixture was ideal. This phenomenon is observed in practice, and the curve of vapour pressure and composition no longer obey's Raoult's law. In many cases the deviations are so extreme as to lead to a maximum in the curve (Fig. 14.17(a)).

The point of maximum vapour pressure on this curve will correspond to a composition of *minimum boiling point*. If liquid of this composition is boiled its composition will not alter, and it will continue to boil at constant temperature. It is known as a **constant boiling mixture** or **azeotrope**. Figure 14.17(b) shows the temperature–composition phase diagram. It may be treated as two simple phase diagrams joined together—for instance the left-hand half of Fig. 14.17(b) behaves exactly like the system in Fig. 14.15.

Ethanol and water form a well-known azeotropic system. No matter how efficient the distillation, the product distilling over will be the azeotrope, with composition 95·6% ethanol, 4·4% water at normal atmospheric pressure. If pure ethanol is required from a dilute aqueous

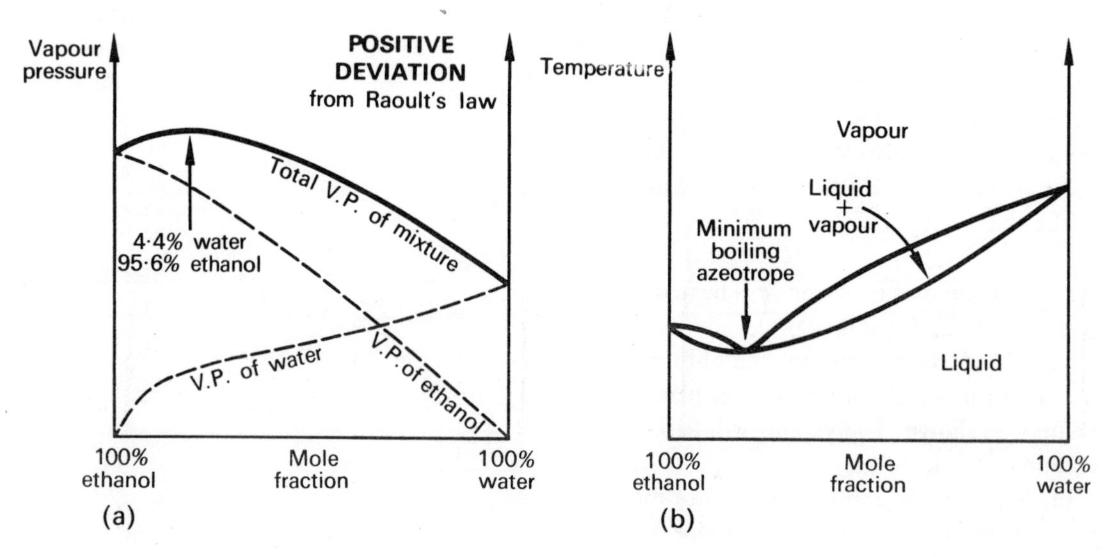

FIG. 14.17.

solution, the final traces of water must either be removed chemically (for instance, with anhydrous calcium oxide), or by adding another component to the system (there is a method of preparing anhydrous alcohol by adding benzene to the system).

14.15 Negative deviations from Raoult's law

Where two miscible liquids have a strong chemical attraction for one another, perhaps showing a tendency towards loose compound formation, evaporation from the mixture will be *less* easy than from an ideal mixture. In extreme cases this commonly leads to another form of azeotrope, this time with a minimum vapour pressure, and maximum boiling point. Figures 14.18(a) and (b) illustrate this behaviour.

Systems such as nitric acid and water, or the hydrogen halides and water, exemplify this type of behaviour. Again it is easy to reason out what will happen when a given mixture is distilled, by treating Fig. 14.18(b) as two separate forms of Fig. 14.15 joined together.

As with the minimum boiling point system above, *if the azeotrope is distilled its composition will remain unchanged*, since the two curves for composition of liquid and vapour meet at this point.

14.16 Azeotropes and bond formation

Systems which show a negative deviation from Raoult's law do so because extra chemical bonds form which prevent such ready vaporization of the molecules. For instance, acetone and trichloromethane show a negative deviation due to the formation of hydrogen bonds. This interpretation is borne out by the fact that, when two liquids which show a negative deviation are mixed, heat is evolved. The amount of heat evolved per mole is not very great (of the order 20 kJ mol^{-1}) showing that the bonds formed are much weaker than covalent or ionic forces.

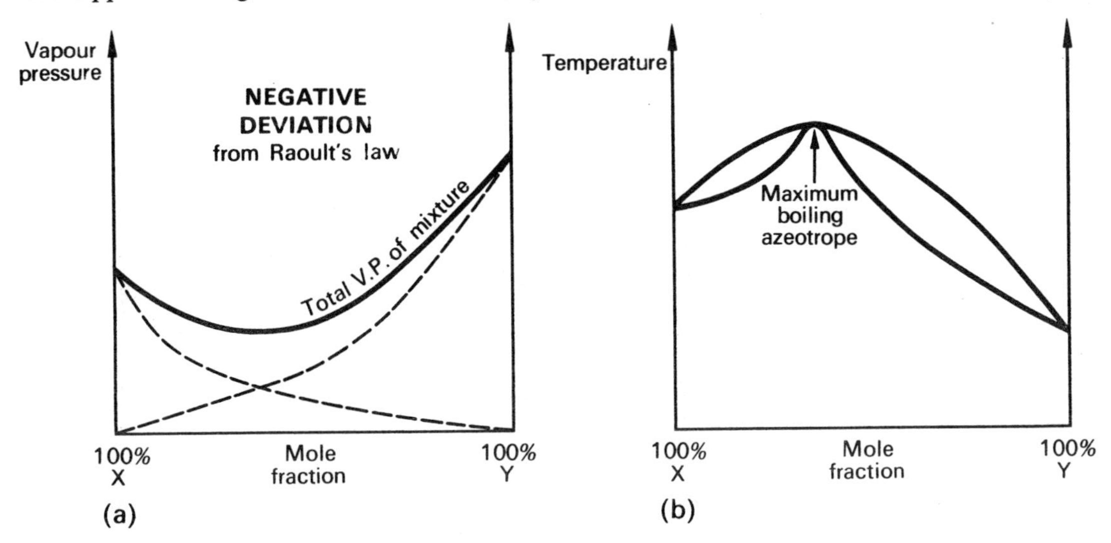

FIG. 14.18.

TABLE 14.1

Name	Formula	Miscibility with water, etc.
Methanol	CH_3OH	Positive deviation from R.L. but no maximum. Miscible with water in all proportions
Ethanol	C_2H_5OH	Positive deviation from R.L., maximum V.P. observed. Miscible with water in all proportions
Propan-1-ol	C_3H_7OH	Pronounced maximum in V.P. curve. Miscible with water in all proportions
Butan-1-ol	C_4H_9OH	Two phases formed on mixing, each component being fairly soluble in the other
Pentan-1-ol	$C_5H_{11}OH$	Two phases formed, solubility being less than for butan-1-ol, etc.

Systems which show a positive deviation from Raoult's law, such as alcohols and benzene, have a positive (endothermic) enthalpy of mixing showing that bonds are being broken.

The formation of two separate phases instead of one, when two liquids are shaken together, is an extreme case of a positive deviation from Raoult's law. The phenomenon can be illustrated by reference to the aliphatic alcohols, Table 14.1.

14.17 Distillation techniques

In section 14.13 the principles underlying fractional distillation were discussed. Even with simple distillation in which no fractionating column is used (Fig. 14.19(a)), a certain amount of condensation and re-evaporation must occur, so that a degree of fractionation must take place. If the distillate is a solid at room temperature, it is sometimes essential to use an **air condenser** rather than a water condenser to avoid blocking the tube (Fig. 14.19(b)).

A preparative chemist often has to heat the reactants to start a reaction. If one of the reactants is volatile, the reaction is heated under **reflux**; the volatile component is condensed and returned to the reaction flask (Fig. 14.19(c)). Once the reaction is complete, the apparatus is allowed to cool and then rearranged for conventional distillation.

If the product being distilled is thermally unstable, the process has to be modified and the distillation is either carried out under reduced pressure (**vacuum distillation**) (Fig. 14.19(d)) or it is distilled with steam (section 14.18). The principle underlying vacuum distillation is to lower the boiling point of a substance by reducing the pressure of the atmosphere above it, either by applying a vacuum or by using a filter pump, which can give a pressure as low as 15 mmHg. A thin capillary tube admits a slow stream of bubbles and prevents serious "bumping" but, even so, it is essential to use safety glasses when carrying out such a distillation.

14.18 Steam distillation

If a mixture of two immiscible liquids is boiled, each phase will exert its own vapour pressure separately, since neither phase is diluting the other. The boiling point of a system of two immiscible liquids is therefore always below that of both the pure liquids.

In organic preparations (for example, the preparation of phenylamine, $C_6H_5NH_2$) volatile components immiscible with water are readily

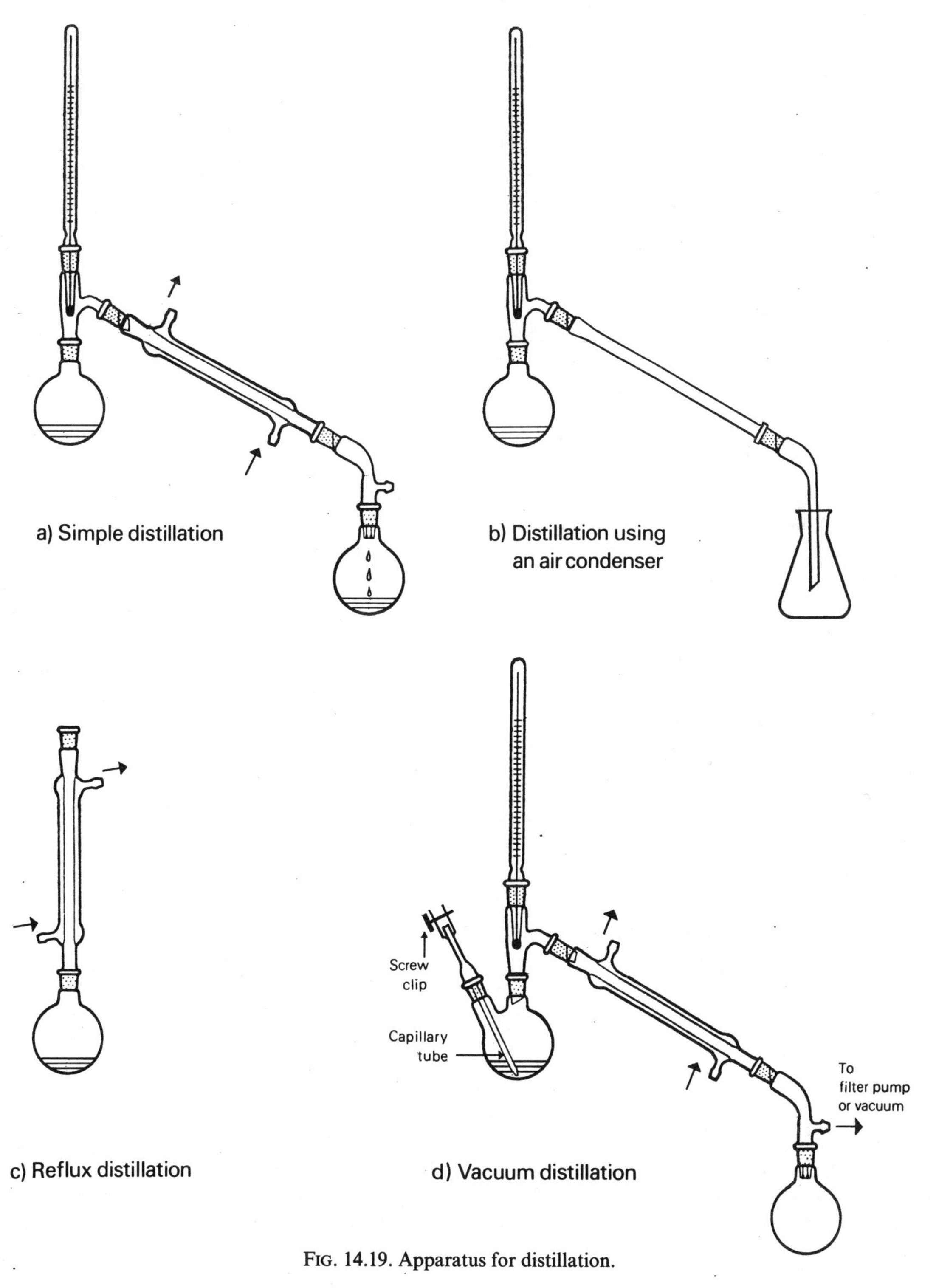

FIG. 14.19. Apparatus for distillation.

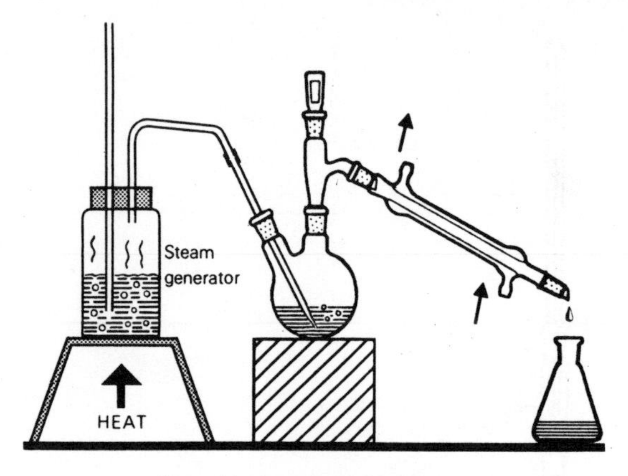

FIG. 14.20. Steam distillation.

removed from the tarry mixture in the preparation flask by blowing steam through the mixture (Fig. 14.20). Distillation will proceed because the temperature of the steam is higher than the boiling point of the mixture. The weight of each component distilled will be proportional to its molar concentration in the vapour. Provided the components can be assumed *completely* immiscible, we can say

$$\frac{\text{number of moles of X in distillate}}{\text{number of moles of water in distillate}} = \frac{\text{s.v.p. of X}}{\text{s.v.p. of water at the distillation temperature}}$$

The distillate will consist of an oily emulsion, which is generally difficult to separate by mechanical means such as with a separating funnel. The usual procedure is to add diethyl ether (or some other solvent immiscible with water but miscible with the non-aqueous phase) and separate the ether layer in a separating funnel. The ether layer may then be stood over a desiccant such as anhydrous calcium chloride to remove traces of dissolved water, and then redistilled.

In theory the method of steam distillation can be used to determine the molecular weight of X (see Study Question 7), but in practice the method is rather inaccurate. This is partly due to the fact that many so-called "immiscible" liquids are in fact partially miscible, and partly because true equilibrium between the phases may not be established.

14.19 Solid-liquid equilibria

The melting point of a pure solid is lowered by adding an impurity. For instance the melting point of pure camphor could be lowered by adding another solute such as naphthalene. The converse is also true: the melting point of pure naphthalene will be lowered by adding small quantities of camphor. Similar behaviour is observed with mixtures of molten metals, and these provide convenient examples for study. One of the simplest is a mixture of lead and tin. It is found that each lowers the melting point of the other, and if a graph of freezing point against composition is plotted, the result is like Fig. 14.21. Careful plotting reveals a sharp minimum in the curve. The alloy at this composition is called a **eutectic** and the point on the curve is called a **eutectic point**.

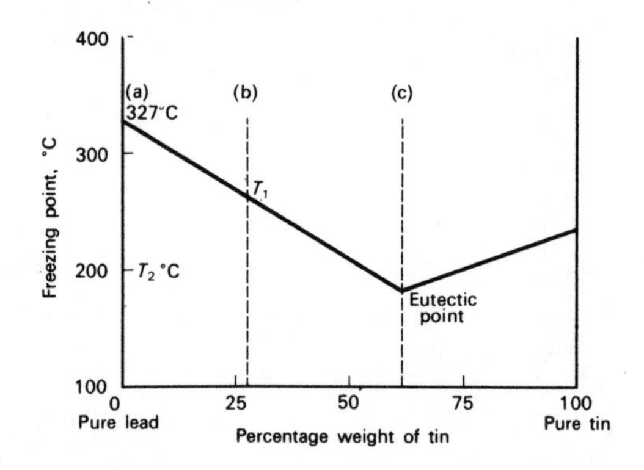

FIG. 14.21. Eutectic formation.

A close examination of the mixtures as they solidify on cooling shows that solidification is gradual, and this can be further verified by plotting a cooling curve for the various compositions. Figure 14.22(a) shows the sharp melting

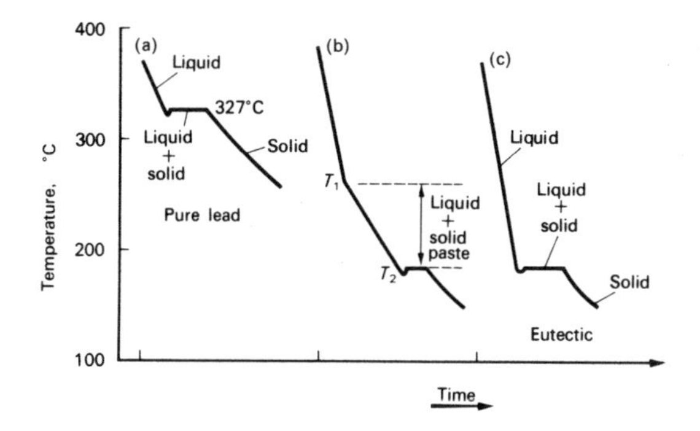

FIG. 14.22.

point associated with pure lead (or pure tin). Figure 14.22(b) shows what happens with an alloy other than the eutectic. Crystallization commences at a temperature T_1, and cooling then slows down. When T_2, the freezing point of the eutectic, is reached the temperature remains constant until the whole has become solid. Thereafter normal cooling of the solid continues.

The explanation of this behaviour is as follows: at T_1, pure lead crystallizes out. The melt becomes richer in tin and its freezing point is correspondingly lowered. This continues until the eutectic composition is reached. The eutectic has the lowest possible freezing point, and thereafter both metals separate out together as two distinct solid phases.

If the eutectic itself is cooled (line (c) in Fig. 14.21) then the freezing point is as sharp as for a pure substance (Fig. 14.22(c)). Sharpness of melting point or freezing point is generally taken to be one of the tests for deciding whether a substance is pure or a mixture, but here the test breaks down. A eutectic is not a pure substance, i.e. not a compound of lead and tin in this case, as the following evidence shows:

(a) It is non-stoichiometric; (this alone would not be a sufficient piece of evidence).

(b) Separate phases of the components can be seen by microscopic examination. The arguments against eutectics being compounds are similar to those for azeotropes. Both eutectics and azeotropes are phenomena which result from the peculiar behaviour of phases in equilibrium.

The reverse of freezing and crystallization, melting, is often used as the main criterion of a solid's purity. In this case the time axis in Fig. 14.22 is reversed. A sharp **melting point** at a particular temperature (Curve (a)) shows a pure solid, whereas gradual melting over a range of temperature (Curve (b)) shows an impure solid. The real identity of a solid can be confirmed by performing a **mixed melting-point** test. The solid is mixed with a pure sample of the compound that it is suspected to be. If the solids are the same, the melting point will again be sharp and at the same temperature. But if the solids turn out to be different, then they will melt gradually over a range of temperature.

14.20 Freezing-points

When a solution is cooled to freezing point, crystals of *pure* solvent nearly always separate first, provided the solution is dilute. The freezing point of the system is that temperature at which the vapour pressure curve of the liquid phase (the solution) intersects that of the solid phase (pure solvent)—see section 14.2. Fig. 14.23 illustrates what happens in the case of aqueous solutions. From this diagram, by a similar argument to that used for elevation of boiling point, the triangles *ABC* and *ADE* are similar. Hence

Depression of freezing point $\propto$ lowering of vapour pressure,

$\therefore$ **depression of freezing point of a solution is proportional to the molar concentration of the solute, provided the solution is dilute.**

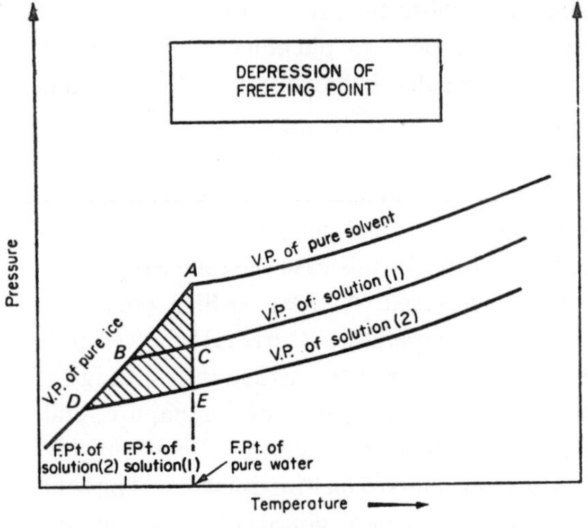

FIG. 14.23.

Various solvents are useful for cryoscopic (freezing point) measurements, and generally they lead to a more accurate determination of molecular weight than ebullioscopic (boiling point) measurements. This is partly because the temperature changes are larger, and partly because the experimental techniques are often simpler.

A typical apparatus for cryoscopic measurements is shown in Fig. 14.24. A Beckmann thermometer is shown, but for quick measurements

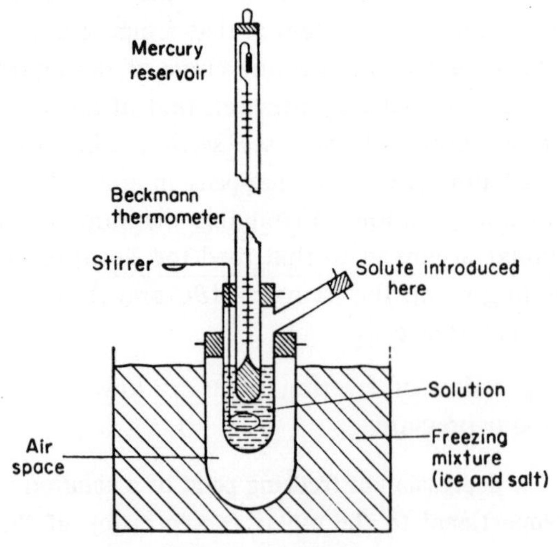

FIG. 14.24. Beckmann's apparatus.

a simple thermometer accurate to one-tenth of a degree will do. The freezing point of a known weight of pure solvent is first taken, with vigorous stirring to prevent supercooling. For maximum accuracy a cooling curve is plotted (Fig. 14.25(a)). After re-warming, a weighed amount of solute is dissolved and a new freezing point taken (Fig. 14.25(b)). The correct freezing point is at X in the figure: once much solvent has crystallized out, the concentration will increase, and the freezing point will gradually drop as shown. For maximum accuracy, a given solution should be warmed and refrozen two or three times, taking the average. Further solute can then be added and the whole process repeated.

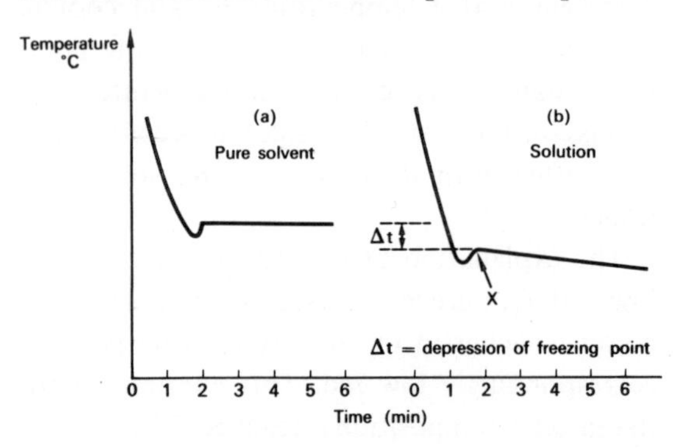

FIG. 14.25. Cooling curves obtained in freezing point determination.

14.21 Determination of molecular weights

The depression of freezing point or elevation of boiling point of a solvent is independent of the nature of solute molecules, and is determined solely by their concentration in moles. For convenience in calculating, tables have been prepared listing:

(1) The elevation of boiling point of typical solvents, in K mol^{-1}, for 1 kg of solvent—this is the **ebullioscopic constant** of the solvent.

TABLE 14.2
EBULLIOSCOPIC AND CRYOSCOPIC CONSTANTS
(K MOL^{-1} FOR 1 KG SOLVENT)

Solvent	Cryoscopic constant	Ebullioscopic constant
Water	1·86	0·52
Ethanol	—	1·15
Trichloromethane	—	3·66
Benzene	5·12	2·67
Ethanoic acid	3·9	2·53
Phenol	7·5	—
Phenylamine	—	3·22
Camphor	40·0	—

(2) The depression of freezing point, in K mol^{-1}—this is the **cryoscopic constant** of the solvent.

Table 14.2 lists some typical constants for well-known solvents. It does *not* follow that solutions of one mole per kg of solvent would have the exact boiling points and freezing points predicted, because solutions of this concentration are too concentrated for ideal behaviour to apply. The constants are purely constants of proportionality, devised to facilitate calculations.

Worked Example 1. A solution of 1·35 g of urea in 72·3 cm^3 of water boiled at 100·162°C (pressure = 760 mm Hg). The ebullioscopic constant of water is 0·52 K mol^{-1} for 1 kg of water. Calculate the molecular weight of urea. Urea may be assumed involatile at 100°C.

1 mole of urea in 1000 g water would theoretically elevate b.p. 0·52 K.

An elevation of b.p. of 0·162 K will be produced by $\frac{0{\cdot}162}{0{\cdot}52}$ mole of urea in 1 kg water, since elevation $\propto$ concentration.

$\therefore$ Number of moles in 72·3 cm^3 of water

$$= \frac{0{\cdot}162}{0{\cdot}52} \times \frac{72{\cdot}3}{1000}.$$

Let molecular weight = M

$$M = \frac{\text{number of grams present}}{\text{number of moles present}}$$

$$= \frac{0{\cdot}52}{0{\cdot}162} \times \frac{1000}{72{\cdot}3} \times 1{\cdot}35 = 60.$$

The large cryoscopic constant of camphor makes it a very useful solvent for determining the molecular weight of organic compounds. In **Rast's method**, the melting point of a small quantity of *pure* camphor (labelled "for molecular weight determinations") is first determined, either in a melting point bath, or alternatively on a heating block. A weighed quantity of solute and a weighed amount of pure camphor are then mixed and melted together, and stirred until homogeneous. The mixture is cooled until solid, ground to a fine powder, introduced into the melting point apparatus, and heated until completely molten. The temperature at which solidification begins on cooling is carefully noted. The whole process is repeated using a standard of known molecular weight, such as naphthalene, $C_{10}H_8$. The molecular weight of solute is calculated as before. Note that the molecular weight of camphor is *not* required, and is not relevant to the problem.

14.22 Colligative properties

Elevation of b.p. and depression of f.p. are colligative properties, and essentially a way of *counting* solute particles, or moles of solute. It frequently happens that a solute *dissociates* completely or partially, and this fact is revealed by freezing point or boiling point data.

Worked Example 2. One-tenth of a mole of the following solutes were each dissolved in 1 kg of solvent with freezing point depressions as quoted in Table 15.2. What conclusions can you draw about the molecular state of the solutes in each case?

(*a*) *sugar in water: f.p.* $-0{\cdot}186°C$;

(*b*) *sodium chloride in water: f.p.* $-0{\cdot}372°C$;

(*c*) *calcium chloride in water: f.p.* $-0{\cdot}558°C$;

(*d*) *ethanoic acid in benzene: depression of f.p. 0·256 K.*

(a) *Sugar in water*. Number of moles of sugar added = 0·1 per kg of water.

Number of moles per kg of water, calculated from cryoscopic constant:

$$= \frac{0{\cdot}186}{1{\cdot}86} = 0{\cdot}1$$

∴ sugar does not dissociate.

(b) *Sodium chloride in water*. Number of moles NaCl(s) added = 0·1 per kg of water.

Number of moles solute per kg of water, calculated from cryoscopic constant:

$$= \frac{0{\cdot}372}{1{\cdot}86} = 0{\cdot}2$$

∴ sodium chloride has dissociated, one mole of NaCl(s) giving two moles of ions.

$$aq + NaCl(s) \rightarrow \underset{\text{one mole}}{Na^+(aq)} + \underset{\text{one mole}}{Cl^-(aq)}.$$

(c) *Calcium chloride in water*. Number of moles of solid added = 0·1 per kg of water. Number of moles of solute per kg of water, calculated from cryoscopic constant

$$= \frac{0{\cdot}558}{1{\cdot}86} = 0{\cdot}3$$

∴ calcium chloride has dissociated, one mole of $CaCl_2$(s) giving three moles of ions.

$$aq + CaCl_2(s) \rightarrow \underset{\text{one mole}}{Ca^{2+}(aq)} + \underset{\text{two moles}}{2Cl^-(aq)}$$

(total: three moles)

(d) *Ethanoic acid in benzene*. Again there is 0·1 mole added per kg of solvent. Number of moles per kg of benzene, calculated from cryoscopic constant

$$= \frac{0{\cdot}256}{5{\cdot}12} = 0{\cdot}05 \text{ mole.}$$

This result is interpreted by postulating *association*. Double molecules, called *dimers*, are formed:

$$\underbrace{CH_3COOH + CH_3COOH}_{\text{two moles}} \longrightarrow$$

```
          O....HO
         //       \
CH3—C             C—CH3
         \       //
          OH....O
```

one mole of dimer

The **osmotic pressure** of a solution is another colligative property that can be used to count moles and determine molecular weights.

14.23 Partition of a solute between two phases

Section 10.12 showed that, if a solute is added to two immiscible solvents, it will obey the equilibrium law at low concentrations. This fact is put to practical use in a number of ways, for example **solvent extraction** and **partition chromatography**.

Ether (ethoxyethane), $C_2H_5OC_2H_5$, is immiscible with water, and organic substances can be extracted from an aqueous emulsion by shaking the emulsion up with ether. For example, in the steam distillation of phenylamine (section 14.18) the distillate contains a small quantity of phenylamine together with a relatively large amount of water in the form of an emulsion which only partly separates out. On adding ether the phenylamine partitions itself between the ether layer (less dense upper layer) and the aqueous layer (lower layer), being very much more soluble in the ether. On passing through a separating funnel, containing most of the phenylamine the lower layer can be obtained separately. A further quantity of ether can then be added to the remaining aqueous layer, enabling most of the remaining phenylamine to be extracted from it.

Worked Example 3. The partition coefficient of a solute X between ether and water is 3, the substance being more soluble in ether. 100 cm^3 of an aqueous solution containing 10 g of X is shaken with 100 cm^3 of ether. Calculate the weight of X left in the aqueous solution.

Since the volumes of the layers are the same,

$$\frac{\text{weight of X in ether}}{\text{weight of X in water}} = \frac{\text{concentration of X in ether}}{\text{concentration of X in water}} = 3$$

If w_1 = weight of X remaining in the aqueous layer, then $(10-w_1)$ = weight of X in the ether layer

$$\therefore \frac{10-w_1}{w_1} = 3$$

$$\text{Whence } w_1 = 2{\cdot}5 \text{ g.}$$

A further 50 cm^3 of ether are added to the aqueous layer, after the first separation. Calculate the weight of X which now remains in the aqueous layer.

Here the two volumes are unequal, and

$$\frac{\text{weight of X in ether}}{\text{weight of X in water}} = \frac{\text{vol. of ether} \times \text{conc. of X in ether}}{\text{vol. of water} \times \text{conc. of X in water}} = \frac{50}{100} \times \frac{3}{1} = \frac{3}{2}$$

Let w_2 = weight of X now remaining in the aqueous layer, then $(2{\cdot}5-w_2)$ = weight of X in the ether layer

$$\therefore \frac{2{\cdot}5-w_2}{w_2} = \frac{3}{2}$$

$$\therefore w_2 = 1{\cdot}0 \text{ g.}$$

A succession of extractions will enable a solute to be extracted from one solvent into another, provided the partition coefficient is reasonably favourable. In the worked example above, with a partition coefficient of only 3, several extractions would be needed to effect a good separation, but in the ether extraction of phenylamine two or three extractions are quite sufficient.

14.24 Countercurrent distribution

Countercurrent distribution is an automatic procedure, used industrially and in research laboratories, for separating two closely related solutes by making use of their slightly different partition coefficients in two immiscible liquid solvents. The technique is used when other methods such as distillation are inadequate.

The apparatus consists of a number of identical tubes (about 100–200) and the lower heavier solvent is placed in each of them. The sample is dissolved in a portion of the upper, lighter, solvent and placed in tube 1, which is shaken, and allowed to settle. The upper solvent is then automatically transferred to tube 2, while a new portion of upper solvent is added to tube 1. The process is repeated with tubes 2 and 3, and so on.

The effect of this is to cause both the solutes in the sample to be gradually transferred along the sequence of tubes, but they will undergo a gradual separation. Suppose that component A favoured the upper solvent more than component B did. Component A would then be transferred from tube to tube in slightly greater amounts than B, and after a number of operations would leave component B behind.

Countercurrent distribution can handle involatile mixtures readily, and can cope with up to 20 g of solute. It can be operated as a closed cycle to save on the number of tubes used: for instance, the sample from tube 50 of a fifty-tube unit can be returned to tube 1, and the whole operation continued.

14.25 Partition chromatography

Paper chromatography operates by a mechanism analogous to countercurrent distribution. We will consider first **partition chromatography** in which we use a strip of chromatography paper (specially prepared absorbent paper rather like filter paper) and place the solute mixture, generally dissolved in water, at one end of the strip. A second solvent is now allowed to travel along the strip, and it will extract the solute in a manner analogous to countercurrent distribution. Figure 14.26 shows two typical arrangements for separation by paper chromatography.

The analogy with countercurrent distribution is as follows: the paper itself holds a quantity of moisture, which can act as solvent for the sample. Moreover, the paper fibres themselves will attract the solute molecules by dipolar forces. The paper is referred to as the **stationary phase**. It is analogous to the lower layer of heavy solvent in countercurrent distribution. The solvent which travels across the paper is known as the **moving phase**, or mobile phase. It is analogous to the upper liquid in countercurrent distribution. Strictly the solvent in the two phases should not mix, though in practice some diffusion of solvent between the phases is bound to occur. Indeed the moving phase used in paper chromatography is very often water-miscible and the "partition" which occurs is really between the liquid moving phase and the polar surface of the paper fibres. In fact elementary experiments, such as the separation of inks, can be carried out on blotting paper or filter paper without a non-aqueous liquid being used: the solutes in ink distribute themselves between the paper fibres and the moving water phase.

Partition chromatography was invented by Martin and Synge in 1941, so it is a relatively recent technique. **Adsorption chromatography** has been known for much longer, however, the first systematic studies being done by Tswett, a Polish botanist whose main work was done at the beginning of this century, in connection with the components of the green pigmentation in plants (chlorophylls and xanthophylls). Adsorption chromatography uses a solid stationary

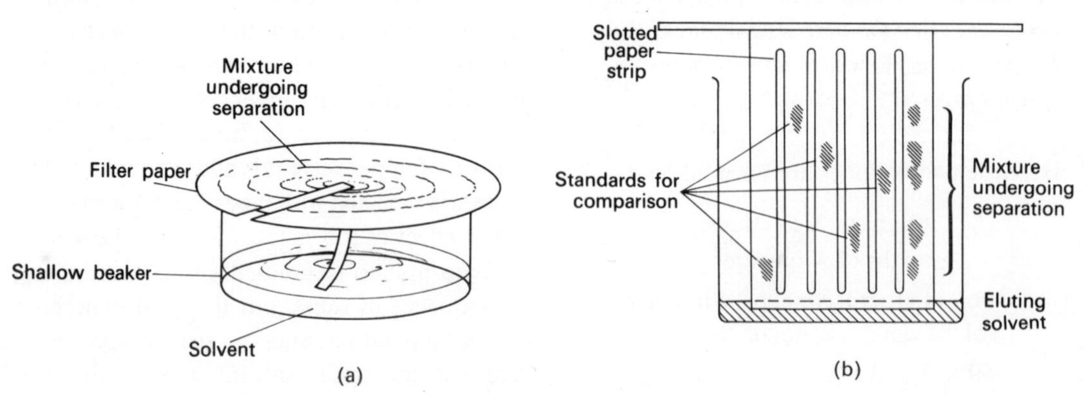

FIG. 14.26. Paper chromatography.

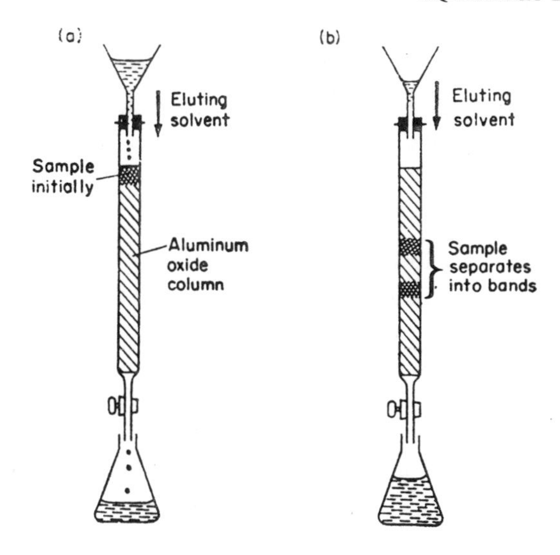

FIG. 14.27. Column chromatography.

phase, such as alumina, and a moving solvent which may or may not be aqueous. Martin and Synge suggested partition chromatography using the following:

stationary phase: water, adsorbed on silica gel;
moving phase: a liquid immiscible with water.

Paper chromatography proceeds by a mechanism which is partly partition and partly adsorption.

Adsorption chromatography, and partition chromatography of the type used by Martin and Synge, can be carried out using a column (Fig. 14.27). The advantage of a column over paper is that it can handle greater weights of solute, but the disadvantages are cost and slowness of operation.

For rapid separations where identification of the components of a mixture is all that is required, paper chromatography is usually the most convenient. For instance the sugars obtained in the hydrolysis of a polysaccharide (Chapter 33) can be identified in this way. The solvent generally used is a mixture of ethyl ethanoate and pyridine. The lower molecular-weight sugars travel further, being more soluble in the mixed solvent and less firmly adsorbed by the paper. When the chromatogram has been "run", the spots are still invisible; the paper is therefore treated with a **locating agent**, in this case phenylammonium phthalate, which reacts with sugars to give coloured products. Individual substances in the chromatogram are characterized by their **R_f values**. R_f value is defined as

$$\frac{\text{distance travelled by solute spot}}{\text{distance travelled by solvent front}}.$$

Since R_f values tend to depend on the actual conditions during an experiment, it is common practice to "run" *known* sugars on the chromatogram at the same time as the unknown mixture; in this way the chromatogram can readily be calibrated (Fig. 14.26(b)). Because of the similarity in the structures of the sugars and the paper, sugars tend to have low R_f values.

Similar separations can be carried out on the solutions of aminoacids obtained by the hydrolysis of proteins. In this case the solvent is often a mixture of butan-1-ol, ethanoic acid and

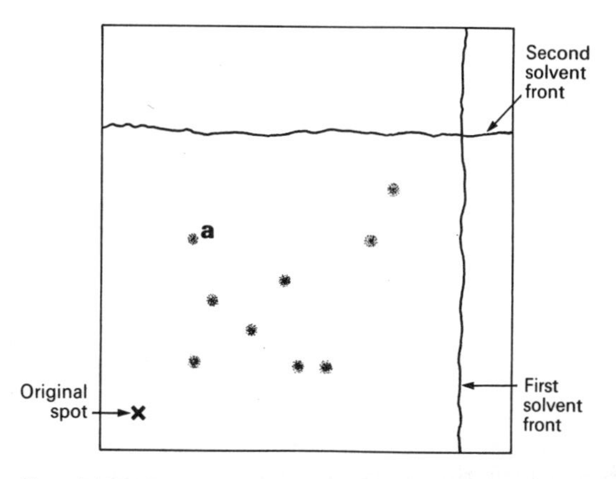

FIG. 14.28. Two-way chromatography. The R_f values for spot *a* are 0·18 in the first solvent and 0·64 in the second solvent.

water, and the locating agent is ninhydrin. Sometimes a **two-way chromatogram** is run in order to obtain a more complete separation. This involves running a normal chromatogram starting in the corner of a square piece of paper, turning the paper through 90° and running the chromatogram again using a different solvent (Fig. 14.28).

For mixtures of lipids (which are biochemical compounds soluble in organic solvents) a thin layer of activated silica gel is the stationary phase, the solvent is a non-aqueous liquid such as benzene and the spots are developed with an indicator such as alkaline bromothymol blue. R_f values approaching 1·0 are quite common in lipid chromatography.

14.26 Gas–liquid partition chromatography

Martin and Synge, in their 1941 paper on partition chromatography, mentioned the possibility of using a gas as the moving phase instead of a liquid. The idea was not developed by anyone until James and Martin published the first paper on gas–liquid chromatography in 1952. The stationary phase is a liquid, such as molten vacuum grease, or silicone oil, impregnated on a powder support and packed in a column. The stationary liquid must be involatile at the temperature used. The moving phase is a carrier gas, such as nitrogen, argon or helium. It is important that the moving phase does not react with the sample, so a carrier gas such as oxygen would generally be unsatisfactory. The whole column is placed in a thermostatically controlled oven or vapour jacket (Fig. 14.29).

The sample may be gaseous, liquid or solid, provided that it is appreciably volatile at the column temperature. For convenience most samples are introduced into the column as vapour or liquid, with a syringe. The emerging sample is passed through a detector whose electrical response depends on the concentration of sample in the gas stream. Various types of detector are in use, two typical arrangements being:

(a) **A katharometer**, shown in Fig. 14.29. The temperature of a heated platinum resistance wire varies with the thermal conductivity of the gas which surrounds it, and will thus alter if the carrier gas contains sample. The wire is connected to a Wheatstone's bridge, and any change in its temperature will alter the resistance, throwing the bridge out of balance. The out-of-balance voltage is fed to an automatic chart recorder.

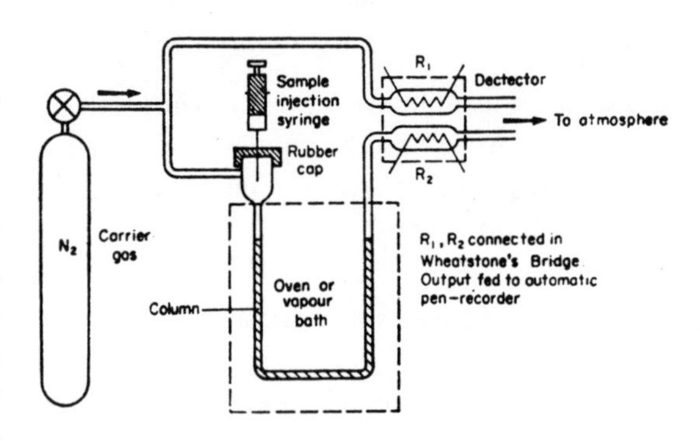

FIG. 14.29. Gas–liquid chromatography.

(b) **An ionization detector.** Some detectors use a hydrogen flame, and a carrier gas of nitrogen. The ionization energy of nitrogen is high and it is not ionized as it passes into the flame, but the ionization energy of organic molecules is generally low. The flame in the detector becomes a weak electrical conductor when organic molecules are passed through it. The current is suitably amplified and fed to a chart recorder as before.

The trace produced by the pen recorder is known as a chromatogram. A typical gas chromatogram is shown in Fig. 14.30. Gas chromatography is used largely as an analytical tool, where it can handle samples as small as 10^{-9} g and effect remarkable separations. Substances which fail to separate even in the most elaborate fractionating column will generally yield to gas chromatographic separation. For instance, a typical, "theoretical plate number" (section 14.13) for fractional distillation is 20. For gas–liquid chromatography it is more like 1000, and by using columns constructed from fine capillary tubing, plate numbers of more than 100 000 have been achieved.

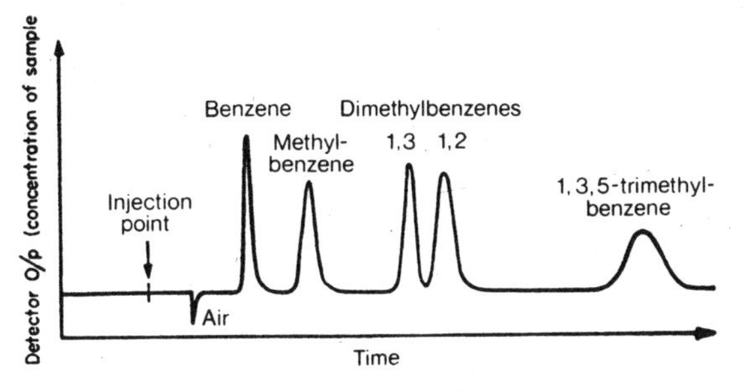

FIG. 14.30. A typical gas chromatogram.

Gas chromatography cannot handle such large masses of sample as fractional distillation or countercurrent distribution, though samples of the order 2–3 g can be separated on special, wide-bore **preparative scale columns**. The method is restricted to samples which can be volatilized.

Factors affecting separation. If the column is a non-polar one, the components of the solute mixture will emerge roughly in the order of their boiling points. The more volatile ones will remain in the vapour phase preferentially and are therefore swept through faster.

If the column is polar it will tend to retard polar samples preferentially, by dipolar attraction. Components in a mixture can be roughly characterized by chromatographing them on two contrasting columns, one polar and one non-polar.

Applications of gas chromatography. The technique has grown enormously since its invention in 1952, and already there are special journals dealing exclusively with published work on the subject. Some industrial applications are:

(1) Petroleum mixtures, which are far too complex for complete chemical analysis, are readily separated by gas–liquid chromatography. The various components are identified by comparing their *retention times* on the column with known standards.

(2) The complex chemical mixtures present in foodstuffs can, when volatile, be identified. Much research has been done into such problems as the odour of coffee and the "bouquet" of wine, presumably to develop ways of improving on the natural product.

(3) Reaction products of industrial processes can be rapidly monitored, and automatic control can be applied to chemical plant.

(4) Alcohol in blood and urine is measured by gas chromatography.

Study Questions

1. Are the following systems homogeneous or heterogeneous?

(a) Sand and water.
(b) Salt solution.
(c) Mist.
(d) Petrol.
(e) Salad cream.
(f) Smoke.
(g) Granite.
(h) Jelly.

2. How many phases and components are present in the following systems?

(a) Ether (l) $\rightleftharpoons$ Ether (g)
(b) Solid tin/lead eutectic
(c) $NH_4Cl(s) \rightleftharpoons NH_3(g) + HCl(g)$
(d) $CaCO_3(s) \rightleftharpoons CaO(s) + CO_2(g)$

3. In Fig. 14.3:

(a) What do the lines *TC* and *TY* represent?
(b) What causes the slope of the line *TY*?
(c) Why does the line *TC* stop at C?
(d) Is it possible to obtain ice above 0·1°C at pressures of less than 220 atm?
(e) Water, obtained below 0°C by supercooling is metastable. What would you expect to happen if a small crystal of ice was added to the water?
(f) Can you suggest how water below 0°C could be obtained, other than by supercooling?

4. The vapour pressure of some salt hydrates and their saturated solutions at 20°C are as follows:

Substance	v.p. of hydrate	v.p. of sat. soln.
$CaCl_2.H_2O$	2·5 mm	7·5 mm
$CuSO_4.5H_2O$	5·0 mm	16·0 mm
$Na_2SO_4.10H_2O$	16·2 mm	16·6 mm

What will happen to these three salts if they are exposed to an atmospheric water vapour pressure of (a) 4 mm, (b) 10 mm, (c) 17 mm?

5. (a) Ethanol boils at 78·4°C; ethyl ethanoate at 77·2°C; an azeotropic mixture containing 31% of the alcohol boils at 71·8°C. How can you account for this behaviour? What will happen if a 50% solution is distilled?

(b) Water boils at 100°C; HCl at −85°C; a maximum boiling mixture of these substances containing 20·2% HCl boils at 108·6°C. How can you account for this behaviour? What will happen if a 50% solution is distilled?

6. Explain the following observations as far as you can:

(a) Above 66°C, phenol and water are miscible in all proportions. Below this temperature, they are miscible only in certain proportions.
(b) Nicotine and water are miscible in all proportions only below 61°C and above 208°C.

7. At 98·5°C, the vapour pressures of phenylamine and water are 42 and 718 mm Hg respectively. In an actual experiment at this temperature, the steam distillate was found to contain 11·02 g of phenylamine and 38·81 g of water. (H_2O = 18.)

(a) Calculate the molecular weight of phenylamine.
(b) Comment on the accuracy of this method as a way of determining molecular weights.

8. The following results refer to the systems (a) 1-naphthol/naphthalene, (b) 2-naphthol/naphthalene, (c) phenol/phenylamine. The freezing-points (°C) are given for mole fractions of the first mentioned component.

	0·0	0·1	0·2	0·3	0·4	0·5	0·6	0·7	0·8	0·9	1·0
(a)	80°		71°		62°		75°		87°		96°
(b)	80°		89°		99°		107°		114°		121°
(c)	−6°	−11°	−3°	16°	28°	31°	28°	20°	15°	30°	41°

(i) Sketch a phase diagram for each system.
(ii) Comment on the features shown by each diagram.

9. (a) Draw the temperature-composition phase diagram for two solids A and B that form a compound AB, and label each area.

(b) Suggest a pair of substances that might show this behaviour.

(c) How many eutectic points would you expect for two substances that form *n* compounds?

10. The distribution coefficient of iodine between water and CCl_4 is 86, the iodine being more soluble in the latter. A solution of 1 g of iodine in 1 dm^3 of water is to be extracted with 100 cm^3 of CCl_4.

(a) How much will be extracted if all the CCl_4 is used at once?

(b) How much will be extracted if the CCl_4 is used as two 50 cm^3 portions?

(c) How would you use the CCl_4 in order to extract the maximum amount of iodine from the water?

11. Suggest methods for separating the following:

(a) A mixture of amino-acids (the breakdown products of proteins).
(b) The hydrocarbons present in a commercial petrol.
(c) $SOCl_2$ (b.p. 78°C) and $POCl_3$ (b.p. 107°C).
(d) Iodobenzene from the sludge after an organic preparation.

12. (a) Calculate the mole fractions of solvent and solute in the following:

(i) 180 g of glucose ($C_6H_{12}O_6$) in 882 g water.
(ii) 1·8 g glucose in 900 g water.
(iii) 12·8 g naphthalene ($C_{10}H_8$) in 70·2 g of benzene (C_6H_6).
(iv) 0·512 g sulphur in 760 g of CS_2.

(b) Why can the concentrations of these solutions *not* be calculated from these figures?

13. (a) The cryoscopic constant for water is 1·86 K mol^{-1} for 1 kg of the solvent. Calculate the freezing temperature you would expect if the solutes had remained undissociated in each of the following cases:

(i) A solution of 8 g of NH_4NO_3 in 10 dm^3 of water froze at −0·0372°C.

(ii) A solution of 0·49 g of sulphuric acid (H_2SO_4) in 1 kg of water froze at −0·0279°C.

(iii) A solution of 3·32 g of KI in 1 kg of water froze at −0·0744°C.

(iv) 4.54 g of HgI_2 were dissolved in a litre of solution (iii). The freezing-point was then −0·0558°C.

(b) Into how many moles has each solute dissociated?

(c) Suggest an equation for the dissociation in each case.

(d) At what temperature would a solution of 0·45 g of glucose in 500 g of water freeze?

14. What are the advantages and disadvantages of each of the following?

(a) Steam distillation rather than ordinary distillation.

(b) Two-way rather than one-way chromatography.

(c) Mixed melting-point rather than simple melting-point tests.

CHAPTER 15

Intermolecular forces

15.1 Van der Waals forces

Van der Waals forces are the weak forces that hold together the atoms in a liquefied noble gas. Such forces exist between all atoms and all molecules. The forces are polar in nature and occur as a result of the positive nuclei in one molecule attracting the negative electrons in a neighbouring molecule. Because the force of attraction depends on the charges on the nuclei and on the number of electrons involved, an approximate relationship would be expected between attractive force and molecular weight. This relationship has already been observed (Chapter 6) where the molecular weights of members of a homologous series were plotted against their boiling-points.

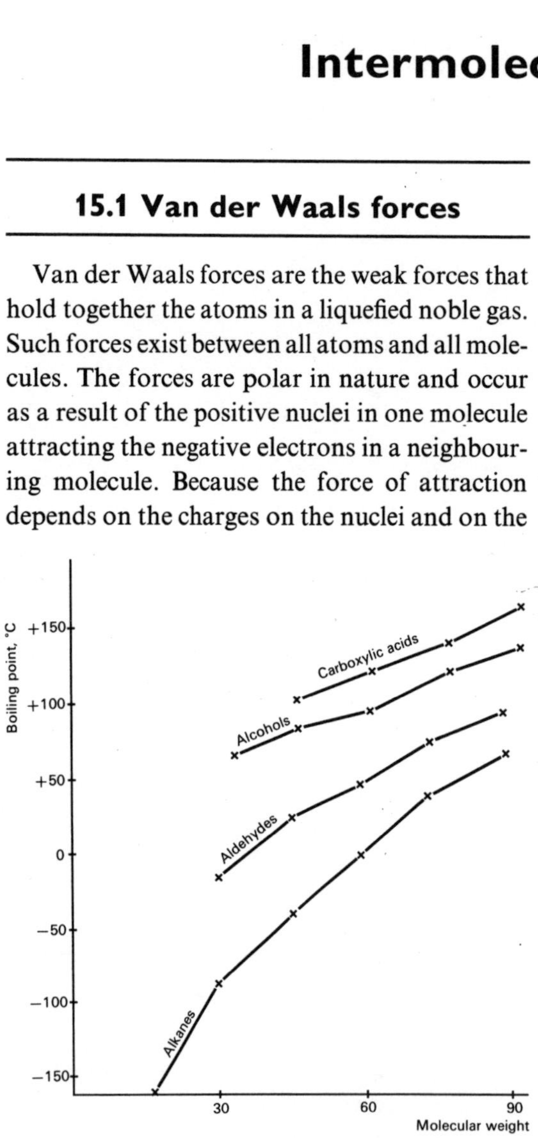

FIG. 15.1. The boiling points of members of four homologous series.

Van der Waals forces can be divided into two types: induced forces such as those that exist between noble gas atoms (section 5.2) and permanent dipole forces such as those that exist between molecules of HCl (section 4.3(b)). The latter are the stronger and this is illustrated in Fig. 15.1 where the boiling points of alkanes, held together by induced dipolar forces, are seen to be much less than those of the aldehydes, which are held together by permanent dipolar forces:

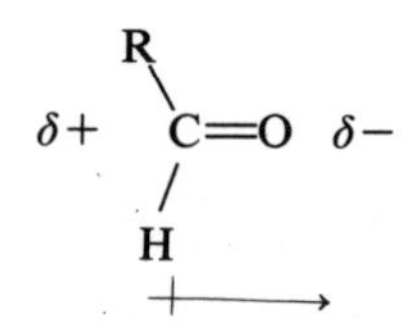

$$\delta+ \quad \overset{\displaystyle R}{\underset{\displaystyle H}{C}}{=}O \quad \delta-$$

$$\longmapsto$$

The shape of a molecule also determines intermolecular forces since the force of attraction between molecules falls off rapidly as the distance between their centres increases. "Spherical" molecules will have lower boiling points than "rod"-shaped molecules (Table 15.1).

TABLE 15.1

Compound	Formula	M. Wt.	B. pt. (K)
1-bromobutane	$CH_3.CH_2.CH_2.CH_2Br$	137	375
2-bromobutane	$CH_3.CH_2.CHBr.CH_3$	137	364
2-bromo-2-methylpropane	$(CH_3)_3CBr$	137	346

15.2 Hydrogen bonding

In Fig. 15.1 the boiling points of the alcohols and acids are shown to be even higher than those of the aldehydes. This is slightly surprising since the aldehydes have larger dipole moments than either the alcohols or the acids; some intermolecular force stronger than permanent dipole attraction must therefore be responsible. This effect is not restricted to organic molecules: for instance if the boiling points of the simplest binary hydrides of each element are plotted against molecular weight (Fig. 15.2) the pattern is as one would expect for van der Waals forces, with the exception that NH_3, H_2O and HF are startlingly high.

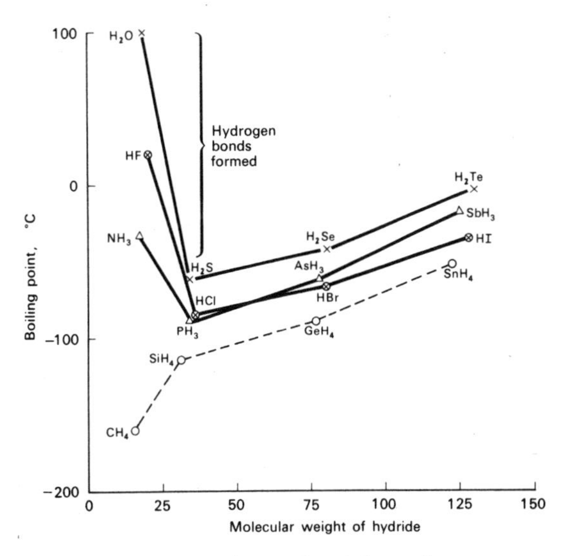

FIG. 15.2. Hydrogen bond formation.

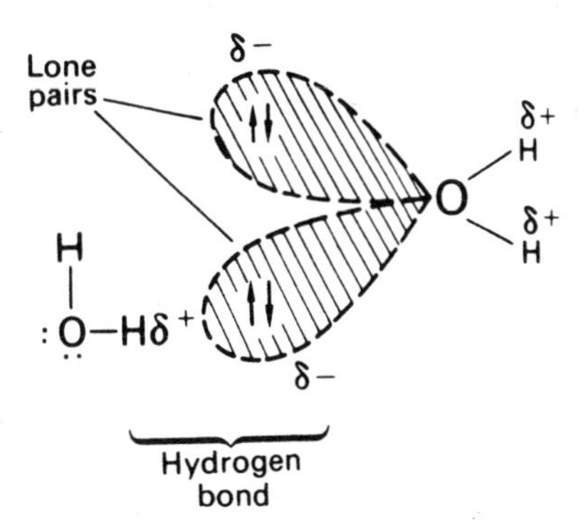

FIG. 15.3.

The explanation of this effect is that in certain circumstances a hydrogen atom can be attracted to strongly electronegative atoms. The force of attraction is strong enough to be thought of as a bond and is called a **hydrogen bond**.

Although the hydrogen bond is essentially electrostatic, it is still directional. It may be illustrated by examining the structure of ice. The water molecule has, as well as two lone pairs, bonds which are strongly polar. The lone pairs occupy regions of negative charge which can attract hydrogen atoms (from other molecules) which are themselves positively charged due to attachment to an electronegative atom. Fig. 15.3 illustrates this.

It is found that hydrogen bonds are strongest where the atoms are attached to the most electronegative first-row elements, N, O, and F. Here the dipolar force will be largest, giving the H atoms a strong charge $\delta+$. Note that a hydrogen atom attached to, say, carbon or silicon is not sufficiently $\delta+$ to be attracted to molecules such as H_2O, NH_3 or HF.

It thus appears that the commonest hydrogen bonds are *bridges* between any two of the atoms N, O, and F.

Hydrogen bond formation also accounts for many solubility effects. For instance methane is almost insoluble in water, while ammonia is highly soluble (one volume of water can dissolve about 1000 volumes of gas) and liquid hydrogen fluoride is miscible in all proportions.

Water and hydrogen fluoride and, to a lesser extent, ammonia, show strong evidence of **association** in the liquid state. Association is the joining together of two or more molecules, and is really a similar effect to the behaviour of H_2O molecules in ice. Liquids are best regarded as *broken-down solids* and liquid water, for instance, consists of small associated units, $(H_2O)_x$ where x = about 6, which have structures akin to ice. Hydrogen fluoride also associates in the liquid state, and $(HF)_x$ chains exist in the vapour just above the boiling point.

Hydrogen bonding between hydrogen and fluorine leads to the formation of an ion (HF_2^-) in solid salts, and trace amounts in liquid hydrogen fluoride. This ion is found to be linear, with equal H—F bond lengths:

$$[F \ldots\ldots H \ldots\ldots F]^-$$

This can be regarded as a structure in which a single electron pair is shared between three atoms. The bond order of these bonds is therefore only one-half. Salts such as KHF_2(s) contain this ion.

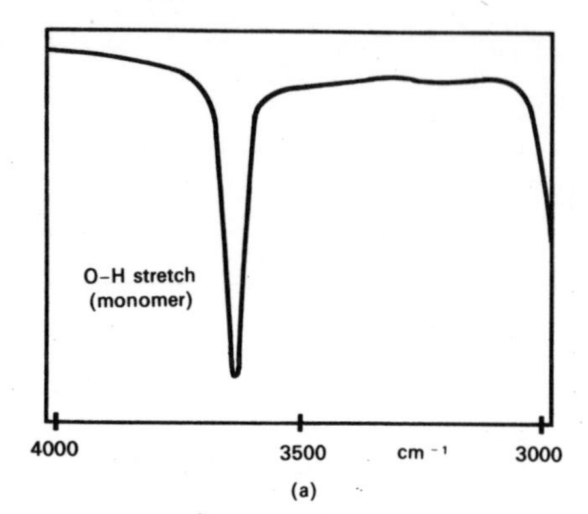

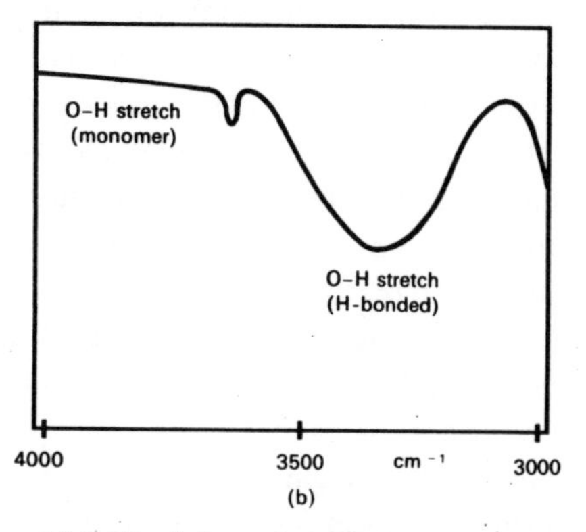

FIG. 15.4. The infra-red spectrum of an alcohol; (a) in the gas phase; (b) as a solution in CCl_4.

The presence of hydrogen bonds can often be detected using infra-red spectroscopy (section 4.3(c)). When a hydrogen bond is formed by an —OH group for instance, the stretching frequency of the O—H bond is lowered and the peak becomes very broad (Fig. 15.4).

Table 15.2 summarizes the forces that can exist between molecules and atoms.

TABLE 15.2
THE FORCES BETWEEN ATOMS AND MOLECULES

Relative strength	Type	Energy (kJ mol^{-1})
Strong	Metallic, ionic, covalent	>150
Fairly weak	Hydrogen bond	10 to 40
Weak	Permanent dipolar	4 to 10
Very weak	Induced dipolar	<4

15.3 Solubility

Dissolving is a process that is accompanied by a considerable increase in entropy. In the expression

$$\Delta G = \Delta H - T\Delta S \text{ (section 12.5)}$$

ΔS is positive for dissolving and hence ΔG will be negative unless ΔH is more positive than $T\Delta S$ is negative. Increasing the temperature will favour dissolving still more. It is really not surprising that so many heats of solution are endothermic.

Since bond-breaking is endothermic and bond-making is exothermic, it is reasonable to suppose that a positive value for the enthalpy of solution, ΔH_{sol}, probably involves the breaking down of a strong intermolecular interaction, and that the sum of the intermolecular forces in the solution is less than the sum of the intermolecular forces that existed in the two pure substances. For example ethanol and cyclohexane mix endothermically: energy is required to break the hydrogen bonds that existed in ethanol and, since only weak forces will exist between ethanol and cyclohexane molecules, this energy has to be obtained from the kinetic energy of the molecules, causing a drop in temperature. On the other hand, propanone (a polar liquid) mixes exothermically with trichloromethane (another polar liquid); this is explained by the formation of a hydrogen bond between the two molecules, an unusual case since the bridge involves a carbon atom:

$$Cl_3C{-}H \cdots O{=}C(CH_3)_2$$

energy is released, therefore the temperature rises. The reason that the hydrogen atom in trichloromethane forms a hydrogen bond is that it acquires a positive charge due to the inductive effect of the three electronegative chlorine atoms.

An appreciably exothermic heat of solution corresponds to a negative deviation from Raoult's law, and a positive ΔH to a positive deviation. It must always be remembered that it is the *difference* in bonding between the solution and the two liquids from which it is formed that determines the solution's behaviour. Thus water and ethanol form hydrogen bonds with each other and yet the water/ethanol system shows a positive deviation from Raoult's law (Fig. 14.17) because the hydrogen bonds in water

TABLE 15.3

Intermolecular forces in first component	Intermolecular forces in second component	Intermolecular forces in solution	Sign of ΔH
Induced dipolar *Hexane*	Induced dipolar *Heptane*	Induced dipolar	0
Induced dipolar *CCl_4*	Polar *Propanone*	Induced dipolar	+
Induced dipolar *Cyclohexane*	H-bonded *Ethanol*	Induced dipolar	+
Induced dipolar *Cyclohexane*	H-bonded *Water*	Immiscible $\Delta H > T\Delta S$	
Polar *$CHCl_3$*	Polar *Propanone*	H-bonded	—
H-bonded *Water*	H-bonded *Ethanol*	H-bonded	Just + see text
H-bonded *Water*	Polar *$CHCl_3$*	Immiscible $\Delta H > T\Delta S$	

itself are stronger than they are either in ethanol or in the solution. Other water and alcohol systems are discussed in section 14.16. Table 15.3 summarizes some of the possibilities that can occur when two liquids are shaken together.

In two of the examples in the table, the components fail to mix. This occurs when the enthalpy term outweighs the entropy term. Normally one of the components will have very strong intermolecular bonding when immiscibility occurs. In these cases the increase in entropy cannot compensate for the energy that must be supplied to separate the water molecules from each other.

Although only one exothermic solution is included in the table, there are other ways in which liquids can mix exothermically. For example H_2SO_4 and H_2O evolve a great deal of heat when they are mixed because of the ionization of the H_2SO_4 molecules.

Study Questions

1. Place the compounds in each of the following sets in increasing order of boiling points:

(a) $CH_3.CH_2.CH_2.CH_3$; $CH_3.CH_2.CHO$; $CH_3.CH_2.COOH$;
(b) CH_3NH_2; CH_3PH_2; CH_3AsH_2; CH_3SbH_2;
(c) $(CH_3)_2CHOH$; $(CH_3)_2CO$; $(CH_3)_2CH.CH_3$;
(d) $(CH_3)_4C$; $(CH_3)_2CH.CH_2.CH_3$; $CH_3.CH_2.CH_2.CH_2.CH_3$;
(e) BCl_3; NCl_3;
(f) C_2H_2; HCN; CO.

2. Why, on first sight, are the following facts surprising? Try to suggest a reason for each of them.

(a) The boiling point of water (373 K) does not lie between that of ammonia (240 K) and that of HF (292 K).

(b) The boiling point of $AlCl_3$ (696 K) is higher than that of PCl_3 (349 K).

(c) The boiling point of nitrogen (77 K) is much less than the isoelectronic ethyne (190 K).

(d) The boiling point of CCl_4 (350 K) is higher than that of $SiCl_4$ (330 K).

3. What intermolecular attractions occur in (a) pentane; (b) pentanal; (c) pentan-1-ol?

What intermolecular attractions would you expect if (d) pentane was mixed with pentanol; (e) pentane was mixed with pentanal; (f) pentane was mixed with hexane? In each of these three cases, state whether you would expect the mixing process to be exothermic, endothermic or thermoneutral.

4. How would the physical properties of water alter if there were no hydrogen bonds between the molecules?

CHAPTER 16

Surfaces

16.1 Adsorption

When a substance becomes concentrated at the surface dividing two phases, it is said to undergo adsorption. There is a distinction between *ad*sorption and *ab*sorption (Fig. 16.1).

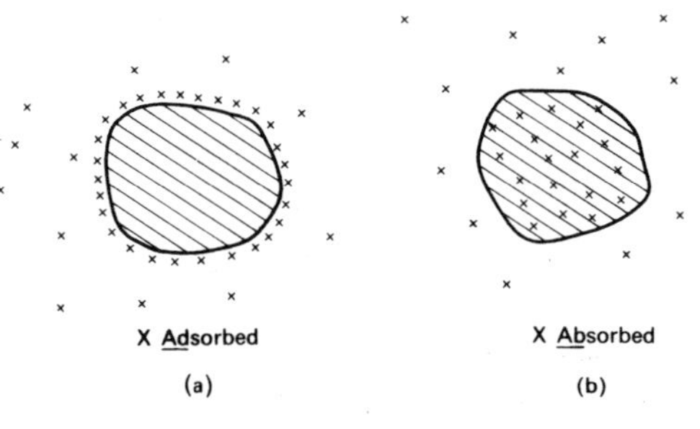

FIG. 16.1.

Adsorption implies that the substance is concentrated at a surface, whereas absorption implies penetration of a solute into the bulk of a phase. A great many examples of adsorption could be quoted, for instance:

(a) *Adsorption of gases by solids.* One of the most powerful adsorbents known is charcoal, which is thought to be a microcrystalline form of graphite with a very high surface area. The large surface area is responsible for the adsorbent properties to a great extent. Gases are readily adsorbed—a fact that was made use of in manufacturing gas-masks for protection against poison gas—and the readiness of a gas to be adsorbed seems to be very approximately in proportion to its readiness to condense to a liquid phase (Table 16.1).

TABLE 16.1
ADSORPTION OF GASES BY CHARCOAL

Gas	Volume adsorbed by 1 g charcoal at 15°	Boiling point (°C)
Sulphur dioxide	380 cm^3	−10
Chlorine	285 cm^3	−35
Ammonia	181 cm^3	−33
Hydrogen sulphide	99 cm^3	−62
Hydrogen chloride	72 cm^3	−85
Dinitrogen oxide	54 cm^3	−88
Carbon dioxide	48 cm^3	−78
Methane	16 cm^3	−160
Carbon monoxide	9·3 cm^3	−190
Oxygen	8·2 cm^3	−183
Nitrogen	8·0 cm^3	−196
Hydrogen	4·7 cm^3	−253

Heats of adsorption of gases on charcoal are low, of the order 20 kJ mol^{-1}. This is comparable to latent heats of vaporizing rather than to the energy needed for forming chemical bonds. It is thought that adsorption generally involves weak attachment of molecules to a surface by van der Waals forces.

Some naturally occurring inorganic solids, for instance the *zeolites* which are complex silicates, have very porous structures favourable to adsorption. These and other substances prepared artificially can be used as **molecular sieves**, since the channels in their structures are of molecular dimensions (0·3–0·5 nm). Different grades of molecular sieve are available commercially as adsorbents for removing impurities from gas streams. For instance, it is possible to obtain a grade which will allow straight chain hydrocarbons through while impeding and adsorbing the more bulky branched chains.

Adsorption of gases in charcoal is generally reversible. The gas can be desorbed by heating. With some adsorbents, a more powerful form of adsorption takes place, where chemical bonds are formed between the adsorbed substances and the surface. This is **chemisorption**, which is generally less easy to reverse. Chemisorption is characterized by a higher heat of adsorption.

The adsorption of gases on solids has recently been applied to chromatography, in **gas–solid chromatography**. Components of the gas mixture are separated by virtue of their differing affinities for the mobile gas phase and the stationary adsorbent surface.

(b) *Adsorption of solutes from solution.* Charcoal is also effective for adsorbing solutes from solutions. Activated charcoal (charcoal which has been rid of adsorbed gases by heating) will extract coloured impurities from a discoloured solution of an organic compound. This is often necessary in the final stages of purification of an organic substance.

(c) *Ion exchange.* A special form of adsorption occurs with some solids which have ionic groups in their structure. The phenomenon was first noticed with the zeolites, and is called **ion exchange**. Ion exchange has important technical uses, such as water softening, and nowadays synthetic resins have replaced the zeolites as ion exchangers. A typical modern ion exchange resin is an organic macromolecule, such as polystyrene, which has been treated chemically to embody sulphonic acid groups, $—SO_3^-$, in the lattice. The resultant structure, known as sulphonated polystyrene, is a sort of "macro-anion" which must have a corresponding number of cations present to neutralize the charge. Most cation exchange resins are sold as their sodium salts (Fig. 16.2).

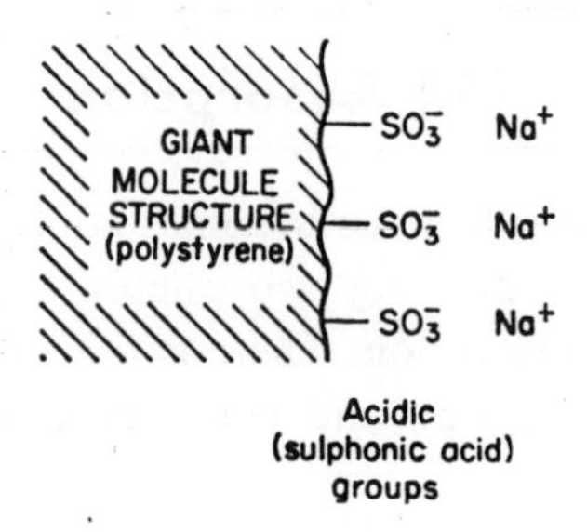

FIG. 16.2.

The resin may be converted quantitatively and reversibly into another salt by allowing a solution of some other cation to flow over it. For instance, if a solution of calcium ions is passed down a column containing beads of the sodium salt of a resin, ion exchange occurs as follows:

$$\underset{\text{mobile}}{Ca^{2+}(aq)}+\underset{\text{stationary}}{2NaR\text{ (resin)}} \rightleftharpoons \underset{\text{stationary}}{CaR_2\text{ (resin)}}+\underset{\text{mobile}}{2Na^+(aq)}$$

The adsorption is reversible, as the sign $\rightleftharpoons$ indicates. If a solution of $Na^+(aq)$ is now passed over the converted resin calcium salt, the sodium salt is regenerated and calcium ions are washed out, or *eluted*.

A resin for exchanging anions can be made, containing basic groups such as quaternary ammonium groups, NR_3^+. Such a resin will exchange quantitatively anions such as OH^-, NO_3^- and SO_4^{2-} (Fig. 16.3).

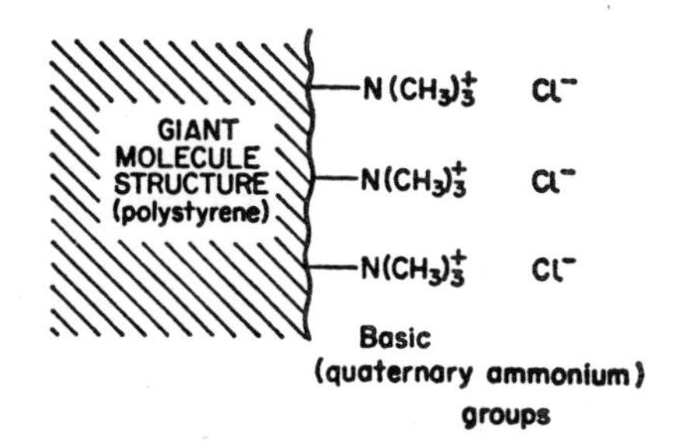

FIG. 16.3.

Simple applications of ion exchange adsorption are:

(i) *Deionization of water*. A mixture of resins containing the hydrogen "salt" of a cation exchanger plus the hydroxide "salt" of an anion exchanger, will remove all the ions present in tap water, substituting instead H^+(aq) and OH^-(aq) ions. Water of such extremely high purity has very low electrical conductivity, and is known as **conductivity water**.

(ii) *Water softening*. Hard water contains magnesium and calcium ions which form insoluble compounds with soaps (section 30.6) and as a result the soap can no longer produce a lather. Hard water is softened by using a sodium cation exchange resin in which the calcium and magnesium ions are replaced by sodium ions which do not affect the detergent properties of the soap. The resin can be regenerated by using concentrated sodium chloride solution.

(iii) *Estimation of the hardness of water*. If tap water is passed through a cation exchanger in its hydrogen form (acid form), it will quantitatively displace hydrogen ions equivalent to the cations it contains (mainly Ca^{2+} and Mg^{2+}). The solution which emerges can be titrated with standard alkali.

Number of moles of Ca^{2+} $= \frac{1}{2} \times$ number of moles of H^+ displaced
$= \frac{1}{2} \times$ number of moles of OH^- required for titration.

16.2 Ion exchange chromatography

A column of ion exchange resin can be used for the chromatographic separation of mixtures of anions or cations, making use of the selective adsorption properties of the resin for the different ions. Any ion-pair will come to equilibrium, concentrations being related by an equilibrium constant. Provided the column has sufficient theoretical plates (section 14.13), ions may be separated even when the equilibrium constants are very close; the process is analogous to a sort of countercurrent adsorption. Two typical applications of ion exchange chromatography are:

(i) A mixture of Cl^-(aq), Br^-(aq) and I^-(aq) can be separated on a column containing the nitrate salt of an anion exchanger, eluting with a concentrated solution of NO_3^-(aq). The separation depends upon the different equilibrium constants of the halide equilibria:

$$NO_3^-(aq) + RX\ (resin) \rightleftharpoons X^-(aq) + RNO_3\ (resin).$$

The effluent is collected in 5 or 10 cm^3 portions, and these are titrated with silver nitrate solution from a burette:

$$Ag^+(aq) + X^-(aq) \rightarrow AgX(s).$$

The results are plotted graphically (Fig. 16.4).

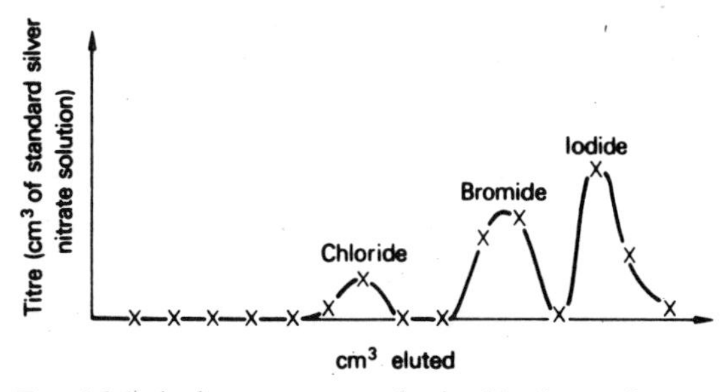

FIG. 16.4. A chromatogram obtained by ion exchange.

(ii) The first straightforward process for separating the lanthanide metals was by ion exchange. Before this, laborious procedures of fractional crystallization had to be resorted to, and even then it was difficult to obtain a pure product. A mixture of lanthanide ions, $M^{3+}(aq)$, is placed on an *anion* exchanger, and eluted with citrate ions. Anionic complex ions are formed, the stabilities of which vary along the series of metals. The separation depends upon these different stabilities: the more stable citrate complexes are formed by the smaller ions at the end of the series, and these are therefore retarded least by the column. Pure samples of the lanthanide elements are made industrially in this way. It is interesting to note that the element promethium, Pm, does not occur in nature and was first made artificially. It was identified by the fact that it fitted into the vacant space in the series on an ion exchange column (Fig. 16.5). Elements 99 (einsteinium) and 100 (fermium) were similarly detected in the dust from a thermonuclear explosion in the Pacific in 1953—on the 100-atom scale!

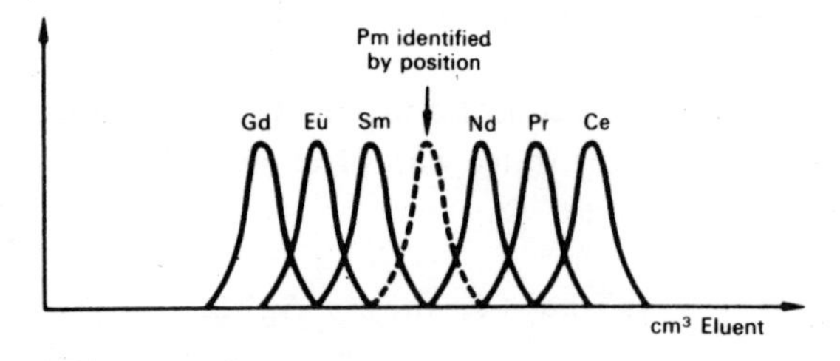

FIG. 16.5. Chromatography of the lanthanide elements.

16.3 Colloids

A **colloid** is a heterogeneous system intermediate between a suspension and a true solution. A true solution contains solute particles less than about 10^{-9} m in diameter (single molecules or ions) dispersed in a solvent. In a colloid the particle size is greater (up to 10^{-6} m), though not so great as to allow the particles to settle out under gravity. A system in which the particles settle slowly out is known as a **suspension**.

The colloidal particles constitute what is called the **disperse phase**, and the "solvent" is known as the **dispersion medium**. Generally the dispersion medium is a liquid, though it would be quite correct to describe a gaseous system such as smoke as a colloid, in which the dispersion medium is a gas. The colloidal particles are usually clumps of molecules or ions, though in the case of substances of very high molecular weight, such as starch and proteins, each colloidal particle may be a single molecule.

CLASSIFICATION OF COLLOIDS

Disregarding colloids in which the dispersion medium is not a liquid, there are various ways in which colloids may be classified.

(a) *Sols and gels*. One way of classifying colloids is by their physical properties, notably viscosity. Take as an example a colloidal solution of gelatin in water: in very dilute solutions the viscosity approaches that of water itself, and the particles of disperse phase are effectively single gelatin molecules. Such a system is called a **sol**, the term **hydrosol** being sometimes used to indicate that the dispersion medium is water. If the concentration of disperse phase is increased, cross-linking between particles begins to occur, and the structure begins to become more rigid —in fact jelly-like. Such a system is called a **gel**. If a gel is heated it melts over a short temperature range, in a manner analogous to an ordinary solid.

(b) *Lyophobic and lyophilic colloids*. Colloids may also be classified in terms of their stability. In lyophobic (literally, solvent hating) colloids there is a tendency for the particles to coagulate. The system is therefore not stable, but tends to revert to a suspension. A number of factors can accelerate this. For instance metals like gold and platinum form lyophobic hydrosols which tend to precipitate in the presence of aqueous metal ions.

Lyophilic (solvent-liking) colloids are stable and do not tend to cluster into particles which settle out. Gelatin forms a lyophilic hydrosol—the particles may cross link, with the occlusion of water molecules, but a jelly will not settle out. Lyophobic colloids on the other hand do not form gels, and their viscosity is always very close to that of the pure dispersion medium. Colloids in which the particles are macro-molecules, such as hydrated silica and albumen, are generally lyophilic.

16.4 Properties of colloids

(a) *Optical properties.* If the particles are comparable in size to the wavelength of light, a beam of light passed through the colloid will show up rather like the beam of light in a dusty atmosphere. This is known as the **Tyndall effect**. Many lyophobic colloids have extremely small particles which fail to show the Tyndall effect, although most colloids do show **Brownian motion** (section 8.1). It is Brownian motion which prevents sedimentation of the particles.

(b) *Colligative properties.* Colligative properties depend upon the number of moles of particles present, and if the particles have a very high average "molecular weight", of the order 1 000 000 as is generally the case, such effects as elevation of boiling point and depression of freezing point will be very slight indeed. Some colloids do have a measurable osmotic pressure however, and this has been used to determine the molecular weight of colloidal particles (section 14.22).

(c) *Electrical properties.* Colloidal particles have a large area in proportion to their volume, so that the properties of their surfaces are often more important than their bulk properties. One important surface phenomenon is the *adsorption* of ions on the particles of a disperse phase. Such adsorption is generally accompanied by the secondary adsorption of further particles of the opposite charge, forming what is known as an **electrical double layer** (Fig. 16.6).

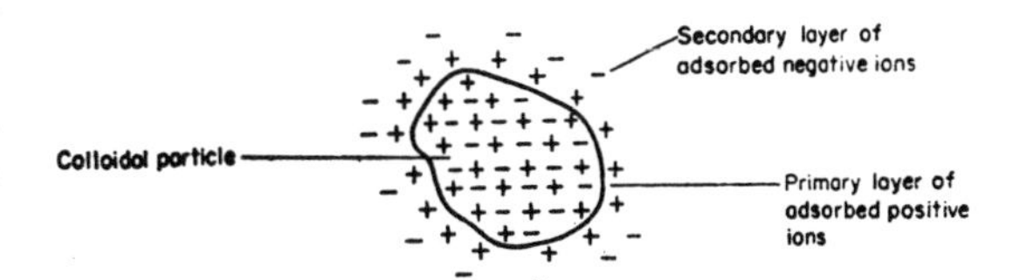

FIG. 16.6. Electrical double layer on a colloid particle.

The particles in a colloid carry an overall charge, either positive or negative due to adsorbed ions, and repulsion between like charges stabilizes the colloid and prevents coagulation. If an electric field is applied, a process analogous to electrolysis occurs, called **electrophoresis**. The colloidal particles are deposited on the electrode and lose their charge. This technique has recently been applied to painting: the article forms an electrode in a colloidal paint bath.

The addition of a quantity of highly charged ions to a hydrosol may cause coagulation to occur by destroying the electrical double layer and hence the charge on the particles. For instance, blood is coagulated by Al^{3+}(aq) (this is made use of in styptic pencils) and colloidal mud at the mouth of a river is coagulated by the salts in sea water (the formation of river deltas relies on this effect). If a negatively charged sol is added to a positively charged one, both will precipitate out.

16.5 Dialysis

The process of separating colloidal particles from the smaller particles which form the solute in a true solution is called **dialysis**. Dialysis is necessary when preparing many lyophobic colloids—it is necessary to remove electrolyte which would otherwise coagulate the particles. (A small amount of electrolyte is necessary to stabilize the colloid.) Dialysis is really a form of filtration, where the filter pores are extremely small. A cellophane membrane will serve, and the arrangement is shown schematically in Fig. 16.7.

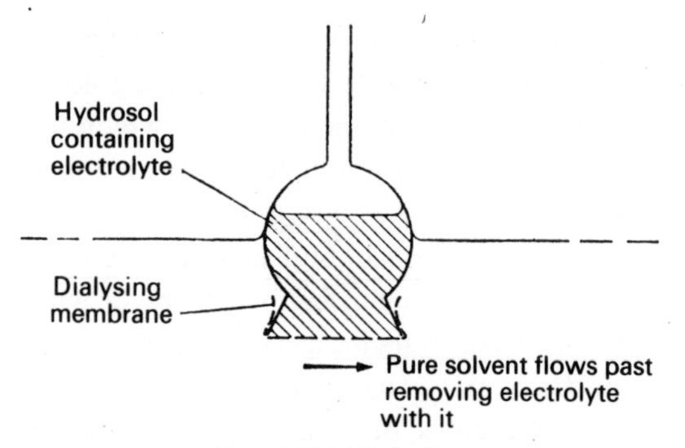

FIG. 16.7. Dialysis.

Dialysis has recently assumed great importance as a means of treating kidney failure. Blood is a colloidal system, in which the disperse phase consists of cells—red and white corpuscles—suspended in an aqueous medium called plasma. Plasma contains dissolved salts, and also traces of waste products such as urea which are normally removed by the kidneys and excreted as urine. In cases of kidney failure, death can follow rapidly due to the build-up of impurities in the blood. An "artificial kidney machine" enables the patient's blood to be dialysed. The high cost of these machines is largely caused by the complex instrumentation needed to monitor all the components in the bloodstream, and keep them in correct balance.

Study Questions

1. (a) Why does a beam of light passing through smoke show the Tyndall effect?

(b) Colloidal As_2S_3, a negative sol, is readily precipitated by a small amount of $AlCl_3$(aq, M): it is also precipitated by about seven times the amount of M $BaCl_2$ or M $ZnCl_2$, but only by several hundred times as much M HCl or M NaCl. Can you see any significance in these results?

(c) Is colloidal As_2S_3 a lyophobic or a lyophilic colloid?

(d) Fe_2O_3 forms a positive sol. Use part (b) to predict whether K_2SO_4 or $MgCl_2$ would be the more effective in precipitating Fe_2O_3 from a colloidal solution.

(e) What would happen if colloidal solutions of Fe_2O_3 and As_2S_3 were mixed?

2. (a) Suggest a method for separating compounds of Am, Cm, Bk and Cf.

(b) If you only had very small quantities, how would you show that a separation of these elements had been achieved?

3. (a) Of the halide ions, which one would be most strongly absorbed by an anion exchange resin? Give a reason for your answer.

(b) Of the halide ions, which will be most readily eluted with water?

(c) Use your answers to (a) and (b) to account for Fig. 16.4.

CHAPTER 17

Reaction rates

17.1 Introduction

This chapter is concerned with the various factors which affect the rate of chemical reactions. So far in this book we have not considered this problem very much, and previous chapters have mainly dealt with equilibria. It was seen that measurement of ΔG could tell us *how far* a reaction was capable of going, though it was incapable of saying *how fast*.

The kinetic theory of matter (Chapter 7) is consistent with experiment insofar as it predicts that, in general, raising the temperature will increase the number of collisions and hence raise the reaction rate. It can be shown that, for molecules in a gas, the number of collisions between molecules rises in proportion to $\sqrt{T}$, provided volume is kept constant. This means that for a rise in temperature of 10 K, the number of collisions only increases a few per cent.

However, it is found by experiment that the rate of most chemical reactions increases *by a factor of two or three* for a 10 K rise in temperature. Evidently the rate of a reaction does not depend simply upon the number of collisions which occur.

We shall return to this important theoretical point in a later section, but first we shall present some experimental facts.

17.2 Definition of reaction rate

Reaction rate, or reaction velocity may be defined as the rate of change of concentration of a stated reactant or product. It is generally convenient to express concentration in mol dm^{-3}, and the time unit might be seconds, minutes or even hours or days. Alternatively, rate may be expressed in terms of *number* of moles, or even number of grams, rather than in terms of concentration of a reactant: such a procedure is more realistic if the total volume of the reactants is changing.

17.3 Factors which can influence reaction rate

Simple experiments show that the following factors must be considered:

(a) **Accessibility of reactants.** By this is meant the ease with which two or more species of reacting molecule can come together in order to react. In the case of a heterogeneous system, in which the reactants occupy different phases the surface area of the interface is an important factor. In a homogeneous reaction system there is no such limitation on reaction rate.

(b) **Temperature.** Reactions increase in rate as the temperature is increased.

(c) **Concentration.** Increasing the concentration of the reactants can have a variety of effects on reaction velocity. Generally an increase in concentration of a given reactant produces an increase in overall reaction velocity, though this is not always the case. In some cases the rate is found to be directly proportional to the concentration of a given reactant, and in other cases rate may be proportional to the square of concentration, or to concentration raised to some other power. The effect of reactant concentration on rate cannot be predicted theoretically (except sometimes by analogy with other known reactions of a similar nature), but must be determined *experimentally*.

(d) **Presence of catalysts.** Many reactions proceed at an increased rate in the presence of another substance called a catalyst. A catalyst is not consumed in a reaction, though it may take part in the formation of intermediate compounds, and certainly forms temporary chemical bonds with reactant molecules at some stage in the process.

17.4 Experimental measurement of reaction velocity

The rate of a chemical reaction is determined by analysis of the reaction mixture at suitable time intervals. The choice of analytical method depends on the reaction being considered.

(a) **Reactions in solution.** These are very often studied most conveniently by titration. Consider for instance the rate of hydrolysis of an ester like methyl ethanoate by $OH^-(aq)$ ions. The equation is:

$$\underset{\text{methyl ethanoate}}{CH_3COOCH_3(aq)} + OH^-(aq)$$

$$\rightarrow \underset{\text{methanol}}{CH_3OH(aq)} + \underset{\text{ethanoate ions}}{CH_3COO^-(aq)};$$

This reaction may be followed by extracting small portions of the reaction mixture by pipette at intervals, and running them into an excess of dilute acid. This effectively stops the reaction proceeding further, and as a further precaution the reaction mixture may be cooled in ice. The excess acid is then titrated with standard alkali to give the amount of alkali that has been consumed.

(b) **Reactions in the gas phase.** These are often more conveniently studied by observing some physical property which does not involve taking samples from the reaction mixture. For instance, the gas phase decomposition of 2-methyl-2-iodopropane could be followed by sampling the mixture at intervals and titrating the liberated hydrogen iodide with $OH^-(aq)$, but it is much more easily followed by plotting a graph of pressure against time, at constant volume:

$$\underset{\text{1 mole}}{(CH_3)_3CI(g)} \rightarrow \underbrace{(CH_3)_2C{=}CH_2(g) + HI(g)}_{\text{total 2 moles}}$$

(c) **Heterogeneous reactions.** The experimental difficulties involved in making theoretical studies of heterogeneous reactions are often considerable, and the results complex, particularly when the reaction takes place at a catalytic surface. Thus a reaction like

$$C_2H_4(g) + H_2(g) \rightarrow C_2H_6(g)$$

which takes place readily at the surface of finely divided nickel is heterogeneous, but very many variables—state of surface, area of surface and method of preparation as well as temperature, concentration and pressure—affect the overall rate.

Even in a non-catalytic heterogeneous process such as

$$CaCO_3(s) \rightarrow CaO(s) + CO_2(g)$$

the structure and nature of the solid surface can have a great effect on the rate.

Measurements do not give the rate of reaction directly: they can only give the concentration of a reactant at a particular time. A graph of concentration against time is then plotted. The rate of reaction is the gradient of the tangent at a given point (Fig. 17.1).

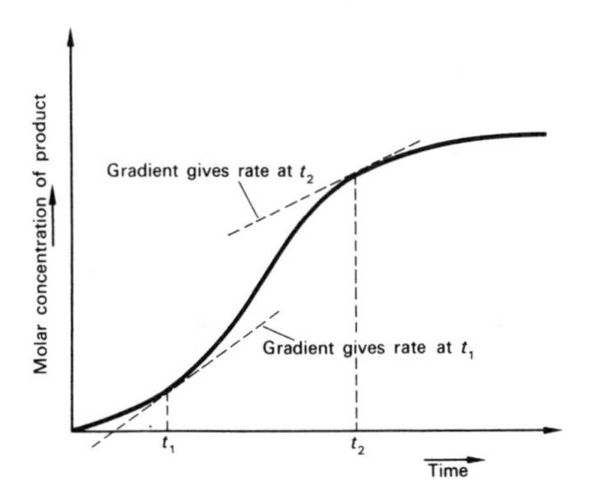

FIG. 17.1.

From these data further graphs may be plotted, for instance rate against time, or rate against concentration. Experiments show that the rates of reactions depend upon the concentration of the reactants, though they are not necessarily directly proportional (Fig. 17.2).

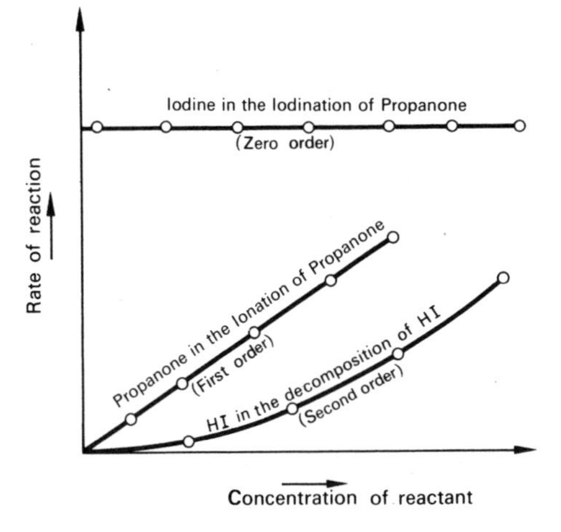

FIG. 17.2.

17.5 Order of reaction

The dependence of rate of a reaction upon concentration of a given reactant can be expressed by the experimentally determined relationship

$$\text{rate} \propto [\text{reactant}]^n$$

where n is defined as the order of reaction with respect to that particular reactant, and we say the reaction follows *nth order kinetics*. The order of a reaction can be determined by plotting log[reactant] against log(rate). The gradient will give n, the order. Where two reactant concentrations affect the rate, we may write

$$\text{rate} \propto [\text{A}]^m[\text{B}]^n$$

and m = order with respect to reactant A, and n = order with respect to reactant B. The overall order of reaction is $(m+n)$.

Order of reaction may be defined as the number of concentration terms in the experimentally determined rate equation.

First order reactions. The thermal decomposition of 2-methyl-2-iodopropane, section 17.4(b), is first order. Another first order reaction is the isomerization of cyclopropane to propene:

$$\text{CH}_2\text{(—CH}_2\text{—CH}_2\text{—, ring)} \rightarrow \text{CH}_3\text{—CH=CH}_2$$

Rate = k[cyclopropane]
where [ ] denotes concentration of reactant, and
k = the **rate constant** of the reaction.

The decay of a radioactive element, though not strictly a chemical process, is first order since the rate of decay, measured by counting the intensity of emitted α- or β-particles, depends upon the amount of the radioactive element present.

Second order reactions. These are much more common. One of the first to be thoroughly studied was the gaseous decomposition of hydrogen iodide, and the reverse reaction (combination of hydrogen and iodine vapour) by Bodenstein in 1896. The reaction proceeds to equilibrium and the results for the equilibrium constant were quoted in Chapter 10. Bodenstein's procedure was to seal the reactants in glass bulbs and place them in a vapour bath at constant temperature. After a given time interval the bulbs were removed, cooled rapidly—**quenched**—to prevent further reaction, and their contents analysed.

We may express the results of the reaction between hydrogen and iodine by writing

$$-R(H_2) = -R(I_2) = k_1[H_2][I_2]$$

where the notation $R(H_2)$ denotes "rate of reaction with respect to hydrogen concentration" and the minus sign denotes rate of *disappearance* of hydrogen. Using this notation the rate could equally well have been expressed in terms of the rate of formation of the product HI(g):

$$R(HI) = k_2[H_2][I_2]$$

Note that $k_2 = 2k_1$, because two moles of hydrogen iodide are formed for every mole of hydrogen or iodine consumed.

However the rate is expressed, the reaction is second order overall, and first order with respect to each separate reactant.

The backward reaction is also second order:

$$-R(HI) = k_3[HI]^2,$$

where $-R(HI)$ denotes rate of disappearance of hydrogen iodide.

Of all the hydrogen halides, Bodenstein was fortunate in choosing hydrogen iodide. The rate equations for the formation and decomposition of HBr are considerably more complex, while HCl and HF do not decompose according to any clear mathematical law.

It is not possible to *deduce* the kinetics of a reaction from its stoichiometric equation, or from its equilibrium constant: the hydrogen halides all follow the same stoichiometry,

$$H_2+X_2 = 2HX$$

but their rate equations are entirely different.

If one of the reactants in a chemical reaction is present in great excess, its concentration may be almost unaffected by the progress of the reaction. In this case its concentration term may be omitted from the rate equation, and the experimentally determined order will be lower. For instance, in the above case, if the reaction were carried out with a small concentration of hydrogen and a very large excess of iodine vapour it would have an overall order of one—first order with respect to hydrogen, and *zero order* with respect to iodine.

Order of reaction then is not an absolutely fundamental property of a reaction system, it is simply a convenient way of describing the rate equation which best fits the facts.

Some reactions, under certain conditions, are found to be of *fractional order*. For instance, an order of 3/2:

$$CH_3CHO(g) \rightarrow CO(g)+CH_4(g);$$
$$-R(CH_3CHO) = [CH_3CHO]^{\frac{3}{2}}$$

17.6 First order rate equation

Most chemical changes gradually slow down as they proceed, since the molar concentration of the reactants falls off as they are consumed. Consider a first order reaction A → products, and let the initial value of $[A] = a$ mol dm^{-3}. Suppose that after time t, x mol dm^{-3} of A have reacted.

$$\therefore \text{ after time } t, [A] = a-x$$

Using the differential notation, we express rate of change of [A] with time (rate of reaction) as follows:

$$-\frac{d[A]}{dt} = -\frac{d(a-x)}{dt}$$

The minus sign indicates that, although rate is a positive quantity, [A] decreases as t increases.

For a first order reaction,

$$-\frac{\mathrm{d}(a-x)}{\mathrm{d}t} = +\frac{\mathrm{d}x}{\mathrm{d}t} = k(a-x)$$

$$\therefore \frac{\mathrm{d}x}{a-x} = k\,\mathrm{d}t$$

$$\therefore \int \frac{\mathrm{d}x}{a-x} = k\int \mathrm{d}t \quad (+\text{integration constant})$$

At the beginning of the reaction, $x = 0$ and $t = 0$, and we may therefore write in limits to the integrals, and eliminate the constant of integration:

$$\int_0^x \frac{\mathrm{d}x}{a-x} = k\int_0^t \mathrm{d}t$$

$$\therefore \log_e\left(\frac{a}{a-x}\right) = kt$$

$$\therefore k = \frac{1}{t}\log_e\left(\frac{a}{a-x}\right)$$

or,

$$k = \frac{2{\cdot}303}{t}\log_{10}\left(\frac{a}{a-x}\right).$$

This is the equation for the graph shown in Fig. 17.3, and enables the rate constant k to be evaluated.

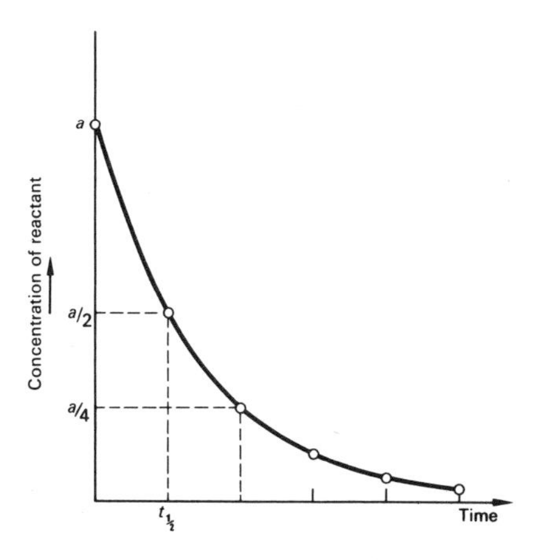

FIG. 17.3. First order reaction.

17.7 Half-life for a first order change

Let $t_{0{\cdot}5}$ = the time taken for exactly half of the original reactant to decompose. That is, when $t = t_{0{\cdot}5}$, $x = a/2$.

Substituting,

$$k = \frac{2{\cdot}303}{t_{0{\cdot}5}}\log_{10} 2$$

Note that this equation does not include the concentration of the reactant. Therefore, for a first order reaction, the time taken for exactly half the reactant to decompose is independent of the initial concentration. In other words, a first order process has a definite **half-life** period. This fact has already been commented on for radioactive decay. The form of curve shown in Fig. 17.3 is known as an **exponential decay** curve.

The time for any other definite fraction of reactant to decompose is also constant. For instance, a first order reaction has a constant "quarter-life" period.

17.8 Rate equations for other orders

For a second order reaction in which rate $\propto [A]^2$, we may write

$$\frac{\mathrm{d}x}{\mathrm{d}t} = k(a-x)^2$$

$$\therefore \int_0^x \frac{\mathrm{d}x}{(a-x)^2} = k\int_0^t \mathrm{d}t$$

$$\therefore \frac{x}{a(a-x)} = kt$$

$$\therefore k = \frac{1}{t}\cdot\frac{x}{a(a-x)}$$

The half-time for a second order change does depend upon the initial concentration, as the following shows:

$$t = \frac{1}{k}\cdot\frac{x}{a(a-x)}$$

$$t_{0\cdot5} = \frac{1}{k}\cdot\frac{a/2}{(a-a/2)a}$$

$$= \frac{1}{k}\cdot\frac{1}{a} = \frac{1}{ka}.$$

A second order process of the type rate $\propto k[A]^2$ therefore has a half-life period inversely proportional to the initial concentration of A. Examination of the half-life period of a reaction, after plotting a graph, is often a suitable way of determining the order of reaction. In general, for an nth order reaction,

$$\text{half-life} \propto \frac{1}{a^{n-1}}.$$

If the rate of reaction is constant, and independent of concentration [A], this can be expressed mathematically as rate $\propto [A]^0$, and the reaction is of zero order. A zero order reaction will produce a straight line plot for concentration against time (Fig. 17.4).

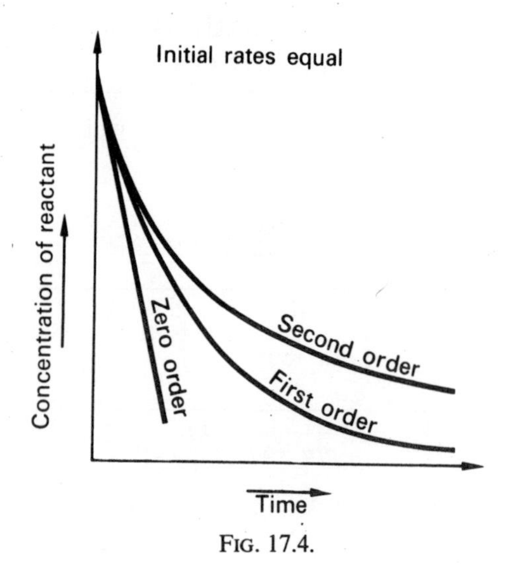

FIG. 17.4.

17.9 Effect of temperature on rate of reaction

Increasing the temperature has the effect of increasing the rate of most reactions. A typical organic reaction shows a doubling or trebling of the rate constant for a rise of 10 degrees. It is found by experiment that the logarithm of the rate constant is inversely proportional to the absolute temperature T, the relationship being

$$\log_e k = \text{constant} - \frac{E}{RT}$$

or, $$E = -2{\cdot}303\, RT \log_{10} k + \text{constant}$$

where R = the gas constant, in J mol^{-1} K^{-1}. E has the dimensions of energy, and is termed the **energy of activation** of the reaction. This equation is strikingly similar to that previously given for the relation between the standard free energy change $\Delta G^\ominus$ and the equilibrium constant K:

$$(\Delta G^\ominus = -2{\cdot}303\, RT \log_{10} K).$$

If $\log_{10} k$ is plotted against $1/T$, a straight line is obtained, Fig. 17.5. The intercept at $1/T = 0$ is the constant above, which may be written

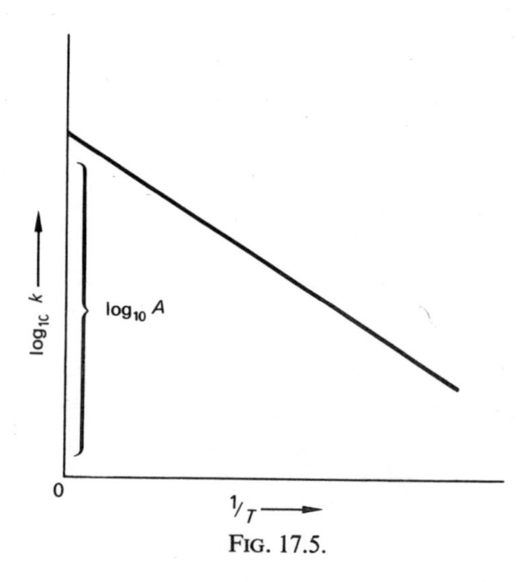

FIG. 17.5.

$\log_{10} A$. The above equations are collectively referred to as the Arrhenius equation, and the constant A is termed the Arrhenius constant or the "A factor".

$$k = Ae^{-E/RT}.$$

17.10 Energy of activation

The energy term E in the Arrhenius equation is the minimum energy which molecules need to acquire before they can react by collision. For instance, in the reaction between H_2(g) and I_2(g) only a very tiny fraction of the collisions actually lead to the production of 2HI(g). In most cases the collision energy is insufficient and the molecules simply rebound elastically. However, if they collide with sufficient energy they may coalesce to form a **transition state**, otherwise known as an activated complex, in which H—H and I—I bonds are breaking and H—I bonds are forming.

$$\begin{matrix} \text{H—H} \\ \text{I—I} \end{matrix} \rightleftharpoons \begin{matrix} \text{H} \cdots \text{H} \\ \vdots \quad \vdots \\ \text{I} \cdots \text{I} \end{matrix} \rightleftharpoons \begin{matrix} \text{H} \quad \text{H} \\ | \quad | \\ \text{I} \quad \text{I} \end{matrix}$$

The energy required to form this transition state is the activation energy. This is depicted graphically in Fig. 17.6.

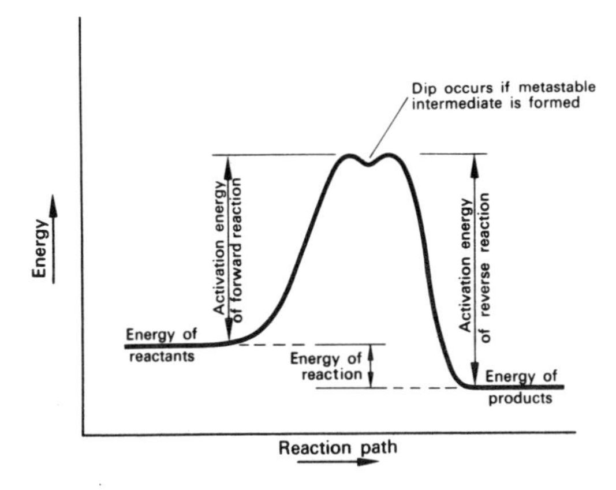

FIG. 17.6. Activation energy.

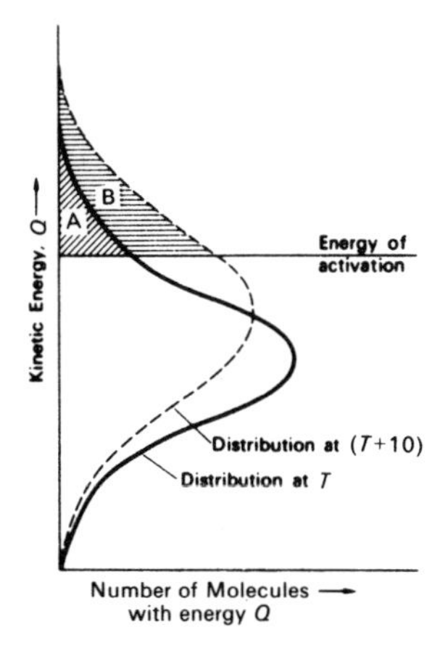

FIG. 17.7. Boltzmann distribution curves at two different temperatures.

At a given temperature the number of molecules possessing a given energy is shown by a **Boltzmann distribution curve**; at higher temperatures the "spread" of energies is larger and far more molecules possess very high energies (Fig. 17.7).

The figure also shows how the total number of molecules whose energy exceeds E varies with temperature. If area A represents the number of molecules with energy greater than E at temperature T, then area B represents the number of additional molecules with energy greater than E at $(T+10)$. When E has a value of about 40–60 kJ mol^{-1}, the number of molecules with energies greater than E is approximately doubled for a temperature rise of 10 K.

Complex reactions which proceed in a series of steps often do not obey the Arrhenius equation. Others may do, by a combination of mathematical factors, though the actual value of E does not have any very real physical significance. Nevertheless the concept of energy of activation is a great help to the chemist in learning how to control reactions and reaction rates.

Energy of activation is comparable in magnitude with the values of bond energy terms

(Chapter 9). This is to be expected, since formation of the transition state involves breaking bonds.

17.11 Molecularity of a reaction step

Some confusion has existed in school text books concerning the meaning of the terms *unimolecular*, *bimolecular*, etc., with regard to chemical reactions. We shall define **molecularity** *as the number of molecules (or atoms or ions) which form the transition state in a reaction.* For instance the molecularity of the reaction

$$H_2(g)+I_2(g) \rightarrow 2HI(g)$$

is two (bimolecular) because the transition state is formed from two molecules.

The confusion probably arises because very many bimolecular processes follow second order kinetics, and unimolecular reactions often follow first order kinetics. This is by no means always true however, and the concepts of order and molecularity should be carefully distinguished:

(a) *Order of reaction* is an experimentally determined quantity, and is nothing more than a mathematical convenience—a number in a rate equation which can have all sorts of values from zero upwards, including fractional values;

(b) *Molecularity* is a theoretical concept—the number of molecules which are *postulated* as participating in formation of the activated complex of a particular *reaction step*. We can refer to the order of a complex reaction, as the observed kinetics of its many steps. We can only refer to the molecularity of each individual stage in a reaction proceeding as a series of steps. Molecularity can only have integral values, usually 1 or 2. A molecularity of 3 would be very rare indeed since the chances of a three-body collision of a suitable type occurring are very remote.

An example of a **bimolecular** reaction is the attack on an alkyl halide such as CH_3Cl by $OH^-(aq)$ (Fig. 17.8). Because the molecularity of the slow, rate-determining step is 2, this is known as an **S_N2 reaction**. The *S* stands for substitution and the N for nucleophilic.

Other alkyl halides are hydrolysed by a different mechanism. The presence of the comparatively weak C—I bond combined with three electron-donating methyl groups on the carbon atom causes 2-methyl-2-iodopropane to dissociate into ions in the slow, rate-determining first step.

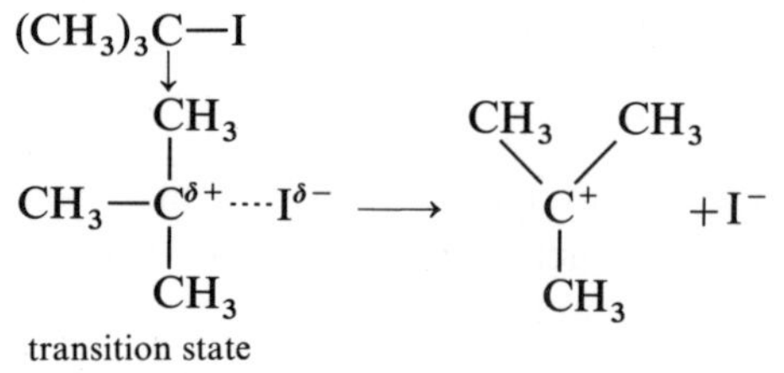

$$(CH_3)_3C\text{—}I \downarrow \quad (CH_3)_3C^{\delta+}\cdots I^{\delta-} \longrightarrow (CH_3)_3C^+ + I^-$$

transition state

The reaction is found to be first order with respect to the alkyl halide, but zero order with respect to the hydroxide ion, and occurs at the rate at which the C—I bonds, activated by

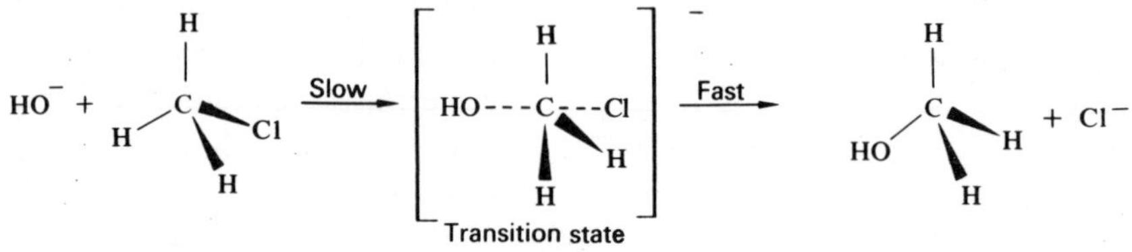

$$HO^- + CH_3Cl \xrightarrow{\text{Slow}} [HO\text{---}CH_3\text{---}Cl]^- \xrightarrow{\text{Fast}} HOCH_3 + Cl^-$$

Transition state

FIG. 17.8.

random collision, break up. The alkyl halide decomposes in a **unimolecular** slow first step in what is therefore known as an **S_N1 reaction**. The second step of the reaction is very rapid since the positively charged carbocation reacts readily with the negatively charged hydroxide ion:

$$(CH_3)_3C^+ + OH^- \rightarrow (CH_3)_3COH.$$

The S_N1 and S_N2 mechanisms are identical from the point of view of stoichiometry, but differ kinetically because of the different structures of the reactants. Many alkyl halides in fact are hydrolysed using a mechanism that it is a hybrid of the S_N1 and S_N2 processes.

The fact that the slowest step determines the overall rate is readily appreciated by considering the "washing up" analogy. Imagine three people washing up plates, one washing, one drying, and one stacking them into piles in the cupboard. The rate determining step in the process will undoubtedly be the drying stage. The third stage cannot proceed any faster than the supply of plates allows it to!

17.12 The effect of radiation on reaction velocity

If radiation is to affect the velocity of a reaction, its wavelength must be such that it is absorbed by one of the components of the reaction mixture. Ultraviolet radiation generally has the most effect, and indeed the results are often very dramatic. For instance, at room temperature in the dark, little observable reaction takes place between $H_2(g)$ and $Cl_2(g)$. In subdued sunlight smooth reaction occurs, forming hydrogen chloride, while in strong ultraviolet light, the mixture will explode.

The presence of radiation is essential to the process of photosynthesis, since here many stages in the process involve an increase in ΔG, and will not take place at all without light.

One of the most important consequences of the absorption of radiation in the visible and ultraviolet regions of the electromagnetic spectrum is the breaking of bonds. The reason why the hydrogen–chlorine reaction becomes explosive in the presence of ultraviolet light is that free atoms of chlorine are formed in the process

$$Cl_2(g) \rightarrow 2Cl(g); \quad \Delta H = +240 \text{ kJ mol}^{-1} \qquad (1)$$

From the relationship, $E = h\nu$

$$E = \frac{Lhc}{\lambda} = \frac{1{\cdot}19 \times 10^{-4}}{\lambda} \text{ kJ mol}^{-1}$$

where h = Planck constant = $6{\cdot}63 \times 10^{-34}$ N s
c = velocity of light = $2{\cdot}998 \times 10^8$ m s^{-1}
L = Avogadro constant = $6{\cdot}022 \times 10^{23}$,

the energy of a quantum of visible light at 400 nm will just break a bond of 300 kJ mol^{-1} energy—a relatively weak bond.

The process by which bonds are artificially ruptured by radiant energy is called **photolysis**. A schematic apparatus is shown in Fig. 17.9. The reaction vessel is generally quartz, which is transparent to ultraviolet radiation, and a mercury lamp is the energy source.

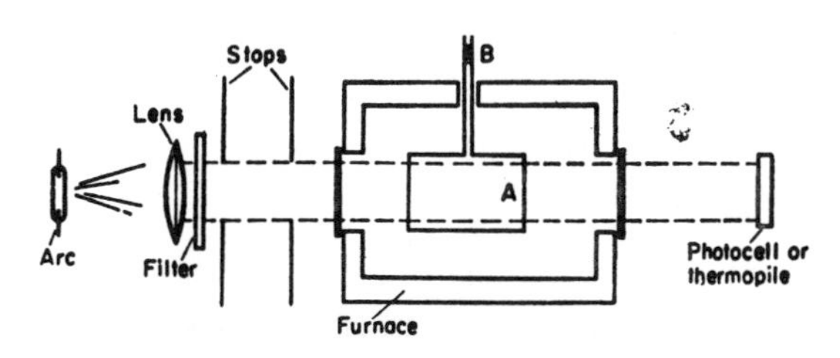

FIG. 17.9. Photolysis apparatus. A, quartz reaction vessel; B, connection to gas handling apparatus.

17.13 Chain reactions

The remarkable thing about the hydrogen–chlorine explosion reaction is its remarkably high **quantum yield**. We may define quantum yield as

$$\frac{\text{number of molecules transformed by reaction}}{\text{number of quanta of radiation absorbed}}.$$

In the above case, quantum yields of up to 10^7 have been observed. This suggests that absorption of a single quantum of energy "triggers off" a whole process, rather in the manner that a small spark can trigger off a combustion reaction.

The process is thought to occur by a whole sequence of reactions, starting with (1) above. Further reactions can occur, such as

$$Cl^{\cdot} + H_2 \rightarrow HCl + H^{\cdot} \qquad (2)$$

$$H^{\cdot} + Cl_2 \rightarrow HCl + Cl^{\cdot} \qquad (3)$$

Reactions (2) and (3) can proceed indefinitely, as long as the free atoms produced are able to react further.

The presence of free atoms and free radicals is often important in chemical reactions. Polymerizations to form "Perspex", polyvinyl chloride and polystyrene are examples.

17.14 Catalysis

A catalyst may be defined as a substance which can increase the rate of a reaction, without itself undergoing permanent chemical change. Catalysts have the following general properties:

(a) A small number of moles of catalyst can catalyse a very large number of moles of reactant. For instance, hydrogen peroxide can be decomposed into water and oxygen by as little as one part per million of colloidal platinum.

(b) A catalyst may undergo intermediate physical changes, and it may form temporary chemical bonds with the reactants, but it will be unchanged in amount and composition at the end of the reaction.

(c) A catalyst cannot affect the ΔG of a reaction, and hence it cannot change the equilibrium position. It may accelerate the attainment of an equilibrium, and its function can be likened roughly to "lubricating the mechanism". Since it is chemically unchanged overall, it cannot add energy to, or remove energy from, the system.

Catalysts may be homogeneous (same phase as the reactants) or heterogeneous (different phase, usually solid), but heterogeneous catalysts have wider applications in practice. Transition metals and their compounds are very often effective. This is generally due to their ability to form intermediate bonds and assist the formation of transition states (activated complexes), though the exact mechanism of a particular catalysis is often obscure. The following are some examples of transition metals catalysts used industrially:

(i) Finely divided nickel catalyses the hydrogenation of the alkene bond C=C. This is used in the conversion of vegetable oils to margarine.

(ii) A finely divided mixture of iron and aluminium oxide is the catalyst for the Haber process for ammonia manufacture.

(iii) Platinized asbestos, and vanadium(V) oxide (vanadium pentoxide) are both used for the reaction $SO_2(g) + \frac{1}{2}O_2(g) \rightarrow SO_3(g)$.

Catalysts work by lowering the energy of activation of the reaction, by providing an alternative path with a lower "hump" to be overcome (Fig. 17.10). Consequently more molecules are present with sufficient energy to react.

Very often a different choice of catalyst will lead to different products from a given reaction

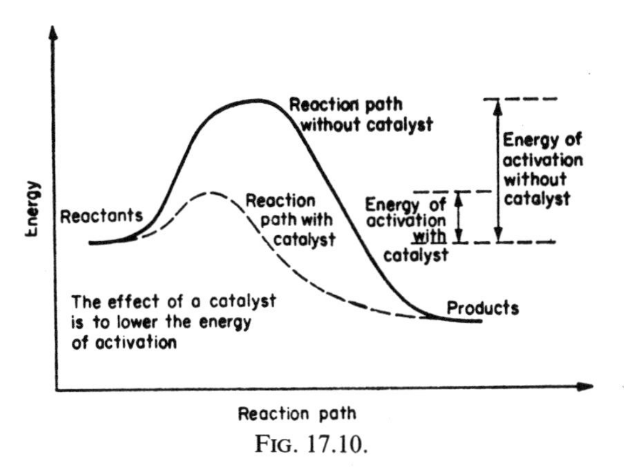

FIG. 17.10.

mixture. Thus the Fischer–Tropsch synthesis of hydrocarbons uses CO(g) and H_2(g) over a mixture of oxides of cobalt and thorium. The same gases passed over zinc oxide yield methanol. Catalysts are highly specific, and the principles underlying their operation are even today far from being understood. The choice of a catalyst for an industrial organic synthesis may often be a combination of science and "cookery", or inspired guesswork, and the recipe for a good catalyst is often a closely guarded industrial secret.

Study Questions

1. Suggest how the rates of the following reactions might be followed experimentally:

(a) The reaction of iodine solution with propanone in the presence of acid.
(b) The thermal decomposition of ethanal vapour to give methane and carbon monoxide.
(c) The action of hydrochloric acid on sodium thiosulphate solution. (One of the products is sulphur.)
(d) The catalytic decomposition of hydrogen peroxide solution into water and oxygen.

2. The decomposition of N_2O_5(g) into NO_2(g) and O_2(g) was followed by observing the change in the pressure. The partial pressure of N_2O_5 in the mixture was as follows:

Time	*Pressure (mm)*	*Time*	*Pressure (mm)*
0 min	348	40 min	105
10 min	247	50 min	78
20 min	185	60 min	58
30 min	140		

Calculate (a) the order of the reaction, (b) the rate constant for this reaction.

3. The thermal decomposition of ethanal was followed at 770 K ($CH_3CHO(g) \rightarrow CH_4(g)+CO(g)$), the overall pressure varying as follows:

Time (seconds)	*Pressure (mm)*
0	363
42	397
105	437
242	497
480	557
840	607

Calculate (a) the order of the reaction, and (b) its rate constant.

4. Comment on the following observations:
(a) Hydrogen and oxygen do not normally react at room temperature, but if a trace of platinum black is present, an explosion occurs.
(b) The enzyme urease causes urea, $CO(NH_2)_2$ to be hydrolysed, yet it has no effect on methyl urea, $NH_2CO(NHCH_3)$.

5. (a) A chemical reaction goes to completion at constant temperature. If you were to plot the rate against time for this reaction, what would you expect the graph to look like?
(b) An "autocatalytic" reaction is one in which one of the products is able to act as a catalyst for the reaction. How would the rate against time graph for an autocatalytic reaction compare with the graph in (a)?

6. Classify the industrial catalysts mentioned in section 17.14 as homogeneous or heterogeneous.

7. D-$C_2H_5(CH_3)CHBr$ can be hydrolysed using NaOH solution.
What shape and configuration will the product possess if the reaction follows (a) first-order, (b) second-order, kinetics?

8. The reaction of NO(g) with H_2(g) is thought to proceed by the following mechanism:

$$2NO \rightarrow (NO)_2; \quad \text{(Fast)}$$
$$(NO)_2 + H_2 \rightarrow \text{Initial products}; \quad \text{(Slow)}$$
$$\text{Initial products} \rightarrow \text{Final products}; \quad \text{(Fast)}$$

What will be

(a) The reaction orders with respect to NO and H_2?
(b) The overall reaction order?
(c) The molecularity of the rate-determining step?

Other questions on reaction rates can be found on pages 211, 229, 381 and 416.

CHAPTER 18

Hydrogen

18.1 Occurrence

Of all the atoms in nature, the hydrogen atom is the simplest. Its lighter isotope consists only of a single proton surrounded by a single electron. Isolated hydrogen atoms, H(g), do not exist under normal conditions at the Earth's surface, though they are present in the Sun and in interstellar space, and can form at high temperatures, for example, in flames. The free element does not occur in the atmosphere except in trace quantities, though large quantities are manufactured for industrial use. The element hydrogen occurs commonly in the following forms:

(a) **Positive ions.** The simplest positive ion is the proton itself, which can exist in a discharge tube but not as a free chemical species in solution. The hydrogen ion in aqueous solution is a proton combined with one or more molecules of water, written for convenience H^+(aq). Crystalline solids containing the ion H_3O^+, the hydroxonium ion, have been prepared, but in an aqueous solution it is difficult to establish the exact composition of H^+(aq), and species such as H_3O^+, $H_5O_2^+$, etc., may all be present in varying concentrations.

The hydrogen ion may exist in other solvents as the solvated proton. For instance, in liquid ammonia we might write H^+(am) for the various species present. The ammonium ion, NH_4^+, is one possible species.

(b) **Negative ions.** Hydrogen occasionally forms negative ions, H^-, and this ion is isoelectronic with He and Li^+, containing a filled 1*s*-orbital. In fact the alkali metals combine with hydrogen gas to form salt-like structures. Sodium hydride, Na^+H^-(s), is in many ways analogous to sodium chloride, Na^+Cl^-(s), and has a simple cubic (6:6) structure. Evidence for the H^- ion is provided by the fact that hydrogen is evolved at the *anode* when a molten alkali metal hydride is electrolysed.

(c) **Covalently bonded atoms.** Hydrogen can also solve its energy level requirements by forming a single covalent bond with other atoms. Most non-metals (but not the noble gases) and some metals will form such bonds.

TABLE 18.1

ENERGIES OF SOME BONDS WITH HYDROGEN ATOMS (kJ mol^{-1})

			H—H 435
C—H 415	N—H 390	O—H 465	F—H 565
Si—H 293	P—H 318	S—H 338	Cl—H 430
	As—H 247	Se—H 276	Br—H 338
		Te—H 238	I—H 297

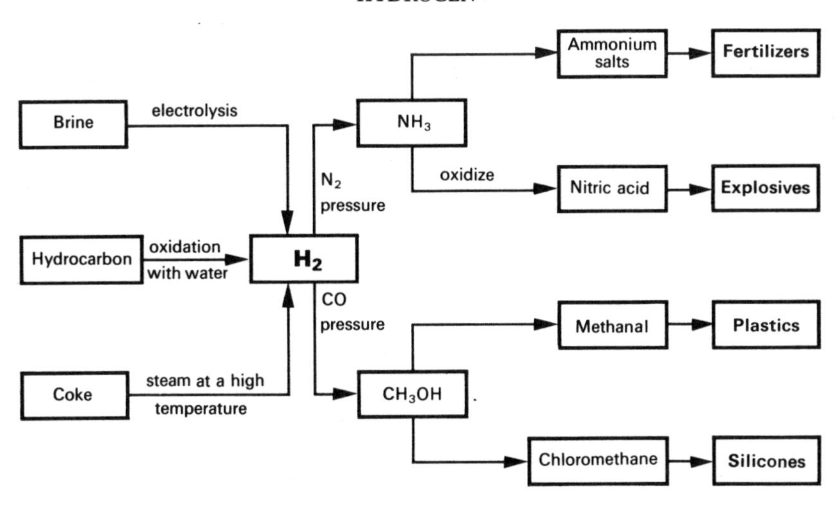

FIG. 18.1.

The simplest bond is that formed in the hydrogen molecule itself, $H_2(g)$, (Chapter 3). Table 18.1 shows a selection of some bonds between elements and hydrogen, together with their bond energies. Metal–hydrogen bonds have low bond energies, and hence hydrides like SnH_4 and SbH_3 are thermally unstable.

18.2 Properties and uses of hydrogen gas

Though a small amount of hydrogen is used for filling balloons, and some pure hydrogen is consumed in the use of the oxy-hydrogen flame for cutting and welding metals, the major bulk of commercially produced hydrogen is used in chemical reactions. Figure 18.1 is a flow-sheet illustrating some of the uses. Impure hydrogen, as a constituent of town gas and water gas, is used extensively as a fuel.

The naturally occurring isotope ^2H, deuterium (often given the symbol D) is used when labelled hydrogen atoms are required. The isotope ^3H, tritium, was first made in 1934 by bombarding deuterium compounds with high energy deuterium nuclei:

$$^2_1D + ^2_1D \rightarrow ^3_1T + ^1_1H; \quad \Delta H = -3{\cdot}8 \times 10^8 \text{ kJ}.$$

As tritium emits β-particles, it is also a useful form of labelled hydrogen. It has been used in studying reaction mechanisms. This nuclear reaction liberates a very large amount of energy, and is an example of **nuclear fusion**. It is a possible source of power in future nuclear reactors using controlled fusion processes though the technical obstacles to obtaining continuous energy in this way are proving enormous. Table 18.2 summarizes some data for hydrogen.

TABLE 18.2

Ionization energy	$H \rightarrow H^+ + e^-$;	$\Delta H = +1310 \text{ kJ mol}^{-1}$
Electron affinity	$H + e^- \rightarrow H^-$;	$\Delta H = -75 \text{ kJ mol}^{-1}$
Bond energy	$H_2 \rightarrow 2H$;	$\Delta H = +435 \text{ kJ mol}^{-1}$
Internuclear distance		0·074 nm
Half-cell potential	$H^+(aq) + e^- \rightleftharpoons \frac{1}{2}H_2(g)$;	$\mathcal{E}^\ominus = 0{\cdot}000 \text{ V}$

18.3 The production of hydrogen gas

From the flow-sheet (Fig. 18.1) it is clear that considerable quantities of hydrogen are required industrially. This hydrogen comes from three main sources:

(i) **Electrolysis.** It is rarely necessary to set up special plant for making hydrogen electrolytically, since it is a by-product of the process for making sodium hydroxide and chlorine from brine, NaCl(aq). Chlorine is liberated at the anode, while at the cathode

$$H^+(aq) + e^- \rightarrow \tfrac{1}{2}H_2(g).$$

Deuterium is manufactured in countries where electrical power is cheap, such as Norway. Its separation from ordinary hydrogen depends on the fact that the rate of liberation of $D_2(g)$ is lower than that of $H_2(g)$. A solution of sodium hydroxide is electrolysed until its relative bulk has been considerably reduced and it will then be enriched in the heavier isotope.

(ii) **Water gas.** This somewhat misleading name refers to the mixture of carbon monoxide and hydrogen produced by blowing steam over white-hot coke:

$$C(s) + H_2O(g) \rightarrow CO(g) + H_2(g); \quad \Delta H = +130 \text{ kJ}.$$

In a process called the *Bosch process*, water gas is mixed with more steam and passed over a catalyst consisting mainly of iron(III) oxide at about 500°C:

$$CO(g) + H_2O(g) \rightarrow CO_2(g) + H_2(g); \quad \Delta H = -42 \text{ kJ}.$$

The carbon dioxide is absorbed by dissolving it in water under pressure. The resultant hydrogen is not all that pure but is pure enough for making ammonia.

(iii) **Petroleum.** When high molecular weight hydrocarbons are *cracked* (broken up catalytically into smaller molecules) some hydrogen is formed. In countries where natural gas is abundant, methane is converted into hydrogen:

$$CH_4(g) + H_2O(g) \xrightarrow{\text{Ni } 900°C} CO(g) + 3H_2(g); \quad \Delta H = +205 \text{ kJ}.$$

Some hydrogen is present in the mixture called *coal gas* evolved when coal is heated in the absence of air. The domestic gas popularly referred to as coal gas, is often derived from petroleum as well as from coal, and contains rather less hydrogen. Its more correct name is *town gas*.

Small laboratories which are not equipped with hydrogen cylinders generally use the following reaction to generate hydrogen, hydrochloric acid being dropped on to impure granulated zinc:

$$Zn(s) + 2H^+(aq) \rightarrow H_2(g) + Zn^{2+}(aq).$$

Hydrogen is produced in a great many reactions involving the reduction of hydrogen ion or water, generally by a metal, for instance:

oxidized (electron removed)

$$Na(s) + H_2O(l) \rightarrow Na^+(aq) + OH^-(aq) + \tfrac{1}{2}H_2(g).$$

reduced (electron added)

Miscellaneous reactions which produce hydrogen are:

(i) the action of alkali on some elements:

$$Al(s) + 3OH^-(aq) + 3H_2O \rightarrow \tfrac{3}{2}H_2(g) + Al(OH)_6^{3-}(aq)$$

(ii) the action of water on saline (salt-like) hydrides:

$$CaH_2(s) + 2H_2O \rightarrow Ca(OH)_2(aq) + H_2(g)$$

18.4 Chemical properties of hydrogen gas

Despite the fact that hydrogen burns, and mixtures with oxygen explode, $H_2(g)$ is not as reactive a gas as one might suppose. Its bond energy is fairly high (435 kJ) which makes it difficult to break. Its readiness to give electrons and form hydrogen ions such as $H^+(aq)$ make it a fairly powerful *reducing agent*. Its reactions may be classified under the following headings:

(a) **Reductions in aqueous solution.** Half-cell potential data predict that hydrogen gas will reduce any couple with a positive $\mathcal{E}^\ominus$ value. Frequently, however, reductions do not occur in the absence of a catalyst, due to the high energy required to break the H—H bond. For instance, $H_2(g)$ fails to reduce $Fe^{3+}(aq)$ to $Fe^{2+}(aq)$ ($\mathcal{E}^\ominus = +0{\cdot}771$ V), and fails to reduce $Ag^+(aq)$ to Ag(s) ($\mathcal{E}^\ominus = +0{\cdot}799$ V).

(b) **Reduction of metal oxides.** The conditions under which it becomes possible for a metal oxide to be reduced by gaseous hydrogen can be determined from studying a free energy diagram like Fig. 13.5. However, a simple aid to memory is that metals below iron (approximately) in the electrochemical series are readily produced by passing hydrogen gas over the heated oxide.

$$CuO(s)+H_2(g) \rightleftharpoons Cu(s)+H_2O(g); \quad \Delta G \text{ negative at } 500°C.$$

In the above case of copper(II) oxide, the equilibrium lies heavily to the right. Iron and steam tend to form an equilibrium with hydrogen and iron(II) diiron(III) oxide:

$$Fe_3O_4(s)+4H_2(g) \rightleftharpoons 3Fe(s)+4H_2O(g); \quad \Delta G \simeq \text{zero at } 500°C.$$

The reaction may be made to proceed in either direction. This behaviour is quite common with equilibria: an equilibrium is readily "driven to completion" by sweeping the products away from the reaction site so that balance is never achieved.

Hydrogen is rarely used for extracting metals on any scale, as it is too expensive compared with, say, coke. Tungsten and germanium are extracted using hydrogen, where quantities are not large and the cost of reductant not an overriding factor.

(c) **Direct combination with elements.** Hydrogen has never, to date, been persuaded to combine with a noble gas, nor does this seem likely from energy considerations. Direct combination has been observed between hydrogen and all other non-metals, however, though in some cases the conditions are rather extreme. For instance at one end of the scale, ethyne is formed in small amounts by blowing hydrogen gas through an arc struck between carbon electrodes. This reaction probably involves the dissociation of hydrogen into atoms which are able to attack the heated carbon.

$$2C(s)+H_2(g) \xrightarrow{\text{high temperature}} C_2H_2(g); \quad \Delta H = +343 \text{ kJ at } 25°C.$$

At the other end of the reactivity scale, hydrogen and fluorine explode spontaneously even in the dark at very low temperatures:

$$H_2(g)+F_2(g) \rightarrow 2HF(g); \quad \Delta H = -535 \text{ kJ at } 25°C.$$

The equilibrium between hydrogen and iodine was considered in Chapter 10.

Metals of Groups IA and IIA combine directly with hydrogen. With the very electropositive metals in Group IA, the formation of the H^- ion is important.

Most transition metals form non-stoichiometric hydrides described as **interstitial** since the hydrogen atoms can fit into the interstices (gaps) in the original metal lattice without much

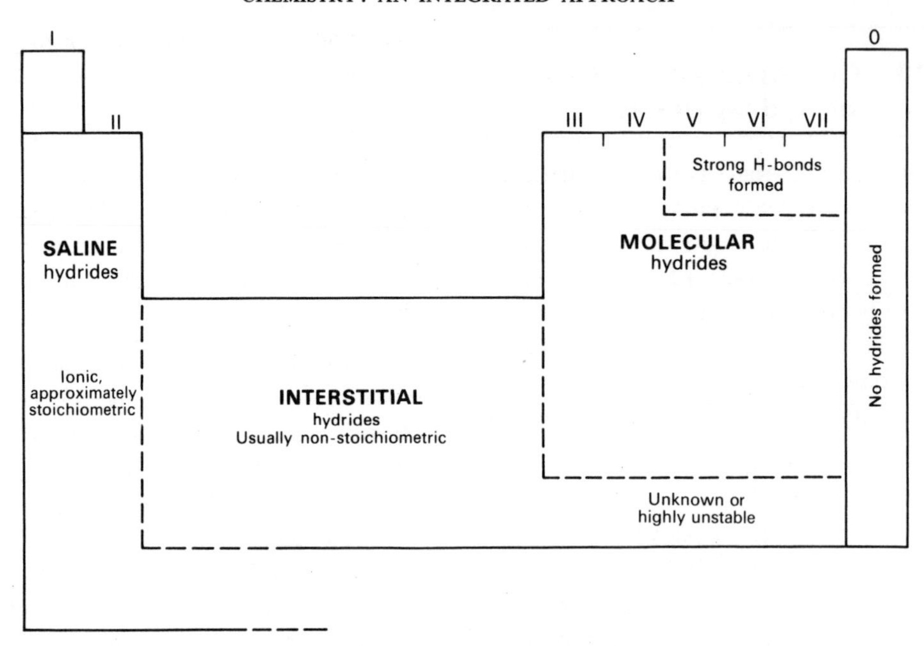

FIG. 18.2. The properties of hydrides.

alteration to the metallic structure. Such hydrides are often more akin to solid solutions than to true chemical compounds, and they are electrical conductors, though their conductivity is usually impaired. A remarkable example is the palladium–hydrogen system, of variable composition PdH_x, where $x = 0$ to 0·75. Palladium powder is used for selectively "mopping up" traces of hydrogen in a gas mixture, and can absorb several hundred times its own volume at room temperature.

The formulae of simple hydrides provide a good illustration of the periodic law (Fig. 1.10), though the heavier hydrides have empirical formulae which do not fit the pattern. Figure 18.2 shows how the properties of hydrides depend upon position in the periodic table.

18.5 Chemical properties of the aqueous hydrogen ion

When we say that an aqueous solution is *acidic*, we are in fact making rather loose use of the term. In the next chapter, where the theory of acids and bases will be developed in more detail, the word *acid* will be given a more specialized theoretical meaning. For practical purposes, however, as long as we consider aqueous solutions, a solution is said to be acidic when the concentration of $H^+(aq)$ is greater than in pure water.

Pure water dissociates slightly into ions by a reaction known as auto-ionization. At room temperature, the product $[H^+] \times [OH^-]$ is about 10^{-14} mol^2 dm^{-6}, and hence the concentration of $H^+(aq)$ is equal to 10^{-7} mol dm^{-3}.

$$H_2O \rightleftharpoons H^+(aq) + OH^-(aq);$$

Concentrations: (mol dm^{-3})	10^{-7}	10^{-7}
Activities:	10^{-7}	10^{-7}

Note that, at a concentration of 10^{-7} mol dm^{-3}, concentration and activity can be taken as numerically equal.

The reactions of dilute aqueous acids, such as hydrochloric acid, HCl(aq), nitric acid, HNO_3(aq) and sulphuric acid, H_2SO_4(aq), are all essentially reactions of the hydrogen ion. Similar reactions are shown by other substances which produce a relatively high concentration of H^+(aq) when added to water. Thus solutions of metal ions low in the electrochemical series show the reaction of acids:

$$Al^{3+}(aq) + H_2O \rightleftharpoons H^+(aq) + \underset{\text{hydroxo-complex}}{Al(OH)^{2+}(aq)}$$

The reactions of H^+(aq) are closely bound up with equilibria, and a proper treatment of them will have to be deferred to the next chapter, but a few of them are summarized here for convenience.

(a) **Indicators.** Indicators are coloured substances which undergo changes of colour depending upon the concentration of H^+(aq) present. Typical reactions of indicators with H^+(aq) are shown in Table 18.3, though it should be borne in mind that the exact concentration of H^+(aq) required to effect the colour change depends upon the indicator chosen (Chapter 19).

TABLE 18.3

Indicator	Colour in acid	Colour in alkali
litmus	red	blue
methyl orange	red	yellow
phenolphthalein	colourless	red

The concentration of H^+(aq) in a solution can be roughly measured by adding **universal indicator**, which is a mixture of several acid-sensitive dyes.

(b) **Metals.** Many metals can reduce hydrogen ions to H_2(g), being themselves oxidized to an aqueous ion. A typical example of this reaction type is

$$Mg(s) + 2H^+(aq) \rightarrow Mg^{2+}(aq) + H_2(g).$$

Metals whose couples with aqueous ions have positive values—those below hydrogen in the electrochemical series—cannot liberate hydrogen from acids.

(c) **Alkalis and bases.** Alkalis are essentially aqueous solutions of OH^- ion, and the neutralization of an alkali like calcium hydroxide, $Ca(OH)_2$, with an acid is essentially

$$H^+(aq) + OH^-(aq) \rightarrow H_2O; \quad \Delta H = -58 \text{ kJ}.$$

Many metal oxides and hydroxides are insoluble in water but nevertheless dissolve in acids with the evolution of heat, by reactions analogous to the above:

$$2H^+(aq) + Cu(OH)_2(s) \rightarrow Cu^{2+}(aq) + 2H_2O(1).$$
$$6H^+(aq) + Fe_2O_3(s) \rightarrow 2Fe^{3+}(aq) + 3H_2O(1).$$

Some metal oxides, for instance chromium(III) oxide, Cr_2O_3, are such strongly-bound structures that they are not attacked by acids appreciably, except with prolonged boiling.

Substances which react with hydrogen ions are known as **bases**, though the term base, like acid, has acquired a fairly special meaning (Chapter 19).

(d) **Catalysis.** Reactions are frequently catalysed by H^+(aq), for instance the hydrolysis of esters.

$$\underset{\text{ethyl ethanoate}}{CH_3COOEt} + H_2O \overset{H^+}{\rightleftharpoons} \underset{\text{ethanoic acid}}{CH_3COOH} + \underset{\text{ethanol}}{EtOH}$$

$$(Et = C_2H_5, \quad \text{ethyl})$$

18.6 Preparation of non-metal hydrides

(a) DIRECT COMBINATION OF ELEMENTS

Although most non-metal hydrides *can* be made by direct combination of the elements this is not always the most convenient way. Indeed some hydrides are of little more than theoretical interest and are rarely prepared. Two very important direct combinations ought to be mentioned:

(i) Ammonia, formed from nitrogen and hydrogen in the Haber process. Details are given in Chapter 23.

(ii) Hydrogen chloride, HCl(g) is manufactured largely from electrolytically produced hydrogen and chlorine which react together by burning. Hydrogen chloride gas is converted into hydrochloric acid by simply dissolving in water.

$$HCl(g) + aq \rightarrow H^+(aq) + Cl^-(aq).$$

(b) HYDROLYSIS OF A SUITABLE "X-IDE"

The hydride of an element X can often be made by hydrolysing suitable "X-ide". For instance, hydrogen sulphide can be made by the hydrolysis, either by water or dilute acid, of a suitably chosen metal sulphide. Similarly phosphine, $PH_3(g)$, can be made by the hydrolysis of a metal phosphide. The method is fairly general, in the sense that suitable reactants can usually be found, e.g.

many metal sulphides + dilute acid → hydrogen sulphide, H_2S.

calcium nitride + water or dilute acid → ammonia, NH_3.

calcium carbide + water or dilute acid → ethyne, C_2H_2, and other hydrocarbons.

magnesium silicide + dilute acid → silane mixture, SiH_4, Si_2H_6, Si_3H_8, etc.

(c) REDUCTION OF HALIDES WITH LITHIUM TETRAHYDROALUMINATE

Lithium tetrahydroaluminate, $LiAlH_4(s)$, is a very vigorous reducing agent used particularly for specific organic reductions. It can be used to prepare fairly pure hydrides, though it would not be used where cheaper, large-scale, methods are available. An example of its use is:

$$SiCl_4(l) + LiAlH_4(s) \rightarrow SiH_4(g) + LiCl + AlCl_3.$$

(d) MISCELLANEOUS REACTIONS

Where small laboratory supplies of certain gaseous hydrogen compounds are required, the common methods do not fall into the above categories. The following is worth mentioning:

$$Ca(OH_2(s) + NH_4Cl(s) \xrightarrow{heat} CaCl_2(s) + H_2O(g) + NH_3(g)$$

Any solid ammonium salt will do here. The gas is alkaline and can be dried over any basic drying agent, e.g. CaO(s), quicklime.

Study Questions

1. Starting with heavy water, D_2O, suggest how the following compounds might be prepared

(a) NaD, (b) ND_3, (c) C_2D_2, (d) DCl, (e) $LiAlD_4$.

2. What would you expect the formulae and properties of the simplest hydrides of the following elements to be?

(a) K, (b) Ni, (c) Ge, (d) Se, (e) Ne.

3. Using Chapter 13 and this chapter, suggest whether molecular hydrogen might be an effective reducing agent for the oxides of

(a) silver, (b) sodium, (c) zinc, (d) nickel.

4. Discuss the relative advantages of the three methods available for the production of hydrogen on an industrial scale (section 18.3).

5. Account for the following observations:

(a) When fused calcium hydride is electrolysed, hydrogen is obtained at the anode.
(b) Calcium hydride reacts with water, giving off a gas.
(c) HCl(l) is a poor conductor of electricity, yet HCl(aq) is an excellent electrical conductor.

6. Like the alkali metals (Chapter 20), hydrogen only has one outer electron. In what ways are the chemistries of hydrogen and the alkali metals (a) similar, (b) different?

7. The element boron forms a number of hydrides: one of these contains 21·8% H and 78·2% B and has a molecular weight of 27·65.

(a) What is the empirical formula of the hydride?
(b) What is the molecular formula of the hydride?
(c) Can you suggest a structure for this compound that is consistent with your theories of bonding?

8. When monosilane, SiH_4, is treated with strong alkali, a gas that burns with a slight explosion is formed. 0·8 g of silane gave 2240 cm^3 of the gas (at s.t.p.).

(a) How many moles of silane gave how many moles of the gas?
(b) What is the gas?
(c) Write an equation for the reaction between SiH_4 and NaOH(aq).

9. Write down equations for the formation of:

(a) H_2S from calcium sulphide, CaS.
(b) NH_3 from magnesium nitride, Mg_3N_2.
(c) Ethyne from calcium dicarbide, CaC_2.
(d) Si_3H_8 from magnesium silicide, Mg_2Si.

10. When hydrogen was passed over a heated red powder, the powder turned into a liquid looking like mercury, which however solidified as it was cooled. 6·85 g of the powder gave 6·21 g of the metal. As the reaction took place, condensation was observed on a cold surface of the reaction tube.

(a) Suggest what the metal was. How many moles of it were formed?
(b) Suggest what the condensate was.
(c) What is the empirical formula of the red powder?
(d) What is the systematic name of the red powder?
(e) What is the trivial name of the red powder?

11. Ammonia is made by passing nitrogen and hydrogen over an iron catalyst. Will the addition of iron speed up, or retard, the decomposition of ammonia into its elements?

CHAPTER 19

Acids

19.1 Acids as solutions of $H^+(aq)$

In the previous chapter an aqueous solution was said to be acidic if its hydrogen ion concentration, $[H^+(aq)]$, was greater than that in pure water, (> about 10^{-7} mol dm^{-3} at room temperature). Such as definition of *acidic* is commonly accepted, but a more general definition is required in order to deal with non-aqueous and heterogeneous systems. The purpose of this chapter is first to examine some of the properties of acids, and second to consider some of the theories of acid–base behaviour.

The familiar dilute acids in the laboratory have one factor in common: they are all the result of adding a substance to water, giving rise to an equilibrium involving hydrogen ions. They vary in equilibrium position however, the so-called strong acids having an equilibrium composition consisting almost entirely (> about 99·9%) of ions. The weak acids, on the other hand, may be less than about 1% ionized yet still give rise to a concentration of $H^+(aq)$ noticeably higher than that of pure water. Table 19.1 gives a selection of familiar acids in aqueous solution.

A substance which dissolves in water to give a high percentage of ions is termed a **strong electrolyte**. A substance which only ionizes to a

TABLE 19.1

Name	Formula of acid	Equation for reaction with water	Approximate value of K_a in mol dm^{-3} at room temperature	Approximate % present as ions in M solution	Classified as
Nitric acid	$HNO_3(l)$	$HNO_3(l)+aq \rightleftharpoons H^+(aq)+NO_3^-(aq)$	large	100	strong
Hydrogen chloride	$HCl(g)$	$HCl(g)+aq \rightleftharpoons H^+(aq)+Cl^-(aq)$	large	100	strong
Sulphuric acid	$H_2SO_4(l)$	$H_2SO_4(l)+aq \rightleftharpoons H^+(aq)+HSO_4^-(aq)$	large	100	fairly strong
Hydrogen fluoride	$HF(g)$ or (l)	$HF(g)+aq \rightleftharpoons H^+(aq)+F^-(aq)$	$6{\cdot}7\times10^{-4}$	2·7	weak
Ethanoic acid	CH_3COOH (l) or (s)	$CH_3COOH(l)+aq \rightleftharpoons H^+(aq)+CH_3COO^-(aq)$	$1{\cdot}8\times10^{-5}$	0·4	weak
Hydrogen sulphide	$H_2S(g)$	$H_2S(g)+aq \rightleftharpoons H^+(aq)+HS^-(aq)$	$1{\cdot}0\times10^{-7}$	0·03	very weak

small extent is termed a **weak electrolyte**. The equilibrium constant here is termed the **dissociation constant**. For acids, it is generally given the symbol K_a.

19.2 Ionic product of water

Water auto-ionizes (Chapter 18) to an extent which varies with temperature. The product $[H^+] \times [OH^-]$ is a constant termed the ionic product of water, and is given the symbol K_w.

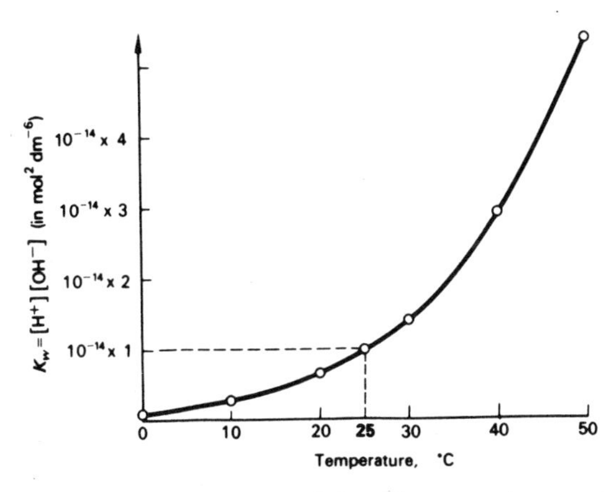

FIG. 19.1. Variation of K_w with temperature.

Figure 19.1 is a plot which shows how K_w varies with temperature. The variation is consistent with le Chatelier's principle, the ionization being endothermic.

$$H_2O(1) \rightleftharpoons H^+(aq) + OH^-(aq); \qquad \Delta H = +58 \text{ kJ}.$$

$[H^+]$ denotes the *activity* of $H^+(aq)$ ions in this expression, but this is numerically equal to the concentration provided the latter is low. At higher concentrations, it is only permissible to treat the two quantities as numerically equal to within an order of magnitude. A few worked examples will make this point clearer.

Worked Example 1. Calculate $[H^+]$ in a solution of M $HF(aq)$, at room temperature.

From the data in Table 19.1,

$$K_a = \frac{[H^+][F^-]}{[HF]} = 6{\cdot}7 \times 10^{-4} \text{ mol dm}^{-3}$$

Inspection of this expression shows the numerator to be far smaller than the denominator, indicating that the hydrogen fluoride is present mainly as molecules rather than ions. We may therefore write $[HF] \simeq 1 \text{ mol dm}^{-3}$.

$$\therefore [H^+][F^-] = 6{\cdot}7 \times 10^{-4} \text{ mol}^2 \text{ dm}^{-6}$$

For every positive ion formed, there will be one negative ion, and therefore

$$\begin{aligned} [H^+] &= [F^-] \\ \therefore [H^+]^2 &= 6{\cdot}7 \times 10^{-4} \\ [H^+] &= \sqrt{6{\cdot}7 \times 10^{-4}} \\ &= 2{\cdot}6 \times 10^{-2} \text{ mol dm}^{-3}. \end{aligned}$$

Notes:

(a) Although hydrogen fluoride is a weak acid, $[H^+]$ has a considerably greater value than in pure water.

(b) When substituting concentration values in the expression for K_a, we have made the assumption that activity = concentration. The calculation is therefore only approximate.

(c) The fact that the calculation is in any case approximate, is some justification for the initial assumption that $[HF] = 1$. Working back, we now see that a value of $[HF] = 1-(2{\cdot}6 \times 10^{-2}) = 0{\cdot}974 \text{ mol dm}^{-3}$ would have been more valid.

Worked Example 2. Calculate $[H^+]$ for a solution of M $H_2SO_4(aq)$, assuming that both H_2SO_4 and HSO_4^- are strong acids.

The above assumptions state that the following equilibria lie mainly to the right:

$$H_2SO_4 + aq \rightleftharpoons H^+(aq) + HSO_4^-(aq)$$
$$HSO_4^-(aq) \rightleftharpoons H^+(aq) + SO_4^{2-}(aq)$$

Hence one mole of H_2SO_4 produces *two* moles of $H^+(aq)$ in the equation. Such an acid is called a **dibasic acid**.

Since the equilibria are almost 100% to the right,

$$[H^+] = 2 \text{ mol dm}^{-3}$$

Note: the **basicity** of an acid is defined as the number of moles of hydrogen ion given by one mole of the acid in the equation.

19.3 Hydrogen ion concentration and activity

It is not easy to obtain a direct measurement of the hydrogen ion *concentration* in a solution, but it is easy to measure its *activity* by means of a cell. If two hydrogen electrodes are coupled together as in Fig. 19.2, the e.m.f. $\mathcal{E}$ is related to the ratio of activities of hydrogen ion, a_1 and a_2 (section 11.7).

$$\mathcal{E} = \frac{RT}{nF}\log_e\frac{a_1}{a_2} = \frac{2{\cdot}303\,RT}{nF}\log_{10}\frac{a_1}{a_2}$$

In conditions where the ionic concentration is very low, we may write $a_1 = c_1$ and $a_2 = c_2$ (where c_1, c_2 are the concentrations of hydrogen ion). Provided the activity or concentration of one dilute solution is known, a value for the other solution may be calculated from e.m.f. measurement.

If the concentration of solution in one cell is kept constant, and the other varied, a graph of e.m.f. against logarithm of concentration can be plotted. This is a straight line, on account of the expression above, provided concentrations are kept low.

19.4 The standard hydrogen electrode and the glass electrode

The standard hydrogen electrode was referred to in Chapter 11. It is the reference electrode to which all quoted values for $\mathcal{E}^\ominus$ are related. Measurements of e.m.f. show that for a solution of dilute hydrochloric acid to have a hydrogen ion activity of 1, the concentration of $[H^+] = 1{\cdot}18$ mol dm^{-3}. This is a good illustration of how widely numerical values of concentrations and activities can differ. The divergence here is 18 per cent.

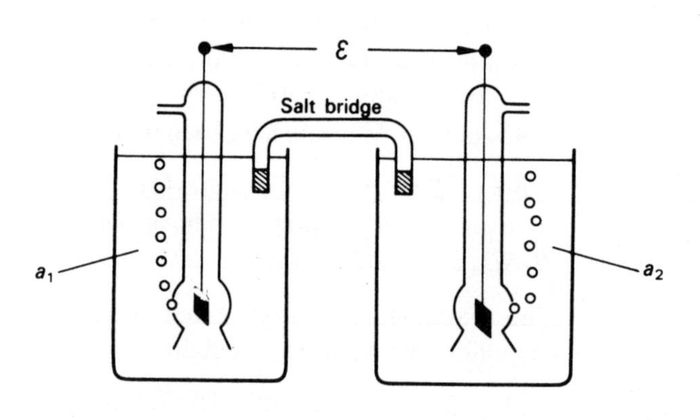

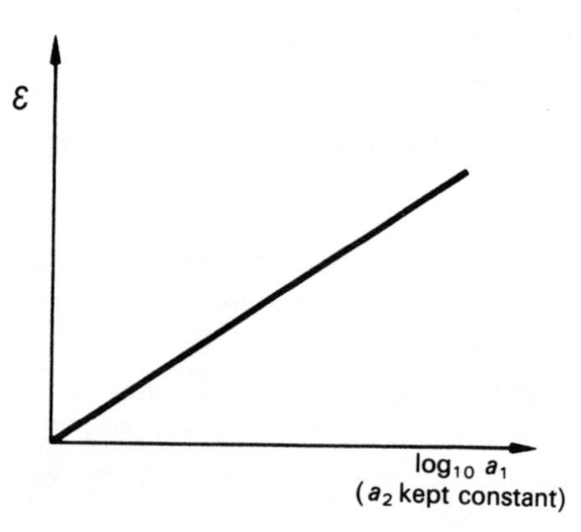

FIG. 19.2.

Hydrogen electrodes are inconvenient to operate for routine measurements of $[H^+]$, and a much simpler alternative is the **glass electrode** (Fig. 19.3).

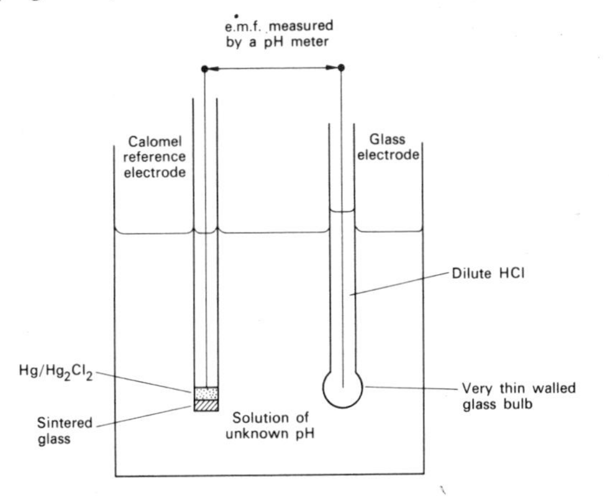

FIG. 19.3. pH measurement with a glass electrode.

Glass has the property of being slightly permeable to hydrogen ions, and the potential difference between the inner and outer surfaces depends upon the relative concentrations of hydrogen ion in contact with these surfaces. It is found experimentally that if a glass electrode is coupled up with a reference electrode unaffected by hydrogen ions, a cell is formed whose e.m.f. is directly proportional to $\log_{10}[H^+]$, just as with an ordinary hydrogen electrode. This is very convenient, for it means that a glass electrode can be used in place of a hydrogen electrode for routine measurements of $[H^+]$. It does however need calibrating before use, and this is described in a later section.

19.5 The pH scale

Since e.m.f. measurements are related to $\log_{10}[H^+]$, and not directly to $[H^+]$ itself, hydrogen ion concentrations are customarily expressed as pH values, where

$$pH = -\log_{10}[H^+].$$

The term pH was coined by Sørensen in 1909, and the notation originally stood for "potential of hydrogen". The "p" notation can be applied to other quantities and always means "minus $\log_{10}$ of". For instance,

$$pOH = -\log_{10}[OH^-]$$
$$pK_w = -\log_{10}K_w, \quad \text{and so on.}$$

Worked Example 3. Express the following values of $[H^+]$ as pH.

(a) $[H^+] = 1 \text{ mol dm}^{-3}$
$pH = -\log_{10}1 = 0$

(b) $[H^+] = 10^{-14} \text{ mol dm}^{-3}$
$pH = -\log_{10}10^{-14} = -(-14)$
$= +14$

(c) $[H^+] = 2 \text{ mol dm}^{-3}$
$pH = -\log_{10}2 = -0{\cdot}3010$

(d) $[H^+] = 2\times 10^{-5} \text{ mol dm}^{-3}$
$pH = -(\bar{5}{\cdot}3010)$
$= +5-0{\cdot}3010$
$= +4{\cdot}699.$

Worked Example 4. Express the following in "p" notation.

(a) $[OH^-] = 10^{-3} \text{ mol dm}^{-3}$
$pOH = -\log_{10}[OH^-] = +3.$

(b) $K_w = 10^{-14}$ at room temperature.
$pK_w = -\log_{10}(10^{-14}) = +14.$

(c) K_a for acetic acid $= 2\times 10^{-5} \text{ mol dm}^{-3}$ at room temperature.
$pK_a = 4{\cdot}699$ (cf. Example 3(d) above).

Worked Example 5. Calculate the pH of the following solutions at room temperature.

(a) $1{\cdot}0$ M $H_2SO_4(aq)$.
From Worked Example 2,
$[H^+] = 2 \text{ mol dm}^{-3}$.
From Worked Example 3(c),
$pH = -0{\cdot}301.$

(b) Pure water.
$K_w = [H^+][OH^-] = 10^{-14}$, and
$[H^+] = [OH^-]$.
$\therefore [H^+] = \sqrt{K_w} = 10^{-7}$
$\therefore$ pH = +7.

(c) 1·0 M NaOH(aq).
Sodium hydroxide is a strong electrolyte,
$\therefore [OH^-] = 1$ mol dm^{-3}
Using the "p" notation,
$pK_w = pH + pOH$
Now, pOH = 0 in this example, and
$pK_w = 14$.
$\therefore$ pH = 14. Note that even this solution, which is strongly alkaline, has still some H^+(aq) ions present, although $[H^+] = 10^{-14}$ mol dm^{-3}.

(d) 0·01 M $Ca(OH)_2$(aq).
Calcium hydroxide is a strong electrolyte, and from its formula, will give two moles of OH^-(aq) per mole of $Ca(OH)_2$.
$[OH^-] = 0{\cdot}02$ mol dm^{-3}
$pOH = -(\bar{2}{\cdot}301) = +1{\cdot}699$
$pH = pK_w - pOH = 14 - 1{\cdot}699$
$= 12{\cdot}301$.

(e) 1·0 M CH_3COOH(aq).
($K_a = 2 \times 10^{-5}$ mol dm^{-3} approximately, at room temperature).

$$K_a = \frac{[H^+][CH_3COO^-]}{[CH_3COOH]}$$
$$= 2 \times 10^{-5} \text{ mol dm}^{-3}$$

$[H^+]^2 = 2 \times 10^{-5}$ mol^2 dm^{-6} (cf. the same argument in Worked Example 1).
$\therefore pH = \frac{1}{2}pK_a = \frac{1}{2}(4{\cdot}699) = 2{\cdot}35$.

19.6 Buffer solutions

A **buffer solution** is one whose pH is relatively insensitive to the addition of small quantities of acid or alkali. Solutions vary widely in their pH sensitivity. Pure water, and pure NaCl(aq), are very sensitive: a litre of pure water would show a marked change (±2 pH units) on adding only 0·01 cm^3 of 1·0 M H^+(aq) or OH^-(aq). In some chemical reactions it is necessary to stabilize the pH, and this can be done by adding substances which can absorb variations in $[H^+]$.

A typical acidic buffer solution is a mixture of a weak acid with a solution of its anion, for example ethanoic acid plus sodium ethanoate. The concentrations of $[CH_3COOH]$ and $[CH_3COO^-]$ are relatively high compared to $[H^+]$, the three concentrations being related by the expression for the dissociation constant of the acid:

$$K_a = \frac{\overbrace{[CH_3COO^-]}^{\text{relatively large}}\overbrace{[H^+]}^{\text{small}}}{\underbrace{[CH_3COOH]}_{\text{relatively large}}} = 2 \times 10^{-5} \text{ mol dm}^{-3}$$
approximately, at room temperature.

Small additions of H^+(aq) are immediately taken up by a shift of the equilibrium

$$CH_3COOH(aq) \rightleftharpoons CH_3COO^-(aq) + H^+(aq)$$

to the left, and small additions of OH^-(aq) by a shift to the right. As long as $[CH_3COOH]$ and $[CH_3COO^-]$ are relatively large, their magnitude will not be much affected by the added acid or alkali, and hence when equilibrium is re-established, pH will be almost the same as it was initially.

Worked Example 6. Calculate the pH of the following buffer solutions.

(a) A mixture which is 1·0 M with respect to both CH_3COOH(aq) and CH_3COONa(aq).
Sodium ethanoate is a strong electrolyte, and therefore $[CH_3COO^-] = 1$. Ethanoic acid is a weak electrolyte, and therefore $[CH_3COOH] = 1$. Substituting,

$$\frac{1 \times [H^+]}{1} = K_a$$

$$\therefore \text{pH} = \text{p}K_a$$
$$= 4{\cdot}699 \text{ (cf. Worked Example 4).}$$

(b) A mixture which is 0·1 M with respect to $CH_3COOH(aq)$ and molar with respect to CH_3COONa.

$$\frac{1 \times [H^+]}{0{\cdot}1} = K_a$$

$$\therefore \text{pH} = \text{p}K_a + 1 = 5{\cdot}699.$$

A buffer which has an alkaline pH can be made up by mixing a weak alkali such as ammonia with one of its salts, such as ammonium chloride (section 19.8).

The pH of any buffer solution can be calculated quickly using a modified expression for the acid dissociation constant.

$$\text{Since for any acid, } K_a = \frac{[H^+][A^-]}{[HA]}$$

$$\text{then } \log K_a = \log\,[H^+] + \log\frac{[A^-]}{[HA]}$$

If an acid is weak, then [HA] will be approximately the same as the acid concentration, assuming that only a small percentage of the molecules dissociate, and $[A^-]$ will be approximately the same as the salt concentration, since the salt will be completely dissociated.

$$\text{Hence pH} = \text{p}K_a + \log_{10}\frac{[\text{salt}]}{[\text{acid}]}.$$

The pH of a solution containing equal concentrations of a weak acid and one of its salts will be the same as the pK_a value of the acid.

19.7 Methods of measuring pH

(a) By pH meter

The most precise way of measuring pH is by means of a glass electrode. This is set up with a reference electrode whose half-cell potential is not affected by pH. One of the features of a glass electrode is its very high electrical resistance ($\sim 10^8\ \Omega$), rendering measurements with a conventional voltmeter useless. A d.c. amplifier is needed, embodying also a potentiometer which enables the e.m.f. to be checked against that of a standard cell (usually a mercury cell).

The e.m.f. of the cell is amplified and fed to the voltmeter. Voltage is related linearly to pH. Simple meters, accurate to about $\pm 0{\cdot}1$ pH unit, have a direct readout of pH on the scale, while other systems use a null method which can give an accuracy of three places of decimals.

The procedure for determining pH with a pH meter is as follows:

(i) Place the pH electrodes in a standard buffer solution, and standardize its e.m.f. against the standard cell provided. The exact procedure varies from instrument to instrument, but the purpose of this operation is to allow for day to day variations in the characteristics of the glass electrode. The meter is set so that the pH it reads corresponds to that of the standard buffer.

(ii) The meter is now ready for use. The electrodes can now be placed in a solution of unknown pH.

(b) Indicator method

This method is generally only accurate to about $\pm 0{\cdot}5$ pH unit. With special precautions, about $\pm 0{\cdot}1$ pH unit can be achieved. The simplest application is the use of **universal indicator** which is a mixture of indicators of varying pK_a values and different colours. A universal indicator shows a different colour for each pH value over a wide range.

The action of indicators is best understood by taking a familiar example, methyl orange, which is yellow in alkaline solution but turns red when $H^+(aq)$ is added*, due to the equilibrium

$$\underset{\text{red}}{HX(aq)} \rightleftharpoons H^+(aq) + \underset{\text{yellow}}{X^-(aq)};$$

$$K_a = \frac{[H^+][X^-]}{[HX]} = 2\times10^{-4}\,\text{mol dm}^{-3}\text{ approx.}$$

Substituting in the expression for K, we see that, to obtain a neutral colour, where $[HX] \simeq [X^-]$, it is necessary for $pH = pK_a$ (cf. Worked Example 6(a)). Methyl orange shows its half-way tint at about pH = 3·7.

If the pH is altered by one unit in the acidic direction, we now have $[HX] = 10\times[X^-]$ if the expression for K_a is to be satisfied (cf. Worked Example 6(b)). Greater change in pH than this will not result in much visible change as the eye cannot detect changes of one coloured species in the presence of a large excess of the other. A similar argument applies to pH change in the alkaline direction. At a pH of 4·7, $[X^-] = 10\times[HX]$ and the solution will be almost completely yellow. The pH range of methyl orange is therefore approximately from 2·7 to 4·7.

TABLE 19.2

Indicator	Colour		pK_a	pH range approximately
	acid	alkali		
Methyl orange	red	yellow	3·7	2·7–4·7
Methyl red	red	yellow	5·1	4·1–6·1
Bromothymol blue	yellow	blue	7·0	6·0–8·0
Phenol-phthalein	colourless	red	9·4	8·4–10·4

Table 19.2 shows the data for some commonly used indicators.

* In the laboratory it is more usual to use screened methyl orange, which is a mixture of two dyes, methyl orange and xylene cyanol FF. This mixture is designed to give a neutral grey at the point where $pH = pK_a$.

19.8 Alkalis and bases

In just the same way as a solution with pH < 7 is defined loosely as acidic, a solution with pH > 7 is said to be **alkaline**. Such a solution will have a concentration of $OH^-(aq)$ greater than the value for pure water at room temperature, namely 10^{-7} mol dm^{-3}. That is, pOH < 7. (Remember that $pH + pOH = pK_w = 14$.)

Hydroxides of the *s*-block metals, Group IA and IIA, are soluble in water forming metal ions and hydroxide ions (with the exception of beryllium hydroxide which is almost insoluble), and hence their solutions are alkaline.

Group IA: $MOH(s) \rightleftharpoons M^+(aq) + OH^-(aq)$
(M = Li, Na, K, Rb, Cs or Fr)

Group IIA: $M(OH)_2(s) \rightleftharpoons M^{2+}(aq) + 2OH^-(aq)$
(M = Mg, Ca, Sr, Ba or Ra)

In fact magnesium hydroxide is only sparingly soluble, with a solubility product, K_s (section 10.12), of only 2×10^{-11} mol^3 dm^{-9} at 25°C. This means that a saturated solution of magnesium hydroxide has a relatively low pH at room temperature, and it may be classed as a weak alkali.

If an alkali is added to an aqueous acid, the reaction which takes place is termed **neutralization**. It is essentially

$$H^+(aq) + OH^-(aq) \rightarrow H_2O(l)$$

This reaction is commonly used to produce salts, for when it occurs the *spectator ions* remain in solution and can be crystallized out. For example:

(i) $HCl(aq) + NaOH(aq)$
$\equiv \underbrace{H^+(aq) + OH^-(aq)}$
$\downarrow$
H_2O

$+ \underbrace{Na^+(aq) + Cl^-(aq)}$
spectator ions
$\downarrow$
$NaCl(s)$ on crystallization

(ii) $2HNO_3(aq) + Ba(OH)_2(aq)$
$\equiv \underbrace{2H^+(aq) + 2OH^-(aq)}$
$\downarrow$
$2H_2O$

$+ \underbrace{Ba^{2+}(aq) + 2NO_3^-(aq)}$
$\downarrow$
$Ba(NO_3)_2(s)$
on crystallization

Many metal hydroxides and oxides are insoluble in water, but can nevertheless neutralize acids in a similar way. Such substances are termed **bases** (basic hydroxides and basic oxides). Alkalis are therefore soluble bases. An example of a neutralization with an insoluble base would be

(iii) $H_2SO_4(aq) + Cu(OH)_2(s) \rightarrow$
$2H_2O(l) + \underbrace{Cu^{2+}(aq) + SO_4^{2-}(aq)}$
$\downarrow$
$CuSO_4.5H_2O(s)$
on crystallization

The only true spectator ion here is SO_4^{2-}—it is the only one which does not actually participate in the reaction—though it is conveniently written into the equation. Any aqueous acid will react with a base producing the ions of the metal. The above can be written

$$2H^+(aq) + Cu(OH)_2(s) \rightarrow 2H_2O(l) + Cu^{2+}(aq)$$

Worked Example 7. Calculate the pH of the following aqueous solutions of bases:

(a) M NaOH(aq)

$$[OH^-] = 1 \text{ mol dm}^{-3}$$
$$\therefore pOH = 0$$
$$\therefore pH = 14 - 0 = 14$$

(b) 0·1 M $Sr(OH)_2(aq)$.
One mole of $Sr(OH)_2$ gives two moles of $OH^-(aq)$, therefore

$$[OH^-] = 2 \times 10^{-1} \text{ mol dm}^{-3}$$
$$\therefore \log_{10}[OH^-] = \bar{1}{\cdot}301$$
$$\therefore pOH = 1 - 0{\cdot}301 = 0{\cdot}699$$
$$\therefore pH = 14 - 0{\cdot}699 = 13{\cdot}301.$$

Worked Example 8. Calculate the pH of the saturated solution formed by adding MgO(s) to water at 25°C.

When the hydroxide is appreciably soluble in water, the oxide will react with water thus:

$$MgO(s) + H_2O(l) \rightarrow \underset{x \text{ mol dm}^{-3}}{Mg^{2+}(aq)} + \underset{2x \text{ mol dm}^{-3}}{2OH^-(aq)}$$

From the data given,

$$K_s = [Mg^{2+}][OH^-]^2 = x.(2x)^2 = 4x^3$$
$$= 2 \times 10^{-11} \text{ mol}^3 \text{ dm}^{-9}$$

$$\text{Hence } x = 1{\cdot}7 \times 10^{-4} \text{ mol dm}^{-3}$$
$$\text{Hence pOH} = -\log_{10}[OH^-]$$
$$= -\log_{10} 2 \times 1{\cdot}7 \times 10^{-4} = 3{\cdot}77$$
$$\text{Hence pH} = 14 - \text{pOH}$$
$$= 10{\cdot}23.$$

Many other substances react with water to give solutions which are alkaline, for instance ammonia.

$$\underset{\text{ammonia}}{NH_3(aq)} + H_2O(l) \rightleftharpoons \underset{\text{ammonium ions}}{NH_4^+(aq)} + OH^-(aq)$$

"Household ammonia" is in fact an aqueous solution of ammonia, in which the above species are present at equilibrium. *Ammonium hydroxide* is the name often given to this solution, but strictly this name should only apply to the species on the right-hand side of the equation.

Worked Example 9. Calculate the pH of a 0.1 M solution of $NH_3(aq)$, given that the dissociation constant, K_a, for NH_4^+ is 5×10^{-10} mol dm^{-3} at room temperature.

$$K_a = \frac{[NH_3][H^+]}{[NH_4^+]} = 5\times10^{-10}.$$

Since the reaction of NH_3 with water produces equal quantities of NH_4^+ and OH^-, then

$$[NH_4^+] = [OH^-] = \frac{10^{-14}}{[H^+]}.$$

The expression may be rewritten

$$K_a = \frac{0{\cdot}1\times[H^+]^2}{10^{-14}} = 5\times10^{-10}$$

$$\begin{aligned}\text{Taking logs; } 1+2\text{pH}-14 &= 9{\cdot}3\\ 2\text{pH} &= 22{\cdot}3\\ \text{pH} &= 11{\cdot}15\end{aligned}$$

A 0·1 M solution of $NH_3(aq)$ is therefore less alkaline than a 0·1 M solution of NaOH(aq) (Worked Example 7).

Worked Example 10. Calculate the pH of a solution that is 0·1 M with respect to NH_3, and 0·1 M with respect to NH_4Cl.

$$K_a = \frac{[NH_3][H^+]}{[NH_4^+]} = \frac{0{\cdot}1\times[H^+]}{0{\cdot}1}$$

Therefore pH $= pK_a = 9{\cdot}3$ (see Worked Example 9).

This is an example of an alkaline buffer solution.

19.9 Applications of pH measurement

(a) **Acid–base titrations.** The end-point of an acid–base titration occurs when there is an abrupt change of pH on adding solution from the burette. An indicator registers the colour change. It is essential to choose an indicator with the correct pK value, except when titrating a strong acid with a strong alkali, when the pH swing is large. The theory of indicators for titrations is dealt with further in section 19.14.

(b) **Soil testing.** Most plants will only grow in soils with pH values between about 6 and 8. Within these limits, however, pH can have a marked effect on the type of plant which is favoured. Soil testing is most conveniently carried out using universal indicator paper or solution, with a colour range specially chosen for the pH values encountered. A portable pH meter, incorporating a glass electrode, is sometimes used, and the soil is shaken with distilled water before testing.

(c) **Reactions in solution.** For some aqueous reactions the pH must be adjusted to within certain limits, such as in the preparation of some complex salts. Universal indicator paper is usually quite accurate enough for this.

(d) **Measurement of dissociation constants.** If the pH of a solution of a weak acid or base is accurately measured with a pH meter, the dissociation constant can be calculated.

Worked Example 11. A 0·1 M solution of methanoic acid, HCOOH(aq), is found to have a pH of 2·4 at 25°C. Calculate its dissociation constant at that temperature.

$$K_a = \frac{[H^+][HCOO^-]}{[HCOOH]} = \frac{[H^+]^2}{[HCOOH]}$$

$$\begin{aligned}\therefore\ pK_a &= 2\text{pH}-\text{p}[HCOOH]\\ \therefore\ pK_a &= 4{\cdot}8-1 = 3{\cdot}8\\ \therefore\ K_a &= 1{\cdot}6\times10^{-4}\ \text{mol dm}^{-3}\end{aligned}$$

This method is not practicable for measuring the dissociation constant of a strong acid, beyond establishing that the value is large. For instance it makes very little difference to the pH of a solution of an acid HX(aq) whether $K_a = 10^{10}$ or 10^{20}. In both cases, for practical purposes the acid is fully (>99·99%) ionized in solution.

TABLE 19.3

	Group III	Group IV	Group V	Group VI	Group VII
First short period	boric(III) acid $HO{-}B(OH)_2$ planar	carbonic acid $O{=}C(OH)_2$ planar	nitric acid $O{=}N(\rightarrow O)OH$ planar	—	—
Second short period	—	silicic acid SiO_2, xH_2O non-stoichio-metric, complex structure	phosphoric acid $O{=}P(OH)_3$ tetrahedral	sulphuric acid $(O)_2S(OH)_2$ tetrahedral	chloric(VII) acid $(O)_3Cl(OH)$ tetrahedral

19.10 Oxo-acids and oxo-anions

Most of the non-metals form **oxo-acids**. With some elements, such as sulphur and phosphorus, a large number of different oxo-acids can form with various structures and oxidation numbers, but in other cases fewer species are known. The well-known mineral acids HNO_3, nitric acid and H_2SO_4, sulphuric acid, are oxo-acids of nitrogen and sulphur respectively.

Table 19.3 shows oxo-acids in which the oxidation number of the central non-metallic element is equal to the number of the group in the periodic table to which that element belongs.

When an element forms oxo-acids of different oxidation states, the oxidation state must be included in the name. Table 19.4 illustrates this for the oxo-acids of chlorine. Certain well-known acids are exceptions to this rule, see Appendix 5.

The strengths of oxo-acids appear to be governed by an empirical rule. If the formula is written out structurally, as $(HO)_nXO_m$, then

$m = 0$: weak acid, pK_a generally about 10;

$m = 1$: fairly weak acid, pK_a generally between 0 and 5;

$m = 2$: moderately strong acid, pK_a generally between -3 and 0;

$m = 3$: a strong acid, practically fully ionized in aqueous solution.

These predictions fit the data for the oxo-acids of chlorine—chloric(I) acid ($m = 0$) is a weak acid, while chloric(VII) acid ($m = 3$) is one of the strongest acids known. Sulphuric acid ($m = 2$) is only moderately strong: although almost fully ionized in aqueous solution, it behaves as a weak acid when dissolved in acetic acid.

TABLE 19.4 For common names; see also Table 11.2

Formula of acid	Systematic name of acid	Formula of anion	Systematic name of anion
$HClO$	Chloric(I)	ClO^-	Chlorate(I)
$HClO_2$	Chloric(III)	ClO_2^-	Chlorate(III)
$HClO_3$	Chloric(V)	ClO_3^-	Chlorate(V)
$HClO_4$	Chloric(VII)	ClO_4^-	Chlorate(VII)

TABLE 19.5

Name of acid	Formula of anion	Name of anion
Manganic(VII) acid (stable in solution only)	MnO_4^-	Manganate(VII)
Chromic(VI) acid	$Cr_2O_7^{2-}$	Dichromate(VI)
	CrO_4^{2-}	Chromate(VI)

Oxo-acids and oxo-ions are also formed by some metals, notably the higher oxidation states of the transition metals. Table 19.5 gives some instances of this.

19.11 Non-aqueous systems

Acid-base reactions are readily carried out in solvents other than water, and the results are often interesting. For instance, ethanoic acid, CH_3COOH (m.p. 16·6°C) is a solvent which auto-ionizes like water, giving solvated protons and solvated anions. Compare the two processes:

$$CH_3COOH(l) \rightleftharpoons H^+(ac) + CH_3COO^-(ac)$$
$$H_2O(l) \rightleftharpoons H^+(aq) + OH^-(aq)$$

Evidence for this is provided by the fact that pure ethanoic acid, like water, has a slight electrical conductivity. Chloric(VII) acid, $HClO_4$, is a strong electrolyte when dissolved in ethanoic acid, and so also is sodium ethanoate.*

$$HClO_4(l) + ac \rightarrow H^+(ac) + ClO_4^-(ac)$$
$$CH_3COONa(s) + ac \rightarrow Na^+(ac) + CH_3COO^-(ac)$$

These solutions can be titrated in a manner exactly analogous to an aqueous titration. Methyl violet is a suitable indicator.

* The state symbol (ac) here denotes ions *solvated* with ethanoic acid, in the same way as (aq) denotes ions *hydrated* with water molecules.

$$H^+(ac) + CH_3COO^-(ac) \rightarrow CH_3COOH(l)$$
(spectator ions, $Na^+(ac)$, $ClO_4^-(ac)$)

In general, acids ionize less strongly in anhydrous ethanoic acid than they would in water. For instance, sulphuric acid can be classed as a strong acid in aqueous solution, but is relatively weak in ethanoic acid.

Another solvent which can be used in place of water is liquid ammonia, b.p. −33°C. Liquid ammonia auto-ionizes thus:

$$2NH_3 \rightleftharpoons NH_4^+ + NH_2^-.$$

An ammonium salt is a typical acid in liquid ammonia, as the ammonium ion is effectively a solvated proton, $H^+ + NH_3$. A typical base would be a metal *amide*, for instance sodium amide, $NaNH_2(s)$.

Non-aqueous reactions are often carried out when the reactants are attacked by water, or fail to dissolve in it, and non-aqueous titrations are often used industrially.

19.12 The Brønsted–Lowry definition of acid-base behaviour

The fact that reactions analogous to aqueous acid–base neutralization can also take place in non-aqueous media such as ethanoic acid or liquid ammonia, means that a more general definition of acid and base is required. Brønsted in 1923 defined an acid as **any substance with a tendency to release a proton**. Conversely a species which tends to accept a proton is defined as a base. When an acid releases a proton, the **conjugate base** is said to be formed, for instance:

$$\underset{\text{acid}}{CH_3COOH} \rightleftharpoons H^+ + \underset{\text{conjugate base}}{CH_3COO^-}$$

An acid with a strong tendency to release protons will have a weak conjugate base. Conversely if a base has a strong tendency to accept

TABLE 19.6

	Acid	⇌ Conjugate base	+ proton	
decreasing strength of acid ↓	$HClO_4$	⇌ ClO_4^-	$+H^+$	increasing strength of base ↑
	HNO_3	⇌ NO_3^-	$+H^+$	
	H_2SO_4	⇌ HSO_4^-	$+H^+$	
	H_3O^+*	⇌ H_2O	$+H^+$	
	CH_3COOH	⇌ CH_3COO^-	$+H^+$	
	NH_4^+	⇌ NH_3	$+H^+$	
	H_2O	⇌ OH^-	$+H^+$	
	NH_3	⇌ NH_2^-	$+H^+$	

* It should be noted that we are writing H_3O^+ here where previously we have been writing $H^+(aq)$. $H^+(aq)$ is more correct, but H_3O^+ is more convenient in balancing equations when applying the Brønsted–Lowry definition. The student should be familiar with both methods.

protons, its conjugate acid must be weak. Thus, on the Brønsted definition ethanoate ions are relatively strongly basic since ethanoic acid itself is not a very strong acid.

It is possible to list acids in order of decreasing strength. Table 19.6 shows a list of half-equations arranged empirically in such an order. These are *not* complete chemical reactions but are analogous to electronic half-equations used to describe redox processes. The equation for an acid–base reaction is derived by combining two half-equations.

The position of equilibrium is determined by the relative strengths of the species involved. For instance chloric(VII) acid has a stronger tendency than H_3O^+ ($= H^+(aq)$) to give up a proton, and H_2O has a stronger tendency to accept protons than ClO_4^-. Consequently the following equilibrium lies to the right:

$$\underset{\text{strong acid}}{HClO_4} + \underset{\text{base}}{H_2O} \rightleftharpoons \underset{\text{weaker acid than } ClO_4^-}{H_3O^+} + \underset{\text{weaker base than } H_2O}{ClO_4^-}$$

This equation is nothing more than a slightly elaborate form of the equation usually written for the ionization of a strong acid when added to water:

$$HClO_4 + aq \rightleftharpoons H^+(aq) + ClO_4^-(aq)$$

The Brønsted theory accounts for what is known as the **levelling** action of water. The species $HClO_4$, HNO_3 and H_2SO_4, in Table 19.6 are all stronger acids than H_3O^+, and hence all react almost completely to give that ion. The same is true of many other acids, for instance the hydrogen halides (except HF). It makes little difference whether the equilibrium constant is 10^{10} or 10^{20}, since almost complete formation of H_3O^+ will be the result in every case (see section 19.1).

Water exerts a similar levelling action at the basic end of the scale. For instance if sodium amide is added to water, ammonia is produced:

$$NaNH_2(s) + H_2O(l) \rightarrow NH_3(aq) + NaOH(aq)$$

Brønsted theory interprets this by saying that NH_2^- is a stronger base than OH^-, and that H_2O is a stronger acid than NH_3. The equation may be rewritten ionically thus:

$$\underset{\text{base}_1}{NH_2^-} + \underset{\text{acid}_2}{H_2O} \rightarrow \underset{\text{base}_2}{NH_3} + \underset{\text{acid}_1}{OH^-};$$

(spectator ions, Na^+)

Note that species like H_2O and NH_3 can act as both acids and bases, since they can both donate and accept protons.

19.13 Hydrolysis of ions as an instance of acid-base behaviour

The Brønsted–Lowry theory is very successful in rationalizing many of the experimental effects observed with pH of aqueous solutions. One such effect is that commonly referred to as **hydrolysis** of salts. Hydrolysis literally means reaction with water, and two main types of effect are observed:

(i) Salts of weak acids dissolve in water to give solutions with alkaline pH. For instance sodium sulphide is quite strongly alkaline, and the solution smells noticeably of hydrogen sulphide:

$$\underset{\text{base}_2}{S^{2-}} + \underset{\text{acid}_2}{2H_2O} \rightleftharpoons \underset{\text{base}_1}{2OH^-} + \underset{\text{acid}_1}{H_2S};$$

(spectator ions, Na^+).

Since H_2S is a weak acid, the sulphide ion is a relatively strong base, which can react with water by removing protons from it.

(ii) Solutions of ions of the weaker metals dissolve in water to give acidic solutions. This again is interpreted by postulating that the aqueous metal ions are themselves weak acids. It may seem strange at first that a metal ion can donate protons, but in fact these ions have a definite number of water molecules more or less firmly attached, and are in effect complex ions. For instance, $Al^{3+}(aq)$ consists mainly of the species $Al(H_2O)_6^{3+}$.

$$\underset{\text{weak acid}}{Al(H_2O)_6^{3+}} \rightleftharpoons \underset{\text{conjugate base}}{[Al(H_2O)_5(OH)]^{2+}} + \underset{\text{pH} < 7}{H^+(aq)}$$

Hydrolysis of salts can be summed up using AB to represent a typical salt which gives two ions, A^+ and B^-, in solution. Two equilibria exist between the ions and the water:

$$A^+ + H_2O \rightleftharpoons AOH + H^+$$

and $$B^- + H_2O \rightleftharpoons BH^+ + OH^-$$

The two equilibrium constants will determine the extent of hydrolysis and hence the pH of the solution.

19.14 Choice of indicators for titrations

Hydrolysis of ions has a profound effect on the suitability of an indicator for a titration. The purpose of a titration is to carry out accurately a desired chemical reaction, and the indicator must be capable of detecting accurately the pH which exists at the endpoint. For instance if ethanoic acid is to be titrated with sodium hydroxide, the end-point will be reached when the pH is that of a solution of sodium ethanoate. Since ethanoate ions are hydrolysed, the end-point will occur at an alkaline pH. Phenolphthalein (pK = 9·6) is a suitable indicator for this titration. Conversely if a weak base is titrated with a strong acid the end-point will be acidic and methyl orange or methyl red will probably be chosen.

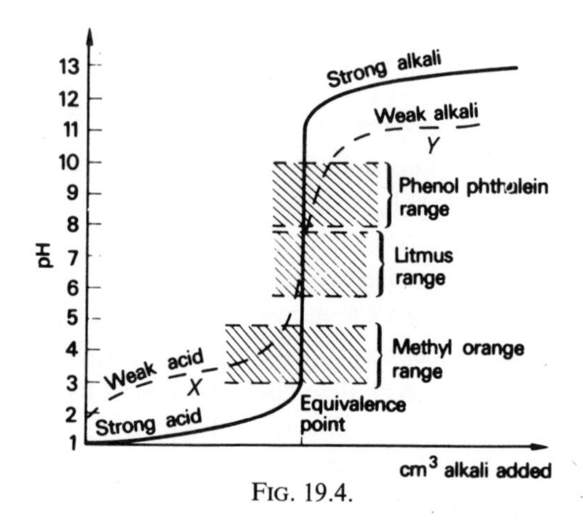

FIG. 19.4.

Figure 19.4 shows how the pH varies during different titrations. Table 19.7 summarizes the choice of indicator.

TABLE 19.7

	pH range of end-point	Example
Strong acid + strong base	3–10 approx.	screened methyl orange clearest, and is unaffected by atmospheric CO_2
Strong acid + weak base	3–6 approx.	methyl orange or methyl red
Weak acid + strong base	8–10 approx.	phenolphthalein

When a polybasic acid is titrated, it is often possible to detect successive stages of the reaction, due to the different acidic strengths of the ions formed. For instance phosphoric(V) acid can form a series of salts NaH_2PO_4 (pH of aqueous solution $\simeq$ 4·7), Na_2HPO_4 (pH ~ 9·7), and Na_3PO_4 (pH ~ 13). An indicator such as methyl orange will change colour at the formation of NaH_2PO_4, while phenolphthalein will detect the formation of Na_2HPO_4. Figure 19.5 shows how pH varies when sodium hydroxide is added to a solution of phosphoric(V) acid.

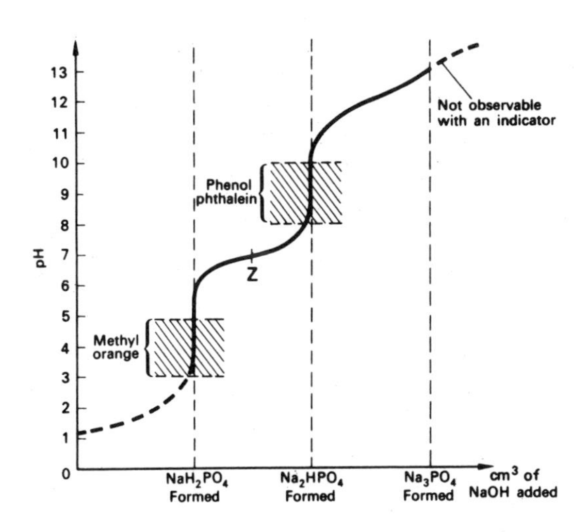

FIG. 19.5.

Carbonates can be titrated with acids if methyl orange is used as an indicator. Carbon dioxide is formed, which reacts to form a weakly acidic solution of pH approximately 4.

$$CO_2 + H_2O \rightleftharpoons H^+(aq) + HCO_3^-(aq)$$

When a weak acid is titrated with a weak base, the change in pH at the end-point is not abrupt enough to allow a colour indicator to be used (Fig. 19.6). In this case titrations can sometimes be carried out using a recording pH meter.

Buffer solutions are solutions with compositions represented by the near-horizontal portions of the curves in Fig. 19.4 and 19.5. Examples of buffer solutions are:

(a) $CH_3COOH(aq) + CH_3COONa(aq)$; e.g. point *X* in Fig. 19.4.

(b) $NH_3(aq) + NH_4Cl(aq)$; e.g. point *Y* in Fig. 19.4.

(c) $Na_2HPO_4(aq) + NaH_2PO_4(aq)$; e.g. point *Z* Fig. 19.5.

Another way of detecting end-points of acid–base reactions is by a **conductometric titration**. If a strong alkali and strong acid are titrated, for instance, the electrical conductivity will fall to a minimum at equivalence point. On either side of the end-point there will be an excess of $H^+(aq)$ or $OH^-(aq)$ and the solution will have a greater conductivity. It is necessary to use an a.c. supply to avoid electrolysis occurring. Figure 19.6 shows the plot of current against cm^3 of alkali added, for a typical acid–alkali titration. If either the acid or base is weak, or a precipitate is formed, a differently shaped graph will be obtained. Normally slightly curved lines, rather than straight lines, are obtained because of the increasing volume of the solution during the titration.

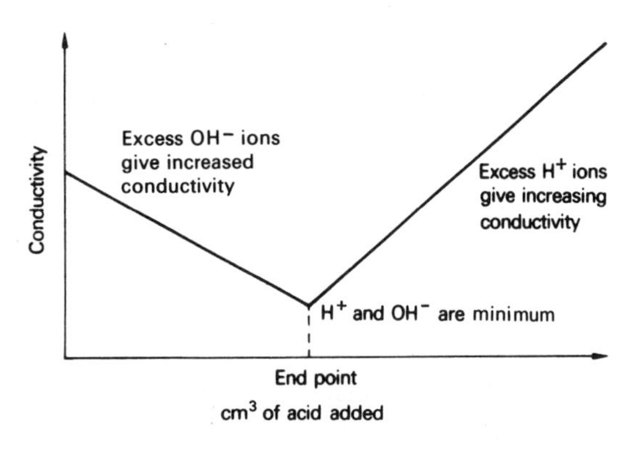

FIG. 19.6. Conductometric titration.

19.15 Limitations of the Brønsted–Lowry concept of acids

The Brønsted–Lowry theory is admirable for rationalizing acid–base reactions which occur in solution in protonic solvents such as water, ammonia and ethanoic acid, though it fails to include a whole series of reactions which, for convenience, are often considered to be acid–base neutralizations.

Consider for instance the reactions of **acidic oxides**. On the Brønsted–Lowry theory, one can hardly call an oxide such as $SO_3(s)$ an acid, since it does not possess any protons which it can donate. Nevertheless it reacts vigorously with water to give a solution of pH $\ll 7$, sulphuric acid. Most non-metal oxides behave likewise.

Furthermore, these acidic oxides will react directly with alkalis in the absence of water. For instance, a tube containing calcium oxide or sodium hydroxide (frequently soda-lime, a mixture of $Ca(OH)_2$ and NaOH) will absorb carbon dioxide with the evolution of heat:

$$CO_2(g) + CaO(s) \rightarrow CaCO_3(s); \quad \Delta H = -178 \text{ kJ}$$

$$CO_2(g) + NaOH(s) \rightarrow NaHCO_3(s); \quad \Delta H = -128 \text{ kJ}$$

An alternative concept of acid–base behaviour was put forward by Lewis, in 1938. Lewis defined a base as a substance capable of donating an electron-pair in the form of a donor bond. A Lewis acid is a substance which can accept an electron-pair by bond formation. Ammonia is a typical Lewis base, for it contains a lone-pair which can donate to a substance with vacant energy levels. (It is also a Brønsted base.) The simplest Lewis acid is the proton, for the proton readily accepts electron pairs, for example:

$$\underset{\substack{\text{proton} \\ \text{electron-pair acceptor}}}{H^+} + \underset{\substack{\text{Lewis base} \\ \text{electron-pair donor}}}{:NH_3} \rightarrow \underset{\text{acid–base complex}}{NH_4^+}$$

It follows that all Brønsted acids and bases are also Lewis acids and bases respectively. However, the term Lewis acid can be applied to substances which do not contain protons and are not therefore Brønsted acids. Another typical Lewis acid is boron trifluoride:

$$\underset{\text{electron-pair acceptor}}{\begin{matrix} F \\ F:\ddot{\underset{..}{B}} \\ F \end{matrix}} + \underset{\text{electron-pair donor}}{\begin{matrix} H \\ :\ddot{\underset{..}{N}}:H \\ H \end{matrix}} \longrightarrow \underset{\text{addition compound (adduct)}}{\begin{matrix} F\,H \\ F:\ddot{\underset{..}{B}}:\ddot{\underset{..}{N}}:H \\ F\,H \end{matrix}}$$

It is usual to reserve the term *Lewis acid*, or *electron-acceptor*, for substances such as BF_3. When the word *acid* is used loosely, it can generally be taken to mean *Brønsted acid* or *proton donor*. Acidic oxides are Lewis acids, but they are not Brønsted acids.

19.16 Acid strength and chemical structure

An acid is described as strong if it is able to ionize almost completely in water. The anions of strong acids (their conjugate bases) must therefore have stable structures. A very weak acid on the other hand is unlikely to form a particularly stable conjugate base. Acid strengths are closely related to the stabilities of the anions that are formed; the ion has to be able to absorb a negative charge and the presence of electronegative atoms or groups will help in this. Table 19.8 shows how the substitution of electronegative halogen atoms for hydrogens increases the strength of an acid.

TABLE 19.8

pK_a VALUES OF ORGANIC ACIDS

$CH_3.COOH$ 4·8	$CH_2F.COOH$ 2·7	$CH_2Br.COOH$ 2·9	$CH_2I.COOH$ 3·1
	$CH_2Cl.COOH$ 2·9	$CHCl_2.COOH$ 1·3	$CCl_3.COOH$ 0·7

In the halogen acids, an electron-pulling ($-I$) effect is operating; the methyl group exerts an electron pushing ($+I$) effect and the strengths of the acids in Table 19.9 show how the methyl acids are much weaker than the halogen acids. Section 30.3 contains a more detailed explanation of the way in which the anions of organic acids are stabilized.

Similar effects are observed with bases. pK_a for NH_4^+ is 9.2, but the $+I$ effect of the methyl group gives $CH_3NH_3^+$ a pK_a value of 10.6 and it is consequently a weaker acid, though of course CH_3NH_2 will be a stronger base than NH_3. On the other hand the $-I$ effect of the carbonyl group, $\gt C{=}O$, means that $CH_3.CONH_3^+$ is a strong acid (pK_a = $-0{\cdot}5$) and $CH_3.CONH_2$ is a very weak base.

TABLE 19.9

pK_a VALUES OF ORGANIC ACIDS

$HCOOH$ 3·8	$CH_3.COOH$ 4·8	$CH_3.CH_2.COOH$ 4·9	$(CH_3)_3C.COOH$ 5·0

Study Questions

1. (a) Calculate the concentration of H^+(aq) in water at 0°, 25° and 100°C if the ionic product, K_w, at these temperatures is 10^{-15}, 10^{-14} and 5×10^{-13} mol^2 dm^{-6} respectively.

(b) Calculate the pH at each temperature.

2. The pK_a values of ethanoic acid, chloroethanoic acid, dichloroethanoic acid and trichloroethanoic acid are 4·8, 2·9, 1·3 and 0·7 respectively, at 25°C.

(a) Which is the strongest acid?

(b) Calculate the approximate pH of a 1·0 M solution of each acid.

(c) Can you suggest a reason for the differences between the acids?

3. Calculate the pH values of the following solutions at 25°C. (At 25°C, $K_w = 10^{-14}$ mol^2 dm^{-6}, HCl and NaOH are strong electrolytes, K_a for the monobasic methanoic acid = 2×10^{-4} mol dm^{-3}, K_a for NH_4^+ = 5×10^{-10} mol dm^{-3}.)

(a) M HCl.
(b) 0·1 M HCl.
(c) 0·2 M HCl.
(d) 5 M HCl.
(e) M HCOOH.
(f) 0·1 M HCOOH.
(g) 0·2 M HCOOH.
(h) 5 M HCOOH.
(i) M NaOH.
(j) 0·1 M NaOH.
(k) 0·2 M NaOH.
(l) 5 M NaOH.
(m) M NH_3.
(n) 0·1 M NH_3.
(o) 0·2 M NH_3.
(p) 5 M NH_3.

4. (a) Calculate the pH values of the following at 25°C ($K_w = 10^{-14}$ mol^2 dm^{-6}, HCl and sodium ethanoate are strong electrolytes, K_a for ethanoic acid = $1{\cdot}8 \times 10^{-5}$ mol dm^{-3}.)

(i) 1 dm^3 of M HCl.
(ii) 1 dm^3 of M ethanoic acid.
(iii) 1 dm^3 of water.
(iv) 1 dm^3 of a solution containing one mole of ethanoic acid and one mole of sodium ethanoate.

(b) What would the pH become in each case if 4 g of NaOH were added to each?
(NaOH = 40)

(c) What would the pH become in each case if 3·65 g of HCl(g) were added to each?
(HCl = 36·5)

(d) What name is given to such solutions as (iv)?

(e) Give an example of a solution like (iv), but with a pH greater than 7.

5. (a) Calculate the pH values of

(i) 0·1 M H_2SO_4.
(ii) 0·02 M H_2SO_4.

(Sulphuric acid is a dibasic acid, totally dissociated in solution.)

(b) Calculate the pOH values, and hence the pH values of

(iii) 0·1 M $Ba(OH)_2$.
(iv) 0·02 M $Ba(OH)_2$.

6. (a) Suggest suitable indicators for the following acid–base titrations:

(i) HCl + NaOH.
(ii) Ethanoic acid + NaOH.
(iii) HCl + ammonia.
(iv) Ethanoic acid + ammonia.
(v) Phosphonic acid, H_2PHO_3 (dibasic acid, pK_a values 1·8 and 6·2 mol dm^{-3}) + NaOH.

(b) Draw curves to show how the pH varies with the amount of base added to a given amount of acid in each case.

7. What reagents would be needed to make the following salts?

(a) Potassium bromide.
(b) Ammonium nitrate.
(c) Cobalt(II) chloride.
(d) Calcium sulphate.
(e) Sodium hydrogensulphate.
(f) Calcium hydrogensulphite.

8. What salts will form when

(a) NaOH neutralizes sulphuric acid.
(b) Magnesium reacts with nitric acid.
(c) Lead(II) nitrate reacts with sodium chloride.
(d) Sodium carbonate reacts with excess HCl(aq).
(e) Equal quantities (moles) of sodium carbonate and HCl(aq) react.

9. Will the following form acidic, neutral or alkaline aqueous solutions? Give reasons for your answers.

(a) Sodium ethanoate.
(b) Sodium chloride.
(c) Ammonium ethanoate.
(d) Ammonium chloride.
(e) Iron(III) chloride.
(f) Aluminium chloride.

10. Label the Brønsted acids and bases in the following equilibria:

(a) $HCl + H_2O \rightleftharpoons H_3O^+ + Cl^-$.
(b) $CH_3NH_2 + H_2O \rightleftharpoons CH_3NH_3^+ + OH^-$.
(c) $CH_3COONa + HNO_3 \rightleftharpoons CH_3COOH + NaNO_3$.
(d) $HSO_4^- + H_2O \rightleftharpoons H_3O^+ + SO_4^{2-}$.
(e) $N^{3-} + 3H_2O \rightleftharpoons NH_3 + 3OH^-$.

11. Classify the following substances as either electron donors or electron acceptors (Lewis bases and acids).

(a) Ammonia, NH_3.
(b) Sulphur dioxide, SO_2.
(c) Ethoxyethane, $C_2H_5OC_2H_5$.
(d) Silicon tetrafluoride, SiF_4.
(e) Cr^{3+} ions.
(f) I^- ions.

12. The dissociation constant for the hydrolysis of a salt such as sodium ethanoate is given the symbol K_h.

$$CH_3COO^- + H_2O \rightleftharpoons CH_3COOH + OH^-$$

Find the relationship between K_h, K_w, and K_a (the dissociation constant of the acid).

13. Borax, $Na_2B_4O_7.10H_2O$, can be titrated as a weak base. A solution containing 1·91 g is exactly neutralized by 20 cm^3 of 0·5 M HCl(aq).

(a) Suggest a suitable indicator for the titration.
(b) How many moles of borax react with how many moles of HCl?
(c) Write down an equation for this reaction.

14. H_3PO_2 is a monobasic, H_3PO_3 a dibasic, and H_3PO_4 a tribasic acid. A hydrogen atom bound to phosphorus is not easily ionized, unlike a hydrogen atom bound to oxygen. On the basis of these data suggest possible structures for the three acids.

15. Certain non-aqueous solvents ionize as follows:

$$2NH_3(l) \rightleftharpoons NH_4^+ + NH_2^-.$$
$$2HCl(l) \rightleftharpoons H^+ + HCl_2^-.$$
$$2BrF_3(l) \rightleftharpoons BrF_2^+ + BrF_4^-.$$

(All the ionic species are solvated).

(a) What are typical acids and bases for each solvent?

(b) Give an example of a neutralization reaction in liquid ammonia.

(c) Are ammonia and hydrogen chloride differentiating or levelling solvents for aqueous acids?

(d) To what extent is it fair to refer to acids and bases when bromine trifluoride is the solvent?

16. When $SnBr_2F_{10}$ is titrated conductometrically against $KBrF_4$ in BrF_3(l), it is found that the conductivity is at a minimum when 0·390 g of $KBrF_4$ have reacted with 0·469 g of $SnBr_2F_{10}$.

(a) What sort of reaction is taking place?

(b) How many moles of $KBrF_4$ react with how many moles of $SnBr_2F_{10}$?

(c) Write down an equation for the reaction.

(d) What is the nature of the substance $SnBr_2F_{10}$?

17. Place the acids in each of the following sets in increasing order of acidity.

(a) HNO_2, HNO_3.
(b) H_3PO_4, H_2SO_4, $HClO_4$.
(c) $CH_3.CH_2.COOH$, $CH_3.CO.COOH$, $(CH_3)_3C.COOH$.
(d) $CH_3.CH_2.COOH$, $CH_3.CO.COOH$, $(CH_3)_3C.COOH$.
(d) $CH_3.CH_2.COOH$, $CH_3.CHCl.COOH$, $CH_2Cl.CH_2.COOH$.
(e) H_3PO_4, $H_2PO_4^-$, HPO_4^{2-}.

18. Acids are often used as catalysts; for example, dilute acid catalyses the reaction of iodine with propanone:

$$I_2(aq) + CH_3COCH_3(aq) \xrightarrow{H^+(aq)} CH_3COCH_2I(aq) + H^+(aq) + I^-(aq)$$

The time taken for a small fixed amount of iodine to react can be measured using a colorimeter. The following results were obtained:

	Initial Concentrations			Time taken for
	$[CH_3COCH_3]$	$[I_2]$	$[H^+]$	iodine to react
A	0.5	0.01	0.5	115 s
B	1.0	0.01	0.5	59 s
C	0.5	0.01	1.0	56 s
D	1.0	0.02	0.5	60 s

(a) Allowing for experimental error, use *A* and *B* to work out the order of reaction with respect to propanone.

(b) Choose suitable experiments and work out the order with respect to (i) iodine, (ii) hydrogen ions.

(c) What is the overall rate expression for this reaction?

(d) Starting with 1·0 M CH_3COCH_3, 1.0 M H^+ and 0.01M I_2, how long would it take for the same amount of iodine to react?

(e) Why is it important to measure the time taken for only a very small amount of the iodine to react?

(f) This is an "autocatalytic" reaction (one of the reaction products acts as a catalyst). What effect will the H^+ that is formed in the reaction have on the results?

CHAPTER 20

The *s*-block metals and aluminium

These final chapters are not intended to fulfil the function of a reference book, but rather to provide a guide to general principles.

20.1 A note on the classification of metals

The term *s*-block refers to that block of elements in the periodic table following the noble gases where only the *s*-electron orbitals are being filled, namely Group IA (the alkali metals) and Group IIA (the alkaline earth metals). Although aluminium is not an *s*-block element, since its valence shell has the configuration $3s^2p^1$, its properties are quite closely related to the *s*-block elements adjacent to it in the periodic table.

We define a B-metal as a metal which follows a transition series in a given period.* By this definition, a number of similar metals are grouped together and they will be considered in Chapter 34. Aluminium forms the cation Al^{3+} with a noble-gas structure like the cations of *s*-block metals, and many of its properties are similar to the properties of beryllium. It is convenient, therefore, to treat it in this chapter, while bearing in mind that many of its properties are more akin to those of the B-metals, especially gallium, indium, and thallium.

* Some older textbooks use the A and B classification differently, viz. B for transition elements and A for "typical" elements. Modern forms of the periodic table, however, use A to denote the elements on the left-hand side and B for the right-hand side (see p. 486).

TABLE 20.1

Principal quantum number of valence shell	Electronic structure of noble-gas core	Group			
		0	I s^1	II s^2	III s^2p^1
2	$1s^2$	He	Li	Be	B
3	$1s^2$; $2s^2p^6$	Ne	Na	Mg	Al
4	$1s^2$; $2s^2p^6$; $3s^2p^6$	Ar	K	Ca	
5	$1s^2$; $2s^2p^6$; $3s^2p^6d^{10}$; $4s^2p^6$	Kr	Rb	Sr	
6	$1s^2$; $2s^2p^6$; $3s^2p^6d^{10}$; $4s^2p^6d^{10}$; $5s^2p^6$	Xe	Cs	Ba	

TABLE 20.2

	Group IA		Group IIA
Lithium	Occurs mainly as complex silicates	Beryllium	Occurs mainly as complex silicates; fairly rare
Sodium	Extensive underground deposits. Approx. 0·5 mol dm^{-3} of $Na^+(aq)$ in sea	Magnesium	Principally as sulphate, carbonate and silicates, and *Carnallite*, $MgCl_2.KCl.6H_2O$
Potassium	Occurs in association with anions such as Cl^-. Extractable as KCl	Calcium	Occurs as carbonate and sulphate
Rubidium Caesium	Very rare; traces occur in association with potassium	Strontium Barium	Occur as carbonate and sulphate
Francium	Radioactive; does not occur naturally on Earth	Radium	Radioactive; found in Th and U ores

Group IIIA—Aluminium—occurs mainly as complex silicates. *Bauxite*, $Al_2O_3.2H_2O$, also common.

20.2 Occurrence of the *s*-block elements

None of the *s*-block elements occurs native (that is, as the free element) since all are far too reactive. The most electropositive elements are found in nature as cations combined with the most electronegative anions. The elements occur as shown in Table 20.2.

20.3 A general survey of properties and uses of *s*-block metals

s-block metals have the following general properties in common, though there are exceptions and some of the properties referred to are shared with other classes of elements:

(a) They are all metals, most of them being markedly reactive with non-metals.

(b) They have relatively low binding energies, due to the fact that there are only one (Group IA) or two (Group IIA) electrons per atom available for bond formation. Moreover, the binding *s*-electrons are present outside a noble gas core and are shielded from the direct attraction of the charge on the atomic nucleus. As consequences of this:

(i) They have melting and boiling points which are lower than those of the transition metals to their right. (See Figs. 1.6 and 1.7.)

(ii) They have relatively low densities, when compared with the metals which are to their right in the periodic table, as a result of their relatively large atomic radii and large atomic volumes (Fig. 1.9). Lithium is so light that it will even float on the oil in which it is customarily stored.

(iii) They have low ionization energies, the outermost *s*-electrons being lost with relative ease. This is the main factor which enables these metals to form positive ions readily. In aqueous solution:

$$M(s) \rightarrow M^+(aq)+e^- \text{ (Group IA)}$$
$$M(s) \rightarrow M^{2+}(aq)+2e^- \text{ (Group IIA)}$$

(c) One of their most characteristic chemical properties is that of constant oxidation number. For this reason, the oxidation number may generally be omitted from the names of compounds. Ions such as Ca^+ and Na^{2+} never occur in chemical compounds. Transition metals and B-metals both show several oxidation states.

(d) They are all powerful reducing agents (electron suppliers) due to (b) above. Reducing power is measured by the half cell potential $\mathcal{E}^{\ominus}$ (Chapter 11), and the data listed in Appendix 1 show that the metals of Group IA and Group IIA appear at the extreme "reducing" end of the list.

(e) With few exceptions their compounds show evidence of the presence of ions. This is in contrast with the transition metals, and with the B-metals, where ionic properties are often less well-defined or where the ions which form are generally complex.

(f) Among the *s*-block elements, the larger ions are not noticeably hydrolysed in aqueous solution. The smaller ions especially Be^{2+}(aq), Li^{+}(aq), and Mg^{2+}(aq), are hydrolysed and behave as weak acids (Chapter 19).

(g) Their hydroxides are soluble in water for the most part, exceptions being $Mg(OH)_2$ and $Be(OH)_2$ which are very sparingly soluble, and $Ca(OH)_2$ which is only slightly soluble. The term *alkali metal* derives from the fact that the metals react with water to form soluble hydroxides, alkalis.

(h) Apart from magnesium and beryllium, both of which have relatively high excitation energies, all the *s*-block elements give characteristic colours when they are introduced into a flame. This property can be used in analysis.

(i) The reactivity of these metals towards non-metals is very marked, though other factors may obscure it. Caesium and rubidium are so reactive that they inflame spontaneously if their surfaces are exposed to the air, and all the alkali metals burn if heated in air. Although magnesium reacts explosively with most non-metals when heated in a finely divided state, large pieces of the metal react only slowly with oxygen owing to the formation of a protective layer of surface oxide.

Uses. Magnesium is a useful structural metal on account of its lightness and is often alloyed with aluminium, for instance in airframe construction. Beryllium ought in theory to be a very attractive proposition as a structural metal, for it is less reactive than magnesium and is even lighter. Unfortunately, however, apart from its cost it possesses two serious snags: in the first place it is rather toxic, and second, it is very brittle except when very pure indeed. It has found uses in nuclear reactors, on account of its capacity to slow down neutrons.

Sodium nowadays finds some rather remarkable applications: formerly most sodium was consumed for chemical purposes, such as special organic reductions. In the laboratory it requires careful handling, owing to its violent reaction with water. It does, however, possess high electrical and thermal conductivity in common with the other Group IA metals, and this has led to its use as a heat transfer agent in nuclear reactors (where it is pumped around as a liquid), and as an electrical conductor. Recent trials have shown that plastic-covered sodium wire, which is very cheap to make by extrusion, is excellent for cables on account of its flexibility and good conductivity. The danger from fire due to chemical

TABLE 20.3

		IA	IIA	IIIA
increasing density, metallic and ionic radii, atomic volume ↓	decreasing m.p., b.p., hardness, ionization energy ↓	Li	Be	
		Na	Mg	Al
		K	Ca	(Sc)
		Rb	Sr	(Y)
		Cs	Ba	(La)
		Fr	Ra	(Ac)

increasing m.p., b.p., hardness, density →

decreasing metallic and ionic radii, atomic volume, thermal and electrical conductivity action is considerably less than that due to electrical short-circuiting.

Apart from sodium and magnesium (and to a lesser extent lithium which finds some structural uses in alloys) only limited quantities of the free *s*-block metals are manufactured and their uses are specialized. Compounds of these metals will be referred to under separate headings below.

20.4 Principal trends in physical properties

Figures 20.1, 20.2, 20.3 show plots of melting points, boiling points, and atomic volumes of these metals. These plots should be compared with Figs. 1.8, 1.9 and 1.10 which show the same properties plotted in sequence for all the elements.

It is a characteristic of a Group of the periodic table that its elements show (a) a **similarity**, (b) a **gradation**, in properties.

(a) Similarities

The similarities within a Group can be traced back to their similar electronic structures. For the isolated atoms in their ground states, these structures are:

Group IA—noble gas plus one outer *s*-electron.

Group IIA—noble gas plus two outer *s*-electrons. The noble gas "core" plays no part at all in the chemistry of these elements, because

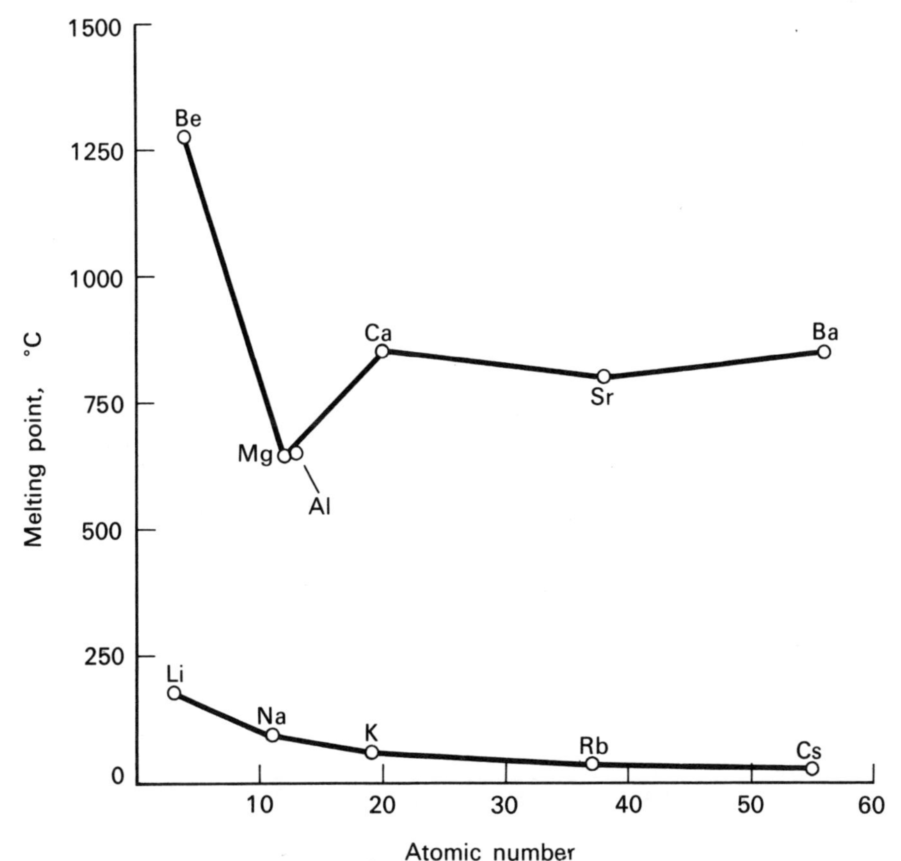

FIG. 20.1. The melting points of the *s*-block elements.

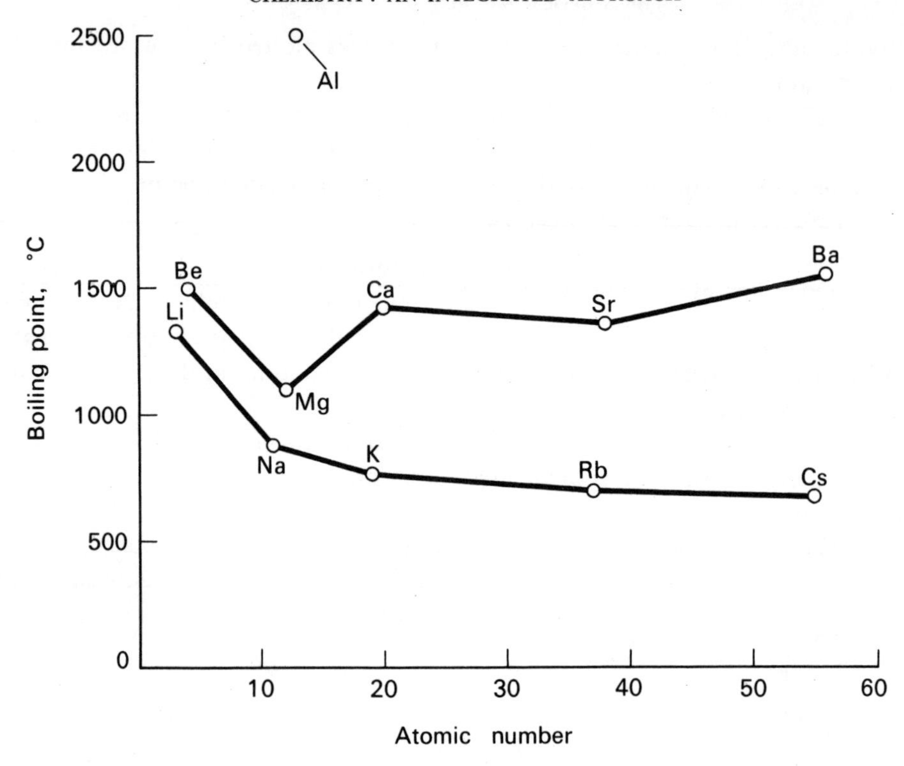

FIG. 20.2. The boiling points of the *s*-block elements.

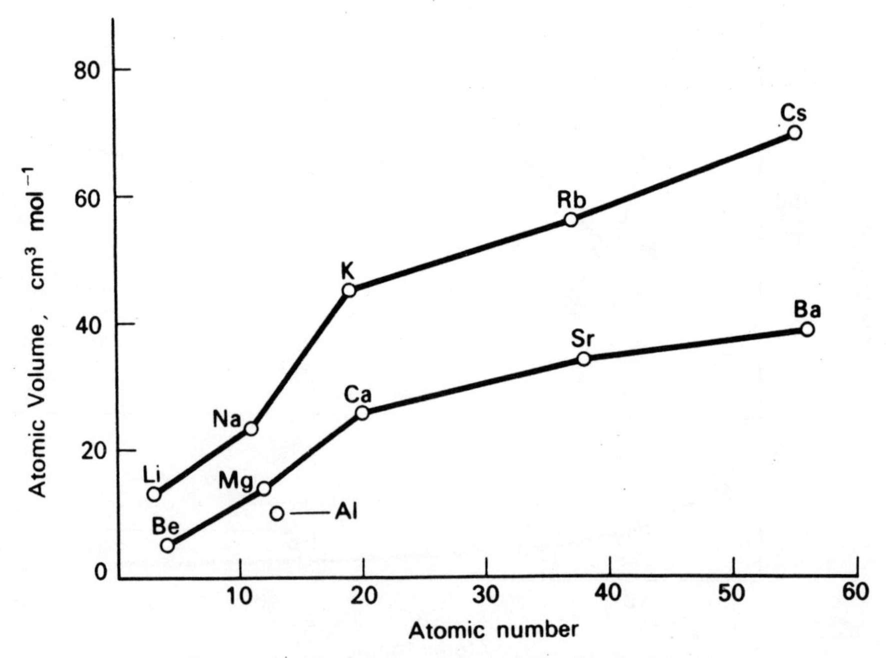

FIG. 20.3. The atomic volumes of the *s*-block elements.

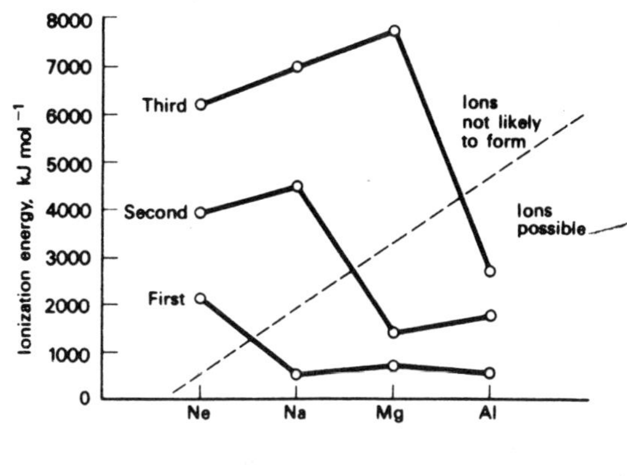

FIG. 20.4.

the extra energy required to remove an electron from it is far too large. Figure 20.4 shows the energy required to remove one, two and three electrons from each of the elements Ne, Na and Mg. The diagram shows clearly that Na^{2+}, Na^{3+} and Mg^{3+} are unlikely to form. Similar behaviour is shown by other sets of elements at the beginning of a period. Figure 20.5 shows the first ionization energy of the alkali metals, and the sum of the first and second ionization energies of the Group IIA metals, plotted against atomic number.

(b) GRADATIONS

The gradual increase in size observed when moving down a given group can be attributed simply to the increasing size of the noble gas "core". Properties such as atomic volume are related closely to metallic radius, because structures are similar within a given group.

Figure 20.6 shows the estimated sizes of atoms and ions of *s*-block elements.

The Be^{2+} ion occurs only rarely, since its very small size causes it to exert an intense electrostatic field at its surface, and it polarizes other ions and forms bonds.

Table 20.4 gives data for the ionic conductances of some ions. One might expect the smallest ions to be more mobile, and to have

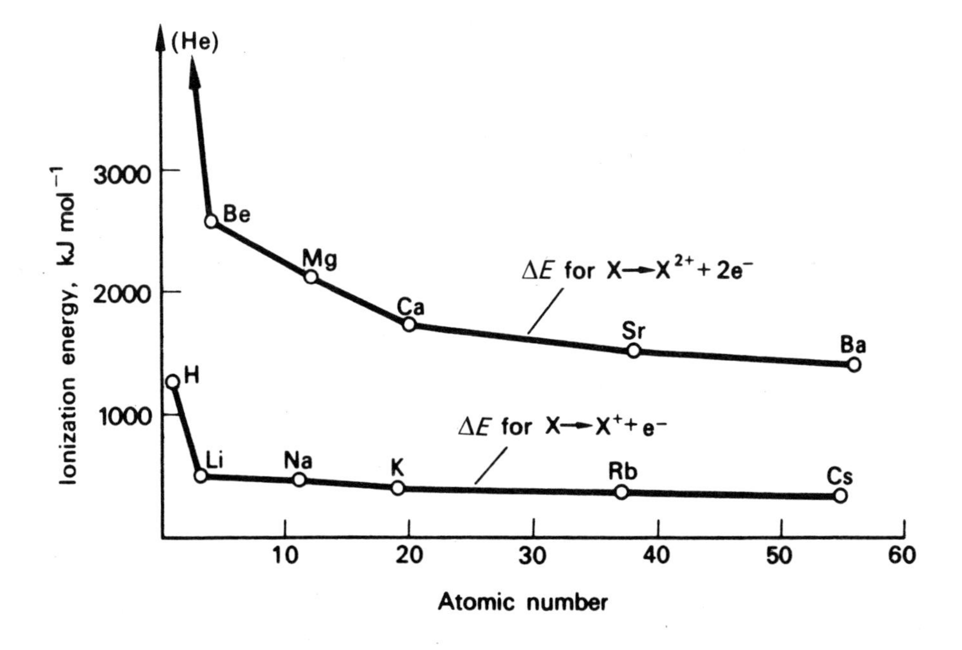

FIG. 20.5.

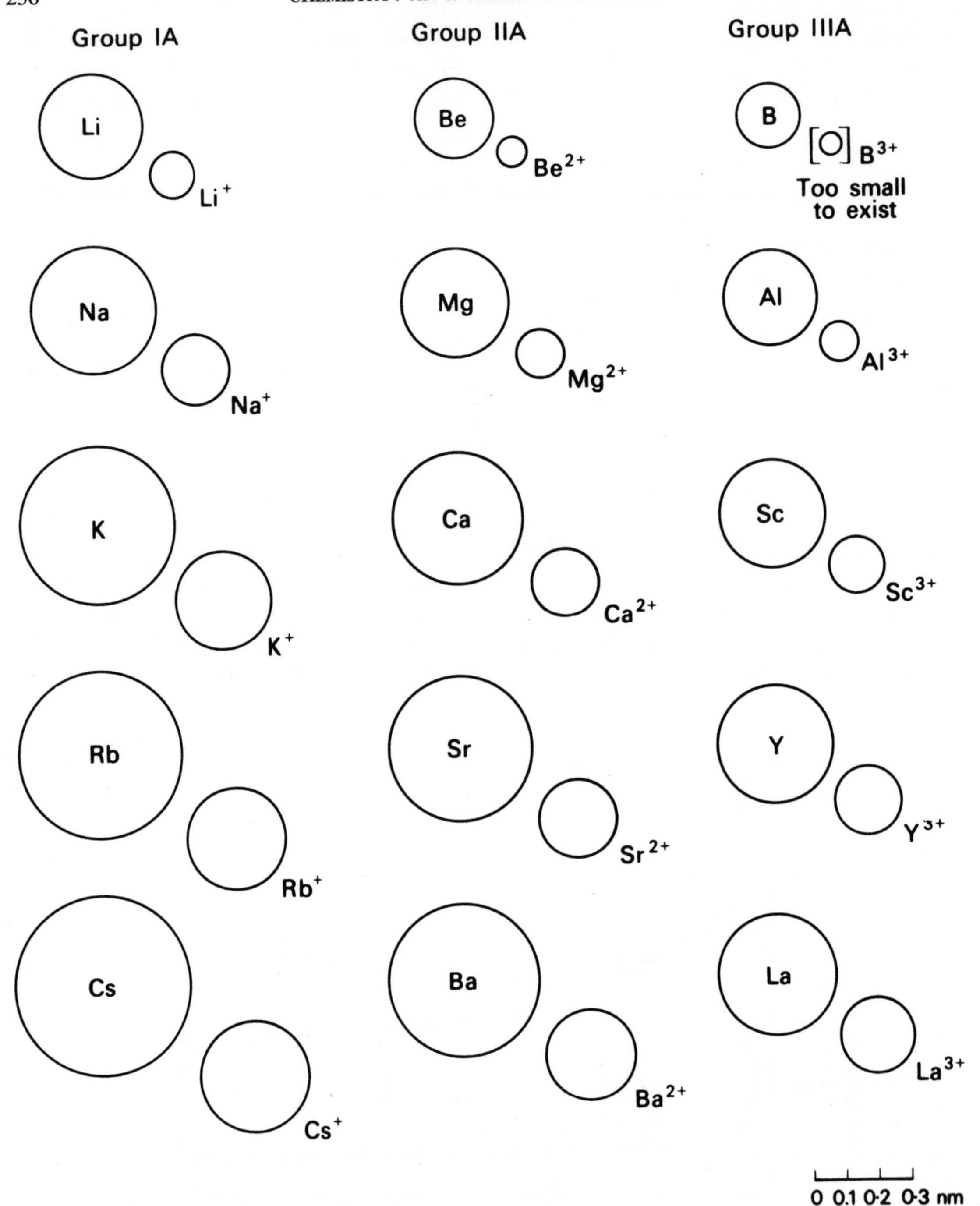

FIG. 20.6. The relative sizes of atoms and ions of *s*-block elements.

TABLE 20.4

IONIC CONDUCTANCES (Ω^{-1} cm^{-2} mol^{-1}) OF GROUP IA METAL IONS

Li^+	Na^+	K^+	Rb^+	Cs^+
33·5	43·5	64·6	67·5	68·0

higher conductances, so that an ion like $Li^+(aq)$ seems to be anomalous; however, the small size of lithium causes a large number of water molecules to be attracted to it. The ion is really $Li(H_2O)_x^+$, relatively large in size. This hypothesis explains other phenomena, such as the abnormal half-cell potential of lithium (section 20.5) and the higher degree of hydration of some of its crystalline salts.

The decrease in ionization energy down a Group (Fig. 20.5) is a result of the increasing distance of the outer electron(s) from the centre of positive charge provided by the core. This is the main factor causing the higher reactivity of the heavier metals.

Larger atoms exert weaker attractions on one another, because electrostatic forces decrease with increasing internuclear distance. The heavier metals within a Group are therefore softer and more volatile.

Heat of sublimation provides a measure of interatomic forces, and this is lower for the heavier elements within a Group (Fig. 20.7).

The metals of Group IIA are harder and less volatile than their Group IA neighbours, showing that their interatomic forces are greater.

This observation is consistent with our notions about the nature of the metallic bond: a metal is regarded as a fairly closely packed assembly of positive ions "welded" together by a mobile electron "gas" (Chapter 5). In the case of the alkali metals there is only one electron per atom available for metallic bonding, whereas in Group IIA there are two electrons per atom available. Moreover, the doubly-charged ions M^{2+} of Group IIA are smaller than their left-hand neighbours in Group IA, M^+, because of the

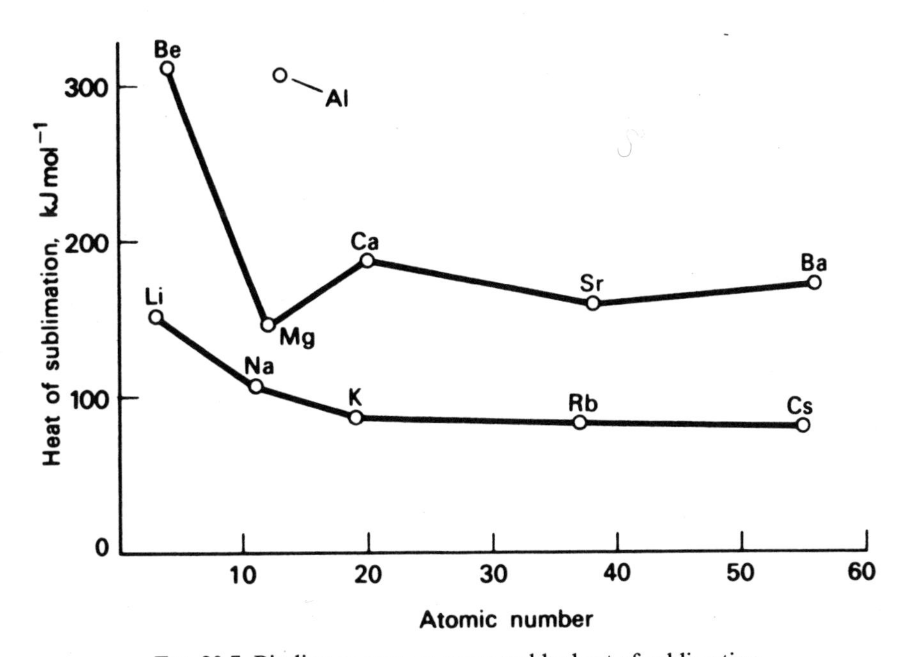

FIG. 20.7. Binding energy, as measured by heat of sublimation.

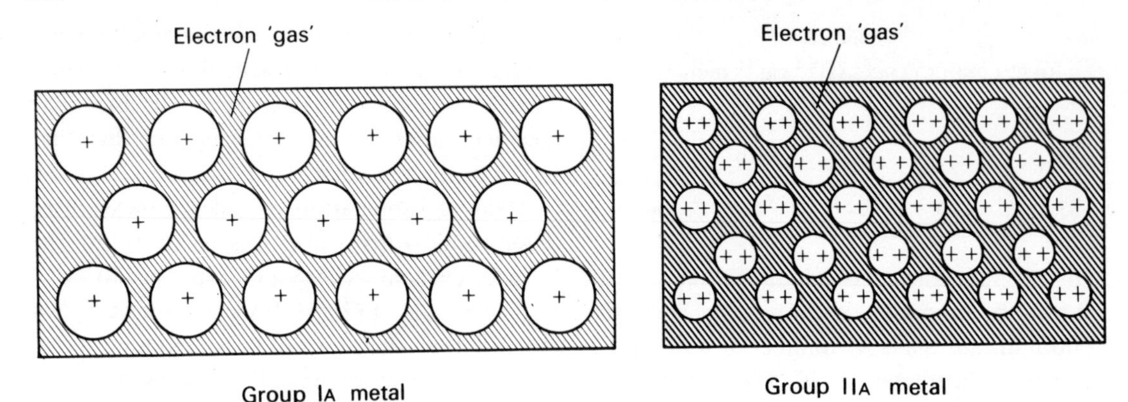

FIG. 20.8.

increased nuclear charge in Group IIA (Fig. 20.8).

20.5 Formation of positive ions by *s*-block elements

The *s*-block elements form positive ions due to the ready loss of their *s*-electrons. One of the important characteristics of *s*-block metals is that only *one* charge is encountered for a given element. Group IA elements form compounds which dissolve to give the ions $M^+(aq)$ (Fig. 20.9), while in Group IIA the characteristic ion is $M^{2+}(aq)$. Thus *s*-block elements have a tendency to lose their outermost *s*-electrons completely. The relative energies needed to form aqueous ions are related to the ionization energies, but depend on other factors as well. Figure 20.10 shows by means of an energy cycle the factors which determine the formation of aqueous ions.

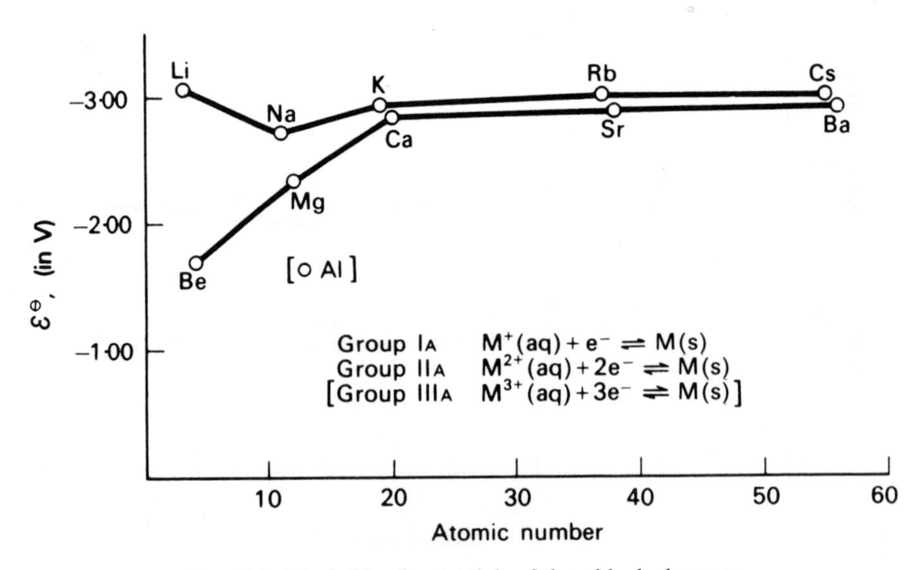

FIG. 20.9. The half-cell potentials of the *s*-block elements.

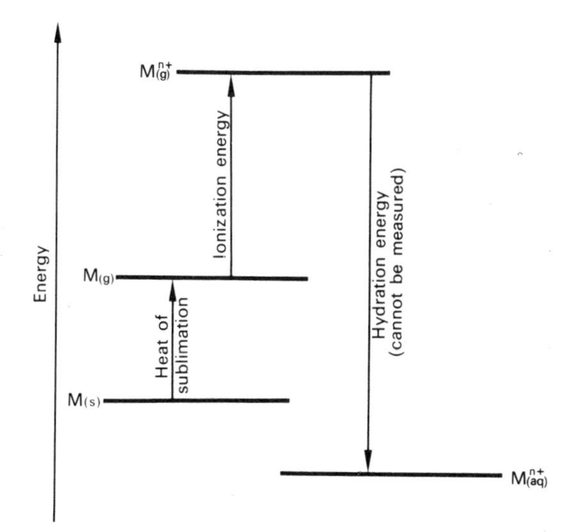

FIG. 20.10. Energy cycle showing the formation of an aqueous ion.

TABLE 20.5

Element	Reactivity	Element	Reactivity
Li	Violent with acids. Moderate with water (cf. barium).	Be	Dissolves rapidly in acids, no reaction with water.
Na	Violent with water and acids. Does not usually catch fire in water in small amounts.	Mg	Slow reaction in water, fast in acids.
K	Violently in water and acids. Usually catches fire.	Ca	Dissolves steadily in cold water, violently in acids.
Rb Cs	Even more violent reaction than potassium.	Sr Ba	About like lithium.

oxidized

$$2Na(s)+2H_2O \rightarrow 2Na^+(aq)+2OH^-(aq)+H_2(g)$$

reduced

The strongly negative $\mathcal{E}^{\ominus}$ values of these elements mean that all react vigorously with dilute acids, $H^+(aq)$. Those with very negative values react vigorously with water also, Table 20.5.

20.6 Halides of the s-block elements

The s-block elements all form simple halides MX (Group IA), or MX_2 (Group IIA), where X is a halogen. These can all be obtained in the anhydrous state, though in cases where the metal ion is small and highly charged the solid which crystallizes from aqueous solution contains water of crystallization, for instance, $CaCl_2.6H_2O$ (Ca^{2+} attracts water molecules), and $BaCl_2.2H_2O$ (Ba^{2+} is larger and has less attraction for water molecules, though more attraction than an alkali metal ion, M^+).

The s-block halides can be regarded as fairly "ideal" ionic compounds—their physical and chemical properties can be interpreted with reasonable accuracy on the assumption that ions

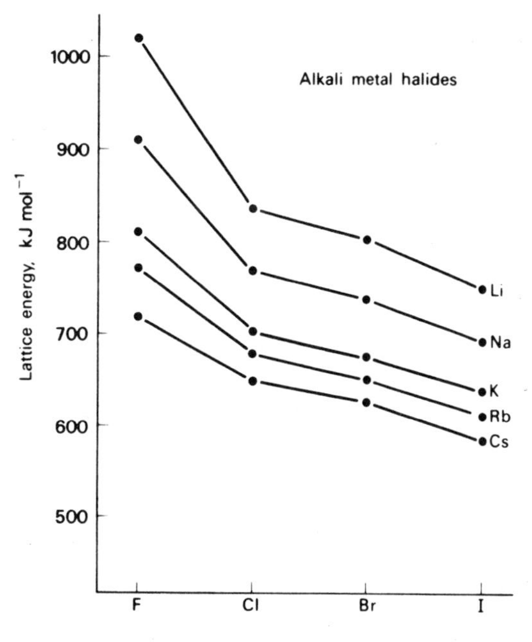

FIG. 20.11. The lattice energies of Group IA halides.

are present in the solid lattice. There is plenty of evidence for the presence of ions in the molten state: on electrolysis the metal appears at the anode, and the halogen at the cathode. Electrolysis of the fused chloride is the method normally chosen for the extraction of the Group IA metals, calcium and magnesium.

The trends in boiling point and lattice energy of the Group IA halides are summarized in Figs. 6.1 and 20.11. Lattice energy may be defined as the energy required to convert one mole of the ionic lattice into gaseous ions: lattices with smaller ions have higher lattice energies.

The halides of *s*-block elements can all be made by the reaction between the metal and the halogen, though this reaction is rarely carried out except as a laboratory exercise. The energetics of this process were considered in Chapter 9, where it was shown that such a reaction is exothermic (Born–Haber cycle, section 9.7).

Most of the alkali metal halides have 6:6 structures, though CsCl, CsBr, and CsI crystallize out with 8:8 coordination. The anhydrous Group IIA halides show a variety of structures depending upon the relative sizes of the ions.

The chlorides, and to a lesser extent bromides and iodides, of sodium and magnesium occur in considerable amounts in sea-water, from which they may be obtained by crystallization.

20.7 Oxides and hydroxides of *s*-block elements

All the *s*-block elements burn vigorously in air, reactivity being greatest in the bottom left-hand corner of the periodic table (caesium and francium). In addition to the expected oxides M_2O (Group IA) and MO (Group IIA), the more reactive members of the groups form peroxides and even superoxides. Peroxides have been shown to contain the ion $[O{-}O]^{2-}$, and superoxides the ion $[O{-}O]^{-}$. On descending the periodic table the tendency is for the element to combine with an increasing amount of oxygen:

$$2Li(s)+\tfrac{1}{2}O_2(g) \rightarrow Li_2O(s)$$
lithium oxide, white solid

$$2Na(s)+O_2(g) \rightarrow Na_2O_2(s)$$
sodium peroxide, white solid

$$Ba(s)+O_2(g) \rightarrow BaO_2(s)$$
barium peroxide, white solid

$$K(s)+O_2(g) \rightarrow KO_2(s)$$
potassium superoxide, yellow solid

This observation is a manifestation of a fairly general rule concerning ionic compounds, namely that the stability of lattices MX, where X is a *large* anion, increases as the size of M increases.

In this case O_2^{2-} is clearly larger than O^{2-}, and it is the larger metal ions at the foot of the periodic table which form stable ionic peroxides. The same effect may be noted for carbonates (section 20.12), and such compounds as polyhalides, e.g. KI_3.

The normal oxides contain the oxide ion, O^{2-}, which is an extremely strong base. Hence they react with water to form the hydroxide by proton exchange:

gains proton

$$O^{2-}+H_2O \rightleftharpoons OH^-+OH^-$$

loses proton

If water is added to calcium oxide (quicklime) CaO, for instance, the above equation can be expressed in full thus:

$$CaO(s)+H_2O(l) \rightarrow Ca(OH)_2(s); \quad \Delta H = -65 \text{ kJ at } 25°C$$

Magnesium and beryllium oxides do not react vigorously with water. If water is added to the product of burning magnesium in oxygen, a weakly alkaline solution (pH about 10) in equilibrium with much undissolved solid is obtained.

Barium peroxide has been used commercially in the manufacture of hydrogen peroxide, since the peroxide ion reacts with water or dilute acid:

$$\overbrace{O_2^{2-}+2H_2O}^{\text{gains protons}} \rightleftharpoons H_2O_2+\underbrace{2OH^-}_{\text{loses protons}}$$

The solubilities of the *s*-block element hydroxides in water are plotted in Fig. 20.12. The two main factors which lead to insolubility of hydroxides are (i) higher cation charge, and (ii) smaller cation size. It is often said that hydroxides like KOH and NaOH are strong bases, while $Ca(OH)_2$ and $Mg(OH)_2$ are weak. The real *base* in aqueous solution is however the ion OH^-, and the "weakness" of an alkali like hydroxide is due to its relative insolubility water.

The Group IA hydroxides will only decompose on very strong heating, but the Group IIA hydroxides are thermally unstable, giving the normal oxide and water vapour:

$$Ca(OH)_2(s) \rightarrow CaO(s)+H_2O(g); \quad \Delta H = +109 \text{ kJ at } 25°C$$

Table 20.6 gives some data for the solubility and thermal stability of the Group IIA hydroxides.

Beryllium hydroxide is exceptional in this Group, since it can behave as an acid as well as a base; such a substance is said to be **amphoteric.** If an excess of $OH^-(aq)$ is added to beryllium hydroxide it will dissolve forming complexes such as tetrahydroxoberyllate(II):

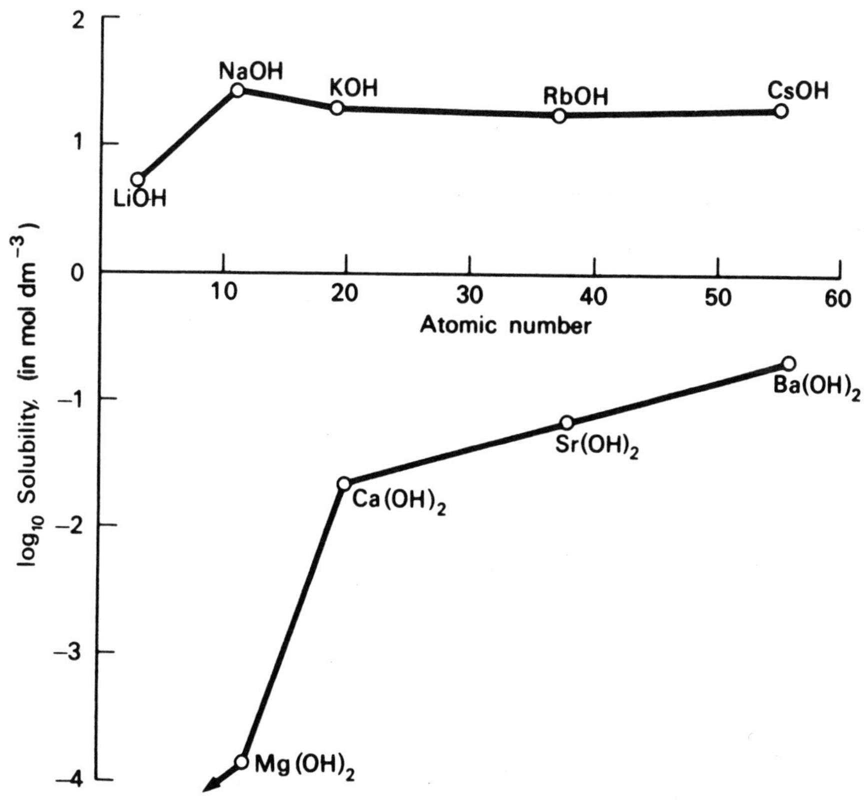

FIG. 20.12. Solubilities of hydroxides.

$$Be(OH)_2(s) + 2OH^-(aq) \rightarrow Be(OH)_4^{2-}(aq)$$

Other amphoteric hydroxides are those of tin, lead, aluminium and chromium; with all these metals there is a strong tendency for the metal atom to occur in the anion in compounds.

TABLE 20.6

Compound	$K_s = [M^{2+}][OH^-]^2$ at 20°C ($mol^3\ dm^{-9}$)	ΔH for $M(OH)_2(s) \rightarrow MO(s) + H_2O(g)$ at 25°C
$Be(OH)_2$	Be^{2+} not formed	+ 54 kJ
$Mg(OH)_2$	$1{\cdot}2 \times 10^{-11}$	+ 81 kJ
$Ca(OH)_2$	$1{\cdot}2 \times 10^{-5}$	+109 kJ
$Sr(OH)_2$	$3{\cdot}2 \times 10^{-4}$	+127 kJ
$Ba(OH)_2$	$1{\cdot}2 \times 10^{-2}$	+146 kJ

20.8 Manufacture and uses of alkalis

Alkalis are used for a variety of purposes in the chemical industry and in the laboratory; among the most common are sodium hydroxide (caustic soda), and calcium hydroxide (slaked lime). Smaller quantities of potassium hydroxide (caustic potash) and other soluble hydroxides are required for special purposes. Calcium oxide, quicklime, is itself used extensively for neutralization of acids; the use of lime on acidic soils is common agricultural practice.

The flowsheet (Fig. 20.13) shows the two main industrial routes to sodium hydroxide.

Reaction (1) occurs on account of the insolubility of calcium carbonate:

$$CO_3^{2-}(aq) + Ca(OH)_2(s, aq) \rightarrow 2OH^-(aq) + CaCO_3(s);$$

spectator ion $Na^+(aq)$ (1)

The route from sodium chloride (2) commences with a concentrated solution of brine, obtained by extraction of underground deposits of sodium chloride by pumping hot water. Various cells are in use for the electrolysis, but all depend upon the following *overall* cathode reaction:

$$H^+(aq) + e^- \rightarrow \tfrac{1}{2}H_2(g).$$

The result of discharging $H^+(aq)$ ions is that Na^+ and OH^- ions are left behind in the cathode region.

The chief technical problem in the design of cells is avoiding interference from the reaction which occurs at the anode:

$$Cl^-(aq) \rightarrow \tfrac{1}{2}Cl_2(g).$$

Steps must be taken to prevent the chlorine diffusing across the cell and being hydrolysed by the OH^- ions;

$$Cl_2(aq) + 2OH^-(aq) \rightarrow Cl^-(aq) + ClO^-(aq).$$

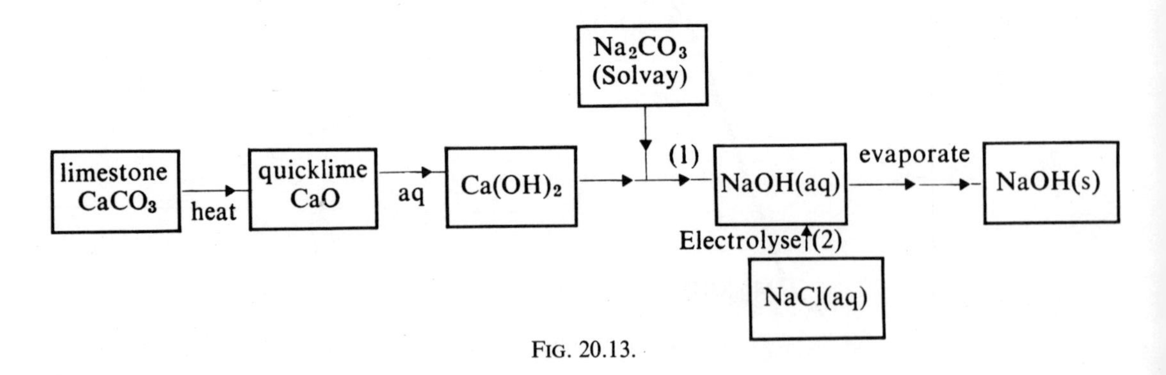

FIG. 20.13.

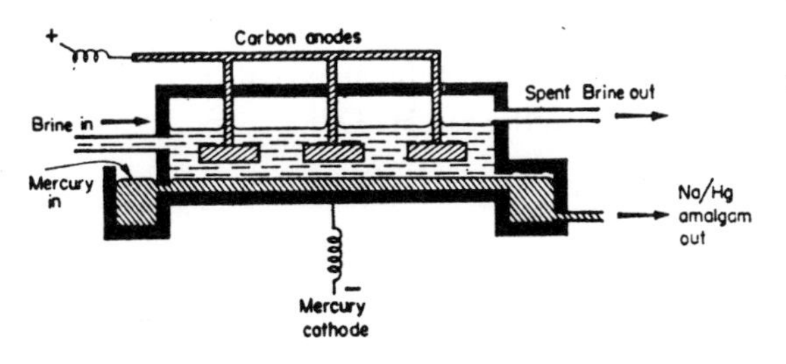

FIG. 20.14. The Kellner–Solvay cell.

One very ingenious design which overcomes this problem is the *Kellner–Solvay cell* (Fig. 20.14). This uses a mercury cathode, which is remarkable in that it allows *sodium* to be discharged in preference to hydrogen. The sodium dissolves in the cathode to form a dilute amalgam, and the half-cell potential for this process is much less negative than for the discharge of pure sodium. Furthermore, the energy of activation (Chapter 17) required to form hydrogen molecules at a mercury surface is very large.

The liquid sodium amalgam is run off, and reacts with water in an iron vessel giving sodium hydroxide solution which is concentrated by evaporation.

20.9 Reactions of alkalis

The reactions of aqueous alkalis are essentially reactions of the hydroxide ion.

(i) Acids and acidic oxides give salts. Excess of a polybasic acid will give rise to one or more acid salts.

(ii) Insoluble hydroxides are precipitated when the ions are mixed.

$$M^{n+}(aq)+nOH^-(aq) \rightarrow M(OH)_n(s).$$

(iii) Amphoteric hydroxides dissolve in excess alkali to give a soluble hydroxo-complex:

$$Be(OH)_2+2OH^-(aq) \rightarrow Be(OH)_4^{2-}(aq);$$

also Al, Sn, Pb, Cr, etc.

(iv) Salts of weak bases react to liberate the free base:

$$NH_4^+(aq)+OH^-(aq) \rightarrow NH_3(aq), g)+H_2O.$$

$$\underset{\text{soluble salt (e.g.: phenylammonium chloride)}}{C_6H_5NH_3^+(aq)}+OH^-(aq) \rightarrow \underset{\text{phenylamine (oil immiscible with water)}}{C_6H_5NH_2(l)}+H_2O.$$

(v) Many non-metals are attacked; sometimes a **disproportionation reaction** occurs when the element, in zero oxidation state, gives one compound containing the element in a negative oxidation state, and another compound in a positive oxidation state. One example is the action of alkali on chlorine described in the previous section (bromine and iodine behave similarly). Another example is the action of alkali on white phosphorus giving phosphine:

$$P_4(s)+3OH^-(aq)+3H_2O \rightarrow 3H_2PO_2^-(aq)+PH_3(g)$$

(vi) Many organic substances are attacked by the OH^- ion which inserts itself in the molecule and displaces a negatively charged ion. Examples are the hydrolyses of esters, alkyl halides and acyl halides.

(vii) Many reactions are catalysed by $OH^-(aq)$, for instance, the polymerization of aldehydes.

20.10 Sulphides of *s*-block elements

All the *s*-block elements combine with sulphur by direct reaction, though the reactions are dangerous and should not be attempted in the laboratory. Group IIA metals combine explosively on warming with powdered sulphur forming MS; Group IA metals form M_2S.

The Group IA sulphides are soluble in water, and the Group IIA sulphides sparingly soluble. They are ionic and the sulphide ion is a weak base (proton acceptor, Chapter 19) which is hydrolysed by water to produce an alkaline solution, which smells of hydrogen sulphide:

accepts protons

$$S^{2-} + 2H_2O \rightleftharpoons H_2S + 2OH^-$$

loses protons

The Group IIA sulphides perform this reaction on warming.

20.11 Other binary compounds with non-metals

Most of the *s*-block elements will combine directly with other non-metals, such as phosphorus, silicon and selenium. Nitrogen is a very unreactive element, though it combines fairly readily with lithium, forming Li_3N, and magnesium, forming Mg_3N_2. These compounds behave as if they contain the nitride ion, N^{3-}, which is hydrolysed by water forming ammonia; the reaction is analogous to the hydrolysis of sulphide ion given above. The elements also react directly with hydrogen if heat is supplied, forming ionic hydrides (sometimes called *saline*, or salt-like, hydrides), which contain the ion H^- (Chapter 18). Stable carbides are formed by calcium, strontium and barium:

$$CaO(s) + 3C(\text{graphite}) \rightarrow CaC_2(s) + CO(g); \quad \Delta H = +460 \text{ kJ}.$$

20.12 Carbonates and hydrogencarbonates

Two series of salts characteristic of the *s*-block elements are the *carbonates*, salts of carbonic acid, H_2CO_3, in which the metal cation is in association with the anion CO_3^{2-}, and *hydrogencarbonates*, in which the metal cation is in combination with the ion HCO_3^-:

O=C(O⁻)(O⁻) — *carbonate ion (planar triangle)

O=C(O⁻)(O—H) — hydrogen-carbonate ion

O=C(O—H)(O—H) — carbonic acid

$$\rightleftharpoons CO_2 + H_2O$$

The hydrogencarbonate ion is itself a weak acid with a tendency to release protons. Solid hydrogencarbonates are formed only by the more electropositive Group I metals, Na, K, Rb, Cs and Fr. On heating they decompose readily (below 100°C) giving carbon dioxide and steam:

$$2NaHCO_3(s) \rightarrow \underset{\text{stable to heat}}{Na_2CO_3(s)} + CO_2(g) + H_2O(g)$$

Solid hydrogencarbonates of the alkaline earth metals cannot be obtained though their ions can exist together in solution. Calcium and magnesium hydrogencarbonates are responsible for *temporary hardness* of water. On boiling the solution the hydrogencarbonate ion decomposes, and the insoluble carbonate is removed from the solution by precipitation:

$$Ca^{2+}(aq) + 2HCO_3^-(aq) \rightarrow CO_2(g) + H_2O(l) + CaCO_3(s).$$

The stability of carbonates themselves towards heat is greatest for metal whose ions have large size and low charge (cf. section 20.7). It may be imagined that the carbonate ion decomposes more easily to give $CO_2(g)$ when it is polarized by the cation. The general reactions

*The negative charges are in fact equally shared among the three oxygen atoms.

are illustrated by the following equations:

$$Li_2CO_3(s) \rightarrow Li_2O(s) + CO_2(g)$$

(Other alkali metal carbonates only decompose with great difficulty.)

$$MgCO_3(s) \rightarrow MgO(s) + CO_2(g)$$

(All Group IIA carbonates can be decomposed by Bunsen heat, but $BaCO_3$ only with great difficulty.)

The carbonate ion is a fairly strong base, the conjugate base of the weakly acidic HCO_3^- ion. Solutions of the carbonate ion behave as alkalis, due to the following hydrolysis reaction:

protons lost

$$CO_3^{2-}(aq) + 2H_2O \rightarrow H_2CO_3(aq) + 2OH^-(aq)$$

protons gained

The carbonates of the alkaline earth Group are all insoluble in water, and occur in the Earth's crust. Although carbon-containing compounds are comparatively rare in the Earth's crust, deposits containing $CaCO_3$(chalk, limestone, marble and Iceland spar), and $MgCO_3$ (such as dolomite, $MgCO_3.CaCO_3$) are extensive. These are derived originally from microscopic organisms.

The most important carbonate industrially is sodium carbonate, which is a cheap alkali used for glass making, soap manufacture, and a large number of industrial processes. Many million tons are produced annually by the Solvay process, developed by the Belgian chemist Ernest Solvay in 1872, using the reaction:

$$CaCO_3 + 2NaCl \rightarrow Na_2CO_3 + CaCl_2.$$

20.13 Nitrates of *s*-block elements

Nitrates are salts of the strong acid, nitric acid, HNO_3. The nitrate ion, NO_3^-, is stable in aqueous solution and is not hydrolysed.

O=N(→O)–O⁻

nitrate ion
planar triangle
(The bonds are in fact equivalent cf. carbonate)

O=N(→O)–O–H

nitric acid
a strong acid
(contrast carbonic acid)

Nitrates, like carbonates, are decomposed by heating. The nitrates of Group IA elements decompose losing oxygen only, forming the nitr*ite** (the ending *-ite* indicates less oxygen):

$$NaNO_3(s) \rightarrow NaNO_2(s) + \tfrac{1}{2}O_2(g);$$

white crystals; sodium nitrite stable to heat; glowing splint test

$$\Delta H = +107 \text{ kJ at } 25°C.$$

In Group IIA the nitrites themselves are unstable to heat, and the overall reaction is the evolution of oxygen and nitrogen dioxide, leaving the metal oxide residue. This latter reaction is typical of most metal nitrates:

$$Ba(NO_3)_2(s) \rightarrow BaO(s) + 2NO_2(g) + \tfrac{1}{2}O_2(g);$$

white crystals; white solid; brown fumes; glowing splint test

$$\Delta H = +500 \text{ kJ at } 25°C.$$

Nitrates, either made synthetically or derived from sources like Chile saltpetre, $NaNO_3$, are an important source of nitrogen for plants.

All common metal nitrates are soluble in water. Neither the anion nor the cation is hydrolysed appreciably for the more electropositive metals, so the solutions have pH values near to 7 at room temperature. Since the ions $Li^+(aq)$, $Be^{2+}(aq)$, and $Mg^{2+}(aq)$ are appreciably hydrolysed by water, nitrates of these metals will be appreciably acidic in water. For instance, magnesium nitrate has a pH of about 6.

*The IUPAC name for nitrite is nitrate(III), and for nitrate, nitrate(V). The common names seem likely to persist for some time and are employed in this book.

$$Mg(H_2O)_x^{2+} + aq \rightleftharpoons \left[Mg\begin{matrix}(H_2O)_{x-1}\\(OH)\end{matrix}\right]^+ + H^+(aq)$$

$[= Mg^{2+}(aq)]$ hydrolysed aqueous ion

pH = 6.

The phenomenon of hydrolysis of metal ions is closely linked to the formation of insoluble hydroxides. The ions which are readily hydrolysed form precipitates if further $OH^-(aq)$ is added, and this is a further stage in the hydrolysis process. For instance, if sodium hydroxide solution is added to a solution of magnesium nitrate, a precipitate of magnesium hydroxide will result. The overall equation is

$$2NaOH(aq) + Mg(NO_3)_2(aq) \rightarrow Mg(OH)_2(s) + 2NaNO_3(aq)$$

though this may be written ionically as a continuation of the hydrolysis equation above:

$$\left[Mg\begin{matrix}(H_2O)_{x-1}\\(OH)\end{matrix}\right]^+ + OH^-(aq) \rightarrow Mg(OH)_2(s) + (x-1)H_2O$$

(spectator ions, Na^+ and NO_3^-).

20.14 Other oxo-salts of *s*-block elements

These are considered under a common heading since they are derived from the oxo-acids of the second-row non-metals, and have certain features in common. It was shown in Chapter 19 that the acids increase in strength across the row. The weaker the oxo-acid, the greater is its tendency to "polymerize", as Table 20.7 shows.

Acid salts can exist for all except chloric(VII) acid. The main *s*-block elements which occur in nature as silicates are lithium, beryllium and occasionally magnesium. The phosphate ion is found in nature as calcium phosphate, and the sulphate ion occurs as magnesium sulphate (*Epsom salt*, $MgSO_4.7H_2O$), gypsum, $CaSO_4.2H_2O$, and barytes, $BaSO_4$. (Note the decreasing amount of water of crystallization in these salts, cf. section 20.6.)

Chlorates(VII) do not occur in nature and their properties will not be considered further here.

Sulphates, phosphates and silicates are markedly stable to heat. This is in contrast to the salts containing the first-row anions, nitrate and carbonate. Very strong heat indeed will cause sulphates to give off a mixture of $SO_2(g)$ and $SO_3(g)$ but such temperatures are not usually achieved in the laboratory.

Calcium silicate is one of the principal components of cement. Cement is made by heating powdered *marl*, a natural mixture of limestone and clay, in an oil-fired furnace. On adding water to the product, a rigid macromolecular structure is obtained, and heat is evolved in the hydration reaction. Concrete consists of cement mixed with gravel or ballast.

Calcium sulphate, which occurs as *gypsum* is of importance as a source of sulphur dioxide in the manufacture of sulphuric acid:

$$2CaSO_4(s) + C(coke) \rightarrow 2CaO(s) + 2SO_2(g) + CO_2(g)$$

TABLE 20.7

acids	$Si(OH)_4$ very weak acid	$O{=}P(OH)_3$ fairly weak acid	$O_2S(OH)_2$ fairly strong acid	$O_3Cl(OH)$
anions formed	SiO_3^{2-} and many complex "polymer" species	PO_4^{3-}, HPO_4^{2-}, $H_2PO_4^-$ and various "polymer" species, e.g. $P_2O_7^{4-}$	SO_4^{2-}, HSO_4^- and a few "polymers" such as $S_2O_7^{2-}$	ClO_4^- only

Alumina, Al_2O_3, and silica, SiO_2, are added to a powdered mixture of gypsum and coke, and calcium silicate and aluminate form a by-product which is a useful cement clinker:

$$CaO + SiO_2 \rightarrow CaSiO_3.$$

Barium sulphate is one of the most important salts of barium; it is highly insoluble and unreactive. The heavy barium nuclei make it opaque to X-rays, and a suspension of it is administered to patients requiring X-ray examination of the digestive tract. It is used as a white pigment, and occasionally as a paper filler.

An important distinction between Group IA and Group IIA is observed in this series of salts: almost all the Group IA salts are soluble in water, whereas nearly all the Group IIA salts are insoluble. Thus the sulphates, phosphates and silicates of Ca, Sr, Ba and Ra are insoluble ($CaSO_4$ sparingly soluble in water); beryllium and magnesium sulphates are however soluble.

20.15 Aluminium and its relationship to *s*-block elements

Aluminium shows a number of similarities to the *s*-block metals; it shows a strong resemblance to the element diagonally above it, beryllium. This is often referred to as a **diagonal relationship**. The reason for the similarity can be seen by examining Figs. 20.3, 20.6 and 20.7, where the trends in atomic and ionic radius, interatomic forces, ionization energy, etc., are seen to cancel out. The metals become more electropositive down a Group, but less electropositive from left to right:

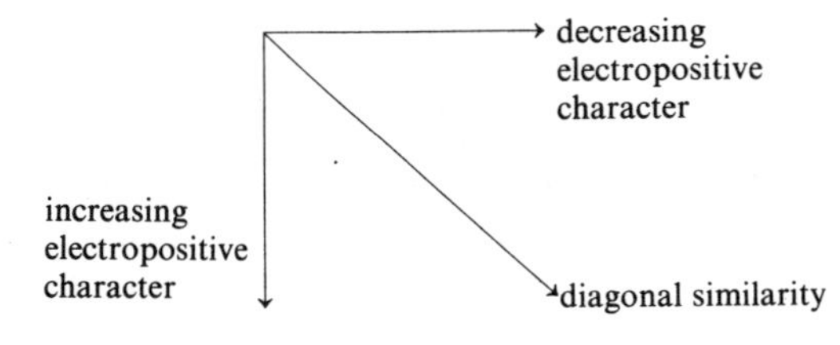

The aluminium atom has the ground state electronic structure $1s^2; 2s^2p^6; \underline{3s^2p^1}$. The underlined electrons are present in the valence shell and are available for bond formation. Figures 1.6 to 1.10 show how the physical properties of aluminium fit in with periodic trends.

Aluminium is one of the commonest elements in nature. Being a relatively "weak" metal, it is not found as chloride or nitrate, or even as carbonate or sulphate, but it occurs widely as complex aluminosilicates such as those present in clay. Such minerals are too stable chemically to be an economic source for extraction of the metal, but deposits of the hydrated oxide, bauxite, $Al_2O_3xH_2O$, also occur and this can be used to obtain the metal. Aluminium oxide is an extremely stable substance (free energy of formation $\Delta G^\ominus = -1580$ kJ mol^{-1}), since the ions Al^{3+} and O^{2-} are both quite small.

Despite the high energy of ionization needed to remove the three outer electrons to form the ion Al^{3+}, this is the stable ion in aqueous solution since considerable energy is regained as hydration energy. The reducing power, as measured by the half-cell potential $\mathcal{E}^\ominus$, is lower than for most *s*-block elements as far as aqueous solutions are concerned (Fig. 20.9).

$$Al(s) \rightleftharpoons Al^{3+}(aq) + 3e^-; \qquad \mathcal{E}^\ominus = -1{\cdot}66\ V$$

Like beryllium, aluminium shows a tendency to form molecular rather than ionic compounds. The ion $Al^{3+}(aq)$ is quite strongly hydrolysed, and behaves as an acid in aqueous solution.

$$[Al(H_2O)_6]^{3+} \rightleftharpoons H^+(aq) + [Al^{(OH)}_{(H_2O)_5}]^{2+}.$$

Aluminium hydroxide, $Al(OH)_3$, is formed as a white gelatinous precipitate when alkali is added to a solution of $Al^{3+}(aq)$ ions. Like beryllium hydroxide, it is amphoteric:

$$\underset{\text{white ppt}}{Al(OH)_3(s)} + \underset{\text{excess alkali}}{OH^-(aq)} \rightarrow \underset{\text{tetrahydroxo-aluminate}}{Al(OH)_4^-(aq)}$$

Addition of more alkali gives species such as $Al(OH)_5^{2-}$ and $Al(OH)_6^{3-}$.

Aluminium is widely used as a light structural metal, either alone or alloyed with other metals such as magnesium. Duralumin is an alloy containing about 4% copper, which has superior strength and corrosion resistance. A hard oxide coat on its surface protects it from corrosion by the atmosphere. The formation of oxide makes aluminium difficult to weld. Typical uses of aluminium are aircraft construction, roofing, liquid containers, domestic utensils and packaging. Its electrical conductivity, though inferior to that of copper, is good enough to be useful when light-weight cables are required.

20.16 Compounds of aluminium

Aluminium reacts directly with most non-metals, including nitrogen, to form compounds in which it has an oxidation number of +3, for instance $Al_2O_3(s)$, $Al_2Cl_6(s)$, $AlN(s)$, and $Al_2S_3(s)$. It burns vigorously in air when finely divided to form a hard refractory solid, m.p. 2050°C. This oxide occurs in various crystalline forms, such as *corundum* which is used in the manufacture of artificial gemstones. The oxide is also used for making refractory crucibles on account of its good stability towards heat.

Aluminium burns readily in chlorine with the formation of a volatile white smoke, Al_2Cl_6, which sublimes at 180°C. This is a molecular substance. Its configuration may be written

```
Cl  Cl  Cl
  \ / \ /
  Al  Al
  / \ / \
Cl  Cl  Cl
```

On heating to high temperatures this molecule dissociates into $AlCl_3(g)$ molecules. Aluminium bromide and iodide have similar structures, but the fluoride crystallizes as an ionic lattice.

Aluminium dissolves in dilute hydrochloric acid to form an ionic solution of aluminium chloride, but the anhydrous compound cannot be obtained by evaporation owing to hydrolysis of $Al^{3+}(aq)$. $AlCl_3$ itself reacts vigorously with water. In either case aluminium hydroxide is the result:

$$Al(s) + 3HCl(aq) \rightarrow AlCl_3(aq) + \tfrac{3}{2}H_2(g)$$

$$AlCl_3(aq) + 3H_2O \xrightarrow{\text{evaporate}} Al(OH)_3(s) + 3HCl(g).$$

Aluminium reacts explosively with sulphur, and this reaction should not be attempted. Al_2S_3 is attacked slowly by water, rapidly by $H^+(aq)$, producing gaseous hydrogen sulphide.

Aluminium forms normal salts with most acids: their formulae can be calculated by balancing the charges on the ions. For instance Al^{3+} and SO_4^{2-} form aluminium sulphate, $Al_2(SO_4)_3$, which crystallizes out as the hydrate $Al_2(SO_4)_3.18H_2O$. This can be made by dissolving aluminium, or aluminium hydroxide, in dilute sulphuric acid. When dilute nitric acid is used, the metal itself is rendered passive due to an impervious layer of oxide, though the hydroxide reacts to give aluminium nitrate.

Aluminium sulphate is less stable to heat than most sulphates:

$$Al_2(SO_4)_3(s) \rightarrow Al_2O_3(s) + 3SO_3(g); \quad \Delta G^\ominus = +408 \text{ kJ at } 25°\text{C}.$$

The very negative free energy of formation of Al_2O_3 undoubtedly has a part to play here.

An important salt is the double salt *potash alum*, of empirical formula $KAl(SO_4)_2.12H_2O$. This is less soluble in water than aluminium sulphate itself, and is often used preferentially for applications such as the above. It is used as a *mordant* in dyeing: aluminium hydroxide is precipitated and adsorbed on the fibres of a fabric, and the dye bonds to the hydroxide more easily than to the fabric alone. On a small scale it is used in styptic pencils for stopping bleeding.

Here the high charge on Al^{3+} causes the colloidal particles in blood to coagulate (Chapter 16).

Aluminium differs from the *s*-block metals in that it reacts readily with warm alkalis:

$$Al(s)+3OH^-(aq)+3H_2O \rightarrow Al(OH)_6^{3-}+\tfrac{3}{2}H_2(g).$$

In this situation it is behaving more like a non-metal than a metal.

20.17 Summary

(1) PHYSICAL PROPERTIES

Soft, relatively light, easily fusible metals.

Group II . Harder, denser, less fusible than Group IA neighbours.

Aluminium. Similar to beryllium.

(2) CHEMICAL PROPERTIES

Ionization. Only one oxidation state; M^+ (Group I) and M^{2+} (Group II) formed, with noble-gas structures; ion formation occurs readily.

Reactivity.

(a) All burn in air, and combine directly with most non-metals.

(b) Heavier members of both Groups react vigorously with water; Be and Al hardly react at all, and Mg very slowly.

(c) They are vigorous reducing agents, due to ready formation of positive ions.

COMPOUNDS.

(a) Only the smallest ions (Be^{2+} and Al^{3+}) have a marked tendency to form covalent bonds.

(b) The largest ions form the most stable compounds with large anions. Hence carbonates, nitrates, peroxides and tri-iodides for instance are most stable in Group 1 and at the bottom of a given Group.

(c) The largest ions form the most soluble hydroxides, and hence the most basic oxides.

(3) TRENDS IN THE S-BLOCK

(a) Ionic radii increase, and hence ionization energies decrease, down a Group.

(b) Ionic radii decrease from left to right.

Trends (a) and (b) tend to cancel along a diagonal, making Li and Mg similar, also Be and Al; (diagonal relationship, Chapter 24).

(c) Stabilities and solubilities of compounds usually follow well-marked trends within a Group (see above).

Although quite a wide range of properties is encountered over the entire range of the *s*-block, extremes being represented by beryllium and caesium, the block of elements forms a closely related set. Group trends are readily observed, and within a Group the elements appear to be closely related. In later Groups, for instance Group V (Chapter 23), the range of chemical behaviour is often wider, and these straightforward relationships are often obscured.

Study Questions

1. (a) Why do ionic radii (i) increase down a Group? (ii) decrease across a period?

(b) Why do ionization energies decrease down a Group, but tend to increase across a period?

2. What are the formulae of the simplest binary compounds formed by the following pairs of elements?

(a) Li and Cl.	(d) Ca and Br.	(g) Al and I.
(b) Li and O.	(e) Ca and S.	(h) Al and O.
(c) Li and N.	(f) Ca and P.	(i) Al and N.

3. In the text it is noted that lithium floats on the oil in which it is stored. (a) Why is it stored in oil? (b) Which other *s*-block elements would you expect to be stored in this way?

4. Calcium forms a fluoride CaF_2. (a) Why does CaF_3 not form with an excess of fluorine? (b) Can you suggest why CaF does not form?

5. The following chlorides are all readily soluble in water. Arrange the solutions in order of increasing pH: NaCl, LiCl, $BeCl_2$, $MgCl_2$ and $AlCl_3$.

6. The properties of elements in the same Group of the periodic table show both similarities and a gradation. Show how this applies to (a) the sulphates of the Group IIA metals (b) the reactivity of the Group IA metals with water.

7. Classify the following as acid–base or redox reactions. Label each species to show how the transfer of electrons or protons takes place.

(a) $2Na+Cl_2 \rightleftharpoons 2NaCl$.
(b) $BaO_2+2H_2O \rightleftharpoons Ba(OH)_2+H_2O_2$.
(c) $Sr+2H_2O \rightleftharpoons Sr(OH)_2+H_2$.
(d) $2Ca_3P_2+6H_2O \rightleftharpoons 2PH_3+3Ca(OH)_2$.
(e) $2NaBr+Cl_2 \rightleftharpoons 2NaCl+Br_2$.

8. Write equations for the following reactions of NaOH(aq):

(i) When it is added to a solution of a zinc salt (Zn^{2+}), a white precipitate forms which dissolves in excess alkali.
(ii) On adding it to a white crystalline solid X, ammonia is evolved.
(iii) On adding it hot to iodine, a solution containing both I^- and IO_3^- forms.
(iv) On refluxing with ethyl ethanoate, ethanol is formed.

9. Why are Group IIA salts generally much less soluble than Group IA salts? The solubilities of the Group IIA sulphates are as follows: $BeSO_4$, 440; $MgSO_4$, 260; $CaSO_4$, 2·0; $SrSO_4$, 0·11; $BaSO_4$ $2{\cdot}5\times10^{-3}$ g dm^{-3}.

(i) Calculate the solubility of each salt in mol dm^{-3}.
(ii) Give the solubility products $[M^{2+}][SO_4^{2-}]$ of each sulphate.
(iii) What trend do you observe? How does this compare with the data for the hydroxides? (Fig. 20.12.)

10. Why is magnesium used

(i) to extract uranium from UF_4?
(ii) in aircraft alloys?
(iii) in fireworks?

11. What have NaF, MgO and AlN in common? What variations would you expect in the properties of these compounds?

12. The lattice energies of LiF, NaF, KF, RbF and CsF are 1000, 890, 795, 760 and 725 kJ mol^{-1}.

(a) How can you account for the gradually diminishing values?
(b) On this basis, which of the Group IA fluorides would you expect to be least soluble in water?
(c) What other factor is important in determining these solubilities? (Draw an energy level diagram.)

13. Account for the following observations:

(a) Molten aluminium bromide is a poor conductor of electricity.
(b) An aqueous solution of the same substance is a good conductor.
(c) This aqueous solution is acidic.
(d) In the gas phase, the molecular weight of aluminium bromide is approximately 530.

14. (a) List as many similarities in the chemistries of Al and Be as you can.
(b) It is believed that, in strong alkali, aluminium forms the complex ions $Al(OH)_4^-$, $Al(OH)_5^{2-}$ and $Al(OH)_6^{3-}$. Yet in strong alkali, BeO forms only $Be(OH)_4^-$, but no $Be(OH)_5^{2-}$ nor $Be(OH)_6^{3-}$. Why is this?

15. Write equations for the following reactions:

(a) Calcium oxide reacts with carbon to give calcium dicarbide, CaC_2, which reacts with water to give ethyne.
(b) Aluminium combines directly with carbon at 1600°C to give Al_4C_3. On hydrolysis, the carbide gives an inflammable gas of m.wt. 16.
(c) A carbide of magnesium, Mg_2C_3, gives propyne, $CH_3{-}C{\equiv}CH$, on hydrolysis.

16. (a) 8·8 g of an element reacted with excess dilute hydrochloric acid to give 2250 cm^3 of hydrogen at s.t.p. Assuming that the element forms ions with only one, two or three positive charges, what possible values are there for the atomic weight of the element?
(b) The element was then ignited; it burned with a red flame to give a white solid that dissolved in water to give a basic solution. What sort of element is it?
(c) When CO_2 was passed into this solution, a white precipitate formed. Which element is it?

17. The elements francium and radium are radioactive and therefore their chemistries (especially that of Fr) have not been investigated in great detail.

(a) How would you expect Fr to react with (i) air (ii) water and (iii) chlorine? What method would you use to try and obtain a sample of metallic francium from one of its compounds?
(b) How would radium react with (i) air, (ii) water, (iii) chlorine?

(c) How soluble would you expect the hydroxide and the sulphate of radium to be? What would the formulae of these compounds be?

(d) Use the diagrams in the chapter to predict values of the boiling points, the $\mathcal{E}^\ominus$ values, the atomic volumes and the ionization energies of these elements.

18. Selenium is a non-metal in the same group as sulphur, Group VIB. Predict what will happen when powdered magnesium and powdered selenium are heated together. What will be the formula of the product? How might you expect the product to react with water?

19. Which ion in Group IIA would you expect to have the highest ionic mobility? Refer to Fig. 20.9 and Table 20.4.

20. Which of the following compounds will be paramagnetic: K_2O, K_2O_2, KO_2, K_3N? Name all these compounds.

Commercial samples of Na_2O_2 are often coloured yellow because of NaO_2 impurities. The addition of water to such samples causes a little oxygen to be evolved and the salt becomes colourless. When a trace of MnO_2 is added, a lot of oxygen is evolved. How do you account for these observations?

21. The hydration energies of the Group IA ions have been estimated to be: Li^+ 515, Na^+ 405, K^+ 322, Rb^+ 293, and Cs^+ 263 kJ mol^{-1}.

(a) What is meant by the hydration energy?

(b) Can you give a reason for the trend that is observed?

(c) How would the hydration energies of the Group IIA ions compare with these values?

22. Make a list of similarities and differences between lithium and magnesium. Explain these relationships as far as possible on the basis of physical principles.

CHAPTER 21

The halogens

A TYPICAL NON-METAL GROUP

In this chapter an important family of non-metals, the halogens, will be surveyed. Many of their properties are characteristic of non-metals in general. The Group as a whole, Group VIIB of the periodic table, affords another good example of the kind of trends observed for the *s*-block metals (Chapter 20), namely a *similarity* and a *gradation* in properties.

21.1 Occurrence and extraction of the halogens

Table 21.1 lists the halogens, with their ground state electronic structures.

The halogens are far too reactive to occur in nature as the free elements, and all are therefore found in combination with metals, generally as simple halides. This is due to the strong tendency of the halogens to gain an electron and form a negative ion, either in aqueous solution or in a crystal lattice. The tendency to form the ion X^- is weakest for astatine, and is relatively weak for iodine—iodine occurs in nature largely as sodium iodate(V), $NaIO_3$. Astatine is radioactive, and has not been found in nature except in very low concentrations. Fluorine occurs in *cryolite*, Na_3AlF_6, and in *fluorite*, CaF_2*. Chlorine occurs mainly as chloride ion in sea water, though deposits of rock salt, *halite*, NaCl, are common. The chief source of bromine is the

* Fluorite is used as a flux, and it is this property which gives rise to the name fluorine (Latin, *fluo*, flow). The term fluorescence comes from the same source, as fluorite has been known for many years to exhibit this property.

TABLE 21.1

Name	Principal quantum number of valence shell	Electronic structure in ground state (abbreviated form)	Electronic structure (full form)
Fluorine	2	2, 7	$1s^2$; $2s^2p^5$
Chlorine	3	2, 8, 7	$1s^2$; $2s^2p^6$; $3s^2p^5$
Bromine	4	2, 8, 18, 7	$1s^2$; $2s^2p^6$; $3s^2p^6d^{10}$; $4s^2p^5$
Iodine	5	2, 8, 18, 18, 7	$1s^2$; $2s^2p^6$; $3s^2p^6d^{10}$; $4s^2p^6d^{10}$; $5s^2p^5$
Astatine	6	2, 8, 18, 32, 18, 7	$1s^2$; $2s^2p^6$; $3s^2p^6d^{10}$; $4s^2p^6d^{10}f^{14}$; $5s^2p^6d^{10}$; $6s^2p^5$

bromide ion in sea water. Iodine occurs in low concentrations in sea water, but is absorbed by certain seaweeds. Seaweed ash contains considerable quantities of iodide ion. Table 21.2 shows the relative abundance of the halogens in nature.

TABLE 21.2

Element	% by weight in Earth's crust	concentration in sea water (average)
Fluorine	0·08	10^{-4} M
Chlorine	0·19	0·56 M
Bromine	0·01	8×10^{-4} M
Iodine	10^{-4}	very small
Astatine	negligible	negligible

The extraction of the free elements from their simple ions is essentially a problem of electron removal, that is, oxidation. The extraction method chosen depends on the gradation of properties down the group. Figure 21.1 shows that electron removal is going to be most difficult for fluorine, slightly easier for chlorine, and so on. In fact the half-cell potential for the reaction $F_2(g) \rightleftharpoons 2F^-(aq)+2e^-$ is more positive than for almost any other known couple, making it impossible to extract fluorine with a chemical oxidizing agent. If electrolysis is used instead, there is no limit to the potential which can be applied to effect the oxidation. In practice chlorine and fluorine are both generally extracted by electrolysis of a fused salt or suitable solution (see below). With bromine and iodine, chemical oxidation is feasible and generally more convenient. A good oxidizing agent is chlorine, which has the merit of being cheap, since its relative abundance is much greater. The extraction of bromine from sea-water depends upon the reaction:

$$Cl_2(g)+2Br^-(aq) \rightarrow Br_2(aq)+2Cl^-(aq);$$
$$\Delta\mathcal{E}^\ominus = 1{\cdot}36-1{\cdot}07\text{ V}$$
$$= +0{\cdot}29\text{ V}$$
$$\Delta G^\ominus = -28\text{ kJ at }25^\circ\text{C}.$$

Extraction of fluorine. Fluorine is extracted by electrolysing a solution of potassium fluoride in anhydrous hydrogen fluoride, at a fairly high temperature (between 100° and 270°, depending upon the electrolyte composition). The chief

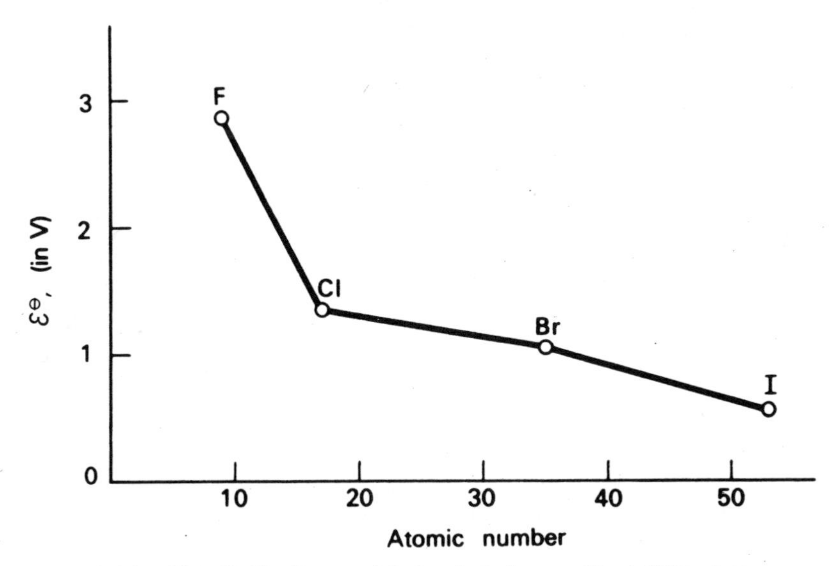

FIG. 21.1. Half-cell potentials for the halogens, $X_2 \rightleftharpoons 2X^-+2e^-$.

problem is to prevent the fluorine from attacking the cell or the anode. All metals are attacked by fluorine, but a protective coat of fluoride often forms, and special steel is used in modern cells, together with a special non-graphite carbon anode.

Extraction of chlorine. An early process (invented by Deacon in 1868) for obtaining chlorine was the oxidation of hydrogen chloride by oxygen in the presence of a catalyst of $Cu^{2+}(aq)$.

$$4HCl(g)+O_2(g) \longrightarrow 2H_2O(g)+2Cl_2(g);$$

In the laboratory, chlorine is made by oxidizing concentrated HCl(aq) with a convenient oxidizing agent such as potassium manganate(VII) (no heat required) or manganese(IV) oxide (gentle heat needed). The latter oxidizing agent was once of industrial importance, but the process using it is now obsolete.

In the United Kingdom, the production of chlorine is closely tied up with the electrolysis of brine for making sodium hydroxide, and this is the main source of the gas.

Extraction of bromine. The usual commercial method for bromine extraction is to evaporate sea water; chlorides, mainly sodium chloride, crystallize out leaving the "mother liquor" relatively rich in bromide ion. Chlorine gas is then bubbled through, and bromine obtained from the solution by distillation. A less common commercial method, but one commonly carried out as a laboratory reaction, is to heat a bromide with concentrated sulphuric acid and manganese(IV) oxide. The reaction can be regarded as two distinct processes:

(i) $KBr(s)+H_2SO_4(l) \rightarrow KHSO_4(s)+HBr(g)$.

Notice that the acid salt is formed, and that the "driving force" of the reaction is the formation of a gaseous product.

(ii) $MnO_2+4H^++2Br^- \rightarrow Mn^{2+}+Br_2+2H_2O$.

The overall equation becomes:

$$MnO_2+2KBr+3H_2SO_4 \rightarrow MnSO_4+2KHSO_4+Br_2+2H_2O.$$

Extraction of iodine. The majority of the world's iodine comes from the small percentage of sodium iodate(V) present in *caliche* (*Chile saltpetre*). Iodate(V) ions, $IO_3^-(aq)$ are reduced with sodium hydrogensulphite,* which provides $HSO_3^-(aq)$. The reaction occurs in two stages:

(i) $3HSO_3^-+IO_3^- \rightarrow I^-+3HSO_4^-$

(ii) $IO_3^-+5I^-+6H^+ \rightarrow 3I_2+3H_2O$.

Reaction (ii) involves two different oxidation states of iodine becoming one (the reverse of disproportionation).

Smaller amounts of iodine are obtained from certain natural brines containing $I^-(aq)$, or from seaweed ash.

Astatine. Astatine does not occur in nature in appreciable amounts. It is made by bombarding bismuth with α-particles, and little is known of its chemistry and physical properties. Its behaviour appears to resemble that of a non-metal more than that of a metal. For instance, it is fairly volatile, and can be dissolved in benzene.

21.2 General survey of reactions and uses of the halogens

The reactions of the halogens are dominated by the strong tendency for the half-reaction:

$$X_2+2e^- \rightarrow 2X^-.$$

Fluorine reacts with all metals, even the noble metals (platinum, gold etc.), and with most non-

*In this book, oxo-ions of sulphur(IV) are termed sulphites.

metals. Chlorine attacks nearly all metals though it is less reactive with non-metals. Bromine and iodine show progressively less reactivity towards metals.

Fluorine gas finds few uses, and it is generally converted into a substance such as SbF_3 which is a milder fluorinating agent. A range of organic compounds containing fluorine in place of hydrogen has been developed in recent years, and a number of such compounds are of commercial importance, for instance the refrigerant gas *freon* (containing CCl_2F_2, $CClF_3$, etc.) and the polymer PTFE (polytetrafluoroethene), $(C_2F_4)_x$ (section 24.6).

Most of the uses of chlorine depend upon its oxidizing power. It is used as a disinfectant, and as a wood-pulp bleach in the manufacture of paper. A large amount of chlorine is used in the organic chemical industry, for instance in the making of dyes, plastics such as PVC, insecticides such as DDT, and explosives.

Bromine is used chiefly in the manufacture of silver bromide for photography, and in drug manufacture (potassium bromide is a widely used sedative).

Iodine is an essential element to health. A deficiency of iodine leads to abnormal swelling of the thyroid gland in the neck—a condition known as *goitre* which is now prevented by controlled addition of iodide to water supplies which require it. Small quantities of iodine are consumed in a miscellany of different ways such as for antiseptics, organic catalysts, and in the manufacture of "polaroid".

21.3 Trends in physical properties of the halogens

The halogen elements all form diatomic molecules X_2, due to the sharing of one electron pair from the uppermost energy level. The atoms in X_2 therefore all have completed octets, and can be regarded as being isoelectronic with their noble-gas neighbours. Figure 21.2 shows how the sizes of atoms and ions, and the bond lengths, increase down the group.

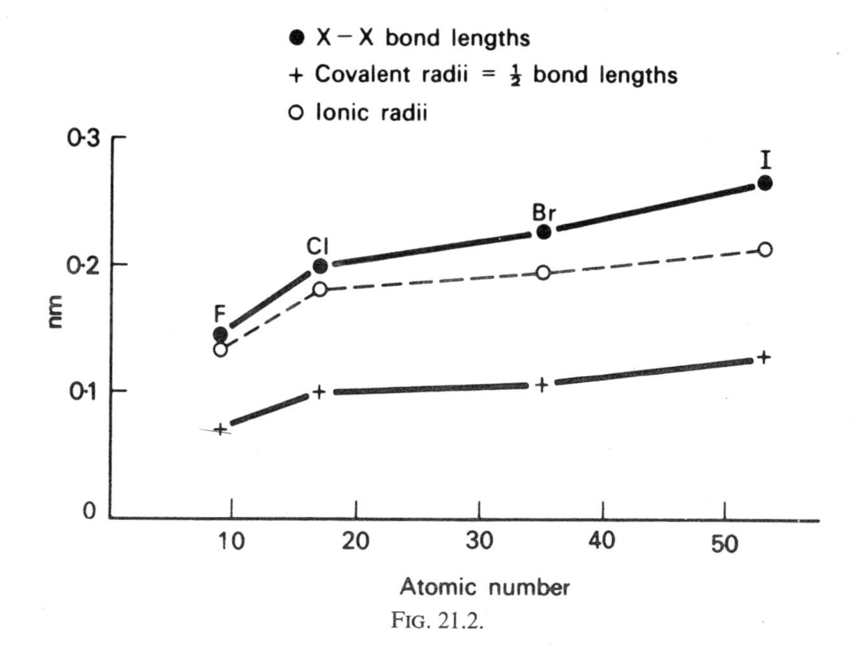

FIG. 21.2.

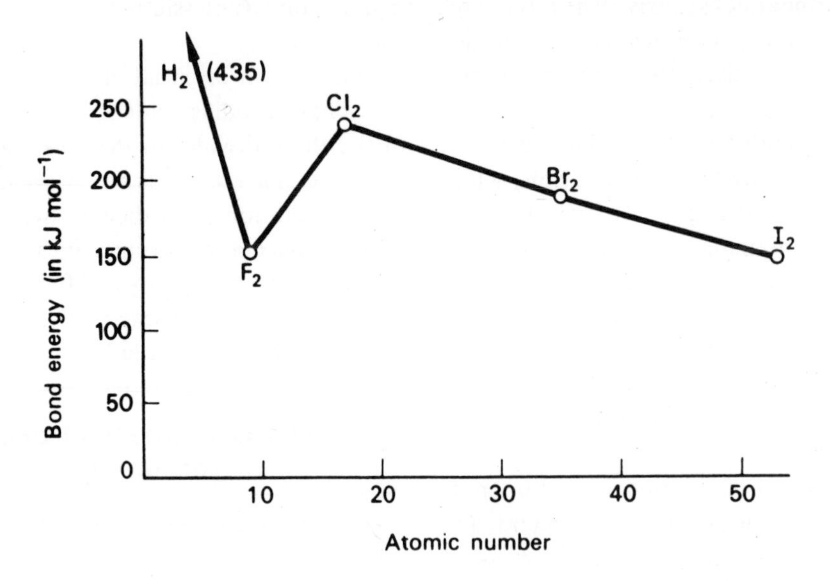

FIG. 21.3. Bond energies in Group VIIB.

Figure 21.3 shows a plot of bond dissociation energy against atomic number of X. Fluorine has an unexpectedly weak bond, and this is part of the reason for fluorine's great reactivity. In contrast the bonds formed between fluorine atoms and other atoms are generally fairly strong. Table 21.3 shows how the strengths of bonds H–X and C–X vary, in comparison with X–X.

TABLE 21.3
BOND ENERGIES (kJ mol^{-1})

	X—X	H—X	C—X
X = fluorine	155	565	438
X = chlorine	242	430	330
X = bromine	192	363	276
X = iodine	150	297	238

The melting points and boiling points of the halogens both increase with molecular weight. Iodine is a solid at room temperature (sublimes at 183°C) and bromine is a liquid.

Refer back to Fig. 3.2 and you will see that the halogens all have high ionization energies, the lowest being iodine (excluding astatine). Hence we would not expect the halogens to form positive ions, X^+, in chemical compounds.*

The tendency for the atoms to form negative ions, X^-, on the other hand, is very marked. The halogens all possess large negative values of **electron affinity.** The electron affinity of an atom X is the opposite of ionization energy of the negative ion X^-, that is:

$$X(g) \rightarrow X^-(g); \quad \text{electron affinity} = \Delta E.$$

Electronegativity. The electronegativity of an element may be loosely defined as "the power of an atom in a molecule to attract electrons". Various attempts have been made to assign numerical values to electronegativity coefficients but none is entirely satisfactory. One such attempt, by Mulliken (Fig. 21.4), is based on

* Binary compounds of iodine containing the element in an oxidation state of +1, for instance ICl, frequently exist, but they are not ionic.

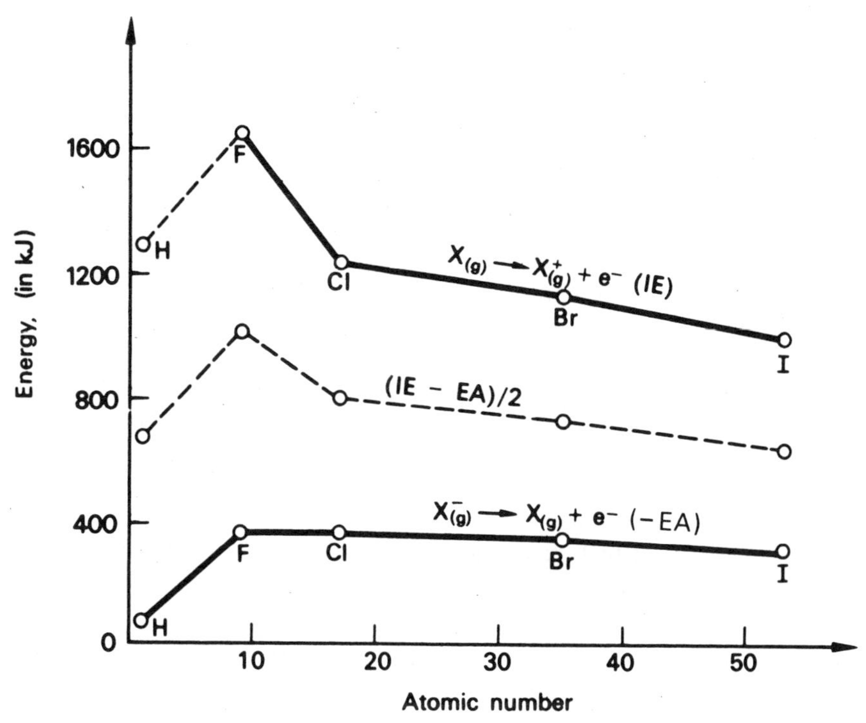

FIG. 21.4. Mulliken's electronegativity = (I.E. − E.A.)/2.

the relationship

$$\text{electronegativity} \propto \frac{\overset{\text{(I.E.)}}{\text{ionization energy}} - \overset{\text{(E.A.)}}{\text{electron affinity}}}{2}$$

In this book the concept is only used qualitatively.

The halogen molecules are non-polar and consequently are not very soluble in water, except where chemical reaction plays a part. Fluorine attacks water and "dissolves" in it in the same sense that sodium "dissolves" in water—a variety of products including ozone and the fluoride ion are formed—while chlorine is moderately soluble due to slow chemical reaction:

$$Cl_2(g) + H_2O(l) \rightarrow HCl(aq) + HOCl(aq).$$

Bromine is sparingly soluble in water giving a red solution. Bromine water kept in the light on the laboratory shelf slowly decolorizes due to hydrolysis. Iodine is practically insoluble in water. All the halogens dissolve readily in non-polar solvents such as carbon tetrachloride.

21.4 The hydrogen halides

Certain of the properties of the hydrogen halides have already been dealt with under different headings in the book. They are summarized here together with the chapter references:

(i) Hydrogen fluoride has a high boiling point (19°C) due to hydrogen bond formation, but the remainder are gaseous substances consisting of molecules HX.

(ii) They are all acids (proton donors) which react with water as follows:

$$HX + aq \rightleftharpoons H^+(aq) + X^-(aq).$$

(iii) They are all highly soluble in water, forming maximum boiling azeotropes. This again indicates bond formation with water.

(iv) Their bond energies decrease down the group (section 21.3 above). For this reason hydrogen iodide is fairly easily dissociated on heating, hydrogen bromide only with difficulty, and hydrogen chloride and fluoride hardly at all.

(v) They are all colourless gases which fume in moist air. This property of fuming is due to reaction between HX and water vapour to form a "fog" of HX(aq).

General methods of making hydrogen halides. Industrially, large amounts of hydrogen chloride, for conversion into hydrochloric acid, are made by burning hydrogen and chlorine together. This method is not suitable for hydrogen fluoride as hydrogen and fluorine explode on contact, even at $-200°C$ in the dark.

In the laboratory any hydrogen halide can be made by hydrolysis of a suitable halide of a metal or non-metal. Taking hydrogen chloride as an example, most non-metal chlorides fume in moist air due to the formation of hydrogen chloride:

$$BCl_3(g) + 3H_2O \rightarrow H_3BO_3(aq) + 3HCl(aq, g)$$

$$PCl_5(s) + H_2O \rightarrow POCl_3(l) + 2HCl(aq, g)$$

Some of the weaker metals have chlorides which resemble non-metal chlorides in their readiness to react with water: aluminium chloride is an example (section 20.16). Other metal halides come to an equilibrium when dissolved in water, without any violent reaction: the only effect observed is an acidic pH value due to hydrolysis. Note that halides of carbon, such as CCl_4, do not react with water (section 24.6).

A hydrogen halide is produced whenever an organic hydroxy-compound is attacked by a suitable non-metal halide. A common test for organic OH-groups is the reaction of the compound with phosphorus pentachloride:

$$ROH + PCl_5(s) \rightarrow RCl + HCl(g) + POCl_3.$$

Frequently a hydrogen halide can be produced by displacement with a stronger, less volatile acid. Concentrated sulphuric acid is suitable for making HF and HCl, and this is a common method for obtaining these compounds. Notice that the *acid* salt is formed in this displacement reaction.

$$NaCl(s) + H_2SO_4(l) \rightarrow NaHSO_4(s) + HCl(g).$$

$$CaF_2(s) + 2H_2SO_4(l) \rightarrow Ca(HSO_4)_2(s) + 2HF(g).$$

Sulphuric acid is not a suitable reagent for making HBr and HI, however, since these are oxidized; phosphoric(V) acid is used instead. The reaction with sulphuric acid can be used as a test for the halide ions (Table 21.4).

TABLE 21.4

Halide	Reaction with concentrated sulphuric acid
Fluoride	Bubbles of HF evolved which stick to the glass test-tube (HF attacks glass).
Chloride	Colourless fuming gas, HCl, evolved.
Bromide	Brown acidic gas evolved (some HBr, some bromine vapour, and some reduction products such as SO_2).
Iodide	Violet vapour, I_2, evolved, and reduction products such as SO_2, H_2S and sulphur.

Hydrogen iodide is a powerful reducing agent, being so readily oxidized to iodine, and it is often used for reducing organic compounds.

Hydrochloric acid finds extensive industrial use where an acid is required, and it is used for cleaning metal surfaces. Hydrofluoric acid is a relatively weak acid, but possesses the unusual property of attacking glass:

$$6HF(aq) + \underset{\text{in glass}}{SiO_2} \rightarrow 2H^+(aq) + \underset{\text{water soluble}}{SiF_6^{2-}(aq)} + 2H_2O.$$

All the aqueous hydrogen halides show typical "acidic" properties, reacting with metals to form hydrogen, with carbonates to form carbon dioxide, and with oxides and hydroxides to form salts (halides). The anhydrous substances or their solutions in dry benzene do not show these acidic properties; for instance anhydrous hydrogen chloride dissolved in benzene will not attack zinc or calcium carbonate. This shows that the "acid" in aqueous solution is $H^+(aq)$ and not HX.

21.5 Interhalogen compounds

A compound consisting of two elements, both halogens, is called an **interhalogen**. Table 21.5 shows the formulae of some interhalogens. Only one or two are of practical importance, and the best known is probably iodine monochloride, ICl(1). The physical properties depend on the molecular weights and polarities of the molecules. The shape of these molecules is of great interest: it will be left to the reader to deduce from arguments of electron-pair repulsions why these molecules have the shapes they do (refer back to section 6.6).

TABLE 21.5

SOME INTERHALOGENS

Linear	T-shaped	Square pyramid	Pentagonal bipyramid
ClF BrF BrCl ICl	ClF_3 BrF_3 IF_3 ICl_3	ClF_5 BrF_5 IF_5	IF_7

Iodine has a strong tendency to aggregate other halogen atoms around itself, and in addition to molecular compounds, **polyhalide** ions are formed in aqueous solution. The best known of these is the ion $I_3^-(aq)$, formed when iodine is dissolved in a solution already containing iodide ions:

$$I_2 + I^-(aq) \rightleftharpoons I_3^-(aq); \quad K = 725 \text{ mol}^{-1} \text{ dm}^3 \text{ at } 25°C$$

This complex ion is relatively easily decomposed, and a solution of tri-iodide ions behaves in effect like a solution of free iodine. Another example of a polyhalide ion is the ion $ICl_4^-(aq)$. Salts of polyhalides are stable when the metal cation is large (cf. section 20.7). Among the alkali metals for instance, KI_3, RbI_3 and CsI_3 are formed though not LiI_3 and NaI_3.

21.6 Oxides and oxo-acids

Fluorine forms a series of oxides at low temperatures, but $F_2O(g)$ is the only one stable at room temperature. They are more correctly regarded as oxygen fluorides, since oxygen is less electronegative than fluorine.

Chlorine forms $Cl_2O(g)$, $ClO_2(g)$, $Cl_2O_6(l)$, and $Cl_2O_7(l)$. Cl_2O_7, dichlorine heptoxide is made by powerfully dehydrating chloric(VII) acid with phosphorus(V) oxide, and it reacts vigorously with water to re-form the acid:

$$2HClO_4(l) + \tfrac{1}{2}P_4O_{10}(s) \rightarrow Cl_2O_7(l) + 2HPO_3(l)$$

$$Cl_2O_7(l) + H_2O \rightarrow 2HClO_4(l).$$

The oxides of chlorine are in general unstable and explosive.

Bromine forms a series of oxides, all of which are unstable.

Iodine forms various oxides, among them iodine(V) oxide, $I_2O_5(s)$. This is made by heating iodic(V) acid crystals above 100°C. At about 300° it dissociates into its elements, but it is more stable thermally than most halogen oxides:

$$2HIO_3(s) \xrightarrow{100^\circ} I_2O_5(s)+H_2O(g)$$

$$I_2O_5(s) \xrightarrow{300^\circ} I_2(g)+\tfrac{5}{2}O_2(g)$$

I_2O_5 is an oxidizing agent, and can be used in the quantitative determination of carbon monoxide:

$$I_2O_5(s)+5CO(g) \longrightarrow I_2(g)+5CO_2(g).$$

Although there does not appear to be a clearly discernible pattern of behaviour among the halogen oxides, the oxo-acids follow a distinct pattern. Fluorine is too electronegative to assume a positive oxidation number, so it does not form true oxo-acids, but among the remaining halogens oxidation numbers of +1, +3, +5 and +7 are observed. Table 21.6 shows the names and formulae of the halogen oxo-ions. Bromine oxo-acids are in general rather unstable.

TABLE 21.6

Oxidation number	Fluorine	Chlorine	Bromine	Iodine
+1	—	ClO^-	BrO^-	IO^-
+3	—	ClO_2^-	—	IO_2^-
+5	—	ClO_3^-	BrO_3^-	IO_3^-
+7	—	ClO_4^-	BrO_4^-	IO_4^-

Chloric(I) acid. Sodium chlorate(I), commercially known as sodium hypochlorite, is the product when cold aqueous sodium hydroxide is reacted with chlorine. This can be regarded as hydrolysis of the element, which disproportionates into the oxidation states +1 and −1:

disproportionation

$$\overset{0}{Cl_2}+2OH^- \rightarrow \overset{-1}{Cl^-}+\overset{+1}{OCl^-}+H_2O$$

A solution of sodium chlorate(I), mixed with sodium chloride, is made by electrolysing brine and allowing the products, OH^- and Cl_2, to react together. In an electrolytic cell for making sodium hydroxide it is necessary to devise some means of keeping the chlorine separate from the sodium hydroxide.

It is not possible to isolate a solid chlorate(I) salt. Any attempt to recover a solid product by evaporation leads to decomposition and loss of oxygen and chlorine oxides. Aqueous sodium chlorate(I) is used as a household bleach, sold commercially under names such as "Domestos". The bleaching action of the chlorate(I) ion is due to its being a powerful oxidizing agent which can oxidize a dye to a colourless product.

Chloric(I) acid can only exist in dilute solution, and the action of dilute acid on the chlorate(I) ion causes decomposition in which chlorine is evolved:

$$2H^++OCl^-+Cl^- \rightarrow Cl_2+H_2O.$$

Chlorate(V). On heating a dilute solution of a chlorate(I) it disproportionates giving the oxidation states +5 and −1:

disproportionation

$$\underset{\text{chlorate(I)}}{\overset{+1}{3ClO^-(aq)}} \longrightarrow \overset{-1}{2Cl^-(aq)}+\underset{\text{chlorate(V)}}{\overset{+5}{ClO_3^-(aq)}}$$

The same overall reaction occurs on reacting chlorine with *hot* alkali:

$$3Cl_2+6OH^- \rightarrow 5Cl^-+ClO_3^-+3H_2O.$$

Sodium and potassium chlorate(V) are white solids, soluble in water, and quite powerful oxidizing agents. Potassium chlorate(V) is a weed-killer.

Chlorate(VII). Heating potassium chlorate(V) to just above its melting point leads to further disproportionation, forming potassium chlorate (VII):

$$\overset{+5}{4KClO_3} \rightarrow \overset{+7}{3KClO_4} + \overset{-1}{KCl}$$

If potassium chlorate(VII) is distilled with concentrated sulphuric acid, anhydrous chloric(VII) acid can be distilled. This is a very powerful oxidizing agent, which is liable to react explosively with organic matter such as dust.

$$H_2SO_4 + KClO_4 \rightarrow HClO_4 + KHSO_4$$

distil under reduced pressure

Iodic(V) acid and iodate(V). Iodine oxo-acids and oxo-ions are relatively stable, and deposits of sodium iodate(V) occur naturally as a constituent of *caliche* (Chile saltpetre, mainly $NaNO_3$). Iodic(V) acid is the product when iodine is oxidized with a powerful oxidizing agent such as concentrated nitric acid. The equation may be derived as follows:

(i) $I_2 + 6H_2O \rightarrow 2IO_3^- + 12H^+ + 10e^-$
(ii) $2HNO_3 + 6H^+ + 6e^- \rightarrow 2NO + 4H_2O$

Take five times (ii) added to three times (i):

$$3I_2 + 10HNO_3 \rightarrow 6H^+ + 6IO_3^- + 10NO + 2H_2O$$

Iodic(V) acid, and the iodate(V) ion, are fairly powerful oxidizing agents. A common reaction is that between iodate(V) and iodide:

$$\underset{+5}{HIO_3} + \underset{-1}{5HI} \rightarrow \underset{0}{3I_2} + 3H_2O$$

21.7 Titrations involving iodine

A solution of iodine will oxidize a wide variety of reducing agents quantitatively, and a standard solution can be used for titration purposes. Iodine is almost insoluble in water, but it dissolves readily in potassium iodide solution, forming the tri-iodide ion (section 21.5). For practical purposes such a solution can be regarded as containing free iodine, since the I_3^- readily decomposes.

Relatively weak reducing agents are oxidized by iodine solution, and a common example used in practice is the thiosulphate ion, $S_2O_3^{2-}(aq)$. Sodium thiosulphate, $Na_2S_2O_3.5H_2O(s)$, can be obtained pure and is used to standardize a solution of iodine. The reaction is:

oxidized

$$2S_2O_3^{2-} + I_2 \rightarrow S_4O_6^{2-} + 2I^-$$

reduced

If a strong oxidizing agent is treated in neutral or acid solution with a large excess of iodide ion, the latter acts as a reducing agent and the oxidant is quantitatively reduced. An equivalent amount of iodine is liberated and this is then titrated with a standard solution of a reducing agent, usually sodium thiosulphate. Below are three examples of equations which can be employed in titrations. Study Question 6 gives practice in balancing a few more.

$$H_2O_2 + 2H^+ + 2I^- \rightarrow 2H_2O + I_2$$
$$2HNO_2 + 2H^+ + 2I^- \rightarrow 2NO + I_2 + 2H_2O$$
$$Br_2 + 2I^- \rightarrow 2Br^- + I_2$$

Iodine is quite strongly brown-coloured in potassium iodide solution, and to some extent acts as its own indicator. The colour can be strongly accentuated by adding starch solution indicator, which reacts with iodine to form a dark blue substance of indefinite composition.

21.8 A general survey of chlorides

In order to understand the properties of halides of the elements, we will first consider the

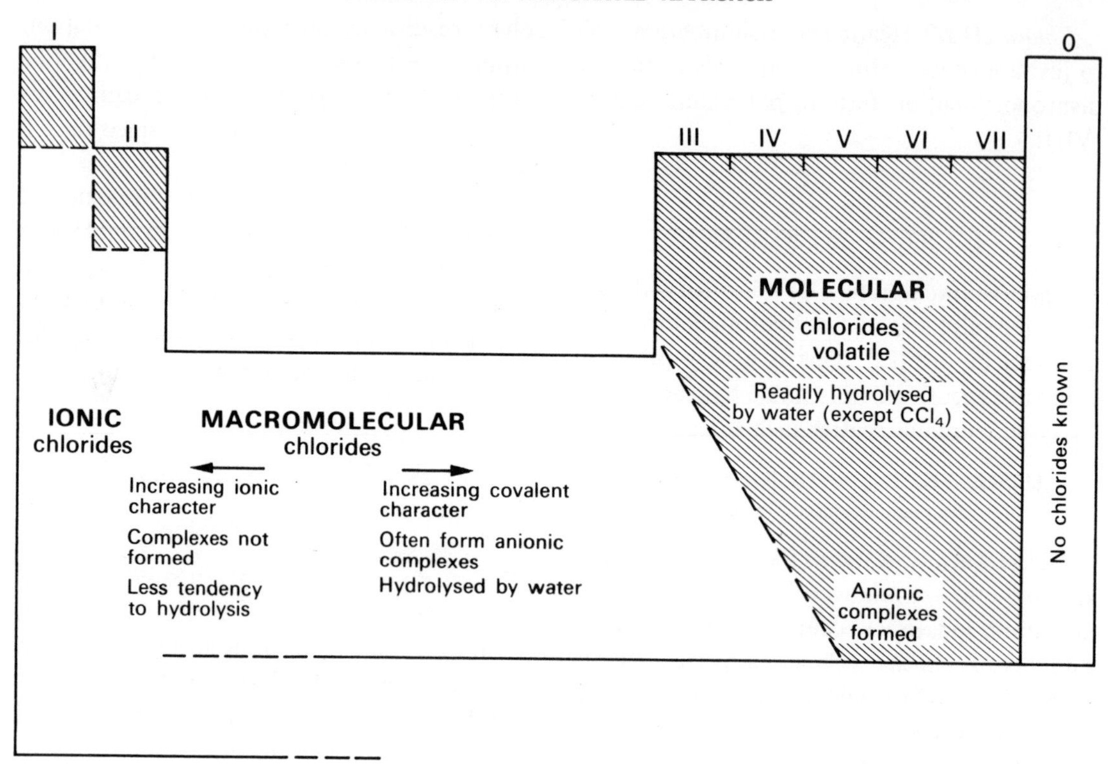

FIG. 21.5.

chlorides. Figure 21.5 shows diagrammatically how the properties of chlorides depend upon the position of the element in the periodic table. Note the following points:

(a) Chlorides of the *s*-block metals, except for $BeCl_2$ and $MgCl_2$, are more or less ideally ionic: they are soluble in water and their ions are hardly hydrolysed at all (pH of aqueous solutions near 7). $MgCl_2$(s) and $BeCl_2$(s) do not fit the ionic model so well, and indeed beryllium chloride has a structure analogous to Al_2Cl_6, which is the result of directional bonds between atoms rather than that due to the packing together of oppositely charged ions:

```
Cl  Cl  Cl  Cl
/ \ / \ / \ / \
  Be  Be  Be
\ / \ / \ / \ /
Cl  Cl  Cl  Cl
```

(b) Chlorides of the non-metals are molecular. Apart from CCl_4 they generally react with water forming hydrogen chloride.

(c) Chlorides of the B-metals and transition metals have giant structures that are neither truly ionic nor truly macromolecular. Those of higher oxidation state are usually molecular (Table 21.7). Most are soluble in water, but silver chloride, AgCl, mercury(I) chloride, Hg_2Cl_2, and lead(II) chloride are rather insoluble. Transition metals and B-metals have a tendency to form anionic complexes, for instance:

$PbCl_4^{2-}$(aq)	tetrachloroplumbate(II)
$CuCl_4^{2-}$(aq)	tetrachlorocuprate(II)
$BiCl_4^-$(aq)	tetrachlorobismuthate(III).

In naming a complex ion, the ending -ate is used to indicate an anion, followed by the oxidation number of the metal. The oxidation number of the metal is easily derived by imagining the ion to dissociate into its constituents:

$$PbCl_4^{2-} \rightarrow \underset{\text{lead(II)}}{Pb^{2+}} + 4Cl^-$$

$$PbCl_6^{2-} \rightarrow \underset{\text{lead(IV)}}{Pb^{4+}} + 6Cl^-$$

Apart from a few chlorides such as those mentioned above, most metal chlorides are soluble in water. If a solution of a metal ion in water gives a precipitate on adding dilute hydrochloric acid, the following metal ions can be suspected: silver(I), lead(II) or mercury(I).

The chloride ion is colourless, and therefore if a solution of a chloride is coloured, the colour is that due to the metal ion present; only transition metals give coloured ions in solution.

(d) Apart from one or two exceptions, chlorides of transition metals and B-metals hydrolyse with water giving HCl(g). It is not generally possible to prepare an anhydrous chloride by evaporating an aqueous solution to dryness.

TABLE 21.7

Giant structures		Molecular structures	
Formula	m.p.	Formula	m.p.
$FeCl_2$	670°C	Fe_2Cl_6	319°C
$PbCl_2$	498°C	$PbCl_4$	−15°C
$SnCl_2$	247°C	$SnCl_4$	−33°C

Zinc chloride for instance gives a basic chloride, and aluminium chloride hydrolyses completely to hydroxide:

$$ZnCl_2(aq) + H_2O \xrightarrow{\text{evaporate}} Zn(OH)Cl(s) + HCl(g)$$

$$AlCl_3(aq) + 3H_2O \rightarrow Al(OH)_3(s) + 3HCl(g)$$

The replacement of chlorine atoms by hydroxy-groups is not a phenomenon peculiar to inorganic chemistry. Organic chlorides hydrolyse in an exactly analogous way:

$$\underset{\text{ethanoyl chloride}}{CH_3COCl} + H_2O \rightarrow \underset{\text{ethanoic acid}}{CH_3COOH} + HCl$$

$$\underset{\text{1-chlorobutane}}{C_4H_9Cl} + H_2O \rightarrow \underset{\text{butan-1-ol}}{C_4H_9OH} + HCl$$

Anhydrous chlorides of the metals which hydrolyse can be prepared by using a method which avoids the use of water. The following methods are available:

(i) Pass a stream of *dry* chlorine gas over the heated metal. Where a metal can exist in more than one oxidation state, this reaction produces a higher state because chlorine is an oxidizing agent. Aluminium chloride, $Al_2Cl_6(s)$, and iron(III) chloride, $Fe_2Cl_6(s)$, can be made in this way (Figure 21.6).

(ii) Pass a stream of *dry* hydrogen chloride gas over the heated metal. This produces a chloride of lower oxidation state, because hydrogen chloride is not such a powerful oxidizing agent. For instance iron filings give mainly iron(II) chloride, $FeCl_2(s)$.

(iii) Heat the hydrated chloride in a stream of dry hydrogen chloride gas. This reaction works by driving the equilibrium composition over to the right:

$$Zn(OH)Cl + \underset{\text{passed into reaction}}{HCl} \rightleftharpoons \underset{\text{residue}}{ZnCl_2} + \underset{\text{driven from reaction}}{H_2O}$$

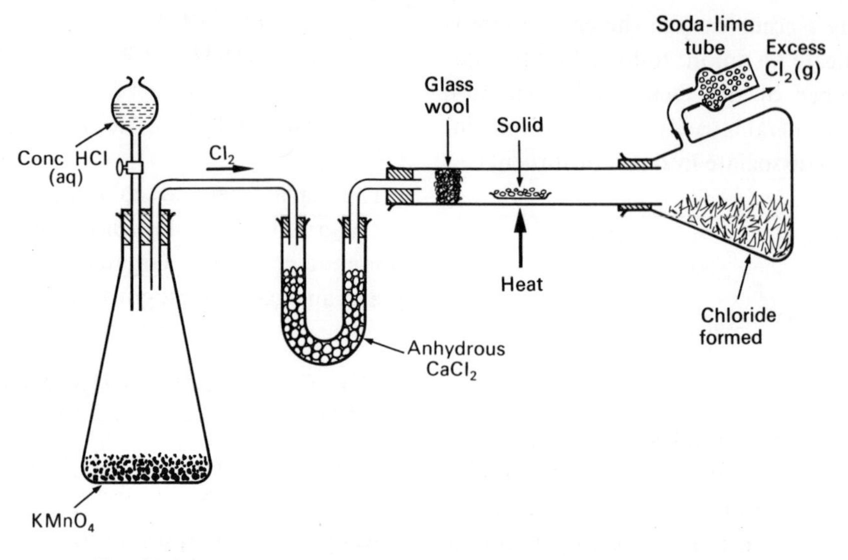

FIG. 21.6. Preparation of the volatile chloride of a metal, e.g. Al_2Cl_6.

21.9 Reactions of the chloride ion

Certain reactions of the chloride ion are quite commonly met with in the laboratory, and they are summarized here for convenience:

(1) $Cl^-(aq) + Ag^+(aq) \rightarrow AgCl(s)$. This reaction forms the basis of a common test for chloride ions; silver nitrate is added, and a white precipitate of silver chloride appears. In order to make the test specific for chloride, the reaction is carried out in acid solution (dilute nitric acid is added first), and the resultant precipitate is confirmed by adding an excess of dilute ammonia solution. AgBr(s) (cream coloured) and AgI(s) (yellow) both form precipitates under the same conditions but these will not dissolve in dilute ammonia solution.

(2) Chloride ions are oxidized to chlorine by any powerful chemical oxidant ($\mathcal{E}^\ominus$ must be greater than +1·36 V), or at the anode in electrolysis. Suitable chemical oxidants include manganese(IV) oxide, MnO_2, and potassium manganate(VII), $KMnO_4$.

21.10 Fluorides compared with chlorides

In many respects fluorides resemble chlorides in their properties. The major point of difference is that fluorides generally behave as more ideally ionic compounds. Their structures are those to be expected from the packing together of oppositely charged ions, whereas chlorides frequently have layer structures due to directional bond formation. Moreover, the solubility of fluorides in water is generally what would be expected from ionic substances. Silver fluoride for instance, is soluble in water.

Fluorine has the property of bringing out the maximum oxidation state in the elements with which it is combined due to a combination of exceptionally high oxidizing power and small size. In fact it is often difficult to prepare a fluoride of low oxidation number. Among the non-metals, compare the formula of SF_6 with SCl_4, and IF_7 with ICl_3. Among the metals, note that uranium forms UF_6 (used in isotope separation) but not UCl_6.

21.11 Noble gas compounds

An example of the great reactivity of fluorine is the formation of fluorides by the **noble gases.** Until 1962, no noble gas compound had been made and some chemists were of the opinion that the noble gases were truly *inert* elements and they were often known as the "inert gases". The first noble gas compound to be synthesised was $Xe^+PtF_6^-$, but before long the binary compounds XeF_2, XeF_4 and XeF_6 had been made by the direct combination of xenon and fluorine.

Oxygen shares with fluorine the property of bringing out high oxidation states in the elements with which it combines. The molecules XeO_3 and $XeOF_4$, and the ion XeO_6^{4-} are known and are prepared from the reaction of the xenon fluorides with water. It is just possible that xenon may be persuaded to form a very unstable chloride, but the noble gases are unlikely to form compounds with any other elements.

Although xenon forms most compounds, both krypton and radon have been shown to combine with fluorine. It is extremely unlikely that neon or argon will ever be persuaded to form compounds, since their formation would be most unfavourable energetically.

The shapes of molecules of noble gas compounds can be predicted using the usual rules. (Chapter 6 and Study Question 5.)

21.12 Bromides and iodides compared with chlorides

We have already seen that whereas fluorine tends to form ideally ionic compounds with metals, chlorine has a tendency to form compounds which depart from the ionic model (layer structures, insolubility in water). The same trend is observed in even more marked degree with bromides and iodides; thus many iodides are insoluble in water. Complex ions readily form, and many iodides dissolve in an excess of iodide ion, for instance:

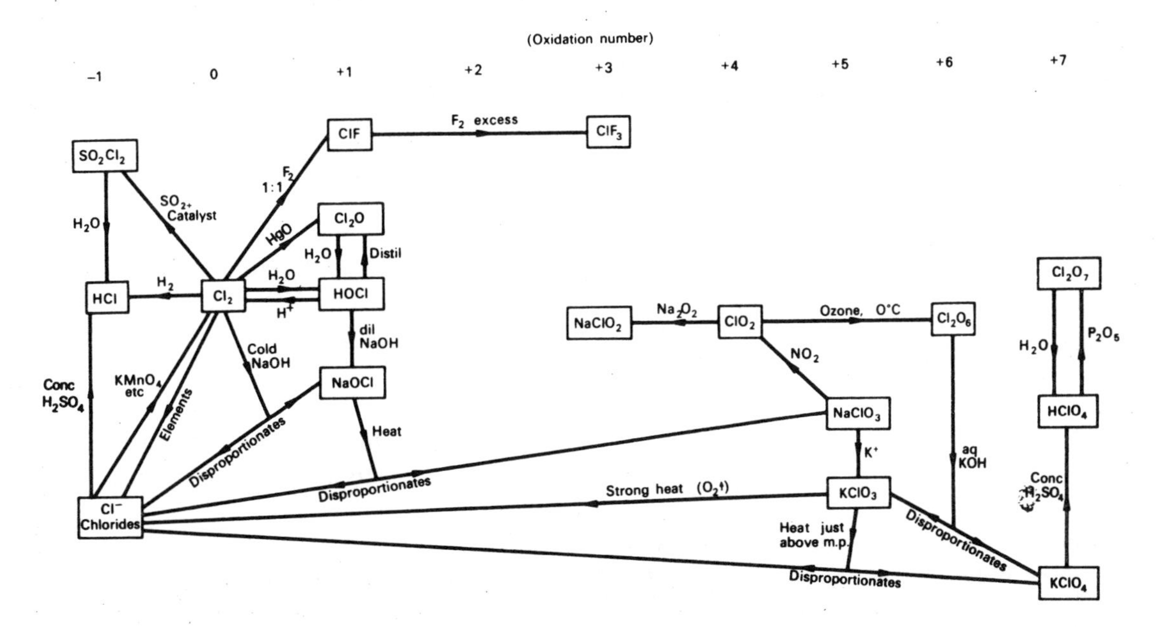

FIG. 21.7. Reactions of chlorine.

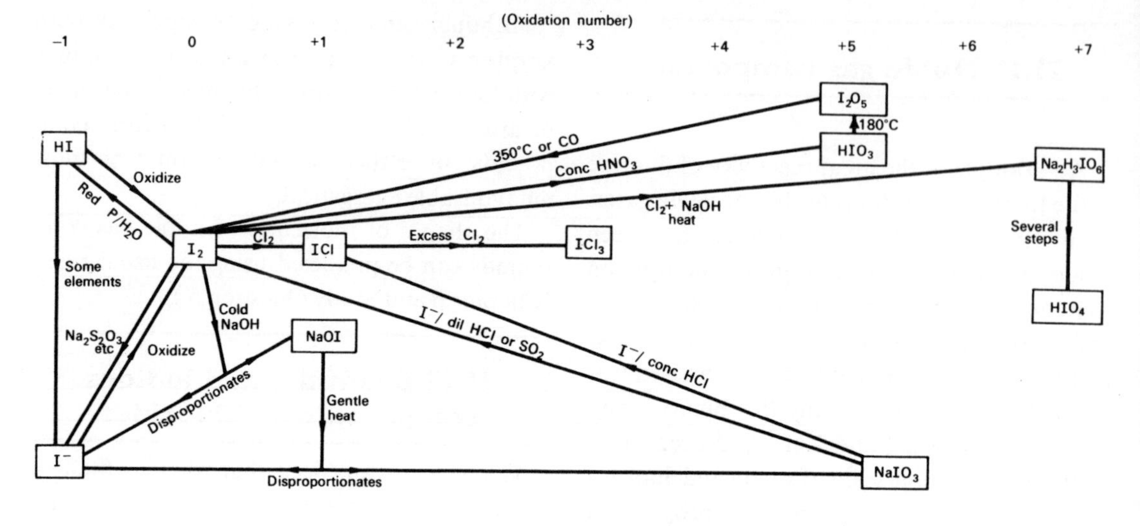

FIG. 21.8. Reactions of iodine.

$$Hg^{2+}(aq)+2I^-(aq) \rightarrow HgI_2(s)$$

$$HgI_2(s)+2I^-(aq) \rightarrow HgI_4^{2-}(aq)$$

The formation of complex ions is just another manifestation of the tendency of iodine to form bonds rather than simple ions.

Bromides are intermediate in properties between chlorides and iodides. The trend down the group is well summarized by considering the halides of silver (Table 21.8).

Figures 21.7 and 21.8 are summary charts giving some of the reactions of chlorine and iodine and their important compounds.

TABLE 21.8

Halide	Structure	Solubility product in water mol² dm⁻⁶	Solubility in ammonia(aq)
AgF	Simple cubic (6:6)	soluble	soluble
AgCl	Simple cubic (6:6)	$K_s \simeq 10^{-10}$	soluble
AgBr	Simple cubic (6:6)	$K_s \simeq 10^{-12}$	soluble only in conc NH_3(aq)
AgI	Tetrahedral (4:4)	$K_s \simeq 10^{-16}$	insoluble even in conc NH_3(aq)

21.13 Summary

(1) PHYSICAL PROPERTIES

X_2 molecules; increasing intermolecular forces down Group, hence Br_2(1) and I_2(s), at room temperature.

(2) CHEMICAL PROPERTIES

IONIZATION

(a) X^- (noble-gas structure) readily formed.

(b) Some oxo-ions XO^-, XO_2^-, XO_3^- and XO_4^- formed (not fluorine).

REACTIVITY

(a) They combine vigorously with metals.

(b) The elements are oxidizing agents.

COMPOUNDS

(a) Ionic lattices containing X^- readily formed with electropositive metals, e.g. in the *s*-block.

(b) Covalent lattices, or complex ions, readily formed with weaker metals, e.g. transition metals and B-metals.

(c) Molecular compounds formed with the non-metals, including hydrogen.

(3) TRENDS DOWN THE GROUP

(a) Ionic radii increase down the Group: the larger ions are readily polarized.

(b) Ionization energies decrease down the group (cf. trends in *s*-block, section 20.17).

(c) Order of oxidizing power is $F_2 > Cl_2 > Br_2 > I_2$ (relatively weak).

(d) Order of reducing power is $I^- > Br^- > Cl^- > F^-$ (last two almost negligible).

(e) Increasing tendency away from non-metallic behaviour down the Group. (cf. trends in Group V, Chapter 23).

The halogens, like the alkali metal and alkaline earth metal Groups, form a closely related family, though the range of chemical behaviour is wider. Fluorine, the first member of the series, departs from the regular Group trends in some respects. Note in particular:

(a) The acid HF is relatively weak.

(b) Fluorine does not form oxo-ions.

(c) Fluorine is exceptionally reactive, due primarily to the unexpectedly small F—F bond dissociation energy.

Study Questions

1. Use Table 21.3 to calculate the energy change that accompanies each of the following reactions.(C—H = 415 kJ mol^{-1}; X is any halogen.)

(a) $X_2+2HI \rightarrow 2HX+I_2$.
(b) $X_2+CH_4 \rightarrow CH_3X+HX$.

2. Describe what happens when dilute hydrochloric acid is boiled.

3. (a) How will the following react with water?

(i) NaCl(s).
(ii) $FeCl_3$(s).
(iii) $SiCl_4$(l).
(iv) ICl(l).
(v) CCl_4(l).

(b) NCl_3 reacts very slowly with water while PCl_3 reacts very rapidly. Can you account for this difference?

4. Write equations for the preparation of hydrogen bromide from:

(a) The hydrolysis of silicon tetrabromide.
(b) The burning of hydrogen in bromine at 300°C.
(c) The reaction of phosphorus tribromide with ethanol.
(d) Distilling potassium bromide with phosphoric(V) acid, (H_3PO_4).

5. Predict the shapes of the following molecules and ions: (a) PBr_3; (b) SF_4; (c) XeF_4; (d) I_3^-; (e) ClO_3^-; (f) ClO_4^-; (g) XeO_3.

6. Balance the following equations. Show the nature of the oxidation-reduction process in each case.

(a) $Ce^{4+}+I^- \rightarrow Ce^{3+}+I_2$.
(b) $BrO_3^-+I^-+H^+ \rightarrow Br^-+I_2+H_2O$.
(c) $MnO_4^-+I^-+H^+ \rightarrow Mn^{2+}+I_2+H_2O$.
(d) $Cr_2O_7^{2-}+I^-+H^+ \rightarrow Cr^{3+}+I_2+H_2O$.
(e) $OCl^-+I^-+H^+ \rightarrow Cl^-+I_2+H_2O$.
(f) $Cu^{2+}+I^- \rightarrow CuI+I_2$.
(g) $I_2+KIO_3+HCl \rightarrow KCl+ICl+H_2O$.
(h) $BrO_3^-+N_2H_4 \rightarrow Br^-+N_2+H_2O$.

7. Chromium forms chlorides in oxidation states 2 and 3. Give equations for the reactions that will probably occur when chromium reacts with (a) HCl(g), (b) Cl_2(g).

8. When the sodium nitrate has been extracted from caliche, an iodate(V) solution containing 10 g of iodine per dm^3 remains. How many (a) moles, (b) grams, of sodium hydrogensulphite must be added to the solution to reduce exactly the iodate(V) to iodine?

9. Classify the following halides as molecular, ionic or giant structures. Which of them will be hydrolysed by water?

(a) BCl_3, (b) $AlBr_3$, (c) NaCl, (d) BaI_2, (e) $TiCl_2$, (f) $TiCl_4$, (g) CHI_3, (h) IBr.

10. (a) What would you expect a sample of astatine to be like at room temperature?
(b) Will HAt be a strong or a weak acid?
(c) How stable would you expect At_3^- to be?
(d) How soluble would AgAt be?
(e) How would conc H_2SO_4 react with an astatide?
(f) What other physical and chemical properties would you expect astatine to show?

11. (a) Give an account of the ways in which the chemistry of fluorine differs from the other halogens.
(b) What physical factors are responsible for the differences?

12. (a) Excess iodide, $I^-(aq)$, in weak acid, was added to 25 cm^3 of 0·1 M iodate(V) solution, $IO_3^-(aq)$. The iodine that was liberated was titrated against 0·5 M sodium thiosulphate solution: 30 cm^3 of the latter was required.
(b) The same iodate(V) solution was titrated into 25 cm^3 of 0·1 M KI solution in very strong hydrochloric acid, in the presence of a little CCl_4. Initially, the layer of CCl_4 became violet, but after 12·5 dm^3 of the iodate(V) solution had been added, it became colourless.

Explain these reactions and give equations for them.

13. Show how the reactions of the halogens can be interpreted in terms of their half-cell potentials.

14. In a number of diagrams in this chapter, hydrogen is included. In what ways does hydrogen (a) resemble; (b) differ from, a halogen?

15. (a) Assuming that the F—F bond energy is 158 kJ mol^{-1} and that the heat of formation of $XeF_4(g)$ is -212 kJ mol^{-1}, calculate the bond energy of an Xe—F bond.
(b) How will the free energy of formation of $XeF_4(g)$ compare with its heat of formation?
(c) What will happen to a sample of XeF_4 if it is heated?
(d) What reasons can you propose for the much greater stability of xenon fluorides as compared with xenon chlorides?

CHAPTER 22

Oxygen and sulphur

SHOWING DIFFERENCES BETWEEN FIRST AND SECOND ROW ELEMENTS

Oxygen and sulphur are the two lightest elements of Group VIB, and this chapter is concerned mainly with the relationship between them. The remaining elements, selenium, tellurium and polonium, are relatively rare and they will not be considered in detail.

A number of differences are observed between oxygen, the "parent" element of the Group, and the remainder. It is in fact a characteristic of any Group of the periodic table that the first member often differs from the rest. Although this chapter stresses the differences between oxygen and sulphur, Group VIB is not exceptional in this respect.

22.1 Occurrence and extraction of oxygen and sulphur

Table 22.1 lists the Group VIB elements, with their ground state electronic structures. Oxygen and sulphur are encountered native—oxygen as a constituent of the atmosphere, and sulphur in certain rocks—whereas the halogens are only found in chemical combination.

Oxygen is present in the atmosphere as diatomic molecules $O_2(g)$, forming approximately one-fifth of the atmosphere by weight. In the upper atmosphere, a certain amount of atomic oxygen is formed due to the action of ultraviolet energy and cosmic rays causing dissociation:

$$O_2(g) + h\nu \rightarrow O(g) + O(g); \quad \Delta E = +497 \text{ kJ}.$$

Free oxygen atoms can combine with ordinary oxygen leading to triatomic molecules of *ozone*, trioxygen, $O_3(g)$:

$$O_2(g) + O(g) \rightarrow O_3(g);$$

TABLE 22.1

Name	Principal quantum number of valence shell	Electronic structure in ground state (abbreviated form)	Electronic structure (full form)
Oxygen	2	2, 6	$1s^2$; $2s^2p^4$
Sulphur	3	2, 8, 6	$1s^2$; $2s^2p^6$; $3s^2p^4$
Selenium	4	2, 8, 18, 6	$1s^2$; $2s^2p^6$; $3s^2p^6d^{10}$; $4s^2p^4$
Tellurium	5	2, 8, 18, 18, 6	$1s^2$; $2s^2p^6$; $3s^2p^6d^{10}$; $4s^2p^6d^{10}$; $5s^2p^4$
Polonium	6	2, 8, 18, 32, 18, 6	$1s^2$; $2s^2p^6$; $3s^2p^6d^{10}$; $4s^2p^6d^{10}f^{14}$; $5s^2p^6d^{10}$; $6s^2p^4$

Oxygen exists in chemical combination in two main forms:

(i) as oxides;
(ii) as oxo-anions.

Examples of compounds in which oxygen is present in the oxo-anions are nitrate, NO_3^-, carbonate, CO_3^{2-}, sulphate, SO_4^{2-}, and phosphate(V), PO_4^{3-}.

Oxygen is extracted from the atmosphere on a large scale by the fractional distillation of liquid air. Air is liquefied by making use of the Joule–Thomson effect (cooling by expansion of the gas), and water vapour and carbon dioxide are removed by solidification. Oxygen boils at a higher temperature than nitrogen, and the two gases form a near-ideal mixture. The principles of the separation are discussed in section 14.13.

Sulphur is fairly widespread in nature, constituting about 0·1 per cent of the Earth's crust. It occurs mainly as sulphides of metals, but also as sulphates and the native element. The chief sources of free sulphur are the following:

(a) *Frasch process*. Sulphur is extracted from underground sulphur-bearing rock in Texas by sinking a shaft about a foot in diameter containing three concentric tubes. Compressed air and superheated water (160°C and about 10 atm) are forced down, and this melts the sulphur which comes up as a foam mixed with air and water. Earthy material is left behind and sulphur of high purity (about 99·5%) is obtained (Fig. 22.1).

(b) *From petroleum*. Most deposits of crude petroleum contain sulphur, and some are economical for extraction when other sources of the element are scarce. The element is usually obtained as hydrogen sulphide, $H_2S(g)$, which can be converted to sulphur by burning with limited air:

$$2H_2S + O_2 \rightarrow 2H_2O + 2S.$$

The chief use of sulphur is for the conversion into sulphur dioxide and thence to sulphuric acid, and so other methods of extraction are employed to obtain sulphur dioxide directly. Two important methods are:

(a) *From iron pyrites, and other sulphide ores*. Iron pyrites, FeS_2, is a common sulphide ore, and when it is heated in air (roasted) it is converted into $Fe_2O_3(s)$ and sulphur dioxide. Other sulphide ores are often roasted as a means of obtaining the metal, and in these cases sulphur dioxide is a useful by-product (section 13.6).

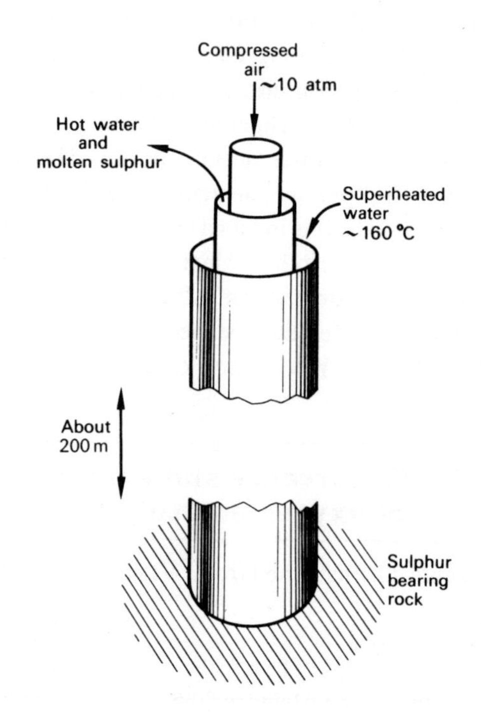

FIG. 22.1. The Frasch process.

(b) *From sulphates*. In the United Kingdom, *anhydrite*, $CaSO_4$, is an important raw material, and this is converted to sulphur dioxide for sulphuric acid manufacture by reduction with carbon.

22.2 Reactions and uses of oxygen and sulphur

Both elements combine directly with most metals. Only the noble metals are totally immune to attack by oxygen. Frequently a protective coating of oxide prevents further reaction from taking place. This is true of aluminium for instance.

The relative stabilities of oxides and sulphides play an important part in the extraction of metals from their ores, and this subject was discussed in some detail in Chapter 13.

The reactions of other elements with oxygen and sulphur differ largely because of the different structures and physical states of the two elements. Sulphur has a marked tendency to **catenate** (it forms chains and S_8 rings) whereas oxygen forms diatomic molecules. The structure of sulphur was discussed in section 5.8, and its allotropy further discussed in section 14.4.

Both elements will combine with most non-metals, though not always by direct reaction. All non-metals will burn in oxygen except the halogens, noble gases, nitrogen and selenium. Even nitrogen will combine directly with oxygen above 2000°C in an endothermic reaction, giving nitrogen monoxide (section 23.9).

In general, oxygen combines more readily with electropositive elements than does sulphur but sulphur combines more easily with the weaker metals such as copper, silver and lead. In nature the weaker metals tend to occur as sulphides rather than as oxides, for example Ag_2S, HgS, PbS and ZnS.

Considerable quantities of oxygen are consumed in the use of oxy-acetylene (oxygen-ethyne) welding equipment and in metallurgical processes. A large amount is used in the petrochemical industry in the manufacture of epoxyethane, and the gas is used extensively in industry for a wide variety of other oxidizing reactions. Smaller amounts of highly purified oxygen are required for medical use.

Sulphur is mainly used for conversion into sulphur dioxide, and thence to sulphuric acid (section 22.7). Smaller amounts are used for vulcanizing (hardening) rubber; the tangled molecules of rubber become cross-linked with —S—S— bridges.

22.3 Physical trends down Group VIB

Figure 22.2 shows how the sizes of atoms and ions vary down Group VIB. All the ions X^{2-} are iso-electronic with the nearest noble gas. Mere consideration of size however is not adequate in explaining the marked differences between

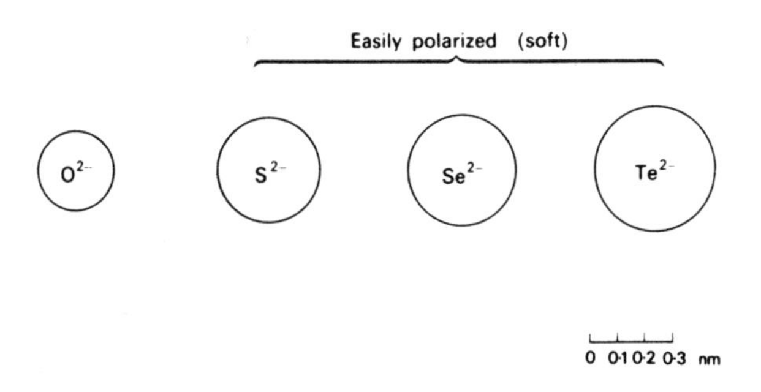

FIG. 22.2.

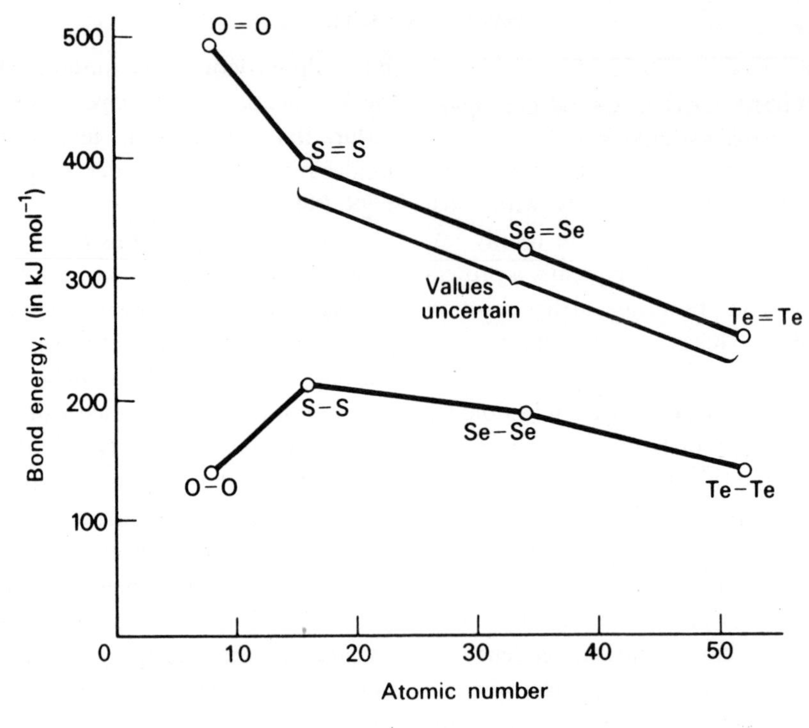

FIG. 22.3.

oxides and sulphides. The ion O^{2-} is far less easily distorted—less *polarizable*—than the remaining ions of the group. This is sometimes expressed another way by saying that O^{2-} is a *hard* ion, while the remainder are soft ions.

Figure 22.3 is a plot of bond energies for the bonds X—X and X=X. It is seen from this that only oxygen has a preference for double bonds O=O*.

The melting and boiling points of the Group VIB elements show marked differences which reflect the different structures of the elements. It will be left to the reader to plot these out and to compare them with other Groups.

Like the halogens, the Group VIB elements all have quite high ionization energies, and consequently positive ion formation is not common. Polonium is, however, a metal, and tellurium does show some metallic properties, these being the elements of the group with the lowest ionization energies.

Those forms of sulphur which contain S_8 molecules are soluble in non-polar solvents like benzene and carbon disulphide. "Polymeric" forms of sulphur, such as plastic and amorphous sulphur, do not dissolve readily in any solvent.

22.4 Compounds with hydrogen

The Group VIB elements all form simple hydrides, H_2X. Water is exceptional in the following respects:

* The double bond in O_2 is unusual in that two of the bonding electrons are unpaired.

(a) it has a higher melting point and boiling point;

(b) it is a weaker acid.

These effects are attributed to hydrogen bonding (section 15.2). Hydrogen bond formation involving sulphur atoms is rare, and is always very weak. Hence $H_2S(g)$ is relatively insoluble in water (a saturated solution at room temperature and pressure is only about 0·1 M).

Hydrogen and sulphur combine together to form *polysulphides*, H_2S_x, where x can be as large as five. This is another manifestation of the property of catenation—sulphur atoms can link to form chains. Oxygen has a weaker tendency to catenate, and the highest known oxide of hydrogen is hydrogen peroxide, H_2O_2.

22.5 Hydrogen peroxide

Hydrogen peroxide, H_2O_2 (m.p. $-0{\cdot}5°C$, b.p. 158°C), is a colourless viscous liquid when pure. It decomposes violently if heated, the reaction $H_2O \rightarrow H_2O+\frac{1}{2}O_2$ being exothermic. It has a bent structure

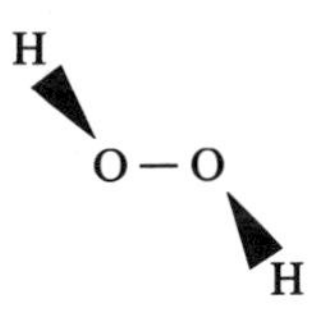

Redox reactions of hydrogen peroxide. Hydrogen peroxide is capable of acting both as an oxidizing and as a reducing agent.

As an oxidizing agent:

$$H_2O_2(aq)+2H^+(aq)+2e^- \rightarrow 2H_2O \quad (1)$$

As a reducing agent:

$$H_2O_2(aq) \rightarrow O_2(g)+2H^+(aq)+2e^- \quad (2)$$

The reducing reactions are generally favoured by alkaline solution (addition of OH^- removes H^+), while oxidizing actions generally proceed in an acidic medium (addition of H^+ favours equation (1)). The following equations illustrate both these processes:

(a) Reducing reactions:

(i) $H_2O_2+OCl^- \rightarrow H_2O+O_2+Cl^-$ (H_2O_2 oxidized to O_2; OCl^- reduced to Cl^-)

(ii) $2[Fe(CN)_6^{3-}]+H_2O_2 \rightarrow 2[Fe(CN)_6^{4-}]+O_2+2H^+$

hexacyano-ferrate(II) → hexacyano-ferrate(III); oxidized; reduced (note drop in oxidation number)

(b) Oxidizing reactions:

(i) $4H_2O_2+PbS \rightarrow 4H_2O+PbSO_4$ (PbS black, $PbSO_4$ white)

(ii) $2Fe^{2+}+H_2O_2+2H^+ \rightarrow 2Fe^{3+}+2H_2O$

(iii) $2I^-+H_2O_2+2H^+ \rightarrow I_2+2H_2O$.

Uses of hydrogen peroxide. Many dyes are bleached by oxidation with hydrogen peroxide; textiles and wood pulp are bleached in this way. Since the only by-product is water, elaborate washing after bleaching is unnecessary.

Preparation of hydrogen peroxide. Hydrogen peroxide can be made by the action of $H^+(aq)$ on a metal peroxide. Sodium or barium peroxide (made by burning the metals in oxygen) and dilute sulphuric acid have been employed:

$$Na_2O_2+2H^+ \rightarrow H_2O_2+2Na^+.$$

A *per*oxide is an oxide which contains structural units $[O—O]^{2-}$.

A *peroxo-salt* is one containing an —O—O— linkage in the anion. The peroxodisulphate(VI) ion provides an example. It is prepared by the anodic oxidation of sulphuric acid under conditions where sulphate ions are discharged preferentially (60% solution, high current density, low temperature).

$$2\left[\begin{array}{c} O \\ \uparrow \\ O\leftarrow S\rightarrow O \\ \downarrow \\ O \end{array}\right]^{2-} \rightarrow \left[\begin{array}{c} O \qquad\qquad O \\ \uparrow \qquad\qquad \uparrow \\ O\leftarrow S-O-O-S\rightarrow O \\ \downarrow \qquad\qquad \downarrow \\ O \qquad\qquad O \end{array}\right]^{2-} + 2e^-$$

Action of dilute mineral acid on a peroxoacid or its salt produces hydrogen peroxide.

Volume concentration of hydrogen peroxide. The concentration of hydrogen peroxide is often expressed as "10-volume", or "20-volume", etc. An *n*-volume solution is defined as one of such concentration that 1 volume of solution gives *n* volumes of oxygen at s.t.p.

$$2H_2O_2 \rightarrow 2H_2O + O_2.$$

2 moles of $H_2O_2 \rightarrow$ 1 mole of O_2 = 22 400 cm^3 of O_2 at s.t.p.

$\therefore$ 2 M $H_2O_2(aq)$ = 22·4-volume solution = 68 g dm^{-3}.

Similarly, a 10-volume solution of H_2O_2 is $\left(2\times\frac{10}{22{\cdot}4}\right)$ M.

22.6 Hydrogen sulphide

Hydrogen sulphide is a colourless gas, highly poisonous, with a strong smell resembling rotten eggs. Sulphur, unlike oxygen, does not react readily with hydrogen, though some hydrogen sulphide is produced if hydrogen is bubbled through boiling sulphur. The following equations show that oxygen has a greater affinity for hydrogen than sulphur has.

$$H_2(g) + \tfrac{1}{2}O_2(g) \rightarrow H_2O(g); \quad \Delta H = -240 \text{ kJ at } 25°C$$

$$H_2(g) + S(\text{rhombic}) \rightarrow H_2S(g); \quad \Delta H = -19{\cdot}6 \text{ kJ at } 25°C.$$

The gas is usually obtained by the action of dilute acid on the sulphide of a metal.

$$FeS + 2H^+(aq) \rightarrow H_2S(g) + Fe^{2+}(aq).$$

For making the pure gas, antimony(III) sulphide and hydrochloric acid are very suitable reagents.

Hydrogen sulphide burns in excess oxygen forming water and sulphur dioxide; in limited air, some free sulphur is produced.

Hydrogen sulphide is a weak acid:

$$H_2S(aq) \rightleftharpoons H^+(aq) + HS^-(aq);$$
$$K_a = 10^{-7} \text{ mol dm}^{-3} \text{ at } 25°C.$$

It reacts as a dibasic acid, though the concentration of free ions $S^{2-}(aq)$ in a solution of hydrogen sulphide is very low indeed.

Hydrogen sulphide is a powerful reducing agent, and sulphur is formed:

$$S + 2H^+ + 2e^- \rightleftharpoons H_2S; \quad \mathcal{E}^\ominus = +0{\cdot}141 \text{ V}$$

Examples of species which hydrogen sulphide will reduce are iron(III) ions to iron(II), chlorine to chloride ion, sulphuric acid to sulphur dioxide and further to sulphur, and manganate (VII) to manganese(II). Equations may be readily balanced using the half-cell reactions.

Worked Example. Obtain an equation for the reduction of iron(III) chloride to iron(II) chloride by hydrogen sulphide.

$$2Fe^{3+}(aq) + 2e^- \rightarrow 2Fe^{2+}(aq)$$

Subtracting the half-cell equation for hydrogen sulphide, we have,

$$2Fe^{3+} + H_2S \rightarrow 2Fe^{2+} + 2H^+ + S.$$

Adding the *spectator ions*, $6Cl^-$, the full equation becomes:

$$2FeCl_3 + H_2S \rightarrow 2FeCl_2 + 2HCl + S.$$

The properties of some sulphides are considered in section 22.13.

OXIDES AND OXO-ACIDS OF SULPHUR

Sulphur forms a range of oxides, among which sulphur dioxide, SO_2 (m.p. −73°C, b.p. −10°C) and sulphur trioxide, SO_3* (m.p. 17°C, b.p. 45°C) are the most important. A large number of oxo-acids also exist, the most important being sulphuric acid, H_2SO_4.

22.7 Sulphur dioxide and sulphurous acid

Sulphur dioxide, SO_2, is a colourless gas with a sharp, choking smell, which liquefies readily under pressure (cf. boiling point above). It is produced industrially either by burning sulphur in air, or as a by-product of *roasting* of a sulphide ore:

$$S + O_2 \rightarrow SO_2; \quad \Delta H = -297 \text{ kJ at } 25°C$$

$$\underset{\text{iron pyrites}}{4FeS_2} + 11O_2 \rightarrow 8SO_2 + 2Fe_2O_3; \quad \Delta H = -3300 \text{ kJ at } 25°C.$$

Sulphur dioxide is also made in the United Kingdom by a process in which anhydrite, $CaSO_4(s)$, is reduced with carbon.

Sulphur dioxide is readily soluble in water (one volume of water dissolves 45 of the gas at 15°C), and the solution contains the dibasic acid, sulphurous acid*, $H_2SO_3(aq)$:

$$SO_2(g) + H_2O \rightleftharpoons H_2SO_3(aq)$$
$$\rightleftharpoons H^+(aq) + HSO_3^-(aq); \ pk_1 = 1{\cdot}8 \text{ at } 25°C.$$
$$\upharpoonleft\downharpoonright$$
$$H^+ + SO_3^{2-}; \ pk_2 = 6{\cdot}2 \text{ at } 25°C.$$

Pure sulphurous acid cannot be isolated: in this respect it resembles carbonic acid, nitrous acid and hypochlorous acid.

* Sulphur trioxide is polymeric (Fig. 22.4) and should strictly be called sulphur(VI) oxide. Sulphurous acid is strictly sulphuric(IV) acid. The common names have been retained in this book. Similarly, the ion SO_3^{2-} has been called sulphite rather than sulphate(IV).

Sulphur dioxide is not inflammable, though it will combine with oxygen in the presence of a catalyst. This reaction is put to good use in the **contact process** for manufacture of sulphuric acid:

$$SO_2(g) + \tfrac{1}{2}O_2(g) \rightarrow \underset{\text{converted to } H_2SO_4}{SO_3(g)}; \quad \Delta H = -98 \text{ kJ at } 25°C.$$

It will only support combustion in the case of metals whose free energy of oxide formation exceeds that of sulphur dioxide itself. Aluminium and magnesium, for instance, will burn in the gas liberating sulphur and forming $Al_2O_3(s)$ and $MgO(s)$ respectively.

Sulphur dioxide is reduced to sulphur by hydrogen sulphide, and hence sulphur is precipitated when $H_2S(g)$ is bubbled into a solution of a sulphite or hydrogensulphite.

$$SO_2(g) + 2H_2S(g) \rightarrow 3S(\text{amorphous}) + 2H_2O(g).$$

This is one of the rare reactions in which sulphur dioxide is acting as an oxidizing agent: in general it is a fairly powerful reducing agent:

$$SO_4^{2-}(aq) + 4H^+(aq) + 2e^- \rightleftharpoons SO_2(aq) + 2H_2O; \quad \mathcal{E}^\ominus = +0{\cdot}17 \text{ V}$$

(cf. hydrogen sulphide above).

The action of sulphur dioxide on aqueous solutions of oxidizing agents is similar to that of hydrogen sulphide, except that sulphur is not precipitated. For instance:

$$2MnO_4^- + 5SO_2 + 2H_2O \rightarrow 5SO_4^{2-} + 2Mn^{2+} + 4H^+$$

Adding the spectator ions for potassium manganate(VII), this equation becomes:

$$2KMnO_4+5SO_2+2H_2O$$
$$\rightarrow K_2SO_4+2MnSO_4+2H_2SO_4.$$

Sulphur dioxide is used industrially as a bleaching agent, a fungicide, and as a refrigerant gas, in addition to its uses as an intermediate in the manufacture of sulphuric acid. In the laboratory it is usually obtained from a syphon. Alternatively small quantities can be made by the action of $H^+(aq)$ on a sulphite, or by reducing concentrated sulphuric acid with a metal such as copper.

The sulphur dioxide molecule is bent, with the sulphur atom in the middle, due to the lone-pair on the sulphur atom. The sulphite ion is pyramidal for the same reason.

22.8 Sulphur trioxide

Sulphur trioxide is the product of passing a mixture of sulphur dioxide and oxygen over a heated catalyst, vanadium(V) oxide, V_2O_5. Sulphur trioxide has a very strong affinity for water, and consequently forms dense white fumes in moist air:

$$SO_3(g)+H_2O(g) \rightarrow H_2SO_4(l); \quad \Delta H = -175 \text{ kJ at } 25°C.$$

On cooling to room temperature, a white solid is formed which contains different crystalline modifications, which are polymeric forms of the monomer SO_3 (Fig. 22.4).

Sulphur trioxide is formed as white fumes when sulphates of some metals are strongly heated, although sulphates are fairly resistant to thermal decomposition, for instance,

$$\underset{\text{white}}{CuSO_4(s)} \rightarrow \underset{\text{black}}{CuO(s)}+SO_3(g)$$

22.9 Sulphuric acid

Sulphuric acid is a very important industrial chemical, and indeed it has been remarked that a country's consumption of it is a fair indication of that country's economic prosperity, for very many processes require its use at some stage or another. It is used in the manufacture of fertilizers such as ammonium sulphate and calcium "superphosphate", and in making explosives, detergents, accumulators, varnishes and artificial fibres. A large amount is used in the preparation of other acids by displacement, and for cleaning metal surfaces.

It is a heavy, oily liquid (density 1·834 g cm^{-3} at 18°C, b.p. 338°C with decomposition), made

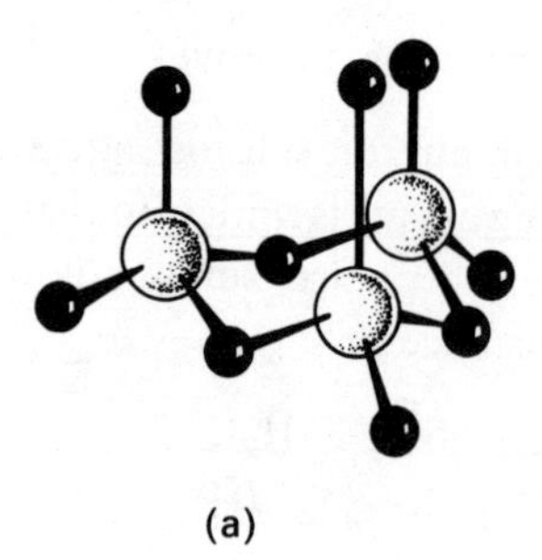
(a)

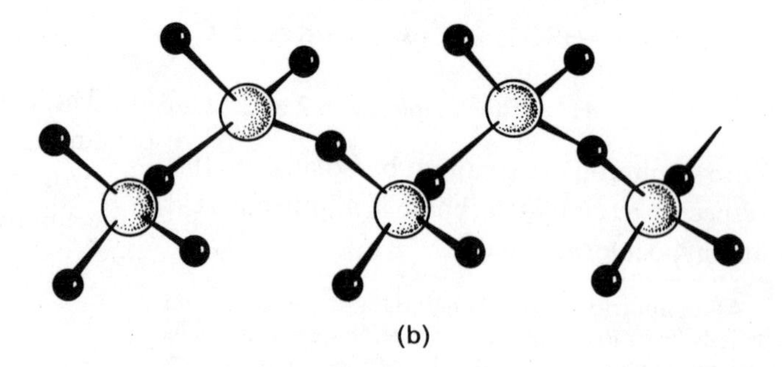
(b)

FIG. 22.4. Forms of sulphur trioxide.

principally by two processes (a) the **chamber process**, which has been in operation for about two hundred years in some form, and which gives an impure acid of moderate concentration, and (b) the **contact process**, which can lead to acid of 100% purity.

(a) *The chamber process*. The chemistry of this process is complex but it involves the oxidation of sulphur dioxide by oxygen in the presence of NO(g) and NO_2(g). The oxides of nitrogen can be regarded as *homogeneous catalysts*, which operate by intermediate compound formation. Acid of about 78% purity is produced, and this is generally converted into fertilizer.

(b) *The contact process*. Sulphur trioxide is hydrated, by absorbing it in fairly concentrated sulphuric acid, previously prepared. If sulphur trioxide were added directly to water, the evolution of heat would be too great and a fine spray of acid would be formed. Anhydrous acid (100%) is readily made by the contact process, as is *oleum*:

$$H_2SO_4(l)+SO_3(g) \rightarrow \underset{\text{oleum}}{H_2S_2O_7(l)}$$

The structures of sulphuric acid and the sulphate ion are based on the tetrahedron (Fig. 22.5). Sulphuric acid has chemical properties as follows:

(a) *As an acid*. It is dibasic, and ionizes in dilute aqueous solution thus:

$$H_2SO_4+aq \rightleftharpoons H^+(aq)+HSO_4^-(aq)$$
$$HSO_4^-(aq) \rightleftharpoons H^+(aq)+SO_4^{2-}(aq)$$

The ions SO_4^{2-}(aq) and HSO_4^-(aq) are not oxidizing agents, and therefore dilute sulphuric acid liberates hydrogen when added to metals with negative half-cell potentials.

The concentrated acid is also an oxidizing agent, and this effect often obscures its simple acidic reactions. For instance metals react to evolve sulphur dioxide, but give no hydrogen.

(b) *As an oxidizing agent*. Concentrated sulphuric acid is readily reduced by many substances, giving sulphur dioxide. For example consider the reaction between copper and concentrated sulphuric acid:

$$\text{half-reaction: } 2e^- + 2H^+ + H_2SO_4 \rightarrow SO_2 + 2H_2O \text{ (reduction)}$$

$$\text{half-reaction: } H_2O + Cu \rightarrow CuO + 2H^+ + 2e^- \text{ (oxidation)}$$

$$H_2SO_4+Cu \rightarrow SO_2+H_2O+CuO \text{ (incomplete)}$$

The equation is incomplete, because we know that CuO is a basic oxide:

$$CuO+H_2SO_4 \rightarrow CuSO_4+H_2O$$

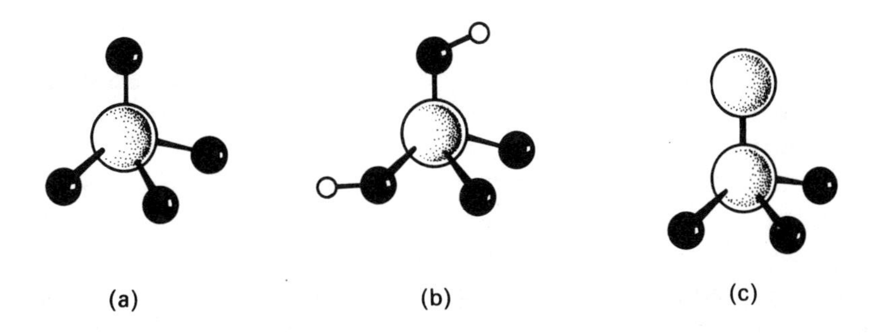

FIG. 22.5. Structures of (a) the sulphate ion; (b) sulphuric acid; (c) the thiosulphate ion.

The final equation is therefore:

$$Cu(s)+2H_2SO_4(1) \rightarrow SO_2(g)+CuSO_4+2H_2O.$$

Concentrated sulphuric acid will oxidize a number of other substances apart from metals, for instance carbon ($\rightarrow CO_2$), sulphur ($\rightarrow SO_2$), bromide ion ($\rightarrow$ bromine) and iodide ion ($\rightarrow$ iodine).

(c) *As a dehydrating agent.* Dehydration differs from drying: drying is simply the physical removal of absorbed water from a substance, while dehydration is the chemical removal of the elements hydrogen and oxygen in the ratio corresponding to H_2O. Sulphuric acid will dehydrate a number of organic substances, notably carbohydrates (of general formula $C_x(H_2O)_y$) which form carbon.

(d) *As a sulphonating agent.* Concentrated sulphuric acid and oleum are used extensively in the organic chemical industry for replacing hydrogen atoms with the sulphonic acid group $-SO_2OH$ (section 27.6). For instance benzene forms benzene-sulphonic acid:

$$C_6H_6+H_2SO_4 \rightarrow C_6H_5SO_2OH+H_2O.$$

Sulphonation is an important reaction in the manufacture of detergents, which are generally salts of an alkylsulphonic acid. Sulphonate detergents are structurally very similar to soapy detergents (soaps) which are discussed in section 30.6.

22.10 Other oxo-ions of sulphur

A large number of oxo-ions of sulphur exist, and there would be little point in cataloguing all their properties. Certain ones are of special interest, if only because they frequently occur in chemical processes.

Sodium thiosulphate $Na_2S_2O_3.5H_2O(s)$ is an important salt though its parent acid thiosulphuric acid, $H_2S_2O_3$, does not exist. The prefix *thio* means "containing sulphur in place of oxygen", and the thiosulphate ion is structurally analogous to sulphate (Fig. 22.5).

It is prepared by boiling sodium sulphite with sulphur:

$$Na_2SO_3(aq)+S(rhombic) \rightarrow Na_2S_2O_3(aq).$$

Sodium thiosulphate is used in volumetric analysis for quantitatively reducing solutions of iodine. It forms the *tetrathionate* ion, $S_4O_6^{2-}$.

Half-cell equation: $S_4O_6^{2-}+2e^- \rightleftharpoons 2S_2O_3^{2-}$

Half-cell equation: $I_2+2e^- \rightleftharpoons 2I^-$

Hence one mole of sodium thiosulphate will reduce half a mole of iodine in the complete equation:

$$I_2+2S_2O_3^{2-} \rightleftharpoons S_4O_6^{2-}+2I^-.$$

One curious property of sodium thiosulphate is its remarkable tendency to supersaturate in aqueous solution. If the solid crystals are gently warmed, they "melt", or dissolve in their own water of crystallization.

This solution can be cooled to room temperature without nucleation taking place to allow crystals to form. Seeding with a tiny crystal, or scratching the inside of the test-tube, causes rapid crystallization to form the solid again, with noticeable evolution of heat.

22.11 Acid-base character of oxides

Almost all the elements form compounds with oxygen (the only exceptions are the lighter noble gases) and these oxides may be classified in a variety of different ways: according to acid–base character, according to structure, according to bond type, and so on. One of the most useful classifications is according to **acid–base character**. The three main categories are:

(i) *Acidic oxides;*
(ii) *Basic oxides;*
(iii) *Oxides which do not display acid–base properties at all.*

This third category is very small and unimportant. Practically all oxides show *some* tendency to undergo acid–base reactions, though the rate of reaction may be slow. At high temperatures for instance, carbon monoxide reacts as an acidic oxide with sodium hydroxide, forming sodium formate.

A number of oxides exist which do not fit precisely into the above categories. For instance *amphoteric oxides* exist, which fall into categories (i) and (ii) simultaneously. Some oxides have a stoichiometric composition corresponding to the presence of two oxidation numbers in the same oxide. For instance red lead oxide, $Pb_3O_4(s)$, can be regarded as being a *compound oxide* made up of $2PbO + PbO_2$. Its systematic name is therefore dilead(II) lead(IV) oxide and it overlaps categories (i) and (ii) above.

The acid–base character of an oxide depends upon two main factors (Fig. 22.6).

(a) The position of the element in the periodic table. Elements on the right (non-metals) form acidic oxides while those on the left form basic oxides.

(b) The oxidation state of the element. The higher the oxidation state, the more strongly acidic the oxide. This is especially true of some metals where a complete range can often be observed, e.g.

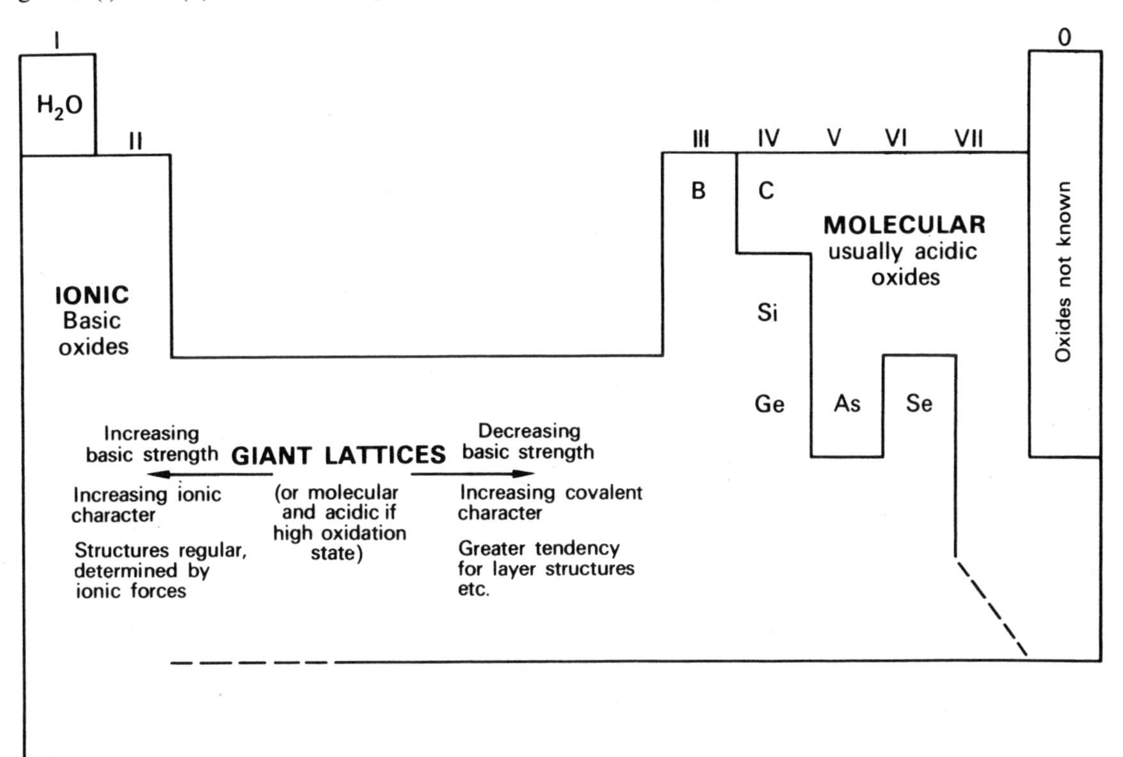

FIG. 22.6. The properties of oxides.

chromium(II) oxide, CrO, basic;
chromium(III) oxide, Cr_2O_3, amphoteric;
chromium(VI) oxide, CrO_3, acidic.

Acidic oxides

Most non-metals form acidic oxides, that is, oxides which will dissolve in alkali to form salts. Most acidic oxides react with water to form $H^+(aq)$, but this is not the case if the oxide is macromolecular. Thus $B_2O_3(s)$ and $SiO_2(s)$ are classed as acidic oxides since they react with alkalis, though they do not react with water.

Table 22.2
A Selection of Acidic Oxides of Non-metals

IIIB	IVB	VB	VIB	VIIB
B_2O_3	CO_2	N_2O_3 N_2O_4 N_2O_5	—	—
	SiO_2	P_4O_6 P_4O_{10}	SO_2 SO_3	Cl_2O_7

Basic oxides

Basic oxides may be sub-divided into two classes:

(a) those which are insoluble in, and not attacked by, water;
(b) those which react with water producing $OH^-(aq)$.

Oxides in category (a) are generally macromolecular structures in which the bonding is essentially covalent. Their structures are frequently layered, and do not correspond to those expected for the packing together of ionic particles. Oxides in category (b) are generally simple ionic lattices containing the O^{2-} ion, though some elements of Group IA and IIA form *peroxides* (containing the $[O{-}O]^{2-}$ ion) and occasionally *superoxides* (containing the $[O{-}O]^-$ ion). These were considered in Chapter 20.

Amphoteric oxides

These are oxides of metals which have the property of dissolving in acids as well as in alkalis, due to the ability of the metal to form complex ions of the type $M(OH)_x^{n-}$:

$$Al_2O_3(s)+6H^+(aq) \rightarrow 2Al^{3+}(aq)+3H_2O.$$
$$Al_2O_3(s)+2OH^-(aq)+3H_2O \rightarrow 2Al(OH)_4^-.$$

Other examples of amphoteric oxides are ZnO, SnO, SnO_2, PbO, PbO_2, and Cr_2O_3.

"Neutral" oxides

A few oxides exist which seem to show no acid–base properties. Examples are dinitrogen oxide, N_2O, nitrogen oxide, NO, and oxygen difluoride, OF_2.

Carbon monoxide is sometimes classed as a neutral oxide, but it will react with alkali at high temperatures, and this reaction is used industrially:

$$CO(g)+OH^-(aq) \rightarrow \underset{\text{methanoate ion}}{H.COO^-(aq)}$$

22.12 Structure of oxides

An alternative way of classifying the oxides of elements is by their structure. This form of classification is in some ways similar to the acid–base classification given above, for structure does have an important bearing upon chemical properties. The main categories which may be distinguished are as follows:

(a) *Essentially ionic structures*. These oxides have regular structures of the form expected from the packing together of oppositely charged ions, usually O^{2-}. Peroxides and superoxides come into this category. These essentially ionic structures are usually soluble in water, though in cases where the lattice energy is high the solubility may be small (as for instance with MgO). Ionic oxides occur on the left-hand side of the periodic table (Fig. 22.6).

(b) *Essentially covalent, macromolecular structures*. Where an oxide has a distorted structure, or a layer structure, it may be concluded that simple ionic forces are not operating but that directional bonds exist between the atoms. Such oxides are common among the transition metals and the B-metals. They are insoluble in water, and often only react with alkali on prolonged boiling. The acidic oxides B_2O_3 and SiO_2 come into this category, though most macromolecular oxides are basic or amphoteric. Many covalent macromolecules of this type are markedly non-stoichiometric.

(c) *Molecular structures*. Molecular oxides are usually acidic; even if they are not gaseous at room temperature, they are nevertheless volatile. They may be oxides either of non-metals or of metals in high oxidation states. Examples are P_4O_6(s) and Mn_2O_7(1).

22.13 Sulphides

It was observed in section 22.3 that the sulphide ion is "softer" than the oxide ion. In other words, the former is more easily distorted (polarized) by the proximity of adjacent positive ions. Most metal ions, except those of the *s*-block metals, are also "soft" ions, and when one attempts to form a compound between two soft ions, what actually happens is that bonds form instead. For this reason, although sulphides may be classified structurally in a similar way to oxides, the ionic category is very small (limited to the metals of Group IA in effect) while the covalent category is very large.

A typical ionic sulphide is sodium sulphide: it is soluble in water and it can be electrolysed when molten. Its physical properties (melting and boiling points, structure) are those expected or simple ions. Most other metal sulphides are

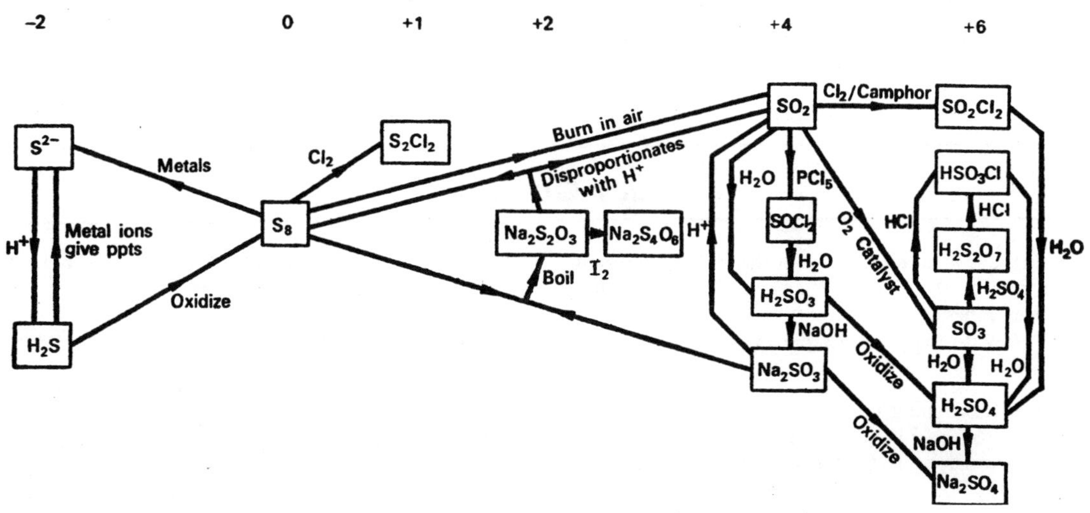

FIG. 22.7. Some reactions of sulphur and its compounds.

almost totally insoluble in water, and they form layer structures or distorted structures typical of bond formation. They are often highly coloured, the colours being different from those of the hydrated metal ions. The colours are due to energy absorbed ($\Delta E = h\nu$) when **charge transfer** occurs. For example mercury(II) sulphide is black because the energy of transfer of charge from S^{2-} to Hg^{2+} is relatively low, and corresponds to the visible region of the electromagnetic spectrum. In contrast, barium sulphide, BaS, is colourless because the charge transfer energy corresponds to absorption in the ultra-violet. The more polarizable the ions, the lower will be the charge transfer energy as a rule, and the greater the likelihood of the compound being coloured.

Most non-metals form sulphides, but only a few have achieved any importance practically. Carbon disulphide, CS_2, is used as a solvent and P_4S_3 finds uses in the match industry. Figure 22.7 summarizes some of the reactions of sulphur.

22.14 Summary

The trends down Group VI are similar to those down the halogen Group (see Study Question 15). The differences between oxygen and sulphur highlight an important feature of the periodic table, namely the differences between first-row and second-row elements.

(a) The relative increase in atomic radius (and hence decrease in ionization energy) is most marked between the first two members of a Group; hence the first-row elements are markedly more electronegative. For instance, N, O and F form *much* stronger hydrogen bonds than P, S and Cl.

(b) The valence shell of first-row elements is limited to eight electrons (Chapter 3) whereas second-row elements have 3*d* levels available for bond formation. Thus:

(i) SF_6 (but OF_2); PCl_5 (but NCl_3); SiF_6^{2-} (but no CF_6^{2-}); AlF_6^{3-} (but BF_4^-).
(ii) $SiCl_4$ hydrolyses but CCl_4 is kinetically stable.

(c) Second-row elements have less tendency to form multiple bonds. Thus:

(i) $O{=}C{=}O$ is a gas, but SiO_2 is a polymeric solid, with single Si—O bonds.
(ii) Compare S_8 with O_2.
(iii) Compare the various allotropic forms of phosphorus with N_2.

Study Questions

1. (a) Use a data book to plot the melting and boiling points of the Group VIB elements against their atomic numbers.
(b) How do the plots compare with those obtained for Group IA?

2. (a) The boiling points of CH_4, NH_3, H_2O and HF are $-160°$, $-33°$, $+100°$ and $+19°C$ respectively. How can you account for the maximum at water?
(b) H_2O_2 boils at 158°C: What can you say about the structure of this substance?

3. Suggest how the molecular formulae of (a) H_2S, (b) SO_2 could be proved experimentally.

4. (a) How many moles of sulphuric acid are there in 1 dm^3 of pure H_2SO_4 at room temperature?
(b) Concentrated sulphuric acid reacts with carbohydrates to give carbon. What would you expect it to give with (i) methanoic acid, HCOOH, (ii) ethanedioic acid, $(COOH)_2$?
(c) Why does concentrated sulphuric acid decolorize blue copper sulphate crystals?

5. (a) Calculate the concentration of a 5-volume solution of hydrogen peroxide.

(b) The decomposition of hydrogen peroxide is a redox process. Label the equation to show the nature of the oxidation and reduction.

6. Which oxide in the following pairs will be the more acidic?

(a) CaO and CO_2.
(b) Cr_2O_3 and CrO_3.
(c) MnO and Mn_2O_7.
(d) N_2O and NO_2.
(e) SeO_2 and TeO.

7. (a) Classify the following oxides in terms of structure:

(i) CaO, (ii) Fe_2O_3, (iii) SiO_2, (iv) Cl_2O_7.

(b) What relationship is there between the structure of oxides and their acid–base character?

8. Write equations for the reactions of:

(a) ZnO with (i) conc HCl, (ii) conc $NaOH(aq)$.
(b) SO_2 with (i) $Fe^{3+}(aq)$, (ii) $Cr_2O_7^{2-}$ (both in acid).

9. Balance the following equations, and show clearly the nature of the oxidation-reduction process in each case:

(a) $MnO_4^- + H_2O_2 + H^+ \rightarrow Mn^{2+} + H_2O + O_2$.
(b) $Fe(CN)_6^{4-} + H_2O_2 + H^+ \rightarrow Fe(CN)_6^{3-} + H_2O$.
(c) $Fe(CN)_6^{3-} + H_2O_2 + OH^-$
$\rightarrow Fe(CN)_6^{4-} + H_2O + O_2$.
(d) $Cl_2 + H_2O_2 \rightarrow HCl + O_2$.
(e) $(NH_4)_2S_2O_8 + H_2O \rightarrow NH_4HSO_4 + H_2O_2$.
(f) $PbS + H_2O_2 \rightarrow PbSO_4 + H_2O$.

10. When SO_2 and chlorine are passed together over a camphor catalyst, a volatile liquid *A* can be obtained. When treated with water, the liquid slowly hydrolyses to a mixture of hydrochloric and sulphuric acids.

(a) 1·35 g of *A* were hydrolysed and the solution treated with excess silver nitrate solution. This precipitated all the chlorine as $AgCl$ (2·87 g). How many moles is this?

(b) The same solution that was obtained in (a) was then treated with barium nitrate solution, which precipitated all the sulphur as $BaSO_4$ (2·33 g). How many moles is this?

(c) Work out the relative numbers of moles of Cl, S and O to obtain the empirical formula of *A*.

(d) Suggest a structural formula for *A*, given that the molecular formula is the same as the empirical formula.

(e) Write down the equation for the reaction of *A* with water.

11. When SO_2 is passed over PCl_5, a mixture of two liquids, *B* and *C*, is obtained: these can be separated by fractional distillation.

(a) 2·38 g of *B* hydrolysed violently with water to give 5·74 g of AgCl after treatment with silver nitrate solution. How many moles is this?

(b) Corrected to s.t.p., 224 cm^3 of *B*'s vapour weighed 1·19 g. What is the molecular weight of *B*?

(c) Suggest a molecular formula for *B*.

(d) Suggest a structural formula for *B*.

(e) What is *C*? Give the equation for the reaction of SO_2 with PCl_5.

12. A solution of $Na_2S_2O_3$ was prepared from Na_2SO_3 and radioactive sulphur. When the sodium thiosulphate was hydrolysed with dilute HCl to give SO_2 and sulphur, it was found that none of the SO_2 contained radioactive sulphur, but that all the radioactivity remained with the precipitated elemental sulphur. Can you suggest an explanation for this?

13. In Chapter 21, it was noted that compounds of the tri-iodide ion, I_3^-, were thermally more stable when the cation was also large. Comment on the relative thermal stabilities you would expect for the following compounds of *s*-block elements:

(a) nitrates; (b) sulphates; (c) oxides and peroxides.

14. Summarize the ways in which the properties of sulphur *differ* from those of oxygen. Find examples of analogous behaviour in Groups III, IV and V.

15. Use your knowledge of the trends in Groups V and VII to predict the properties of Se, Te and Po.

CHAPTER 23

The Group V elements

THE TRENDS DOWN A GROUP

In this chapter, as in Chapter 21, a complete family of elements will be examined. The gradiation in properties is here even more strongly marked; the first member, nitrogen, is a typical non-metal, while the last member bismuth has the characteristic properties of a metal.

23.1 Occurrence and extraction of Group V elements

Table 23.1 lists the elements of Group VB, together with their ground state electronic structures.

The first member, nitrogen, occurs native where it constitutes approximately four-fifths of the Earth's atmosphere, and in chemical combination mainly as nitrates. Nitrogen is an essential element to living matter, being an important constituent of proteins and nucleic acids. Phosphorus is also necessary for life, while the remaining elements, especially arsenic, are toxic to human beings.

Phosphorus occurs in phosphate ores such as *apatite*, $3Ca_3(PO_4)_2.CaF_2$. The remaining elements of the Group behave essentially as weak metals, and accordingly occur chiefly as the sulphides; this is in accord with the trend noted in Chapter 13. Bismuth, being more electropositive than the other elements, occurs also with the more electronegative anion O^{2-}

TABLE 23.1

Name	Principal quantum number of valence shell	Ground state electronic structure (abbreviated form)	Electronic structure (long form)
N	2	2, 5	$1s^2$; $2s^2p^3$
P	3	2, 8, 5	$1s^2$; $2s^2p^6$; $3s^2p^3$
As	4	2, 8, 18, 5	$1s^2$; $2s^2p^6$; $3s^2p^6d^{10}$; $4s^2p^3$
Sb	5	2, 8, 18, 18, 5	$1s^2$; $2s^2p^6$; $3s^2p^6d^{10}$; $4s^2p^6d^{10}$; $5s^2p^3$
Bi	6	2, 8, 18, 32, 18, 5	$1s^2$; $2s^2p^6$; $3s^2p^6d^{10}$; $4s^2p^6d^{10}f^{14}$; $5s^2p^6d^{10}$; $6s^2p^3$

(as Bi_2O_3) but the chief ores of these elements contain FeAsS(s), As_2S_3(s), Sb_2S_3(s), and Bi_2S_3(s). Trace quantities of arsenic occur in many metal ores and their removal (essential on account of the toxicity) presents many problems.

Nitrogen is extracted on a large scale by the fractionation of liquid air (section 14.13). Large amounts are consumed in the manufacture of ammonia and nitric acid, and many metallurgical processes use it as an inert atmosphere where oxidation must be prevented.

Phosphorus is normally obtained as the red allotrope (Chapter 5) for the production of the mixture used on the sides of safety match boxes, or as the white allotrope for conversion to phosphoric acid and phosphates. It is extracted by heating, in an electric furnace, a mixture of silica, coke and a phosphate ore such as *apatite*. The process may be regarded as consisting of two essential reactions:

(1) slag formation: $Ca_3(PO_4)_2(s) + 3SiO_2(s) \xrightarrow{2000°C} 3CaSiO_3(l) + \frac{1}{2}P_4O_{10}$

(2) reduction: $P_4O_{10} + 10C(s) \longrightarrow P_4(g) + 10CO(g)$

Note that in stage (1) the more volatile acidic oxide, phosphorus(V) oxide, is displaced by the less volatile silicon(IV) oxide, and that in stage (2), P_4(g) *molecules* are produced. Upon cooling, white phosphorus (the volatile, molecular allotrope, section 5.7) is the product, and this is precipitated electrostatically. Conversion from white phosphorus to red requires a catalyst, usually iodine, when done on a small laboratory scale, but for larger scale industrial conversion no catalyst is required, and white phosphorus is simply heated to about 270°C for 4 or 5 days until the transition is complete.

Demand for arsenic and its compounds is nowadays very low, on account of its highly toxic nature; in former years arsenic was used medicinally, and arsenic(III) sulphide as a yellow pigment. Demand for metallic antimony and its compounds is far greater. Soft metals such as tin and lead are hardened by alloying with antimony, and type-metal alloys often contain antimony in varying proportions according to hardness required. Antimony is not appreciably attacked by sulphuric acid, and it is therefore used in the manufacture of accumulator grids. The metal is obtained by roasting the sulphide to give the oxide, followed by carbon reduction. This is the usual method for all but the most electropositive metals (Chapter 13):

$$Sb_2S_3(s) + 4\tfrac{1}{2}O_2 = Sb_2O_3(s) + 3SO_2(g);$$
$$Sb_2O_3(s) + 3C(s) = 2Sb(l) + 3CO(g).$$

The same method is applied to the extraction of bismuth metal, which is important in the manufacture of low melting alloys. Some bismuth occurs native and is extracted by melting.

23.2 General survey of reactions and trends in Group V

There is a strong gradation from non-metallic to metallic properties down the Group. It has been noted in other chapters that the first member of a Group often differs in properties from later members, and this is certainly true of nitrogen which is the only gaseous element in Group V at room temperature; this property can be attributed to its strong tendency to form triple bonds N≡N in preference to single bonds N—N. Diatomic molecules are therefore stable for nitrogen but not for other members of the group. Phosphorus, arsenic and antimony all form molecules X_4 in the vapour state, in which single bonds X—X are present. Figure 23.1 is a plot comparing the relative bond energies of these elements. Bismuth and

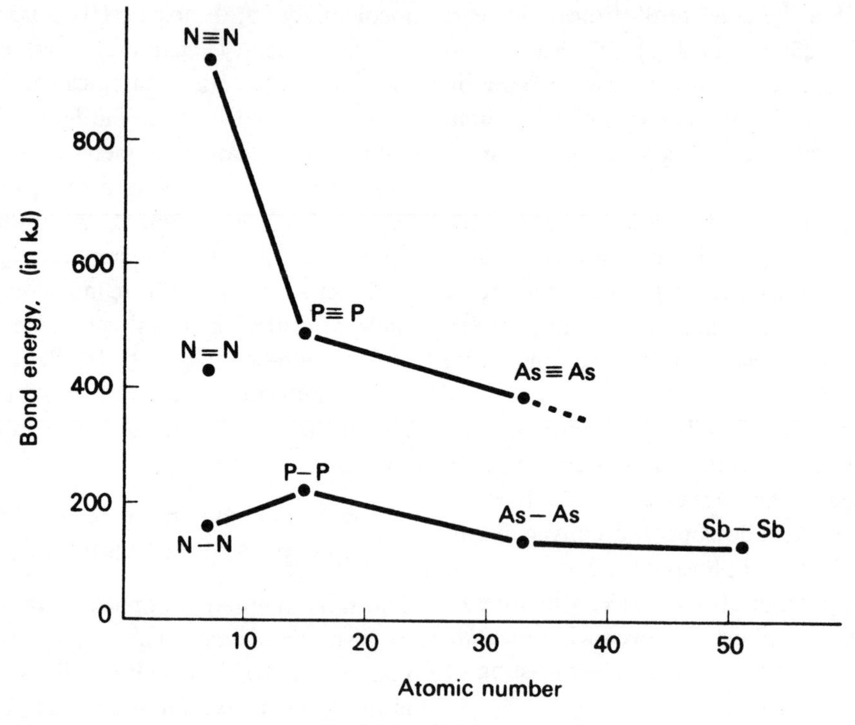

FIG. 23.1

antimony both conduct electricity in the solid state, the conductivity decreasing when temperature is increased. They are therefore both metals. The N≡N bond in nitrogen is exceptionally strong (heat of dissociation 940 kJ mol^{-1}) and this makes it rather unreactive. It does not react with many metals directly, though some of the *s*-block metals react when heated in the gas.

$$3Mg(s)+N_2(g) = \underset{\text{magnesium nitride}}{Mg_3N_2(s)}; \quad \Delta H = -430\,kJ$$

$$3Li(s)+\tfrac{1}{2}N_2(g) = \underset{\text{lithium nitride}}{Li_3N(s)}; \quad \Delta H = -197\,kJ \text{ at } 25°C.$$

Phosphorus, although less electronegative than nitrogen, combines more readily with metals to form phosphides as it is more reactive. The other elements of the Group also combine with metals, though in the case of antimony and bismuth the process is probably more correctly regarded as alloy formation.

The structures of the Group V elements present an interesting pattern. Nitrogen is very simple, forming diatomic molecules; the other elements form diatomic molecules at high temperature in the vapour. Phosphorus, arsenic and antimony all form tetrahedral molecules X_4 (Fig. 5.13), and this molecular structure persists in the solid state when the vapour is condensed.

In addition to these forms, all the elements form giant lattices with a gradual trend from covalent to metallic bonding and structure. At one end of the scale, violet phosphorus and red phosphorus form layered structures with a co-ordination number of three (Fig. 5.13). With

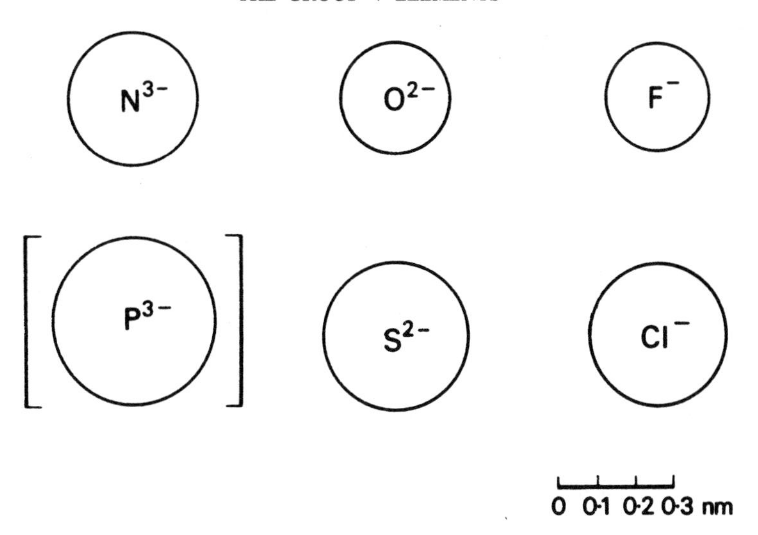

FIG. 23.2. The sizes of simple negative ions. The ion P^{3-} cannot exist, owing to polarization.

arsenic and antimony the layers are becoming more closely packed—these are intermediate covalent-metallic structures—and with bismuth the packing of atoms, and the bonding, is truly metallic.

23.3 Ion formation in Group V

The ground state electronic structures of the atoms of Group V elements are all *noble gas minus three electrons*. An examination of the energy levels suggests that an ion X^{3-} ought to be formed in favourable cases: in fact only nitrogen appears to form this ion, and even then only in certain nitrides of electropositive metals. Two factors weigh against the existence of stable ions of this type:

(i) It is difficult to form an ion of high negative charge, because each additional electron added to the atom is repelled by the ones already there.

(ii) The more negative charges an ion possesses, the *larger* it becomes (Figure 23.2) and a large ion is easily polarized (that is, it is "soft", cf. sulphide ion in Chapter 22). Moreover, a very large ion cannot fit easily into a lattice with smaller ions. For this reason, true ions P^{3-}, As^{3-}, etc., do not appear to exist. Compounds such as Li_3P, lithium phosphide, may *appear* to contain P^{3-} ion, but their structure is not in accord with the ionic model.

All the elements in this Group can exist in oxidation states of +3 or +5; in general negatively-charged oxo-ions are formed. Only the metals antimony and bismuth can form positive ions.

Table 23.2 summarizes the main oxo-ions of +5 oxidation state formed by Group V elements.

TABLE 23.2

Element	Name of anion	Formula of anion
Nitrogen	Nitrate	NO_3^-
Phosphorus	Trioxophosphate(V) and	PO_3^-
	phosphate(V)*	PO_4^{3-}
Arsenic	Arsenate(V)	AsO_4^{3-}
Antimony	Antimonate(V)	SbO_3^-
Bismuth	Bismuthate(V)	BiO_3^-

* Strictly, tetraoxophosphate(V).

Bismuth is able to form a cation $Bi^{3+}(aq)$, though this is readily hydrolysed to the oxo-cation $BiO^+(aq)$:

$$Bi^{3+}(aq)+H_2O \rightleftharpoons BiO^+(aq)+2H^+(aq)$$

$Sb^{3+}(aq)$ is even more strongly hydrolysed to $SbO^+(aq)$. The two metals may be regarded as characteristic examples of B-metals, and the following factors determine the nature of the ions formed:

(i) The heavy B-metals, such as bismuth, show a tendency to have an oxidation state equal to the *Group number minus two*. This effect is known as the **inert pair effect** (Chapter 34).

(ii) The B-metals show a stronger tendency than *s*-block metals to form complex ions; such complexes are frequently formed in such a way as to neutralize some of the charge originally present. The ions $BiO^+(aq)$ and $SbO^+(aq)$ are of lower charge than the simple species from which they are derived, and ions such as $BiCl_4^-(aq)$ are also examples. B-metal cations are more easily polarized—they are softer—than *s*-block cations and for this reason a large number of B-metal compounds are essentially covalent in their bonding.

23.4 Group V hydrides

All the Group V elements probably form hydrides XH_3, though the existence of BiH_3 has been questioned—certainly if it exists it is exceedingly unstable. The lighter elements also form hydrides X_2H_4, and nitrogen forms hydrogen azide, HN_3. These compounds are summarized in Table 23.3.

TABLE 23.3

Element	Name of hydride XH_3	Formula of XH_3 (state at room temp)	Other hydrides
Nitrogen	Ammonia	$NH_3(g)$	Hydrazine, $N_2H_4(l)$ Hydrogen azide, HN_3
Phosphorus	Phosphine	$PH_3(g)$	Diphosphane, $P_2H_4(l)$
Arsenic	Arsine	$AsH_3(g)$	
Antimony	Stibine	$SbH_3(g)$	
Bismuth	(Bismuthine)	(BiH_3, very unstable)	
	↓ decreasing thermal stability		

Hydrolysis of nitrides gives ammonia, and hydrolysis of phosphides gives phosphine. This is an example of the general reaction mentioned in section 18.6: where an "X-ide" is hydrolysed, a hydride of X is the result.

$$Mg_3N_2+6H_2O \rightarrow 2NH_3+3Mg(OH)_2$$
$$Ca_3P_2+6H_2O \rightarrow 2PH_3+3Ca(OH)_2$$

This method is of limited practical application.

23.5 Ammonia

Considerable amounts of nitrogen are converted into ammonia by the Haber synthesis, $N_2+3H_2 \rightleftharpoons 2NH_3$. This reaction has been used in this book to illustrate equilibrium constant and le Chatelier's principle, and the reader should refer to the relevant chapter (sections 10.3 and 10.4). The precise experimental conditions vary from one chemical plant to another, but all processes use (a) a high pressure (between 250 and 1000 atm), (b) the lowest temperature compatible with a reasonable rate of reaction (generally around 500°C), and (c) a catalyst based on iron and aluminium oxide (the precise details of catalysts are generally closely guarded industrial secrets).

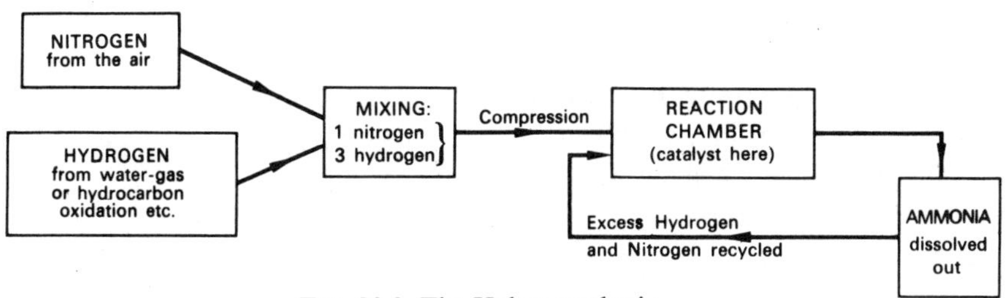

FIG. 23.3. The Haber synthesis.

The flow-sheet (Fig. 23.3) shows the essential details of the process. Despite the fact that the yield of a given equilibrium process may be only about one-tenth, almost total conversion is finally achieved since the gases are recycled after dissolving out the ammonia formed.

Ammonia is gaseous at room temperature but readily liquified (b.p. -33°C). It is highly soluble in water (at room temperature and pressure a solution of 35% by weight, density $0{\cdot}880\ \mathrm{g\,cm^{-3}}$, can be obtained). The high solubility and relatively low volatility of ammonia are attributed to hydrogen bond formation (section 15.2). The ammonia molecule is pyramidal, with one lone pair, and its structure and electronic configuration are shown in Fig. 23.4. In aqueous solution a small amount of ionization takes place, causing ammonia to behave as a weak base:

$$NH_3(g)+H_2O(l) \rightleftharpoons \underset{\text{hydrogen-bonded}}{NH_3 \ldots H_2O(aq)}$$
$$\rightleftharpoons NH_4^+(aq)+OH^-(aq);$$
$$K = 1{\cdot}8\times10^{-5}\ \mathrm{mol\,dm^{-3}}\ \text{at } 25°\text{C}.$$

Liquid ammonia is itself a liquid rather like water, even to the extent of autoionizing (section 19.11). It is described as an ionizing solvent because many ionic compounds dissolve in it, and reactions analogous to aqueous reactions can be carried out in it.

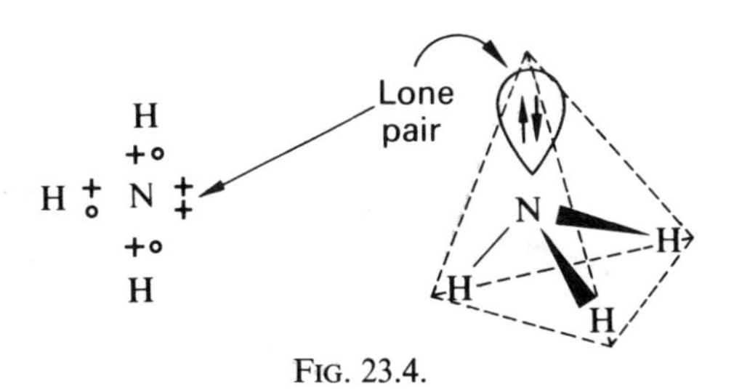

FIG. 23.4.

Ammonia will burn in air with difficulty, forming mainly nitrogen and water vapour. In the presence of a hot platinum wire, ammonia can be oxidized catalytically by the exothermic reaction

$$4NH_3(g)+5O_2(g) = 4NO(g)+6H_2O(g);$$
$$\Delta H = -900\ \text{kJ at } 25°\text{C}.$$

The gas has an irritating smell even in low concentrations. It is evolved from decayed living matter, and may be detected with Nessler's reagent.

The ammonia molecule is an important **ligand** in complex ion formation. Many metal ions (mainly transition metals) form **ammine** complexes, generally with a change of colour; the well-known colour change of copper(II) ions from pale blue to dark blue on adding ammonia is due to ammine formation:

$$\underset{\text{pale blue}}{Cu^{2+}(aq)}+4NH_3(aq) \rightleftharpoons \underset{\text{tetraammine-copper(II) ion dark blue}}{Cu(NH_3)_4^{2+}}$$

The basic nature of ammonia causes it to form the ammonium ion, NH_4^+. The presence of this ion in aqueous solution has been mentioned above, but the ion forms readily in the presence of any proton donor (Brønsted acid):

$$\underbrace{H^+ + NH_3 = NH_4^+}_{\substack{\text{acid-base pair} \\ \text{(conjugate pair)}}}$$

The ammonium ion is present in a series of salts, for instance in ammonium nitrate, $NH_4NO_3(s)$, and in ammonium sulphate, $(NH_4)_2SO_4(s)$. Ammonium chloride, $NH_4Cl(s)$, is formed as a white smoke when ammonia and hydrogen chloride gases are mixed:

$$NH_3(g)+HCl(g) \rightleftharpoons NH_4Cl(s); \quad \Delta H = -177\,kJ \text{ at } 25°C.$$

On heating, the solid readily dissociates by the reverse reaction.

Another example of the basic nature of ammonia is its ability to donate its lone pair of electrons to any other electron-accepting compound. For instance, boron trifluoride, $BF_3(g)$, combines with ammonia to form an acid-base complex (Section 19.15).

Ammonia is the parent compound of an important series of organic substances called *amines**, derived from ammonia by the substitution of one or more hydrogen atoms by hydrocarbon groups.

The formula of ammonia has been established by completely decomposing it into its elements, and showing that the product is one-quarter nitrogen by volume and three-quarters hydrogen:

2 vols. ammonia ⇌ 3 vols. hydrogen + 1 vol. nitrogen

∴ by Avogadro's law,

2 moles ammonia ⇌ 3 moles hydrogen + 1 mole nitrogen.

$$\therefore 2NH_3 \rightleftharpoons 3H_2 + N_2.$$

A large amount of manufactured ammonia is converted into nitric acid by catalytic oxidation, and thence to explosives and fertilizers. Fertilizer manufacture is an important application: it was first realized at the end of the nineteenth century that the supplies of nitrogen essential to life would ultimately give out (when the Chile deposits of sodium nitrate became exhausted), and this led chemists to tackle the problem of making nitrogen compounds from the nitrogen in the air, known as the **fixation** of nitrogen. Nowadays most nitrogen is "fixed" by conversion to ammonia, but the problem has always been made difficult by the high bond energy, and consequent low reactivity, of the element. There exist nitrogen-fixing bacteria capable of combining with elementary nitrogen at room temperature, but the precise mechanism by which these work remains a mystery. Much current work is devoted to the development of catalysts which might mimic the behaviour of these organisms, and future variants of the

* The terms *amine* and *ammine* must be carefully distinguished.

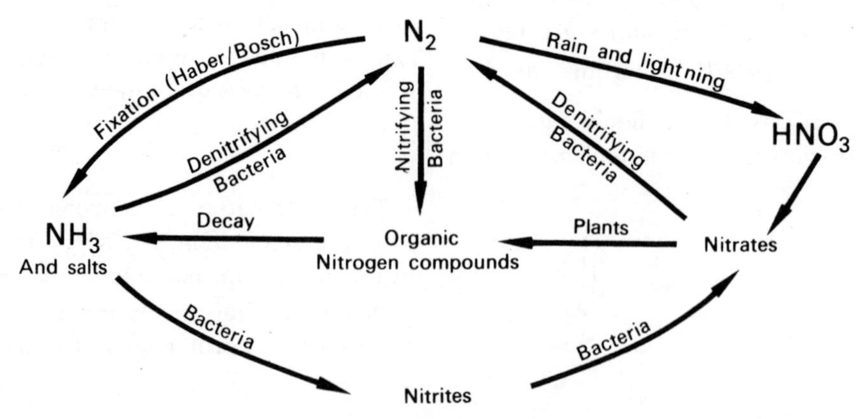

FIG. 23.5. The nitrogen cycle.

Haber process might ultimately work at near room temperature in almost 100% yield. In this context it is interesting to note that complexes containing the N_2 molecule as ligand have recently been prepared.

Figure 23.5 is a diagram of the **nitrogen cycle**. It illustrates how nitrogen is exchanged between the atmosphere, the Earth's crust and living matter.

23.6 Phosphine

Phosphine is a gas, but unlike ammonia it does not exhibit hydrogen bonding; consequently it has a lower boiling point and is almost insoluble in water. It is generally made by the hydrolysis of white phosphorus by sodium hydroxide solution—an example of a disproportionation reaction:

oxidized (O.N. increases, 0 to +1)

$$P_4 + 3OH^- + 3H_2O \rightarrow 3H_2PO_2^- + PH_3$$

reduced (O.N. decreases, 0 to −3)

Phosphine is a much weaker base than ammonia, and the remaining Group V hydrides are weaker still, although the phosphonium ion, PH_4^+, and its salts, do exist.

23.7 Arsine and stibine

These gases are obtained when compounds of arsenic and antimony are subjected to vigorous reduction in aqueous solution. These reactions form the basis of the Marsh test for arsenic, and later modifications such as Gutzeit's test. In Gutzeit's test arsine (AsH_3) or stibine (SbH_3), evolved from the reaction with granulated zinc in dilute sulphuric acid, is allowed to react with $Ag^+(aq)$, and black metallic silver is deposited.

The thermal stability of the Group V hydrides varies as follows:

$$NH_3 > PH_3 > AsH_3 > SbH_3 > BiH_3$$

In the older Marsh test, arsine was passed down a heated tube, and a mirror of metallic arsenic deposited on the cooler part of the tube. The lower stability of stibine means that metallic antimony is deposited even before the very hot portion of the tube is reached (Fig. 23.6).

23.8 Halides of Group V elements

All the Group V elements form halides XHa_3, where X = Group V element and Ha = halogen; in addition, some halides XHa_5 exist, such as PCl_5. Phosphorus forms a mixture of PCl_3 and PCl_5 when the element phosphorus is burned in chlorine.

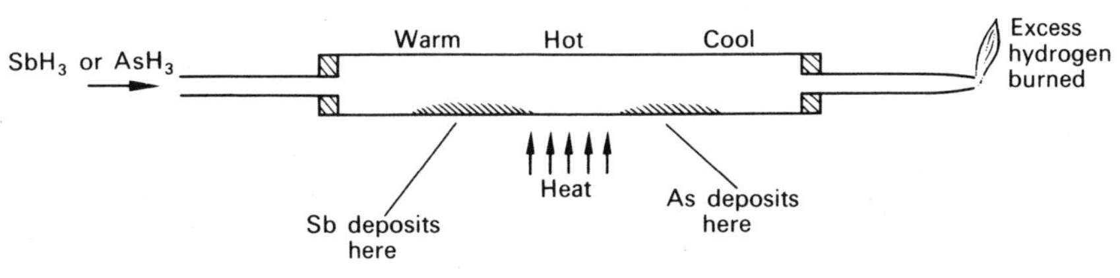

FIG. 23.6. The Marsh test.

Phosphorus trichloride is a colourless liquid, and is a typical non-metal chloride, fuming in moist air, and reacting with water to form phosph*onic* acid, H_2PHO_3, formerly called phosphorous acid, H_3PO_3.

$$PCl_3+3H_2O = H_2PHO_3+3HCl; \quad \Delta H \text{ negative.}$$

Phosphorus pentachloride has a trigonal bipyramid shape in the vapour phase (Figure 6.16), but in the solid this is replaced by what appears to be an ionic structure made up of PCl_4^+ (tetrahedral) and PCl_6^- (octahedral). It reacts with water (and with organic compounds containing hydroxyl groups) to give first phosphorus trichloride monoxide:

$$PCl_3+H_2O = POCl_3+2HCl; \quad \Delta H = -92\,\text{kJ at } 25°\text{C.}$$

Phosphorus trichloride monoxide is a liquid consisting of tetrahedral molecules, this observation being in accord with the rules for molecular shape (section 6.1). It is relatively stable towards hydrolysis, but boiling water converts it to phosphoric acid. In these hydrolysis reactions the oxidation number of phosphorus remains +5 throughout:

$$^{-1}Cl—\overset{+5}{P}(Cl^{-1})_4 \qquad ^{-1}Cl—\overset{+5}{P}(Cl^{-1})_2{=}O^{-2}$$

$$^{-1}HO—\overset{+5}{P}(OH^{-1})_2{=}O^{-2}$$

Antimony and bismuth form chloride oxides where the oxidation number is +3.

$$SbCl_3(aq)+H_2O \rightleftharpoons \underset{\text{white precipitate}}{SbOCl(s)}+2HCl(aq); \quad \text{(also Bi.)}$$

The above reaction affords a good illustration of the principle of equilibrium and le Chatelier: addition of $H^+(aq)$ ions readily dissolves

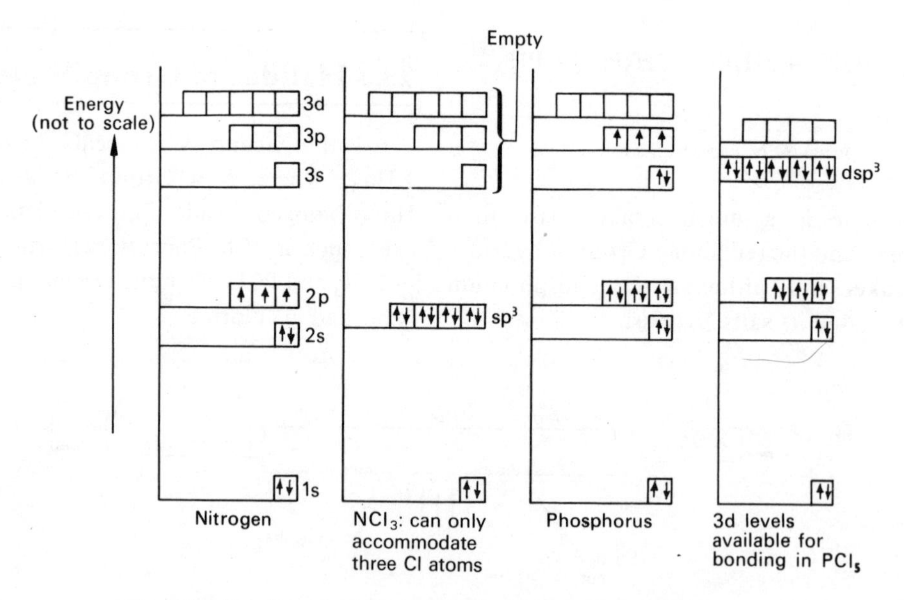

FIG. 23.7. To show that nitrogen cannot form a chloride NCl_5.

the antimony chloride oxide precipitate, and dilution of this solution leads to reprecipitation.

Nitrogen is exceptional in that it forms only one chloride, NCl_3. There are insufficient energy levels to accommodate five halogen atoms and form NCl_5, as the energy group of principal quantum number 2 does not have *d*-levels.

23.9 Oxides and oxo-acids of Group V elements

The pattern presented by the oxides of Group V elements is more complicated than that of the chlorides, due to the appearance of oxidation states other than +3 and +5. Table 23.4 lists the main ones, together with their molecular formulae and physical states at room temperature.

Dinitrogen oxide, N_2O, and nitrogen oxide, NO, do not display acidic or basic tendencies—they were classed as *neutral* oxides in section 22.11. With the remainder, a distinct trend from acidic to basic is observed:

N	P	As	Sb	Bi
Strongly acidic oxides	⟶	weakly acidic oxides	⟶ amphoteric oxides (Sb_2O_5 acidic) ⟶	weakly basic oxide

The acidic tendency can be observed with nitrogen oxides, where nitrous and/or nitric acids are formed:

$$N_2O_3 + H_2O = 2HNO_2 \text{ (nitrous acid)}$$
$$N_2O_5 + H_2O = 2HNO_3 \text{ (nitric acid)}$$

$$\overset{+4}{N_2O_4} + H_2O = \overset{+3}{HNO_2} + \overset{+5}{HNO_3}$$
(disproportionation)

Oxides of phosphorus react with water to form the corresponding oxo-acids (Figs. 23.8, 23.9).

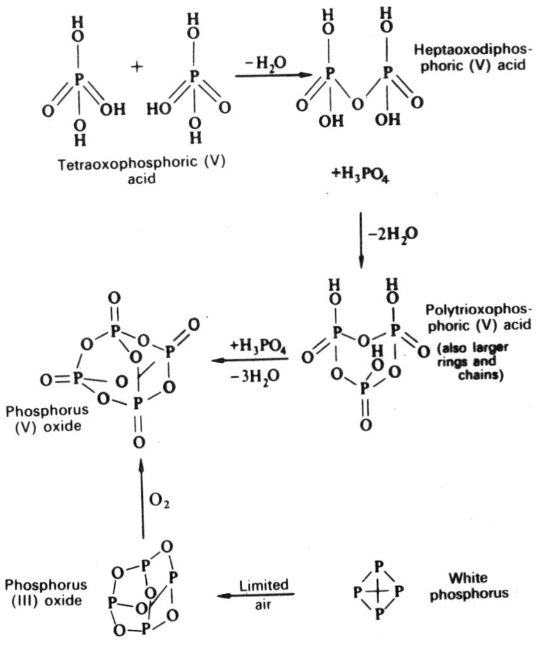

FIG. 23.8.

TABLE 23.4

Element	Oxidation number of element in oxide				
	+1	+2	+3	+4	+5
Nitrogen	$N_2O(g)$	$NO(g)$	$N_2O_3(g)$	NO_2 $N_2O_4(g)$	$N_2O_3(s)$
Phosphorus			$P_4O_6(s)$	$PO_2(s)$*	$P_4O_{10}(s)$
Arsenic			$As_4O_6(s)$		$As_4O_{10}(s)$
Antimony			$Sb_2O_3(s)$*	$SbO_2(s)$*	$Sb_2O_5(s)$*
Bismuth			$Bi_2O_3(s)$*		

* Indicates formula to be empirical: structure consists of larger molecules or giant lattice.

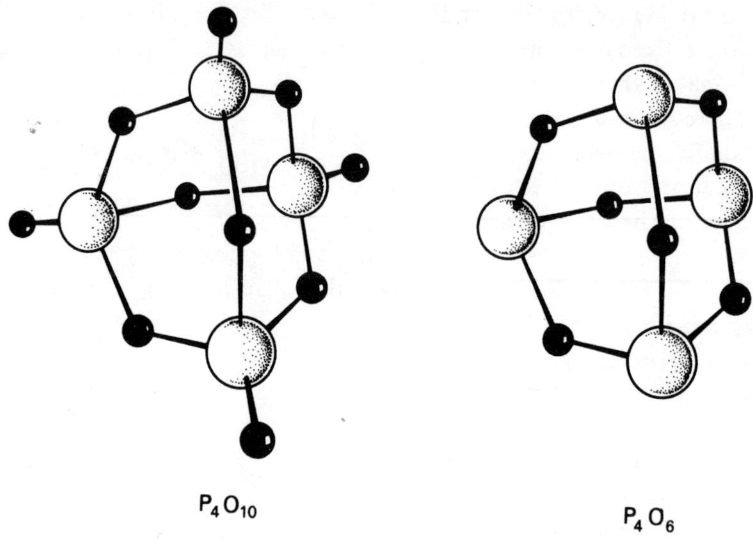

FIG. 23.9.

$$\tfrac{1}{2}P_4O_{10}+H_2O = 2HPO_3$$
$$HPO_3+H_2O = H_3PO_4$$

An oxide which can react either with an acid or a base is classed as amphoteric, and antimony(III) oxide comes into this category:

$$Sb_2O_3(s)+6H^+(aq) = 2Sb^{3+}(aq)+3H_2O.$$
$$Sb_2O_3(s)+6OH^-(aq) = 2SbO_3^{3-}(aq)+3H_2O.$$

Bismuth(III) oxide dissolves readily in dilute acids, but does not dissolve in alkali; it is therefore a true basic oxide although weak compared with, say, the oxide of a Group IA metal. Figure 23.10 is a plot of half-cell potentials of the oxides and oxo-acids in the +5 oxidation state. Note the following points:

(i) Phosphoric acid is not a strong oxidizing agent.

(ii) Bismuthic(V) acid is an extremely strong oxidizing agent, and will even oxidize manganese(II) directly to manganese(VII).

Some of the oxo-acids of Group VB, and their salts, are of special importance, and they are considered in the sections which follow.

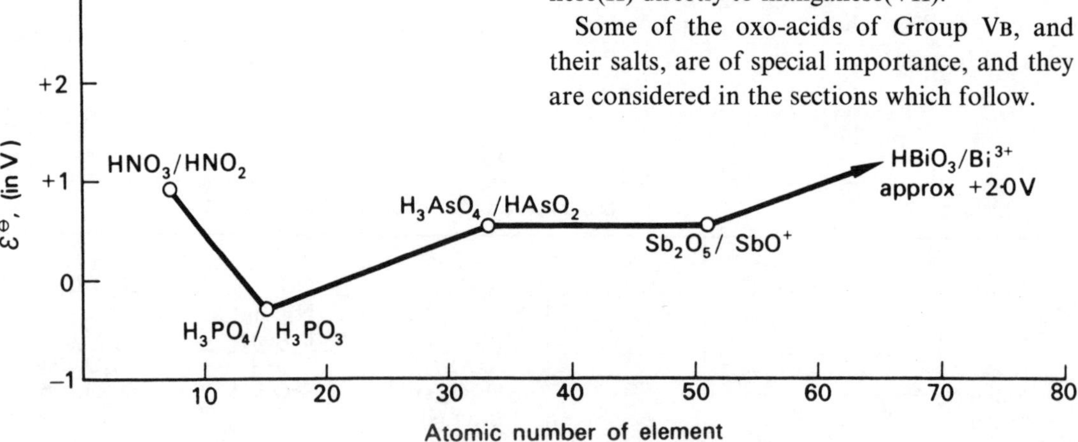

FIG. 23.10. Half-cell potentials for $X(V) \rightleftharpoons X(III)$.

23.10 Nitric acid

This is the most important oxo-acid of nitrogen, and very large amounts are manufactured for the chemical industry using the oxidation of ammonia. In dilute aqueous solution it ionizes forming $H^+(aq)$ and the planar ion, $NO_3^-(aq)$. Nitric acid is obtained by the following sequence of reactions:

$$\text{ammonia} \xrightarrow[\text{oxidation}]{\text{catalytic}} \text{nitrogen oxide} \xrightarrow{\text{mix with air}}$$

$$\text{nitrogen dioxide} \xrightarrow[\text{oxygen}]{\text{water and}} \text{nitric acid}$$

Nitrogen oxide, NO, is a colourless gas which reacts instantly with the oxygen in air to form brown $N_2O_4 \rightleftharpoons NO_2$. In the presence of excess oxygen and water in an absorption tower, nitric acid is produced.

$$3NO_2 + H_2O = 2HNO_3 + NO.$$

Nitric acid is a colourless liquid when pure, but is normally yellow due to dissolved oxides of nitrogen produced by its decomposition. So-called "concentrated" nitric acid used in the laboratory contains about 30% water, and the true anhydrous acid has markedly different properties (for instance it will not react with copper). The reactions of dilute nitric acid are again different from those of laboratory "conc" acid. Concentrated nitric acid is a very powerful oxidizing agent while dilute nitric acid is only mildly so.

Reactions of dilute nitric acid

Although most dilute acids react with metals to give hydrogen gas, due to the reduction of $H^+(aq)$ by the metal, this is not the case with dilute nitric acid which gives mainly oxides of nitrogen such as NO and NO_2. This shows that the metal has reduced the nitrate ion instead of hydrogen ion. A very dilute solution of acid will, however, give hydrogen with a reactive metal such as magnesium. Apart from this, dilute nitric acid behaves as a normal acid in its reactions with oxides, hydroxides and carbonates, forming nitrates.

Reactions of concentrated nitric acid

These may be considered under the following headings:

(i) *Acidic*. Concentrated nitric acid forms nitrates when added to a basic substance (oxide, carbonate, or hydroxide for instance).

(ii) *Oxidizing*. Concentrated nitric acid will oxidize very powerfully, generally giving nitrogen oxide as its reduction product. It is usually easier to treat the oxidation reactions as simple oxygen transfer when balancing equations, even though they are also examples of electron transfer. Consider, for instance, the oxidation of hydrogen sulphide to sulphur:

$$2HNO_3 = 2NO + H_2O + 3[O]$$
$$3H_2S + 3[O] = 3S + 3H_2O$$
$$\overline{2HNO_3 + 3H_2S \rightarrow 3S + 2NO + 4H_2O}$$

Although concentrated nitric acid attacks all metals except the noble metals (gold, platinum, etc.) its reaction with many metals is slowed down because it renders the surface *passive*. This effect is very marked with iron and chromium.

(iii) *Nitrating*. Many organic compounds, particularly aromatic compounds, are nitrated by a mixture of concentrated nitric acid and concentrated sulphuric acid, e.g.

$$\underset{\text{benzene}}{C_6H_6} + HNO_3 = \underset{\text{nitrobenzene}}{C_6H_5NO_2} + H_2O$$

The reaction takes place because of the presence of nitryl cations, NO_2^+, produced by proton transfer between the two substances:

$$HNO_3 + 2H_2SO_4 = NO_2^+ + H_3O^+ + 2HSO_4^-$$

(protons gained: HNO_3 → H_3O^+; protons lost: H_2SO_4 → HSO_4^-)

Nitration in the manufacture of dyes and explosives is the main industrial use of nitric acid.

23.11 Nitrates

Since nitric acid is strong acid, most metals form nitrates. All of these are soluble in water; they are less stable to heat than other commonly occurring salts such as sulphates, chlorides and phosphates. The usual mode of decomposition is by loss of oxygen and oxides of nitrogen (mainly NO_2), leaving a residue of the metal oxide:

$$Pb(NO_3)_2(s) \rightarrow PbO(s) + NO_2(g) + \tfrac{1}{2}O_2(g).$$

With the very electropositive metals however (metals below sodium in Group IA) the only gas given off is oxygen, for these metals form stable nitr*ites*:

$$NaNO_3(s) \rightarrow NaNO_2(s) + \tfrac{1}{2}O_2(g).$$

Where the oxide is itself unstable with respect to heating (where its free energy of formation, $\Delta G_f^{\ominus}$, becomes positive, Chapter 13) the residue is the metal itself. This is only true of the metals near the bottom of the electrochemical series:

$$AgNO_3(s) \rightarrow Ag(s) + NO_2(g) + \tfrac{1}{2}O_2(g).$$

Since all nitrates are soluble in water, it is not possible to devise a precipitation test for the nitrate ion: instead some other reaction has to be used. Two tests are commonly employed:

(i) Devarda's alloy (Cu 50, Al 45, Zn 5%) will rapidly reduce nitrate ions to ammonia in the presence of aqueous alkali.

(ii) Take the suspected solution of $NO_3^-(aq)$ in a test tube, make acid with dilute sulphuric acid and add $Fe^{2+}(aq)$ ions, as iron(II) sulphate. Slowly add concentrated sulphuric acid so that it does not mix, but forms a separate layer beneath the aqueous layer. A brown ring forms at the interface between the two liquids if a nitrate is present. The colour is thought to be due to nitrosyl-complexes of iron(II) such as the following:

$$[Fe(NO)(H_2O)_5]^{2+} \quad \text{(octahedral)}$$

Nitrates are oxidizing agents on account of their readiness to give up oxygen. Sodium nitrate occurs naturally in Chile, and other important nitrates include ammonium nitrate —a nitrogen-rich fertilizer—and potassium nitrate.

23.12 Nitrous acid

Nitrous acid, HNO_2, is thermally unstable, and exists in dilute solution only; it is a weak monobasic acid ($K_a = 4 \times 10^{-4}$ mol dm^{-3} at 25°C) and its salts are called nitrites. Nitrous acid reacts with all compounds containing the NH_2-group, giving nitrogen and an OH-group. The acid is made *in situ* from sodium nitrite and hydrochloric acid. Nitrous acid contains nitrogen in a lower oxidation state (+3), and is not such a powerful oxidizing agent as nitric acid: it can in fact act both as an oxidant and as a reductant in different reactions.

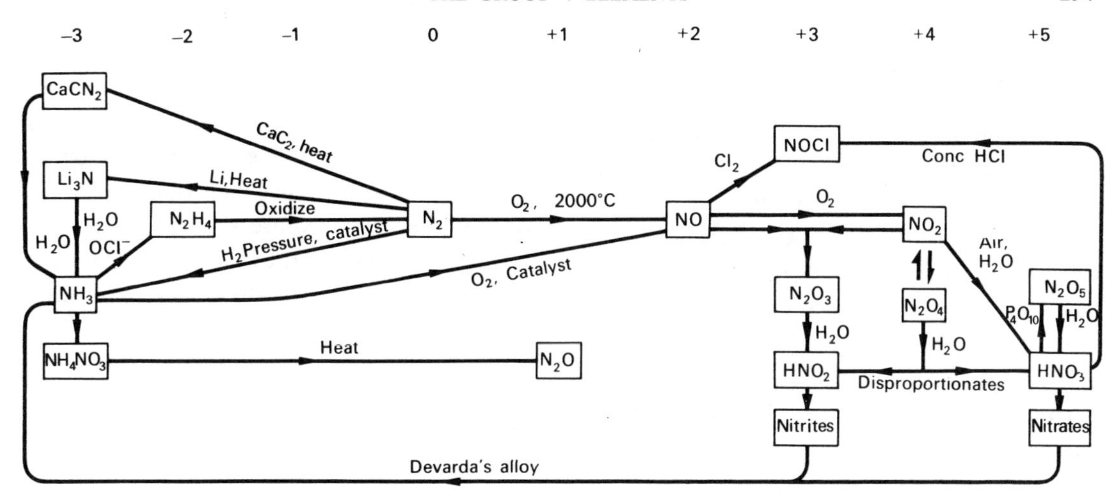

FIG. 23.11. Some reactions of nitrogen and its compounds.

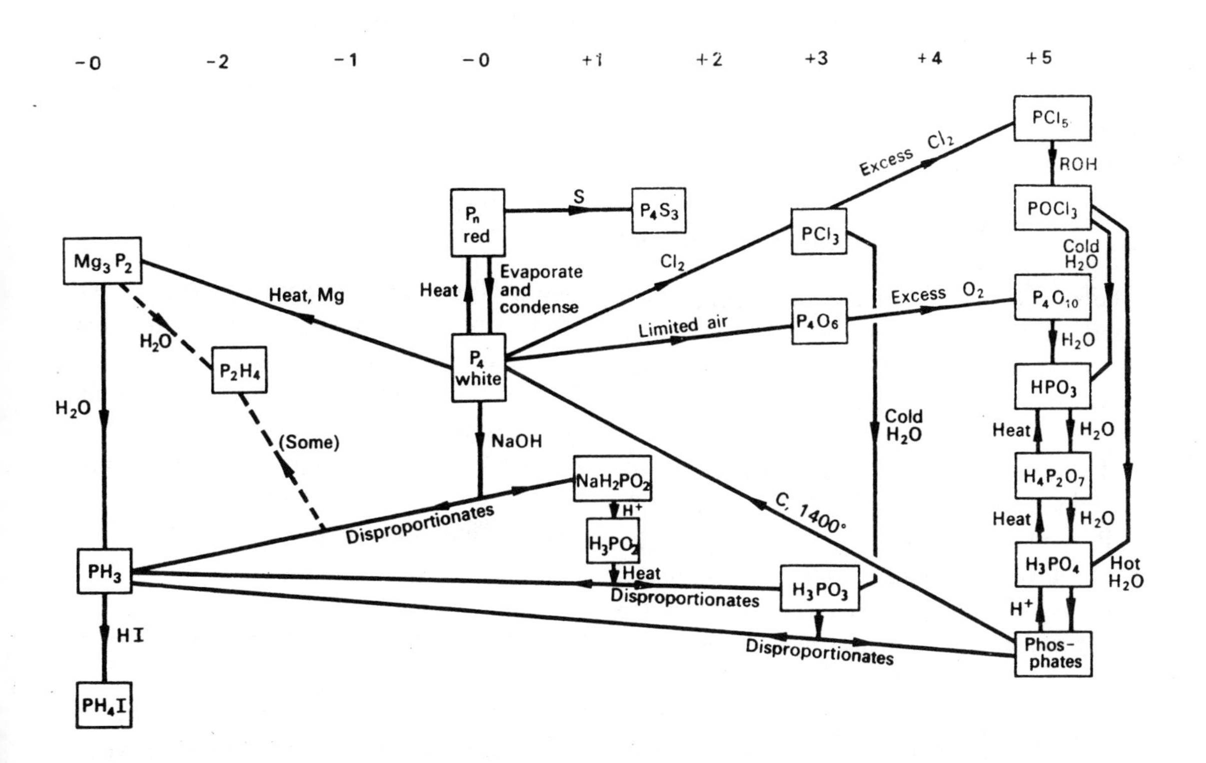

FIG. 23.12. Some reactions of phosphorus and its compounds.

23.13 Phosphoric acid

The commonest oxo-acid of phosphorus is phosphoric(V) acid, H_3PO_4. It is a member of the iso-electronic series H_3PO_4, H_2SO_4 and $HClO_4$, and is the weakest member of this series (section 19.10). It is classed as a tribasic acid—that is, it has three replaceable hydrogen atoms—though the replacement of the third atom is not normally carried out in titration reactions. It was observed by Pauling that, where an acid is polybasic, its successive dissociation constants, in mol dm^{-3}, are in the rough ratio

$$k_1 : k_2 : k_3 = 1 : 10^{-5} : 10^{-10}.$$

Phosphoric(V) acid provides a good illustration of Pauling's rule:

$$H_3PO_4 \rightleftharpoons H^+ + H_2PO_4^-; \quad k_1 = 0{\cdot}75 \times 10^{-2}; \text{ (not very strong)}$$

$$H_2PO_4^- \rightleftharpoons H^+ + HPO_4^{2-}; \quad k_2 = 0{\cdot}67 \times 10^{-7}; \text{ (weaker than ethanoic acid)}$$

$$HPO_4^{2-} \rightleftharpoons H^+ + PO_4^{3-}; \quad k_3 = 1{\cdot}0 \times 10^{-12}; \text{ (extremely weak acid)}$$

The reason for this progressive weakening is partly electrostatic: it is less easy for $H_2PO_4^-$ to lose a proton than for H_3PO_4 since the former has an overall negative charge which can attract the proton.

The titration of phosphoric(V) acid with sodium hydroxide is dealt with in section 19.13.

Pure phosphoric(V) acid is a syrupy, relatively involatile, liquid, which decomposes below its boiling point. The syrupy nature is due to hydrogen bonding in which O—H . . . O bridges form (sulphuric acid exhibits the same effect). It does not show the oxidizing properties associated with sulphuric and nitric acids: it can be used for instance to prepare hydrogen iodide by distillation with an iodide:

$$KI(s) + H_3PO_4(l) \rightarrow KH_2PO_4(s) + HI(g).$$

Under comparable conditions using sulphuric acid, the hydrogen iodide would be oxidized to iodine. In writing this equation, it is important to write the most acid salt since the conditions of the reactions are strongly acidic and Na_2HPO_4 and Na_3PO_4 cannot form.

Careful heating of phosphoric(V) acid leads to dehydration:

$$2H_3PO_4 - H_2O \rightarrow H_4P_2O_7;$$

$$2H_3PO_4 - 2H_2O \rightarrow 2HPO_3;$$

$$2H_3PO_4 - 3H_2O \rightarrow \tfrac{1}{2}P_4O_{10}.$$

Phosphoric(V) acid is used for treating metal surfaces and as a food flavouring. A large amount is converted into phosphate fertilizers.

Calcium phosphate(V), $Ca_3(PO_4)_2$, is made by the action of lime on phosphoric(V) acid, but unfortunately it is not a good fertilizer since it is insoluble in water and cannot be taken up by plants. In recent years it has been customary to convert it to the acid salt which is soluble in water and sold as "*superphosphate*":

$$Ca_3(PO_4)_2 + 4H_3PO_4 = 3Ca(H_2PO_4)_2.$$

Sodium salts of phosphoric(V) acid are manufactured for a variety of purposes which include emulsifying of processed cheese and manufacture of synthetic detergents. A polymer of $NaPO_3$ is sold as "Calgon", a water-softening agent. It forms soluble phosphate complexes with the Ca^{2+}(aq) and Mg^{2+}(aq) in hard water, thereby preventing insoluble stearate precipitates forming when soap is added.

23.14 Summary

The elements in Group V afford a particularly interesting example of Group trends. In particular:

(i) the range of observed chemical behaviour is wider than in Groups I and VII;

(ii) the differences between the first two members nitrogen and phosphorus is quite marked (cf. oxygen and sulphur, Chapter 22).

Trends down the Group

(1) Structure, bonding, and chemical properties show a complete change, from non-metallic N_2(g) to metallic Bi(s) which has a close-packed structure.

(2) The inert-pair effect is shown strongly by bismuth.

(3) The stability to heat, and the basic strength, of hydrides varies $NH_3 \gg PH_3 > AsH_3 > SbH_3 > BiH_3$(?). (Cf. similar trends in hydrogen halides.)

(4) Oxides become more basic down the Group.

Exceptional properties of nitrogen

(a) The N≡N bond is exceptionally strong (cf. section 22.14).

(b) Nitrogen cannot accept more than 8 electrons in its valence shell.

(c) It is the only member of the Group which is a gas at room temperature.

(d) It is the only member of the Group that forms an ion X^{3-}.

(e) It is the only member of the Group that forms hydrogen bonds.

(f) Its oxides, oxo-acids, and chlorides do not conform to the patterns of the remainder of the Group.

Study Questions

1. Describe what you would expect to see when the following compounds are heated. Give balanced equations in each case:

(a) Potassium nitrate. (b) Mercury(II) nitrate.
(c) Copper(II) nitrate. (d) Ammonium nitrite.

2. 25 cm^3 of a gaseous hydride of nitrogen were passed over a heated iron catalyst and completely decomposed into 75 cm^3 of a mixture of nitrogen and hydrogen. On passing this mixture over heated copper oxide, the volume dropped to 25 cm^3. (All volumes at s.t.p.)

(a) How does the nitrogen–hydrogen mixture react with copper oxide?

(b) How many cm^3 of hydrogen were produced in the decomposition?

(c) How many cm^3 of nitrogen were produced in the decomposition?

(d) What is the empirical formula of the nitrogen hydride?

(e) What is the molecular formula of the nitrogen hydride?

3. On heating, a white solid A sublimed as two gases, B and C. B had a powerful obnoxious smell, while C was an acid gas. At s.t.p., 448 cm^3 of B weighed 0·68 g, and 112 cm^3 of C weighed 0·64 g. The gases recombined when condensed on a cold surface, reforming A. Identify A, B and C.

4. (a) Write down the oxidation numbers of phosphorus in H_3PO_3, H_3PO_4 and PH_3.

(b) H_3PO_3 disproportionates on heating, giving phosphine. Suggest an equation for this reaction.

5. If a mixture of nitric and hydrochloric acids (aqua regia) is heated, it is possible to distil off a gas, which can be liquified using an ice-salt mixture. The gas has a molecular weight of 65·5 and contains 56% by weight of chlorine.

(a) Suggest a molecular formula for the gas.

(b) Suggest a structural formula for the gas.

(c) Suggest how the molecular weight of the gas could be measured.

6. The half-cell potentials for As, $3H^+/AsH_3$(g) and Ag^+/Ag are −0·60 and +0·80 V respectively.

(a) Could AsH_3 reduce Ag^+(aq) completely to silver, or would there be an appreciable quantity of Ag^+(aq) at equilibrium?

(b) Give a balanced equation for the reaction of AsH_3 with Ag^+(aq).

(c) Do these data tell you anything about the rate of this reaction?

7. Give equations for the following reactions of nitrous acid:

(a) With $Fe^{2+}(aq)$ to give NO(g).
(b) With ethanamide, CH_3CONH_2.
(c) The spontaneous slow decomposition into nitric acid and nitrogen oxide.

8. Some bismuth(III) nitrate was dissolved in concentrated hydrochloric acid. On diluting this solution, a white precipitate appeared, which redissolved on adding more concentrated hydrochloric acid. This cycle could be repeated. Write down the equation for this equilibrium.

9. (a) Of the elements in Group VB, which will:

(i) have the lowest first ionization energy?
(ii) form an ionic compound with calcium?
(iii) form the most basic oxide?
(iv) form the hydride, XH_3, with the lowest boiling point?

(b) With sulphuric acid, arsenic gives an oxide, while antimony and bismuth form sulphates; with nitric acid, arsenic and antimony form oxides, but bismuth forms a nitrate. How can you account for these differences?

10. Elements in the same Group show (a) similarities, (b) gradations in their physical and chemical properties. Discuss the truth of this statement when applied to the elements of Group VB.

11. "The properties of the first element of a Group often differ considerably from those of the remainder." Show the extent to which this is true for nitrogen.

12. The standard heats of formation of $NH_4NO_3(s)$, $H_2O(g)$ and $N_2O(g)$ are -365, -242 and $+83$ kJ mol^{-1} respectively. The corresponding free energies of formation are -184, -228, and $+104$ kJ mol^{-1}.

(a) Write down the equations for the decomposition of ammonium nitrate into (i) nitrogen, oxygen and water (ii) dinitrogen oxide and water.
(b) What are the enthalpy and free energy changes associated with each equation?
(c) Is it possible to use these data to forecast which reaction will occur?

13. When 3·45 g of a white solid, X, were warmed gently with alkali, 0·51 g of ammonia were evolved. The resulting solution was acidified with nitric acid and the addition of excess ammonium molybdate(VI) solution caused a yellow precipitate to form in the cold.

(a) Suggest a formula for X.
(b) What would you expect the pH of an aqueous solution of X to be?

14. A white solid, Y, was dissolved in hydrochloric acid. When hydrogen sulphide was passed into the solution, a brown precipitate formed. When the original solution in hydrochloric acid was diluted, a white precipitate formed. The addition of concentrated sulphuric acid to solid Y caused acid brown fumes to be evolved.

(a) Identify Y.
(b) What would you expect the action of heat on Y to be?

15. The nitryl cation, NO_2^+, is formed when concentrated nitric and sulphuric acids are mixed.

(a) Which substance is the Brønsted base in this reaction?
(b) Predict the shape of NO_2^+.

16. (a) Polytrioxophosphoric(V) acid, $(HPO_3)_n$, exists as both rings and chains. Draw diagrams of these structures.

(b) What is the structure of nitric acid, HNO_3?
(c) Can you suggest why HNO_3 and HPO_3 have such different structures?

17. For each of the following reactions, (a) give an equation, (b) explain the nature of the reaction, and (c) give the equation for the corresponding reaction where nitrogen replaces oxygen, and liquid ammonia replaces water as the solvent:

(i) Sodium hydroxide solution reacting with hydrochloric acid.
(ii) Anhydrous copper sulphate turning blue when water is added.
(iii) Sodium reacting with water and giving off hydrogen.
(iv) A solution of zinc ions, Zn^{2+}, giving a white precipitate with sodium hydroxide solution.
(v) The precipitate in (iv) redissolving on the addition of excess sodium hydroxide solution.

18. (a) Use Appendix I to predict whether the following disproportionation reactions can occur in acid solution:

(i) N_2O_4 to NO_3^- and NO.
(ii) N_2O_4 to NO_3^- and HNO_2.
(iii) $HAsO_2$ to As and H_3AsO_4.

(b) Why are NO_3^- and NH_4^+ unable to disproportionate?

CHAPTER 24

Boron, carbon and silicon

SHOWING DIAGONAL RELATIONSHIPS

At this point the book departs from the previous pattern of treating a complete Group of the periodic table in a single chapter. The non-metals boron, carbon and silicon form a closely related set and it is profitable to consider them together. Carbon is the building element of life—all organic molecules are based on skeletons of carbon atoms—and at first sight silicon and boron appear to behave differently. A closer investigation, however, reveals that silicon and boron have many properties in common with carbon. Carbon is often exceptional, for it is the only element apart from hydrogen where the number of valence electrons is numerically equal to the number of valence orbitals. The similarities in the chemistries of boron and silicon are often referred to as a **diagonal relationship**.

TABLE 24.1

Element	Principal quantum number of valence shell	Ground state electronic structure (abbreviated form)	Ground state electronic structure (long form)
B	2	2, 3	$1s^2; 2s^2p^1$
C	2	2, 4	$1s^2; 2s^2p^2$
Si	3	2, 8, 4	$1s^2; 2s^2p^6; 3s^2p^2$

24.1 Occurrence and extraction of boron, carbon and silicon

Carbon is the only element of the three to occur native. This is a remarkable fact for carbon is quite a strong reducing agent. In the presence of such large quantities of oxygen it may be wondered why all the carbon on Earth has not burned away; as it is, deposits of coal are quite extensive and graphite and diamond occur naturally. The reason for the existence of free carbon is that it is formed from living systems: carbon is, as far as we know, the only element capable of building the vast complexity of molecules which constitute living matter. All carbon compounds are thermodynamically unstable in the presence of oxygen, but they are *kinetically* very stable due to the fact that their energy levels are completely filled. A carbon skeleton does not have electron vacancies or lone pairs.

The remainder of carbon in nature occurs mostly as carbon dioxide, evolved from living matter, or as carbonates, mainly $MgCO_3$ and $CaCO_3$, which constitute sedimentary rocks and are derived from living matter.

While carbon is the "organic" building element, silicon is its inorganic counterpart. The Earth's crust is made up very largely of complex silicates—giant lattices containing

—Si—O—Si—O—Si— linkages—and, after oxygen, silicon is the most abundant element in the Earth's crust. Boron by comparison is relatively rare though the borates in which it occurs are very similar to silicates in structure and properties. Neither element is found free in nature.

Graphite. Impure forms of graphite carbon which occur in nature, such as coal and anthracite are first converted to coke by heating in the absence of air. A finely powdered mixture with sand is then heated in an electric furnace, and crystals of graphite are formed, probably via the intermediate formation of silicon carbide, SiC.

Colloidal suspensions of graphite in oil and water are used as lubricants; the layer structure of graphite is partly responsible for its lubricant properties (Chapter 5). A mixture of graphite and clay is used in pencil "lead".

Diamond. Diamond crystals are formed from graphite at extreme pressures and fairly high temperature (1200°C and 100 000 atm) and synthetic diamonds made this way can nowadays compete economically with natural diamonds for industrial purposes. Such diamonds are dark, sometimes being completely black, and are useful as abrasives, but modern technology has not yet succeeded in solving the problem of synthesizing large gem-stones in this way.

Silicon. Silicon is obtained by the reduction of silicon(IV) oxide, though the free energy graph (Fig. 24.1) shows that only an element like magnesium is capable of achieving this. Carbon will also reduce SiO_2 at very high temperatures but the product contains a high proportion of silicon carbide. Very pure silicon (>99·99%) is required in the manufacture of transistors, and the technique of zone refining is employed. Zone refining is a procedure in which impurities are removed from a solid phase by melting and resolidifying. It is therefore essentially a recrystallization process. Figure 24.2 shows the principle, though in practice more than one heater is used and the tube is moved through the stationary heaters.

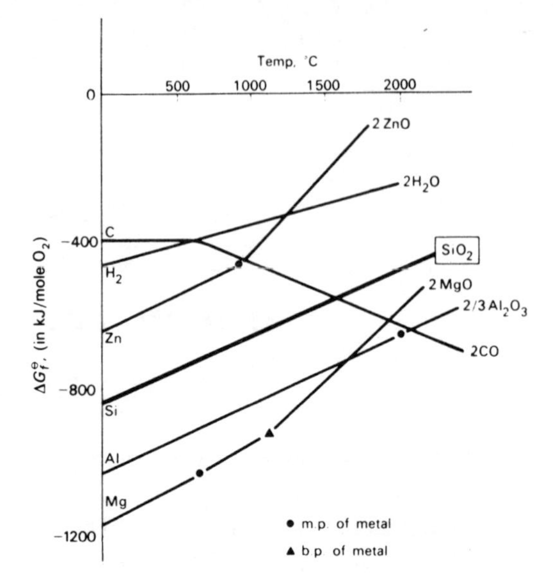

FIG. 24.1.

Boron. Boron is a difficult element to obtain pure, and it is mainly produced by reduction of boron(III) oxide, B_2O_3, with an electropositive element like magnesium. The free element does not find many uses.

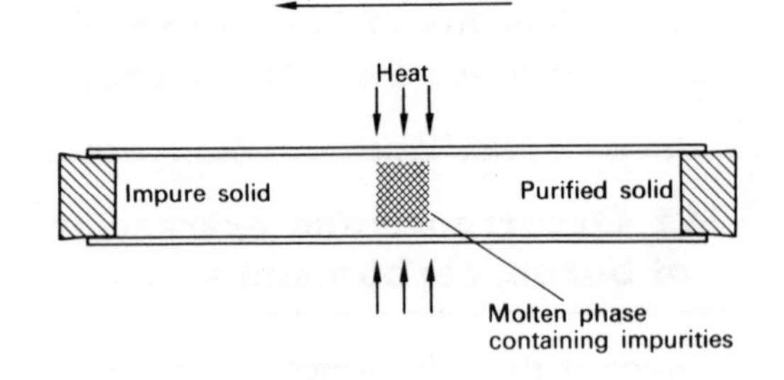

FIG. 24.2. Zone refining.

24.2 General survey of properties

The three elements are all macromolecular in structure, with high binding energies. Carbon and silicon have four electrons per atom available to form bonds while boron has three; these figures are reflected in the high heats of sublimation of the elements.

The structure of boron is complicated, the atoms being arranged in eicosahedra (20-faced). One would expect boron to be a metal, since this is usually the case when there are surplus orbitals with insufficient electrons to fill them. However, boron is a non-conductor and the electrons are not able to move freely through the lattice; a detailed explanation of why this should be so is outside the scope of this book.

Carbon is able to form a tetrahedral structure with a co-ordination number of four. It is also able to form a layer structure (graphite) with a coordination number of three. These structures are described in Chapter 5. The remaining elements of Group IVB do not form layer structures analogous to graphite, though all the elements apart from lead form diamond-like structures.

Table 24.2 is a comparison of some of the physical data for the three elements.

TABLE 24.2

	Boron	Carbon (graphite)	Silicon
Melting point	2300°C	3700°C	1420°C
Boiling point	2600°C	4800°C	2600°C
Density (g cm^{-3})	2·34	2·26	2·35
First ionization energy (kJ)	795	1085	785
Atomic radius (nm)	0·080	0·077	0·117

The elements all burn in air, though boron and silicon do not do so easily unless they are finely divided. Fluorine attacks all three elements, while the other halogens react with boron and silicon but not with carbon. This fact is fortunate, for many metals can be extracted by electrolysis of their fused chlorides using carbon anodes which are immune to attack by the evolved chlorine. Carbon will react directly with sulphur in an electric furnace:

$$C(graphite)+2S(g) = CS_2(g); \quad \Delta H = +115 \text{ kJ at } 25°C.$$

Boron combines directly with nitrogen when heated in the gas, to form a nitride, BN(s), which is structurally similar to graphite and isoelectronic with it; (the ions B^- and $N+$ are isoelectronic with the carbon atom). Recently a diamond-like form called *borazon*, an even harder substance than diamond, has been prepared.

Both boron and silicon form a number of hydrides, though these are nothing like as stable kinetically as those of carbon, and are not formed by direct combination of the elements. They are discussed in section 24.5.

All three elements combine directly with most metals to form compounds of various structures and stoichiometry (section 24.4).

24.3 Bond formation and reactivity

Figure 24.3 compares the energy level occupancy of boron, carbon and silicon, and shows that carbon is the only element which, when it forms electron pair bonds, exactly fills all its valence orbitals. This is the factor which accounts for the kinetic stability of carbon compounds.

The point is well illustrated by comparing the three chlorides BCl_3(l), CCl_4(l) and $SiCl_4$(l). The first and last named fume strongly in moist air and are highly reactive towards attack from negative ions such as OH^- and NH_2^-, and from molecules containing atoms with lone pairs of electrons such as water and ammonia. Carbon tetrachloride in contrast is highly inert and even prolonged boiling with concentrated alkali has almost no effect upon it. In terms of free energy

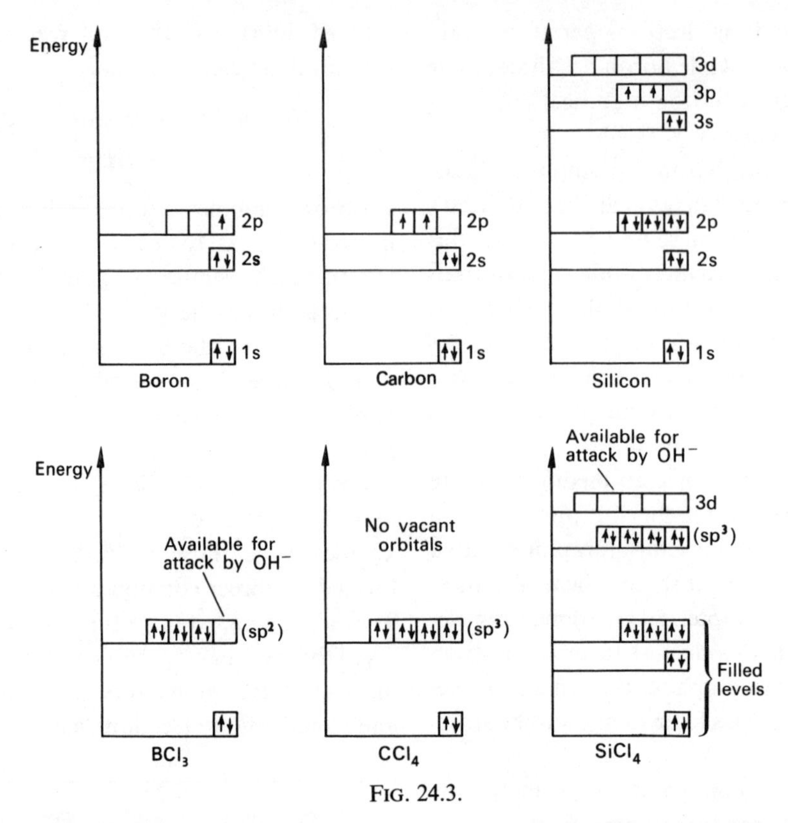

FIG. 24.3.

all three chlorides would be expected to hydrolyse, as the following standard free energy changes at 25°C show:

$$BCl_3 + \tfrac{3}{2}H_2O \rightarrow \tfrac{1}{2}B_2O_3 + 3HCl; \quad \Delta G^\ominus = -142 \text{ kJ}$$

$$SiCl_4 + 2H_2O \rightarrow SiO_2 + 4HCl; \quad \Delta G^\ominus = -142 \text{ kJ}$$

$$CCl_4 + 2H_2O \rightarrow CO_2 + 4HCl; \quad \Delta G^\ominus = -275 \text{ kJ}.$$

Carbon tetrachloride does not react, despite the favourable free energy change. It is not a *thermodynamic* factor which stabilizes carbon tetrachloride, but a *kinetic* factor: the reason that carbon tetrachloride does not react is that the required energy of activation cannot be attained. Figure 24.4 illustrates how the energies of activation differ for the attack of the three molecules by OH^- ion.

A similar explanation applies to the hydrides of carbon and silicon: these are superficially very similar, and both elements form homologous series:

$$CH_4, \quad C_2H_6, \quad C_3H_8, \quad C_4H_{10}, \quad \ldots, C_nH_{2n+2}\text{—the alkanes}$$

$$SiH_4, \quad Si_2H_6, \quad Si_3H_8, \quad Si_4H_{10}, \quad \ldots, Si_nH_{2n+2}\text{—the silanes.}$$

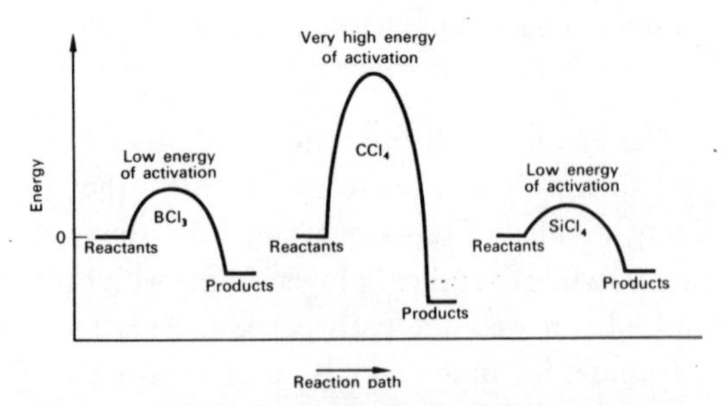

FIG. 24.4. Ease of hydrolysis of chlorides depends on energy of activation.

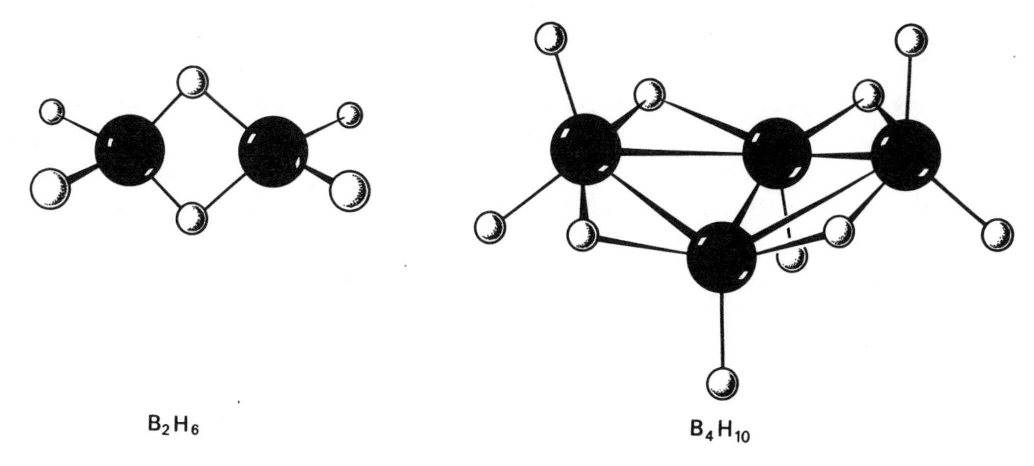

FIG. 24.5. Structures of two boron hydrides.

Whereas hydrocarbons are stable and obtainable in almost unlimited chain lengths, silicon hydrides inflame spontaneously in air and have not been isolated above Si_8H_{18}.

A different set of conditions applies to silicates however. Here the roles are reversed, and carbonate are *less* stable to heat than silicates. Calcium carbonate and calcium silicate have similar heats of formation. Calcium carbonate decomposes because it forms a gas, with consequent increase in disorder (entropy, Chapter 12):

$$CaCO_3(s) = CaO(s)+CO_2(g); \quad \text{increase in entropy}$$

$$CaSiO_3(s) = CaO(s)+SiO_2(s);$$

solid phases throughout,

$\therefore$ negligible entropy change.

Boron hydrides are reactive compounds for the same kind of reason as BCl_3 is reactive: they have unoccupied valence orbitals. In fact the bonding in boron hydrides was not understood until comparatively recently, when it was found that electron pairs in boron hydrides could bind several atoms together instead of simply two. The complex structures of boron hydrides, and their peculiar formulae, were thus explained (Fig. 24.5).

24.4 Ion formation by boron, carbon and silicon

By analogy with aluminium we might expect boron to form an ion B^{3+}, but in fact this ion would be too small to exist, and too much energy of ionization would be needed to form it. For the same reasons, C^{4+} and Si^{4+} do not exist (Fig. 24.6). There is no evidence for positive ion formation by these elements except in a discharge tube, and this is one reason why they are classed as non-metals.

All three elements form oxo-anions. In the

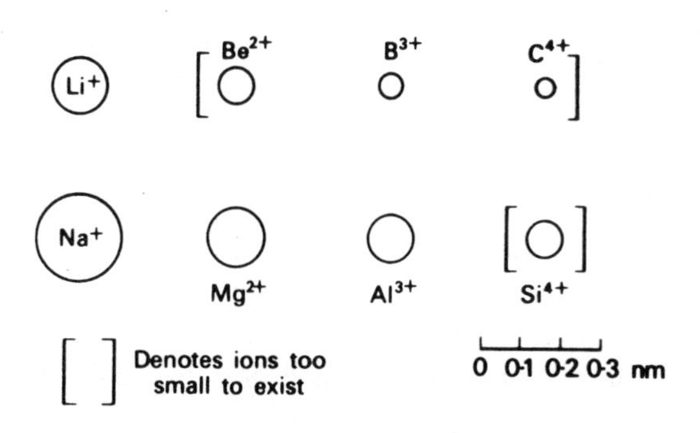

FIG. 24.6.

case of carbon the carbonate ion, CO_3^{2-}, is particularly simple, being a planar equilateral triangle in shape (section 20.12). Boron forms the analogous borate(III) ion, BO_3^{3-}, and also a series of ions representing intermediate degrees of hydration of the parent acid such as the ion $B_4O_7^{2-}$ which occurs in borax, $Na_2B_4O_7 \cdot 10H_2O(s)$, and ions such as $B_3O_6^{3-}$. This tendency to form poly-ions (ions with more than one atom of the element in question) is shared with silicon, where the property is extremely marked. Simple silicate ions SiO_3^{2-}, analogous to carbonate, do not exist but instead silicon tends to form extended polymeric structures.

Carbon forms the rather unexpected dicarbide ion C_2^{2-}, which occurs in compounds such as CaC_2. On adding water to calcium dicarbide, ethyne, $C_2H_2(g)$, is evolved and this observation (together with X-ray data on solid calcium dicarbide) leads to the conclusion that the dicarbide ion contains a triple bond:

$[C{\equiv}C]^{2-}$ dicarbide ion

$H{-}C{\equiv}C{-}H$ ethyne

Boron and silicon do not form analogous ions: the reluctance of second-row elements to enter into multiple-bond formation has already been noted, and in the case of boron there are insufficient valence electrons available to form a triple bond.

24.5 Hydride formation by boron, carbon and silicon

The most striking fact in dealing with the hydrides of these three elements is the vast range of "stable" hydrocarbons. Very many occur naturally on Earth: North Sea gas consists largely of methane, CH_4, and crude oil deposits consist largely of paraffin hydrocarbons of general formula C_nH_{2n+2}, where n = up to about 40. Hydrocarbons are also present in wood and coal. They owe their stability to the fact that reactions such as combustion require a high energy of activation. Silanes, Si_nH_{2n+2}, inflame spontaneously in air but paraffins need to be "triggered off" with a spark.

Structurally the following features may be noted:

(i) The structures of the boron hydrides (boranes) are different from those of carbon and silicon hydrides due to the different bonding (section 24.3).

(ii) Hydrocarbons frequently exhibit structures where there are multiple bonds, C=C and C≡C. For instance:

$H_2C{=}CH_2$

ethene

Silicon can only form single-bonded structures.

A mixture of silanes can be prepared by the action of water on magnesium silicide, $Mg_2Si(s)$. For instance:

$$Mg_2Si(s)+4H_2O(l) = SiH_4(g)+2Mg(OH)_2(aq).$$

Similarly a mixture of boranes can be made by the action of water on magnesium boride. This method is of general use (section 18.6), and can also be applied to the preparation of certain hydrocarbons though it rarely is in practice.

If a pure sample of a hydride is required, the usual procedure is to reduce the corresponding chloride with the powerful reducing agent, lithium tetrahydridoaluminate, $LiAlH_4$. This reagent is frequently employed for reducing organic compounds, and its uses here can be summarized in the form of the following equation:

$$SiCl_4 + LiAlH_4 \rightarrow SiH_4 + LiCl + AlCl_3.$$

Physical properties. The boiling points and melting points of these hydrides increase with molecular weight within a given series. Figure 6.18 shows a plot of boiling point against number of carbon atoms for the paraffin hydrocarbons, and similar trends are observed for boron and silicon hydrides. Hydrogen bonds do not form among these compounds: hence methane is considerably more volatile than NH_3, H_2O and HF, and is considerably less soluble in water (Fig. 15.2). The absence of hydrogen bonds is explained by the fact that these elements are not very electronegative, and also that they do not possess lone pairs in their compounds.

24.6 Halides of boron, carbon, and silicon

Carbon tetrachloride. The exceptional unreactivity of carbon tetrachloride, CCl_4(l), has already been commented upon (section 24.3). This compound has found uses as a non-polar solvent, suitable for dissolving oils and fats; proprietary dry-cleaning fluids were until recently based upon it. Its solvent properties are similar to that of petrol (a hydrocarbon mixture) but it has the added advantage of being non-inflammable. It is now falling into disuse on account of its toxicity; in the presence of a naked flame it reacts with oxygen to give carbon dichloride oxide, $COCl_2$(g), which is highly poisonous. Nowadays, other chlorine-substituted hydrocarbons such as $CH_3.CCl_3$ are taking the place of carbon tetrachloride, and substances such as these also find uses in fire extinguishers, particularly where there is a risk of electrical fires.

PVC. The carbon-chlorine bond is quite strong (bond energy = 330 kJ mol^{-1}) and is in most cases fairly unreactive. The plastic "polyvinyl chloride" (PVC) has a structure analogous to "polythene", but with chlorine atoms in place of some hydrogen atoms:

```
  H   H   H   H   H   H   H   H
  |   |   |   |   |   |   |   |
—C—C—C—C—C—C—C—C—
  |   |   |   |   |   |   |   |
  H   H   H   H   H   H   H   H
```

a section of a molecule of polythene

```
  H   Cl  H   Cl  H   Cl  H   Cl
  |   |   |   |   |   |   |   |
—C—C—C—C—C—C—C—C—
  |   |   |   |   |   |   |   |
  H   H   H   H   H   H   H   H
```

a section of a PVC molecule

PTFE. The abbreviation PTFE stands for polytetrafluoroethene, $(C_2F_4)_x$. It is quite extraordinarily unreactive, due to the high energy of the C—F bond (438 kJ mol^{-1}), and also possesses a very low coefficient of friction with most other substances. It is a difficult material to work and is expensive, but finds uses where frictionless bearings and non-stick surfaces (e.g. cooking pans) are required. Strong heat causes it to depolymerize into smaller structural units of the same empirical formula.

Freons. The general term "Freon" refers to gaseous compounds such as $CClF_3$ and CCl_2F_2 which are used as refrigerant gases. On a large scale ammonia is used, but for domestic units "Freon" is less toxic though more expensive.

Silicon halides. Silicon tetrachloride fumes in moist air, and is hydrolysed by water, depositing hydrated silica in a hydrated form:

$$SiCl_4(l) + 4H_2O \rightarrow [Si(OH)_4] + 4HCl(g)$$

$$\downarrow H_2O$$

$$SiO_2.xH_2O(s).$$

The same hydrolysis reaction is used in the manufacture of *silicones*. If $(CH_3)_2SiCl_2(l)$ is hydrolysed, the product is a mixture of ring and chain polymers:

$$(CH_3)_2SiCl_2 + 2H_2O \rightarrow [(CH_3)_2Si(OH)_2] + 2HCl$$

$$\downarrow -H_2O$$

$$-\underset{CH_3}{\overset{CH_3}{Si}}-O-\underset{CH_3}{\overset{CH_3}{Si}}-O-\underset{CH_3}{\overset{CH_3}{Si}}-O-\underset{CH_3}{\overset{CH_3}{Si}}-O-$$

The physical properties of a silicone can be varied by introducing other compounds: $(CH_3)_3SiCl$ can be included to introduce chain ends, and CH_3SiCl_3 leads to branched chains:

$$Cl-\underset{CH_3}{\overset{CH_3}{Si}}-CH_3 \longrightarrow -O-\underset{CH_3}{\overset{CH_3}{Si}}-CH_3$$

chain end

$$Cl-\underset{Cl}{\overset{Cl}{Si}}-CH_3 \longrightarrow -O-\underset{O}{\overset{O}{Si}}-CH_3$$

branched chain

Silicone polymers can thus be "tailor made" with a wide choice of properties: some are oils, and others with a higher degree of cross linking form useful plastics. Silicone products are strongly water-repellent.

24.7 Oxides and oxo-acids of boron, carbon and silicon

The principal oxides of the three elements which will be considered are $B_2O_3(s)$, $CO(g)$, $CO_2(g)$ and $SiO_2(s)$. All except carbon monoxide are very weakly acidic, dissolving in alkali to form salts:

TABLE 24.3
OXIDES OF BORON AND SILICON

Formula	Name	Melting point
$CO(g)$	carbon monoxide	$-205°C$
$CO_2(g)$	carbon dioxide	$-78°C$ (subl.)
$SiO_2(s)$	silicon(IV) oxide, silica	1728°C
$B_2O_3(s)$	boron(III) oxide	577°C

$$CO_2(g) + 2OH^-(aq) \rightarrow CO_3^{2-}(aq) + H_2O; \quad \text{(rapid)}$$

$$B_2O_3(s) + 6OH^-(aq) \rightarrow 2BO_3^{3-} + 3H_2O; \quad \text{(slow)}$$

$$SiO_2(s) + 2OH^-(aq) \rightarrow SiO_3^{2-}(aq) + H_2O; \quad \text{(slow)}$$

Boron(III) oxide and silicon(IV) oxide are solids with macromolecular structures which only dissolve in alkali when finely ground and heated; carbon dioxide is a gas which is absorbed very rapidly by alkali. Carbon dioxide also forms well-defined acid salts, known as hydrogencarbonates:

$$CO_2(g) + OH^-(aq) \rightarrow HCO_3^-(aq)$$

Carbon monoxide is normally classed as a neutral oxide (section 22.11) though it does react with alkali to form the *methanoate* ion at high temperature and pressure:

$$CO(g) + OH^-(aq) \longrightarrow H-C\begin{matrix} \nearrow O \\ \searrow O^- \end{matrix} \quad (aq)$$

Carbon dioxide is present in the air (about 0·03%) due to its evolution from animals and plants, and by processes such as burning and fermentation. Its concentration remains remarkably constant, and it is continually removed from the air by photosynthesis. The gas is obtained by:

(i) the action of $H^+(aq)$ on a carbonate or hydrogencarbonate;
(ii) action of heat on carbonates and hydrogencarbonates;
(iii) oxidation of carbon, carbon monoxide, or carbon compounds;
(iv) fermentation.

The gas is liquefied by pressure, and the liquid is stored in cylinders. Rapid expansion of the gas leads to pronounced cooling, and solid carbon dioxide (sold as "dry-ice") is thus obtained. The gas is used in refrigeration, "fizzy" drinks, fire extinguishers, fruit preservation, some metallurgical processes, and the manufacture of some carbonates (e.g. the white pigment $PbCO_3(s)$).

When carbon dioxide is dissolved in water, some carbonic acid, $H_2CO_3(aq)$, is formed, but the anhydrous acid cannot be obtained.

Carbonates. The *s*-block elements (except beryllium) form stoichiometric carbonates; the thermal stability of these was discussed in section 20.12.

B-metals and transition metals form non-stoichiometric *basic* carbonates when $CO_3^{2-}(aq)$ is added to an aqueous solution of the metal ion. For instance copper carbonate forms $x\text{Cu(OH)}_2.y\text{CuCO}_3$. Some metals do not form carbonates at all, e.g. aluminium and beryllium.

Apart from the Group IA carbonates, all carbonates are insoluble in water, and decompose on heating giving the oxide and $CO_2(g)$.

Compounds containing the ion HCO_3^- were described in section 20.12.

Carbon monoxide is formed whenever carbon or its compounds react with oxygen at a high temperature, or where the supply of oxygen is insufficient. It also forms when a metal oxide reacts with carbon at high temperature. Essentially its formation depends upon the following endothermic reaction which becomes favoured above about 700°C:

$$\text{C(graphite)} + \text{CO}_2\text{(g)} \rightleftharpoons 2\text{CO(g)}; \quad \Delta H = +172 \text{ kJ}$$

Carbon monoxide can be made in the laboratory in this way but a purer product is obtained by dehydrating methanoic acid with concentrated sulphuric acid:

$$\text{H.COOH} - \text{H}_2\text{O} \rightarrow \text{CO}.$$

Carbon monoxide is very poisonous due to its ability to combine with haemoglobin in the blood to form a stable addition compound; this prevents the normal exchange between haemoglobin and oxygen from taking place. An industrial fuel called *producer gas*, consisting of carbon monoxide diluted with nitrogen, is made by passing air over red-hot coke. A better fuel is *water gas*, a mixture of carbon monoxide and hydrogen made by passing steam over heated coke. The latter reaction is endothermic, but by mixing air and steam the overall process can be made approximately thermoneutral.

$$2\text{C(graphite)} + \text{O}_2\text{(g)} = 2\text{CO(g)}; \quad \Delta H = -222 \text{ kJ}$$

$$\text{C(graphite)} + \text{H}_2\text{O(g)} = \text{CO(g)} + \text{H}_2\text{(g)}; \quad \Delta H = +130 \text{ kJ}$$

Water gas is also used as a source of hydrogen, carbon monoxide, and methanol.

Carbon monoxide forms volatile compounds, which are known as carbonyls, with some transition metals (section 35.13).

Silicon(IV) oxide differs markedly in structure from carbon dioxide, due to the reluctance of silicon to form double bonds (Fig. 24.7). This difference in structure is also reflected in differences between silicates and carbonates. Sand and quartz are naturally occurring forms of silicon(IV) oxide, and a variety of other crystalline forms—polymorphs or allotropes—also

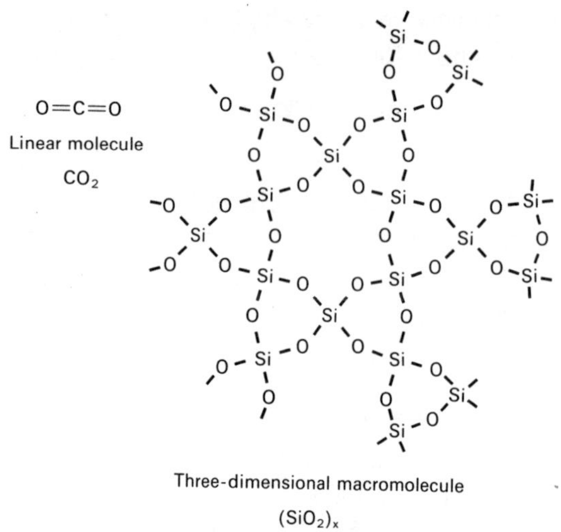

O=C=O

Linear molecule

CO_2

Three-dimensional macromolecule

$(SiO_2)_x$

FIG. 24.7.

occur. If silicon(IV) oxide is melted (m.p. 1728°C) and recooled it becomes **vitreous**, or glass-like; it loses its regular crystalline lattice and becomes amorphous. This glass eventually devitrifies and becomes crystalline once more.

Vitreous silica is used in special laboratory apparatus; it has a low coefficient of thermal expansion and consequently does not shatter like ordinary glass when subject to violent thermal shock. Acids do not attack it, but alkalis slowly react forming silicates. Aqueous hydrogen fluoride dissolves it, forming the ion SiF_6^{2-}:

$$SiO_2(s) + 6HF(aq) \rightarrow 2H^+(aq) + SiF_6^{2-}(aq) + 2H_2O.$$

Silicic acid, $SiO_2.xH_2O$, is the name given to the colloidal gel produced when $H^+(aq)$ is added to a solution of silicate ion. A continuous range of substances of varying degrees of hydration can be obtained by heating this. One such hydrate, a dry powder called silica gel, is widely used as a drying agent.

Sodium silicate, Na_2SiO_3 (approximate formula), is formed as a glass when sodium carbonate is fused with silica:

$$Na_2CO_3 + SiO_2 \rightarrow Na_2SiO_3 + CO_2.$$

Prolonged heating with water under pressure converts it to a treacly aqueous solution known as "water glass", formerly used as an egg preservative; its main uses now are as an additive for detergents, and as a cheap adhesive.

It is probable that the extended lattice formed by silicon and oxygen atoms in silica persists in water glass, thereby giving it a high viscosity.

Silicates in the Earth's crust. The empirical formula of naturally occurring silicates are complex, and their structures fall into three distinct types:

(i) **Fibrous structures**, in which chain molecules exist, e.g. the *amphiboles*.

(ii) **Layer structures**, in which the solid readily splits into flat sheets, e.g. *mica*.

(iii) **Solid structures**, in which the macromolecules are three-dimensional, e.g. *felspars* (components of granite).

In these structures there may be varying proportions of water; specially processed *zeolites* treated to remove the water have been found to act as **molecular sieves**. For instance a sieve is manufactured which will allow straight chain hydrocarbons to pass through it, while blocking branched chains. Other zeolites find uses as ion exchangers (section 16.1) though they have been largely superseded by synthetic polystyrene resins.

Glass. Glass has been known for many centuries, and ordinary cheap varieties are made by fusing together sand, limestone and an alkali such as sodium hydroxide or sodium carbonate. The resultant non-stoichiometric compound is glass, in effect a mixture of silicates. Soda-glass, made by using sodium carbonate, is described as soft glass as it has a low softening point on heating; by substituting potassium carbonate a harder glass is obtained. Glasses never have sharp melting points, but always soften over a range of temperature due to their amorphous nature.

Coloured glasses are made by adding small traces of transition metal oxides which form coloured ions; the green colour of cheap glass is due to traces of iron impurity. Glass of high refractive index is made by using lead oxide in place of calcium oxide. Modern laboratory glassware uses oxides of boron and aluminium in place of calcium oxide, and this is known as *borosilicate glass* ("Pyrex" is a well-known example).

24.8 Borides, carbides and silicides

Metals form binary compounds with boron, carbon and silicon, the two main classes being **ionic** (*saline* or salt-like), and **interstitial** (where the non-metal atoms occur within a metal lattice).

Ionic. (a) The carbides of the alkali and alkaline-earth metals contain discrete C_2^{2-} units. These give ethyne, C_2H_2, on hydrolysis.

(b) Be_2C and Al_4C_3 contain discrete carbon atoms or ions, and give methane, CH_4, on hydrolysis.

The difference in property can be understood by considering the sizes of the ions which might be expected. It has previously been remarked for carbonates (section 20.12) and polyhalides (section 21.5), that large anions form more stable lattices with large cations. Here the larger ions of Groups I and II favour the larger C_2^{2-} ion.

Interstitial borides, carbides and silicides (and also nitrides and phosphides) are formed by many transition metals. These are metal-like (they conduct electricity) and are usually non-stoichiometric. The lattice is based upon the metallic element. A reverse state of affairs occurs with certain compounds based on the graphite structure, in which metal atoms are "sandwiched" interstitially between the layers, e.g. KC_8, KC_{24} and KC_{36}. Borides such as CaB_6 occur also.

24.9 Cyanides

Cyanides are salts of the acid hydrogen cyanide, HCN (b.p. 26°C). This gas dissolves in water giving a very weak acid:

$$HCN(g) + aq \rightleftharpoons \underbrace{H^+(aq) + CN^-(aq)}_{\text{hydrocyanic acid}};$$

$$K_a = 7{\cdot}2 \times 10^{-10}\ \text{mol dm}^{-3} \text{ at } 25°C.$$

Hydrogen cyanide is in some respects similar to the hydrogen halides, and indeed the cyanide ion has properties similar to the halide ions in general. The gas *cyanogen*, $(CN)_2$, bears a formal resemblance to the halogens and is often referred to as a *pseudo-halogen*.

Hydrogen cyanide and cyanides are extremely poisonous substances which require careful handling. The gas smells of almonds. The antidote is a solution of iron(II) ions which reacts with $CN^-(aq)$ to form the harmless complex ion $Fe(CN)_6^{4-}$ (see below).

The HCN molecule is linear, and it is isoelectronic with ethyne, possessing a triple bond:

$$H-C\equiv N \quad \text{or} \quad H\overset{\times}{\bullet}C\overset{\times\circ}{\underset{\times\circ}{\times\circ}}N\,\overset{\circ}{\circ}$$

The cyanides of the alkali metals are probably ionic in the solid state, but the structure of most others suggests that they are covalent. The majority are insoluble in water, but dissolve in excess $CN^-(aq)$ forming complex ions.

Sodium cyanide is manufactured by the following reactions:

$$Na + NH_3 \rightarrow NaNH_2(s) + \tfrac{1}{2}H_2$$
sodium amide (sodamide)

$$NaNH_2 + C \rightarrow NaCN + H_2.$$

Since hydrogen cyanide is very weak, the conjugate base $CN^-(aq)$ is fairly strong, and can react with water to produce an alkaline solution:

$$CN^-(aq) + H_2O \rightleftharpoons HCN(aq) + OH^-(aq)$$
proton accepted; proton lost

Cyano complexes. The $CN^-(aq)$ ion is a powerful ligand, that is to say, it can join to metal ions forming stable complexes. In many cases the ion is so stable that the properties of the free ions cannot be detected in solution; this is true of the $Fe(CN)_6^{4-}$ ion mentioned above:

$$Fe^{2+}(aq) + 6CN^-(aq) \rightleftharpoons Fe(CN)_6^{4-}(aq);$$
iron(II) ion; hexacyanoferrate(II) ion

Transition metals and B-metals form stable complexes with CN^-; *s*-block metals do not, and neither does aluminium.

24.10 Thiocyanates

Another pseudohalide ion, closely related to cyanide, is the thiocyanate ion, CNS^-. This is also a ligand, and one of its special uses is in the test for the presence of iron(III). When $Fe^{3+}(aq)$ and $CNS^-(aq)$ are mixed, a variety of species can be obtained depending on the relative ratios of the simple ions present, but all have a deep blood red colour. This is used to estimate iron(III) colorimetrically: the intensity of absorption of light can be compared with a known standard in a colorimeter. Alternatively, photoelectric measurements can be made.

24.11 Summary

Boron and silicon afford one of the clearest examples of a diagonal relationship. Hence:

(a) The elements are similar in physical properties, existing in both amorphous and crystalline forms.

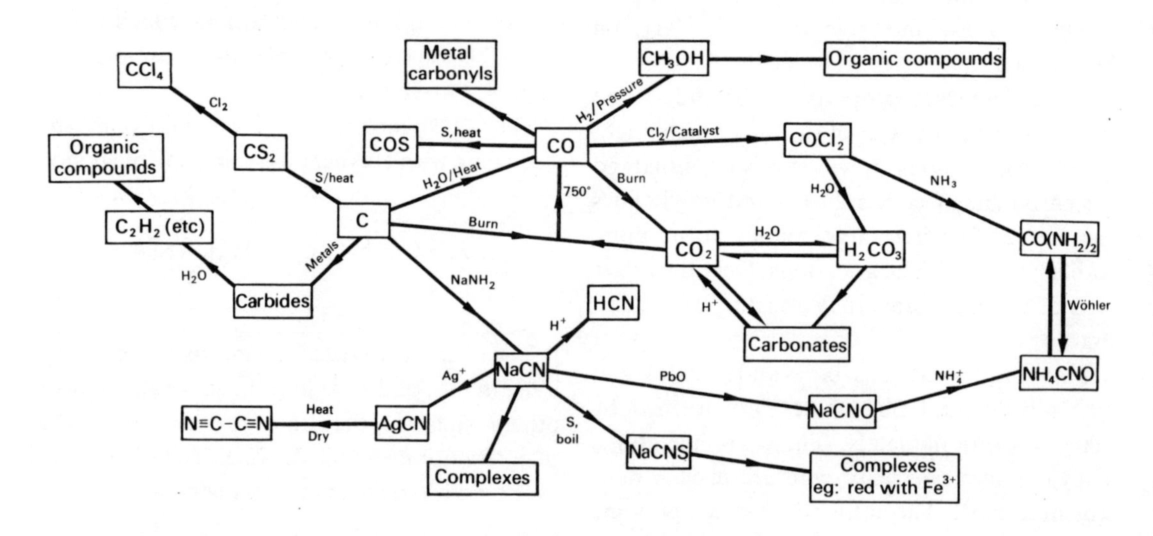

FIG. 24.8. A summary of reactions of carbon.

(b) Their oxides are similar—weakly acidic, macromolecular, giving rise to similar "polymeric" oxo-ions and readily forming glasses.

(c) Both elements form a series of hydrides which are kinetically unstable in the presence of oxygen (i.e. they ignite spontaneously).

(d) Both form readily hydrolysable molecular chlorides.

Other examples of diagonal relationships

Lithium and magnesium react similarly with air, water and non-metals. Lithium resembles Group II rather than Group I in having a thermally unstable carbonate, in forming a hydrated chloride when the aqueous solution is evaporated, and in forming an ion $Li^+(aq)$ which hydrolyses in aqueous solution.

Beryllium and aluminium form similar hydroxides which are amphoteric; the metals themselves dissolve in alkali giving hydrogen. Similar complexes are formed, e.g. $BeCl_4^{2-}$ and $AlCl_4^-$. (But note BeF_4^{2-} and AlF_6^{3-}, due to expansion of aluminium valence shell.) Beryllium and aluminium chlorides have analogous structures.

Table 24.4

Chemical Trends in the First Row (Li—Ne)

Property	Li	Be	B	C	N	O	F
Elements							
B.p. (in K)	1600	2750	4200	4300	77	90	85
Coordination number	8	12	*see* Chap. 5	3 or 4	1	1	1
	Metallic character decreases across the Period ⟶						
Hydrides							
Empirical formula	LiH	BeH_2	BH_3	CH_4	NH_3	H_2O	HF
Structure	Ionic	Polymeric	Molecular	Molecular	Molecular	Molecular	Molecular
Chlorides							
Empirical formula	$LiCl$	$BeCl_2$	BCl_3	CCl_4	NCl_3	Cl_2O	ClF
Structure	Ionic	Intermediate	Molecular	Molecular	Molecular	Molecular	Molecular
Oxides							
Empirical formula	Li_2O	BeO	B_2O_3	CO_2	N_2O_3 etc.	—	F_2O
Structure	Ionic	Giant	Giant	Molecular	Molecular	—	Molecular
Acid–base character	Alkaline	Amphoteric	Weakly acidic	Weakly acidic	Acidic	—	Does not react
	Acid strength increases across the Period ⟶						
Ions							
Simple ions	Li^+	Be^{2+}	—	—	N^{3-}	O^{2-}	F^-
Oxo-ions	—	BeO_2^{2-}	BO_3^{2-}	CO_3^{2-}	NO_3^-, NO_2^-	—	—

24.12 Chemical periodicity

Chapters 19 to 24 dealt with the chemistry of all the elements in the first (Li—Ne) and second (Na—Ar) rows. Chapter 1 concentrated on the periodicity of physical properties, and the Tables in this section summarize the trends in chemical behaviour of elements in these two Periods.

For further details relating to Tables 24.4 and 24.5, see Chapter 5 (the elements), Section 18.4 (the hydrides), Section 21.8 (the halides) and Sections 22.11 and 22.12 (the oxides).

TABLE 24.5

CHEMICAL TRENDS IN THE SECOND ROW (Na—Ar)

Property	Na	Mg	Al	Si	P	S	Cl
Elements							
B.p. (in K)	1160	1390	2720	2950	554	718	239
Coordination number	8	12	12	4	3	2	1
			Metallic character decreases across the Period ⟶				
Hydrides							
Empirical formula	NaH	MgH_2	AlH_3	SiH_4	PH_3	H_2S	HCl
Structure	Ionic	Ionic	Polymeric	Molecular	Molecular	Molecular	Molecular
Chlorides							
Empirical formula	NaCl	$MgCl_2$	$AlCl_3$	$SiCl_4$	PCl_3, PCl_5	S_2Cl_2 etc.	—
Structure	Ionic	Ionic	Molecular (Al_2Cl_6)	Molecular	Molecular	Molecular	—
B.p. (in K)	1740	1690	696	330	349	409	—
Oxides							
Empirical formula	Na_2O	MgO	Al_2O_3	SiO_2	P_2O_3, P_2O_5	SO_2, SO_3	Cl_2O etc.
Structure	Ionic	Ionic	Giant	Giant	Molecular (dimeric)	Molecular	Molecular
Acid–base character	Alkaline	Weakly alkaline	Amphoteric	Weakly acidic	Acidic	Acidic	Acidic
		Covalent nature and acidic strength increase across the Period ⟶					
Ions							
Simple ions	Na^+	Mg^{2+}	Al^{3+}	—	—	S^{2-}	Cl^-
Oxo-ions	—	—	AlO_2^-	SiO_3^{2-}	PO_4^{3-}, PO_3^{3-}	SO_4^{2-}, SO_3^{2-} etc.	ClO_4^- etc.

Study Questions

1. List the ways in which the chemistry of aluminium is (a) similar to, and (b) different from the chemistry of boron. How can you account for the differences?

2. (a) Aluminium forms an ion AlF_6^{3-}. Why is BF_6^{3-} not formed?

(b) Suggest why $BCl_3(g)$ is monomeric, while the chloride of aluminium in the gas phase consists of Al_2Cl_6 dimers.

3. (a) Compare and contrast

(i) CO_2 and SiO_2.
(ii) $CaSiO_3$ and $CaCO_3$.
(iii) $SiCl_4$ and CCl_4.
(iv) $(CH_3)_2CO$ and $(CH_3)_2SiO$.

(b) What reasons can you suggest for the differences that you mention?

4. (a) List as many similarities as you can in the chemistry of boron and silicon.

(b) Explain, in terms of fundamental principles, why the chemistries of boron and silicon are similar.

(c) Is there any evidence for a diagonal relationship between carbon and phosphorus?

5. (a) List the typical properties of a halogen.

(b) Cyanogen has been described as a "pseudo-halogen". How many of the halogen properties does cyanogen possess?

6. When 10·01 g of a white solid, X, were heated, 2·2 g of an acid gas, A, that turned lime water milky, were given off, together with 0·9 g of a gas, B, which condensed to a colourless liquid. The solid that remained, Y, weighing 6·91 g, dissolved in water to give an alkaline solution, which with excess barium chloride solution gave 9·85 g of a white precipitate, Z. Z effervesced with acid, giving off carbon dioxide.

(a) Identify Z, A and B.

(b) How many moles of Z, A and B were produced?

(c) Deduce the nature of Y.

(d) Deduce the nature of X and write down an equation for its thermal decomposition.

7. A litre of M sodium carbonate was evaporated and 286 g of white crystals were obtained. These were left in a dry laboratory and were later found to weigh 124 g.

(a) What are the formulae of the two solids?

(b) What is the name given to the process by which the crystals lost weight?

8. When a solution of sodium hydrogencarbonate is added to a solution of zinc ions, zinc carbonate, $ZnCO_3$, is precipitated.

(a) What else is formed in this reaction?

(b) Give a balanced equation for the reaction.

(c) Zinc carbonate is readily decomposed by heat. Suggest what the products of the thermal decomposition might be.

9. Metal oxides and salts are sometimes heated on a charcoal block in a blowpipe flame.

(a) Which metal oxides would be reduced to the metal in this way?

(b) How would you expect a nitrate or chlorate(V) to react in this test?

10. Give as many reasons as you can why carbon is often preferred to silicon or boron as a reducing agent.

11. When HF(g) reacts with glass in the presence of water, hexafluorosilicic(IV) acid, "H_2SiF_6", is formed.

(a) Suggest an equation for this reaction.

(b) What shape would you expect the SiF_6^{2-} ion to be?

(c) Can you suggest why $SiCl_6^{2-}$ has yet to be made?

(d) When barium hexafluorosilicate(IV) is heated, SiF_4 is evolved. Suggest an equation for this reaction.

12. (a) The four heavy atoms in $(SiH_3)_3N$ are coplanar. Why is this surprising?

(b) Would you expect $(SiH_3)_3N$ to be an electron donor?

CHAPTER 25

Saturated hydrocarbons

25.1 Organic chemistry

The term **organic chemistry** originally meant simply "the chemistry of living things", but nowadays it has been extended to include the study of all compounds containing C—H bonds. The special study of the chemistry of *living* systems is called **biochemistry**.

Despite the relatively low abundance of carbon on earth, the number of known carbon compounds exceeds the combined total of known compounds of all other elements. Carbon is able to form long *kinetically stable** chains of atoms, and these form the skeletons of organic molecules. Throughout the majority of organic reactions the **carbon skeleton** remains intact, and such changes as do occur affect only those atoms or groups of atoms that are attached to the skeleton.

25.2 The stability of carbon skeletons

Carbon and hydrogen are the only two elements in which the number of outer electrons (valence electrons) equals the number of outer-most electron orbitals. Hence whenever carbon forms single electron-pair bonds with other atoms, there are no vacant energy levels to be attacked by electron pairs from another reacting species, nor does the carbon atom have a lone pair with which to attack another species that does possess vacant energy levels. (Figure 25.1.) This restricts the available mechanisms by which bonds can be broken, and accounts for the kinetic stability of carbon skeletons. Moreover, when the bond is a C—C bond it is non-polar and will not attract potential attacking-reagents electrostatically.

C—C bonds are however *thermodynamically* unstable in the presence of the halogens, oxygen, and many other reagents, though they do not break unless the conditions are unusual or severe.

In addition to C—C bonds, organic chemistry is also concerned with C—H bonds, which are just as stable kinetically as C—C bonds, and

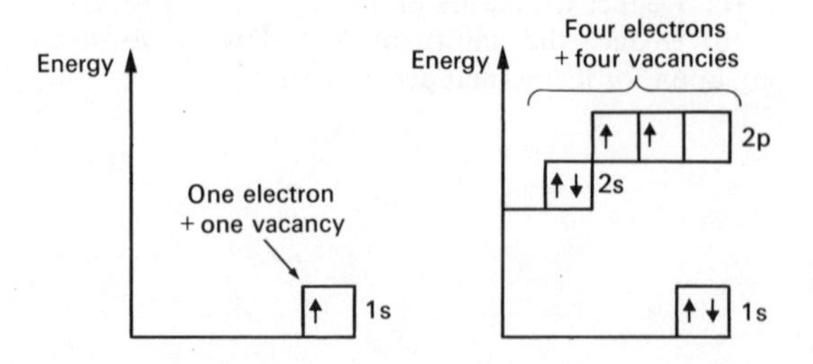

FIG. 25.1. The occupancy of energy levels in carbon and hydrogen.

* See chapter 24 for a discussion of thermodynamic and kinetic stability.

for many of the same reasons. The electronegativities of carbon and hydrogen are similar (Fig. 4.8) so that C—H bonds are of relatively low polarity and do not attract attacking species electrostatically.

Compounds that contain only C—C single bonds and C—H bonds are known as **saturated hydrocarbons**. The term "saturated" implies that no further atoms can add to the carbon skeleton.

25.3 Saturated hydrocarbons (alkanes)

The simplest organic compounds are the **alkanes**, which are all hydrocarbons composed entirely of C—C and C—H bonds, with no lone pairs or vacant orbitals. It can readily be shown that if there are n carbon atoms in a given alkane molecule, there must be $2n+2$ hydrogen atoms. The **general formula** of the alkanes is therefore C_nH_{2n+2}. A series of compounds with the same general formula, each representing an increase of one carbon atom on the previous compound, is called a **homologous series** (section 6.9). If there are several carbon atoms, the molecule can be either "straight-chain" or "branched chain" in structure, though it must be remembered that in reality the so-called straight-chain structures are zig-zag due to the tetrahedral angle between carbon atoms.

The alkanes occur extensively in the earth's crust in the form of crude petroleum, which consists mainly of a mixture of straight-chain alkanes up to about C_{30}. Table 25.1 summarizes some information on some of the simple alkanes.

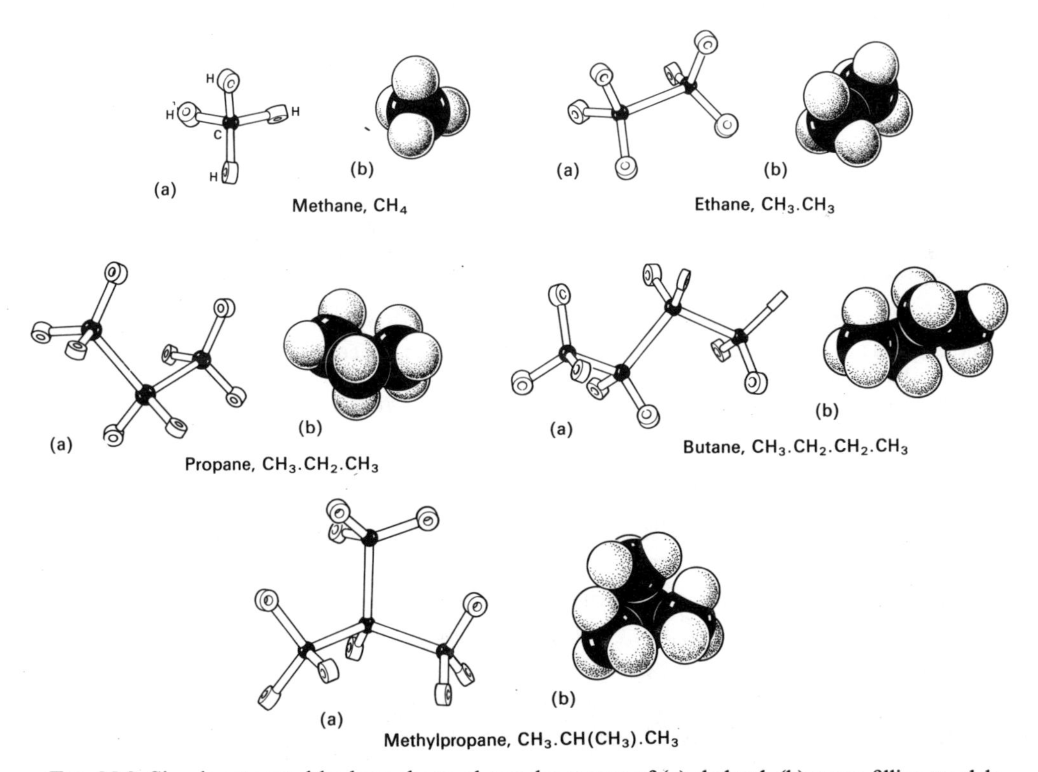

FIG. 25.2. Simple saturated-hydrocarbons, shown by means of (a) skeletal, (b) space-filling models. (The skeletal models depicted are made with the "Orbit" molecular building kit, see page xiii.)

TABLE 25.1 SIMPLE ALKANES

Molecular formula and name	Abbreviated structural formula*	Full structural formula	Skeletal notation‡
CH_4 methane	CH_4	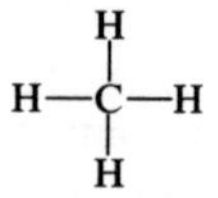 H H—C—H H	(not used)
C_2H_6 ethane	$CH_3.CH_3$	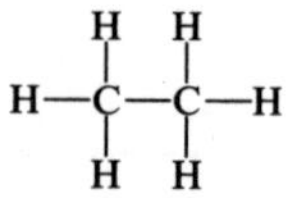 H H H—C—C—H H H	(not used)
C_3H_8 propane	$CH_3.CH_2.CH_3$	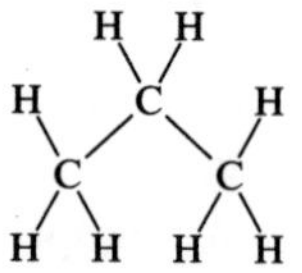 H H H C H C C H H H H	
C_4H_{10} butane	$CH_3.CH_2.CH_2.CH_3$	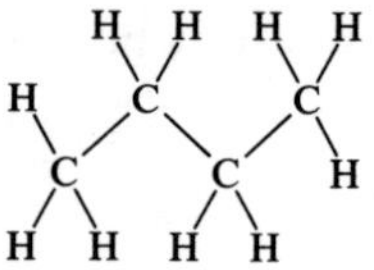 H H H H H C C C C H H H H H	
C_4H_{10} methylpropane	$CH_3.CH(CH_3).CH_3$	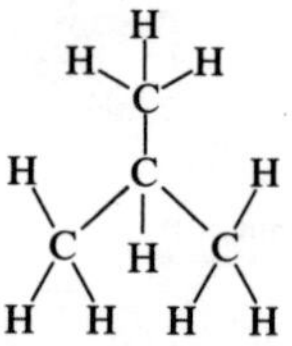 H H C H H C H C H C H H H H	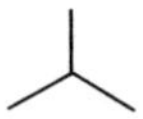

* Where a dot is used to show a bond, it means a link between carbon atoms of the main chain. Lines are not used in these formulae, as they would imply direct links between the atoms at each end of the line.

‡ The skeletal notation shows all C—C single bonds by a line, and omits all other bonds entirely. Similarly, double bonds may be shown by double lines (Table 26.1). This abbreviated notation is useful when writing complex formulae.

25.4 Nomenclature of alkanes

Until a few years ago the naming of organic compounds was rather unsystematic, but in recent years the much more logical IUPAC nomenclature has been developed and is largely replacing the older "trivial" names. The process continues, and in this book systematic names have been used wherever possible. Once the IUPAC system is understood, it is possible to write the formula of a compound correctly from the name. In cases of doubt, the reader is referred to Appendix V. Alternative names of compounds are listed in the index.

The basis of IUPAC nomenclature of organic compounds is that the name consists essentially of

(prefix) – stem – suffix.

TABLE 25.2

STEMS USED IN IUPAC NAMES OF CARBON CHAIN COMPOUNDS

Number of carbon atoms	Stem	Number of carbon atoms	Stem	Number of carbon atoms	Stem
1	meth-	6	hex-	11	hendec-
2	eth-	7	hept-	12	dodec-
3	prop-	8	oct-	13	tridec-
4	but-	9	non-	14	tetradec-
5	pent-	10	dec-	*etc.*	
				n	alk-

The stem gives the number of carbon atoms in the longest "straight chain" in the carbon skeleton. The suffix gives the principal functional group or groups in the compound. The prefix is optional, and is used to denote extra functional groups of atoms attached to the carbon skeleton where these exist.

The saturated hydrocarbons contain in effect no functional group at all, and the suffix used to denote this situation is **-ane**. Up to *four* carbon atoms the stems are rather arbitrary, hence **meth-** denotes only *one* carbon atom in methane. Similarly, **but-** denotes *four* carbon atoms in butane. For *five* or more carbon atoms, prefixes are derived from the Greek numbers. The stem **alk-** is used as a general stem to denote *n* carbon atoms. Table 25.2 illustrates this. The name alkane refers to any saturated hydrocarbon of general formula C_nH_{2n+2}.

25.5 Nomenclature of branched-chain compounds

Branched-chain alkanes are readily named by regarding them as consisting of a straight chain with one or more side chains attached. A hydrocarbon side chain must have the general formula C_nH_{2n+1}, since it will have one spare bond for attachment to the main chain. The suffix to denote this is **-yl**, and hence the simplest hydrocarbon group is methyl, $CH_3—$. Similarly we have propyl, $C_3H_7—$, hexyl, $C_6H_{13}—$, and so on. The general name for a saturated hydrocarbon group with one spare bond for attachment to the main skeleton is **alkyl**. The names of side chains are then used as prefixes. Table 25.1 illustrated the simplest example, methylpropane, $CH_3.CH(CH_3).CH_3$.

A complication arises with longer chains, in that it is necessary to identify in some way the carbon atoms to which any side chains may be attached. This is done by numbering the carbon atoms in the main chain of the carbon skeleton, starting at one end, and then incorporating these numbers in the name of the compound. For instance:

$$^{1}CH_3.^{2}CH_2.^{3}CH_2.^{4}CH_2.^{5}CH_2.^{6}CH_3$$

hexane

$$CH_3.\underset{\underset{CH_3}{|}}{CH}.CH_2.CH_2.CH_2.CH_3$$

2-methylhexane

$$CH_3.CH_2.\underset{\underset{CH_3}{|}}{CH}.CH_2.CH_2.CH_3$$

3-methylhexane

The main carbon chain is numbered in such a way as to keep the numbers as low as possible, hence the names 4- and 5-methylhexane would be incorrect as they are the same as 3- and 2-methylhexane respectively.

If there is more than one side chain, then separate numbers must appear in the name for each of them, for instance:

$$\begin{array}{l} CH_3.CH.CH.CH_2.CH_2.CH_3 \\ \quad\;\;\; | \quad\;\; | \\ \quad H_3C \;\; CH_3 \end{array}$$

2,3-dimethylhexane

$$\begin{array}{l} \quad\;\;\; CH_3 \\ \quad\;\;\; | \\ CH_3.C.CH_2.CH_2.CH_2.CH_3 \\ \quad\;\;\; | \\ \quad\;\;\; CH_3 \end{array}$$

2,2-dimethylhexane

$$\begin{array}{l} \quad\;\;\; CH_3 \\ \quad\;\;\; | \\ CH_3.C.CH_2.CH.CH_2.CH_3 \\ \quad\;\;\; | \qquad\;\; | \\ \quad\;\;\; CH_3 \;\;\; CH_3 \end{array}$$

2,2,4-trimethylhexane

Remember to find the *longest possible* continuous chain of carbon atoms when naming a compound: this must always determine the stem of the name. Hence the name 2-butylhexane is wrong: its correct name turns out to be 5-methylnonane:

$$\begin{array}{l} \qquad\qquad\quad CH_3 \\ \qquad\qquad\quad | \\ CH_3.(CH_2)_3.CH.(CH_2)_3.CH_3 \end{array}$$

5-methylnonane

Study questions 1 and 2 give practice in the use of nomenclature; further rules will be introduced in later chapters of the book.

25.6 Cycloalkanes

Compounds considered so far have all consisted of saturated chains of carbon atoms, either "straight" (zig-zag because of electron-pair repulsion) or "branched". It is also possible for carbon atoms to form rings, and such compounds are named by using the prefix **cyclo-**. Cycloalkanes are therefore saturated ring hydrocarbons. Simple cycloalkanes, where only one ring exists, have the general formula C_nH_{2n}. A common example is cyclohexane, C_6H_{12}.

It is instructive to make models of different alkanes, both chain and cyclic, using a skeletal-model kit. The angle between adjacent carbon bonds in a stable, unstrained saturated molecule is tetrahedral, i.e. about 109°. Cyclopropane, C_3H_6, has as its carbon skeleton an equilateral triangle with bond angles of 60°. The degree of strain is therefore $(109-60) = 49°$ and it is not surprising therefore that cyclopropane is unstable and reactive compared to alkanes in general. Cyclobutane, C_4H_8, has a square skeleton, with $(109-90) = 19°$ of strain, and is also reactive. Cyclopentane, C_5H_{10}, is practically strain-free since the internal angle of the regular pentagon is 108°. With six or more carbon atoms in the ring, the molecule can relieve strain by buckling up, and so all cycloalkanes above C_6 are stable and unreactive (Fig. 25.3).

If a molecular model is made of cyclohexane, it will be noted that it can be twisted into different shapes, and that a *slight* strain has to be applied in order to convert from one shape to another. The two most stable shapes, or **conformations** as they are called, are the "chair" and "boat" forms illustrated in Fig. 25.3. It is not possible to isolate either conformation in the case of cyclohexane—they are in dynamic equilibrium with an activation energy of conversion of rather less than the mean thermal energy of the molecules at normal temperatures. The study of the conformations of molecules is, however, of importance to chemists in elucidating reaction-mechanisms.

25.7 Physical properties of alkanes

The physical properties of a given alkane are determined by two factors:

(i) the size and shape of the molecule;
(ii) the very low overall polarity of all alkane molecules.

Factor (ii) is common to all alkanes, so they are

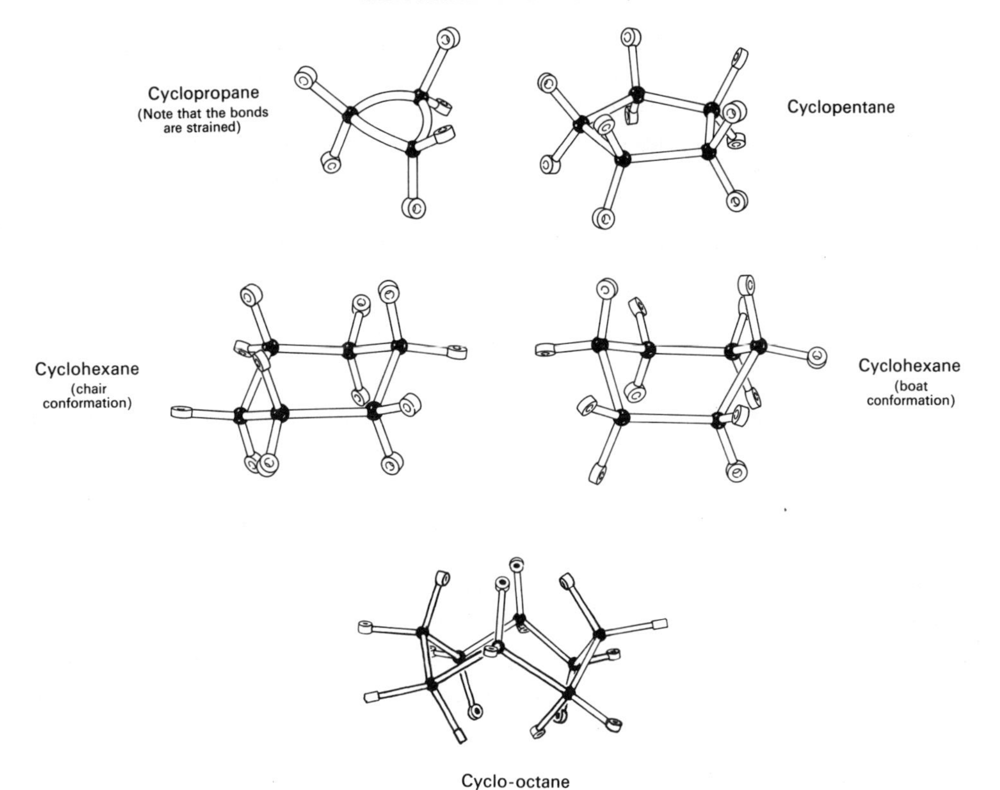

FIG. 25.3. Simple cycloalkanes shown by means of skeletal models, illustrating strain in C_3H_6, and distortion of the ring in C_6H_{12} and C_8H_{16}.

all insoluble in water and have low melting and boiling points when compared to other compounds of similar molecular weight. All have densities less than 1 g cm^{-3} in the solid or liquid state, and hence float on water. They dissolve readily in other non-polar liquids, and in each other.

The way in which physical properties of compounds varies along a homologous series was illustrated in Chapter 6.

TABLE 25.3

THE PHYSICAL PROPERTIES OF COMPOUNDS OF MOLECULAR FORMULA C_5H_{12}

Name and abbreviated structural formula	Skeletal formula	Density (in g cm^{-3})	Boiling point (in K)	Melting point (in K)	Viscosity (in centipoise)
Pentane $CH_3.CH_2.CH_2.CH_2.CH_3$		0·626	309	143	0·240
Methylbutane $CH_3.CH(CH_3).CH_2.CH_3$		0·620	301	113	0·223
Dimethylpropane $CH_3.C(CH_3)_2.CH_3$		0·591	283	257	0·206

Some interesting properties are observed when comparing isomers. In general, molecules with many side chains tend to give rise to low melting points: this is because a regular crystalline state is difficult to achieve with an irregularly shaped molecule. However, a few molecules with side chains are very regularly shaped and compact; these may have higher melting points than their "straight-chain" isomers. Straight-chain isomers can pack more closely together and tend to have higher densities (Table 25.3). Alkanes are colourless, and viscosity of the liquid increases with molecular weight.

25.8 The reactions of alkanes

The very existence of organic compounds on this earth is a remarkable fact. In the presence of atmospheric oxygen, all organic compounds are thermodynamically unstable. Fortunately for our survival on this planet, the saturated carbon atom is kinetically stable. Nevertheless, the most striking chemical reaction of alkanes in everyday life is their ability to burn in oxygen or air to give, ultimately, carbon dioxide and water. In limited oxygen, carbon monoxide may be formed or soot may be deposited, but it is a characteristic of alkanes not to deposit much soot (in contrast to unsaturated and aromatic hydrocarbons, see section 26.5). The use of petroleum-derived alkanes as fuels is well known.

In the organic chemistry laboratory, the reactions of alkanes are not of great importance, and only two will be mentioned.

1. In the presence of ultraviolet light, alkanes are attacked by halogens, especially chlorine and bromine, forming substituted alkanes. Hence methane forms initially chloromethane:

$$Cl_2(g) + CH_4(g) \xrightarrow{h\nu} CH_3Cl(g) + HCl(g); \quad \Delta H = -98\ \text{kJ mol}^{-1}$$

The reaction is a **substitution reaction** (section 17.11). It is a photochemical reaction: although exothermic, ultraviolet light energy is required to trigger it off. Chlorine absorbs this energy and dissociates into chlorine atoms:

$$Cl_2(g) \rightarrow {}^{\cdot}Cl(g) + .Cl(g); \quad \Delta H = +242\ \text{kJ mol}^{-1} \qquad (1)$$

An atom or group of atoms which has an unpaired electron is called a free radical. The covalent bond is broken in such a way that each atom retains one electron from the electron pair —this is called **homolytic fission***.

The reaction step (1) above is called the **initiating step**, and it is followed by collision processes which enable further reaction to proceed; these stages are called **propagation steps**:

$$\left.\begin{array}{l} CH_4 + {}^{\cdot}Cl \rightarrow CH_3{}^{\cdot} + HCl \quad \Delta H = +4\ \text{kJ mol}^{-1} \\ Cl_2 + {}^{\cdot}CH_3 \rightarrow CH_3Cl + {}^{\cdot}Cl; \quad \Delta H = -97\ \text{kJ mol}^{-1} \end{array}\right\} \qquad (2)$$

The ˙Cl atom formed continues the reaction. Stage (3) of the reaction mechanism is the **termination** of the process by the removal of free radicals from the system:

$$\left.\begin{array}{l} Cl^{\cdot} + .Cl \rightarrow Cl_2; \quad \Delta H = -242\ \text{kJ mol}^{-1} \\ Cl^{\cdot} + .CH_3 \rightarrow CH_3Cl; \quad \Delta H = -339\ \text{kJ mol}^{-1} \\ CH_3. + .CH_3 \rightarrow C_2H_6; \quad \Delta H = -346\ \text{kJ mol}^{-1} \end{array}\right\} \qquad (3)$$

* If a bond is broken so that one atom retains both electrons from the electron pair, the process is called **heterolytic fission**, and the products are ions, e.g.

$$Cl_2(g) \rightarrow :Cl^-(g) + Cl^+(g); \quad \Delta H = +1140\ \text{kJ mol}^{-1}$$

Free radical reactions that proceed in this manner are known as **chain reactions**. Note that three stages are always necessary: (1) initiation; (2) propagation; (3) termination. In the above case the overall effect is the formation of $CH_3Cl(g)$ and $HCl(g)$ from chlorine and methane, with the possibility of side reactions such as that giving $C_2H_6(g)$. Further substitution can also proceed giving $CH_2Cl_2(g)$, $CHCl_3(g)$ and finally $CCl_4(g)$.

Since bond formation in the termination steps is very exothermic, the bonds will only form if collision takes place on the walls of the reaction vessel, which can then absorb the energy that is released.

2. When subjected to a high temperature, an alkane molecule vibrates strongly enough to break up into smaller fragments, some at least of which must be unsaturated. This process is termed **thermal cracking** and may be illustrated by an equation such as the following:

$$C_{18}H_{38} \rightarrow \underbrace{C_8H_{18}}_{\text{saturated}} + \underbrace{C_8H_{16} + C_2H_4}_{\text{unsaturated}}$$

In industry, *the importance of alkane chemistry cannot be overemphasized*: crude petroleum is one of the major raw materials of the world's economy, and it is increasingly being applied to uses which traditionally have been provided by coal, e.g. the manufacture of aromatic compounds.

25.9 Petroleum refining

The term petroleum means "rock-oil". The exact processes by which it was formed are still the subject of much discussion, but deposits always occur in association with sedimentary rock. It is possible that the vegetable matter in the waters from which the sediments were laid

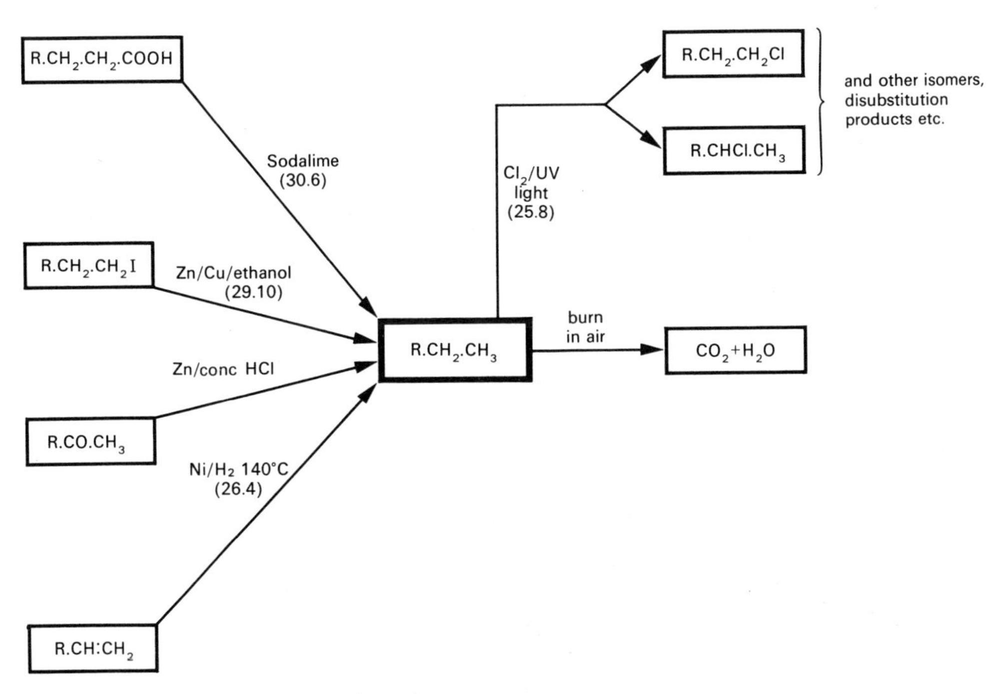

FIG. 25.4. Summary of laboratory reactions involving a typical alkane $R.CH_2.CH_3$ ($R = C_nH_{2n+1}$). Numbers refer to sections in the book where these reactions are described.

down, became entrapped with the mineral deposits, and underwent bacterial decay to form residues that were subsequently transformed to petroleum by some mechanism as yet unknown.

The first stage in obtaining petroleum products is **fractional distillation**, to obtain five or six fractions as shown in Fig. 25.5. Since the boiling point of a substance tends to be in rough proportion to its molecular weight, this initial separation is roughly according to molecular weight. Further distillation, together with some chemical methods of purifying individual products, was originally the only route available for processing petroleum. Nowadays, two further processes are of great importance, namely **cracking** and **reforming**. These enable the molecules present in the original petroleum to be changed into more useful ones. Cracking (section 25.8) is needed in order to supply the demand for sufficient low-molecular-weight hydrocarbon material for gasoline and naphtha; both thermal cracking and catalytic cracking processes are used.

Reforming resembles cracking, in that large molecules are broken into fragments, but a higher pressure is employed together with an appropriate catalyst to enable the fragments to recombine to form different molecules. In this way, branched-chain isomers may be made from the predominantly straight-chain material in crude petroleum, and cyclisation reactions can occur producing aromatic compounds such as benzene, C_6H_6, and its derivatives.

The principal use of alkane products from petroleum is in the manufacture of fuels, naphtha—for processing into other chemicals—

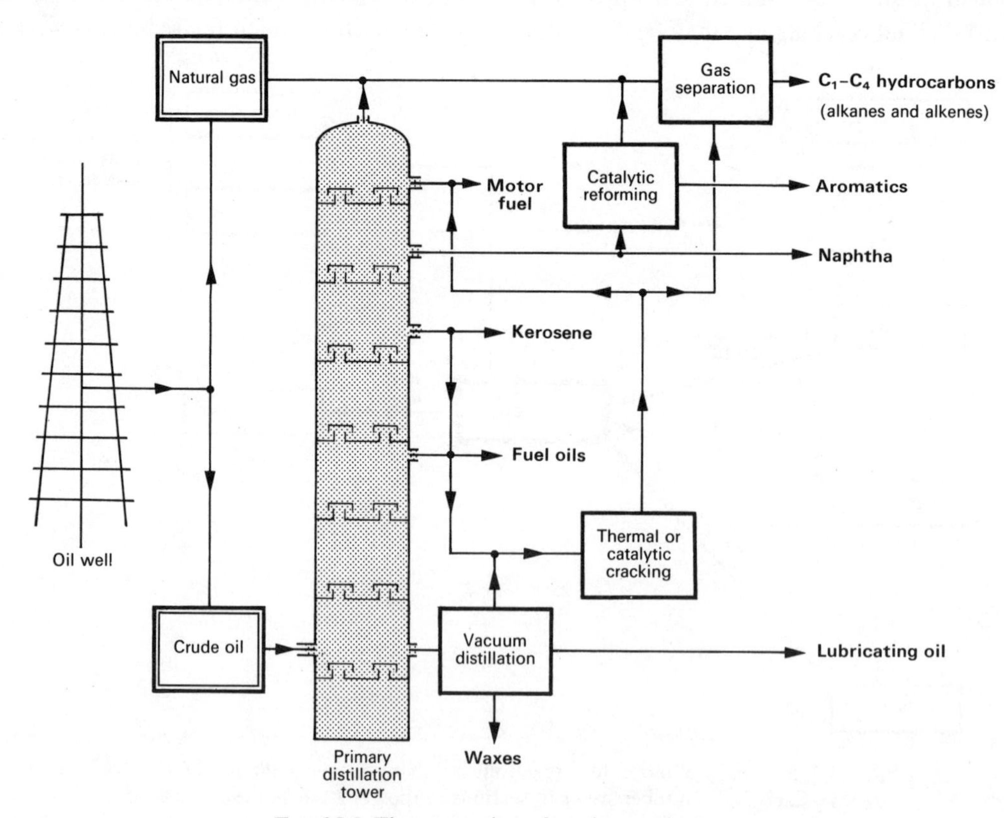

FIG. 25.5. The processing of crude petroleum.

and lubricating oils. A detailed discussion is outside the scope of this book, but a summary of the essential processes is given in Fig. 25.5.*

The most important single product of petroleum refining is ethene (still known industrially as ethylene). This and other products are described in later chapters of this book.

* For a more detailed description see for instance '*Chemicals from Oil*' by R. T. Nye (Pergamon Press Ltd.).

Study Questions

1. Write (a) full structural formulae, (b) formulae using "dot" notation, (c) "skeletal" formulae, for the following alkanes:

(i) 2-methylhexane;
(ii) 2,3-dimethylpentane;
(iii) cyclo-octane;
(iv) 1,3-diethylcyclopentane.

2. Give the names of the following alkanes:

(a) $CH_3.CH_2.C(CH_3)_2.CH_2.CH_3$;
(b) $CH_3.CH_2.CH(CH_2.CH_3).CH_3$;
(c) $C(CH_3)_4$;

(d) 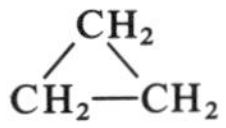 CH_2 / CH_2—CH_2 (three-membered ring)

(e)

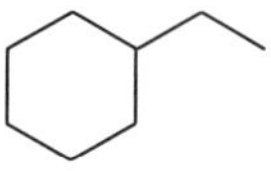

3. Use section 25.7 to place the compounds in each of the following sets in order of increasing boiling point:

(a) 2-methylheptane, 2-methylhexane, 2-methyloctane;
(b) 2,3-dimethylbutane, hexane, 2-methylpentane;
(c) cyclohexane, hexane, methylcyclopentane.

4. Dichloromethane can be formed photochemically from methane and chlorine.

(a) Write step-by-step equations to show how this might occur.
(b) How would you attempt to "trigger off" this reaction in the dark?

5. (a) Explain why high molecular-weight fractions from petroleum need to be cracked and reformed.
(b) Explain clearly the difference between cracking and reforming.
(c) Chemists are nowadays able to identify the oil from a tanker by matching it with that from the oilfield from which it originated. Suggest what techniques would be used to do this.

6. Explain clearly the meaning of the following terms: (a) homologous series; (b) general formula; (c) substitution reaction; (d) chain reaction.

CHAPTER 26

Unsaturated hydrocarbons

26.1 Occurrence and uses of alkenes

Alkenes are hydrocarbons which contain a C=C double bond. We can think of an alk*ene* as an alk*ane* which has lost two atoms of hydrogen from adjacent carbon atoms. Such a compound can readily undergo addition-reactions, and is said to be **unsaturated**.

The C=C bond is too reactive for more than traces of it to be present in crude petroleum. Nevertheless, enormous quantities of alkenes are manufactured from crude oil by cracking. The most important of these are ethene (ethylene) $CH_2{:}CH_2$, propene (propylene) $CH_2{:}CH.CH_3$, and buta-1,3-diene (butadiene) $CH_2{:}CH.CH{:}CH_2$. These compounds are used mainly for conversion into polymers such as polythene, polypropylene, polystyrene and synthetic rubbers (section 26.8).

Many natural products contain the C=C bond, for instance the so-called "unsaturated" vegetable oils. These oils are converted into saturated compounds in the manufacture of margarine.

Further treatment of the alkenes from petroleum refining leads to a variety of materials important to modern life—solvents, detergents and insecticides are a few examples.

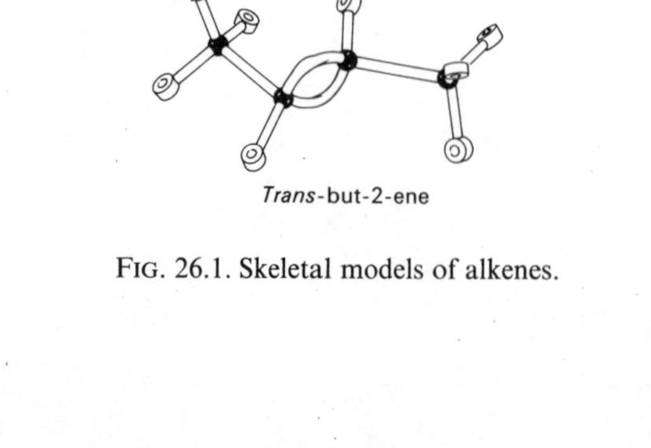

Ethene, $CH_2{:}CH_2$, showing the double bond

Propene $CH_2{:}CH.CH_3$

Trans-but-2-ene

FIG. 26.1. Skeletal models of alkenes.

26.2 Nomenclature of alkenes

In IUPAC nomenclature the suffix **-ene** denotes the functional group C=C, i.e. the presence of a double bond somewhere in the chain. It may be necessary to insert a number in the name to denote the position of the double bond; this is done in a manner similar to the numbering of side chains. The longest chain of carbon atoms in the molecule is sought, and the bonds in it are numbered in such a way that the numbers are lowest (section 25.4). Table 26.1 illustrates the principles with respect to some simple alkenes.

TABLE 26.1

NOMENCLATURE OF ALKENES

Molecular formula and name	Abbreviated structural formula*	Full structural formula	Skeletal notation
C_2H_4 ethene	$CH_2{:}CH_2$	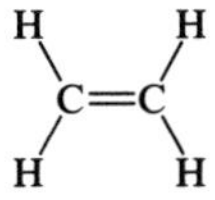	(not used)
C_3H_6 propene	$CH_2{:}CH.CH_3$	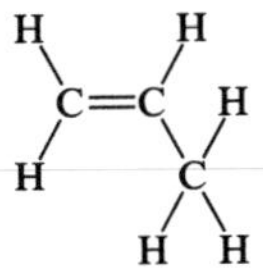	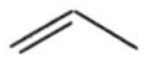
C_4H_8 but-1-ene	$CH_2{:}CH.CH_2.CH_3$	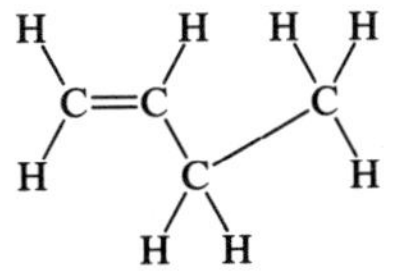	
C_4H_8 but-2-ene	$CH_3.CH{:}CH.CH_3$	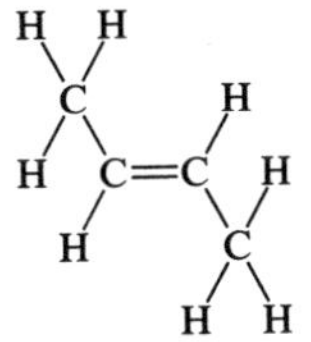	 (trans)
C_4H_6 buta-1,3-diene	$CH_2{:}CH.CH{:}CH_2$	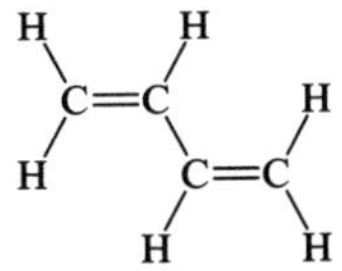	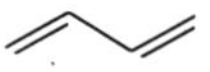
C_6H_{10} cyclohexene	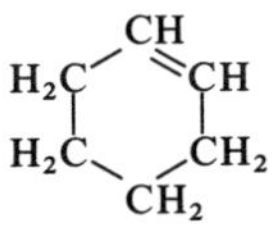	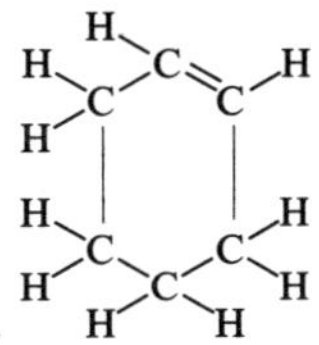	

*A single dot . is here used conventionally to denote an electron-pair bond between the two *carbon* atoms; a double dot : shows a double bond between two carbon atoms.

26.3 The shapes of alkene molecules

A carbon atom forming an alkene bond has three co-planar bonds, with bond angles of approximately 120°. This is in accordance with the principles of molecular shape (section 6.6); free rotation cannot occur about a double bond. In the simplest alkene, ethene C_2H_4, all six atoms therefore lie in one plane and cannot move out of this plane, except to a slight degree by vibration of the molecule (Fig. 26.2). The configuration of the bonds in ethene also exists

FIG. 26.2. The ethene molecule represented by means of a space-filling model. Note that all the atom centres are co-planar.

TABLE 26.2

COMPARISON OF PHYSICAL DATA FOR C—C AND C=C

Property	C—C	C=C
Length (nm)	0·154	0·135
Wavelength of stretching mode in I–R spectrum (cm^{-1})	Approx. 750	1650
Bond energy term (in kJ) (see Chapter 9)	346	610

in more complex molecules containing C=C bonds, giving rise to the possibility of **geometric isomerism**. The concept was introduced in section 6.10. But-2-ene is the simplest example: it can exist as *cis* and *trans* isomers. Table 26.1 illustrates the *trans* isomer: the *cis* isomer is

H H
C=C or
CH_3 CH_3

Not all alkenes can exhibit *cis-trans* isomerism, for instance, but-1-ene cannot because one of its carbon atoms has two identical groups (in this case hydrogen atoms) attached to it.

Cis and trans isomers often have markedly different physical properties, but differences in their chemical behaviour are restricted to those cases where molecular shape is important. Note that it is the lack of free rotation about the double bond axis that makes it possible for isomers to exist. On the other hand there is only one straight-chain form of but*ane*: many shapes (conformations) exist but all are readily interconverted, so all represent forms of the same compound.

26.4 The nature of the alkene double bond

The C=C double bond, like the C—C single bond, is non-polar because it links identical atoms. It is stronger than the single bond yet paradoxically at first sight, it is more reactive chemically. X-ray analysis shows it to be shorter, and the vibrational frequency of its stretching mode is higher than that of the single bond (section 4.3).

The apparent paradox of the greater reactivity of C=C and its greater strength can be resolved by considering how the difference between its bond energy and that of the C—C bond arises. The predominant type of reaction of the C=C bond is the **addition reaction**. The overall result of addition to the double bond is generally energy release, or exothermic reaction. Hydrogen for instance adds on readily in the presence of a catalyst of finely divided nickel at about 140°C:

$$C_2H_4(g) + H_2(g) \rightarrow C_2H_6(g); \quad \Delta H = -126 \text{ kJ mol}^{-1}$$

The hydrogenation reaction can be examined in terms of an energy diagram, using known values for the bond energies of H—H (436 kJ mol^{-1}) and C—H (413 kJ mol^{-1}) as well as the values for C—C and C=C quoted above (Fig. 26.3).

The alkene double bond is thought to consist essentially of two parts:

(a) an axially symmetrical charge cloud containing an electron pair, resulting from the overlap of two 2*s* orbitals—this is known as a **sigma-bond** (σ-bond);

H H
C C
H H

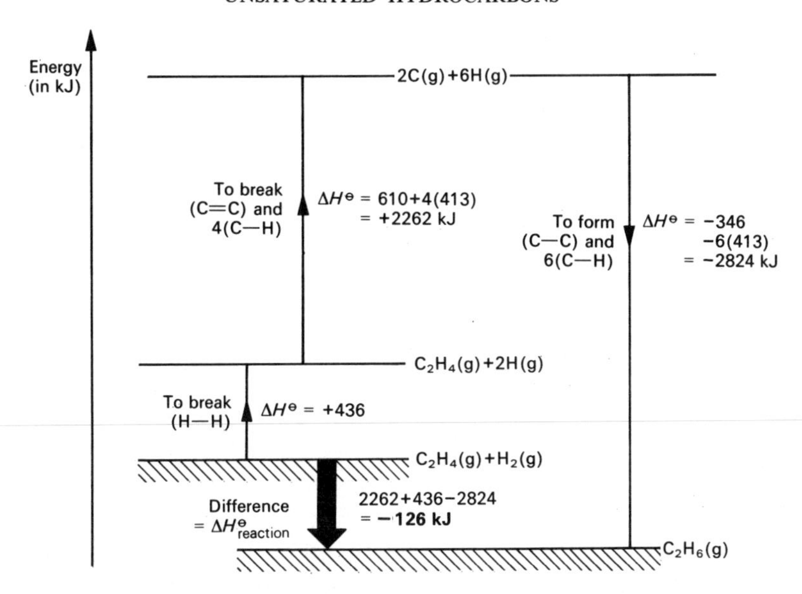

FIG. 26.3. Factors determining the enthalpy of hydrogenation of the C=C bond in ethene.

(b) two charge clouds, one above and one below the plane of the bond system, resulting from the overlap of two $2p$ orbitals; these two charge clouds together comprise the other electron pair, and constitute the **pi-bond** (π-bond).

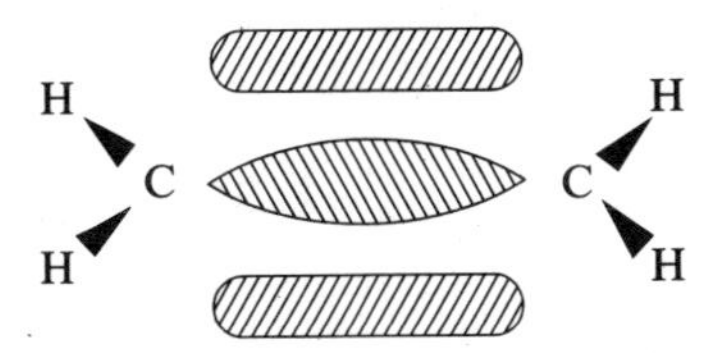

Enthalpy data show us that the π-bond is *thermodynamically* less stable than the σ-bond. The above charge-cloud picture shows us that it is also *kinetically* less stable. The negative-charge clouds of the π-bond are readily attacked by positively-charged species or **electrophiles**. The overall result of such attack is generally an addition reaction of some sort. Before considering the mechanism of such reactions, let us examine them from a practical point of view.

26.5 The chemical reactions of alkenes

Physically, alkenes closely resemble alkanes of corresponding molecular weight and shape, as would be expected from the composition and non-polar nature of their molecules. The volatile alkenes are recognizable by a characteristically unpleasant smell. When burned in air they give a smokier flame than alkanes, on account of their higher carbon content.

Chemically, alkenes are far more reactive than alkanes, and their main type of reaction is **addition**, in contrast to the substitution reactions of alkanes. An addition reaction is one in which one species combines with another; the product is sometimes referred to as an **adduct** (*add*ition prod*uct*). As well as the hydrogenation reaction mentioned in section 26.4, the following addition reactions are readily observable in the laboratory.

(a) Hydrogen halides *add on* when their concentrated aqueous solution is added to the alkene in the cold; the reaction may be generalized thus:

$$\rangle C{=}C\langle + H{-}X \longrightarrow -\underset{H}{\overset{|}{\underset{|}{C}}}-\underset{X}{\overset{|}{\underset{|}{C}}}-$$

the hydrogen atom attaches itself to that carbon atom which already contains the greater number of hydrogen atoms: this rule is often known as **Markownikov's rule**. For instance but-1-ene reacts with hydrogen bromide to give mainly 2-bromobutane:

$$CH_2{:}CH.CH_2.CH_3 + HBr$$
$$\rightarrow CH_3.CHBr.CH_2.CH_3$$

(b) Among the halogens themselves, fluorine is so reactive that an explosion results. Iodine on the other hand is rather too unreactive. Chlorine and bromine react readily at room temperature to form dihalogenoalkanes:

$$\rangle C{=}C\langle + X{-}X \longrightarrow -\underset{X}{\overset{|}{\underset{|}{C}}}-\underset{X}{\overset{|}{\underset{|}{C}}}-$$

hex-1-ene, for example, would react with liquid bromine rapidly to form 1,2-dibromohexane.

(c) Aqueous chlorine and bromine do not give a high yield of dihalogenoadduct. Instead, the solution behaves as HOCl or HOBr, and adds on to form a colourless product called a "chlorhydrin" (or bromhydrin):

$$\rangle C{=}C\langle + HO{-}Br \longrightarrow -\underset{HO}{\overset{|}{\underset{|}{C}}}-\underset{Br}{\overset{|}{\underset{|}{C}}}-$$

The reaction with bromine water forms a very convenient and rapid laboratory test for the C=C bond, resulting in immediate decolorization of the bromine water, since $Br_2(aq) + H_2O \rightleftharpoons HOBr(aq) + H^+(aq) + Br^-(aq)$.

(d) Dilute alkaline potassium manganate(VII) is readily reduced by C=C. In effect, the solution can be regarded as providing a mole of H_2O and a mole of [O] atoms, the overall process being the formation of a **diol** (a compound containing two hydroxy-groups):

$$\rangle C{=}C\langle + H_2O + [O] \longrightarrow -\underset{HO}{\overset{|}{\underset{|}{C}}}-\underset{OH}{\overset{|}{\underset{|}{C}}}-$$

This reaction, like (c), is useful for detecting the C=C bond, though it must be remembered that other compounds are also capable of reducing purple $MnO_4^-(aq)$ to green MnO_4^{2-} in these conditions. In acid conditions the MnO_4^- is reduced to the virtually colourless $Mn^{2+}(aq)$ ion.

(e) Ozone (trioxygen, O_3) reacts rapidly to form an adduct (an ozonide) which can be decomposed by $H^+(aq)$ with complete fission of the C=C bond. The overall process is termed **ozonolysis**:

$$RR'C{=}CR''R''' + O_3 \longrightarrow \text{ozonide (ring: } C\text{–}O\text{–}C \text{ and } C\text{–}O\text{–}O\text{–}C\text{; } R, R' \text{ on one C, } R'', R''' \text{ on the other)}$$
$$\xrightarrow{H^+} RR'C{=}O + O{=}CR''R''' + H_2O_2$$

The importance of this reaction is in locating the exact position of a double bond in a carbon chain: the molecule is subjected to ozonolysis and the fragments are then analysed. Alternative reagents, easier to use but more expensive, are sometimes employed for the same task.

(f) An important addition reaction of alkenes is with concentrated sulphuric acid, which reacts to form an alkyl hydrogensulphate ($RHSO_4$,

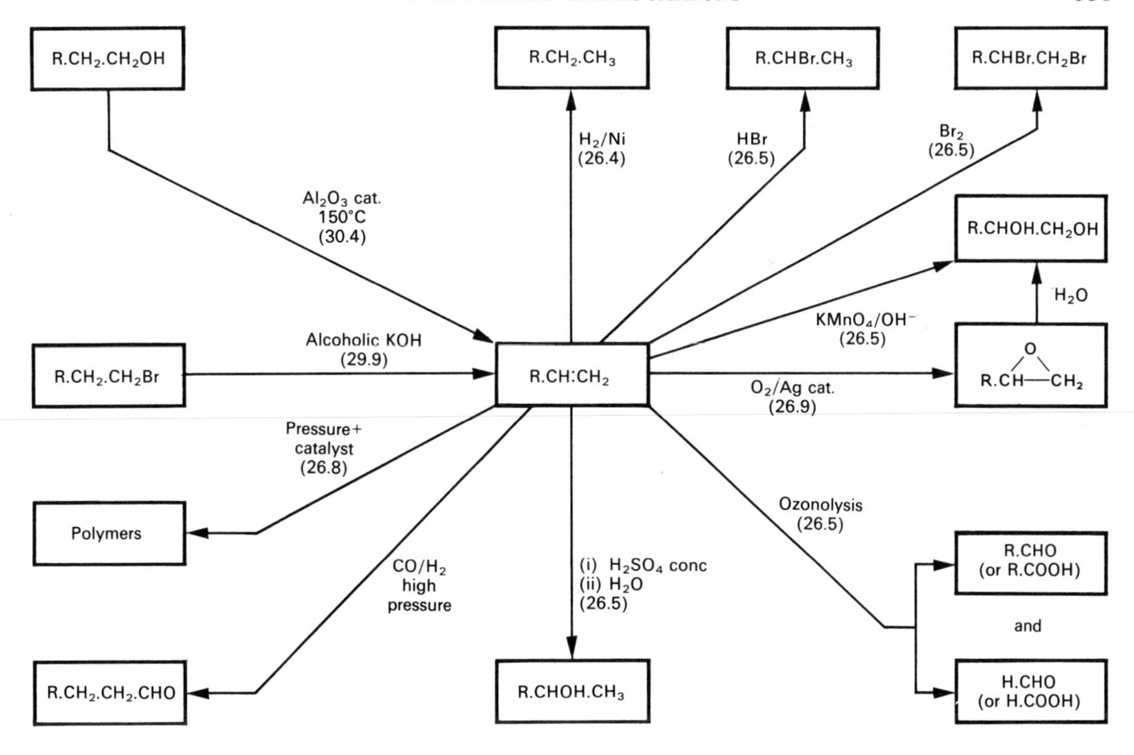

FIG. 26.4. Reactions involving a typical alkene R.CH:CH$_2$.

where R = alkyl). Using ethene itself as an example:

$$\mathrm{H_2C{=}CH_2 + H_2SO_4 \longrightarrow H{-}CH_2{-}CH_2{-}O{-}SO_2OH}$$

Ethyl hydrogensulphate is an example of an **ester**, i.e. a substance derived from an acid by the replacement of an acidic hydrogen atom by a hydrocarbon group such as an alkyl group. Alkyl hydrogensulphates, in common with other esters, can be hydrolysed by dilute $H^+(aq)$ to form the corresponding alcohol. This and other reactions are summarized in Fig. 26.4.

(g) A special case of the addition reaction is when the molecule adds on to itself: this is **polymerization**. Many of the higher alkenes polymerize on storage to form resin-like substances of higher molecular weight. The controlled polymerization of many of the lower alkenes is of immense importance in the plastics industry, and is considered in section 26.8.

26.6 Mechanism of addition reactions to C=C

Essentially we believe that the first step in all the addition reactions considered above is attack by an electrophile on the π-bond. This may be illustrated by considering bromination:

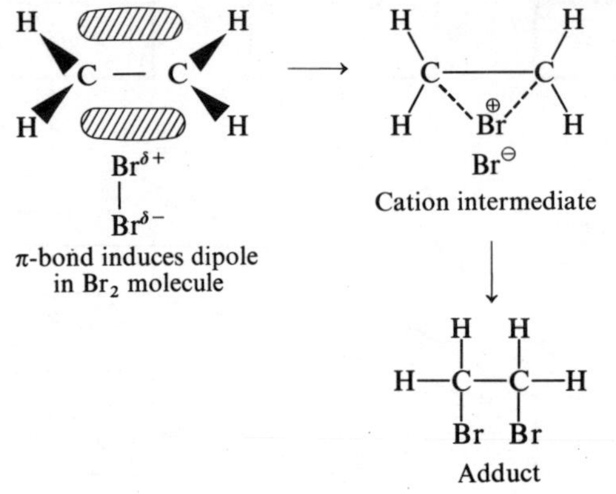

π-bond induces dipole in Br_2 molecule

Cation intermediate

Adduct

the positively charged intermediate species is called a **carbocation** or **carbonium** ion. Its exact structure is still the subject of some discussion among organic chemists. Presumably before the final stage, the Br atom in the carbocation must transfer entirely to one carbon atom, and electrons flow from the other carbon atom to leave that electron-deficient, thus allowing the Br^- to become attached. The fact that other negative ions, such as Cl^- and NO_3^-, can attach themselves in place of one of the bromine atoms (but not both) is supporting evidence for this mechanism.

26.7 Alkenes from petroleum

Table 26.3 summarizes the main alkenes important to the petroleum and petrochemicals industry. Perhaps the most important single compound is ethene, used not only for polymerization to polythene, but for a host of other uses.

26.8 Manufacture of polymers

Polymers made from alkenes are all **addition polymers**, that is, the monomers link by addition reactions without elimination taking place. Whereas some of the higher alkenes polymerize spontaneously, the lower ones require higher temperatures and catalysts to overcome the activation energy barriers. The first polythene was made by a process still in use today. Patented by I.C.I. in 1933, it involves the use of pressures of up to 2500 atmospheres, and under these conditions not surprisingly a somewhat irregularly shaped polymer molecule, with a degree of cross-

TABLE 26.3

ALKENES FROM PETROLEUM

Name and formula	Source	Uses
CH_2:CH_2 ethene (ethylene)	Cracking of liquid hydrocarbons, and refinery gases.	Polymerization to polythene, PVC, polystyrene. Hydration to ethanol. Bromination to $BrCH_2.CH_2Br$ for motor fuel anti-knock. Conversion to "ethylene oxide" and hence glycols (section 30.9).
CH_2:$CH.CH_3$ propene (propylene)	Cracking processes similar to those which produce ethene.	Polymerization to polypropylene and various rubber-like co-polymers. Hydration to propan-2-ol (section 30.9). Conversion to "propylene oxide" and hence glycols and polyurethane polymers. Synthesis of glycerine (section 30.9).
CH_2:$CH.CH$:CH_2 buta-1,3-diene (butadiene)	Cracking followed by extraction from C_4 alkanes in refinery gases.	Making poly-butadiene rubber; styrene-butadiene co-polymers. Manufacture of *nylon*.

linking, is the result. A better product is obtained using catalysts such as those developed by Ziegler, at pressures of only a few atmospheres. The overall process is essentially the formation of a very-long-chain alkane:

$$n\ \mathrm{H_2C{=}CH_2} \longrightarrow \left(\begin{array}{c}\mathrm{H\ \ H}\\ |\ \ \ |\\ -\mathrm{C{-}C}-\\ |\ \ \ |\\ \mathrm{H\ \ H}\end{array}\right)_n$$

The product of the Ziegler process is more crystalline, with better-packed molecules, and is harder with a higher melting point. It is called "high density" polythene.

Propene is nowadays polymerized by catalysts developed by Natta and Ziegler, which gives a hard, semi-transparent rigid product with a high degree of crystallinity called **isotactic** polypropylene. The catalytic control of the polymerization process causes the molecules all to link in the same orientation, with the side-chain methyl groups all pointing the same way (Fig. 26.5). Uncontrolled polymerization on the other hand gives a randomly arranged polymer with a soft rubbery consistency and few useful properties, called **atactic** polypropylene.

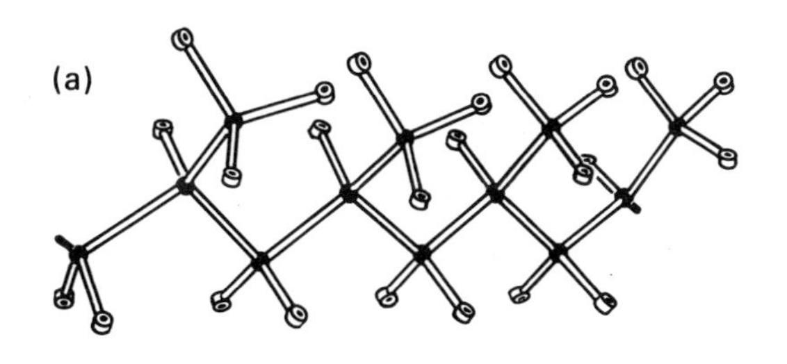

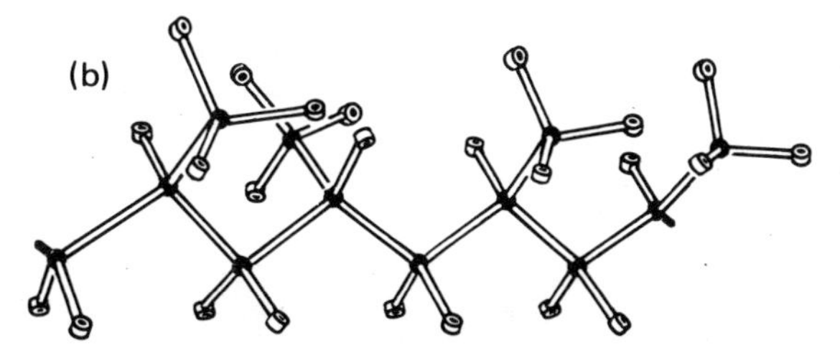

FIG. 26.5. (a) Isotactic and (b) atactic forms of polypropylene.

Where more than one monomer is incorporated in the polymer, the result is called a **co-polymer**. Variation of the ratio of monomers enables polymers to be tailor-made with the desired properties, at economic costs. Ethene-propene co-polymers are assuming economic importance for their rubber-like properties, and may well challenge synthetic rubbers currently in use, which are made from butadiene and similar monomer materials.

26.9 Alkene oxides (epoxyalkanes)

Elimination of HCl from the "chlorhydrin" (section 26.5) produced by reacting propene with aqueous chlorine, results in a cyclic molecule 1,2-epoxypropane, known industrially as propylene oxide:

$$\mathrm{CH_2{:}CH.CH_3} \xrightarrow{\mathrm{HOCl}} \mathrm{CH_2(OH){-}CH(Cl).CH_3} \xrightarrow{\mathrm{Ca(OH)_2}} \mathrm{CH_2{-}CH.CH_3}\ (\text{bridged by O})$$

A similar product, epoxyethane (ethylene oxide), can be made from ethene by direct catalytic oxidation:

$$\mathrm{CH_2{:}CH_2} + \tfrac{1}{2}\mathrm{O_2} \longrightarrow \mathrm{CH_2{-}CH_2}\ (\text{bridged by O})$$

These substances are of great importance as intermediates in the manufacture of solvents, lacquers, anti-freeze, etc.

26.10 Alkynes

The suffix **-yne** in IUPAC nomenclature means a triple bond, containing three electron pairs, between two carbon atoms. The simplest compound is ethyne, C_2H_2, commonly known as acetylene. The link consists of one σ-bond and

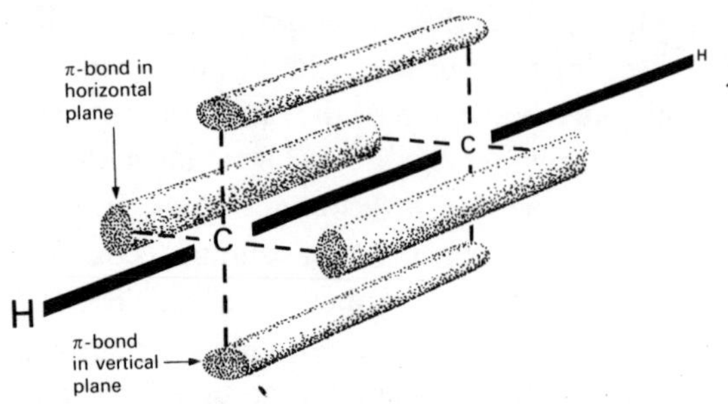

FIG. 26.6. The charge distribution in the alkyne bond.

two π-bonds, the two latter coalescing to form in effect a barrel of negative charge around the σ-bond. Since there are no lone pairs, the result is a linear molecule (see section 6.8).

Alkynes undergo many addition reactions similar to those of alkenes (Fig. 26.7). Although the triple bond is thermodynamically less stable than the double bond, it appears to be kinetically *more* stable, requiring higher activation-energy (section 17.10). Hence, perhaps surprisingly, alkynes are less reactive in some respects than alkenes.

Where there is a hydrogen atom attached directly to a triply-bonded carbon, it is found to be slightly acidic and can be replaced by certain metals to give unstable salts. Hence ethyne itself, and all alk-1-ynes, can give a yellow-white precipitate when passed through ammoniacal silver(I) nitrate:

$$2Ag^+(aq) + H{-}C{\equiv}C{-}H(g) \rightarrow Ag{-}C{\equiv}C{-}Ag(s) + 2H^+(aq)$$

Such salts, systematically named dicarbides, occur widely, though not all can be made directly from the alkyne. Calcium dicarbide, $CaC_2(s)$, is made by reacting calcium oxide with coke at high temperatures. It is ionic, containing the C_2^{2-} ion, and hydrolyses, slowly with water and rapidly with acids, to form ethyne:

$$2H^+(aq) + C_2^{2-} \rightarrow C_2H_2(g).$$

Nowadays an increasingly large amount of ethyne is also made by the pyrolysis of methane from petroleum gases. Ethyne thus made has a

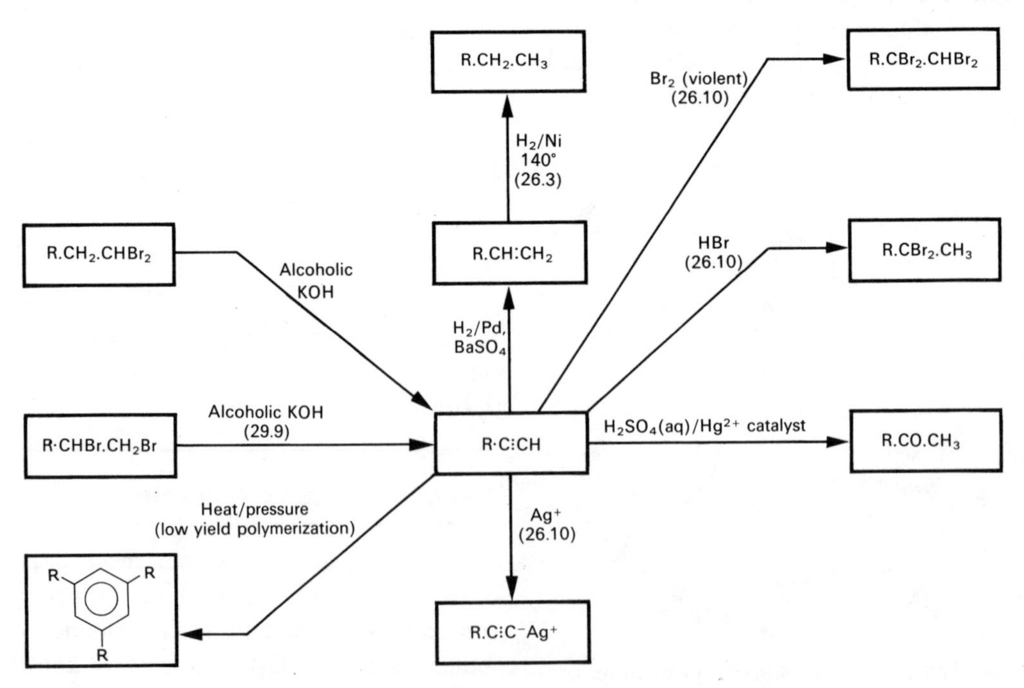

FIG. 26.7. Reactions involving a typical alkyne R.C:CH.

number of important uses. Perhaps the most important is in the manufacture of "vinyl chloride", chloroethene, for polymerization to "polyvinyl chloride", PVC:

$$HC{\equiv}CH + HCl \rightarrow H_2C{=}CHCl.$$

The reaction occurs by the controlled addition of *one* mole of hydrogen chloride. Addition of a second mole of HCl gives $CH_3.CHCl_2$ rather than $CH_2Cl.CH_2Cl$.

One addition reaction used industrially with ethyne does not occur so readily with ethene: this is the addition of hydrogen cyanide to form "acrylonitrile", cyanoethene:

$$CHCH{:}CH + HCN \rightarrow CH_2{:}CH.CN$$

Acrylonitrile is the basis of acrylic fibres such as *Acrilan*. Another use of ethyne is in the manufacture of chlorinated solvents. For instance, direct addition of two moles of chlorine produces 1,1,2,2-tetrachloroethane:

$$HC{\equiv}CH + 2Cl_2 \rightarrow CHCl_2.CHCl_2$$

Figure 26.7 summarizes the important reactions of a typical alkyne.

Study Questions

1. Write (a) full structural formulae, (b) formulae using "dot" notation, (c) "skeletal" formulae, for the following unsaturated hydrocarbons:

(i) propadiene;
(ii) propyne;
(iii) hexa-1,3,5-triene;
(iv) cyclopentene.

2. Give the names of the following alkenes:

(a) $CH_2{:}C(CH_3).CH{:}CH_2$;

(b) CH, H_2C, CH, H_2C, CH_2, CH_2 (ring)

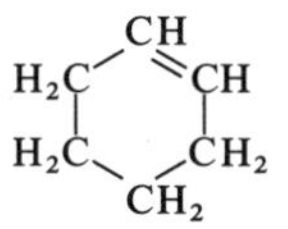

(c)

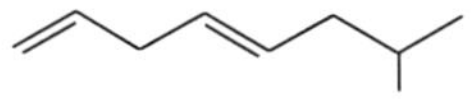

3. (a) State the number of geometric isomers for each of the following substances:

(i) but-1-ene;
(ii) but-2-ene;
(iii) 2-methylpenta-1,3-diene.

(b) What condition must be fulfilled for an alkene to show geometric isomerism?

(c) Are there any circumstances in which a saturated compound could show geometric isomerism?

4. A compound **A**, of empirical formula CH_2, decolorized acidified $KMnO_4$(aq). **A** reacted with bromine to give **B**, $C_3H_6Br_2$, which reacted with an ethanolic solution of potassium hydroxide to give **C**, C_3H_4. **C** was a gas that gave a yellow precipitate with ammoniacal silver nitrate. **C** gave **D**, C_3H_8, when reacted with excess hydrogen in the presence of a nickel catalyst at 140°C.

(a) Identify the compounds **A** to **D**.
(b) Write balanced equations for the reactions.

5. A compound **E** contained 87·8% carbon and 12·2% hydrogen by mass. **E** decolorized bromine water, and with excess hydrogen in the presence of finely divided nickel at 140° gave **F** (C 85·7%; H 14·3%).

(a) Write structural formulae for at least three possible isomers of **E**.

(b) How could the actual structure of **E** be established?

6. Upon ozonolysis, a hydrocarbon **G** gave propanone, $CH_3.CO.CH_3$, as the only organic product. What is the structural formula of **G**?

7. A sample from an unlabelled cylinder of gas burns to give carbon dioxide and water. 100 cm^3 of the gas weighs $0{\cdot}12 \pm 0{\cdot}02$ g at 20°C and 760 mmHg.

(a) Calculate the molecular weight of the gas.

(b) Assuming the gas to be a hydrocarbon, write down as many structural formulae as you can which fit this value for the molecular weight.

(c) Describe simple chemical tests you would apply to establish the identity of the gas.

8. "Perspex" is a transparent colourless solid with empirical formula $C_5H_8O_2$. It does not react with bromine water or potassium permanganate. When heated it forms a liquid of molecular formula $C_5H_8O_2$. This liquid decolorizes both bromine water and acidified potassium permanganate.

(a) Explain how the structure of the liquid differs from that of solid "perspex".

(b) When the liquid is gently heated in the presence of a catalyst it solidifies. What is the name given to this type of reaction?

(c) What is the solid product obtained in (b)?

9. (a) Why would you expect cyclo-octyne to be an *unstable* liquid?

(b) Write an equation for the reaction of cyclo-octyne with hydrogen bromide.

(c) What would you expect to observe when cyclo-octyne is added to ammoniacal silver nitrate?

CHAPTER 27

Aromatic hydrocarbons

27.1 Aromatic hydrocarbons

The compounds considered in Chapters 25 and 26 are termed **aliphatic** hydrocarbons: another important class of compounds are the so-called **aromatics**, of which the simplest is benzene, C_6H_6. Benzene and its derivatives have been important industrially for many years, and occur naturally as constituents of coal tar. The gradual decline of the coal industry, and corresponding growth of the oil industry, have meant that an increasing number of aromatic compounds are nowadays being manufactured from petroleum by catalytic reforming (section 25.9).

The general terms **arene** and **aryl** are used to denote aromatic derivatives of benzene. It will be seen later in the chapter that, although the suffix -ene occurs in the name, true C=C double bonds are not present in the ring structure, and the chemical properties are quite different. Table 27.1 summarizes the properties of some simple arenes.

The physical properties of the arenes are very similar to those of alkanes or alkenes of corresponding molecular weight. Hence they are colourless, less dense than water, insoluble in

TABLE 27.1
SOME SIMPLE ARENES

Molecular formula and systematic name	Trivial name	Abbreviated structural formula	Skeletal formula (see section 27.3)
C_6H_6 benzene	benzene	C_6H_6	
C_7H_8 methylbenzene or phenylmethane	toluene	$C_6H_5.CH_3$	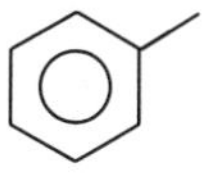
C_8H_{10} 1,2-dimethylbenzene	orthoxylene	$CH_3.C_6H_4.CH_3$	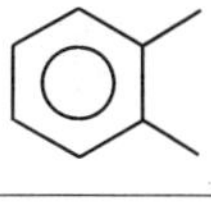

polar liquids such as water, and soluble in non-polar solvents. Benzene is worthy of note on account of its exceptionally high melting point, 5·5°C, in contrast to its simple derivatives, e.g. methyl benzene (−95·0°C). The extremely regular shape and compact nature of the benzene molecule enables a stable crystal lattice to form.

Aromatic hydrocarbons burn with a very sooty flame, which enables them to be distinguished from saturated aliphatic compounds. The uses of aromatic compounds are described in section 27.11.

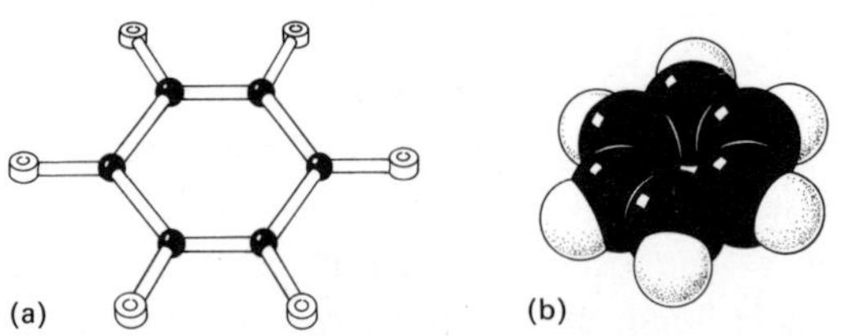

FIG. 27.1. Benzene shown as (a) a skeletal model, (b) a space-filling model.

27.2 Aromaticity

One of the commonest hydrocarbons is benzene, C_6H_6. It has a far lower hydrogen content than a corresponding alkene (e.g. hexene, C_6H_{12}) or alkane (e.g. hexane, C_6H_{14}), which implies the presence of a large number of C=C double bonds. Despite this, it does not show the property of unsaturation to any marked degree. Its high carbon-hydrogen ratio causes it to burn with a very smoky flame, in contrast to saturated hydrocarbons, yet it does not decolorize bromine water, nor does it turn dilute alkaline potassium manganate(VII) solution green. In other words, chemical evidence suggests that ordinary C=C double bonds are absent.

For many years the structure of benzene presented a problem, and the question was largely resolved by Kekulé in 1865, who took the significant step of proposing a hexagonal formula thus:

```
          H
          |
   H    //C\     H
     \ C     C /
       |     ||
       C     C
     /  \\  /  \
   H      C      H
          |
          H
```

We now know that Kekulé's version cannot be entirely correct, though he was right in proposing a planar hexagonal shape, because:

(i) X-ray analysis of crystalline benzene shows all C—C bond lengths to be identical (0·139 nm), in contrast to the C—C bond length found in ethane (0·154 nm) and the C=C bond length in ethene (0·134 nm);

(ii) the enthalpy of combustion of benzene is about 170 kJ mol^{-1} less than the value calculated from the bond energy terms in the Kekulé formula (see Study Question No 1).

These facts may be explained by assuming that six of the electrons in benzene, one from each carbon atom, are able to move freely around the ring. One electron from each carbon atom occupies a spare *p* orbital. The orbitals of the six *p* electrons can overlap to form ring-shaped charge clouds above and below the ring of carbon atoms. These new orbitals are called π-orbitals, just as in alkenes and alkynes. Within these charge clouds the six electrons can circulate freely, and no one electron belongs to any particular carbon atom. We say that the electrons are **delocalized**.

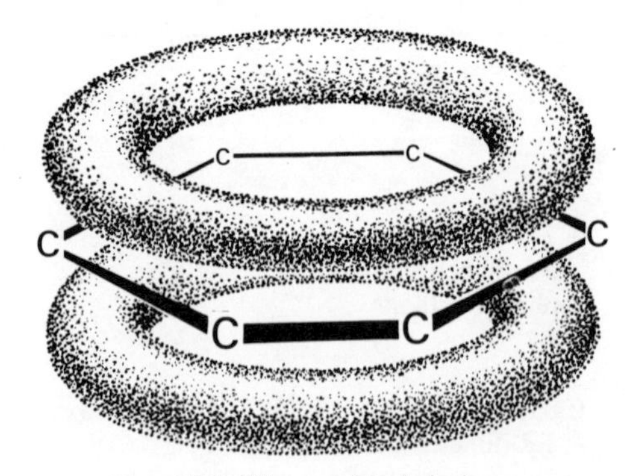

FIG. 27.2. The π-orbitals in benzene.

Delocalization is a common phenomenon in chemistry, and whenever it occurs it will lead to extra stabilization of the molecular structure. This is not surprising—indeed it follows from the fact that electrons repel one another and hence get as far from one another as the energy levels in the system will allow. Where delocalization is possible, the electrons in the system can "spread themselves out" and achieve a more stable configuration.

27.3 Resonance

It is sometimes convenient to describe the delocalized structure of a molecule such as benzene in terms of conventional valence-bond structures. Benzene itself is described as a blend or hybrid of two structures thus:

The two structures illustrated have no physical reality—they are simply imaginary formulae used to aid the description of benzene in valence-bond terms. The imaginary structures are referred to as **canonical forms**, and the actual structure which results from blending them is called a **resonance hybrid.*** Alternatively, the delocalized form is sometimes called mesomeric (*meso* meaning "intermediate") and the phenomenon termed **mesomerism**.

A double-ended arrow ↔ is used to indicate that the concept of resonance is being used to describe a structure. It must not be confused with the equilibrium sign ⇌. The latter denotes the actual existence of two species changing reversibly from one to another, that is, it depicts an actual chemical change. The resonance sign, ↔, simply indicates that two or more canonical forms have been written down in an attempt to describe a structure which cannot be represented in terms of simple electron-pair bonds.

Another aid to understanding delocalized structures is the concept of **resonance energy**, which is the difference between the actual energy content of the resonance hybrid and the energy of the canonical forms. Resonance energy is illustrated in the energy diagram, Fig. 27.3.

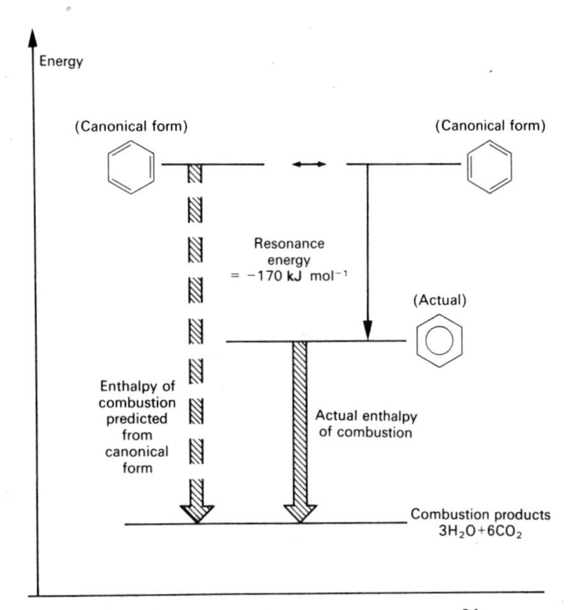

FIG. 27.3. Calculation of resonance energy of benzene (energy axis not to scale).

A curved-arrow notation is sometimes used to denote the movement of electron pairs from one position to another. Using curved arrows‡

* The concept of resonance was first invoked by Kekulé, but he regarded the canonical forms of benzene as oscillating from one to another. Modern theory would regard this as incorrect, though Kekulé is without doubt the founder of present day ideas about the structure of benzene and similar molecules.

‡ The curved arrows are only a convention, to aid the imagination. It must not be thought that the electron pairs actually follow paths around the curves of the arrows!

to denote electron movement, the structure of benzene can therefore be depicted thus:

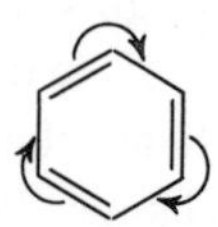

Compounds which have ring systems containing delocalized electrons are said to be **aromatic**, in contrast to **aliphatic** compounds, where a delocalized ring is absent. Until recently, textbooks have tended to represent the benzene ring in skeletal form as if it were a canonical form, but a better convention is to draw a circle to denote the **aromatic sextet** of delocalized electrons thus:

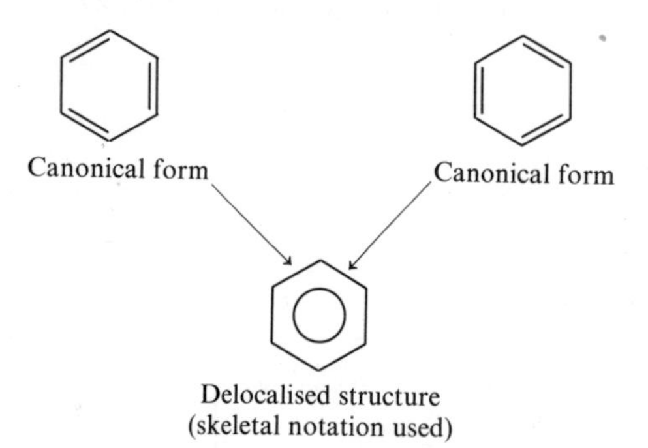

Delocalised structure (skeletal notation used)

Sometimes two or more rings are fused together, for instance in naphthalene, $C_{10}H_8$, and anthracene, $C_{14}H_{10}$:

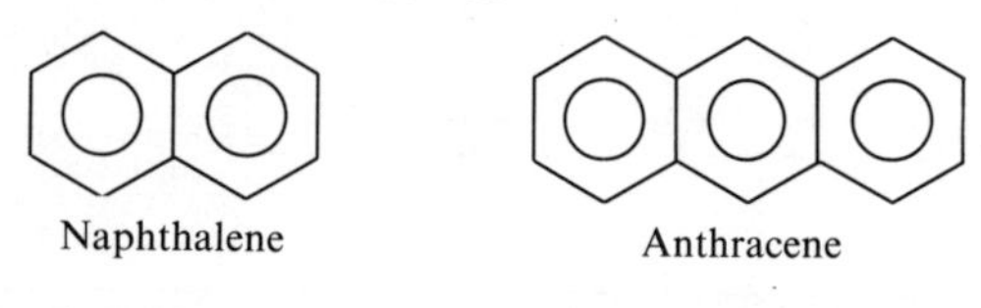
Naphthalene Anthracene

Such hydrocarbons are termed **polycyclic** aromatic hydrocarbons. In other compounds, a carbon atom may be replaced by a nitrogen atom, as for instance in pyridine, C_5H_5N, and pyrimidine, $C_4H_4N_2$:

Pyridine Pyrimidine

Compounds such as these are called **heterocyclic**. Derivatives of pyrimidine are in fact important structural units in DNA (Chapter 33).

27.4 Nomenclature of benzene derivatives

Simple derivatives of benzene are easy to name. For instance replacement of one hydrogen atom by a methyl group gives methylbenzene, $C_6H_5.CH_3$, commonly known as toluene. It is not necessary to include a number in the name, because all carbon atoms in the ring are identical. Other examples of mono-substituted benzene derivatives are listed in Table 27.2.

The group C_6H_5— is termed the **phenyl** group. For instance, an alternative name for methylbenzene would be phenylmethane. The general name for substituted derivatives of the phenyl group is **aryl**.

Disubstituted benzene derivatives require numbers to denote the relative positions of the two attached groups. A given pair of groups can give rise to three isomers, 1,2 (formerly called *ortho*), 1,3 (*meta*) and 1,4 (*para*). For example:

CH_3, CH_3 — 1,2-dimethylbenzene

OH, NO_2 — 3-nitrophenol

CH_3, NO_2 — 1-methyl-4-nitrobenzene

The labelling is done in such a way as to keep the numbers as small as possible, as with aliphatic names; it would, for instance, be incorrect to write 3,4-dimethylbenzene.

TABLE 27.2

SIMPLE DERIVATIVES OF BENZENE

Name of substituent group as IUPAC prefix. (Suffix in brackets)	Structure of substituent group	Systematic name benzene derivative	Trivial name of benzene derivative	See section
		(Recommended name printed **bold**)		
Methyl-	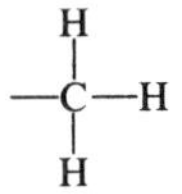 H —C—H H	**Methylbenzene**	(Toluene)	27·9
Nitro-	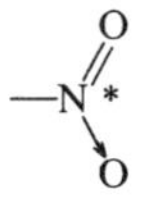 O —N* O	**Nitrobenzene**	—	31·11
Amino- (-amine)	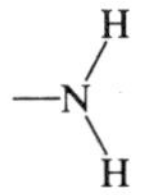 H —N H	**Phenylamine**	(Aniline)	31·3
Hydroxy- (-ol)	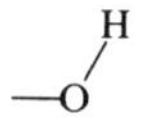 H —O	(Hydroxybenzene)	**Phenol**	30·5
Carboxy- ‡(-carboxylic acid)	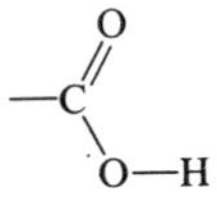 O —C O—H	(Benzenecarboxylic acid)	**Benzoic acid**	30·6
Chloro-	— Cl	**Chlorobenzene**	—	29·6

‡ The termination **-oic acid** cannot be used here as the carbon atom in —COOH does not form part of a carbon chain. * Delocalisation (resonance) can occur.

27.5 The reactivity of benzene

The presence of two π-electron rings would lead us to expect that benzene, like alkenes and alkynes, would be attacked by electrophiles more readily than by nucleophiles. This is indeed the case, but in practice the predominant reactions of benzene turn out to be substitution rather than addition.

The reason that benzene, unlike alkenes and alkynes, does not favour addition reactions is that an addition reaction involves the loss of the aromatic sextet and hence of delocalization energy stabilization. Both benzene and ethene can be hydrogenated by addition of $H_2(g)$ in the presence of finely divided nickel, but benzene requires a temperature about 150 K higher:

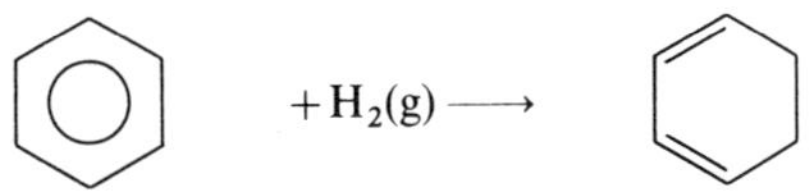

benzene $+ H_2(g) \longrightarrow$ cyclohexa-1,3-diene (first stage)

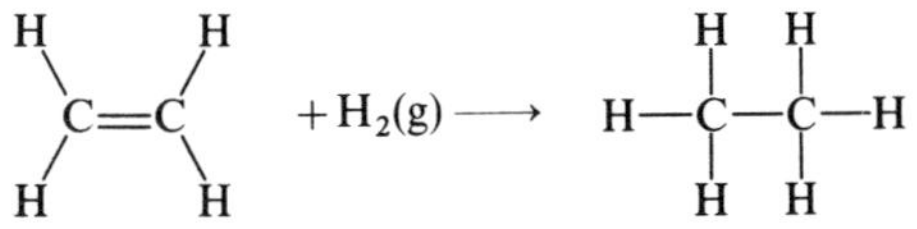

$H_2C{=}CH_2 + H_2(g) \longrightarrow H_3C{-}CH_3$

ethane ethene

Figure 27.4 shows comparative energy diagrams for the two hydrogenations.

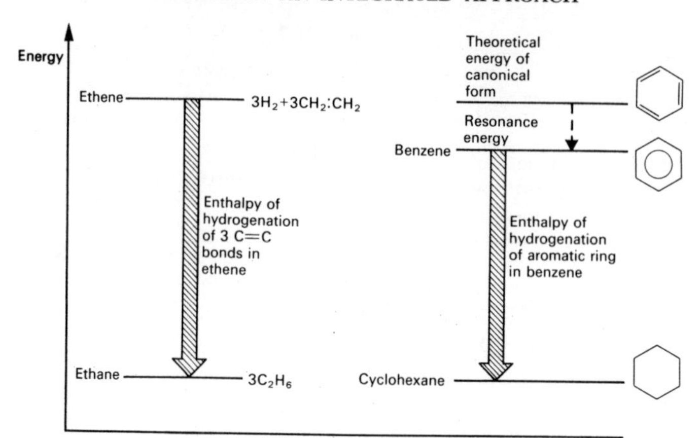

FIG. 27.4. Hydrogenation of benzene compared with that of ethene.

Benzene undergoes substitution reactions with electrophiles, which for the moment we may represent in general as $(X)^+$. Initially the electrophile is attached to the π-electron ring, then it attacks one of the carbon atoms. A transition state (section 17.10) is formed, in which the C—H bond is weakening, and the C—X bond is strengthening. The transition state does not possess aromatic character, and hence sufficient activation energy has to be provided to overcome the delocalization energy of the benzene ring. The transition state is then liable to break down in one of two possible directions:

(a) it might lose a proton, H^+, and form a derivative C_6H_5X, or

(b) it might lose its electrophile $(X)^+$ again, reforming benzene.

The whole process may be summarized thus:

electrophile attacks π-electrons

$+(X)^{\oplus} \longrightarrow$ $(X)^{\oplus}$

$\oplus$ H, X — transition state —(a)→ C_6H_5X; —(b)→ benzene

It might be asked why the transition state does not combine with an anion Y^- in the system, to form a straightforward adduct C_6H_6XY in the same way as alkenes do. The reason however is that the adduct, cyclohexa-1,3-diene, is not stabilized by delocalization.

27.6 Electrophilic substitution reactions of benzene

(a) *Nitration*. A typical electrophilic substitution reaction of benzene is **nitration**. The electrophile here is the nitryl cation (nitronium ion), NO_2^+, produced when concentrated nitric and sulphuric acids are mixed together, the former acting rather surprisingly as a Brønsted base (section 19.12):

protons gained

$$HNO_3 + 2H_2SO_4 \rightarrow NO_2^+ + H_3O^+ + 2HSO_4^-$$

protons lost

Evidence for the existence of NO_2^+ is as follows:

(i) cryoscopic data shows four ions are formed;

(ii) infra-red (and Raman) spectroscopy.

The product of the reaction is nitrobenzene,

$C_6H_5NO_2$, a yellow oil. More vigorous conditions lead to further substitution giving mainly 1,3-dinitrobenzene.

$$C_6H_6 \xrightarrow{NO_2^+} C_6H_5NO_2 \xrightarrow{NO_2^+} C_6H_4(NO_2)_2$$

(b) *Sulphonation*. Another electrophilic substitution reaction of benzene is **sulphonation**, which occurs when benzene is refluxed for several hours with fuming sulphuric acid (H_2SO_4+ SO_3):

$$C_6H_6 + SO_3 \longrightarrow C_6H_5\text{-}SO_2\text{-}O\text{-}H$$

Benzenesulphonic acid

The electrophile here is thought to be sulphur trioxide itself, rather than a positively charged species, so the subsequent elimination of a proton from the transition state need not occur.

The sulphonic acid group, $-SO_2.OH$, is much more acidic than the carboxylic acid group $-COOH$, and benzenesulphonic acid is a strong acid that forms salts such as sodium benzenesulphonate, $C_6H_5.SO_2O^-Na^+$. The grouping can be removed by strong nucleophilic attack, i.e. by fusing sodium benzenesulphonate with sodium hydroxide to produce phenol:

$$C_6H_5SO_2O^{\ominus} + OH^- \longrightarrow C_6H_5OH + SO_3^{2-}$$

(c) *Halogenation*. Among the halogens, bromine and chlorine perform substitution reactions with benzene in the presence of a catalyst. Such a catalyst is usually an electron-deficient compound (i.e. a Lewis acid, section 19.15) which presumably promotes the partial formation of a halogen cation which can act as the electrophile. For instance, using aluminium bromide as the catalyst:

$$Br_2 + AlBr_3 \rightarrow Br^{\delta+} \ldots AlBr_4^{\delta-}$$

Br₂: Lewis base; AlBr₃: Lewis acid; $Br^{\delta+} \ldots AlBr_4^{\delta-}$: Positive end of complex acts as electrophile

The overall equation is:

$$C_6H_6 + Br_2 \longrightarrow C_6H_5Br + HBr$$

The appropriate halogenobenzene, C_6H_5Hal, is formed. Iodine is too unreactive, while fluorine is over-reactive and produces disruption of the entire ring, so that direct substitution is effectively confined to chlorine and bromine.

(d) *Alkylation*. Other reagents that attack benzene are halogen-containing carbon compounds such as alkyl halides, again in the presence of an electron-deficient catalyst:

$$CH_3.CH_2Br + AlBr_3 \rightarrow CH_3.CH_2^{\delta+} \ldots AlBr_4^{\delta-};$$
$$\text{giving } C_6H_5.CH_2.CH_3$$

This is known as the Friedel–Crafts reaction. The electrophile is the positive end of the acid-base complex, in effect a carbocation $CH_3.CH_2^+$. The electrophilic part of the acid-base complex attacks, to give $C_6H_5.CH_2.CH_3$.

The Friedel–Crafts reaction will also allow acyl groups to be introduced into the benzene ring. **Acyl** is the general name for groups R.CO— where R = alkyl (section 25.4).

$$R.COCl + C_6H_6 \longrightarrow C_6H_5CO.R + HCl$$

R.COCl: an acyl chloride

For instance, to prepare phenylethanone, $C_6H_5.CO.CH_3$, equimolar amounts of $AlCl_3$ and benzene are placed in a flask and ethanoyl (acetyl) chloride is added dropwise. The mixture is heated under reflux at 60°C for half an hour. The product is extracted with benzene, washed with alkali, dried and redistilled.

27.7 Other reactions of benzene

Benzene is not a particularly reactive substance, and it finds uses as a solvent for organic reactions. It is, however, rather toxic, and other solvents are therefore often to be preferred.

Like other hydrocarbons, benzene burns readily in oxygen; oxygen behaves as a free radical in this respect. Halogens can undergo reactions of the free-radical type in the presence of ultra-violet light, to give adducts, e.g. hexachlorocyclohexane, $C_6H_6Cl_6$. The attacking species in the latter is thought to be free chlorine atoms produced by photolytic dissociation of the Cl_2 molecule.

27.8 Reactions of simple benzene derivatives

The reactivity of the benzene ring is profoundly modified by the presence of a substituent group already present. For instance, nitrobenzene is relatively unreactive towards electrophiles whereas methylbenzene is appreciably more reactive than benzene itself. Two distinct effects underlie this behaviour:

(a) *Inductive effect*. This has been discussed in Chapter 19, but the main results may be summarized by saying that most substituent groups show what is termed a $-I$ effect, that is, they attract electrons. A few groups, notably alkyl groups, show a $+I$ effect, that is, they are electron-releasing. The C—H bond itself is of low polarity and is assumed to represent zero inductive effect.

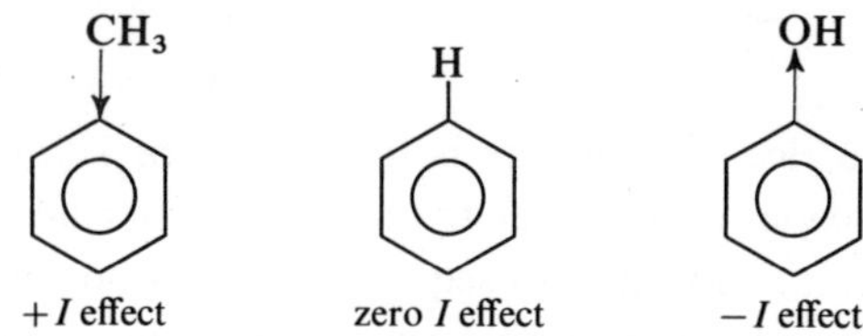

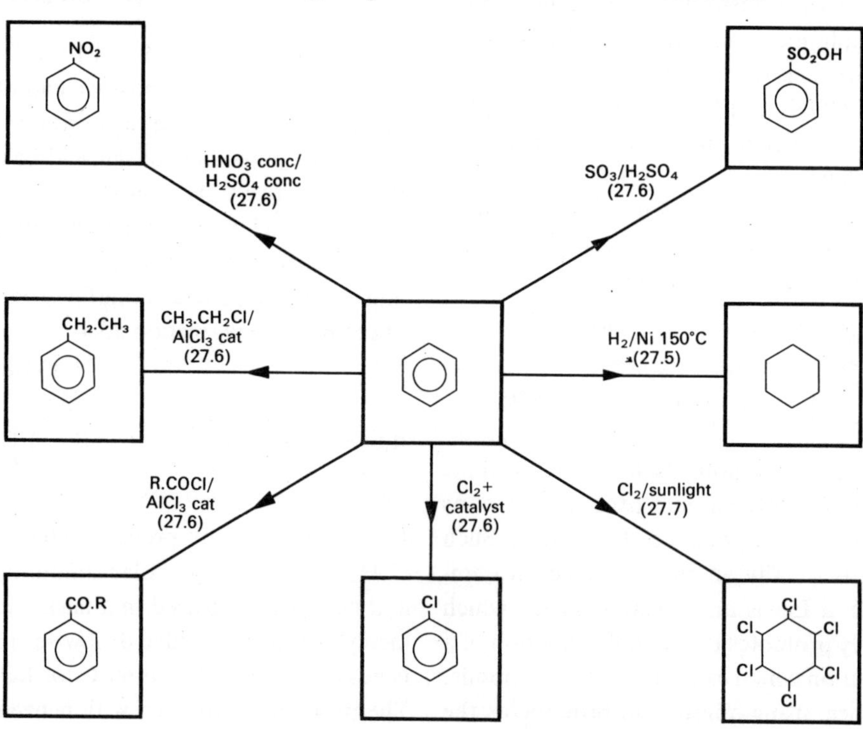

FIG. 27.5. Reactions of benzene.

An approximate order of groups arranged in order of inductive effect is as follows:

$$\begin{array}{cccccccc} C_2H_5 & CH_3 & (H) & I & Br & Cl & OH & F \end{array}$$
$$\longleftarrow (+I) \longleftarrow \text{zero} \qquad \longrightarrow (-I) \longrightarrow$$

The general result of the inductive effect is for the $+I$ effect to activate the benzene ring by pushing more negative charge into the π-electron system, and for the $-I$ effect to deactivate the ring by withdrawing negative charge from the benzene ring.

(b) *Resonance effect*. It is however relatively rare for the inductive effect to operate on its own: where the side-group contains either a multiple bond or a readily-donated lone pair, then the possibility of delocalization of the side-group electrons into the π-electron system of the ring gives rise to another effect, known as the **resonance effect** (mesomeric effect).

It is outside the scope of this book to discuss the details of how the inductive and resonance effects interact to produce overall activation or deactivation, but for reference purposes Table 27.3 gives the main results observed for **nitration**, a typical kind of electrophilic attack. It will be noticed that the side-groups containing a double bond that is separated from the π-electron system by a single bond, all deactivate the ring and all give the 1,3-disubstituted derivative as the predominant isomer. By contrast, atoms or groups possessing lone pairs on the atom attached directly to the ring produce mainly the 1,2 and 1,4 isomers. They usually activate the ring.

It is the resonance effect, not the inductive effect, that is most important in determining the relative *rates* of substitution at different carbon atoms in the ring, and hence the relative amounts of the different isomers produced.

The presence of a side-group can have quite striking effects on the reactivity of the benzene ring. Thus while benzene itself is brominated and nitrated relatively slowly, phenol (hydroxybenzene) reacts immediately under *mild* conditions, to give *tri*-substitution:

$$C_6H_5OH(aq) + Br_2(\text{dil. aq}) \longrightarrow \text{2,4,6-}Br_3C_6H_2OH$$

2,4,6-tribromophenol

$$C_6H_5OH(aq) + HNO_3(\text{dil. aq}) \longrightarrow \text{2,4,6-}(NO_2)_3C_6H_2OH$$

2,4,6-trinitrophenol

In contrast to the strongly activating nature of the hydroxy-group of phenol, the nitro-group of nitrobenzene has a deactivating effect, and a mixture of fuming nitric and sulphuric acid is needed, at 70°C, to convert nitrobenzene to 1,3-dinitrobenzene:

$$C_6H_6 \xrightarrow[\text{conc, 40°C}]{HNO_3/H_2SO_4} C_6H_5NO_2 \xrightarrow[\text{fuming, 70°C}]{HNO_3/H_2SO_4} \text{1,3-}C_6H_4(NO_2)_2$$

1,3-dinitrobenzene

27.9 Methylbenzene (toluene)

Methylbenzene is part alkane and part arene, and shows some of the reactions of both classes of compound. It tends to react with the same reagents as benzene itself, but the $+I$ effect of the methyl side-group makes its ring more reactive. This very fact makes its preparation from benzene difficult: the Friedel–Crafts reaction will not stop readily at the first stage, but will go on to produce isomers of dimethylbenzene. Fortunately, however, methylbenzene occurs naturally in coal and other materials.

TABLE 27.3 REACTIVITY OF SIMPLE BENZENE DERIVATIVES TOWARDS NITRATION

Compound	Reactivity compared with that of benzene	Main nitration products
(a) Compounds of the general type 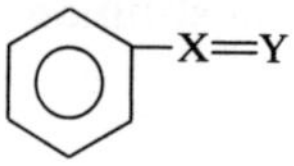 C_6H_5—X=Y	Usually less (resonance produces positive charge at 2, 4 and 6 positions)	Mainly the 3-nitro derivative
Examples: 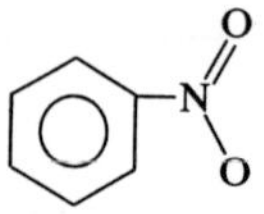 C_6H_5—NO_2 (—N(=O)—O)	Less	1,3-dinitrobenzene 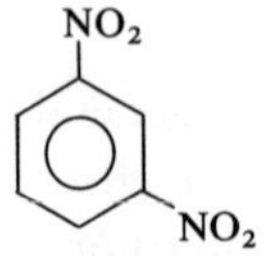 (NO_2, NO_2)
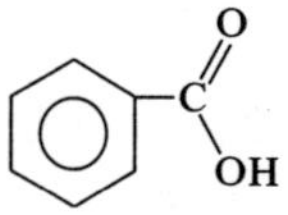 C_6H_5—C(=O)—OH	Less	3-nitrobenzene-carboxylic acid 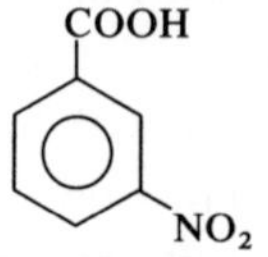 (COOH, NO_2)
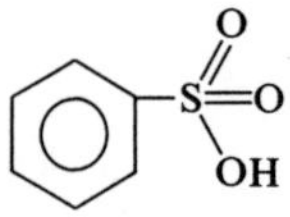 C_6H_5—S(=O)(=O)—OH	Less	3-nitrobenzene-sulphonic acid 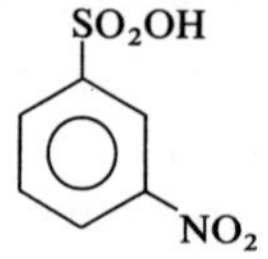 (SO_2OH, NO_2)
(b) Other compounds	Usually greater (resonance produces negative charge at 2, 4 and 6 positions)	Mainly a mixture of 2-nitro and 4-nitro derivatives
Examples: —CH_3	Slightly greater	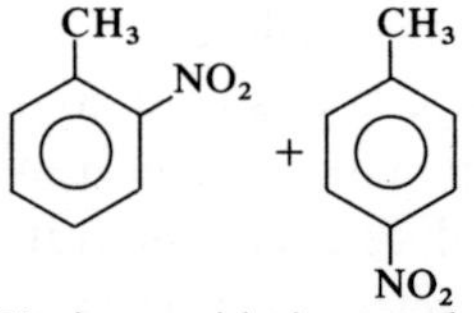 CH_3, NO_2 (2-) + CH_3, NO_2 (4-)
—NH_2	Much greater	Various oxidation products
—NH.CO.R	Slightly greater	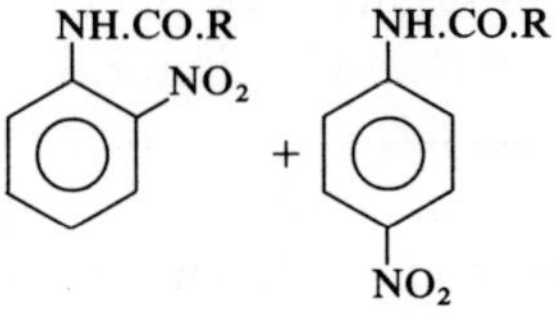 NH.CO.R, NO_2 (2-) + NH.CO.R, NO_2 (4-)
—OH	Much greater	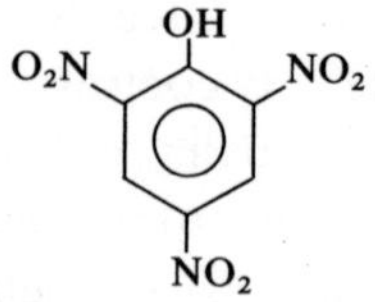 OH, O_2N, NO_2, NO_2 (2,4,6-)
—O.CO.R	Slightly greater	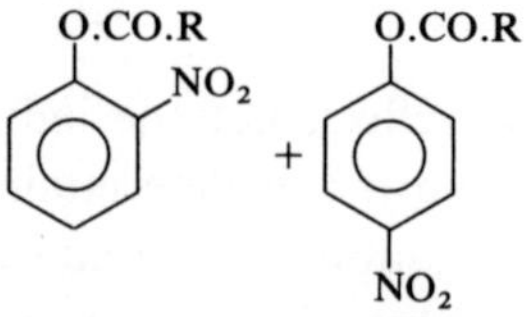 O.CO.R, NO_2 (2-) + O.CO.R, NO_2 (4-)
—Cl	Less (strong inductive effect gives ring +charge)	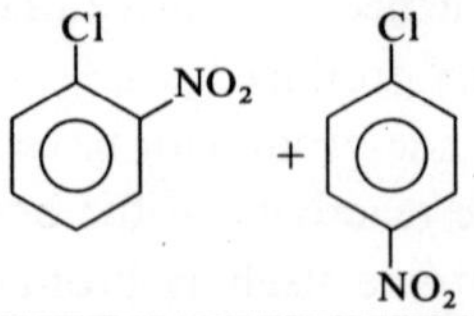 Cl, NO_2 (2-) + Cl, NO_2 (4-)

The alkyl part of the molecule, i.e. the alkyl side-group, is more reactive than in ethane, due to the presence of the benzene ring. Thus it can undergo *controlled* oxidation:

$$C_6H_5CH_3 \xrightarrow{CrO_2Cl_2} C_6H_5CHO \quad \text{(Étard reaction)}$$

Benzaldehyde

$$C_6H_5CH_3 \xrightarrow[MnO_4^-]{\text{Alkaline}} C_6H_5COOH$$

Benzoic acid

The reactions of methylbenzene are summarized in Fig. 27.6.

If chlorine is passed into boiling methylbenzene in the absence of a catalyst, side-chain substitution takes place in stages, each stage being slower than the previous one. The successive products can be isolated when the appropriate weight increase occurs. Note that in naming the products, brackets are used because the methyl side-chain is itself substituted. Trivial names for these substances are still in common use and they are given in square brackets beneath the systematic names.

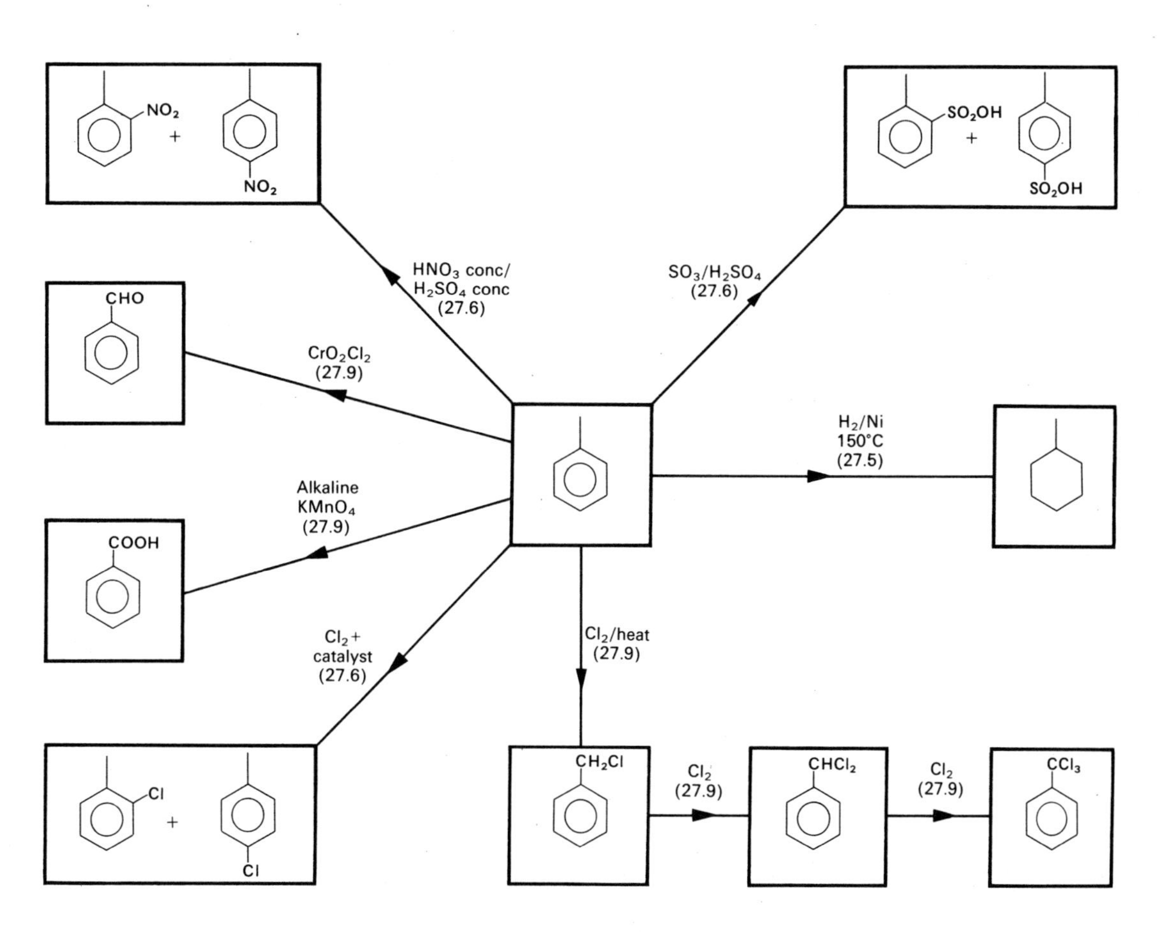

FIG. 27.6. Reactions of methylbenzene.

CH_3 —Cl_2→ CH_2Cl —Cl_2→

Methylbenzene [toluene]   (Chloromethyl)benzene [benzyl chloride]

$CHCl_2$ —Cl_2→ CCl_3

(Dichloromethyl)benzene [benzal chloride]   (Trichloromethyl)benzene [benzotrichloride]

27.10 Nucleophilic attack on the benzene ring

Nucleophiles are generally negatively-charged groups, seeking centres of positive charge. There are two good reasons which prevent them attacking carbon atoms in benzene, and displacing hydrogen:

(a) the π-electron rings repel nucleophiles;

(b) if hydrogen were displaced, it would have to leave as H^- to carry away the negative charge from the transition state, and this would only be possible under extremely basic conditions which do not normally prevail.

Nucleophiles will however occasionally attack carbon atoms to replace a halogen. Although condition (a) still applies, a halogen atom *can* be eliminated as its anion. Hence under vigorous conditions, in the presence of catalysts in industrial processes, chlorobenzene can be attacked by OH^- and NH_3:

Benzene (from coal tar) —Cl_2, 40°C→ chlorobenzene (Cl)

Chlorobenzene —NH_3→ NH_2

Chlorobenzene —10% NaOH(aq), 330°C, 350 atm.→ O^- —H^+(aq)→ OH

It has been found that a strongly electron-withdrawing group in the 2 or 4 position relative to the halogen atom enables the latter to be substituted more easily by nucleophiles. For instance, 1-chloro-2,4-dinitrobenzene is much more readily hydrolysed by alkalis than chlorobenzene itself. The important reagent 2,4-dinitrophenylhydrazine (section 28.3) is made by making use of the nucleophilic properties of hydrazine, N_2H_4:

Cl —Controlled nitration→ Cl, NO_2, NO_2 —N_2H_4→ NH.NH_2, NO_2, NO_2

1-chloro-2,4-dinitrobenzene   2,4-dinitrophenylhydrazine

27.11 Manufacture and uses of arenes

In addition to the arenes (benzene, toluene, naphthalene and xylenes mainly) extracted from coal tar an appreciable quantity can be extracted from the naphtha fraction (of crude petroleum) after subjecting it to catalytic reforming.

One of the main uses of benzene is in the manufacture of styrene, by the following process:

Benzene —C_2H_4→ $CH_2.CH_3$ —600°C, Fe_2O_3 catalyst→ CH:CH_2

ethylbenzene (phenylethane)   styrene (phenylethene)

Styrene is then polymerized, either alone to make polystyrene, or with other monomers such as butadiene to make various co-polymers which are useful plastics.

Benzene is also converted to cyclohexane by catalytic hydrogenation: subsequent oxidation processes lead to hexane-1,6-dioic acid, commonly known as adipic acid, for manufacture of *Nylon*.

Methylbenzene, toluene, is widely used in the manufacture of polyurethane plastics. Other useful applications of toluene are summarized in Fig. 27.6.

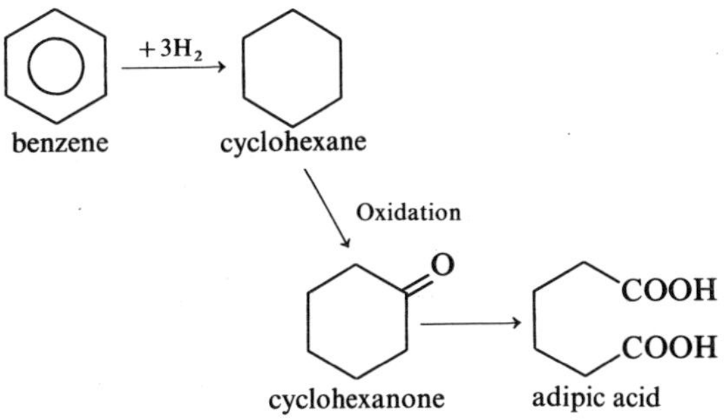

benzene $\xrightarrow{+3H_2}$ cyclohexane $\xrightarrow{\text{Oxidation}}$ cyclohexanone $\rightarrow$ adipic acid (COOH, COOH)

Study Questions

1. The following question is an exercise comparing the stability of benzene with the theoretical stability of the canonical form (section 27.3) used in writing its resonance structure.

Bond energies in $kJ\,mol^{-1}$: C—C 346; C=C 610; C—H 413; C—Cl 336; Cl—Cl 242; H—Cl 431.

$\Delta H^{\ominus}_{at}$ for C(s): +715 kJ per mole of C.

$\Delta H^{\ominus}_{at}$ for H_2(g): +436 kJ per mole of H_2.

$\Delta H^{\ominus}_{f}$ for C_6H_6(g): +82 kJ per mole of benzene.

(a) Use the data to determine the theoretical $\Delta H^{\ominus}_{f}$ of gaseous cyclohexa-1,3,5-triene, the canonical form.

(b) Calculate the difference between your answer to (a) and $\Delta H^{\ominus}_{f}$ for benzene.

(c) What is the name given to the quantity of energy calculated in (b)?

(d) Calculate the theoretical $\Delta H^{\ominus}$ for the following reactions:

(i) 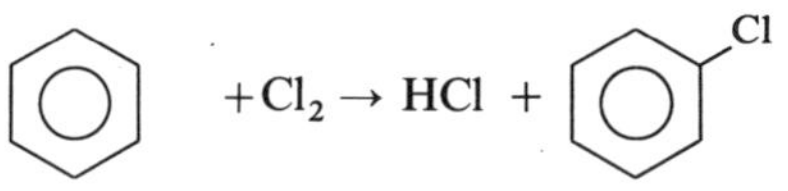 benzene $+ Cl_2 \rightarrow HCl +$ chlorobenzene (Cl)

(ii) 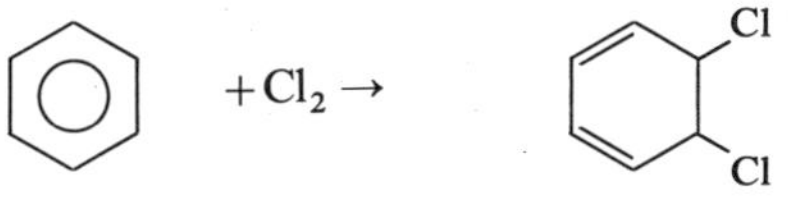 benzene $+ Cl_2 \rightarrow$ dichlorocyclohexadiene (Cl, Cl)

(iii) 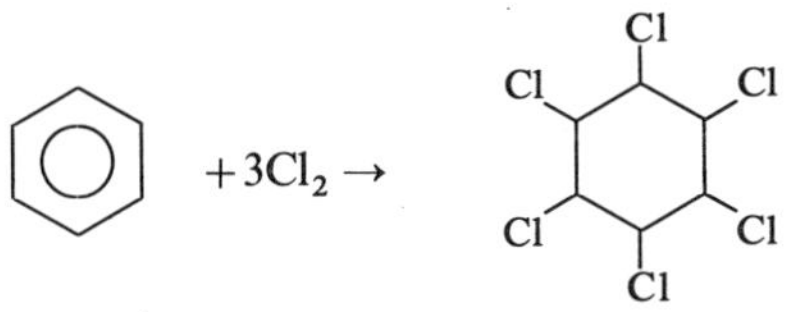 benzene $+ 3Cl_2 \rightarrow$ hexachlorocyclohexane (Cl, Cl, Cl, Cl, Cl, Cl)

(e) How does benzene actually react with chlorine? Is this consistent with your answers to (d)?

2. State, giving your reasons, whether you would expect the nitration of (a) phenylamine, (b) benzoic acid, to be faster or slower than that of benzene.

3. Indicate the position(s) in each of the following where substitution would occur with electrophilic attack:

(a) 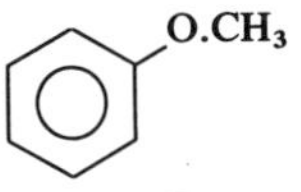 benzene ring with O.CH_3

(b) 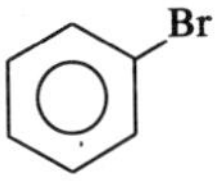 benzene ring with Br

(c) 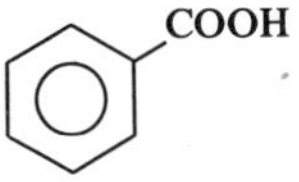 benzene ring with COOH

(d) 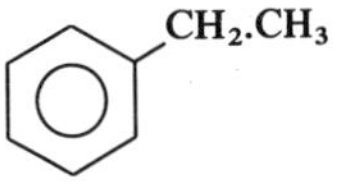 benzene ring with $CH_2.CH_3$

(e) 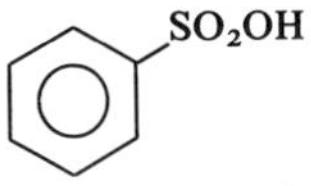 benzene ring with SO_2OH

(f) 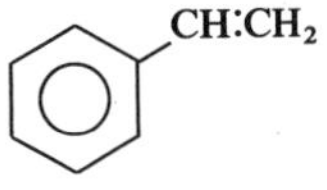 benzene ring with CH:CH_2

4. Starting from methylbenzene, state how you would attempt to prepare:

(a) 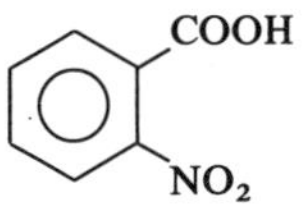 benzene ring with COOH and NO_2 (ortho)

(b) 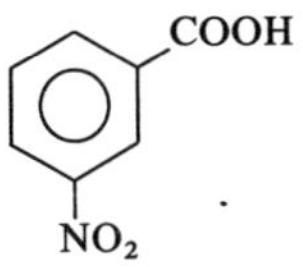 benzene ring with COOH and NO_2 (para)

5. When 1-chloropropane reacts with benzene in the presence of anhydrous aluminium chloride, the product is 2-phenylpropane rather than 1-phenylpropane.

(a) Draw structural formulae for the two products.
(b) How does 1-chloropropane react with aluminium chloride?
(c) What must happen to the cation before it attacks the benzene?

6. A compound **J** (C 90%, H 10% by mass) was oxidized with alkaline potassium manganate(VII) to give benzene-1,2-dicarboxylic acid.

(a) Write the skeletal formula of benzene-1,2-dicarboxylic acid.
(b) Derive the molecular formula of **J**.
(c) Suggest a structural formula for **J**.
(d) Suggest the general reaction of alkylbenzene compounds with alkaline potassium manganate(VII).

CHAPTER 28

Compounds containing the carbonyl group

28.1 The carbonyl group

The carbon atom is capable of forming double bonds with other atoms as well as with itself. When a double bond exists between carbon and oxygen, a **carbonyl group** is said to be present. The chemical properties of this group are rather different from those of the C=C bond (Chapter 26) because carbon and oxygen have markedly different electronegativities. In the carbonyl group, the π-bond is distorted, and the bond is polar.

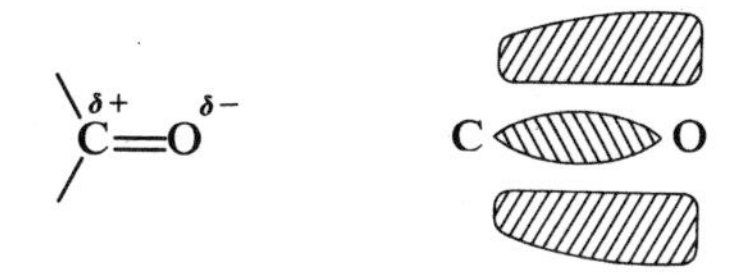

Since the oxygen atom carries a partial negative charge, it is susceptible to attack by electrophiles. The carbon atom on the other hand carries a partial positive charge, and is susceptible to attack by negatively charged, positive-seeking groups. Such attacking groups are known as **nucleophiles**.

When a carbonyl compound reacts chemically therefore, the initial step is often the attachment of a nucleophile to the carbon atom, followed by the addition of an electrophile, usually H^+, to the oxygen atom (Fig. 28.1).

$$\overset{\delta+}{>C}=\overset{\delta-}{O} \xrightarrow{X^-} -\underset{X}{\overset{|}{C}}-O^{\ominus} \xrightarrow{H^+} -\underset{X}{\overset{|}{C}}-OH$$

Nucleophile Electrophile

FIG. 28.1. Mechanism of addition of HX to the carbonyl group.

28.2 Nomenclature of carbonyl compounds

The chemical properties of the carbonyl group are strongly affected by the proximity of other functional groups. The simplest case is when the carbonyl group is attached to two simple alkyl or aryl groups: such compounds are called **ketones**. The suffix for denoting ketones in the IUPAC system is **-one**. Table 28.1 gives some simple examples: note that there is no need for a number to appear when naming propanone and butanone—if the compound is a ketone, there is only one possible place for the carbonyl group to occur.

When a hydrogen atom is attached to the carbonyl group, the compound is an **aldehyde**. The IUPAC suffix is **-al**. In writing formulae, the aldehyde group is usually written —CHO, but this is just a convenient way of representing the group

$$-C\begin{matrix}\diagup H\\ \diagdown\!\!\diagdown O\end{matrix}$$

It is not necessary to number the aldehyde group when naming compounds, since it must be at the end of a carbon chain, and this is generally carbon atom number 1 (Table 28.1). Aldehydes and ketones are very similar in their chemical reactions.

Other functional groups may also join to the carbonyl group, and the properties of the

TABLE 28.1
ALDEHYDES AND KETONES

Molecular formula and name	Abbreviated structural formula	Skeletal notation
C_3H_6O propanone	$CH_3.CO.CH_3$	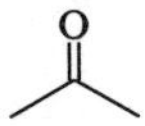 O
C_4H_8O butanone	$CH_3.CO.CH_2.CH_3$	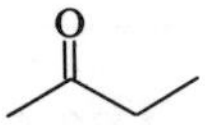 O
$C_5H_{10}O$ pentan-2-one	$CH_3.CO.CH_2.CH_2.CH_3$	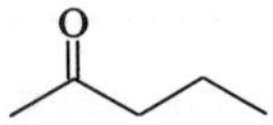 O
$C_5H_{10}O$ pentan-3-one	$CH_3.CH_2.CO.CH_2.CH_3$	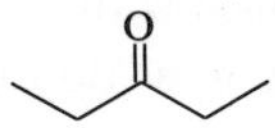 O
$C_5H_{10}O$ pentanal	$CH_3.CH_2.CH_2.CH_2.CHO$	 CHO
C_8H_8O phenylethanone	$C_6H_5.CO.CH_3$	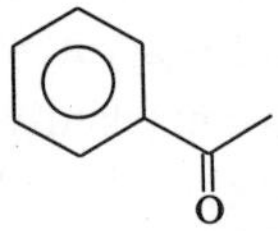 O

TABLE 28.2
OTHER CLASSES OF COMPOUND CONTAINING THE CARBONYL GROUP

Type of Compound	Structure of functional group	IUPAC suffix	Example	Section reference
Carboxylic acid	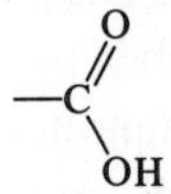 $-C(=O)OH$	-oic acid	butanoic acid $CH_3.CH_2.CH_2.COOH$	30·6
Amide	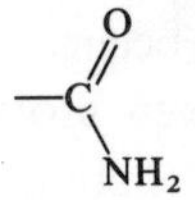 $-C(=O)NH_2$	-amide	butanamide $CH_3.CH_2.CH_2.CONH_2$	31·9
Ester	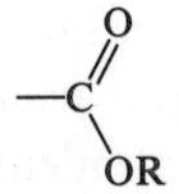 $-C(=O)OR$	(alkyl) -oate	methyl butanoate $CH_3.CH_2.CH_2.COOCH_3$	30·7
Acid	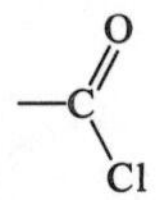 $-C(=O)Cl$	-oyl chloride	butanoyl chloride $CH_3.CH_2.CH_2.COCl$	29·2

resultant compound often differ markedly from simple aldehydes and ketones. Table 28.2 gives examples, together with section references.

28.3 Ketones and aldehydes

Just as with compounds containing the C=C double bond, the predominant behaviour of the C=O bond in aldehydes and ketones is *addition*. Here however the similarity ends. The carbonyl group is polar, and aldehydes and ketones have higher boiling points than their parent hydrocarbons. For instance propanone (commonly known as acetone) is miscible with water and boils at 56°C, in contrast to propane which is a gas insoluble in water (section 25.7).

(a) The carbonyl group in aldehydes and ketones may be hydrogenated with hydrogen gas in the the presence of finely divided nickel at about 140°C; this addition reaction is similar to that of alkenes. The product contains the hydroxy-group, —OH, and is called an **alcohol** (IUPAC suffic **-ol**, section 30.1).

$$>C{=}O \xrightarrow[140°C]{H_2/Ni} -\overset{|}{\underset{H}{C}}-\underset{H}{O}$$

For example, butanone can be converted to butan-2-ol, and butanal to butan-1-ol:

$$CH_3.CH_2.CO.CH_3 + H_2 \rightarrow CH_3.CH_2.\underset{\displaystyle OH}{\underset{|}{CH}}.CH_3$$

$$CH_3.CH_2.CH_2.CHO + H_2 \rightarrow CH_3.CH_2.CH_2.CH_2OH$$

This reduction can also be effected using sodium amalgam and water.

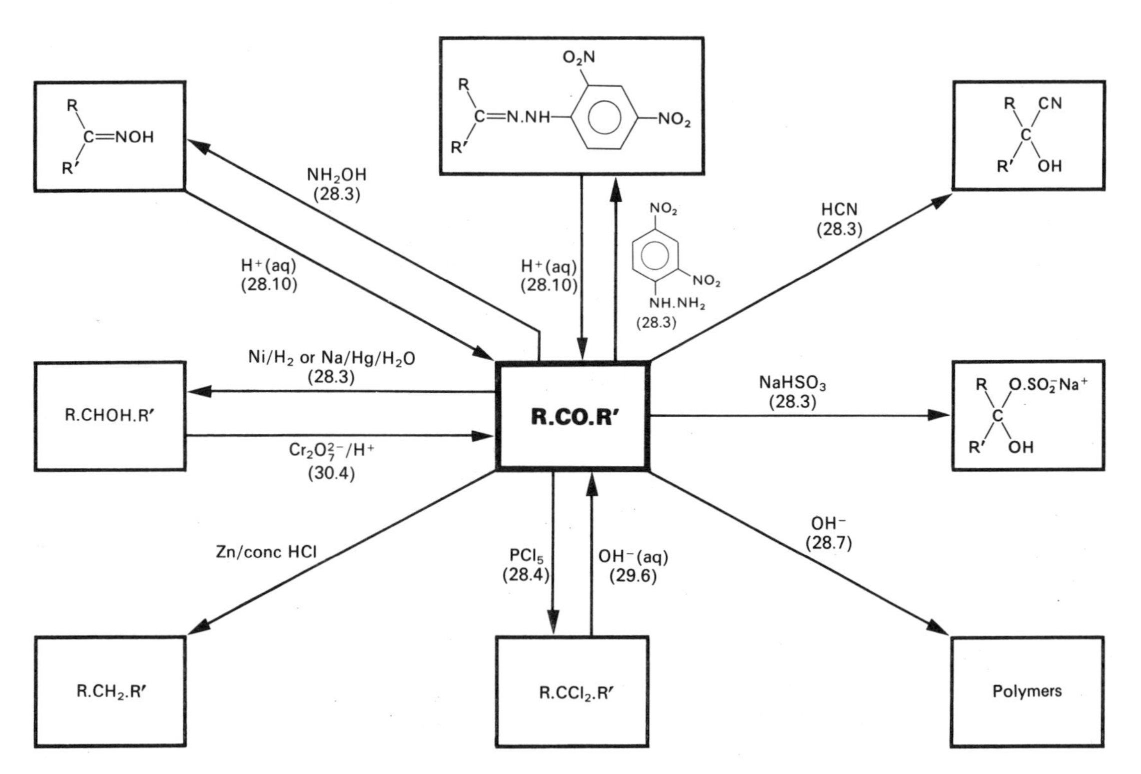

FIG. 28.2. Reactions involving ketones and aldehydes (R' = H for aldehydes).

(b) Other addition reactions of aldehydes and ketones involve attack by a nucleophile containing a lone pair. Water itself sets up an equilibrium, but this generally lies far to the left:

$$R(R')C{=}O + H_2O \rightleftharpoons R(R')C(OH)_2$$

(c) Stronger nucleophilic attack is usually effected with nitrogen compounds containing lone pairs; the initial reaction is often followed by subsequent elimination of water. For instance, propanone reacts with hydroxylamine, NH_2OH, to form an **oxime**:

$$(H_3C)_2C{=}O + NH_2OH \longrightarrow [(H_3C)_2C(OH)NHOH] \longrightarrow (H_3C)_2C{=}N{-}OH$$

(d) Aldehydes and ketones may be detected in analysis by adding "Brady's reagent", 2,4-dinitrophenylhydrazine, which forms a yellow or orange precipitate, called a **phenylhydrazone**, with carbonyl compounds:

$$(H_3C)_2C{=}O + H_2N.NH{-}C_6H_3(NO_2)_2 \longrightarrow (H_3C)_2C{=}N.NH{-}C_6H_3(NO_2)_2$$

propanone phenylhydrazone

Reactions (c) and (d) are examples of what are termed **condensation reactions**. Condensation may be defined as *addition* followed by the *elimination* of a small **leaving group** such as H_2O.

The crystalline adducts formed may be recrystallized from aqueous ethanol, and their melting points determined. This affords a valuable means of identifying the original aldehyde or ketone (section 14.19).

(e) An important synthetic route involves the addition of hydrogen cyanide to the C=O bond. The reaction is sometimes useful for adding a carbon atom to the skeleton:

$$(H_3C)_2C{=}O + HCN \longrightarrow (H_3C)_2C(OH)CN$$

(note no subsequent elimination)

(f) Sodium hydrogensulphite, $NaHSO_3$, also forms crystalline adducts:

$$R.CO.R' + NaHSO_3 \longrightarrow R(R')C(OH)O.SO_2^-Na^+$$

28.4 The reactions of aldehydes ketones with halogens

When iodine reacts with propanone in the presence of hydrogen ion, the reaction is found to be first order with respect to propanone and to hydrogen ion, but zero order with respect to iodine concentration (i.e. independent, see Chapter 17).

$$CH_3.CO.CH_3(aq) + I_2(aq) \rightarrow CH_3.CO.CH_2I(aq) + HI(aq);$$
$$\text{rate} = k[CH_3.CO.CH_3][H^+].$$

This suggests that the rate-determining step involves a propanone molecule and an aqueous hydrogen ion. A possible mechanism for the reaction might be:

$$CH_3.\overset{O}{\overset{\|}{C}}.CH_3 + H^{\oplus} \rightarrow CH_3.\overset{^{\oplus}OH}{\overset{\|}{C}}.CH_3 \xrightarrow[\text{(rate determining)}]{\text{slow}} CH_3.\overset{OH}{\overset{|}{C}}{:}CH_2$$

$$\xrightarrow{I_2,\ \text{fast}} CH_3.\overset{OH}{\overset{|}{C}}\!-\!CH_2 \ (\text{I}\ \text{I}^{\ominus}) \xrightarrow{\text{fast}} CH_3.\overset{O}{\overset{\|}{C}}.CH_2I + H^{\oplus} + I^{\oplus}$$

The substitution of halogen atoms for hydrogen atoms occurs more rapidly in aldehydes and ketones than in simple alkanes (section 25.8) and the reaction is catalyzed by $OH^-(aq)$ as well as by acids.

If an aldehyde or ketone is warmed with phosphorus pentachloride, the $\rangle C{=}O$ group is converted into the group $\rangle CCl_2$:

$$R(R')C{=}O + PCl_5 \longrightarrow R(R')CCl_2 + POCl_3$$

28.5 The iodoform reaction

If an aldehyde or ketone containing a methyl group attached to carbonyl ($CH_3.CO—$) is iodinated under basic conditions, tri-iodination will eventually result, producing the group $CI_3.CO—$. The presence of three iodine atoms on one carbon *weakens* the C—C bond, causing it to break even under these relatively mild alkaline conditions:

$$CI_3.CO.R + OH^- \rightarrow CHI_3 + R.COO^-$$

Tri-iodomethane, CHI_3, is produced as a yellow precipitate. This reaction is used as a test for the group $CH_3.CO—$, and is called the **iodoform test**. The reaction is noteworthy in that it is unusual for the carbon skeleton to be so easily broken.

Some alcohols can also give a positive iodoform test (section 30.4(d)). Such alcohols must be able to produce the group $CH_3.CO—$ when oxidized by the iodine.

The iodoform reaction does not occur with ethanoic acid, $CH_3.COOH$, nor with ethanamide, $CH_3.CONH_2$, although the grouping $CH_3.CO—$ is present in both cases. The properties of the carbonyl group are modified by the adjacent lone pair in each case, and the mechanism described in section 28.4 cannot occur.

28.6 Differences between aldehydes and ketones

Although most of the reactions of ketones are shared also by aldehydes, certain differences occur in aldehydes due to the presence of a hydrogen atom attached directly to the carbonyl group. This C—H bond is quite reactive and is susceptible to oxidation. Gentle refluxing of an aldehyde with acidified potassium dichromate(VI) results in oxidation to a **carboxylic acid**:

$$R{-}C({=}O){-}H + [O] \longrightarrow R{-}C({=}O){-}O{-}H$$

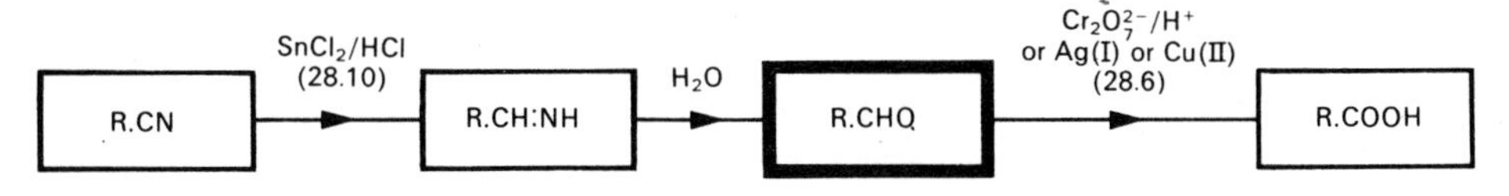

FIG. 28.3. Reactions involving aldehydes but not ketones.

For instance, butanal can be oxidized to butanoic acid:

$$CH_3.CH_2.CH_2.CHO + [O] \rightarrow CH_3.CH_2.CH_2.COOH$$

This reaction can proceed under quite mild conditions: under comparable conditions a ketone fails to oxidize. Very powerful conditions will in fact oxidize a ketone, but only by breaking up the carbon skeleton, giving products, generally acids, with less carbon atoms than the original ketone.

This difference in ease of oxidation is used in chemical tests to distinguish aldehydes from ketones. To put it another way, aldehydes are good reducing agents whereas ketones are not.

Compounds containing the aldehyde group will:

(i) reduce ammoniacal silver(I) nitrate to metallic silver (silver mirror test):

$$2Ag^+ + R.CHO + OH^- \rightarrow 2Ag + R.COOH + H^+;$$

(ii) reduce Fehling's solution (alkaline copper(II) tartrate) to copper(I) oxide, a red precipitate which appears on warming.

Aldehydes can be quite readily reduced to primary alcohols (section 30.4) whereas ketones are more difficult to reduce, and yield secondary alcohols.

Aldehydes and ketones exist also in aromatic chemistry. For instance benzaldehyde (benzenecarbaldehyde, $C_6H_5.CHO$) can be oxidized to benzoic acid (benzenecarboxylic acid, $C_6H_5.COOH$) or reduced to benzyl alcohol (phenylmethanol, $C_6H_5.CH_2OH$).

The reactions of aldehydes and ketones are summarized in chart form in Figs. 28.2 and 28.3.

28.7 Polymerization of carbonyl compounds

Aldehydes and ketones can undergo a variety of polymerization reactions. These are usually addition polymerizations, and aldehydes commonly polymerize more easily than ketones. For instance ethanal polymerizes with dilute alkali to give firstly 3-hydroxybutanal (aldol):

$$CH_3CHO \xrightarrow[\text{slow}]{OH^\ominus} {}^\ominus CH_2CHO \xrightarrow[\text{fast}]{CH_3CHO} CH_3CH(O^\ominus)CH_2CHO \xrightarrow[H^+]{\text{fast}} CH_3CH(OH)CH_2CHO$$

More-concentrated alkali, accompanied by heating, produces further polymerization resulting in the formation of a resin.

Similar reactions occur with other aldehydes and ketones, but it is essential that *one* carbon atom next to the carbonyl group carries a hydrogen atom. If there is no such hydrogen atom, the first step cannot occur and instead disproportionation occurs. The carbonyl compound is simultaneously reduced to an alcohol and oxidized to a carboxylate anion. This reaction is called **Cannizzaro's reaction**, and may be illustrated with reference to (i) methanal (formaldehyde, H.CHO) and (ii) benzenecarbaldehyde (benzaldehyde, C_6H_5CHO):

(i) $2H.CHO + OH^\ominus \rightarrow H.CH_2OH + H.COO^\ominus$

(ii) $2\,C_6H_5CHO + OH^\ominus \longrightarrow C_6H_5CH_2OH + C_6H_5COO^\ominus$

Acid conditions may also result in polymerization. Ethanal can be polymerized to a solid sold as *Meta* fuel for portable stoves. Aqueous methanal (formaldehyde solution or *formalin*) is used for preserving biological specimens, but it tends to throw down a polymer precipitate on storage.

It was noted in Chapter 24 that silicon did not form bonds Si=O formally equivalent to C=O. Methyl **silicones** may be regarded as the silicon analogues of propanone:

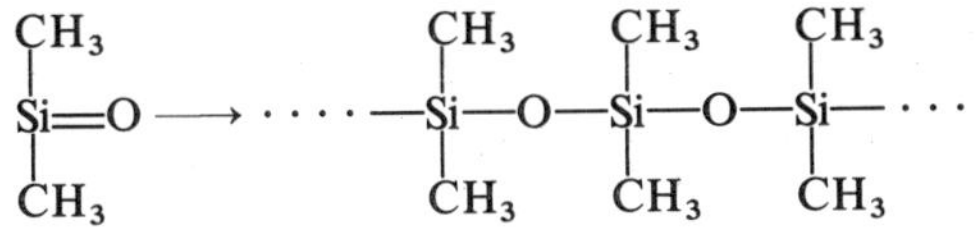

$$(CH_3)_2Si{=}O \longrightarrow \cdots{-}Si(CH_3)_2{-}O{-}Si(CH_3)_2{-}O{-}Si(CH_3)_2{-}\cdots$$

Theoretical "silicone monomer" — Silicone polymer

28.8 Other compounds containing the carbonyl group

Table 28.2 listed other compounds containing the C=O group. In each case an electronegative group is associated with the carbonyl group, and the carbonyl group is thereby prevented from showing its simple properties. None of these compounds forms condensation products with hydroxylamine or Brady's reagent, none undergoes addition with hydrogen cyanide and none gives the iodoform test. All can, however, be reduced but generally only with a very powerful reducing agent such as lithium tetrahydridoaluminate(III), $LiAlH_4$.

The compounds listed in Table 28.2 will be considered in more detail in later sections.

28.9 Manufacture and uses of aldehydes and ketones

Aldehydes and ketones are generally made by the oxidation of the appropriate alcohol.

(a) A primary alcohol, $R.CH_2OH$ oxidizes to an aldehyde, R.CHO.

(b) A secondary alcohol, R.CHOH.R′ oxidizes to a ketone, R.CO.R′.

On the industrial scale, propanone (acetone) is made by atmospheric oxidation of propan-2-ol with a catalyst:

$$\text{petroleum} \xrightarrow{\text{cracking}} CH_2{:}CH.CH_3 \xrightarrow{H_2O(\text{cat})}$$

$$CH_3.CHOH.CH_3 \xrightarrow{O_2(\text{cat})} CH_3.CO.CH_3$$

Another important ketone industrially is cyclohexanone:

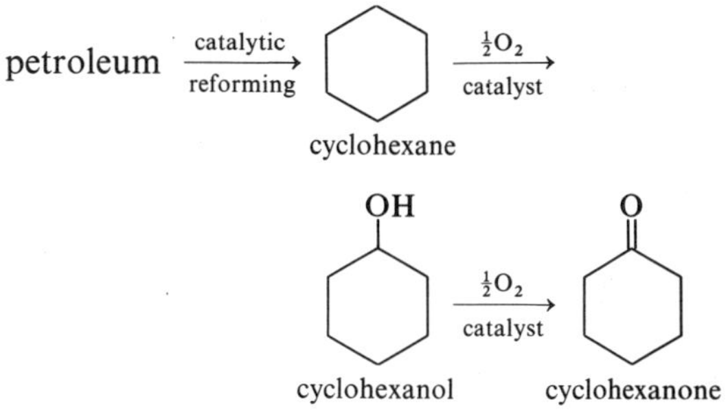

$$\text{petroleum} \xrightarrow[\text{reforming}]{\text{catalytic}} \text{cyclohexane} \xrightarrow[\text{catalyst}]{\frac{1}{2}O_2}$$

$$\text{cyclohexanol} \xrightarrow[\text{catalyst}]{\frac{1}{2}O_2} \text{cyclohexanone}$$

cyclohexane

cyclohexanol cyclohexanone

Propanone is widely used as a solvent, and also in the manufacture of *Perspex*. Cyclohexanone is an intermediate in the manufacture of *Nylon*.

Methanal (formaldehyde) and ethanal (acetaldehyde) are made industrially by the catalytic oxidation of the alcohol using air or oxygen. Methanal is widely used in the manufacture of condensation polymers such as phenol-formaldehyde and urea-formaldehyde, which are hard, thermosetting plastics. Ethanal is used in the manufacture of ethanoic acid (acetic acid).

In the laboratory, aldehydes and ketones are readily prepared by distilling a mixture of the appropriate alcohol with sodium dichromate(VI) and dilute sulphuric acid. The aldehyde or ketone is removed from the system as it forms because it has a lower boiling point than the alcohol (section 6.9).

28.10 Nitriles and related compounds

In this chapter we have considered the formal relationship between the groups C=O and C=C. In a similar manner we may imagine the triply bonded **nitrile group** or **cyano group**, C≡N, to be formally related to the group C≡C.

In the IUPAC system there are two alternative ways of naming nitriles. The group may be named as the prefix, **cyano-**, or as the suffix, **-onitrile**:

$CH_3.CN$	cyanomethane or ethanonitrile
$CH_3.CH_2.CN$	cyanoethane or propanonitrile etc.

Nitriles are hydrolyzed to the corresponding acid by heating with acid, or to the salt of the acid by boiling with alkali. For instance butanonitrile would form butanoic acid:

$$CH_3.CH_2.CH_2.CN + H^+ + 2H_2O \rightarrow CH_3.CH_2.CH_2.COOH + NH_4^+$$

Like the carbonyl group, the nitrile group can be reduced either catalytically (H_2/Ni) or with sodium amalgam, to form an **amine**:

$$CH_3CN + 2H_2 \rightarrow \underset{\substack{\text{aminoethane} \\ \text{(ethylamine)}}}{CH_3.CH_2NH_2}$$

Tin(II) chloride is a less vigorous reducing agent and in the presence of hydrogen chloride reduces a nitrile by adding only one mole of H_2, forming an **imine**:

$$R.C\vdots N + SnCl_2 + 2HCl \rightarrow R.CH\text{:}NH + SnCl_4$$

Imines contain a double bond C=N like oximes and phenylhydrazones (section 28.3), and all these compounds are readily hydrolyzed by acids back to the carbonyl compound from which they are derived (Fig. 28.2):

$$R.CH\text{:}NH + H_2O \rightarrow R.CHO + NH_3$$

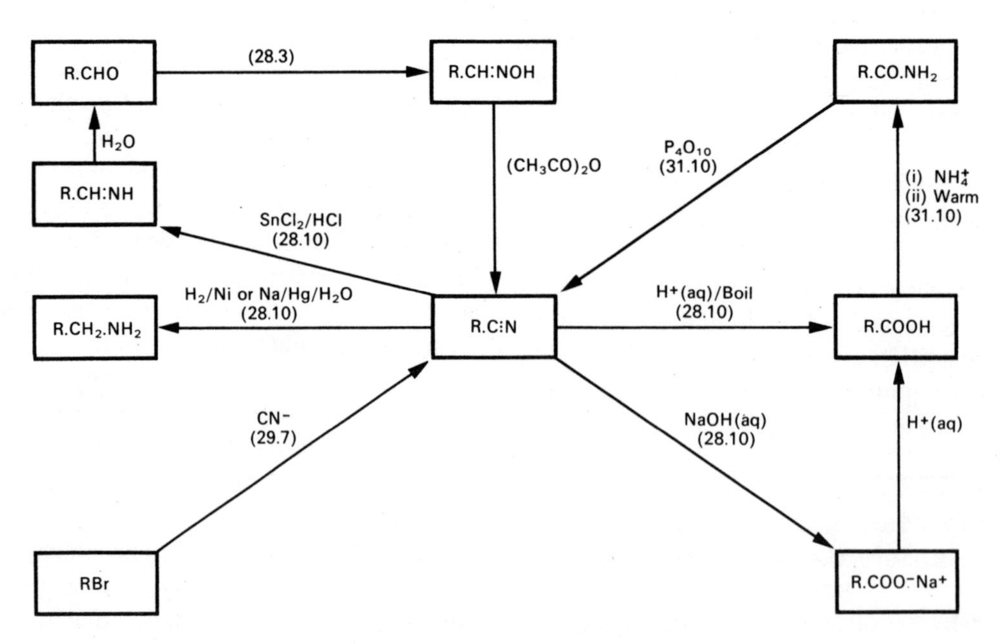

FIG. 28.4. Reactions of nitriles.

Both imines and oximes can be reduced to amines using sodium and ethanol.

Compounds containing the bond $C{=}N$ or $C{\equiv}N$ are similar chemically to those containing the bond $C{=}O$, since the electronegative nitrogen and oxygen atoms leave the adjacent carbon atom positively charged, thereby rendering it liable to attack by nucleophiles; alternatively a proton can attack the nitrogen (or oxygen) atom.

The reactions of nitriles are summarized in Fig. 28.4.

Study Questions

1. (a) How does the electron distribution in $CCl_3.CHO$ differ from that in $CH_3.CHO$?

(b) Suggest a reason why $CCl_3.CHO$ crystallizes from water as $CCl_3.CH(OH)_2$ whereas $CH_3.CHO$ can be recovered from water unchanged.

2. (a) What type of isomerism is shown especially by compounds containing double bonds?

(b) Draw a "dot and cross" electronic structure of $CH_3.CH{:}NOH$.

(c) Draw a structural formula of $CH_3.CH{:}NOH$, showing in particular the bond angle at the nitrogen atom.

(d) $CH_3.CH{:}NOH$ exists as geometric isomers. Explain why this is the case.

(e) Would you expect $(CH_3)_2C{:}NOH$ to exhibit geometric isomerism? Give a reason for your answer.

3. Give equations and reaction conditions for the conversion of ethanal to 2-hydroxypropanoic acid (lactic acid) in only two steps.

4. Which of the following compounds will give a yellow precipitate with iodine and alkali?

(a) Propanal, $CH_3.CH_2.CHO$.
(b) Propanone, $CH_3.CO.CH_3$.
(c) Ethanoic acid, $CH_3.COOH$.
(d) Ethyl ethanoate, $CH_3.COOC_2H_5$.

5. A compound **M**, C_4H_8O, gave a yellow precipitate with 2,4-dinitrophenylhydrazine, and a yellow precipitate with iodine and alkali, but no reaction with Fehling's solution.

(a) Deduce the structural formula of **M**.

(b) Write a balanced equation for the reaction of **M** with iodine and alkali.

6. A white solid **N**, containing 37·5% of carbon by mass, reacted with water to give an inflammable gas **O** containing 92·3% carbon. This gas reacted with dilute sulphuric acid containing mercury(II) sulphate, to give a very volatile liquid **P** containing 54·4% carbon. **P** reduced Fehling's solution to a red solid and in doing so formed **Q**, containing 40% carbon.

(a) Identify the lettered compounds.
(b) Write balanced equations for the reactions.
(c) What would you observe on adding 2,4-dinitrophenylhydrazine to **P**? Give the structural formula of the product.
(d) What would you expect to observe on adding iodine and alkali to **P**?

7. Compounds that react as R^-MgX^+, where R is an alkyl group and X a halogen, are called Grignard reagents.

(a) A Grignard reagent reacts with the carbonyl group to give the group

```
   OMgX
 \ /
  C
 / \
    R
```

Suggest a mechanism for this reaction.

(b) Dilute H^+(aq) converts

```
   OMgX           OH
 \ /            \ /
  C      to      C
 / \            / \
    R              R
```

Predict the structures of the products formed when the following compounds react with RMgX followed by dilute acid: (i) methanal; (ii) ethanal; (iii) propanone; (iv) ethanonitrile, $CH_3.CN$; (v) carbon dioxide.

8. (a) Write an essay comparing and contrasting the reactions of the $C{=}O$ bond with those of the $C{=}C$ bond (see Chapter 26).

(b) In a similar way, compare and contrast the reactions of the $C{\equiv}N$ bond with those of the $C{\equiv}C$ bond.

CHAPTER 29

Compounds containing carbon-halogen bonds

29.1 The carbon-halogen bond

In Chapter 21 it was observed that halogen atoms formed single bonds with other atoms. A large number of organic compounds exist in which a halogen atom can replace a hydrogen atom on the carbon skeleton.

The strength of the carbon-halogen bond decreases in the order

$$C—F > C—Cl > C—Br > C—I$$

Actual values are given in Table 21.3.

The carbon-fluorine bond is remarkably strong (452 kJ mol^{-1}), even stronger than C—H (435 kJ mol^{-1}); this makes organic fluoro-compounds very unreactive, and not typical of halogen derivatives as a whole. The remainder of the halogens form weaker bonds which render them readily susceptible to attack by nucleophiles.

29.2 Nomenclature of organic halogen compounds

In the IUPAC system, a halogen atom substituted in a carbon skeleton is named as a prefix. Where more than one alternative exists, numbering may be necessary just as with other functional groups. For instance:

$CH_3.CHBr.CH_2CH_3$ 2-bromobutane

$CHCl_2.CH_2.CH_2.CH_3$ 1,1-dichlorobutane

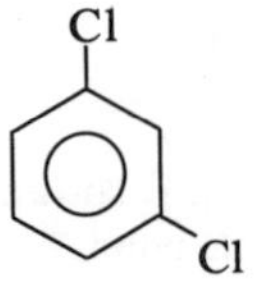

Cl, Cl 1,3-dichlorobenzene

A special case is the functional group —CO.Hal (usually chloride). Such compounds are called **acid halides**, and are named with the termination **-oyl halide**. For instance:

$CH_3.COOH$ ethanoic acid

$CH_3.COCl$ ethanoyl chloride

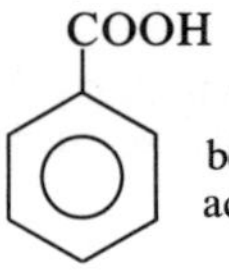

COOH benzoic acid

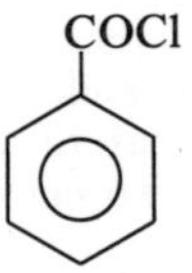

COCl benzoyl chloride

29.3 Physical properties and uses of organic halides

The carbon-halogen bond is more polar than C—H, and roughly equivalent in polarity to C=O. Simple halogeno-alkanes are therefore comparable in boiling point to aldehydes or ketones of corresponding molecular weight. The more halogen atoms there are in the molecule, the higher the molecular weight and the less

TABLE 29.1

SOME TYPICAL ORGANIC HALIDES

Molecular formula and name	Structural formula	Boiling point (in °C)	Uses
C_2H_5Cl chloroethane	$CH_3.CH_2Cl$	12	Synthesis of lead(IV) tetra-ethyl
$C_2H_4Br_2$ 1,2-dibromoethane	$CH_2Br.CH_2Br$	131	Fuel additive
$CHCl_3$ trichloromethane	$CHCl_3$	61	Formerly an anaesthetic (chloroform). Used in synthesis
C_2H_3Cl chloroethene (vinyl chloride)	$CH_2{:}CHCl$	−14	Polymerization to PVC (polyvinyl chloride)
C_2HCl_3 1,1,2-trichloroethene (trichloroethylene)	$CHCl{:}CCl_2$	87	Grease-solvent
C_6H_5Cl chlorobenzene	Cl (benzene ring)	132	Manufacture of DDT and phenol

volatile the resultant compound. Since, as a rule, there is no possibility of hydrogen bonding the halogeno-alkanes are insoluble in water, and soluble in non-polar solvents. The relatively greater mass of Cl, Br and I makes compounds of these halogens generally denser than water. Most halogeno-alkanes are colourless with a sweetish smell: a selection of typical organic halides is listed in Table 29.1.

29.4 Nucleophilic attack

The mechanism of attack on the carbon-halogen bond by nucleophiles was considered in section 17.11. The two important points to remember are:

(a) A nucleophile is a *negatively charged* group that will seek a positively charged centre—in this case the carbon atom carrying the halogen; nucleophiles will be *bases* and will have *lone pairs*; examples of nucleophiles are OH^-, NH_3, H_2O, and CN^-.

(b) The halogen atom itself constitutes a good negative leaving group. If a nucleophile attacks a molecule, it is often necessary for a leaving group to carry away the negative charge.

It is clear therefore that an electrophile cannot attack the carbon-halogen bond. Although there appears to be no reason why it should not directly attack the negatively-charged end of the bond, a halogen atom cannot form a positively charged leaving group to complete the reaction. Hence C—Hal is generally attacked by nucleophiles, in contrast to C—H which is generally attacked by electrophiles.

The two types of substitution, S_N1 and S_N2, described in Chapter 17, are extreme cases of possible reaction-mechanisms. In reality most reactions proceed by a combination of both mechanisms, with various side reactions, such as elimination, occurring to complicate the process and lower the overall yield.

29.5 Notation used in depicting reaction mechanisms

It is convenient when describing the mechanisms of chemical reactions to use a curved arrow to denote the transfer of an electron pair. It is important to draw these arrows carefully to indicate the exact positions of electron transfer. Using this notation, we may depict the attack of a simple halogeno-alkane, $R.CH_2X$ by the nucleophile OH^- as follows:

$$HO{:}^{\ominus} + RCH_2{-}X \longrightarrow HO{-}CH_2R + {:}X^{\ominus}$$

The left-hand curved arrow denotes that a lone pair from OH^- transfers to form a bond between oxygen and carbon. The right-hand arrow indicates that *simultaneously* the electron-pair of the C—X bond is transferring completely to the halogen atom forming an X^- ion.

In depicting reaction mechanisms, it is important to show only *one stage at a time*. In the S_N2 case considered above, both electron-pair transfers occur at once and are shown by one equation. The S_N1 substitution mechanism on the other hand occurs as a two-stage process, and these two stages *must* be shown separately:

$$R_3C{-}X \rightleftharpoons R_3C^{\oplus} + {:}X^{\ominus}$$

carbocation (carbonium ion)

$$R_3C^{\oplus} + {:}OH^{\ominus} \longrightarrow R_3C{-}OH$$

29.6 Reactions of organic halides with OH^- ion

It is readily found by experiment that the susceptibility to attack of the carbon-halogen bond by hydroxide ion varies widely from compound to compound, and the outcome is not always substitution. Within a series of halogen compounds, the reactivity increases in the order fluoride < chloride < bromide < iodide. This is readily explained in terms of bond energy (section 29.1), since the weaker the bond, the greater the ease of release of the halogen atom.

Even wider variations are observed with different compounds of a given halogen, say chlorine. For instance, reactivity increases very markedly along the series:

$$C_6H_5Cl < CH_3.CH_2Cl < CH_3.COCl$$

chlorobenzene < chloroethane < ethanoyl chloride

Chlorobenzene is unaffected by boiling alkali, though industrially it can be converted to phenol, C_6H_5OH, by nucleophilic attack by OH^- in the presence of a catalyst. Chloroethane, at room temperature a gas, is readily converted to ethanol, C_2H_5OH, by cold aqueous alkali, while ethanoyl chloride reacts violently with cold water to form ethanoic acid, $CH_3.COOH$.

The unreactivity of the C—Cl bond in chlorobenzene is explained by the delocalization of a lone pair from the chlorine atom into the π-electron system of the benzene ring. As a result, electron transfer can occur around the ring system, leaving positive charge on the halogen atom and negative charge on the 2, 4, or 6 position of the benzene ring. Using the curved arrow notation and the "Kekulé" notation for the benzene ring:

forming etc.

Hence the actual structure of chlorobenzene may be thought of as a resonance hybrid (section 27.3) of these forms, resulting in a strengthened carbon-chlorine bond with some residual positive charge on the chlorine atom. The actual structure of the compound becomes difficult to depict accurately in terms of simple electron-pair bonds on account of this delocalization, but an approximation to the structure would be

$Cl^{\delta+}$ $\delta-$ $\delta-$ $\delta-$

The very high degree of reactivity of the group —CO.Cl in ethanoyl chloride (commonly known as acetyl chloride) may be explained by the fact that there are *two* electronegative atoms attached to the same carbon atom. As a result the carbon centre becomes positively charged and hence extremely prone to the initial attack by the nucleophile OH^-.

If more than one halogen atom is present in a molecule, all of them will react if the conditions are appropriate. For instance, hydrolysis of 1,2-dibromoethane gives ethane-1,2-diol (ethylene glycol):

$$CH_2Br.CH_2Br + 2OH^- \rightarrow CH_2OH.CH_2OH + 2Br^-$$

If more than one halogen atom is attached to the same carbon atom, the product is however a *carbonyl* compound, not a *diol*. It is not usually possible for two hydroxy-groups to remain attached to the same carbon atom; instead a mole of water is eliminated:

$$RR'CCl_2 + 2OH^- \rightarrow RR'C(OH)_2 + 2Cl^-$$
$$RR'C(OH)_2 \xrightarrow{-H_2O} RR'C{=}O$$

If the group $—CCl_3$ is hydrolysed, the product is the carboxyl group which, in the alkaline conditions present, appears as its anion:

$$R.CCl_3 + 4OH^- \rightarrow R.COO^- + 3Cl^- + 2H_2O.$$

29.7 Reactions of organic halides with other nucleophiles

(a) *Alkyl halides*. OH^- is not the only nucleophile that attacks the carbon-halogen bond.

(i) Water itself is a weaker nucleophile, carrying a partial negative charge on its oxygen atom. It will readily attack carbocations (carbonium ions) produced by the initial stage of the S_N1 reaction:

$$H_2O: + {}^{\oplus}CR_3 \rightarrow H_2O^{\oplus}—CR_3 \rightarrow HO—CR_3 + H^{\oplus}$$

(ii) Alcohols, ROH, can also act as nucleophiles under these conditions, resulting in the overall substitution of the halogen atom by the group —OR. For instance, ethanol would react with $(CH_3)_3CI$ to form $(CH_3)_3C.O.C_2H_5$, an **ether**:

$$CH_3.CH_2(H)O: + {}^{\oplus}C(CH_3)_3 \rightarrow CH_3.CH_2(H)O^{\oplus}—C(CH_3)_3 \rightarrow CH_3.CH_2.O.C(CH_3)_3 + H^{\oplus}$$

(iii) A stronger nucleophile than ethanol is the ethoxide ion, $C_2H_5O^-$. In general, ethers can be prepared by treatment of the appropriate halogen-alkane with sodium alkoxide in the corresponding alcohol, e.g.

$$CH_3Br + C_2H_5O^- \rightarrow C_2H_5.O.CH_3 + Br^-$$

However, with higher homologues an alternative course of reaction, elimination, may occur leading instead to an alkene.

(iv) Ammonia reacts with halogeno-alkanes to give **amines**; the reagents are heated in a sealed tube:

$$RI + 2NH_3 \rightarrow RNH_2 + NH_4I$$

This reaction is complicated by the fact that the product is almost as strong a nucleophile as ammonia itself, and results in further attack to form R_2NH and R_3N and ultimately $NR_4^+I^-$ (section 31.4).

(v) Cyanide ion, CN^-, is an effective nucleophile, usually obtained by dissolving potassium cyanide in ethanol. It leads to the formation of a nitrile (section 28.10) by substitution; the halide is refluxed with ethanolic potassium cyanide:

$$CH_3.CH_2Br + CN^- \rightarrow CH_3.CH_2.CN + Br^-$$

The reaction is an excellent way of increasing the length of a carbon skeleton.

Carboxylate ions, for instance ethanoate, $CH_3.COO^-$ (acetate), are weak nucleophiles. If a halide is warmed with silver ethanoate, an **ester** is formed by nucleophilic substitution:

$$CH_3.\underset{\underset{O}{\|}}{C}-\ddot{O}^{\ominus}\!:\ \ \underset{\underset{CH_3}{|}}{CH_2}-Br \rightarrow CH_3.\underset{\underset{O}{\|}}{C}-O-\underset{\underset{CH_3}{|}}{CH_2} + Br^{\ominus}$$

Use of the silver salt assists this reaction, by removal of the halide ion as the insoluble silver halide.

(b) *Acyl halides*. Compounds of general formula R.COHal (where R = alkyl or aryl) react similarly to alkyl halides but much more vigorously. The following examples are listed for comparison with alkyl halides listed in subsection (a) above.

(i) Water vigorously attacks acyl halides, giving fumes of hydrogen chloride in moist air:

$$H_2O + R.COCl \rightarrow R.COOH + HCl$$

(ii) Alcohols react, more vigorously than with *alkyl* halides, to form esters. Phenols also react with acyl halides though not readily with alkyl halides.

$$R'OH + R.COCl \rightarrow R.COOR' + HCl$$

(iii) The alkoxide ion reacts extremely vigorously to form the ester as with the alcohol.

(iv) Ammonia reacts with acyl halides to form **amides** (functional group $-CONH_2$).

$$NH_3 + R.COCl \rightarrow R.CONH_2 + HCl$$

The liberated HCl forms a salt with the excess ammonia which is present.

(v) Carboxylate ions, for instance ethanoate $CH_3.COO^-$, react to form acid anhydrides. For instance, ethanoyl chloride and sodium ethanoate will form ethanoic (acetic) anhydride:

$$CH_3.COONa + CH_3.COCl \rightarrow CH_3.CO.O.CO.CH_3$$

The presence of the carbonyl group means that there is a considerable positive charge on the carbon atom due to the attraction on electrons exerted by the oxygen. Hence initial attack by the nucleophile is encouraged.

(c) *Aryl halides*. Aryl halides are extremely unreactive—the term "aryl" implies that the halogen atom is attached *directly* to the benzene ring, as in chlorobenzene for example. Nucleophilic attack is hard to achieve in the laboratory, and examples of reactions which occur industrially are mentioned in sections 27.10 and 30.8.

29.8 Factors influencing mechanism of substitution

It may be wondered why some compounds react with nucleophiles via a bimolecular transition state (S_N2) whereas others react via a carbocation (carbonium ion), which represents a unimolecular transition state. To take extreme cases, 1-chlorohexane reacts mainly S_N2 whereas 2-iodo-2-methylpropane, $(CH_3)_3CI$, reacts mainly S_N1. Indeed this is not the end of the story: under other conditions substitution may fail to occur and elimination may occur instead. The factors which decide between S_N1 and S_N2 are as follows:

(a) **Steric factor**. Bulky groups attached to the halogen-carrying carbon atom will tend to favour ionization by S_N1 and inhibit S_N2. S_N2 is inhibited in the attack of $(CH_3)_3CI$ by OH^- because the bulky methyl groups block the initial approach of the nucleophile. S_N1 is favoured by the fact that the resultant ion $(CH_3)C^+$ is planar, enabling the bulky methyl groups to be further apart.

(b) **Bond energy factor**. Iodides will favour S_N1 more strongly than chlorides. This is because the initial ionization step has a relatively low energy of activation, on account of the weak C—I bond. Figure 29.1 depicts the energy changes which are thought to occur in the S_N1 attack of 2-iodo-2-methylpropane by water.

29.9 Elimination reactions

So far in the "organic" chapters of this book we have considered mainly addition reactions and substitution reactions. The converse of addition is **elimination**, and elimination reactions play an important part in the chemistry of some organic halides. Whether a substitution reaction or an elimination reaction occurs depends largely on the conditions chosen. This may be illustrated with respect to 1-bromopropane.

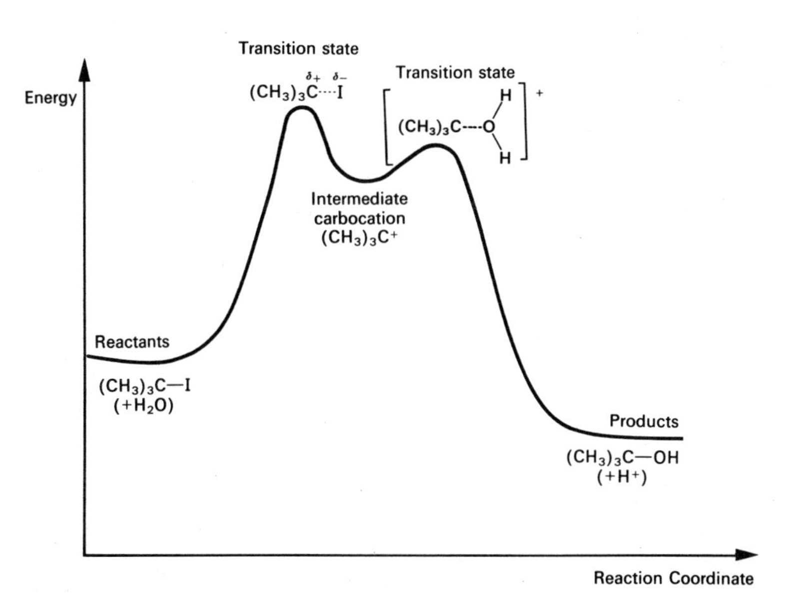

FIG. 29.1. The variation in energy during the S_N1 hydrolysis of 2-iodo-2-methylpropane (see also section 17.11).

(a) *Substitution*. This is effected by warming the compound with dilute aqueous alkali:

$$CH_3.CH_2.CH_2Br+OH^- \rightarrow CH_3.CH_2.CH_2OH+Br^-$$

(b) *Elimination*. This is effected by heating the compound more strongly, with a concentrated ethanolic solution of potassium hydroxide. The overall reaction is the elimination of hydrogen bromide (which, under the alkaline conditions, forms potassium bromide and water).

$$CH_3.CH_2.CH_2Br \rightarrow CH_3.CH{:}CH_2+HBr$$

Where two halogen atoms are present, elimination of two moles of hydrogen halide leads to an alkyne:

$$CH_3.CHBr.CH_2Br \rightarrow CH_3.C{:}CH+2HBr$$

To promote elimination, in preference to substitution, it is necessary to choose a solvent which does not readily stabilise the carbonium ion (i.e. a non-polar solvent). The carbonium ion therefore decomposes before it has a chance to react with the nucleophile. Elimination occurs more readily with higher homologues and is thus a good preparative method for higher alkenes, though not effectively for ethene itself.

The reaction mechanisms described represent extremes of behaviour, and under normal conditions will proceed simultaneously. By careful choice of conditions one of them can be made to predominate, but remember that yields of organic reactions are rarely 100 per cent because of side reactions.

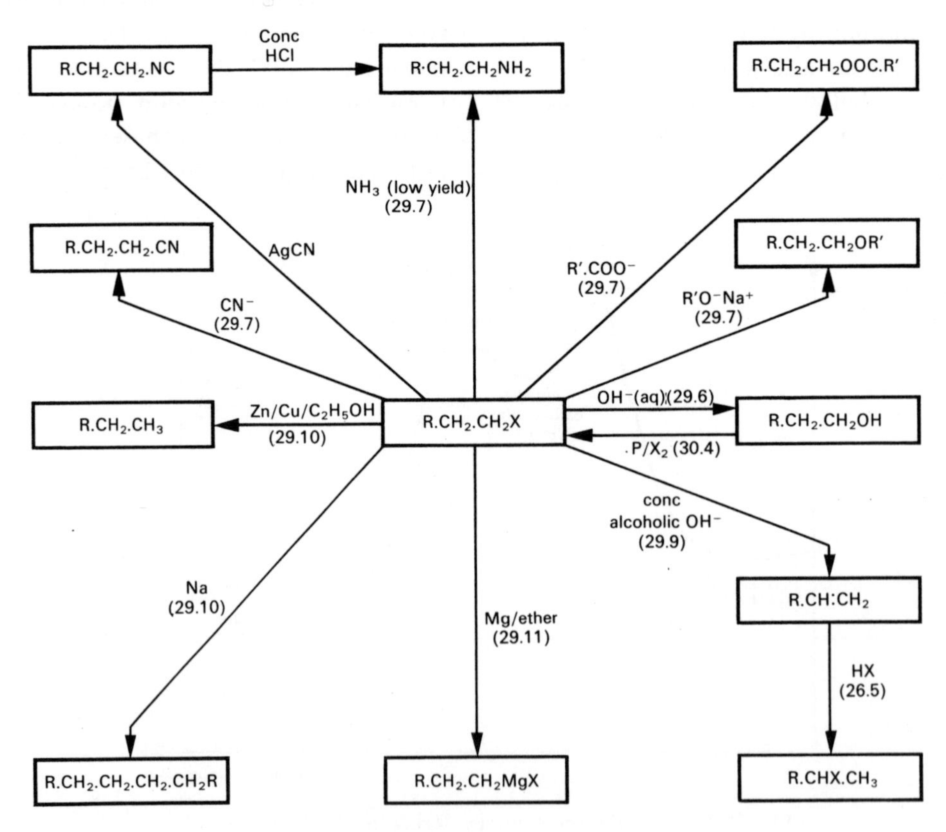

FIG. 29.2. Reaction of alkyl halides.

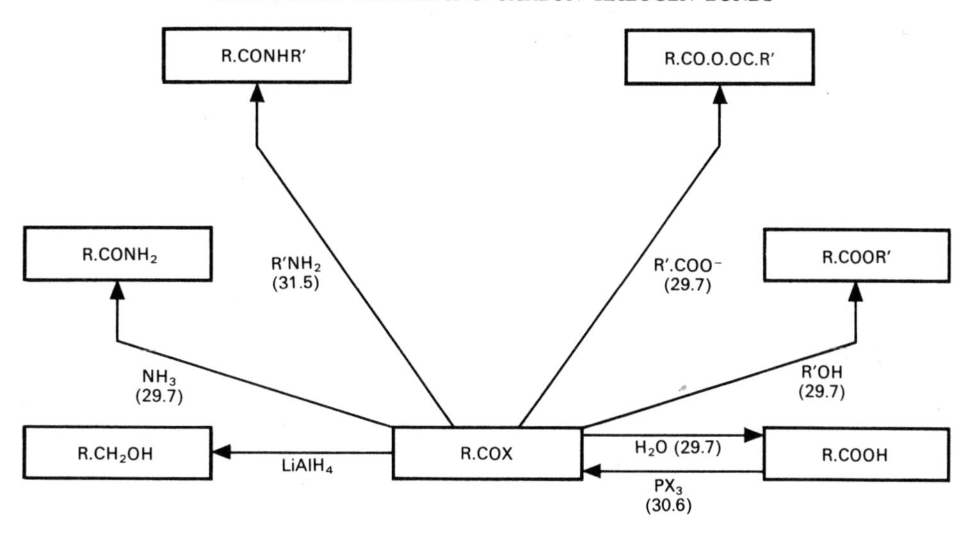

FIG. 29.3. Reactions of acyl halides.

29.10 Reduction of organic halides

Alkyl halides are reduced to the corresponding alkane by a zinc–copper couple in ethanol. For instance bromohexane would be reduced to hexane. In general:

$$RX \rightarrow RH$$

Acyl halides are reduced by lithium tetrahydridoaluminate(III), $LiAlH_4$, to the corresponding alcohol:

$$R.COCl \rightarrow R.CH_2OH$$

Alkyl halides react with sodium, to give alkanes by a condensation reaction. The reaction is carried out in dry ether (ethoxyethane) and the sodium halide is formed as a precipitate. For instance:

$$2CH_3.CH_2I + 2Na \rightarrow CH_3.CH_2.CH_2.CH_3 + 2NaI$$

The reaction is called the **Wurtz reaction**.

29.11 Grignard reagents

If an alkyl or aryl iodide (or bromide) is dissolved in anhydrous ether and refluxed with magnesium turnings, a compound known as a **Grignard reagent** is formed. Grignard reagents are unstable and cannot be isolated, on account of their highly polar, and consequently highly reactive, C—Mg and Mg—Br bonds:

$$R—Br + Mg \rightarrow R—Mg—Br$$

The reagent is kept in ether solution, and used for a wide variety of synthetic routes (see Chapter 28, Study Question 8).

Grignard compounds are members of the class of **organometallic compounds**, which may be defined as compounds which have a direct, covalent link between a metal atom and carbon. Other examples of organometallic compounds are lead(IV) tetraethyl, used as "anti-knock" additive in motor fuel, and aluminium triethyl, used as a catalyst in the Ziegler process for the polymerization of ethene.

29.12 Fluorocarbons

The carbon-fluorine bond is much more stable than the other carbon-halogen bonds so that fluoroalkanes are chemically extremely stable and do not react with oxidizing or reducing agents, alkalis, or acids. Compounds in which all the hydrogen atoms of an alkane are replaced by fluorine are known as **fluorocarbons**. They have physical properties similar to the parent alkane: up to $n = 4$ all the compounds C_nF_{2n+2} are gases.

Fluorocarbons are used industrially for specialized needs where extreme chemical inertness is required. They find uses as oils, coolants and sealing liquids. Tetrafluoroethene, $CF_2{:}CF_2$, is the fluorine analogue of ethene, $CH_2{:}CH_2$. It is polymerized to give a plastic, polytetrafluoroethene, or PTFE. PTFE is far more expensive than polythene but it has remarkable properties. It is resistant to almost all known chemicals, and is marketed as *Teflon* and *Fluon*. It has an exceptionally low coefficient of friction and is used for special bearings. Domestically it is used in non-stick coatings for kitchen utensils.

Mixed fluorochloro derivatives of alkanes are used as refrigerant gases, marketed under the name *Freon*, and as aerosol propellants. Examples are CF_2Cl_2 and $CFCl_3$.

Study Questions

1. Use sections 29.1 and 29.6 to arrange the compounds in each of the following sets in order of increasing reactivity towards nucleophiles:

(a) C_6H_5Br, $C_6H_5.COBr$, $C_6H_5.CH_2Br$;
(b) C_2H_5Br, C_2H_5Cl, C_2H_5F, C_2H_5I;
(c) 1-iodopentane, 2-iodopentane, 2-methyl-2-iodobutane;
(d) C_6H_5F, $CH_3.COCl$, $CH_3.CH_2.CH_2Br$, $CH_3.CHI.CH_3$.

2. (a) Give the formula and name of the main product formed when $C_6H_5.CH_2Cl$ reacts with NaOH(aq).

(b) The major product formed when $C_6H_5.CHCl_2$ reacts with NaOH(aq) is $C_6H_5.CHO$. Account for this.

(c) Predict the major product formed when $C_6H_5.CCl_3$ reacts with NaOH(aq).

3. Give equations and reaction conditions to show how the following conversions could be achieved:

(a) $C_2H_5.COCl$ to $C_2H_5.CONH_2$;
(b) $C_2H_5.COCl$ to $C_2H_5.COOCH_3$;
(c) C_2H_5Br to C_2H_5OH;
(d) C_2H_5Br to $H.CO.OC_2H_5$;
(e) C_2H_5Br to $C_2H_5.COOH$ (two steps).

4. The reaction of ethanolic potassium hydroxide with 1-bromopropane, to form propene, is a low-yield reaction.

(a) Suggest what other organic products might be formed, instead, to reduce the yield.

(b) For each product in (a), suggest a mechanism by which it might form.

5. 0·925 g of **A**, an organic compound containing chlorine, reacted vigorously with water to give a solution which reacted with exactly 50 cm^3 of 0·2 M $AgNO_3$(aq). The solution also contained **B**, an organic compound with no chlorine. **A** also reacted with concentrated ammonia to form a solid **C** which contained 19·2% of nitrogen by mass. Suggest structural formulae for **A**, **B** and **C**.

6. A compound **D** of molecular formula $C_2H_2Cl_2O$, reacted with ethanol to give **E**, $C_4H_7ClO_2$. Boiling **E** with NaOH(aq) gave ethanol and a solution of **F**, $C_2H_3O_3Na$.

(a) Write down the structural formulae of **D**, **E** and **F**.

(b) Write balanced equations for the reactions.

CHAPTER 30

Compounds containing carbon-oxygen single bonds

30.1 Classification and nomenclature

Oxygen forms two covalent bonds per atom. Compounds containing the double bond C=O were dealt with in Chapter 28; this chapter deals with the single bond between carbon and oxygen. The compounds are classified and named as in Table 30.1.

This first part of this chapter deals with category (a) in Table 6.1, i.e. hydroxy-compounds. The remainder of the chapter deals with category (b).

It should be noted that the term "ester" refers strictly to *any* compound in which the replaceable hydrogen atom of an acid is replaced by any hydrocarbon group R; in this chapter, the special case, of esters of carboxylic acids, will be considered.

TABLE 30.1

CLASSIFICATION AND NOMENCLATURE OF COMPOUNDS CONTAINING C—O SINGLE BONDS

Class	Functional group	IUPAC prefix or suffix	Example
(a) *Compounds containing OH-groups*			
Alcohols (30·4)	—OH	**-ol** (hydroxy-)	propan-1-ol $CH_3.CH_2.CH_2OH$
Phenols (30·5)	—OH	**-ol** (hydroxy-)	1-hydroxy-2-methylbenzene $HO.C_6H_4.CH_3$
Carboxylic acids (30·6)	—C(=O)OH	**-oic acid**	butanoic acid $CH_3.CH_2.CH_2.COOH$
(b) *Compounds containing C—O—C links* (section 30.7)			
Ethers	(R)—O—(R′)	**alkoxy-**	ethoxyethane $CH_3.CH_2.O.CH_2.CH_3$
Esters	(R)—C(=O)O(R′)	(alkyl) **-oate**	methyl ethanoate $CH_3.COOCH_3$
Acid anhydrides	(R)—C(=O)—O—C(=O)—(R′)	**-oic anhydride**	ethanoic anhydride $CH_3.CO.O.CO.CH_3$

30.2 Physical properties of hydroxy-compounds

The hydroxy-group, —OH, is a group capable of forming hydrogen bonds (section 15.2). Hence all hydroxy-compounds are soluble in water to a reasonable degree, and all have melting and boiling points which, for their molecular weight, are relatively high. The carboxyl group, —COOH, is more polar than the simple hydroxy-group, so carboxylic acids generally are less volatile and more soluble in water than alcohols of comparable molecular weight. A long-chain acid or alcohol, for instance octadecanoic acid, $C_{17}H_{35}COOH$, or cetyl alcohol (hexadecan-1-ol), $C_{16}H_{33}OH$, will be almost insoluble in water, whereas a compound with several hydroxy-groups such as sucrose (section 32.7) or glycerol (section 30.9) will be highly soluble in water and very involatile. Phenol dissolves in water to form two partially-miscible layers.

As would be expected, simple organic hydroxy-compounds are colourless, and have densities less than unity.

30.3 Acidic properties of the hydroxy-group

It is well-known, and implicit in their names, that compounds containing the group —COOH are acidic. Generally they are weakly ionized in aqueous solution, but produce sufficient $H^+(aq)$ to react with indicators, magnesium, calcium carbonate and so on. Phenols contain an OH-group attached *directly* to a benzene ring, and they show very much weaker acidic properties. They barely affect pH paper, and phenol itself is too weak an acid to react with carbonates, though it does form salts with alkalis. Alcohols contain an OH-group attached to an alkane chain; they do not show any acidic properties when dissolved in water and they do not react with $OH^-(aq)$ to form salts. They do however react directly with electropositive metals such as sodium to form salts containing the **alkoxide ion**, RO^-. In general, the acidic strength of organic hydroxy- compounds varies in the order

carboxylic acids > phenols > alcohols.

Table 30.2 gives some values for pK_a of representative compounds (see Chapter 19 for a treatment of ionization of acids).

TABLE 30.2

pK_a (= $-\log_{10}K_a$) OF SOME ORGANIC HYDROXY COMPOUNDS

Compound	pK_a	Compound	pK_a
Methanol, CH_3OH	16	Methanoic acid, $H.COOH$	3·8
Ethanol, $CH_3.CH_2OH$	18	Ethanoic acid, $CH_3.COOH$	4·8
Phenol, C_6H_5OH	10	Propanoic acid, $CH_3.CH_2.COOH$	4·9
		Benzoic acid, $C_6H_5.COOH$	4·2

The main factors determining the acidic strength of an organic hydroxy- compound are the relative stability of the compound itself and of the ion (conjugate base) which it produces upon loss of a proton. The following are the ions formed in three representative cases:

(a) *Carboxylic acid.*

$$CH_3.COOH \rightleftharpoons CH_3.COO^- + H^+$$

ethanoate ion

(b) *Phenol.*

$$C_6H_5OH \rightleftharpoons C_6H_5O^- + H^+$$

phenoxide ion

(c) *Alcohol.*

$$C_2H_5OH \rightleftharpoons C_2H_5O^- + H^+$$

ethoxide ion

(a) *Stabilization of the carboxylate anion*

Among the three typical ions given above the ethanoate ion is the most strongly stabilized, because simple transfer of electron pairs can give rise to an identical electron arrangement and strong resonance can therefore occur:

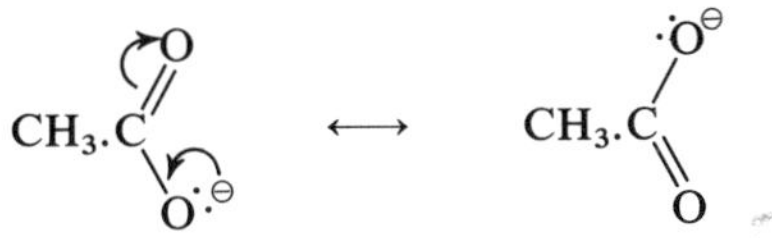

$$CH_3.C(=O)O^{\ominus} \longleftrightarrow CH_3.C(O^{\ominus})=O$$

The actual structure of the ethanoate anion (and of other similar carboxylate ions) can be depicted as follows:

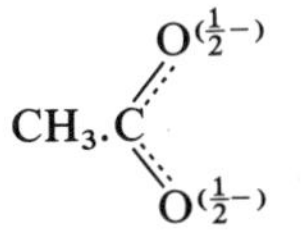

$$CH_3.C\begin{matrix} O^{(\frac{1}{2}-)} \\ O^{(\frac{1}{2}-)} \end{matrix}$$

The high degree of delocalization of the electrons results in some stabilization of the anion. A similar migration of electron pairs may occur in the un-ionized acid, but the structure which results would be strongly dipolar and not equivalent to the first. Consequently it is argued that resonance does not play such an important part in the stabilization of the un-ionized acid:

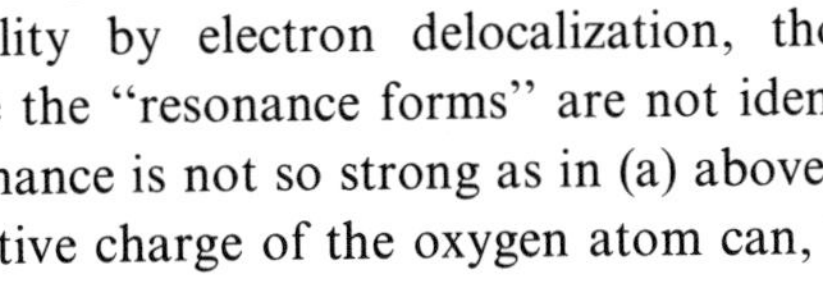

$$CH_3.C(=O)\text{—}O\text{—}H \longleftrightarrow CH_3.C(O^{\ominus})=\overset{\oplus}{O}\text{—}H$$

(b) *Stabilization of the phenoxide ion*

The phenoxide ion can also gain some stability by electron delocalization, though, since the "resonance forms" are not identical, resonance is not so strong as in (a) above. The negative charge of the oxygen atom can, however, distribute itself around the benzene ring in a series of electron-pair movements:

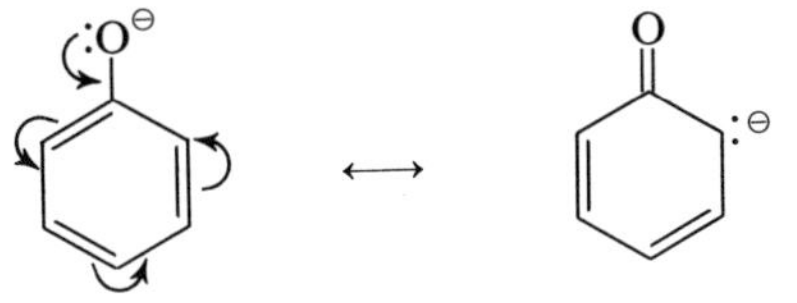

$$C_6H_5O^{\ominus} \longleftrightarrow O{=}C_6H_5{:}^{\ominus} \quad \text{etc.}$$

Hence the resultant structure of the phenoxide ion is approximately:

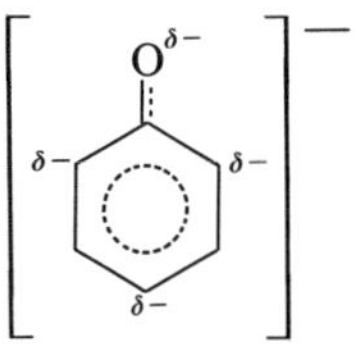

$$\left[C_6H_5O^{\delta-} \;(\delta- \text{ at positions } 2, 4, 6) \right]^{-}$$

Again it is argued that while a similar type of electron delocalization can take place with the un-ionized form of phenol, in this case the effect is not nearly so strong because the resonance forms are dipolar in the same way as with acetic acid:

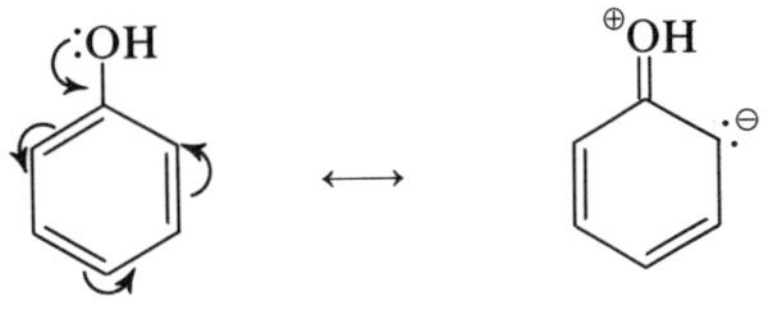

$$C_6H_5OH \longleftrightarrow {}^{\oplus}OH{=}C_6H_5{:}^{\ominus} \quad \text{etc.}$$

Note however that even with phenol itself some slight negative charge does appear at positions 2, 4 and 6 in the benzene ring (cf. chlorobenzene, section 27.8). This factor has an important bearing on the reactions of the aromatic ring in these compounds.

In alcohols there are only saturated carbon atoms; this makes resonance stabilization impossible, both for the alkoxide ion and the un-ionized molecule. This is easily understood if it is remembered that resonance structures must fulfil the normal rules for occupying energy levels, and must not break any bonds.

The acidic strength of a carboxylic acid can be greatly modified by the presence of electron-withdrawing ($-I$) groups or electron-pushing ($+I$) groups. This effect is known as the **inductive effect**, and it was discussed in Chapter 19. Similarly, the acidic strength of phenols may be modified. For instance 2,4,6-trinitrophenol is a sufficiently strong acid to affect indicators. It has a pK_a value of 1·0 and is commonly known as picric acid, an explosive.

30.4 Reactions of alcohols

Alcohols are important in organic chemistry as polar solvents, for use when water is impracticable: methanol, ethanol and propan-2-ol are particularly important industrially. In everyday language the word *alcohol* refers to *ethanol*, the substance produced by fermentation that has been both friend and villain throughout history. Another common compound in nature is glycerol, the systematic name of which is propane-1,2,3-triol, $CH_2OH.CHOH.CH_2OH$. Glycerol is a component of animal and vegetable fats and oils, in which it is present in the form of long-chain carboxylic esters (section 30.7). The main reactions of alcohols are summarized below, and in Fig. 30.1.

(a) *Acidic*. Alcoholic OH-groups react with sodium metal, liberating hydrogen. The alkoxide ion formed is an extremely strong base (stronger than OH^-); hence if water is added, it reacts to re-form the alcohol.

$$ROH + Na \rightarrow RO^-Na^+ + \tfrac{1}{2}H_2$$
$$RO^- + H_2O \rightleftharpoons ROH + OH^-$$

Alcohols will not react with hydroxide ion, the equilibrium lying far to the right in the above equation.

(b) *Oxidation*. Oxidation is in effect the elimination of two moles of hydrogen atoms:

$$-\overset{|}{\underset{|}{C}}-\underset{|}{O} \longrightarrow -\overset{|}{C}=O$$

Alcohols containing the group $-CH_2OH$ are called **primary alcohols** and these oxidize to give the aldehyde, which can undergo further oxidation to the carboxylic acid:

$$R.CH_2OH \longrightarrow R.CHO \longrightarrow R.COOH$$

Alcohols containing the group $>$CHOH attached to two alkyl groups are called **secondary alcohols**, and these oxidize to ketones:

$$\begin{matrix} R \\ & CHOH \\ R' \end{matrix} \xrightarrow{[O]} \begin{matrix} R \\ & C=O \\ R' \end{matrix}$$

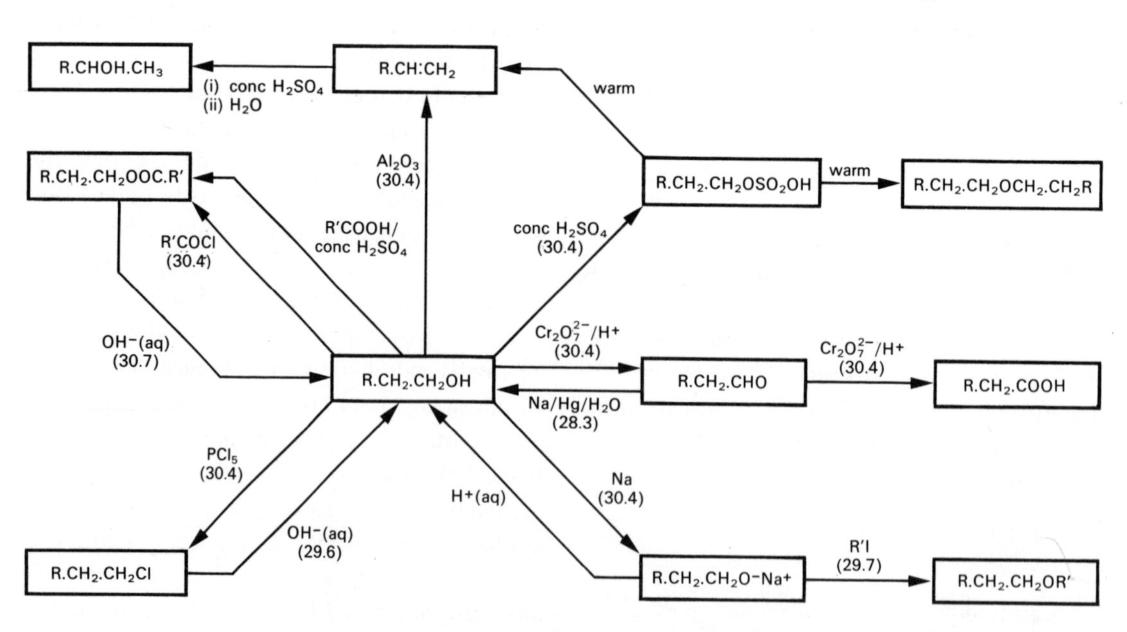

FIG. 30.1. The reactions of primary alcohols.

In the laboratory, a suitable method of oxidation is to distil the alcohol with acidified sodium dichromate(VI). If the carboxylic acid is required, the primary alcohol must be heated under reflux to convert the aldehyde. Industrially, oxidations are often carried out using oxygen in the presence of a catalyst.

So-called **tertiary alcohols**, which contain the group $\geq$COH attached to three alkyl groups, cannot be oxidized as they do not possess a hydrogen atom in the position necessary for the initial elimination reaction to occur. However, very vigorous oxidizing agents—such as potassium manganate(VII)—will oxidize them, by splitting off a carbon chain as CO_2 and leaving a ketone or an acid with fewer carbon atoms than the original alcohol.

(c) *Esterification*. If an alcohol is heated with an acid (organic or inorganic) in the presence of concentrated sulphuric acid, an equilibrium is set up resulting in the formation of an **ester**. The function of the sulphuric acid is partly catalytic and partly the removal of water, to assist the formation of ester by disturbing the equilibrium. For instance, ethanol may be converted to ethyl benzoate:

$$CH_3.CH_2OH + C_6H_5COOH \overset{c.H_2SO_4}{\rightleftharpoons} C_6H_5COOC_2H_5 + H_2O$$

Esters are recognized by their characteristic fruity smells; their reactions are dealt with in section 30.7.

Inorganic acids may also be esterified by alcohols. For instance, sulphuric acid reacts with ethanol exothermically to form the ester ethyl hydrogensulphate:

$$C_2H_5OH + H_2SO_4 \rightarrow C_2H_5.OSO_2OH + H_2O$$

Note that this compound differs from a sulphonate in that it has a C—O—S linkage, not C—S. Strong heat causes alkyl hydrogensulphates to undergo elimination of H_2SO_4 leaving the alkene:

$$C_2H_5.OSO_2OH \longrightarrow C_2H_4 + H_2SO_4.$$

The overall effect, then, of heating an alcohol with concentrated sulphuric acid is dehydration of the alcohol (see Fig. 30.1).

Concentrated aqueous solutions of hydrogen halides may also be esterified, using concentrated sulphuric acid as a catalyst. An alternative catalyst is another Lewis acid, zinc chloride:

$$ROH + HCl \longrightarrow RCl + H_2O$$

(d) *Substitution of the OH-group by halogen* A number of reagents will react directly with OH-groups in alcohols, phenols or carboxylic acids, to replace the OH-group with a halogen atom. The commonest are phosphorus trichloride, PCl_3, phosphorus pentachloride, PCl_5, and sulphur-oxide dichloride (thionyl chloride), $SOCl_2$. From a preparative point of view, the latter has the advantage that the by-products of the reaction are all gaseous.

$$3R.OH + PCl_3 \rightarrow 3R.Cl + H_3PO_3$$
$$R.OH + PCl_5 \rightarrow R.Cl + HCl + POCl_3$$
$$R.OH + SOCl_2 \rightarrow R.Cl + HCl + SO_2$$

In the presence of concentrated sulphuric acid, the hydrogen halide itself can react directly with alcohols. For instance hydrogen bromide, prepared by the reaction of solid potassium-bromide with concentrated sulphuric acid, can convert butan-1-ol to 1-bromobutane:

$$C_4H_9OH + HBr \rightarrow C_4H_9Br + H_2O.$$

The function of the sulphuric acid, partly, is to assist the removal of water. This reaction is in effect the esterification of the hydrogen halide by the alcohol.

An alternative way of replacing —OH by —Br is to use a mixture of red phosphorus and bromine, which behaves in effect as PBr_3 and performs a reaction analogous to that of PCl_3. Similarly, a mixture of phosphorus and iodine can be used to replace —OH by —I.

Note that whereas RBr reacts with CN^- and NH_3, ROH does not.

Iodoform reaction. Alcohols which contain the grouping $CH_3.CHOH.$ give a positive iodoform test (section 28.5).

(e) *Acylation* The general term **acyl** refers to the group R.CO—, that is, an alkyl group plus a carbonyl group. Examples are ethanoyl, $CH_3.CO—$ (commonly known as acetyl), and propanoyl, $CH_3.CH_2.CO—$.

Acylation is the replacement of certain types of hydrogen atom—e.g. that in the hydroxy-group of an alcohol, phenol or carboxylic acid—by the group R.CO—. For instance ethanoyl chloride (acetyl chloride) will **"acetylate"** an alcohol or a phenol to form the ethanoate ester:

$$CH_3OH + CH_3.COCl \rightarrow CH_3.COOCH_3 + HCl$$

$$\text{(OH on benzene ring)} + CH_3.COCl \rightarrow \text{(O.COCH}_3\text{ on benzene ring)} + HCl$$

A similar reaction occurs with benzoyl chloride, $C_6H_5.COCl$, and is referred to as **benzoylation**:

$$C_2H_5OH + \text{(COCl on benzene ring)} \longrightarrow \text{(COOC}_2\text{H}_5\text{ on benzene ring)} + HCl$$

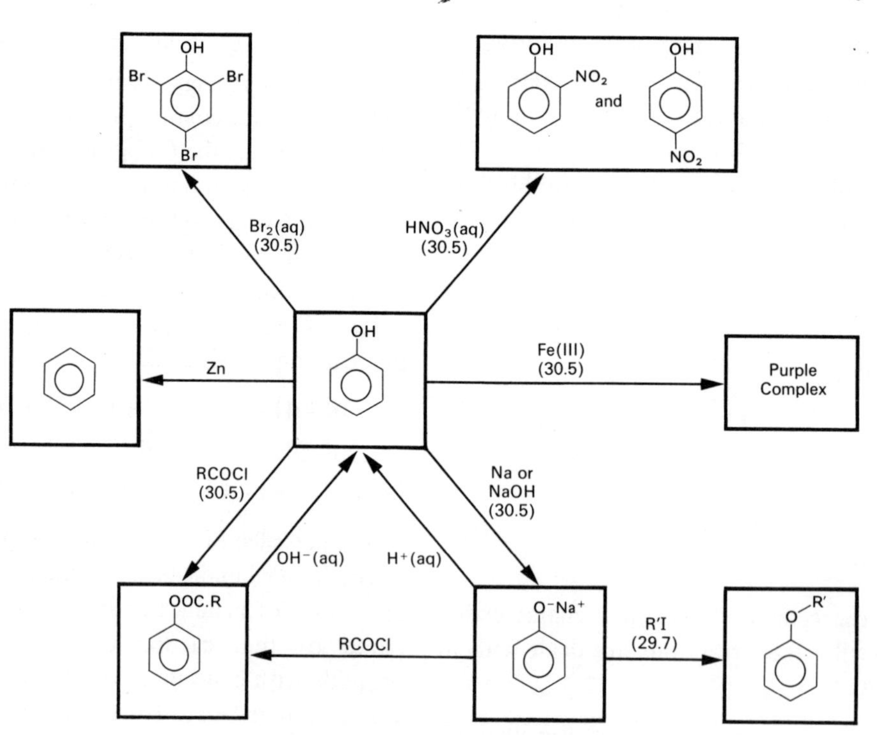

FIG. 30.2. The reactions of phenol.

Acetylation and benzoylation are often important in organic synthesis (see for instance section 31.5).

(f) *Dehydration* Primary and secondary alcohols can be dehydrated by passing the vapour over a heated catalyst of aluminium oxide at about 150°C, e.g.

$$C_2H_5OH(g) \rightarrow C_2H_4(g) + H_2O(g).$$

The same process can be effected by heating an alcohol strongly with concentrated sulphuric acid; the reaction then goes via the alkyl hydrogensulphate ester (section 26.5):

$$C_2H_5OSO_2OH \rightarrow C_2H_4 + H_2SO_4.$$

30.5 Reactions of phenols

Many of the reactions listed above for alcohols occur also for phenols, but important differences occur due to the more acidic nature of phenols.

(a) *Acidic.* Like alcohols, phenols react with sodium and similar metals. Unlike alcohols, they also react with sodium hydroxide to form salts. Phenol itself forms salts called phenoxides. Phenol is, however, too weak an acid to attack carbonates.

(b) *Oxidation.* Phenols cannot oxidize by the elimination of hydrogen, and reagents such as acidified dichromate have no effect. Stronger oxidizing agents produce various polymeric products.

(c) *Esterification.* Direct esterification of a phenol cannot be achieved with an acid in the presence of concentrated sulphuric acid (contrast this behaviour with that of alcohols).

(d) *Acylation.* Phenols *can* be acetylated and benzoylated like alcohols, to form esters.

(e) *Removal of the OH-group.* It is difficult to remove the OH-group from phenols completely. The C—O bond is strengthened by interaction with the π-electrons of the ring, and this makes attack even by strong nucleophiles impossible. Thus whereas PCl_5 will attack alcohols, it has no effect on phenols.

(f) *Reactions of the aromatic ring.* Resonance effects make the ring in phenols extremely sensitive to attack by electrophiles. It was seen in section 30.3 that in phenol itself the main concentrations of negative charge were at the 2, 4 and 6 positions. Hence bromine water produces an immediate precipitate of 2,4,6-tribromophenol:

$$C_6H_5OH + 3Br_2 \longrightarrow C_6H_2Br_3OH + 3HBr$$

(phenol, OH on ring → 2,4,6-tribromophenol: OH; Br, Br; Br)

Dilute nitric acid produces a mixture of 2-nitrophenol and 4-nitrophenol, while concentrated nitric acid produces 2,4,6-trinitrophenol, picric acid. (See also section 30.3).

(g) *Complex formation.* Phenols form complexes with many metal ions. Particularly noteworthy are the coloured complexes formed between phenols and iron(III) ions (Chapter 35).

30.6 Reactions of carboxylic acids

The acidic nature of the group —COOH has already been examined. The group cannot be oxidized. Reaction with an alcohol in the presence of concentrated sulphuric acid results in the formation of an ester. Acylation—e.g. acetylation and benzoylation—can occur, forming an **acid anhydride**:

$$\underset{\text{ethanoic acid}}{CH_3.COOH} + \underset{\text{ethanoyl chloride}}{CH_3.COCl} \rightarrow \underset{\text{ethanoic anhydride}}{CH_3.CO.O.CO.CH_3} + HCl$$

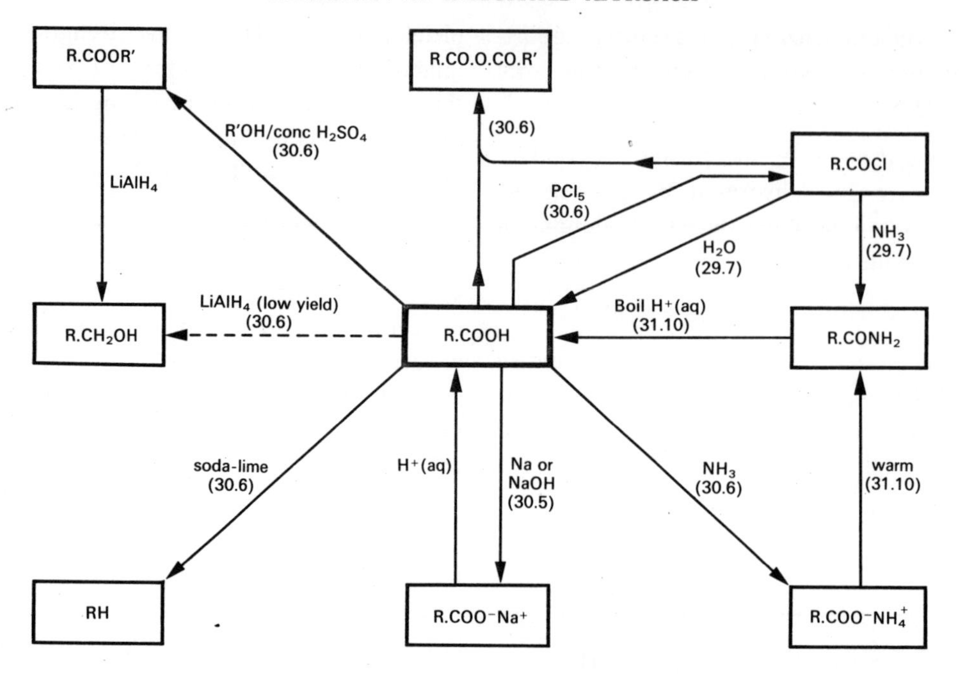

FIG. 30.3. The reactions of carboxylic acids.

Reagents such as PCl_5 remove the OH-group and replace it by Cl, just as with alcohols. The product is an **acid chloride**:

$$CH_3COOH + PCl_5 \rightarrow CH_3.COCl + POCl_3 + HCl$$

Most reducing agents have no effect on the carboxyl group, in contrast to the free, carb*onyl* group of aldehydes and ketones which is readily reduced to the alcohol. Very strong reducing conditions produced by lithium tetrahydridoaluminate(III) will, however, reduce the group $—COOH$ to the primary alcohol $—CH_2OH$.

The proximity of the hydroxyl group to the carbonyl group in $—COOH$ has other effects: the carbonyl group will not undergo addition reactions (e.g. with HCN, NH_2OH or 2,4-dinitrophenylhydrazine, cf. section 28.3).

One important reaction of carboxylic acids is convenient to mention at this stage, namely the removal of the carboxyl group by heating with "soda-lime" (effectively a mixture of sodium and calcium hydroxides, produced by slaking calcium oxide with concentrated sodium hydroxide solution).

$$R.COOH + \underset{\text{soda-lime}}{NaOH} \rightarrow RH + NaHCO_3$$

This reaction is called **decarboxylation**, and it affords a means of reducing the number of carbon atoms in a compound by one.

Long-chain alkyl carboxylic acids are sometimes called fatty acids since their esters with glycerol are present in animal and vegetable fats and oils.

Soap is essentially the mixture of sodium salts of fatty acids produced by hydrolysing an oil or fat with aqueous alkali. Its **detergent** action arises from its ability to lower the surface energy of the interface between water and oil, or dirt, thereby promoting the formation of smaller particles of an emulsion which is held in suspension. A fatty-acid anion has a long-chain **hydrophobic** (non-polar) end and an ionic **hydrophilic** (polar) end, Fig. 30.4.

FIG. 30.4. A detergent molecule.

Dicarboxylic acids are commonly encountered, and these generally have a carboxylic acid group at either end of an alkyl chain, for instance

ethanedioic acid (oxalic acid)
HOOC.COOH

hexane-1,6-dioic acid (adipic acid)
$HOOC.(CH_2)_4.COOH$

benzene-1,4-dicarboxylic acid
(terephthalic acid) $HOOC.C_6H_4.COOH$

Adipic acid is used in the manufacture of *Nylon* (section 32.2), and terephthalic acid in the manufacture of *Terylene*. Oxalic acid is a common reagent in volumetric analysis; it can be used to standardize oxidizing agents such as $KMnO_4$ and cerium(IV) salts. It is oxidized to carbon dioxide:

$$H_2C_2O_4 \rightarrow 2CO_2 + 2H^+ + 2e^-$$

Concentrated sulphuric acid dehydrates oxalic acid, giving a mixture of carbons monoxide and dioxide. This reaction can be used as a test for oxalic acid and oxalates.

30.7 Compounds containing the C–O–C linkage

The three classes of compounds to be considered were listed in Table 30.1. They are ethers, esters, and acid anhydrides. The relationship between them is seen by imagining either alkyl or acyl groups attached to oxygen:

ethers: alkyl-O-alkyl
esters: acyl-O-alkyl
acid anhydrides: acyl-O-acyl

The above classification is used simply for illustration; generally speaking, aryl may be substituted for alkyl, or a group such as benzoyl for acyl, without altering the classification. For instance, phenoxybenzene, $C_6H_5.O.C_6H_5$, is classed as an ether, and phenyl benzoate, $C_6H_5.COOC_6H_5$, as an ester.

The order of reactivity of these compounds, towards nucleophiles, depends on the amount of positive charge carried by the central oxygen atom; this amount is affected by the proximity of a neighbouring carbonyl group. The reactivity order is therefore

acid anhydrides > esters > ethers.

At one end of the scale, acid anhydrides are attacked by water at room temperature; at the other end, ethers are extremely resistant to attack by most reagents. Esters are intermediate in reactivity, but are attacked by $OH^-(aq)$.

(a) *Acid anhydrides.* A common laboratory reagent is acetic anhydride (ethanoic anhydride) which is widely used as an acetylating agent, being less violent for this purpose than acetyl chloride. Acetic anhydride is hydrolysed to the carboxylic acid by water (in effect the water is acetylated). Alcohols are acetylated to esters:

$$CH_3.CO.O.CO.CH_3 + H_2O \rightarrow 2CH_3.COOH$$

$$CH_3.CO.O.CO.CH_3 + 2ROH \rightarrow 2CH_3.COOR + H_2O$$

Another advantage of acetic anhydride over acetyl chloride is that the former does not generate hydrogen chloride fumes.

Another nucleophile which will attack acid anhydrides is ammonia, which reacts to give an **acid amide** (IUPAC suffix **-amide**):

$$\underset{\text{ethanoic anhydride}}{(CH_3.CO)_2O} + NH_3 \rightarrow \underset{\text{ethanamide}}{CH_3.CONH_2} + CH_3.COOH$$

Ethanoic anhydride is used industrially in the manufacture of cellulose acetate and polyvinyl acetate—plastics used extensively by the packaging industry and elsewhere.

(b) *Esters.* The formation of esters has already been discussed (section 30.4(c)). Esters are readily attacked by nucleophiles such as OH^-, but not so rapidly as acid anhydrides:

$$\underset{}{R.COOR'} + OH^- \rightleftharpoons \underset{\text{acid anion}}{R.COO^-} + \underset{\text{alcohol}}{R'OH}$$

Water acts similarly as a nucleophile, the reaction being catalyzed by aqueous hydrogen ion:

$$R.COOR' + H_2O \overset{H^+(aq)}{\rightleftharpoons} R.COOH + R'OH$$

A special case is the formation of soap by the action of alkali on oils or fats. A typical fat is a triester of propane-1,2,3-triol and a long-chain fatty acid ($R = C_{17}H_{35}$ in stearate for instance):

$$\begin{array}{l} CH_2.O.CO.R \\ CH.O.CO.R \\ CH_2.O.CO.R \end{array} + 3OH^- \longrightarrow \underset{\substack{\text{glycerol} \\ \text{(propane-1,2,3-triol)}}}{\begin{array}{l} CH_2OH \\ CHOH \\ CH_2OH \end{array}} + \underset{\text{stearate anion}}{3R.COO^{\ominus}}$$

Ammonia is another nucleophile that attacks esters, though the reaction is slower than the comparable one with acid anhydrides. For instance, concentrated alcoholic ammonia heated with ethyl propanoate gives propanamide:

$$CH_3.CH_2.COOC_2H_5 + NH_3 \rightarrow CH_3.CH_2.CONH_2 + C_2H_5OH$$

Esters are used extensively in artificial food flavourings (naturally occurring esters are largely responsible for fruit flavours), and in the perfumery industry. They also find uses as solvents for medicinal preparations, being less toxic than other alternatives.

(c) *Ethers.* Apart from being highly inflammable, ethers are chemically very inert. Ethoxyethane, $CH_3.CH_2.O.CH_2.CH_3$, commonly known simply as "ether", is a commonly used solvent for fats, oils and resins, and is also used as an anaesthetic. It is moderately polar, and this gives it an ability to dissolve a wide range of materials.

The chemical inertness of ethers is due to the lack of a carbonyl group, or similar electronegative group, adjacent to the oxygen atom. A

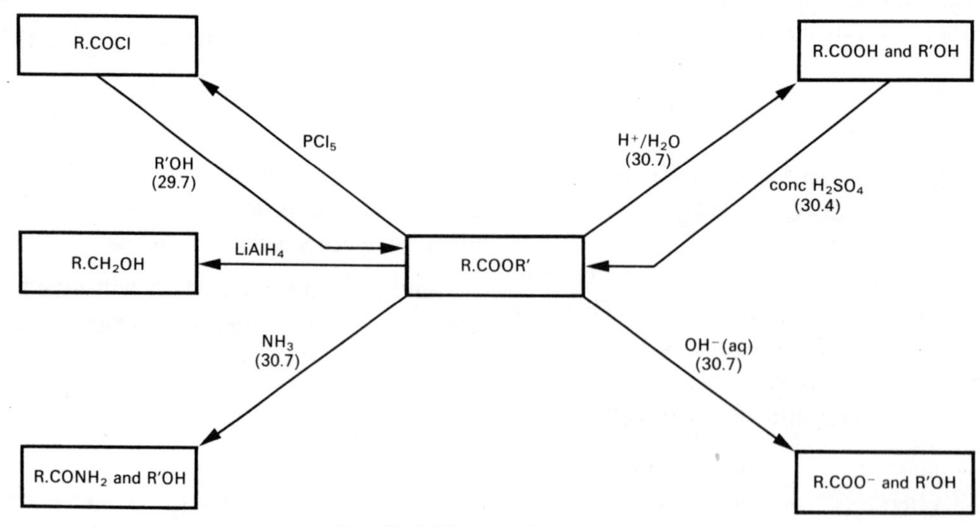

FIG. 30.5. The reactions of esters.

really strong nucleophile such as PCl_5 will attack on heating:

$$R.O.R' + PCl_5 \rightarrow RCl + R'Cl + POCl_3$$

Cyclic ethers are known, and where these are strain-free they resemble their open-chain analogues. A common solvent is tetrahydrofuran:

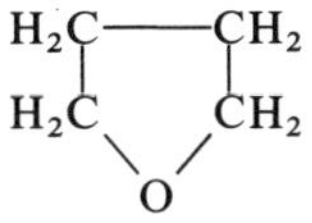

$$\begin{array}{ccc} H_2C & \text{—} & CH_2 \\ | & & | \\ H_2C & & CH_2 \\ & \diagdown O \diagup & \end{array}$$

Where strain exists however, the compound is highly reactive. An example is epoxyethane (ethylene oxide, section 26.9). Although formally a cyclic ether with a three-membered ring, it is highly reactive due to strain, and is for instance readily hydrolysed to ethane-1,2-diol:

$$\underset{\diagdown O \diagup}{CH_2\text{—}CH_2} + H_2O \longrightarrow \underset{OH\quad\ OH}{\overset{|\qquad\ |}{CH_2\text{—}CH_2}}$$

The above reaction is of great industrial importance.

The presence of lone pairs on the oxygen atom of an ether enables them to function as bases. For instance, aluminium chloride can dissolve monomerically in ethoxyethane by the formation of a dative covalent bond:

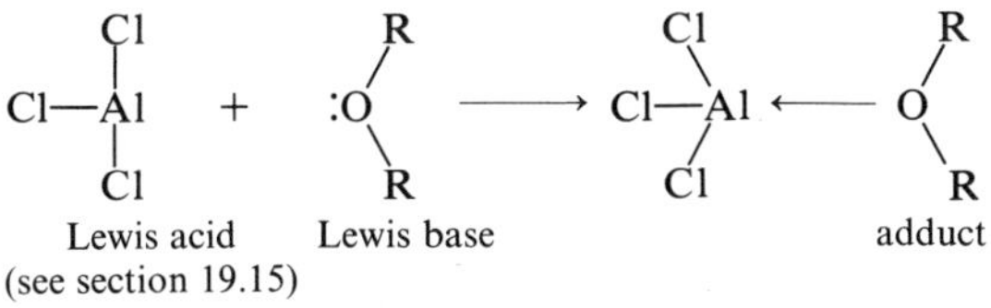

$$\underset{\text{Lewis acid (see section 19.15)}}{Cl\text{—}AlCl_2} + \underset{\text{Lewis base}}{:OR_2} \longrightarrow \underset{\text{adduct}}{Cl\text{—}AlCl_2 \leftarrow OR_2}$$

30.8 Manufacture of alcohols and phenols

From an industrial point of view the most important alcohols are methanol, ethanol, propan-2-ol and ethane-1,2-diol.

Methanol is made by the catalytic reaction

$$CO(g) + 2H_2(g) \xrightarrow{ZnO} CH_3OH(g)$$

The mixture of carbon monoxide and hydrogen is called "synthesis gas" and is nowadays largely obtained from petroleum gas by reacting methane with steam at 900°C with a nickel catalyst:

$$CH_4(g) + H_2O(g) \xrightarrow{Ni} CO(g) + 3H_2(g)$$

A similar gas mixture can be obtained from coke and steam, and is called "water gas", (section 18.3).

Ethanol for industrial use is made no longer by fermentation but, largely, by the direct hydration of ethene with steam and a catalyst of phosphoric acid absorbed on kieselguhr at 300°C:

$$CH_2{:}CH_2(g) + H_2O(g) \rightarrow CH_3.CH_2OH(g)$$

Propan-2-ol is made similarly by the direct hydration of propene. Note that propan-1-ol is not produced (Markownikov's rule, section 26.6).

Ethane-1,2-diol is made mainly by the hydrolysis of "ethylene oxide" (section 26.9).

Phenol. Large quantities of phenol and its homologues are obtained from coal tar, but more has to be manufactured from benzene. The most important route to phenol nowadays is the so-called *Cumene process*. The initial stage is a kind of Friedel–Crafts reaction, to form cumene; the second stage splits off the side-chain by oxidation:

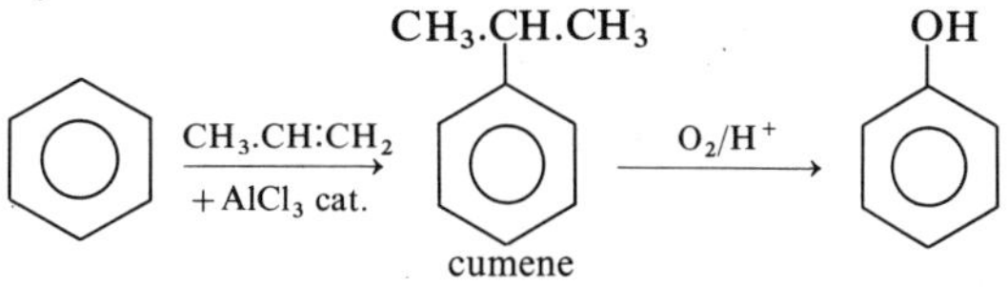

$$C_6H_6 \xrightarrow[+AlCl_3\text{ cat.}]{CH_3.CH{:}CH_2} \underset{\text{cumene}}{C_6H_5\text{—}CH_3.CH.CH_3} \xrightarrow{O_2/H^+} C_6H_5OH$$

An alternative route is via chlorobenzene. The first stage is effectively the direct chlorination of benzene, using hydrogen chloride and oxygen;

the second stage is the nucleophilic removal of the chlorine atom and its replacement by the OH-group under extreme conditions (section 27.10):

$$C_6H_6 \xrightarrow[\text{(Raschig)}]{HCl_2/O_2} C_6H_5Cl \xrightarrow{NaOH/300°C} C_6H_5OH$$

30.9 Uses of alcohols and phenols

Methanol is used to make methanal (formaldehyde) for making thermosetting plastics (see below). It is also used in the manufacture of *Perspex* and as a solvent for varnishes and paints. Its 40% aqueous solution is used in biology as *formalin.*

Ethanol is used largely in the manufacture of ethanal (acetaldehyde) and hence for the manufacture of acetic acid, for cellulose acetate and polyvinyl acetate. It is also used as a solvent.

Propan-2-ol, besides being a useful solvent, is used in the manufacture of propanone (acetone).

Ethane-1,2-diol is widely used as an anti-freeze and de-icing fluid. It is also of use in making **polyester fibres**. Being an alcohol with *two* functional groups, it can be made to react with an acid containing *two* carboxyl groups, benzene-1,4-dicarboxylic acid (terephthalic acid), and hence make an ester which is a condensation co-polymer:

$$n\left[HOOC{-}C_6H_4{-}COOH\right] + n\left[HOCH_2.CH_2OH\right]$$

$$\downarrow$$

$$\left[{-}OC{-}C_6H_4{-}COO.CH_2.CH_2.O{-}\right]_n + (2n-1)H_2O$$

terylene

Propane-1,2,3-triol (glycerol), obtained by hydrolysis of oils and fats, is used as a food flavouring, and in the manufacture of its trinitrate ester, which is the explosive "nitroglycerine".

Study Questions

1. Place the compounds in each of the following sets in order of increasing acidity:

(a) $C_6H_5.CH_2OH$, $C_6H_5.CH_2.COOH$, $CH_3.C_6H_4OH$;

(b) $CH_3.CH_2.COOH$, $(CH_3)_3C.COOH$, $CH_3.CO.COOH$;

(c) $CH_3.CH_2.COOH$, $CH_3.CHCl.COOH$, $CH_2Cl.CH_2.COOH$;

(d) phenol, 2-methylphenol, 2-nitrophenol.

2. Give equations and reaction conditions to show how the following conversions could be achieved:

(a) ethanol to methoxyethane (give two routes, each of two steps);

(b) ethanol to ethanoyl chloride (two or three steps);

(c) phenol to phenyl ethanoate;

(d) propanoic acid to propanamide, $CH_3.CH_2.CONH_2$ (two routes);

(e) ethanol to ethyl ethanoate (two steps, using no other compound containing carbon);

(f) ethanol to propanoic acid (three steps);

(g) phenol to 1,3,5-tribromobenzene.

3. (a) Write structural formulae for (i) ethoxybenzene, (ii) benzoic anhydride, (iii) phenyl ethanoate.

(b) Place these compounds in increasing order of reactivity towards water.

4. Place the compounds in each of the following sets in order of increasing solubility in water:

(a) C_2H_5OH, $C_2H_5OC_6H_5$, C_6H_5OH;
(b) cyclohexane, cyclohexanol, glucose (section 32.5);
(c) $C_3H_7.COOH$, $C_3H_7.COONa$, $C_3H_7.COOC_2H_5$;
(d) butan-1-ol, butan-2-ol, 2-methylpropan-2-ol.

5. Phenol can be acetylated using ethanoyl (acetyl) chloride, but not directly with ethanoic acid.

(a) Suggest why the ester is not readily formed by phenol and ethanoic acid.

(b) Write an equation to show how the ester is formed when ethanoyl chloride is used.

6. 2-nitrophenol can be separated from 4-nitrophenol by steam distillation (section 14.18).

(a) What sort of forces are likely to exist between OH- and NO_2-groups?
(b) Which of the two given nitrophenols will have the stronger intermolecular forces?
(c) Suggest a reason why one nitrophenol is volatile in steam and the other not.

7. (a) Why is sodium chloride added to precipitate a soap after the *saponification* of a fat?
(b) Suggest why a soap solution forms a scum on addition to hard water (hard water contains dissolved Ca^{2+}(aq) ions).
(c) Explain the mechanism by which soap lowers the surface tension of water.

8. What products would you expect when epoxyethane (ethylene oxide) reacts with (a) H_2O, (b) HCl, (c) NH_3, (d) HCN; (e) a Grignard reagent RMgX followed by dilute acid (see Chapter 28, study question 8).

9. Assign the following compounds, which contain C, H and O only, to their correct classes:

(a) a neutral liquid which does not react with sodium, but which forms an alcohol when boiled with NaOH(aq);
(b) a solid which burns with a very smoky flame, melts easily, reacts with sodium giving off hydrogen and decolorizes bromine water;
(c) a neutral liquid which reacts readily with ethanoyl chloride and turns acidified dichromate(VI) solution from orange to green;
(d) a neutral liquid which burns with a clear flame, reacts with sodium giving hydrogen, but is unaffected by acidified dichromate(VI) solution.

10. A compound **A**, C_3H_8O, gave two products when treated with 90% sulphuric acid. One was **B**, $C_6H_{14}O$. The other, **C**, C_3H_6, reacted with dilute sulphuric acid to give some **D**, C_3H_8O, which could be oxidized to **E**, C_3H_6O, but could not be oxidized further without breaking the carbon chain. When **A** was refluxed with acidified dichromate(VI) solution **F**, $C_3H_6O_2$, was obtained. When heated with zinc oxide, **F** gave **G**, $C_6H_{10}O_3$. **A** and **F** together with concentrated sulphuric acid gave **H**, $C_6H_{12}O_2$.

(a) Identify the lettered compounds **A** to **H**.
(b) Which of these compounds would evolve hydrogen upon addition of sodium?
(c) Place **D**, **E** and **F** in increasing order of acidity.
(d) Place **B**, **G** and **H** in increasing order of reactivity towards water.

11. A neutral compound **J**, $C_8H_8O_2$, was hydrolyzed with NaOH(aq). The two products of the reaction were separated, made neutral, and treated with iron(III) chloride. One of the compounds formed a purple complex, the other a deep red colour, when thus treated. What is the structural formula of **J**?

12. Alcohols and ethers can be regarded as compounds in which one or both of the hydrogen atoms in water (Chapter 22) have been replaced by alkyl. Compare and contrast the properties of water, ethanol and ethoxyethane.

13. The hydrolysis of methyl ethanoate with aqueous alkali using equimolar quantities of the reactants follows second order kinetics. In the presence of a large excess of alkali, the kinetics become first order.
(a) What are the products in this reaction? Write a complete equation.
(b) Why does the order change as the concentration of alkali increases?
(c) Will the mechanism of the reaction also alter?

CHAPTER 31

Compounds containing carbon-nitrogen single bonds

31.1 Classification and nomenclature

Nitrogen forms three covalent bonds per atom. Compounds containing the triple bond C≡N were dealt with in Chapter 28, and the double bond C=N was encountered in oximes and similar condensation-products formed by the interaction of nitrogen-containing compounds with the carbonyl group. This chapter deals with the rather diverse range of compounds containing the single bond between carbon and nitrogen. The compounds are classified and named in Table 31.1.

Amines are derivatives of ammonia, in which one or more of the hydrogen atoms have been replaced by a hydrocarbon group. An additional classification to the one given in Table 31.1 is to subdivide them as **alkylamines**, e.g. aminoethane, $C_2H_5NH_2$, and **arylamines**, e.g. aminobenzene, $C_6H_5NH_2$. The relationship between alkylamines and arylamines is analogous to that between alcohols and phenols, and differences in chemical reactions are explained in a similar way.

Amides are related to amines in the same way that carboxylic acids are related to alcohols or phenols: the presence of the group >C=O modifies the reactions of —NH_2 in the same kind of way as it modifies —OH.

Nitro-compounds were introduced in section 27.6, and their reactions are considered in this chapter, though they have only limited relationship with other compounds containing the C—N single bond.

Diazonium salts are unstable compounds which are rarely isolated but are of considerable importance in synthesis, especially that of organic dyestuffs.

31.2 Occurrence and physical properties of amines

The nitrogen atom in amines possesses a lone pair and is capable of forming hydrogen bonds. Hence the melting and boiling points of amines are higher than those of alkanes of similar molecular weight (Chapter 15). Lower homologues are highly soluble in water, solubility decreasing with increasing molecular weight. Phenylamine, commonly known as aniline, is a liquid at room temperature and is sparingly soluble in water.

The smell of alkylamines is characteristic, being ammonia-like but also reminiscent of rotting fish. Indeed amines are frequently formed as products of food spoilage especially of fish and meat. Arylamines have a different

TABLE 31.1

CLASSIFICATION AND NOMENCLATURE OF COMPOUNDS CONTAINING C—N SINGLE BONDS

Class	Functional group	IUPAC prefix or suffix	Example of nomenclature
(a) *Amines*			
Primary amines (salts:	$—NH_2$ $—NH_3^+$)	amino- -amine	$CH_3.CH_2.CH_2NH_2$ aminopropane or propylamine* (propylammonium salts)
Secondary amines	$>NH$	-amine	$(C_2H_5)_2NH$ diethylamine
(salts:	$>NH_2^+$)		
Tertiary amines	$—N<$	-amine	$(CH_3)_3N$ trimethylamine
(salts:	$—NH^+<$)		
Quaternary ammonium salts	$—N^+—$ (four bonds)	-ammonium	$(CH_3)_4N^+ I^-$ tetramethylammonium iodide
(b) *Amides*	$—C(=O)—NH_2$	-amide	$CH_3.CH_2.CH_2.CONH_2$ butanamide
(c) *Nitro-compounds*	$—N(\rightarrow O)=O$	nitro-	$C_6H_5NO_2$ nitrobenzene
(d) *Diazonium salts*	$—N^+\equiv N$	-diazonium	$C_6H_5N\equiv N\ Cl^{\ominus}$ benzenediazonium chloride

* Difunctional primary amines are named, e.g. hexane-1,6-diamine $NH_2(CH_2)_6NH_2$.

type of smell. *Aniline is toxic by skin absorption* and should be handled with care.

The amino-group occurs in amino-acids: these compounds are dealt with in Chapter 32.

31.3 The basic character of the amino-group

The dominant property of the amino group is its *basic* character, due to the presence of the lone pair on the nitrogen atom. Hence, alkylamines are nucleophiles which participate in many reactions in a similar fashion to ammonia itself.

All amines are bases, readily accepting a proton donated from an acid to make the corresponding alkylammonium ion, e.g.

$$\underset{\text{base}}{RNH_2} + \underset{\text{proton}}{H^+} \rightleftharpoons \underset{\text{conjugate acid}}{RNH_3^+}$$

Hence the amines react with acids to form well-characterized crystalline salts, e.g. ethylammonium chloride, $C_2H_5NH_3^+ \ Cl^-$, and phenylammonium hydrogensulphate, $C_6H_5NH_3^+ \ HSO_4^-$.

The basic strength of an amine in aqueous solution may be expressed in terms of the ionization constant, K_a, of its conjugate acid (section 19.12). The stronger the base, the weaker the conjugate acid and the smaller the value of K_a, hence the larger the value of pK_a.

From Table 31.2 it is seen that phenylamine is a somewhat weaker base than ammonia, whereas methylamine is significantly stronger. The increase in strength obtained when substituting alkyl for hydrogen in ammonia is readily explained in terms of the $+I$ inductive effect of the alkyl group. The ability of the nitrogen atom to donate its lone pair to an acid is enhanced by the electron-pushing effect of the alkyl group:

H

$CH_3 \rightarrow N:$ (Base)

$(+I)$ H

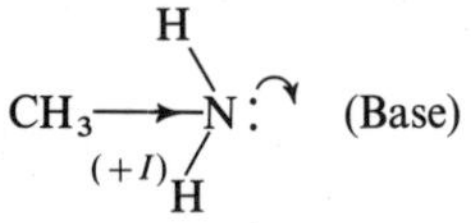

The fact that phenylamine (aniline) is a weaker base than ammonia can be explained in terms of resonance (mesomeric effect). Since the lone pair of nitrogen is, here, less available for donation to an acid, it must be withdrawn into the benzene ring by delocalization. If we draw resonance structures, it is readily seen how this occurs (cf. section 30.3):

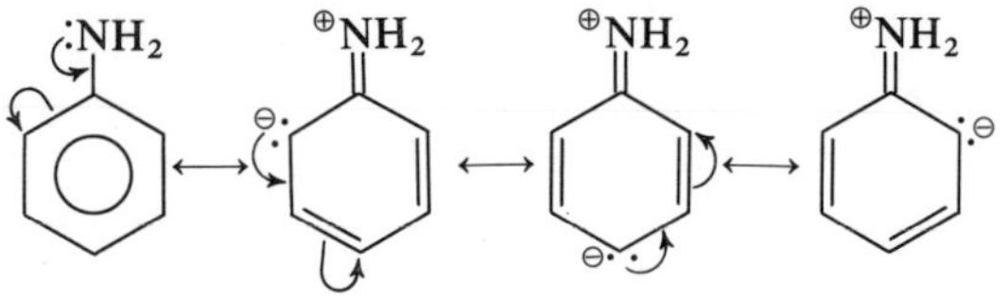

$\ddot{N}H_2 \quad {}^{\oplus}NH_2 \quad {}^{\oplus}NH_2 \quad {}^{\oplus}NH_2$

Electron movements such as those depicted by the above resonance forms will result in an overall charge distribution approximately as follows:

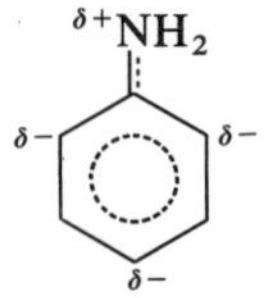

$\delta+NH_2$

$\delta- \quad \delta-$

$\delta-$

From this it follows that phenylamine has relatively little tendency to accept protons. The ion $C_6H_5NH_3^+$ does not have this possibility of resonance stabilization. Phenylamine is sparingly soluble in water, but does dissolve in strong acids, e.g. concentrated hydrochloric acid, to form the ion $C_6H_5NH_3^+$. Addition of alkali causes immediate reprecipitation of the free amine as an emulsified oil which gradually separates from the water, forming two liquid layers.

TABLE 31.2

BASIC STRENGTHS OF SOME AMINES ILLUSTRATED BY VALUES OF IONIZATION CONSTANTS OF THEIR CONJUGATE ACIDS

$$\text{e.g. } K_a = \frac{[H^+(aq)][RNH_2(aq)]}{[RNH_3^+(aq)]}; \ pK_a = -\log_{10} K_a$$

Name of amine	Formula of amine	Name of conjugate acid cation	Formula of conjugate acid cation	pK_a
Phenylamine	$C_6H_5NH_2$	Phenylammonium	$C_6H_5NH_3^+$	4·6
Ammonia	NH_3	Ammonium	NH_4^+	9·2
Methylamine	CH_3NH_2	Methylammonium	$CH_3NH_3^+$	10·6
Dimethylamine	$(CH_3)_2NH$	Dimethylammonium	$(CH_3)_2NH_2^+$	10·8

31.4 Amines as nucleophiles—alkylation reactions

It was stated in Chapter 29 that the action of ammonia on an alkyl iodide produced a complex series of products. The initial reaction is the simple attack of the iodide by the nucleophile NH_3:

$$RI + NH_3 \rightarrow RNH_2 + HI$$

However, RNH_2 is itself a nucleophile like ammonia, and so can react further with the alkyl iodide:

$$RI + RNH_2 \rightarrow \underset{\text{secondary amine}}{R_2NH} + HI$$

and again:

$$RI + R_2NH \rightarrow \underset{\text{tertiary amine}}{R_3N} + HI$$

Since the amines are all bases, all will combine with hydrogen iodide by salt formation, giving rise to $RNH_3^+I^-$, $R_2NH_2^+I^-$ and $R_3NH^+I^-$.

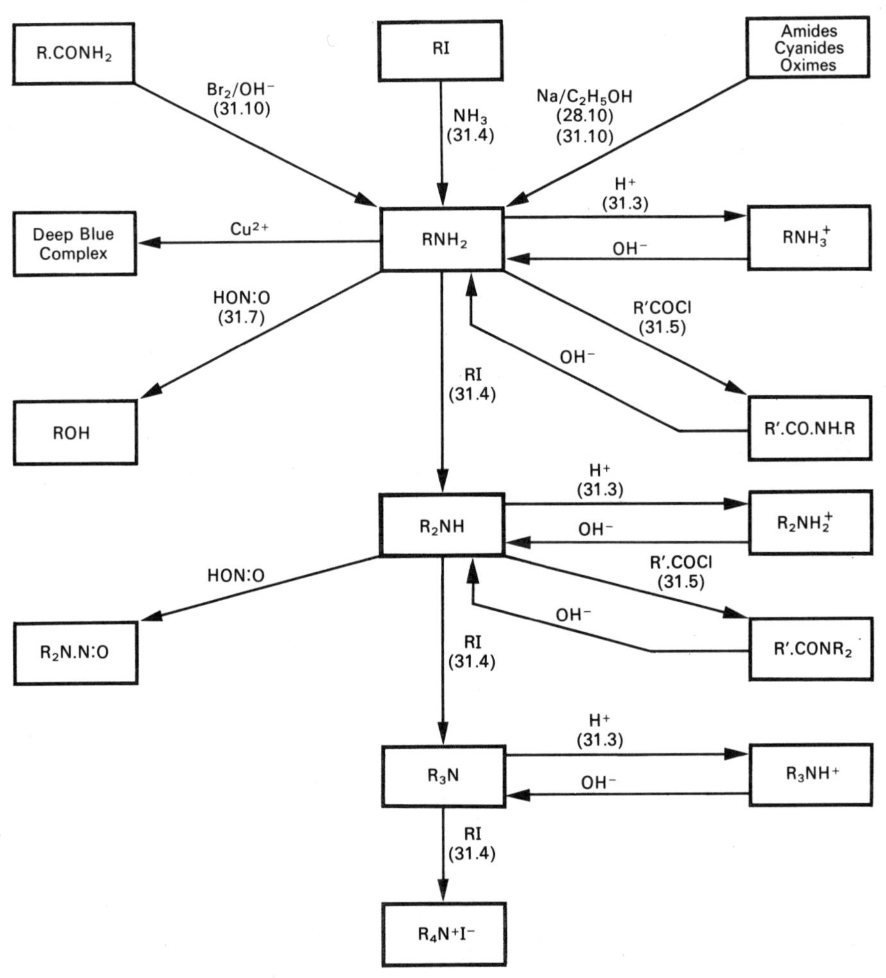

FIG. 31.1. Reactions of amines.

Finally, the tertiary amine itself can attack the alkyl iodide to form a **quaternary ammonium salt**:

$$RI + R_3N \rightarrow R_4N^+I^-$$

In practice therefore, simply heating an alkyl iodide with alcoholic ammonia will yield a mixture of products, and separation may be difficult. Nevertheless, manufacture of secondary and tertiary amines is of technical importance. Mixed derivatives such as

$C_6H_5NH.CH_3$ and $C_6H_5N(CH_3)_2$

are important in the dyestuffs industry (section 31.6(b)).

31.5 Acylation of the amino-group

Primary and secondary amino-groups possess hydrogen atoms which can be replaced by the acyl group. The most common uses of acylation in practice are substitution of $CH_3.CO—$ (acetylation) and $C_6H_5.CO—$ (benzoylation).

Acylation of the groups $—NH_2$ and $>NH$ is very similar to the acylation of the OH-group described in section 30.5, with the important difference that whereas carboxylic acids fail to acylate the OH-group they will, in fact, acylate the more basic amino-groups. In decreasing order of vigour of reaction, acylating agents for the amino-group are:

(i) acyl chlorides, e.g. ethanoyl chloride, $CH_3.COCl$;
(ii) acid anhydrides, e.g. ethanoic anhydride, $(CH_3.CO)_2O$;
(iii) carboxylic acids, e.g. ethanoic acid, $CH_3.COOH$;
(iv) carboxylic esters, e.g. methyl ethanoate, $CH_3.COOCH_3$.

The reactions are exactly parallel to the conversion of the above reagents to amides by ammonia, and the products are substituted amides:

$$C_6H_5NH_2 + CH_3.COCl \rightarrow C_6H_5NH.CO.CH_3 + HCl$$

$$C_6H_5NH_2 + (CH_3.CO)_2O \rightarrow C_6H_5NH.CO.CH_3 + CH_3.COOH$$

$$C_6H_5NH_2 + CH_3.COOH \rightarrow C_6H_5NH.CO.CH_3 + H_2O$$

$$C_6H_5NH_2 + CH_3.COOCH_3 \rightarrow C_6H_5NH.CO.CH_3 + CH_3OH$$

The product may be named either as a substituted amide (*N*-phenylethanamide) or as a substituted amine (*N*-ethanoylphenylamine). The symbol *N* appears in the name to indicate a group attached to a nitrogen atom rather than to a numbered carbon atom in the parent compound.

Acylation reactions, especially acetylation and benzoylation, are important in preparing crystalline derivatives of amines as a means of identification by melting point. They are also important as an aid to synthesis of ring-substituted aromatic amines.

When the acyl group is benzoyl, $C_6H_5.CO—$, reactivity is usually lower, and it is customary to use benzoyl chloride as the benzoylating agent. The reaction is called the **Schotten–Baumann method**, and is very similar to the benzoylation of phenol:

$$C_6H_5NH_2 + C_6H_5COCl \rightarrow C_6H_5NH.CO.C_6H_5 + HCl$$

N-phenylbenzamide
(*N*-benzoylphenylamine)

31.6 Complex formation by the amino-group

In addition to the basic and nucleophilic reactions mentioned above, the electron-donating properties of the NH_2-group make it important in complex formation. Ammonia itself is an important ligand (section 35.10) for forming complex ions with transition metals and B-metals, for instance the well-known complex, tetra-amminocopper(II), $Cu(NH_3)_4^{2+}$. Simple alkyl amines will form complexes in a similar manner, RNH_2 taking the place of NH_3 in the complex ion. Difunctional primary amines can perform as ligands—with two points of attachment (bidentate ligands, Chapter 35)—and form complexes with a ring structure known as **chelates**. For instance, 1,2-diaminoethane (ethane-1,2-diamine) reacts with Cu^{2+}(aq) to form a soluble, deep-blue, complex ion:

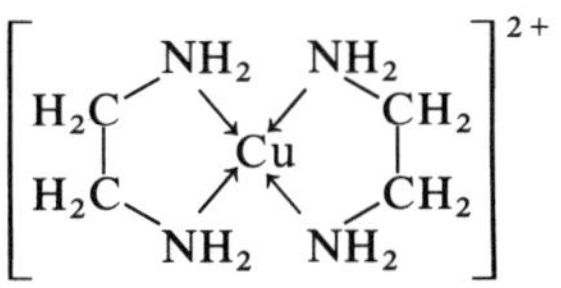

$$\left[\begin{array}{c} H_2C-NH_2 \quad NH_2-CH_2 \\ | \qquad \searrow \swarrow \qquad | \\ \qquad Cu \qquad \\ | \qquad \nearrow \nwarrow \qquad | \\ H_2C-NH_2 \quad NH_2-CH_2 \end{array}\right]^{2+}$$

bis(ethane-1,2-diamine)copper(II) ion

Similarly, the compound

$$H_2NCH_2.CH_2.NH.CH_2.CH_2NH_2$$

can act as a tridentate ligand, with all three nitrogen atoms forming points of attachment with the central metal atom. Polydentate ligands are used in inorganic chemistry on account of their ability to form very stable complexes: for the complex to break up it is necessary for all the points of attachment to become detached simultaneously. One of the most powerful complexing ligands known is the hexadentate ligand known as **edta** which has two tertiary amino-groups and four carboxylic acid groups in its structure:

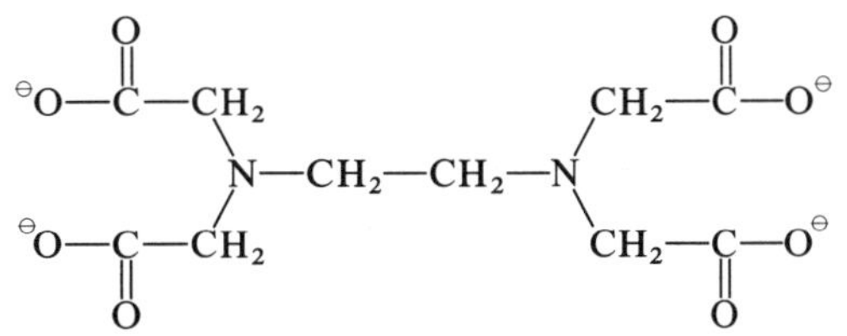

$$(^{\ominus}O-CO-CH_2)_2N-CH_2-CH_2-N(CH_2-CO-O^{\ominus})_2$$

31.7 Diazonium salts

When the salt of a primary aromatic amine is reacted with nitrous acid (made *in situ* by the action of concentrated hydrochloric acid on sodium nitrite) at 0°C, a curious reaction takes place, resulting in the formation of a **diazonium salt**:

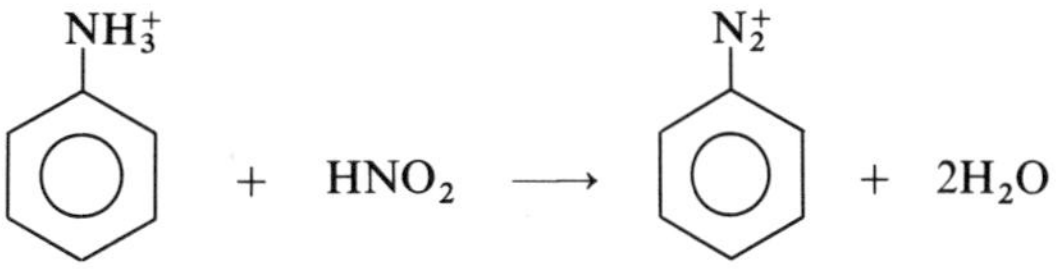

$$C_6H_5NH_3^+ + HNO_2 \longrightarrow C_6H_5N_2^+ + 2H_2O$$

Benzenediazonium ion

Alkylamines fail to perform this reaction directly; instead, nitrogen gas is evolved and an alcohol is formed:

$$C_2H_5NH_3^+ + HNO_2 \rightarrow C_2H_5OH + N_2 + H_3O^+$$

Alkyldiazonium ions do not form in this way. The ion has this structure:

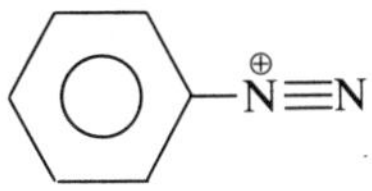

$$C_6H_5-\overset{\oplus}{N}\equiv N$$

The process of converting $ArNH_2$ into $ArN_2^+\ Cl^-$ is described as **diazotization**. Diazonium salts are unstable, explosive compounds when isolated, and are consequently used in aqueous solution. They are important in synthesis in the two types of reaction which follow.

(a) *Reactions evolving nitrogen*, leading to the introduction of other groups on the benzene ring. For instance warming the aqueous solution of the diazonium salt much above 0°C causes it to decompose:

$$C_6H_5-\overset{\oplus}{N}\equiv N + H_2O \longrightarrow C_6H_5-OH + N_2(g) + H^+(aq)$$

The overall result is therefore the same as with an alkylamine and nitrous acid, i.e. the replacement of $-NH_2$ by $-OH$. Other reactions of this type are summarized in Fig. 31.2.

(b) *Coupling reactions*, leading to the formation of dyestuffs. An example of this is the preparation of a dye called Orange II by coupling the anion of 2-naphthol with the ion produced by diazotizing 4-aminobenzene-sulphonic acid:

2-naphthol (anion) $O^{\ominus}$ + $N\equiv\overset{\oplus}{N}-C_6H_4-SO_2O^{\ominus}\,Na^{\oplus}$ ⟶ Orange II ($SO_2O^{\ominus}\,Na^{\oplus}$, $N=N$, OH)

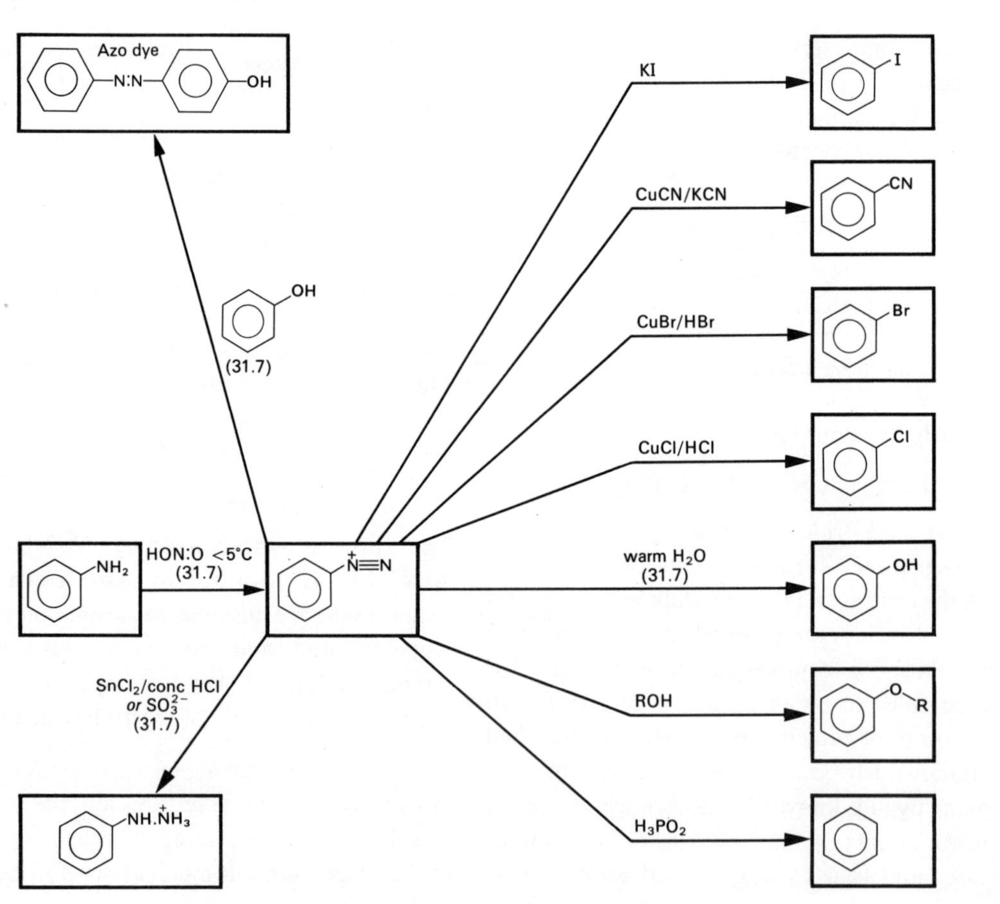

FIG. 31.2. Reactions of diazonium salts.

The coupling reaction is a good instance of electrophilic substitution. The diazonium ion is a weak electrophile which attacks the aromatic nucleus of the phenolic compound in its most strongly activated position. The group —N=N— is known as the **azo** group. It has the ability to absorb light from the visible spectrum: such groups are known as **chromophores**. Methylbenzene (aniline) and *N*-substituted derivatives are the starting point of many commercially important dyes, the so-called aniline dyes.

Reduction of the benzenediazonium ion under mild conditions leads to the conjugate acid of phenylhydrazine:

$$C_6H_5N_2^+\,Cl^- + 2SO_3^{2-} + 2H_2O \rightarrow C_6H_5NH.NH_3^+ + 2SO_4^{2-}$$

31.8 Ring reactions of aromatic amines

Resonance makes the benzene ring in aromatic amines rather susceptible to attack by electrophiles. An additional complication is that the group $—NH_2$ is somewhat susceptible to oxidation, giving a variety of products depending upon the conditions.

It was shown in section 31.3 that the mesomeric effect produces a strong concentration of negative charge at the 2, 4 and 6 positions in phenylamine (cf. the same effect in phenol, section 30.3). Aqueous bromine reacts immediately to give a precipitate of 1-amino-2,4,6-tribromobenzene, by a reaction similar to that of phenol:

$$C_6H_5NH_2 + 3Br_2(aq) \longrightarrow C_6H_2Br_3NH_2 + 3HBr(aq)$$

Direct nitration of aniline cannot be effected because nitric acid is too strong an oxidizing agent. If nitro-derivatives of phenylamine are required the technique of **protection** is used. First the amino-group is acetylated (or benzoylated):

$$C_6H_5NH_2 + (CH_3CO)_2O \longrightarrow C_6H_5NH.CO.CH_3 + CH_3COOH$$

The resultant substituted group does not activate the 2, 4 and 6 positions of the ring so strongly as the free amino-group, moreover, it is more resistant to oxidation. Electrophilic attack by the usual "nitration mixture," concentrated sulphuric and nitric acids, gives a mixture of 2-nitro and 4-nitro derivatives:

$$C_6H_5NH.CO.CH_3 \xrightarrow{HNO_3/H_2SO_4} 2\text{-}NO_2C_6H_4NH.CO.CH_3 + 4\text{-}NO_2C_6H_4NH.CO.CH_3\ \text{(mainly)}$$

It is of interest to note that although statistically one might expect an equal chance of attack at the 2 and 4 positions, resulting in two parts of 2-nitro to one part of 4-nitro derivative (there being two equivalent 2-positions), in fact the 4-isomer predominates. The effect that leads to this is known as a **steric effect**: the bulky side group $CH_3.CO.NH—$ obstructs the adjacent 2-positions but does not hinder attack at the 4-position.

The acyl group is readily removed by warming the compound with alkali, leading to the free

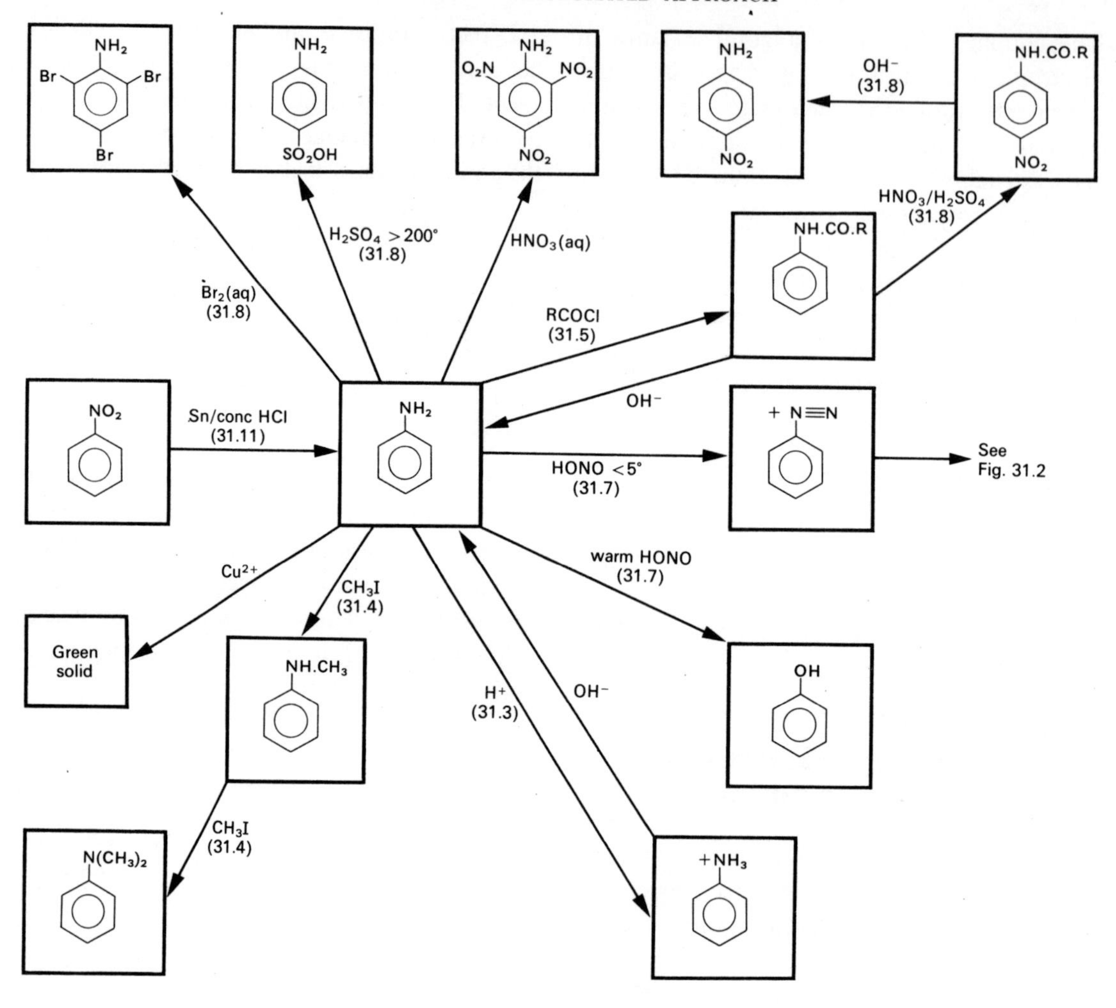

FIG. 31.3. Reactions of aromatic amines.

amino-group once more. Thus the amino-compound was *protected* during nitration by first acylating it.

Phenylamine can be sulphonated to give 4-aminobenzenesulphonic acid. The reaction is essentially the rearrangement of the salt produced by initial attack of the base by the acid. The sulphonic acid produced can form an internal salt—dipolar ion (zwitterion)— with itself:

$$C_6H_5NH_2 \xrightarrow{H_2SO_4} C_6H_5NH_3^{\oplus}HSO_4^{\ominus} \xrightarrow{heat} NH_2C_6H_4SO_2OH \rightleftharpoons NH_3^{\oplus}C_6H_4SO_2O^{\ominus}$$

31.9 Amides—physical properties and classification

When a compound contains an amino-group attached directly to a carbonyl group it is called an **amide** (acid amide). Compounds are named with the suffix **-amide** in the IUPAC system, the carbon atom of the functional group being itself included when giving the stem of the name. Amides may be regarded as being derived from carboxylic acids by the replacement of —OH by —NH_2. Table 31.3 illustrates the nomenclature, and gives the melting and boiling point of simple aliphatic amides.

The acid amides are colourless compounds, relatively involatile for their molecular weight, and the lower members of the series are highly soluble in water. These physical properties suggest strong hydrogen bonding. Intermolecular hydrogen bonding can occur in a manner similar to carboxylic acids.

$$\text{R(O=)C–N(H)–H}^{\delta+}\cdots\text{O}^{\delta-}\text{=C(R)–N(H)–H}^{\delta+}\cdots\text{O}^{\delta-}\text{=C(R)–N(H)–H}$$

31.10 Chemical properties of amides

A number of reactions forming amides have already been described, for instance the action of ammonia on acyl chlorides (Fig. 29.3), esters (section 30.7) and acid anhydrides (section 30.7). A common way of preparing amides however is simply to heat the ammonium salt of the parent carboxylic acid with an excess of the acid itself, for some hours. The reaction which proceeds is dehydration:

$$R.COO^-NH_4^+ \rightarrow R.CONH_2 + H_2O$$

(a) *Acid-base reactions.* Amides are much weaker bases than amines. The lone pair of the nitrogen atom is less available for donation because it is partially used in stabilizing the molecule by resonance:

$$\text{R–C(=O)–}\ddot{N}H_2 \longleftrightarrow \text{R–C(–O}^{\ominus}\text{)=}\overset{\oplus}{N}H_2$$

Once a proton has been added to the amide group, this delocalization cannot occur. Conversely, the amide group can react as a weak acid, by proton loss, giving rise to the ion $RCONH^-$ which is weakly stabilized by resonance:

TABLE 31.3
SIMPLE ALIPHATIC AMIDES

Name and formula of acid amide	Name and formula of parent acid	M.p. of amide (°C)	B.p. of amide (°C)
methanamide $H.CONH_2$	methanoic acid $H.COOH$	2	193
ethanamide $CH_3.CONH_2$	ethanoic acid $CH_3.COOH$	82	222
propanamide $CH_3.CH_2.CONH_2$	propanoic acid $CH_3.CH_2.COOH$	79	213

$$R{-}C({-}O^{\ominus}){=}NH \longleftrightarrow R{-}C({=}O){-}NH^{\ominus}$$

Amides are therefore **amphoteric**, that is, they can function both as acids and bases.

(b) *Reactions breaking the C—N bond.* The C—N bond is broken quite readily when an acid amide is heated with either acid or alkali. The reaction with OH^-(aq) affords a practical means of distinguishing between amides and amines: amides give off ammonia.

$$R.CO.NH_2 + OH^- \rightarrow R.COO^- + NH_3$$
$$R.CO.NH_2 + H_3O^+ \rightarrow R.COOH + NH_4^+$$

Nitrous acid produces a reaction exactly parallel to that with amines, evolving nitrogen gas:

$$R.CO.NH_2 + HNO_2 \rightarrow R.COOH + N_2 + H_2O$$

(c) *Reduction with lithium tetrahydridoaluminate.* The group $-CONH_2$ is as resistant to reduction as its counterpart —COOH. Lithium tetrahydridoaluminate will however reduce it smoothly to a primary amine:

$$R.CO.NH_2 + 4[H] \rightarrow R.CH_2NH_2 + H_2O$$

(d) *Molecular rearrangement—the Hofmann reaction.* Amides react with bromine in a solution of sodium hydroxide to give amines, with the loss of one carbon atom, the overall change being

$$R.CONH_2 \xrightarrow{Br_2/OH^-} R.NH_2$$

This reaction is of interest in being one of the few examples of reaction by **molecular rearrangement** that have been encountered in this book [stage (iv) in the reaction sequence below]. The stages are:

(i) $R.CONH_2 + BrO^- \rightarrow R.CONHBr + OH^-$

(ii) $R.CONHBr + OH^- \rightarrow R.CONBr^- + H_2O$

(iii) $R.C({=}O){-}NBr^- \longrightarrow R.C({=}O){-}N + Br^-$

unstable intermediate

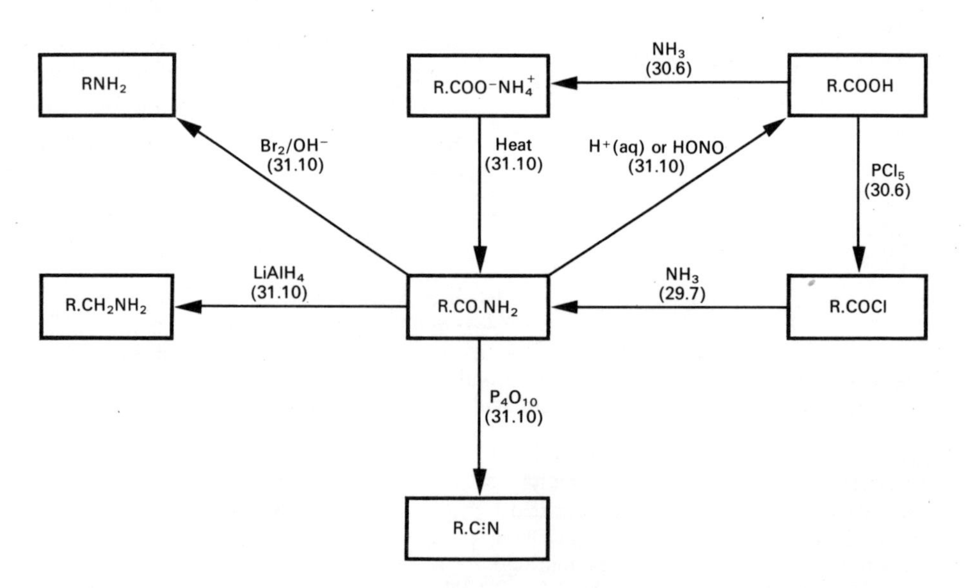

FIG. 31.4. Reactions of amides.

(iv) the intermediate then rearranges thus:

$$R{-}C(=O){-}N \longrightarrow R{-}N{=}C{=}O$$

an alkyl isocyanate

(v) finally the isocyanate hydrolyses in the presence of alkali:

$$R.NCO + 2OH^- \rightarrow R.NH_2 + CO_3^{2-}$$

The Hofmann reaction is of value in synthesis, especially when it is required to shorten a carbon chain by one atom (Fig. 31.6).

(e) *Dehydration.* Phosphorus(V) oxide, P_4O_{10}, is a powerful dehydrating agent, and will convert the group $—CO.NH_2$ to $—CN$. For instance, ethanamide is converted to ethanonitrile:

$$CH_3.CONH_2 - H_2O \rightarrow CH_3.CN$$

Nitriles do not strictly belong to this chapter since they have a multiple bond between the carbon and nitrogen atoms: they were introduced in Chapter 28, and a summary of their reactions was given in Fig. 28.4. A further figure, 31.5, relates nitriles to the compounds mentioned in this chapter, showing the major oxidation and reduction routes in synthetic organic chemistry.

31.11 Nitro-compounds

The prefix **nitro-** denotes the presence of the group $—NO_2$ in a compound. The group is attached to the carbon skeleton via its nitrogen atom, and is stabilized by resonance:

$$C_6H_5{-}N(=O){\rightarrow}O \longleftrightarrow C_6H_5{-}N(\leftarrow O){=}O$$

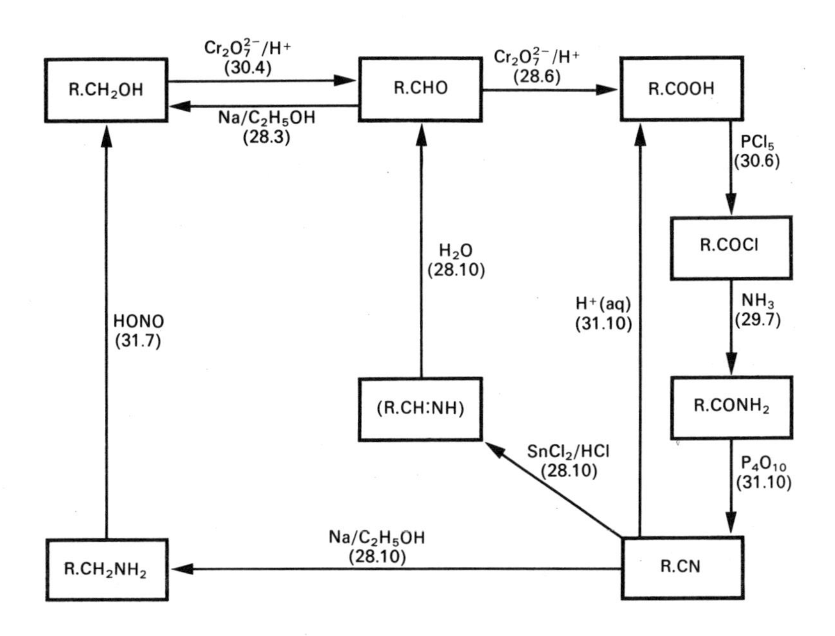

FIG. 31.5. Oxidation and reduction routes.

The nitro-group should not be confused with the isomeric nitr*ite* group —ON=O which occurs in nitrous acid derivatives.

Nitro-compounds are only slightly polar, and are hence fairly volatile and insoluble in water. The nitro-group confers a yellow or red colour on most compounds containing it; hence nitrobenzene is a yellow, oily liquid. Nitration of benzene produces initially nitrobenzene, but more vigorous conditions produce first 1,3-dinitrobenzene and finally 1,3,5-trinitrobenzene. Similar conditions produce the trinitro-derivative of methylbenzene (trinitrotoluene or **TNT**):

CH_3 (methylbenzene) $\xrightarrow[\text{heat}]{HNO_3/H_2SO_4}$ CH_3, O_2N, NO_2, NO_2

1-methyl-2,4,6-trinitrobenzene
(TNT)

Aromatic nitro-compounds are of importance as explosives. Their high oxygen content means that the carbon in the molecule can react to give carbon dioxide by an exothermic reaction without the need for a separate source of oxygen. Detonation is needed to provide the necessary

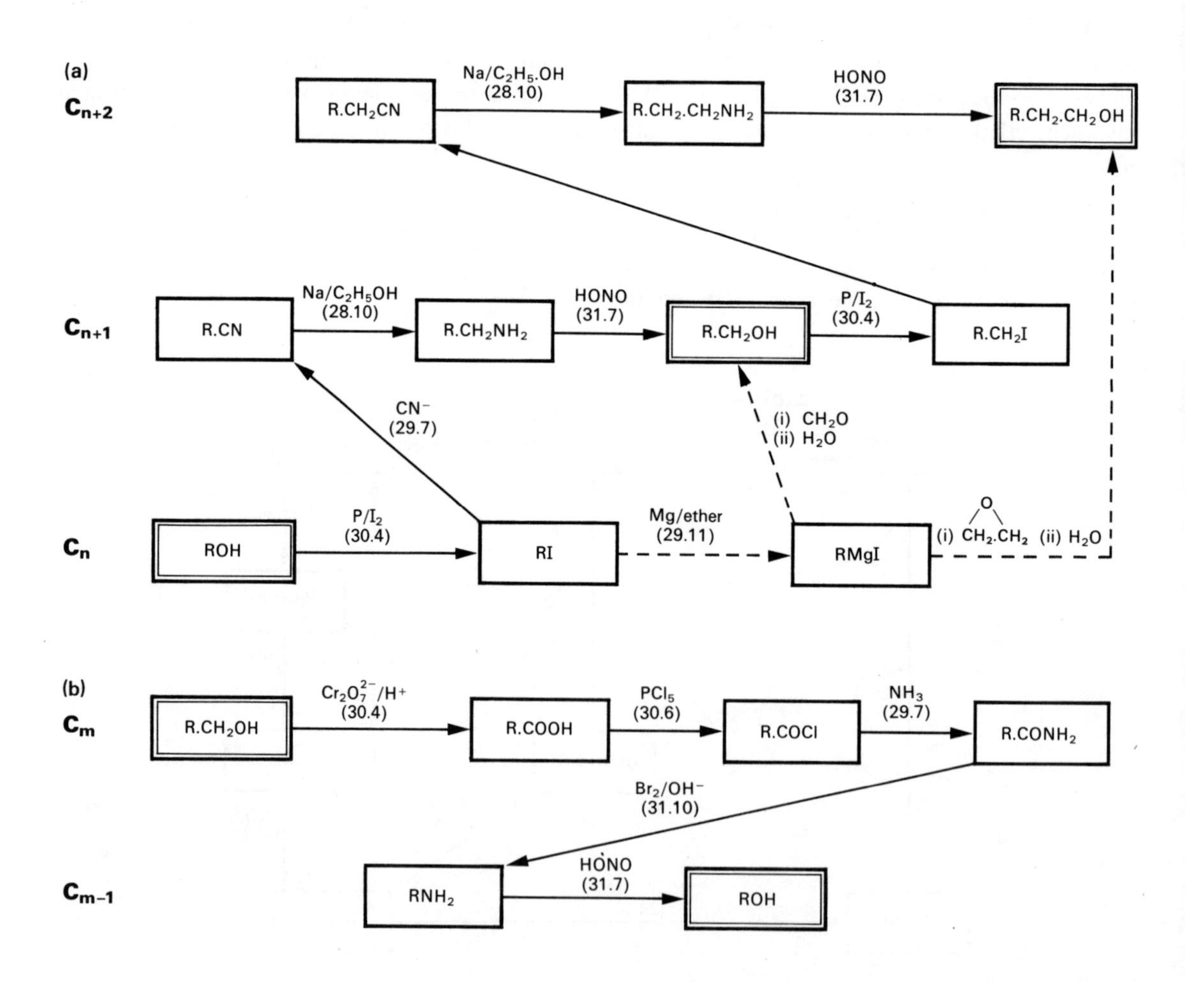

FIG. 31.6. (a) Ascent, (b) descent of a homologous series.

activation energy, and the explosion proceeds by a branching free-radical chain process (cf. section 25.8).

Although aromatic nitro-compounds are more common, nitroalkanes are also encountered. Industrially they are made by the vapour-phase nitration of alkanes with nitric acid, and are used as intermediates in synthesis, as high energy fuels, and as solvents.

Aromatic nitro-compounds are also important as a route to amines, which are made from them by reduction in acidic conditions. Industrially, iron is used as a reducing agent, but in the laboratory a convenient reducing agent is tin in the presence of concentrated hydrochloric acid:

$$R.NO_2+3Sn+7H^+ \rightarrow R.NH_3^+ +3Sn^{2+} +2H_2O$$

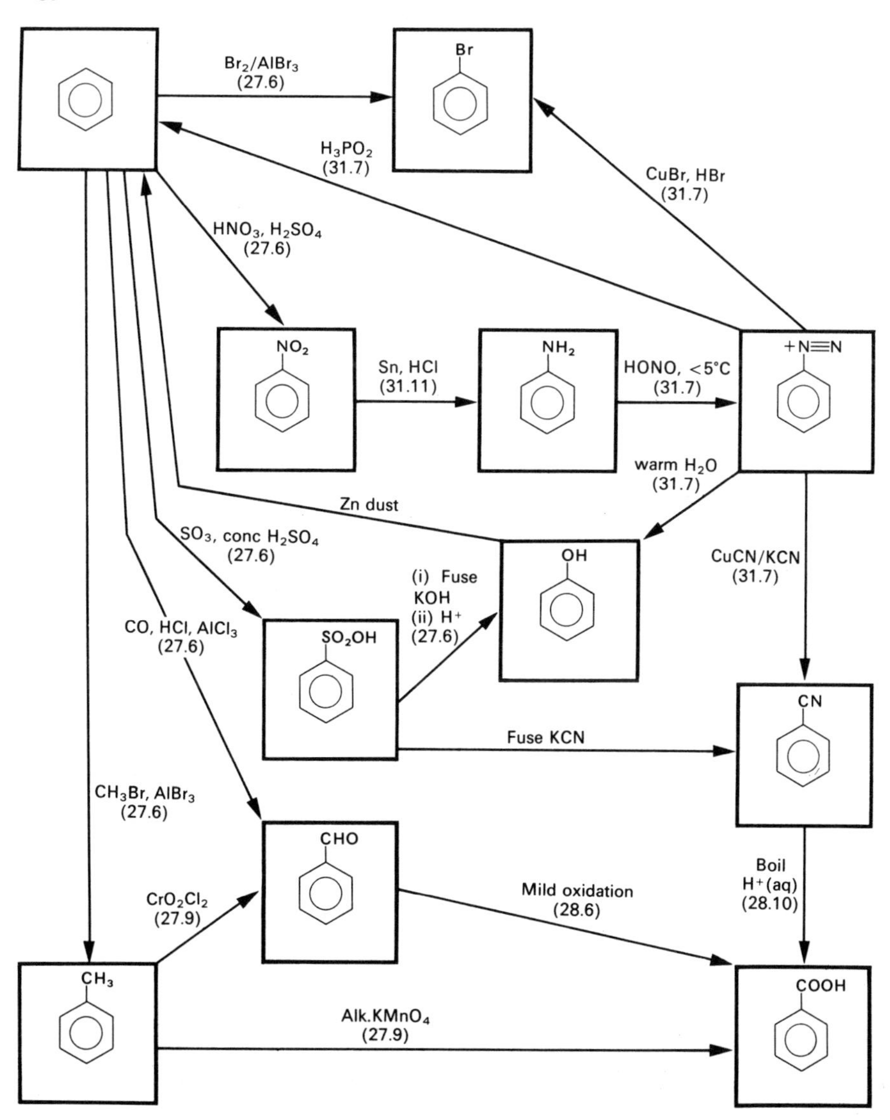

FIG. 31.7. Some major synthetic routes in aromatic chemistry.

Addition of alkali removes a proton from the $R.NH_3^+$ ion, forming the free amine.

Reduction of nitro-compounds in neutral or alkaline conditions gives rise to various condensation products containing groups such as the azo-group, —N=N—, and will not be considered further here.

Figure 31.7 shows the relationship of nitrobenzene to other monosubstituted aromatic compounds. The figure also summarizes some of the major synthetic routes in aromatic chemistry.

Study Questions

1. Use section 31.3 to place the compounds in each of the following sets in order of increasing basic strength:

(a) NH_3, CH_3NH_2, $ClCH_2NH_2$;
(b) $C_6H_5NH_2$, $C_6H_5.CH_2NH_2$, $C_6H_5.CONH_2$;
(c) CH_3NH_2, $(CH_3)_2NH$, $(CH_3)_3N$.

2. (a) Would you expect $C_6H_5N(CH_3)_3^+ OH^-$ to be a strong or a weak base? Give your reasons.

(b) How would you expect the pH of solutions of $C_6H_5NH_3^+ Cl^-$ and $C_6H_5N(CH_3)_3^+ Cl^-$ to differ?

(c) Predict the positions of substitution when $C_6H_5NH_2$ and $C_6H_5N(CH_3)_3^+$ are nitrated. How would the rates of substitution in the two cases compare with the rate of nitration of benzene?

3. Give equations, and reaction conditions, to show how the following conversions could be achieved:

(a) $CH_3.CHO$ to $CH_3.CONH_2$;
(b) $CH_3.CONH_2$ to $CH_3.CHO$;
(c) $CH_3.CH_2OH$ to $CH_3.CH_2.CH_2OH$;
(d) $CH_3.CH_2.CH_2OH$ to $CH_3.CH_2OH$.

4. Give equations to show how the following conversions might be achieved:

(a) benzene to phenol;
(b) phenylamine to benzoic acid;
(c) nitrobenzene to bromobenzene;
(d) nitrobenzene to 1,3-dibromobenzene;
(e) nitrobenzene to 1,3,5-tribromobenzene.

5. When ethylamine, $CH_3.CH_2NH_2$, reacts with nitrous acid, the following products can all be identified. Name them and suggest how each might be formed.

(a) A liquid that reacts with sodium and can be oxidized to ethanoic acid.

(b) A gas that reduces alkaline $MnO_4^-(aq)$ and decolorizes bromine water.

(c) An unreactive liquid with a molecular weight of 74.

6. (a) Urea, $CO(NH_2)_2$, is the amide of carbonic acid, $CO(OH)_2$. How would you expect urea to react with (i) water, (ii) hot alkali, (iii) nitrous acid, (iv) phosphorus(V) oxide; (v) bromine and alkali?

(b) When 1·2 g of urea is heated, 224 cm^3 of ammonia, measured at s.t.p., is evolved. Cryoscopic measurements show that the solid residue has a molecular weight of 103. (i) What is the molecular formula of the solid product? (ii) Suggest a structural formula for the solid product.

7. Assign each of the following compounds to its correct class:

(a) a neutral solid which reacts with nitrous acid to give a carboxylic acid: on warming with bromine and alkali a fishy odour is evolved;

(b) a basic liquid which reacts with ice-cold nitrous acid to give a solution that forms an orange dye with phenol;

(c) an acidic solid which reacts with cold NaOH(aq) giving off an inflammable alkaline gas;

(d) a basic inflammable liquid which does *not* react with acetyl chloride.

8. A white solid **A**, $C_9H_{11}ON$, was reacted with NaOH(aq) to give **B**, C_6H_7N, and **C**, $C_3H_5O_2Na$. On acidification, **C** gave **D**. With phosphorus pentachloride **D** formed **E**, which in turn formed **F**, C_3H_7NO, with concentrated aqueous ammonia. On treatment with bromine and alkali, **F** formed **G**, C_2H_7N.

(a) Write down the structural formulae of the compounds **A** to **G**.

(b) Place **B**, **F** and **G** in order of increasing basic strength.

(c) How could **A** be made from **B**?

9. An ionic solid **H**, $C_4H_{11}NO_2$, was heated with phosphorus(V) oxide to give **I**, C_4H_7N. With sodium and ethanol, **I** gave **J**, $C_4H_{11}N$, which in turn gave **K**, $C_4H_{10}O$, with nitrous acid.

(a) Suggest structural formulae for the compounds **H** to **K**.

(b) How can **K** be reconverted into **H**?

10. A compound **L**, C_3H_5N, was isolated from the reaction between silver cyanide and bromoethane. On reduction with sodium and ethanol, **L** gave **M**, C_3H_9N, which with nitrous acid gave **N**, $C_3H_8N_2O$. **P**, an isomer of **L**, was also isolated from the reaction mixture and it was reduced to **Q**, an isomer of **M**. With nitrous acid **Q** gave **R**, C_3H_8O.

Identify the compounds **L** to **R**.

11. Phenylamine (aniline) can be prepared by reducing nitrobenzene. Nitrobenzene is refluxed with a mixture of tin and hydrochloric acid. The mixture is then made alkaline and steam distilled. The distillate is then shaken in turn with three portions of ether (ethoxyethane) which are separated using a funnel. The combined ether layers are dried with anhydrous sodium sulphate and then distilled (ether boils at 35°C, phenylamine at 183°C).

(a) Write a balanced equation for the reduction.

(b) Why is the mixture made alkaline before the steam distillation?

(c) Why does the phenylamine distil over with the steam?

(d) Why are three separate portions of ether used in turn, when it would have been quicker to have used the same amount of ether all at once?

(e) What does the experimental method given tell you about the solubility of water in ether?

(f) It is claimed that a higher yield of phenylamine is obtained if vacuum distillation is used at the end. What are the advantages of vacuum distillation over distillation at atmospheric pressure?

12. Write an essay comparing and contrasting the properties of ammonia (Chapter 23) with those of a primary alkylamine, RNH_2.

CHAPTER 32

Compounds with more than one functional group

32.1 The interaction of functional groups

When an organic compound contains more than one functional group, the groups can either behave independently or react with one another in such a way as to produce a new type of compound, in which new types of chemical reaction are observed. In cases where the groups in the molecule are far apart, they will usually behave independently, and the behaviour of one group is only slightly modified by the proximity of the other group. For instance, an amino-acid is a compound containing the amino group ($—NH_2$, section 31.1) and the carboxylic acid group ($—COOH$, section 30.1), and possesses all the reactions characteristic of both these groups. In addition however, the amino group can *react* with the carboxylic acid group to give rise to new properties.

This chapter is concerned with **polyfunctional** compounds, such as amino-acids, in which the functional groups modify one another's properties. Many such compounds occur in nature as the building blocks of biological molecules (such as proteins, carbohydrates and nucleic acids) and these will be dealt with in later sections in this chapter. First however, some simple examples of polymerization and cyclization will be considered.

32.2 Condensation polymerization

Polymerization can be of two types:

(a) *addition polymerization*, in which a double bond is opened in the monomer to form the linkages in the polymer (section 26.8);

(b) *condensation polymerization*, in which two different functional groups unite, followed by an elimination reaction. Consider the compound

$$H_2N.(CH_2)_5.COCl.$$

This compound contains the amino group and the acid chloride group, and is named 6-amino-hexanoyl chloride. It can polymerize to form a substance known as *Nylon*-6 which has the formula

$$-\!\!\left[—CO.NH.(CH_2)_5.CO.NH.(CH_2)_5—\right]_n\!\!-.$$

The linkage $—CO.NH—$ is also common in proteins, where it is known as a **peptide link**; it is formed by the elimination of HCl when the acid chloride and amino groups react:

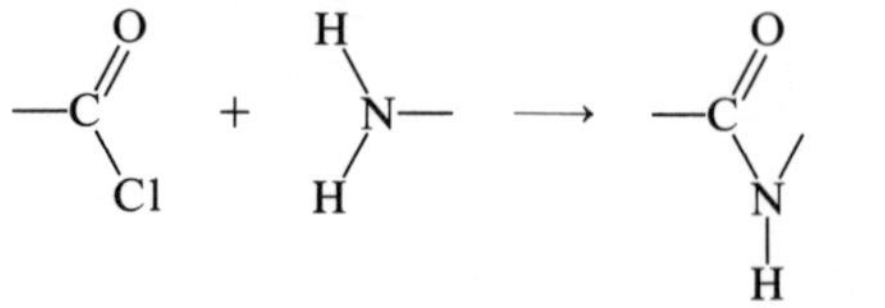

$$—C(=O)—Cl \;+\; H—N(H)— \;\longrightarrow\; —C(=O)—N(H)—$$

peptide link

Polymers formed by condensation reactions are known as **condensation polymers**. Other types of condensation polymer involve *two* different monomers, each with *two* similar functional groups, for example *Nylon* 6,6:

$$H_2N.(CH_2)_6.NH_2 + Cl.CO.(CH_2)_4COCl$$

1,6-diaminohexane | hexane-1,6-dioyl chloride

$$\downarrow$$

$$\left[\!-NH.(CH_2)_6.NH.CO.(CH_2)_4.CO-\right]_n$$

Nylon-6,6

Polymers involving more than one monomer are called **co-polymers**.

Another common condensation co-polymer is *Terylene*. *Terylene* is a *polyester* whereas *Nylon* is a *polyamide*. The monomers are ethane-1,2-diol (ethylene glycol) and benzene-1,4-dicarboxylic acid (terephthalic acid):

$$HO.CH_2.CH_2.OH \quad + \quad HOOC-C_6H_4-COOH$$

ethane-1,2-diol — benzene-1,4-dicarboxylic acid

$$\downarrow$$

$$\left[-CH_2.CH_2.O.CO-C_6H_4-CO.O-\right]_n$$

Terylene

Both *Nylon* and *Terylene* can be moulded or drawn into fibres. They are **thermosetting** plastics, i.e. once formed they cannot be remelted without decomposition. They are chemically inert, unattacked by solvents, and stable to moderate heat. Their properties make them very important commercially, especially in the textile industry.

32.3 Cyclization (ring formation)

When the two groups in a difunctional compound can react with one another, reaction either occurs *between* two molecules (*inter-*molecular reaction, resulting in a chain polymer, e.g. *Nylon*-6) or *within* the molecule (*intra*-molecular reaction), resulting in a ring being formed. For example a molecule of the compound 6-aminohexanoyl chloride, considered in the foregoing section, is capable of undergoing condensation between its two ends. Ring-forming reactions of this type are called **cyclization reactions**.

$$H_2C(CH_2.COCl)-H_2C(CH_2.CH_2.NH_2) \longrightarrow \text{ring }(H_2C, CH_2, C{=}O, NH, CH_2, CH_2, H_2C) + HCl$$

The extent to which cyclization can occur is controlled by various factors:

(a) the strain factor—rings with fewer than five atoms are less readily formed on account of the strained angles between bonds (section 25.6);

(b) the steric factor—cyclization may be prevented by the presence of bulky groups which prevent the skeleton from bending;

(c) the statistical factor—it is difficult to form very large rings, because the probability of the two ends of a very long chain meeting one another is low.

Great interest has been shown in the production of compounds containing large rings, especially by the perfumery industry. Many naturally occurring perfumes are large-ring compounds, e.g. ketones with a 15-membered ring, and such compounds are difficult to synthesize because under normal conditions *inter*molecular reactions tend to occur in preference to *intra*molecular.

The following are examples of cyclization reactions:

(a) *Dehydration of a dicarboxylic acid:*

$$C_6H_4(COOH)_2 \xrightarrow{\text{heat}} C_6H_4(CO)_2O + H_2O$$

benzene-1,2-dicarboxylic acid (phthalic acid) — benzene-1,2-dicarboxylic anhydride (phthalic anhydride)

(b) *Formation of cyclic esters* A compound containing the groups —OH and —COOH can form internal esters. For example, 4-hydroxybutanoic acid loses water when heated, to form a cyclic ester sometimes known as a **lactone**:

$$\underset{\displaystyle CH_2OH}{CH_2}-CH_2-C\begin{matrix}\nearrow O\\ \searrow OH\end{matrix} \longrightarrow \begin{matrix} CH_2-C(=O) \\ | \quad\quad \diagdown \\ | \quad\quad\quad O \\ | \quad\quad \diagup \\ CH_2-CH_2 \end{matrix} + H_2O$$

lactone

32.4 Building blocks for biological molecules

Biological systems, such as living cells, depend for their function on the presence of special polymers, biochemical macromolecules, synthesized by the organism. Now, *Nylon* has already been quoted as an example of a synthetic polymer in which the building blocks are difunctional monomer units; by a similar argument, these biochemical macromolecules can be regarded as condensation polymers: the monomer units all have at least two functional groups. The structures of the resultant polymers are intricate.

Table 32.1 summarizes the types of monomer responsible for building polymers in biological systems. The remainder of this chapter describes some of the properties and chemical reactions of the monomer units themselves (which are often biologically important molecules in their own right) and the next chapter describes some of the properties of the polymers which exist in living systems.

32.5 Monosaccharides

The **monosaccharides** are compounds of general formula $C_x(H_2O)_y$ which form the building blocks, or monomers, from which **polysaccharides** such as starch and cellulose are built up. Compounds of general formula $C_x(H_2O)_y$ are termed **carbohydrates**.

Monosaccharides can be classified according to the number of carbon atoms they contain. For instance, the simplest possible carbohydrate is one in which $x = 3$: this is called a **triose**, glyceraldehyde, $C_3(H_2O)_3$. Figure 32.1 shows the structure of glyceraldehyde. Note that the centre carbon atom is asymmetric because it contains four different attached groups. The diagram shows D-glyceraldehyde: monosaccharides with the L-configuration (section 6.10) are rare in nature.

The commonest monosaccharides are those in which $x = 5$ (**pentoses**) and $x = 6$ (**hexoses**). Table 32.2 summarizes some of the main examples.

TABLE 32.1

A SUMMARY OF SOME BIOLOGICALLY IMPORTANT MOLECULES

Monomer	Functional groups present	Polymer formed from monomer units	Function of polymer
Monosaccharides (section 32.5)	—OH —C=O	Polysaccharides, e.g. starch, cellulose (section 33.1)	Source of energy, or structural
Amino-acids (section 32.9)	$-NH_2$ —COOH	Proteins (section 33.3)	Building tissue, also enzymes
Ribose + phosphate group + organic base	—OH =N—	DNA, RNA	Carrier of genetic information

TABLE 32.2

SOME MONOSACCHARIDES

Class	Number of atoms	Molecular formula	Example	Functional groups present
Triose	3	$C_3H_6O_3$	D-Glyceraldehyde	—OH —CHO
Tetrose	4	$C_4H_8O_4$	(rare in nature)	
Pentose	5	$C_5H_{10}O_5$	Ribose (occurs in RNA, see section 33.4)	—OH (—CHO)*
Hexose	6	$C_6H_{12}O_6$	Glucose	—OH (—CHO)*
			Fructose	—OH $\left(>C=O\right)$*

* The functional group in brackets only occurs in the open-chain form of the compound.

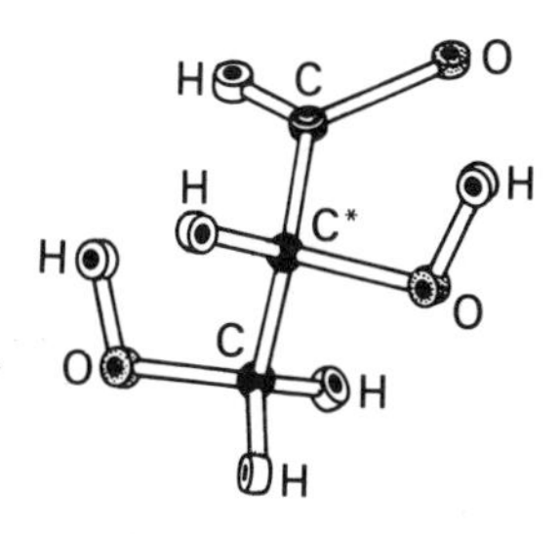

FIG. 32.1. D-glyceraldehyde. [This figure and others in Chapters 32 and 33 represent models made with the "Orbit" molecular building kit, see page xiii.]

Glyceraldehyde, $CH_2OH.\overset{*}{C}HOH.CHO$, is shown in Fig. 32.1 in the D form. The carbon atom marked with an asterisk * is asymmetric. It possesses the characteristic properties of aldehydes as well as those of alcohols. It is important as a biochemical intermediate, but not of great importance as a structural unit, and so will not be considered further here. Tetroses are also of little importance.

Ribose, $CH_2OH.\overset{*}{C}HOH.\overset{*}{C}HOH.\overset{*}{C}HOH.CHO$, is an important pentose, being one of the building units of RNA (ribonucleic acid). The three carbon atoms marked with an asterisk are asymmetric and the groups attached to them must be oriented in a specific way. In the free state, ribose frequently occurs in the straight-chain form indicated, but is able to cyclise into a 5-membered ring (**furanose ring**). Cyclisation occurs because a lone pair of electrons on an HO-group can attack the positive carbon atom in the carbonyl group (sections 28.1 and 28.3) thus forming an ether linkage. Figure 32.2 shows the configuration of the ring structure of D-ribose.

It is helpful to construct molecular models when studying compounds with asymmetric carbon atoms such as monosaccharides: it is difficult to express on paper the exact orientation of every carbon atom, yet a reversal of the configuration around any one carbon atom will result in a molecule, of a different shape, which is a stereoisomer of the original substance.

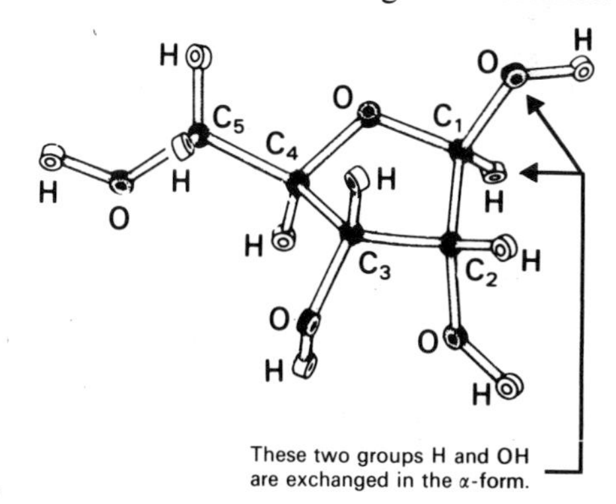

FIG. 32.2. The 5-membered ring (furanose ring) formed by the cyclization of D-ribose. This particular form is called β-D-ribofuranose. An alternative form of furanose ring can be obtained from D-ribose if the cyclization reaction at carbon atom number 1 occurs in such a way as to cause the two arrowed groups to exchange positions (forming the α-isomer).

Ribose contains the functional groups of an aldehyde and of an alcohol, and therefore performs the chemical reactions characteristic of these classes.

Glucose, $CH_2OH.\overset{*}{C}HOH.\overset{*}{C}HOH.\overset{*}{C}HOH.\overset{*}{C}HOH.CHO$, is an important hexose sugar. The four asymmetric carbon atoms are indicated by asterisks. Since each asymmetric carbon atom is capable of existing in two different, mirror-image configurations, there are 2^4 possible isomers of glucose. Most of these are not important biologically.

Like ribose, glucose is capable of undergoing cyclisation, either to a **furanose** (5-membered) ring or, more commonly, to a **pyranose** ring which is 6-membered. Figure 32.3 shows the two possible configurations that can result—they are designated by the prefixes α and β.

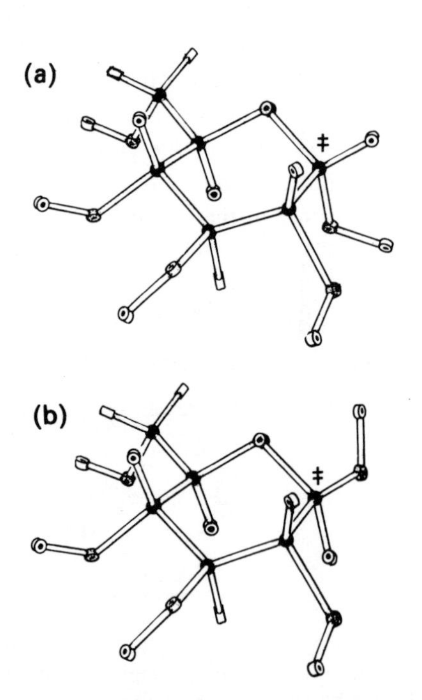

FIG. 32.3. (a) α-D-glucopyranose; (b) β-D-glucopyranose. The carbon atom marked ‡ gives rise to an asymmeric centre when cyclization occurs, hence the two forms α and β are possible.

Although glucose exists commonly in nature in the six-membered, pyranose ring-form, the free compound contains the straight-chain form in solution and hence exhibits the functional group properties of an aldehyde.

Fructose, $CH_2OH.(\overset{*}{C}HOH)_3.CO.CH_2OH$, is another hexose. It contains the ketone group and is therefore called a **ketohexose**, in contrast to the hexoses which contain the aldehyde group which are called **aldohexoses**.

32.6 Occurrence and properties of the monosaccharides

The monosaccharides are all very soluble in water, as would be expected from the presence of a large number of hydroxyl groups (which give rise to hydrogen bonding). They are white, crystalline solids with high melting points; most decompose on stronger heating before the boiling point is reached. They are insoluble in non-polar organic solvents.

Aldoses such as glucose and ribose, contain the aldehyde functional group and consequently possess reducing reactions. For instance:

(a) they reduce Fehling's solution (section 28.6) to copper(I) oxide;

(b) they reduce ammoniacal silver nitrate solution to silver;

(c) strong oxidizing agents oxidize the aldehyde group and the *primary* alcohol group to the carboxylic acid group:

$$\begin{array}{c} CHO \\ | \\ (CHOH)_4 \\ | \\ CH_2OH \end{array} \xrightarrow{3[O]} \begin{array}{c} COOH \\ | \\ (CHOH)_4 \\ | \\ COOH; \end{array}$$

(d) addition of CN^- forms a cyanohydrin;

(e) various other compounds form condensation products analogous to those formed by aldehydes—for instance glucose and phenylhydrazine form a compound called an **osazone**:

$$\begin{array}{l} CHO \\ (CHOH)_4 + 3\ C_6H_5\text{—}NH.NH_2 \longrightarrow \\ CH_2OH \end{array}$$

$$\begin{array}{l} CH{=}N\text{—}NH\text{—}C_6H_5 \\ | \\ C{=}N\text{—}NH\text{—}C_6H_5 + C_6H_5\text{—}NH_2 \\ (CHOH)_3 \\ CH_2OH \qquad\qquad + 2H_2O + NH_3 \end{array}$$

(f) reduction with sodium amalgam and water converts the aldehyde group into an alcohol group. Glucose for instance forms a polyhydric alcohol:

$$\begin{array}{l} CHO \\ (CHOH)_4 \\ CH_2OH \end{array} \xrightarrow{2[H]} \begin{array}{l} CH_2OH \\ (CHOH)_4 \\ CH_2OH \end{array}$$

32.7 Glucose as an energy source

Glucose is the most important monosaccharide, being a source of energy from foodstuffs. Glucose is the monomer from which starch and cellulose are built up. The human body breaks down starch (but not cellulose) into glucose by means of biological catalysts called **enzymes**. Cellulose can be used as an energy source by certain animals, the ruminants. Rumination involves the conversion of cellulose to low molecular-weight acids, carbon dioxide and methane, using micro-organisms.

Glucose is needed by the body as a source of muscular energy; it is broken down by a complex mechanism the overall result of which is oxidation to carbon dioxide and water:

$$C_6H_{12}O_6 + 6O_2 \rightarrow 6CO_2 + 6H_2O; \quad \Delta H = -2820 \text{ kJ mol}^{-1}.$$

The structure of glucose-based polysaccharides is discussed in Chapter 33.

32.8 Disaccharides

Two common carbohydrates in foodstuffs are maltose and sucrose. They are termed **disaccharides** and may be regarded as the product of a condensation reaction between two monosaccharide residues. Maltose, $C_{12}H_{22}O_{11}$, is a disaccharide whose structure consists of two D-glucose residues, joined together (Fig. 32.4) and sucrose (cane sugar, also $C_{12}H_{22}O_{11}$) can be regarded as glucose linked to fructose (Fig. 32.5). Sucrose is one of the main sources of energy for the body—it is first converted into glucose and fructose by the action of digestive enzymes and these are subsequently oxidized by the process mentioned in section 32.7.

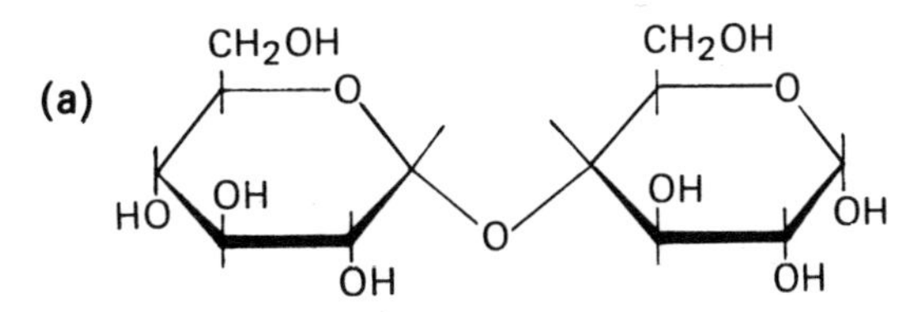

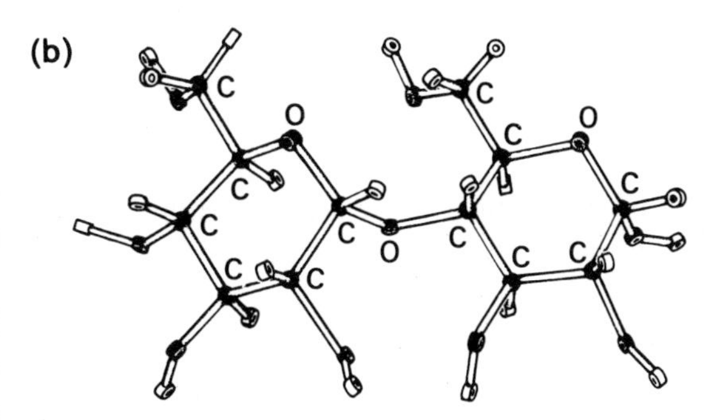

FIG. 32.4. Maltose. (a) Diagrammatic form in which the symbols of ring carbon atoms, and of single hydrogen atoms attached to them, are omitted. (b) Skeletal model.

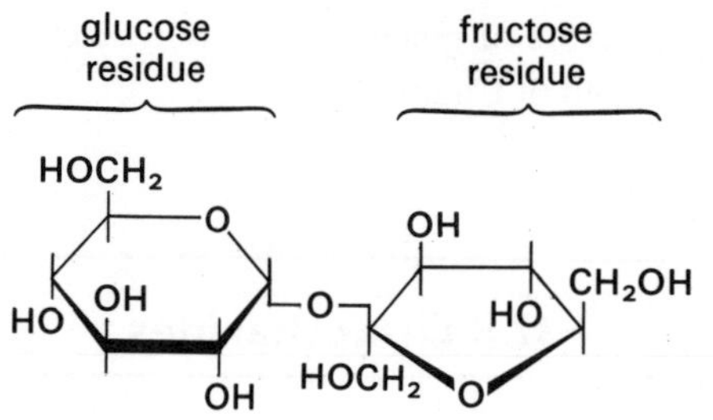

glucose residue

fructose residue

$HOCH_2$ O OH OH HO O HO CH_2OH OH $HOCH_2$ O

FIG. 32.5. Sucrose. For the sake of clarity the rings are shown as planar. Ring carbon atoms, and single hydrogen atoms attached to them, have not been shown

Disaccharides are readily split into their monomers by the action of hot dilute acid, e.g.

$$\text{maltose} \xrightarrow{H^+} \text{glucose} + \text{glucose}$$
$$\text{sucrose} \xrightarrow{H^+} \text{glucose} + \text{fructose}$$

Monosaccharides and disaccharides can be identified either by thin-layer or paper chromatography (section 14.25).

The rate of this reaction may be followed by observing the rate of change in optical rotation in a polarimeter (section 4.3). Sucrose rotates plane-polarized light in the dextro (+) direction; glucose is (+) and fructose is (−), but the specific rotation of fructose is higher than glucose. Hence the sign of the optical rotation changes from (+) to (−) as the reaction nears completion. The hydrolysis of sucrose is accordingly known as **inversion** and the mixture of glucose and fructose produced is called "invert sugar".

Fructose and sucrose reduce Fehling's solution only slowly; glucose and maltose on the other hand are true reducing sugars.

Sucrose is manufactured by extraction either from sugar cane or sugar beet. Maltose is produced by the partial hydrolysis of starch, e.g. in the "malting" of barley for brewing. Another disaccharide which occurs in foodstuffs is lactose, a constituent of milk.

32.9 Amino-acids

Amino-acids are the "monomer" units from which protein molecules are made. Only 20 occur commonly in nature, and these are all α-amino-acids, that is, the amino group is attached to the carbon atom adjacent to the carboxyl group. The general formula of an amino-acid is $H_2N.CHR.COOH$, where R is a side chain.

Proteins may be regarded as complex copolymers of amino-acids. The sequence of amino-acid residues is critically important in determining the physical and chemical characteristics of the protein.

The simplest amino-acid is glycine, aminoethanoic acid, $H_2N.CH_2.COOH$. In aqueous solution it exists largely as a **zwitterion** (or dipolar ion) in which a proton has transferred from the carboxyl group to the amino group, $^+H_3N.CH_2.COO^-$.

Alanine is another simple amino-acid. It is the methyl derivative of glycine in which $R = CH_3$ in the general formula. All amino-acids in nature—except glycine itself—are optically active, belonging to the L series. Figure 32.6 depicts a molecular model of L-alanine.

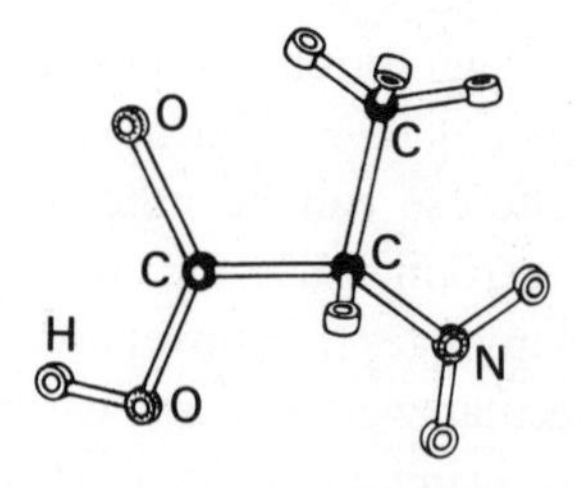

FIG. 32.6. L-alanine, illustrating the "corn" rule for recognizing amino-acids of the L series.

A useful rule employed by biochemists for recognizing amino-acids of the L series is the "corn" rule: view the α-carbon atom along the

H—C bond, with the hydrogen atom towards you. In a clockwise direction the groups read CO—R—N (CO of the carboxyl group, R the substituted side chain, and N the amino group). Figure 32.6 shows L-alanine from this viewpoint.

32.10 Amino-acids with two carboxyl groups

If the group R, of the general formula, itself contains a carboxyl group, the whole molecule will contain two such groups. Examples are aspartic acid (R = $CH_2.COOH$) and glutamic acid (R = $CH_2.CH_2.COOH$). The sodium salt of the latter, monosodium glutamate, is widely used in the food industry as a meat flavour.

Whereas the simple amino-acids described in section 32.8 are approximately neutral in pH, on account of the effects of their two main functional groups cancelling one another out, amino-acids with two carboxyl groups will have an overall acidic reaction to indicators.

Similarly, R can contain an extra amino group, as in lysine [R = $—(CH_2)_4.NH_2$] which is alkaline to indicators.

32.11 General properties of amino-acids

The amino-acids are white solids, soluble in water to a varying degree depending on the nature of the side-chain R. Their pH also depends on the nature of R as discussed in the foregoing two sections.

A mixture of amino-acids results when a protein is hydrolysed by heating with dilute mineral acid, and the identification and quantitative estimation of the components of the mixture is an important step in characterizing a protein. Chromatography (either paper or thin-layer) commends itself as an ideal technique, but

TABLE 32.3
SUMMARY OF THE MAIN AMINO-ACIDS IN NATURE

Name	Abbreviation	Structure of R
(a) *R = aliphatic side chain*		
glycine	gly	—H
alanine	ala	$—CH_3$
cysteine*	cys	$—CH_2SH$
serine	ser	$—CH_2OH$
threonine	thr	$—CH_2.CH_2OH$
methionine	met	$—(CH_2)_2.S.CH_3$
valine	val	$—CH(CH_3)_2$
leucine	leu	$—CH_2.CH(CH_3)_2$
isoleucine	ile	$—CH(CH_3).CH_2.CH_3$
(b) *R = aromatic side chain*		
phenylalanine	phe	$—CH_2.C_6H_5$.
tyrosine	tyr	$—CH_2.C_6H_4OH$
(c) *R = heterocyclic side chain*		
tryptophan	try	$—CH_2$— (indole ring structure, N)
histidine	his	$—CH_2$— (imidazole ring structure, N, N)
(d) *R = acidic side chain*		
aspartic acid	asp	$—CH_2.COOH$
glutamic acid	glu	$—(CH_2)_2.COOH$
(e) *R = basic side chain*		
lysine	lys	$—(CH_2)_4.NH_2$
arginine	arg	$—(CH_2)_3.NH.$ $C(=NH).NH_2$
(f) *Imino-acids*		
proline	pro	ring: CH_2, H_2C, CH_2, N(H)—CH —COOH

* See p. 406 for note on —S—S— bridges.

unfortunately with up to 20 amino-acids present, an extra refinement has to be introduced. Either two-dimensional chromatography (two-way chromatography, section 14.25), or else the technique of paper-electrophoresis has to be employed. In the latter case the "acidic" amino-acids, such as glutamic and aspartic acids, migrate to the anode on account of a predominance of negative COO^- groups, while the "basic" amino-acids, such as lysine, migrate towards the cathode on account of the predominance of NH_3^+ groups.

Table 32.3 summarizes the formulae of the naturally occurring amino-acids. Category (f) consists strictly of *imino*-acids rather than amino-acids, since one of the hydrogen atoms of the amino group is lost in forming a cyclic system, leaving the basic **imino** group, $\rangle NH$.

No attempt is made to name the amino-acids according to IUPAC rules, since such names would be excessively lengthy. It is convenient to use three-letter abbreviations when referring to sequences of amino-acids in more complex molecules, and these shortened forms are given in the table.

32.12 Disulphide bridges

Two amino-acids—cysteine and methionine—contain sulphur, in addition to the other four elements. Thus proteins containing either of these residues will contain sulphur. The presence of cysteine is of particular importance in the structure of proteins because it frequently occurs in the form of two molecules linked together. Two cyst*eine*-residues can link by elimination of two hydrogen atoms forming cyst*ine*, a molecule which contains the linkage S—S, referred to as a **disulphide bridge**:

$$\text{cysteine residues:} \quad -CH_2SH + HSCH_2- \xrightarrow{[O]} \text{disulphide bridge:} \quad -CH_2.S.S.CH_2-$$

32.13 Structural units of DNA and RNA

Ribonucleic acid (RNA) and deoxyribonucleic acid (DNA) are complex polymers whose monomer sequence carries, encoded in a form known popularly as the **genetic code**, the

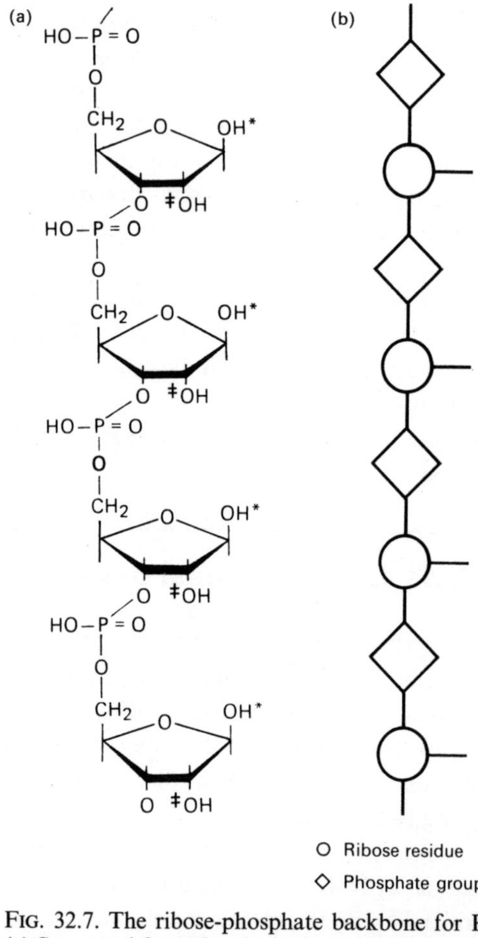

FIG. 32.7. The ribose-phosphate backbone for RNA. (a) Structural formula; the hydroxyl groups marked * are detached when the bases are in position. The hydroxyl groups marked ‡ are replaced by single H atoms in the backbone of DNA. (b) Diagrammatic form, cf. Fig. 33.10.

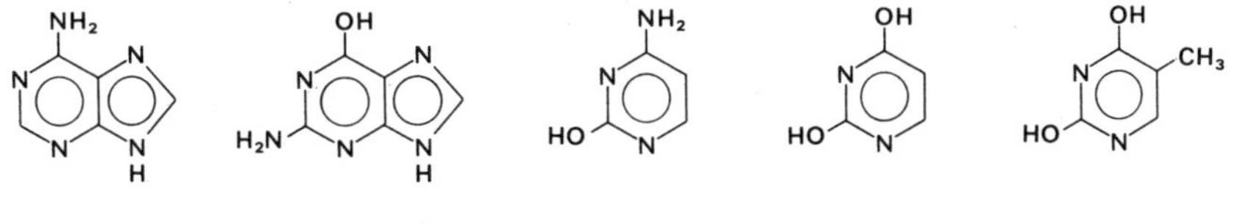

adenine guanine cytosine uracil thymine

FIG. 32.8. The organic bases responsible for carrying genetic information in DNA and RNA.

information which enables both the synthesis, by cells, of proteins and the self reproduction of cells. To understand the structure of RNA and DNA it is necessary first to examine the structure of the building units.

Both RNA and DNA consist essentially of a "backbone", with ribose rings in the furanose-(5-membered)-configuration linked through phosphate groups. The phosphate groups link rings of β-D-ribofuranose via carbon atoms 1 and 5 (Fig. 32.7). The backbone for DNA differs from that for RNA in that every ribose residue becomes deoxyribose, by the substitution of a single hydrogen atom for the hydroxyl group attached to carbon atom No. 2, i.e. the atoms labelled C_2 in Fig. 32.2.

DNA and RNA will be considered in more detail in the chapter which follows, but the essential feature of both structures is the presence of four organic bases attached in sequence to the points marked * in Fig. 32.7. For RNA the bases are adenine, guanine, cytosine and uracil; for DNA the bases are adenine, guanine, cytosine and thymine. In other words there are five bases, three of them being common to both structures and each of the remaining two (uracil and thymine) being present in only one structure.

The sequence of organic bases is highly ordered, and determines the coding of genetic information (Fig. 32.8).

The organic bases are heterocyclic compounds (section 27.3): cytosine, uracil and thymine may be regarded as derivatives of *pyrimidine* which is itself an analogue, of benzene, in which two of the CH-groups have been replaced by nitrogen atoms (Fig. 32.9).

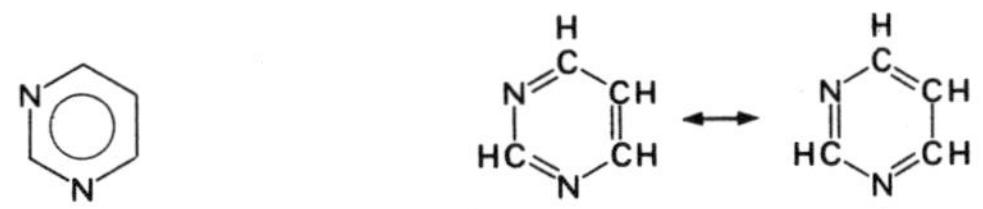

Canonical forms (cf. benzene):

FIG. 32.9. The structure of pyrimidine, a heterocyclic compound similar to benzene in structure.

Adenine and guanine are derivatives of the double-ring compound *purine* (Fig. 32.10).

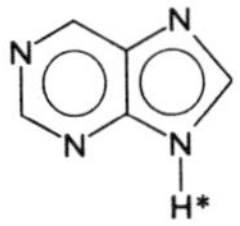

FIG. 32.10. Purine, the heterocyclic compound from which adenine and guanine are derived. The hydrogen atom marked * is lost when a purine derivative is linked in RNA or DNA.

The next chapter (section 33.4) will show how DNA is built by means of hydrogen-bonds linking base pairs. As a result of molecular size and shape, thymine will only pair with adenine, and cytosine only with guanine. Uracil occurs only in RNA, not in DNA; RNA is responsible for carrying genetic information in the cell, but base-pairing does not occur.

Study Questions

1. Explain why:

(a) only one naturally occurring amino-acid is optically inactive;

(b) some simple sugars reduce Fehling's solution rapidly, whereas others do so only slowly;

(c) there are 16 chain-form optical-isomers of glucose;

(d) the ring form of glucose possesses two isomers, α and β.

2. State the difference between:

(a) an addition polymer and a condensation polymer;

(b) a polyester and a polyamide;

(c) a thermoplastic and a thermosetting plastic;

(d) *Nylon*-6 and *Nylon*-6,6;

(e) an intermolecular reaction and an intramolecular reaction.

3. The following reactions each occur at about 200°C. Identify each of the lettered compounds. (**A**, **C** and **E** are the three hydroxybutanoic acids).

(a) **A**, $C_4H_8O_3$, gives **B**, $C_4H_6O_2$. **B** does not decolorize bromine water.

(b) **C**, $C_4H_8O_3$, gives **D**, $C_4H_6O_2$. **D** decolorizes bromine water.

(c) **E**, $C_4H_8O_3$, gives **F**, $C_8H_{12}O_4$. **F** does not decolorize bromine water.

(d) **G**, $C_4H_6O_4$, gives **H**, $C_4H_4O_3$. 25 cm^3 of a 0·1 M solution of **G** is exactly neutralized by 25 cm^3 of 0·2 M NaOH(aq).

(e) Explain why **A**, **C** and **E** behave differently when they are heated.

4. (a) Two dibasic acids, fumaric and maleic, share the formula $C_4H_4O_4$ and have a double bond between their two central carbon atoms. On heating, fumaric acid is unaffected but maleic acid loses a mole of water.

(i) Give the structural formula of each acid.

(ii) Give the systematic name of each acid.

(b) Malonic acid (propane-1,3-dioic acid) loses not water, but carbon dioxide when heated.

(i) Suggest why malonic acid behaves in this way.

(ii) Give the structural formula of the organic product of this reaction.

5. An aromatic compound **A**, $C_9H_{10}O_2$, reacted with sodium hydroxide solution to give (after acidification) an acid **B**, $C_8H_8O_2$. On reacting **B** with hot alkaline $KMnO_4$ and acidifying, a second acid **C**, $C_8H_6O_4$ was isolated. When heated, **C** lost water and formed **D**, $C_8H_4O_3$. Identify the compounds **A** to **D** and summarize your answer in a reaction chart.

6. Pentane-2,4-dione is observed to be acidic: its central hydrogen atoms can be donated to a suitable base. Suggest why this should be the case.

7. A mixture of butan-1-ol, ethanoic acid and water is often used as the eluting solvent in the separation of amino-acids by paper chromatography. The following R_f values (see section 14.25) are observed with a typical eluting mixture:

gly 0·20; ala 0·24; leu 0·58; glu 0·25; lys 0·12; ser 0·19.

(a) Account for the difference in R_f values between gly, ala and leu.

(b) Why is the R_f value of ser less than that of ala?

(c) Why is the R_f value of glu less than that of leu?

When a mixture of phenol and water is used as eluting solvent, the following R_f values are observed:

gly 0·40; ala 0·55; leu 0·82; glu 0·33; lys 0·55; ser 0·34.

(d) Why does lys travel, relatively, much further in this solvent?

(e) Why does ala travel further than glu in this solvent?

(f) Why are these two solvents often used together in the two-way (two-dimensional) chromatography of amino-acids?

CHAPTER 33

Biological macromolecules

This chapter deals with some of the macromolecular structures encountered in living systems, from the standpoint of their structure and physical properties. A detailed consideration of chemical mechanisms involving these substances forms part of the study of **biochemistry**, and is not dealt with in this book.*

33.1 Polysaccharides

Combination of two monosaccharide residues, with the elimination of water, produces a disaccharide (Chapter 32). Further polymerization leads to a **polysaccharide**, for example cellulose or amylose.

The nature of the linkage between monomer residues is important. For instance, maltose consists of two residues of D-glucose, in the pyranose ring configuration, with the hydroxyl group of carbon atom 1 in the α-configuration (α-D-glucose). Maltose is said to contain a 1α-4 link:

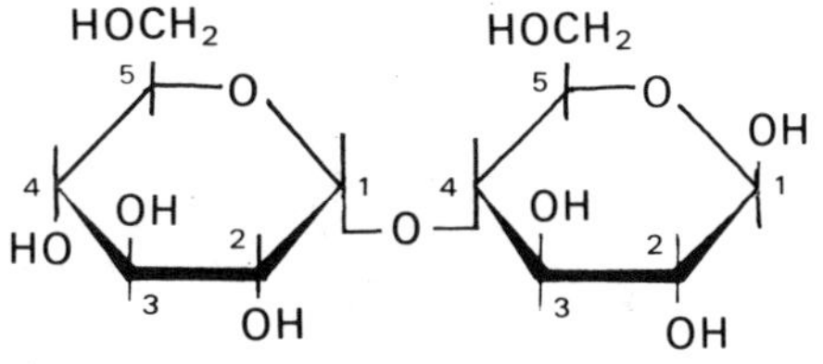

α-link

If on the other hand two glucose molecules link 1-4 with the C_1 atom in the β-configuration, the result is an entirely different disaccharide called cellobiose:

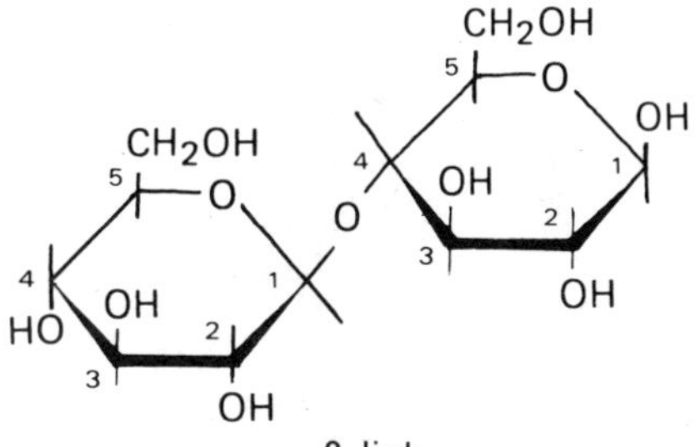

β-link

Cellulose is the main structural component of cell walls of plants, and is fibrous in nature. It consists of β-D-glucose residues linked at the 1-4 position.

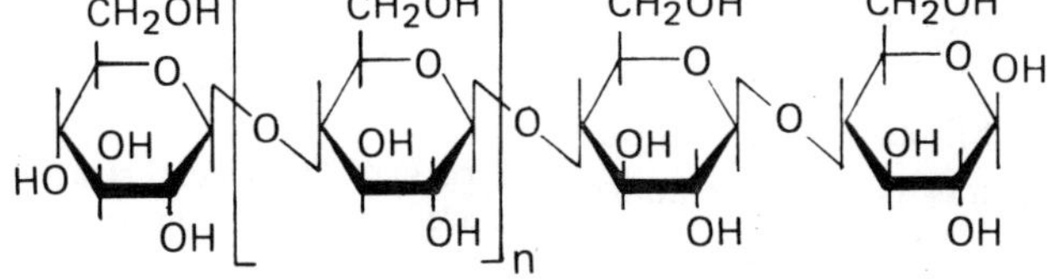

* See *A Background to Biochemistry*, T. J. Jennings, Pergamon Press, for an introductory account.

FIG. 33.1. Cellulose.

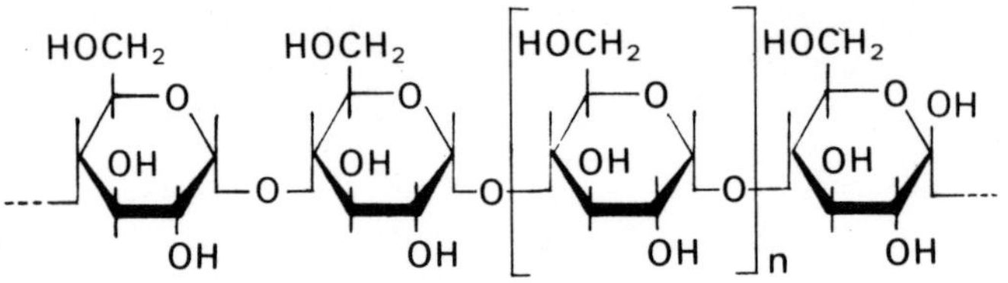

FIG. 33.2. Amylose, a component of starch.

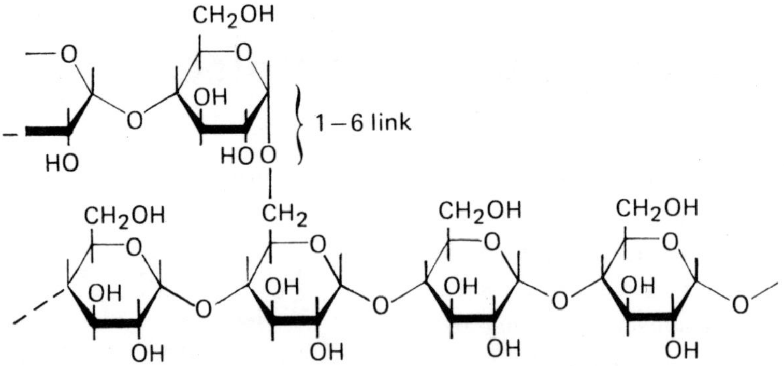

FIG. 33.3. Amylopectin, another component of starch, showing the formation of branches in the polymer chain.

Amylose comprises some 30% of starch, and consists of α-D-glucose residues joined at the 1-4 position. The α-linkages are identical to those in maltose.

The α-linkages in amylose give rise to a coiled configuration of the polymer chain, whereas the β-linkages in cellulose tend to give extended chains. This accounts for the fact that starch is non-fibrous whereas cellulose is fibrous.

The other main constituent of starch is *amylopectin*, which differs from amylose in having branched chains in its structure. Branches involve a 1-6 linkage (Fig. 33.3).

Glycogen is a polysaccharide not unlike amylose but is more highly branched. It is the main energy store in the muscles and liver of animals, and has a molecular weight of up to 10^7.

The presence of α- or β-links between residues is important when considering enzyme action, as enzymes are generally highly specific in the type of linkage they attack. For instance, humans do not possess digestive enzymes capable of breaking the β-linkages of cellulose, whereas the α-linkages in the components of starch are readily dealt with. Small polymers, for instance tri- and tetra-saccharides, are liberated in the digestive tract as intermediates in the digestion of polysaccharides.

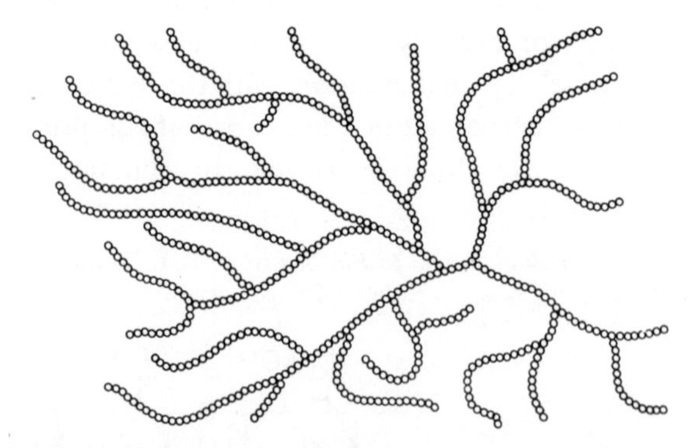

FIG. 33.4. The general structure of amylopectin. Each circle represents a glucose residue. From Baldwin E. (1967) *The Nature of Biochemistry*, Cambridge University Press.

The well-known starch–iodine reaction is the result of the forming of a loosely co-ordinated complex in which a molecule of iodine is surrounded by a coil of glucose residues in amylose. The complex has the familiar dark blue colour.

33.2 Polypeptides

When two amino-acids join together they do so by a condensation reaction involving the amino-group of one unit and the carboxyl-group of the other. The linkage —CO.NH— formed is known as the **peptide link**:

```
  O                 ⊖O
  ‖                  |
  C        ⟷         C
 / \ ..             / \\ ⊕
    N  /                N  /
    |                   |
    H                   H
```

The lone pair on the nitrogen atom is delocalized, and hence the atoms which form the peptide link are co-planar. The nitrogen atom has a planar rather than a pyramidal configuration (cf. section 6.1).

The ordering of amino-acid residues along a polypeptide chain is known as the **primary structure**. Diagrammatically, the primary structure of any polypeptide may be represented thus:

```
     O    R    H    O    R″
     ‖    |    |    ‖    |
     C    CH   N    C    CH
 .../ \  / \  / \  / \  / \...
      N    C    CH   N
      |    ‖    |    |
      H    O    R′   H
```

The backbone of the chain is $-\!\!\left[\!-\overset{|}{C}H.CO.NH-\right]_n\!\!-$. The end groups of the chain are free amino and carboxyl groups respectively, and these can ionize in the same way as in a free amino-acid.

The peptide link may be tested for by making use of its ability to form a co-ordinated complex with copper(II) ion. The material containing the suspected peptide link is dissolved in water and made alkaline with bench sodium hydroxide. A drop of copper(II) sulphate solution will produce a mauve colour if the peptide link is present. This is known as the *biuret test*.

The presence of free amino groups in amino-acids or polypeptides is detected by adding a compound called ninhydrin which produces a complex red or blue dye with NH_2-groups. Ninhydrin is an excellent locating agent for paper chromatography.

33.3 Proteins

Most animal tissues contain nitrogen in the form of proteins. Proteins are complex amino-acid polymers, and perform a great variety of functions. Some are used to build up connective tissue, some act as enzymes to perform biochemical reactions. For instance insulin is a protein which is present in the pancreas and which governs the combustion of sugar by the body; it contains 51 amino-acid residues and has a molecular weight of about 5700.

All proteins are characterized by the amino-acid sequence they contain; the full characterization of a protein must involve three stages:

(i) determination of the amino-acid composition by acid hydrolysis followed by paper chromatography and quantitative estimation of the components (section 14.25);

(ii) determination of the primary structure by techniques such as N-terminal analysis (see below);

(iii) determination of the spatial configuration of the molecule, that is, its **secondary structure**.

N-terminal analysis is a technique by which the end of a polypeptide chain with a free NH_2— group can be characterized. The compound

FDNB (1-fluoro-2,4-din trobenzene) has been found to bond itself to free amino groups; the peptide is then hydrolyzed with acid. The amino-acids are thereby separated from one another, but the residue at the end now has the FDNB attached:

$$O_2N\text{-}C_6H_3(NO_2)\text{-}F + H_2N.CHR.CO.NH.CHR'.CO.NH.CHR''.CO\ldots$$

$$\downarrow OH^-$$

$$O_2N\text{-}C_6H_3(NO_2)\text{-}NH.CHR.CO.NH.CHR'.CO.NH.CHR''.CO\ldots.$$

$$\downarrow H^+$$

$$O_2N\text{-}C_6H_3(NO_2)\text{-}NH.CHR.COOH + H_2N.CHR'COOH + H_2N.CHR''COOH \quad \text{etc.}$$

In this way, the end amino-acid is "labelled" and can be recovered as its FDNB derivative and identified by paper or thin-layer chromatography.

X-ray analysis of proteins reveals an intricate three-dimensional structure, the so-called secondary structure. Two main types of feature which can be observed are:

(a) sheet structures, e.g. the β-pleated sheet (Fig. 33.5);

(b) helical structures, e.g. the α-helix (Fig. 33.6).

The secondary structure taken up by a protein depends on the nature of the side groups R of the amino-acids, in particular on their size and their reaction to changes in pH. Secondary structure is largely determined by the way in

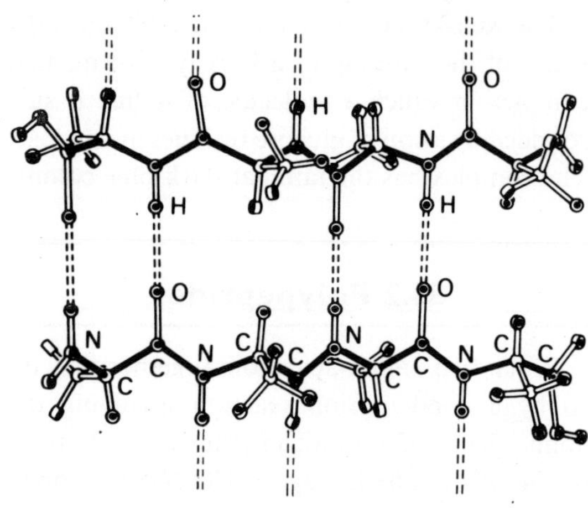

FIG. 33.5. Model of a β-pleated sheet. The dotted lines illustrate the weak hydrogen-bond links between chains. This model only shows two peptide backbones, whereas in practice a large number will occur, arranged in parallel to form a sheet.

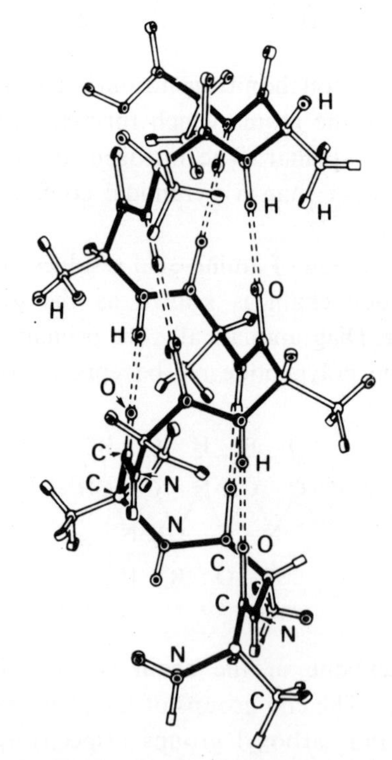

FIG. 33.6. Model of the α-helix. This structure occurs most commonly in non-fibrous proteins. The backbone is shown here by black bonds, and for simplicity the monomer here is L-alanine ($R = CH_3$ throughout). The dotted lines illustrate hydrogen bonds.

which hydrogen bonds can form between neighbouring amino-acid residues in the molecule. Figures 33.5 and 33.6 are drawings of molecular models and it will be seen that although hydrogen bonds are individually weak, a very large number of them together can combine to form a stable structure.

Where two cysteine residues come together a disulphide bridge (section 32.12) can form, giving rise to cross links.

Secondary structure of proteins is readily destroyed by only moderate heat (above about 70°C), by strongly acidic or alkaline conditions, or by breaking up the hydrogen bonds in any other way such as using ultra-violet light or adding a non-polar solvent. If the secondary structure of the protein is lost, the biological activity of the molecule disappears and the protein is said to be **denatured**. Once denatured, the protein will not recover its secondary structure, even though the primary structure may well remain intact.

Superimposed upon the secondary structure of a protein is a higher-order configuration known as **tertiary structure**. For instance the coils of an α-helix may themselves be coiled (in a manner roughly analogous to the "coiled-coil" of an electric light bulb) to form a sort of super-helix, or else a globular structure. Adjacent coils of the super-helix may be held together partly by hydrogen bonds and partly by disulphide bridges. The tertiary structure of a protein may again be of biological importance, especially in the case of enzymes where a molecule of an exact shape is required to perform some specific chemical task (Fig. 33.7).

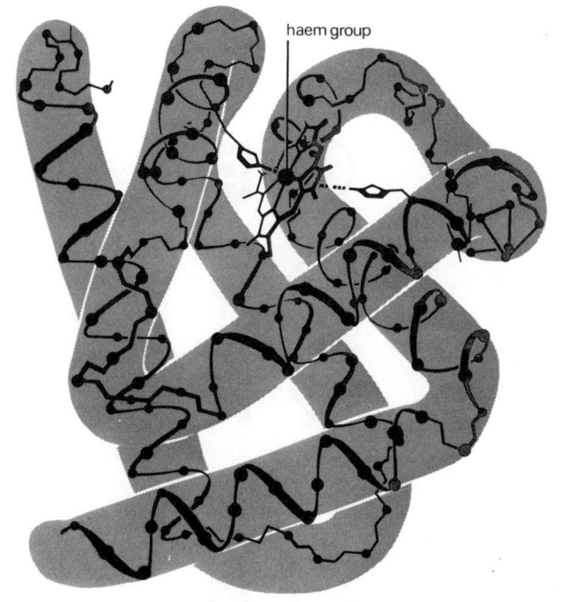

FIG. 33.7. The structure of myoglobin, a complex protein which stores oxygen in blood. It consists of an iron atom surrounded by a complex ligand (haem group) with a superhelix of protein material coiled around it. (After Dickerson R. E. (1964)).

33.4 Nucleic acids

The nucleic acids perform the very important function in cells of controlling the replication of protein molecules, and determining the sequencing of amino-acids in them. Nucleic acids form the "template" which orders the amino-acid sequencing, and hence they contain all the information required for replication.

The nature of the backbone was described in Chapter 32. It was also shown that five organic bases were present. Further analysis has shown that in DNA

$$\frac{\text{number of moles of guanine}}{\text{number of moles of cytosine}} = 1$$

and

$$\frac{\text{number of moles of adenine}}{\text{number of moles of thymine}} = 1$$

Thus the molecules are seen to pair off, and this is explained by the formation of hydrogen bonds

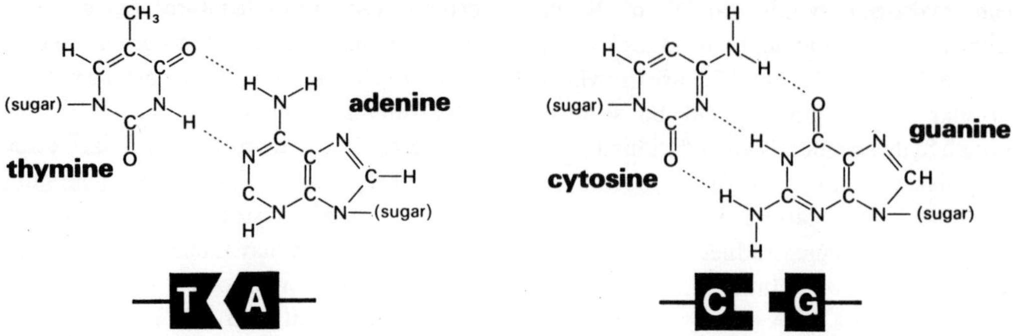

FIG. 33.8. Formation of base pairs in DNA. The pairing is controlled by (a) the number of hydrogen bonds that can form and (b) relative sizes of the two molecules. The block diagram form of showing the bases is used again in Figs. 33.9 and 33.10.

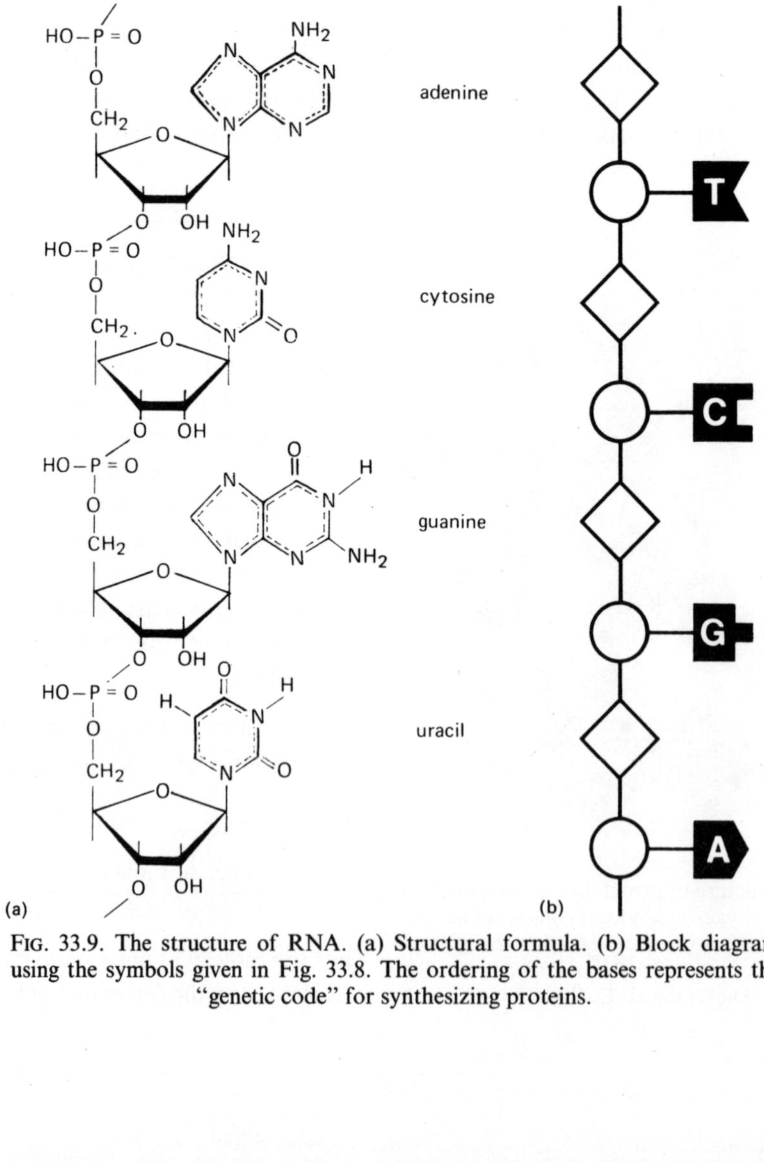

FIG. 33.9. The structure of RNA. (a) Structural formula. (b) Block diagram using the symbols given in Fig. 33.8. The ordering of the bases represents the "genetic code" for synthesizing proteins.

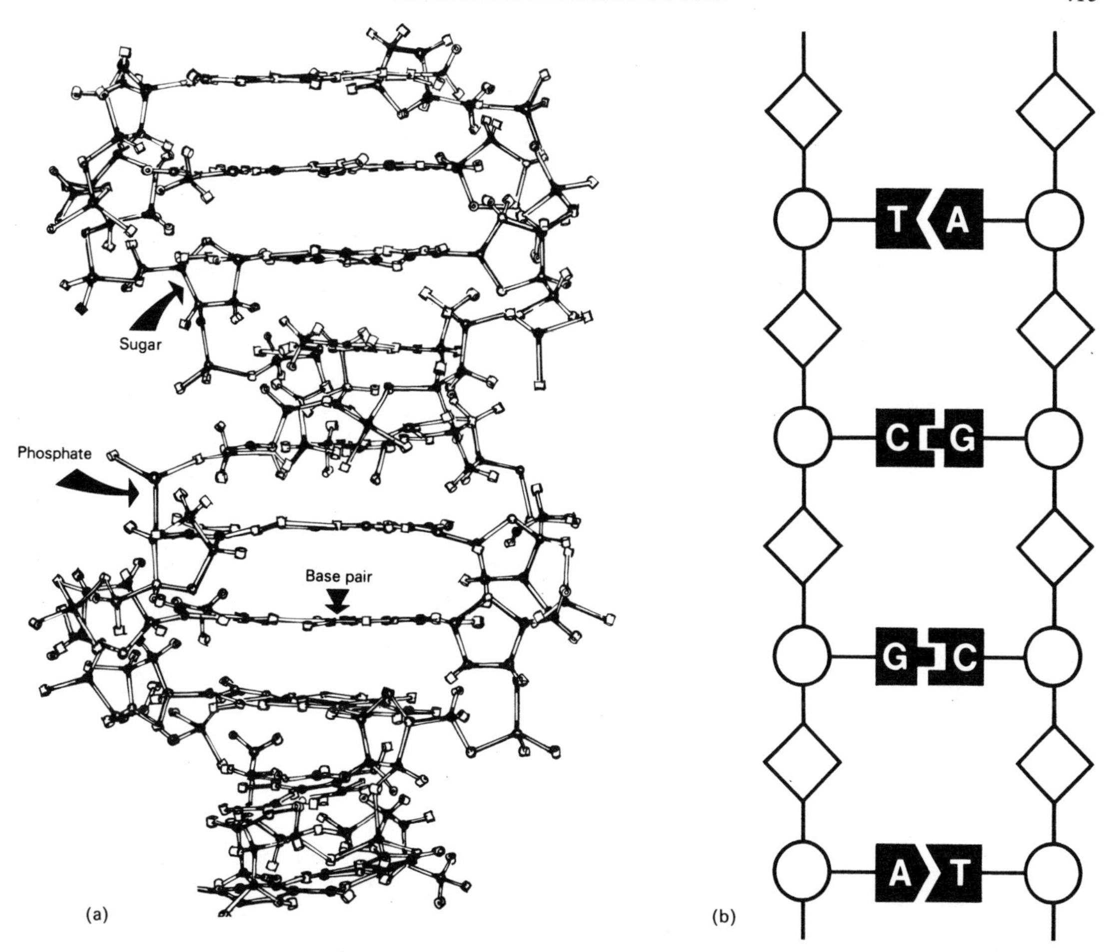

FIG. 33.10. The structure of DNA. (a) A skeletal model. (b) Block diagram showing how one strand can act as a "template" for building up the other strand.

between them in the DNA molecule (Fig. 33.8).

X-ray analysis revealed that DNA was in fact double-stranded, each strand matching up to the other by means of base-pairs (Fig. 33.10). Furthermore, the whole was found to have a secondary structure consisting of a double helix which could be likened to a kind of spiral staircase in which the "treads" were base pairs. Figure 33.10(a) shows a molecular model of a short section of DNA.

Study Questions

1. List (a) similarities, (b) differences between:

(i) amylose and amylopectin;
(ii) amylose and cellulose;
(iii) DNA and RNA.

2. Explain what is meant by (a) primary, (b) secondary, (c) tertiary structure of a protein.

3. Enzymes are biochemical catalysts and all are proteins. Use your knowledge of the structure of proteins to account for the following:

(a) enzymes are more specific than other catalysts, e.g. inorganic catalysts;

(b) enzymes are readily poisoned by the ions of heavy metals;

(c) enzymes work best within a narrow pH range, whereas other catalysts are generally less sensitive to changes in pH;

(d) most enzymes become less efficient at about 50°C and are irreversibly destroyed at 100°C, whereas other catalysts are often much more efficient at 100°C than at room temperature.

4. Partial hydrolysis of a protein leads to smaller polypeptides. Dipeptide structures can be worked out using 1-fluoro-2,4-dinitrobenzene (FDNB, section 33.3), followed by hydrolysis. Thus gly-ala (glycyl-alanine) forms alanine and DNBglycine.

A hexapeptide gave the following dipeptides on hydrolysis: gly-leu, gly-gly, gly-glu, leu-gly and glu-ala. The FDNB experiment on the hexapeptide gave DNBgly.

(a) Which amino-acid is at the NH_2-end of the hexapeptide chain?

(b) What two possibilities exist for the amino-acid sequence within the chain?

5. The hydrolysis of urea, $CO(NH_2)_2$, is catalysed by the enzyme urease. A number of reactions all at 18°C, were carried out. Each was stopped after ten minutes by adding mercury (II) ions and the ammonia that was formed was titrated with 0.02M HCl. In each case, the same amount of urease solution was used, only the urea concentration being different.

Experiment	$[CO(NH_2)_2]$	HCl needed (cm^3)
A	0 (control)	0
B	0.02	1.2
C	0.04	2.1
D	0.06	3.1
E	0.08	3.8
F	0.10	4.0
G	0.20	3.9
H	0.50	3.8

(a) Why is the amount of HCl needed a measure of the rate of reaction?

(b) Plot the rate (amount of HCl) against the urea concentration.

(c) What is the order of reaction with respect to urea when

(i) the urea solution is very dilute (Expts *A* to *E*).
(ii) the urea solution is more concentrated (Expts *F* to *H*)?

(d) Why does the order of reaction alter as the urea concentration increases?

6. It is often said that the rate of many enzyme catalysed reactions approximately doubles for every rise in temperature of 10°C.

(a) Assuming that such a reaction has a rate of "1 unit" at 17°C (290K), work out the rate of reaction at 300, 310, 320, 330 and 340K.

(b) The rate constant, k, is proportional to the rate. Plot $\log_{10}$ rate against $1/T$ and calculate the gradient ($\log_{10}$rate$/1/T$) of your graph (Section 17.9).

(c) Use the expression Gradient $= -E/2\ 3R$ to work out a value for *E*, the activation energy. ($R = 8.31 \times 10^{-3}$ kJ mol^{-1}).

(d) Suggest a reason why your graph is not a straight line.

CHAPTER 34

The B-metals

34.1 Classification and electronic structure

The B-metals are normally taken to be those metals which follow a transition series in a given period. Thus aluminium is not normally classified as a B-metal and it has properties which are generally more akin to those of *s*-block metals. Zinc, cadmium and mercury have properties similar to those of the metals on their right in the periodic table and will be considered as B-metals in this book.†

TABLE 34.1

IB	IIB	IIIB	IVB	VB	VIB	VIIB	0
		Al*					(Ar)
Cu§	Zn‡	Ga	[Ge]	[As]			(Kr)
Ag§	Cd‡	In	Sn	Sb	[Te]		(Xe)
Au§	Hg‡	Tl	Pb	Br	Po	[At]	(Rn)

* Aluminium does not follow a transition series.

§ These elements are more like transition metals than B-metals.

[ ] These elements are borderline between metal and non-metal.

‡ Some books treat these elements as transition metals.

† For a note on the nomenclature adopted in this book, see also section 20.1.

The ground state electronic structures of B-metal atoms may be derived by taking the noble gas structure at the end of the period, and removing an appropriate number of electrons. Table 34.3 shows how this is done. For instance, the configuration of an atom of tin may be written down by taking the configuration for xenon, and writing p^2 instead of p^6;

$$1s^2;\ 2s^2p^6;\ 3s^2p^6d^{10};\ 4s^2p^6d^{10};\ 5s^2p^2$$

or, in the abbreviated form,

2, 8, 18, 18, 4.

It is normally the valence electrons which are of greatest interest to the chemist, and the following table summarizes how these vary in this region of the periodic table, often known as the **p-block**.

TABLE 34.2

Group	No. of electrons in valence shell	Configuration of valence shell
IIIB	3	s^2p^1
IVB	4	s^2p^2
VB	5	s^2p^3
VIB	6	s^2p^4
VIIB	7	s^2p^5
0	8	s^2p^6

TABLE 34.3

II p^0	III p^1	IV p^2	V p^3	VI p^4	VII p^5	VIII p^6	Abbreviated electronic configuration of noble gas	Electronic configuration of noble gas written out in full
Zn	Ga	Ge	As	Se	Br	Kr	2, 8, 18, 8	$1s^2$; $2s^2p^6$; $3s^2p^6d^{10}$; $4s^2p^6$
Cd	In	Sn	Sb	Te	I	Xe	2, 8, 18, 18, 8	$1s^2$; $2s^2p^6$; $3s^2p^6d^{10}$; $4s^2p^6d^{10}$; $5s^2p^6$
Hg	Tl	Pb	Bi	Po	At	Rn	2, 8, 18, 32, 18, 8	$1s^2$; $2s^2p^6$; $3s^2p^6d^{10}$; $4s^2p^6d^{10}f^{14}$; $5s^2p^6d^{10}$; $6s^2p^6$

34.2 Occurrence and extraction of B-metals

Most of the B-metals occur as oxides or sulphides, particularly the latter. It is a characteristic of B-metals that their ions are "soft" (easily polarized or distorted), and they therefore occur with soft anions such as sulphide. Some mercury occurs native. The sulphide ores are converted into the oxide by roasting, and the oxide is then reduced with carbon. The essentials of these extraction processes are described in Chapter 13.

This chapter will be mainly concerned with Group IIB (zinc, cadmium and mercury) and Group IVB (germanium, tin and lead). The Group VB elements have been considered in Chapter 23, and the Group IIIB elements (gallium, indium and thallium) are relatively rare and will not be considered in any detail. Germanium, and many of the elements in this

TABLE 34.4

Element	$\mathcal{E}^\ominus$ value		Name of ore	Formula of ore	Method of extraction
Zn	Zn^{2+}/Zn,	$-0{\cdot}763$ V	Zinc blende	ZnS	Roast; C reduction
Cd	Cd^{2+}/Cd,	$-0{\cdot}403$ V	E.g. greenockite	CdS	C reduction; extracted from Zn ores
Sn	Sn^{2+}/Sn,	$-0{\cdot}136$ V	Cassiterite	SnO_2	C reduction
Pb	Pb^{2+}/Pb,	$-0{\cdot}126$ V	Galena	PbS	Roast; C reduction
Sb	Sb_2O_3/Sb,	$+0{\cdot}152$ V	Stibnite	Sb_2S_3	Roast; C reduction
Bi	Bi^{3+}/Bi,	$+0{\cdot}32$ V	Bi_2O_3, Bi_2S_3 and $(BiO)_2CO_3$		By-product of Sb, etc., extraction
Hg	$Hg_2^{2+}/2Hg$,	$+0{\cdot}789$ V	Cinnabar and native	HgS	Roast in air; Hg vapour condenses

borderline region between metal and non-metal, have assumed technological importance on account of their semi-conductor properties. Germanium itself is widely used in the manufacture of transistors, and compounds between elements of Groups IIIB and VB such as gallium arsenide, GaAs, are also important. The requirements of the electronics industry have forced chemists to manufacture these elements and compounds to exceptional purity standards. Germanium for transistor manufacture needs to be at least 99·99% pure, and the technique known as zone refining is used.

Table 34.4 gives a summary of the methods of extraction used for the main B-metals, the metals being arranged in order of the electrochemical series.

34.3 Properties and uses of the B-metals

The B-metals are generally low melting (compared with the transition metals) and fairly soft. Zinc, cadmium and mercury are the most volatile metals in this part of the periodic table.

Chemically, the B-metals are generally fairly "weak", that is, they have less tendency than the *s*-block metals to combine with non-metals such as oxygen and chlorine, and this is reflected in their $\mathcal{E}^{\ominus}$ values (Table 34.4). One of the most reactive B-metals is zinc, and even this is less reactive with non-metals than any *s*-block metal.

Gallium and germanium are used in the electronics industry. Zinc is used for protecting iron in "galvanized" articles; it is "sacrificially" corroded when an electrolyte bridges the two metals (Chapter 11). Brass is a copper–zinc alloy and zinc is widely used as a building material and in the preparation of pigments. Cadmium is a soft low-melting metal which is used in some anti-friction bearings, and in low melting alloys such as Wood's metal (m.p. between 60° and 70°C). A cadmium–mercury amalgam is used in the Western standard cell. Mercury is the only metal which is liquid at room temperature, and is widely used in thermometers, manometers, and other scientific apparatus.

Metallic tin is used mainly as tin-plate for protecting steel containers from corrosion; food cans are coated in this way. So-called "tin foil" is more likely these days to be made from aluminium, but some tin is used in the manufacture of alloys such as pewter (lead–tin), bronze (copper–tin), bearing metals, type metal and solder (lead–tin).

Most of the lead produced at the present time is used for the manufacture of accumulators, especially in motor vehicles. Other uses are the manufacture of anti-knock agents such as tetramethyl-lead(IV), $Pb(CH_3)_4$, electric cable sheaths and lead pipes. Lead blocks are used in radioactive screening: the high nuclear charge of lead makes it an efficient absorber of radiation.

34.4 Ion formation and the inert-pair effect

The ions of B-metals, although comparable in size with those of *s*-block metals (e.g. Zn^{2+} has a similar radius to Mg^{2+}, Fig. 34.1), are nevertheless very different. The main differences between *s*-block ions and B-metal ions may be summarized as follows:

(a) B-metals enter into chemical combination less readily than *s*-block metals, because B-metals have higher ionization energies. Figure 34.2 shows this effect.

(b) B-metal ions are more readily polarized (softer, more easily distorted) than *s*-block metal ions. Thus when compounds do form, they are more likely to be covalent:

(i) they enter into complex formation readily;

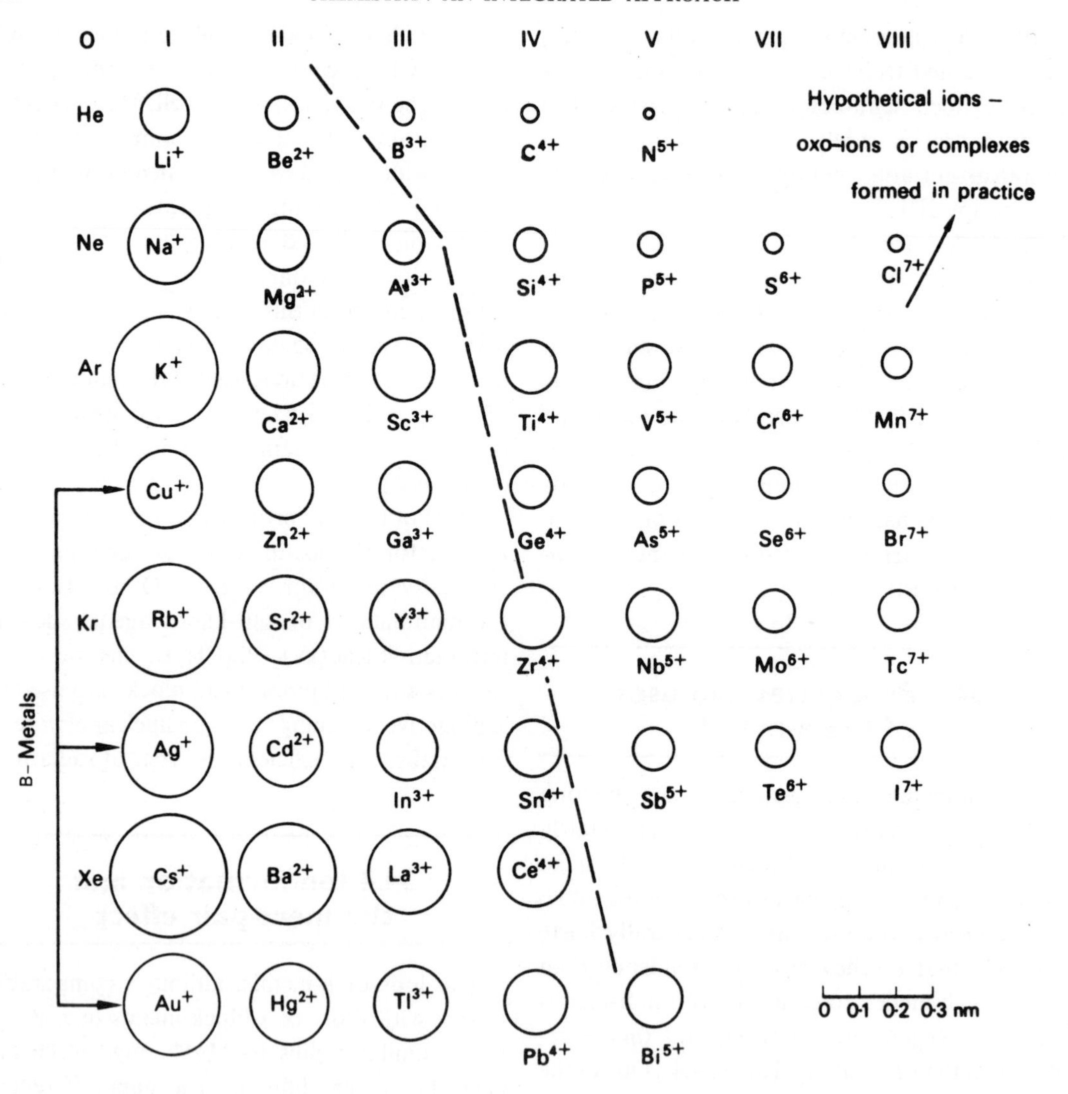

FIG. 34.1.

(ii) the structures of compounds do not fit the ionic model, i.e. they are frequently layered or distorted;

(iii) physical properties of compounds, such as solubility in water and volatility, are often not consistent with the existence of true metal ions.

(c) B-metals show variable valency, exhibiting two states:

(i) oxidation number = Group number.
(ii) oxidation number = Group number minus two.

This is the **inert-pair** effect (see below). Loss of electrons from a B-metal atom can take place in two ways:

(i) All the outer s and p electrons may be lost: in this case a compound exists in which the oxidation number equals the Group number, for

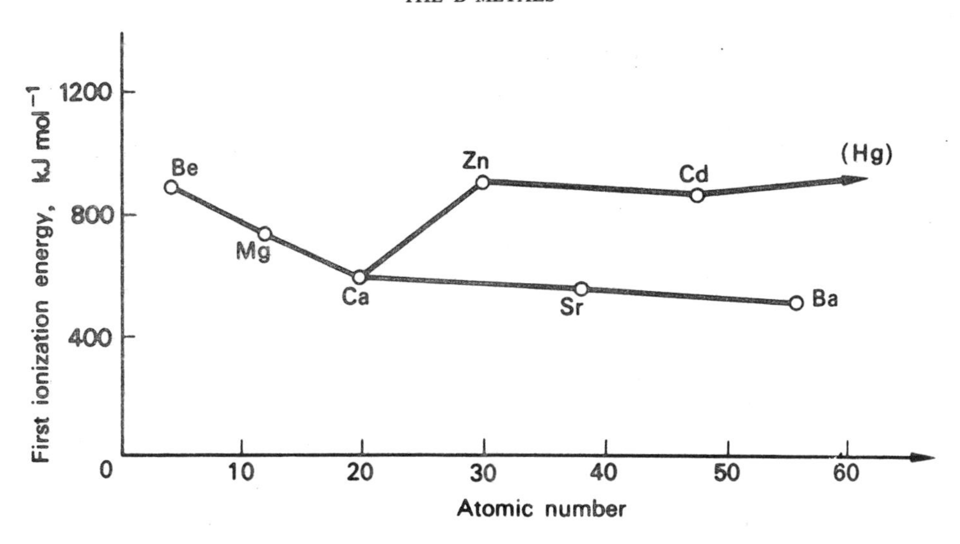

FIG. 34.2. B-metals have higher ionization energies than A-metals.

instance, tin(IV) and antimony(V). Such compounds are not ionic: electron-pair bonds are formed, either in neutral molecules or complex ions.

(ii) The outermost *s* electrons are frequently not lost, and the oxidation number of the element is then equal to *Group number minus two*, for instance, thallium(I), lead(II) and bismuth (III). The pair of *s*-electrons is said to have become inert. The inert-pair effect is most strongly marked for the heavier elements of a Group such as those named above.

The ions formed by B-metals are quite unlike those formed by *s*-block metals: they are very much more polarizable. The "hard-soft concept" has been used to illustrate this: we say that *s*-block metal ions are hard, but B-metal ions are soft. The consequence of this is that B-metals enter into complex formation and electron-pair bond formation much more readily than *s*-block metals of similar atomic radius. Ions of very high charge are not often formed in any case, and one would not expect to observe simple ions such as Bi^{5+}, partly because their size would be too small and partly because their energy of ionization would be too great; in aqueous solution such ions would certainly hydrolyze.

Ligands of negative charge, which can neutralize the charge present on the simple ion, form particularly stable complexes. Hence although the simple ion Pb^{4+} does not occur, complex ions such as $PbCl_6^{2-}$ are stable:

$$\underset{\text{lead(IV)}}{Pb^{4+}} + 6Cl^- \rightleftharpoons \underset{\text{hexachloroplumbate(IV)}}{PbCl_6^{2-}}$$

Figure 34.1 is a diagram summarizing the atomic and ionic radii of s-block and p-block elements. The radii shown to the right of the dotted line are hypothetical only: the ions are too small and highly charged to exist free, and oxo-ions or complexes occur instead.

34.5 Oxides and hydroxides of B-metals

The B-metals react directly with oxygen, though in general they do not occur very high up the electrochemical series and reaction is usually slow. With elements such as mercury reaction only occurs on heating the metal in oxygen.

Oxides in this region of the periodic table often have structures of an irregular kind suggesting that covalent bonds exist between atoms. Symmetrical structures determined by such factors as the radius ratio of the ions are encountered less often. B-metal oxides are insoluble in water, though a very weak equilibrium is set up in which metal ions and hydroxide ions are formed in aqueous solution.

Many of the B-metal oxides and hydroxides are **amphoteric**, that is to say, they react either as acids or as bases. When an amphoteric oxide dissolves in an acid it gives a solution containing cations $M^{n+}(aq)$; when it dissolves in an alkali it gives a solution containing hydroxo-complexes of the type $M(OH)_n^{y-}$.

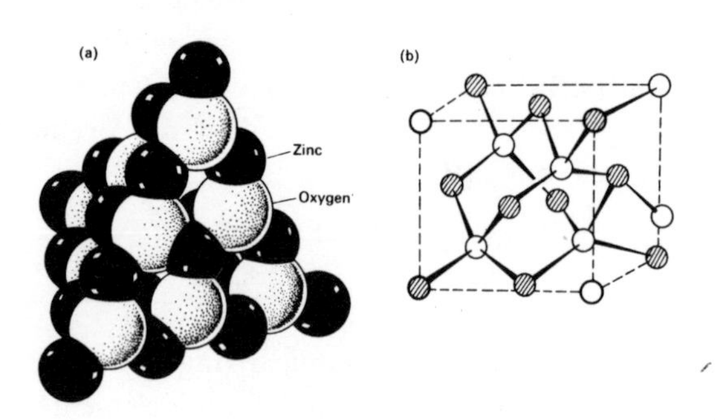

FIG. 34.3.

Zinc. Zinc oxide, ZnO, is a white solid with a tetrahedral structure showing 4:4 co-ordination (Fig. 34.3). It is formed when zinc is heated in oxygen, and is insoluble in water. It dissolves in acids and alkalis:

$$ZnO(s) + 2H^+(aq) \rightarrow Zn^{2+}(aq) + H_2O$$

$$ZnO(s) + 2OH^-(aq) + H_2O \rightarrow \underset{\text{tetrahydroxozincate}}{Zn(OH)_4^{2-}(aq)}$$

Zinc hydroxide is precipitated when alkali is added to a solution of $Zn^{2+}(aq)$, but excess reagent redissolves the precipitate:

$$Zn^{2+}(aq) + 2OH^-(aq) \rightleftharpoons Zn(OH)_2(s);$$

$$K_s = 2 \times 10^{-14}\,mol^3\,dm^{-9};$$

$$Zn(OH)_2(s) + 2OH^-(aq) \rightarrow Zn(OH)_4^{2-}(aq)$$

Zinc oxide and hydroxide are therefore amphoteric. Zinc metal is itself amphoteric in nature, reacting with alkali to give hydrogen:

$$Zn(s) + 2OH^-(aq) + 2H_2O \rightarrow Zn(OH)_4^{2-} + H_2(g)$$

Zinc forms only one oxidation state, with oxidation number $= +2$. It is not necessary to write zinc(II) in the names of compounds because there can be no ambiguity.

Cadmium. Cadmium forms a brown oxide, CdO(s), and a white hydroxide $Cd(OH)_2(s)$ under conditions similar to those for zinc. These compounds are not amphoteric: they dissolve in acids but not in alkalis.

Mercury. Mercury forms two series of compounds containing respectively the +1 and +2 oxidation states of the metal. On heating mercury in oxygen, mercury(II) oxide, HgO, is obtained as a red solid which readily dissociates into its elements on strong heating ($\Delta G_f^\ominus$ positive, Chapter 13). There is also a yellow form of mercury(II) oxide, which differs from the red form only in the size of its constituent particles. Addition of alkali to a solution of $Hg^{2+}(aq)$ gives mercury(II) oxide: the hydroxide does not exist though the oxide is hydrated when first precipitated. Mercury(I) oxide, Hg_2O, does not exist though salts of it are readily obtained.

Tin. Ordinary tin does not oxidize easily, though molten tin will combine with oxygen to form tin(IV) oxide, SnO_2. This same compound is present in *cassiterite*. The same oxidation state of tin is formed when steam is passed over the heated metal, though water itself has a negligible effect on tin:

$$Sn(s) + 2H_2O(g) \rightarrow SnO_2(s) + 2H_2(g)$$

Tin(IV) oxide is more acidic than basic in character; in this respect it is more like the oxide of a non-metal. High oxidation states of a metal behave in a "non-metallic" fashion, and other examples will be encountered, for instance manganese(VII) and chromium(VI). Concentrated sulphuric acid slowly dissolves tin(IV) oxide to give the sulphate:

$$SnO_2 + 2H_2SO_4 \rightarrow Sn(SO_4)_2 + 2H_2O$$

Alkalis dissolve it forming a stannate(IV) ion:

$$SnO_2(s) + 2OH^-(aq) + 2H_2O \rightarrow \underset{\text{hexahydroxo-stannate(IV)}}{Sn(OH)_6^{2-}}$$

Hydroxides of tin do not appear to exist: the addition of alkali to a solution of $Sn(SO_4)_2$ precipitates SnO_2.

Tin shows the inert-pair effect: it forms a series of compounds with an oxidation state +2, which is "Group valency minus two". Tin(II) oxide is precipitated in hydrated form by adding alkali to $Sn^{2+}(aq)$ ions; on heating this precipitate in absence of air a brown powder is obtained, approximating to SnO in composition. On exposure to air this oxide quickly reacts to form SnO_2. It is amphoteric, dissolving in acids to give tin(II) ions, $Sn^{2+}(aq)$, and in alkali to give stannate(II) hydroxo-complexes. $Sn^{2+}(aq)$ is a good reducing agent ($\mathcal{E}^\ominus$ for $Sn^{4+}|Sn^{2+} = +0{\cdot}15$ V).

Lead. Like tin, lead forms two series of compounds, in which the oxidation states are +2 and +4 respectively. Lead shows the inert pair effect more strongly than tin, and consequently lead(II) compounds are the more common while lead(IV) compounds are powerful oxidizing agents ($\mathcal{E}^\ominus$ for $Pb^{4+}|Pb^{2+} = +1{\cdot}5$ V). Lead (II) oxide, *litharge*, PbO(s), is formed as a yellow powder when lead is strongly heated in air, or when a higher oxide of lead is heated strongly. It is also formed when lead(II) nitrate or lead(II) carbonate is heated in air. It is amphoteric and forms $Pb^{2+}(aq)$ when dissolved in acids, and $Pb(OH)_4^{2-}$ when heated with concentrated alkali. True lead hydroxide is not formed by adding $OH^-(aq)$ to $Pb^{2+}(aq)$: instead a hydrated oxide approximating to $2PbO.H_2O$ is precipitated.

The structure of lead(II) oxide is strange, and quite clearly not derived from the ions Pb^{2+} and O^{2-}. The environment of the lead atoms is shown in Fig. 34.4: it is thought that the inert pair of electrons might occupy the apex of the pyramid.

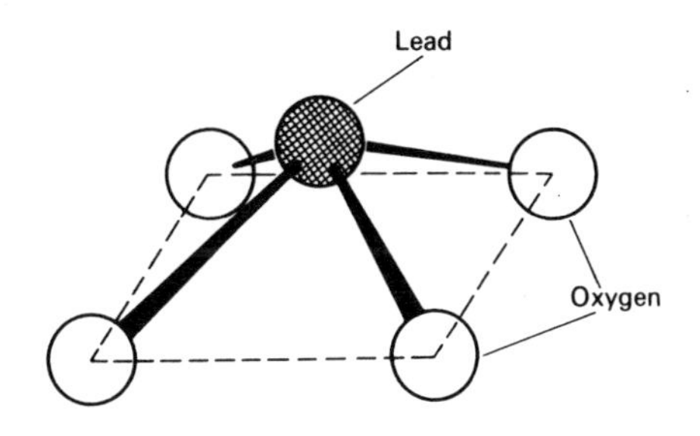

FIG. 34.4. The environment of a lead atom in lead(II) oxide (cf. ZnO, Fig. 34.3).

If lead(II) oxide is heated in air for some time at about 450°C, a red powder of formula Pb_3O_4, *red lead oxide*, is formed. Its constitution suggests that it is a "salt" compounded of the acidic oxide PbO_2 and the basic oxide PbO, i.e. $2PbO.PbO_2$. Its systematic name is therefore dilead(II) lead(IV) oxide. If red lead oxide is warmed with dilute nitric acid dark brown lead(IV) oxide, $PbO_2(s)$, is formed: it is as if the "basic" part of the Pb_3O_4 had been dissolved out:

$$Pb_3O_4(s) + 4HNO_3(aq) \rightarrow 2Pb(NO_3)_2(aq) + PbO_2(s) + 2H_2O$$

cf. $$PbO(s) + 2HNO_3(aq) \rightarrow Pb(NO_3)_2(aq) + H_2O$$

Lead(IV) oxide is an acidic oxide: it will not dissolve in acids except by some other type of reaction, e.g. oxidation:

$$PbO_2(s)+4HCl(aq) \rightarrow PbCl_2(aq)+Cl_2(g)+2H_2O$$

Lead(IV) oxide is used in accumulators, where energy is derived from the process

$$Pb(IV)+Pb(0) \rightleftharpoons 2Pb(II)$$

Antimony and bismuth. The oxides of these metals are discussed in Chapter 23, and they should be compared with the other B-metal oxides discussed in this section.

34.6 Halides of the B-metals

Although they are not so reactive as the *s*-block metals, the B-metals will enter into direct combination with all the halogens and most of the other non-metals. A few of the more important halides are considered in detail below.

Zinc halides. Zinc reacts directly with all the halogens to form $ZnX_2(s)$. These compounds are all white crystalline solids soluble in water; in solution the ions $Zn^{2+}(aq)$ and $2X^-(aq)$ are produced. The hydrated zinc ion is hydrolyzed by water and the solution therefore has an acidic pH (section 19.13). Another method of preparing the halide is by the action of HX(aq) on zinc metal. For instance dilute hydrochloric acid produces a solution of zinc chloride:

$$Zn(s)+2HCl(aq) \rightarrow ZnCl_2(aq)+H_2(g)$$

It is difficult to obtain the anhydrous salt from this solution by evaporation, because of hydrolysis, but dry hydrogen chloride gas produces the same reaction giving the anhydrous salt.

Zinc fluoride is much less soluble in water than the other zinc halides. This is probably a lattice energy effect (cf. lithium fluoride, Chapter 20).

Mercury halides. If chlorine is passed over heated mercury the higher oxidation state, mercury(II) chloride, is produced:

$$Hg(l)+Cl_2(g) \rightarrow HgCl_2(s); \quad \Delta H = -230\,\text{kJ at } 25°\text{C}$$

It is relatively easy to oxidize mercury to the +2 state, and all the halogens are capable of doing this. The +1 state is produced by heating a mixture of metallic mercury and the mercury(II) compound. For instance:

$$Hg(l)+HgCl_2(s) \rightarrow \underset{\text{mercury(I) chloride}}{Hg_2Cl_2(s)}$$

Although the *empirical* formula of mercury(I) chloride is HgCl, the molecular formula is Hg_2Cl_2. In the solid state and in the vapour phase linear molecules exist:

Cl—Hg—Cl and Cl—Hg—Hg—Cl

There is little evidence for the presence of ions in the solid, and the shape of these molecules is that predicted from the repulsion of electron pairs (mercury has a completely filled 5*d* shell and no lone pairs).

The Hg—Hg bond present in mercury(I) chloride persists in aqueous solutions of the mercury(I) ion. Measurements of the e.m.f.s of concentration cells and of ionic conductances point conclusively to the existence of the ion Hg_2^{2+}. Mercury(I) chloride is used in the *calomel electrode*, used as a reference electrode for pH measurement.

Like silver(I) halides, mercury(I) halides are insoluble in water. If dilute hydrochloric acid is added to an aqueous solution of a metal salt, silver(I), mercury(I) and lead(II) are the only common ions which will form precipitates —a fact which is often made use of in analysis. It is interesting that this insolubility occurs despite the fact that the ionic radii of Ag^+ and Pb^{2+} are comparable with K^+ and Ba^{2+}, and indicates that these B-metal halides are not really ionic.

Mercury readily forms complex ions where the ligand is a halide ion. Addition of I^-(aq) to Hg^{2+}(aq) results first in the precipitation of red HgI_2, but excess iodide redissolves the precipitate to form a colourless solution containing a complex ion:

$$HgI_2(s) + 2I^-(aq) \rightarrow HgI_4^{2-}(aq)$$

Tin halides. Tin forms halides with all the halogens, and the oxidation number may be either +2 or +4: direct action of the halogen on the metal produces the +4 state. Except for tin(IV) fluoride, the +4 halides are molecular in structure; the +2 halides have higher melting points but their structures are layer-like, suggesting that they are macromolecular rather than ionic.

Compounds of tin(IV) are very strongly hydrolysed by water forming hydrated tin(IV) oxide. $SnCl_4$ is a fuming liquid which reacts vigorously with water. The +2 halides are not so strongly hydrolysed, though their solutions are acidic.

Tin forms colourless complex ions with the halogens, such as hexachlorostannate(IV), $SnCl_6^{2-}$. The formation of such ions is typical of B-metals.

Lead halides. Lead shows the inert-pair effect more strongly than tin, and hence only fluorine is capable of oxidizing the metal directly up to the +4 state. Chlorine reacts with heated lead, to form lead(II) chloride, $PbCl_2$, and bromine and iodine react similarly though rather more slowly. Lead(IV) chloride is a yellow, unstable oil which fumes in air and reacts with water giving PbO_2(s) (cf. tin(IV) chloride).

The usual way of making the +2 halides is by precipitation: all are practically insoluble in water, the most soluble being the chloride which dissolves quite readily in hot water.

$$Pb^{2+}(aq) + 2Cl^-(aq) \rightarrow PbCl_2(s).$$

Stable halide complexes are formed, and these are soluble in water; concentrated hydrochloric acid will dissolve lead(II) chloride with the formation of tetrachloroplumbate(II), $PbCl_4^{2-}$(aq). In a similar fashion, potassium iodide solution will precipitate lead(II) ions as yellow PbI_2, which redissolve on adding excess reagent with the formation of PbI_4^{2-}(aq).

34.7 Sulphides of the B-metals

Sulphur combines directly with most metals; among the B-metals, the more reactive ones such as zinc combine violently when finely divided. B-metal sulphides are insoluble in water, and their structures are macromolecular rather than ionic. Most of them are best prepared by precipitation reactions, rather than by direct combination of the elements.

Zinc sulphide, ZnS. This compound occurs naturally in two crystalline forms, known as *zinc blende* and *wurtzite*. Both lattices have a co-ordination number of 4:4 (tetrahedral) though they differ in symmetry. A white precipitate of zinc sulphide is obtained on adding sulphide ions (or hydrogen sulphide) to a solution of Zn^{2+}(aq), in neutral or slightly alkaline solution. Zinc sulphide is almost completely insoluble in water:

$$ZnS(s) \rightleftharpoons Zn^{2+}(aq) + S^{2-}(aq);$$
$$K_s \simeq 10^{-24}\ \text{mol}^2\ \text{dm}^{-6} \text{ at } 25°\text{C}.$$
$$S^{2-}(aq) + 2H^+(aq) \rightleftharpoons H_2S(aq);$$
$$pK_a = 20, \text{ at } 25°\text{C}.$$

If $[H_2S] = 0{\cdot}1\ mol\,dm^{-3}$ for a saturated solution, and $[H^+] = 1$, then $[S^{2-}]$ becomes 10^{-21} $mol\,dm^{-3}$. Hence in theory a concentration of $Zn^{2+} > 10^{-24}/10^{-21}\ mol\,dm^{-3}$ ought to cause precipitation if hydrogen sulphide is passed into an acidified solution of $Zn^{2+}(aq)$ ions. In practice such a precipitate takes about a month to form. This is another example of a kinetic factor overriding a thermodynamic factor.

Mercury(II) sulphide, HgS. Like zinc sulphide this compound can exist in two crystalline forms; the form commonly found in nature is a red ore called *cinnabar* which has a macromolecular structure in which each mercury atom has two near sulphur neighbours.

If hydrogen sulphide or sulphide ion is added to a solution of $Hg^{2+}(aq)$, a series of colour changes through orange and brown usually occurs, resulting finally in the formation of a black precipitate of HgS with an amorphous structure. This is one of the least soluble sulphides, with a solubility product quoted as 10^{-54}; it will not dissolve in $H^+(aq)$ and is not attacked even by concentrated nitric acid.

Tin sulphides. Direct action of sulphur on heated, finely divided tin results in the exothermic formation of tin(II) sulphide, SnS. Sulphur is not a powerful enough oxidizing agent to oxidize tin to the +4 state. Tin(IV) sulphide does, however, exist. Both sulphides can be prepared by precipitation:

$$Sn^{2+}(aq) + S^{2-}(aq) \rightarrow SnS(s); \quad \text{(black)}$$

$$Sn(OH)_6^{2-}(aq) + 2S^{2-}(aq) \rightarrow SnS_2(s) + 6OH^-(aq); \quad \text{(yellow)}$$

Tin readily forms complex ions with S^{2-} in the +4 oxidation state. A solution of ammonium sulphide, which can be regarded as containing both sulphur and sulphide ions, will dissolve both the sulphide precipitates as thiostannate complexes:

$$SnS + S + S^{2-} \rightarrow SnS_3^{2-}$$

$$SnS_2 + S^{2-} \rightarrow SnS_3^{2-}$$

Arsenic and antimony sulphides behave in a similar fashion.

Lead(II) sulphide. Lead(II) sulphide forms as a black precipitate when $Pb^{2+}(aq)$ and $S^{2-}(aq)$ ions are mixed, or when the elements are heated together. It occurs naturally as *galena*. The solubility product is less than that of zinc sulphide by a factor of about 1000 ($K_s = 3 \times 10^{-27}$ $mol^2\,dm^{-6}$ at 25°C) and it does not dissolve in dilute acids. Hot dilute nitric acid, being an oxidizing agent, does dissolve lead(II) sulphide:

2 N reduced from +5 to +4

$$PbS(s) + 4HNO_3(aq) \rightarrow Pb(NO_3)_2(aq) + S + 2NO_2(aq) + 2H_2O$$

S oxidized from −2 to 0

Group VB sulphides. The trends observed for Group IVB sulphides (tin and lead) are repeated in a parallel fashion in Group VB (antimony and bismuth). In particular:

(i) Antimony, like tin, forms sulphides which can dissolve in ammonium sulphide as thio-complexes, such as SbS_3^-. These two metals can be identified in analysis by this property. Bismuth, like lead, fails to form thio-complexes.

(ii) The inert-pair effect shows itself in the same sort of way. Bismuth fails to form a sulphide Bi_2S_5 but forms instead the "Group valency minus two" oxidation state in Bi_2S_3. The same effect is observed with lead and thallium.

34.8 Complex ion formation by the B-metals

Whereas it was a characteristic of *s*-block metals that their ions are generally simple, simple ions of the B-metals are rarely encountered. The ions Ba^{2+}(aq) and Zn^{2+}(aq) for instance are not very similar: in the case of the barium ion the state symbol (aq) denotes an indefinite number of water molecules loosely attached around a simple barium ion as a hydration sheath, while in the case of the zinc ion there is evidence for the presence of four water molecules fairly firmly held in tetrahedral positions. The zinc ion is really a complex ion $Zn(H_2O)_4^{2+}$, and the same effect occurs with other B-metal ions (it is also true of aqueous ions of transition metals, Chapter 35).

Stepwise replacement of ligands. Other ligands apart from the water molecule can attach themselves to B-metal ions by displacing the water molecules present. For instance ammines are formed if ammonia is added to an aqueous solution of zinc ions; in the first instance only one ligand is replaced:

$$Zn(H_2O)_4^{2+} + NH_3(aq) \rightarrow [Zn^{(H_2O)_3}_{(NH_3)}]^{2+} + H_2O$$

Successive replacements of H_2O by NH_3 lead finally to $Zn(NH_3)_4^{2+}$—in fact, this is the ion normally present in an aqueous solution containing zinc ions in the presence of excess ammonia.

Negatively charged ligands behave similarly: addition of Cl^-(aq) ions to a solution of Zn^{2+}(aq) leads mainly to $ZnCl_3^-$(aq) and $ZnCl_4^{2-}$(aq).

Ligands which will add to B-*metals.* In addition to H_2O and Cl^- mentioned above, the B-metals will form complexes with other halide ions, such as F^-, Br^- and I^-. Ions such as cyanide, CN^-, and thiocyanate, CNS^- (referred to previously as pseudohalides, section 24.9) will also form stable complexes with some, though not all, B-metals. Zinc for instance forms a stable tetracyanozincate ion, $Zn(CN)_4^{2-}$, but the corresponding complexes of tin and lead are less stable.

Ligands with more than one point of attachment to the central metal atom are known as **polydentate** and polydentate ligands form complexes which often have greater stability than simple ligands. Ethane-1,2-diamine, $H_2N.CH_2.CH_2.NH_2$ is an example: both nitrogen atoms can donate an electron pair to the same metal atom by forming a ring structure (known as a **chelate**):

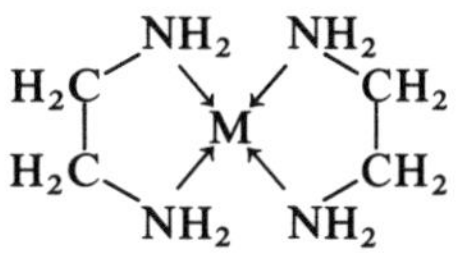

NH_2 NH_2
H_2C CH_2
M
H_2C CH_2
NH_2 NH_2

Such a complex is more stable than a simple ammine complex, because even if one point of attachment comes adrift the other end of the ligand remains attached. Tridentate ligands are even more stable than bidentate, and so on. One of the most powerful chelating agents of all is "edta" (ethylene-diaminetetra-acetic acid), the anion of which has *six* points of attachment. Such a ligand can even form reasonably stable complexes with "non-complexing" ions such as Ca^{2+} and Li^+, and if added to a B-metal it forms a complex species of great stability. The simple ion is literally "wrapped up" in the ligand, and removed from the solution just as effectively as if it had been precipitated out. If edta is added to a solution of zinc ion, it is impossible for instance to precipitate the sulphide of the metal with hydrogen sulphide.

Antimony and tin form soluble thio-complexes SbS_3^- and SnS_3^{2-} when a solution of ammonium sulphide (an alkaline solution containing S^{2-}) is added to a sulphide of antimony or tin. The solubility of these sulphides in ammonium sulphide is used as an analytical test for antimony and tin.

Properties of B*-metal complexes*. Complex ions of B-metals have the following properties:

(i) They are colourless, except when there is colour associated with the ligand.

(ii) They are generally more easily decomposed than complex ions of transition metals, though far more stable than complex ions of *s*-block metals (where these exist).

(iii) They are usually soluble in water, and do not readily form precipitates.

This last property requires some explanation, for a precipitate is in effect a complex of zero charge, which happens to be insoluble in water. In the stepwise loss of protons which occurs on adding alkali to $Zn(H_2O)_4^{2+}$, the following species are obtained with increasing pH:

$$\underset{\text{soluble}}{[Zn^{(H_2O)_3}_{(OH)}]^+} \rightarrow \underset{\text{insoluble}}{Zn(OH)_2.xH_2O(s)} \rightarrow \underset{\text{soluble}}{[Zn^{(H_2O)}_{(OH)_3}]^-} \rightarrow \underset{\text{soluble}}{Zn(OH)_4^{2-}}$$

The solubility of precipitates and the stability of complexes are therefore closely connected.

Stability constants. The tendency of a complex ion to decompose in aqueous solution is expressed quantitatively by its **stability constant** at a given temperature, which is the equilibrium constant for the equilibrium:

$$M + nL \rightleftharpoons M_nL$$

(M = metal; L = ligand; charges omitted)

The reciprocal of K, which is the equilibrium constant for the reaction from right to left, is sometimes termed the instability constant.

B-metal complex ions in general have lower stability constants than those of transition metals, though there are exceptions. (See also section 35.11).

34.9 Some common salts containing B-metals

Simple binary salts (halides and sulphides) have been dealt with and they will not be considered further. Other salts may be classified into

(a) *Those in which the metal is the cation:* e.g: lead nitrate, zinc sulphate etc.

(b) *Those in which the metal is in the anion:* e.g: sodium dioxozincate, Na_2ZnO_2.

Many B-metal oxides (ZnO, SnO, SnO_2, PbO_2, Sb_2O_3) will dissolve in alkalis such as sodium hydroxide, forming hydroxo-complexes. On evaporation of the solution the solid which is obtained is the *oxo*-salt formed by loss of water. B-metal salts, when soluble, form colourless solutions provided there is no other ion present which might confer colour. Many solid binary salts of B-metals are coloured due to charge transfer absorption. Charge transfer absorption is the absorption of electromagnetic energy due to the transfer of an electron from one atom to another. Where this energy of absorption corresponds to a wavelength in the visible region, the substance will be highly coloured. Thus halides, oxides and sulphides are frequently brightly coloured or black.

When a B-metal salt is soluble in water, the solid obtained on crystallization is often hydrated. This is generally due to the hydrated cation of the metal being itself included in the crystal lattice. Zinc sulphate is soluble in water and crystallizes out as $ZnSO_4.7H_2O$. Such behaviour is also observed with *s*-block metals and magnesium, for example, forms $MgSO_4.7H_2O$.

All nitrates of metals are soluble in water. B-metal nitrates decompose on heating to give the oxide, oxygen, and nitrogen dioxide.

Addition of CO_3^{2-}(aq) to a solution of a B-metal ion generally results in the precipitation of a non-stoichiometric basic carbonate such as xPbCO$_3$.yPb(OH)$_2$. In some cases the normal carbonate or something approximating to it can be obtained by using hydrogencarbonate ions, HCO_3^-(aq), as the precipitating agent. All B-metal carbonates, whether normal or basic, are white in colour and insoluble in water.

Most of the B-metal sulphates are soluble in water and crystallize out in hydrated form but lead(II) sulphate forms as an unhydrated white precipitate when the relevant ions are mixed. This reaction is sometimes used as a test for lead ions (among the common metals only barium forms a sulphate whose insolubility is comparable), and for the gravimetric estimation of lead.

34.10 Summary

The B-metals have characteristic properties which distinguish them from *s*-block metals and transition metals.

(1) They are generally fairly soft, dense, with relatively low melting points.

(2) Unlike *s*-block metals, they have little tendency to form truly ionic lattices; this is often interpreted by saying that the ions which they form are more readily polarized.

(3) Like transition metals, they readily form complex ions, but these differ from transition metal complexes in that

(a) they are colourless (unless the ligand causes the colour);

(b) they do not contain unpaired electrons, and are thus diamagnetic.

(4) Their compounds show two oxidation numbers, "Group number" and "Group number minus two", especially at the bottom of the periodic table (inert-pair effect). This is in contrast to transition metals, where adjacent oxidation states are more commonly one unit apart, and *s*-block metals where variable oxidation state is not shown.

For a table summarizing the properties of metals see section 35.19.

34.11 Summary of the properties of Group IV elements

In addition to the two common B-metals tin and lead, Group IV also contains two of the commonest non-metals, carbon and silicon (Chapter 24). Table 34.5 summarizes the similarities and differences that occur within this Group. The lack of available d-levels in carbon leads to important differences from the remaining members of the Group (illustrated here by silicon):

(a) *Formulae*. SiF_6^{2-} forms readily, CF_6^{2-} cannot form.

(b) *Structure*. SiO_2 is a polymeric solid; CO_2 is a monomeric gas.

(c) *Bonding*. Silicon does not form formal double bonds (single bonds between silicon and other atoms do have multiple character due to the overlap of *d*-orbitals); carbon forms double bonds with itself, and with N, O and S.

(d) *Reactivity*. $SiCl_4$ reacts with cold water; CCl_4 is unaffected by boiling water. SiH_4 explodes in air; CH_4 is kinetically stable in air.

TABLE 34.5

THE PROPERTIES OF GROUP IV

Property	Carbon	Silicon	Germanium	Tin	Lead
Element					
B.p. (in K)	4300	2950	3100	2960	2025
Co-ordination number	3 or 4	4	4	4 or 6	12
	Increasing metallic character down the Group →				
Oxidation number	Usually 4	+4	+4, (+2)	+2, +4	+2, (+4)
	Increasing inert-pair effect down the Group				
Hydride, XH_4	Thermal stability decreases markedly down the Group →				
Other hydrides	Very many (section 6.9)	Si_nH_{2n+2} only	Ge_nH_{2n+2} only	None	None
Chlorides, b.p. (in K)					
XCl_2	—	—	—	896	1227
XCl_4	350	330	356	386	Decomposes
	Of the chlorides, only $PbCl_2$ shows any marked ionic character				
Oxides, m.p. of XO_2 (in K)	217	1883	1390	1400	Decomposes (PbO 1160)
	Acidic character decreases down the Group →				
	Ionic character increases down the Group →				

Study Questions

1. In what ways does zinc resemble

(a) A Group IIA metal?
(b) A transition metal?
(c) A typical B-metal such as lead or tin?

2. Show how the properties of lead and tin differ from the corresponding properties of silicon. In what ways do the three elements resemble each other?

3. How do (a) the oxides, (b) the sulphides of antimony and bismuth compare with those of tin and lead?

4. Thallium forms two oxides.

(a) What would you expect their formulae to be?
(b) Would you expect them to be coloured?
(c) Which oxide would be the more basic?

5. Addition of NaOH(aq) to $CdCl_2$(aq) gives a white precipitate insoluble in excess alkali.

(a) Suggest a formula for the precipitate.
(b) How does the behaviour of cadmium differ from that of zinc in this case?
(c) How would you expect the precipitate to react with HCl(aq)?

6. (a) Why does lead(II) sulphide dissolve in hot concentrated hydrochloric acid?
(b) Why does arsenic(III) sulphide dissolve in ammonium sulphide?
(c) Why does lead(II) bromide dissolve in potassium bromide solution?

7. (a) Suggest why lead appears to react only very slightly with hydrochloric acid.

(b) Find out why tetraethyl-lead(IV) is used as an antiknock compound in petrol.

(c) Find out why lead is often used on roofs.

(d) Find out why lead is used in car batteries.

8. (a) Why are mercury(I) salts diamagnetic?

(b) HgF_2 boils at 650°C and $HgCl_2$ boils at 303°C. The former is insoluble, but the latter soluble in organic solvents. Comment on the likely nature of these compounds.

9. (a) Use Appendix 1 to determine whether

(i) $Hg^{2+}(aq)$ will react with $Sn^{2+}(aq)$.

(ii) $Hg^{2+}(aq)$ will react with $Pb^{2+}(aq)$.

(b) What would you expect to *observe* if solutions of mercury(II) chloride and tin(II) chloride were mixed?

(c) What would you expect to *observe* if solutions of mercury(II) chloride and lead(II) nitrate were mixed?

10. What evidence is there for the inert-pair effect in Group IIB?

11. When a brown solid, A, was heated, oxygen was evolved and a red solid, B, and eventually a yellow solid, C, were formed. C dissolved in dilute nitric acid to give a solution of D, which with hydrochloric acid gave a white precipitate of E. Dry E reacted with chlorine to give the thermally unstable liquid, F. Molten E could be electrolyzed to give the metal G at the cathode. When the solution of D was treated with H_2S, a black precipitate, H, formed.

(a) Identify the lettered compounds.

(b) Give equations for the reactions.

(c) How can H be converted into G?

(d) How can B be converted directly into D?

12. When a white solid, J, was heated, carbon dioxide was evolved and the solid became yellow. On cooling, the solid, K, became white. K dissolved in hydrochloric acid to give a solution of L. When L was treated with sodium hydroxide solution, a white precipitate, M, formed, but this then dissolved in excess base to give N.

(a) Identify the lettered compounds.

(b) Give equations for the reactions.

13. Aluminium is a Group III element. In what ways does it resemble (a) an *s*-block metal (b) a B-metal?

CHAPTER 35

The transition metals

The transition metals occupy the central region of the periodic table known as the *d*-block. There are three transition series, each marking the filling up of *d*-orbitals in the atoms, but this chapter will be mainly concerned with the **first transition series**, from scandium to copper. Zinc is treated in some books as a transition metal, being the next member of the series after copper, but as was shown in Chapter 34, it is more like the B-metals to its right in the periodic table.

35.1 Electronic structures of the first transition series

It was stated in Chapter 3 that, except for hydrogen, the sub-levels *s*, *p*, *d*, etc., within a quantum shell differ in energy. Energy level diagrams such as those in Fig. 3.9 illustrate this. As the atomic number increases, and orbitals interact with each other to a greater extent, this splitting becomes even more marked, and before the first transition series is reached, the principal quantum levels actually begin to overlap in energy. It is for this reason (tacitly assumed in earlier chapters) that the 3*d* energy level does not begin to fill with electrons immediately after the 3*p* is full. The element which follows argon in the periodic table is potassium ($1s^2$; $2s^2p^6$; $3s^2p^6$; $\underline{4s^1}$) and the 3*d* level remains unfilled because it is higher in energy than the 4*s*.

Once the 4*s* energy level is filled, the 3*d* can begin to fill; here the pattern of behaviour differs from that in the previous period. The transition metal electronic structures are given in Table 35.1.

The filling up is not entirely regular, due to the fact that the 3*d* and 4*s* levels are very close in energy along this series. Although the 3*d* level normally fills preferentially, chromium and copper are exceptional in that a 4*s* electron is "lost" to the 3*d*. It appears that half-filled and filled *d*-shells are favoured.

35.2 Physical properties and uses of the transition metals

The metals themselves differ from the *s*-block elements and the B-metals, because of the high binding energy which results from incompletely filled *d*-levels. The chief characteristics are:

(i) high melting and boiling points, and high heats of vaporization;

(ii) considerable mechanical strength, especially in alloys.

Figure 35.1 is a plot of melting and boiling points for the metals of the first long period. Boiling point of an element is a more reliable

TABLE 35.1

Element		Electronic structure in ground state (short form)	Electronic structure (full form)	
s-block	Ar	2, 8, 8	$1s^2$; $2s^2p^6$; $3s^2p^6$	
	K	2, 8, 8, 1	[Ar core]; $4s^1$	
	Ca	2, 8, 8, 2	[Ar core]; $4s^2$	
d-block	Sc	2, 8, 9, 2	[Ar core] $3d^1$; $4s^2$	
	Ti	2, 8, 10, 2	[Ar core] $3d^2$; $4s^2$	
	V	2, 8, 11, 2	[Ar core] $3d^3$; $4s^2$	
	Cr	2, 8, 13, 1	[Ar core] $3d^5$; $4s^1$	half-filled *d*-shell
	Mn	2, 8, 13, 2	[Ar core] $3d^5$; $4s^2$	half-filled *d*-shell
	Fe	2, 8, 14, 2	[Ar core] $3d^6$; $4s^2$	
	Co	2, 8, 15, 2	[Ar core] $3d^7$; $4s^2$	
	Ni	2, 8, 16, 2	[Ar core] $3d^8$; $4s^2$	
	Cu	2, 8, 18, 1	[Ar core] $3d^{10}$; $4s^1$	filled *d*-shell
	Zn	2, 8, 18, 2	[Ar core] $3d^{10}$; $4s^2$	filled *d*-shell

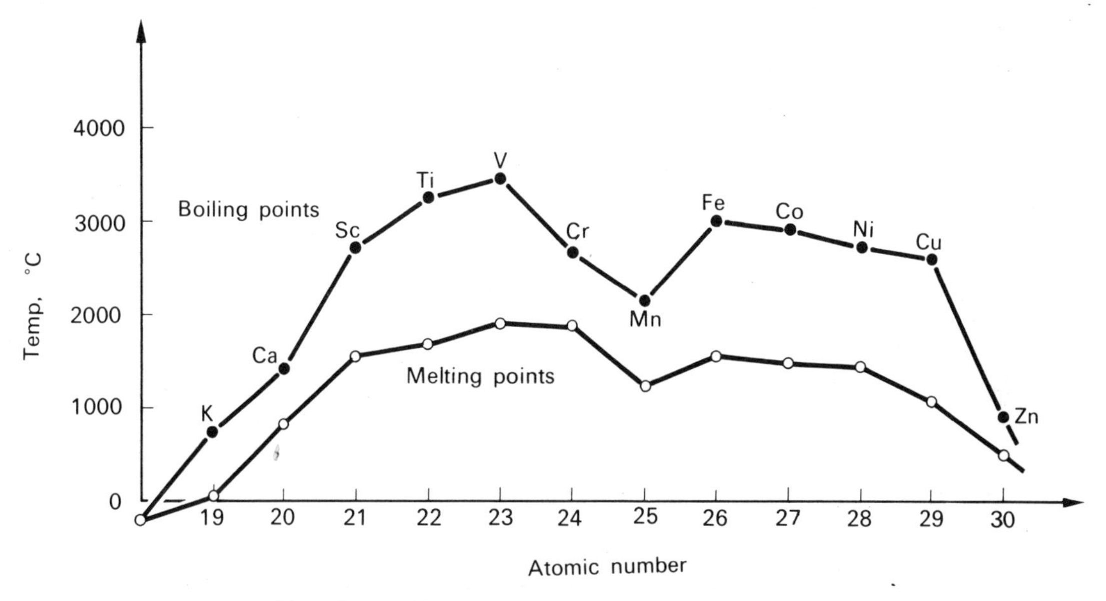

FIG. 35.1. Melting and boiling points of transition metals.

guide to binding energy than melting point, since the latter is rather dependent upon crystal structure, yet both properties are seen to follow the same trend. Figure 35.2 shows how binding energy, as measured by the heat of vaporization, varies along the same series. It is very noticeable that zinc has a low melting point, boiling point and binding energy; all this suggests that relatively weak forces must exist between the atoms in the metal. In Chapter 5 a correlation between binding energy and number of valence electrons available for bond formation was noted: if this conclusion is valid then we must further conclude that the *d*-electrons in zinc play little part in metallic bond formation. A completely filled *d*-shell, $3d^{10}$, is stable and to some extent "inert"

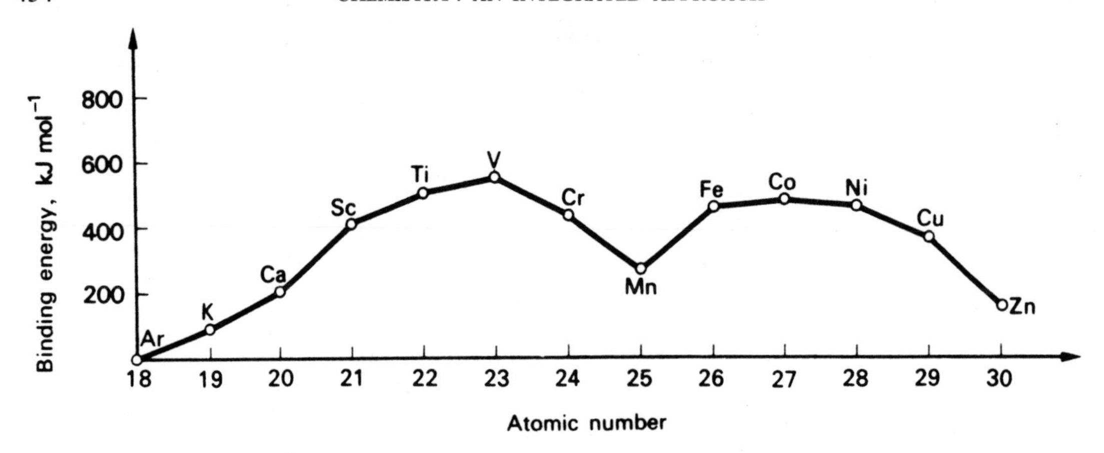

FIG. 35.2. Binding energy, as measured by heat of vaporization.

—it can be treated as part of the inner core of electrons.

The dip in binding energy (also m.p. and b.p.) at manganese coincides with the configuration $3d^5$; half-filled d-shells have some sort of stability associated with them, but this is not so great as that for filled shells.

The boiling points of the first transition series are lower than those of corresponding metals in the second and third transition series. It is evident that stronger interatomic forces operate in the later series; tungsten, for instance, is the least volatile metal known, and because of its involatility it is used in electric light filaments.

The high mechanical strength of the transition metals make some of them valuable structurally: iron is the most abundant and the most widely used transition metal; titanium has in recent years become a structural metal in its own right —it is lighter than steel but superior to aluminium in strength, corrosion resistance and heat resistance. Unfortunately despite its widespread occurrence titanium is very costly to extract. Some of the transition metals find specialized uses—nickel and chromium because of their resistance to chemical attack and copper because of its high electrical conductivity—and the remainder, such as manganese and cobalt, are widely used in special steels. Mechanical properties cannot be explained simply in terms of crystal structure and binding energy, since they depend more upon such factors as grain structure and lattice defects.

35.3 Occurrence and extraction of transition metals

The transition metals are less electropositive than the s-block metals as shown by their higher first ionization energies (Fig. 35.3), and occur mainly as oxides or sulphides. The "noble" metals, gold, platinum, osmium, iridium, etc., are extremely unreactive and usually occur native.

Apart from titanium, for which special methods of extraction are required, the oxides of the metals can be reduced with carbon at high temperatures. Since carbon is a cheap reducing agent, this is the method normally employed, though for special purposes aluminium may be used as reducing agent (Thermit process). The sequence of operations for extracting the metal are therefore:

(i) Mechanical concentration of the ore (flotation, magnetic separation, etc., where appropriate).

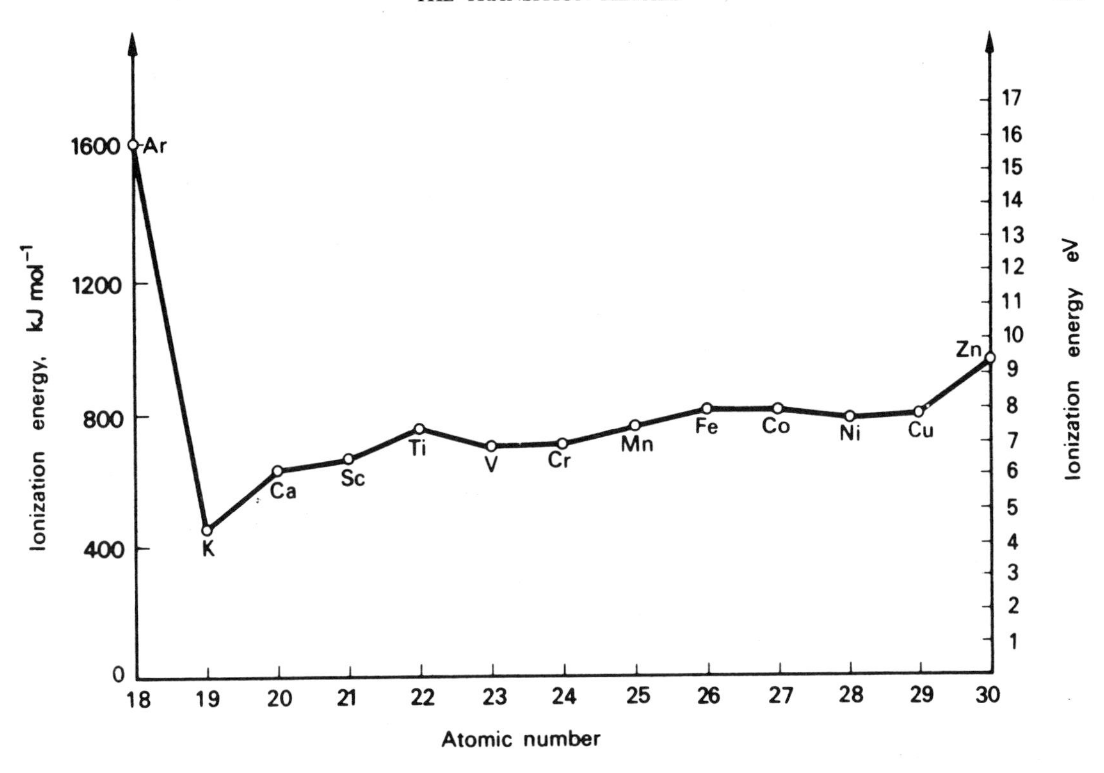

FIG. 35.3. First ionization energies along first transition series.

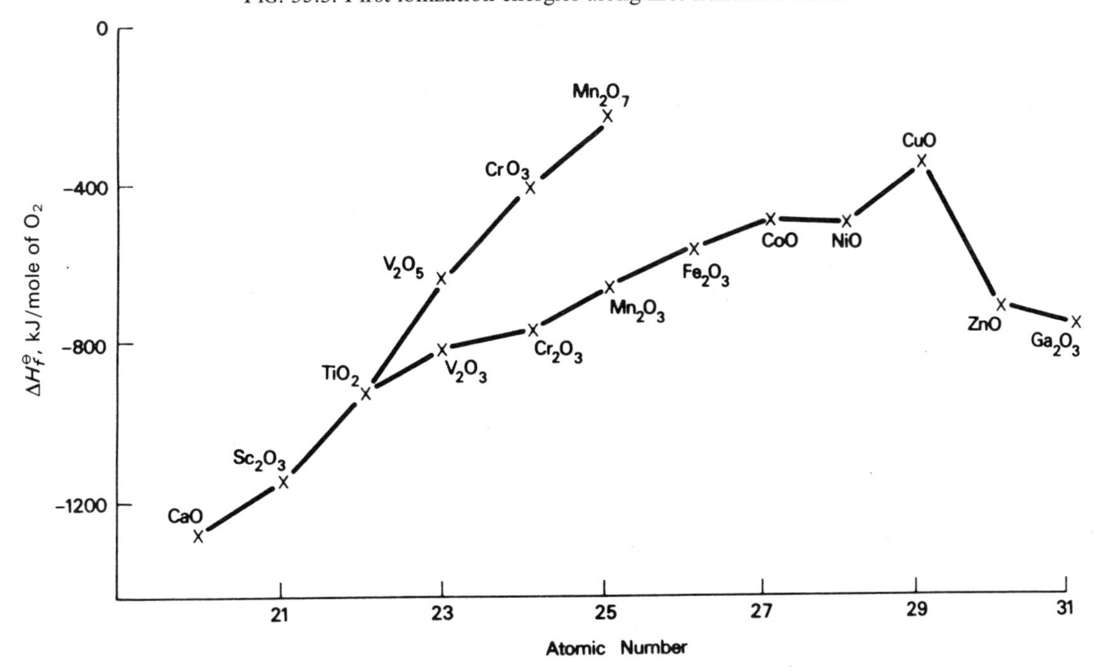

FIG. 35.4. Heats of formation of transition metal oxides.

(ii) Roasting of sulphide ores to convert to the oxide, as sulphides are not readily reduced by carbon directly (see Chapter 13).

(iii) Reduction of the ore with carbon in the form of coke.

(iv) Refining of the metal. This may involve lowering of the carbon content by blowing air through the molten metal (→ CO). In many cases refining is done by electrolysis (copper, chromium, silver).

For many purposes a pure metal is not required. In the case of chromium, for instance, the pure metal is obtained by electrolytic deposition (chromium plate) but the alloy *ferrochrome*, composed of chromium and iron, is obtained by reducing the ore *chromite*, $FeCr_2O_4$, and is quite suitable for making stainless steel.

Figure 35.4 is a plot of the heat of formation of the common oxides of the transition metals against atomic number of the metal. The extraction of some transition metals (iron, chromium, nickel and copper) from their ores is dealt with in Chapter 13.

35.4 Ion formation in the first transition series

One of the main features which distinguishes a transition metal from other metals is the property of **variable oxidation state**. The B-metals are capable of forming compounds in two oxidation states due to the inert-pair effect, but the transition metals can show *several* states, frequently differing from each other by only one unit. Figure 35.5 is an oxidation number diagram for the first transition series: the oxidation states shown are either simple hydrated ions (lower oxidation states) or oxo-ions (higher oxidation states).

The main features shown in Fig. 35.5 are as follows:

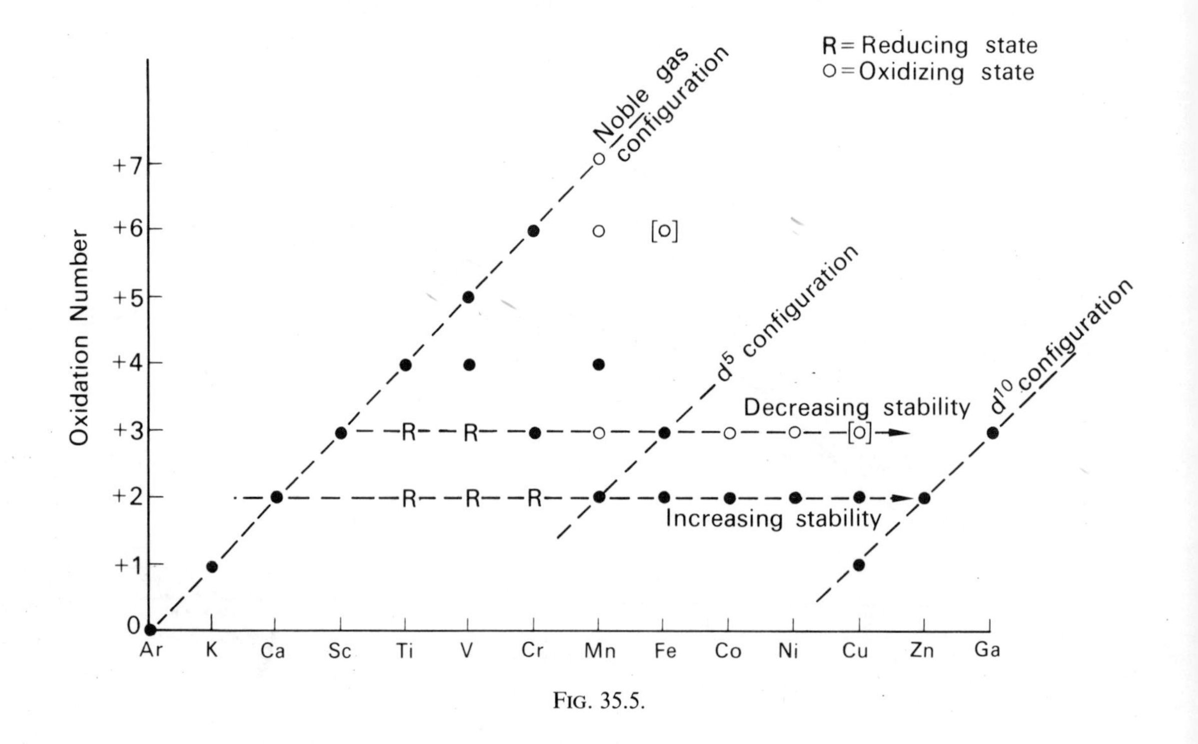

FIG. 35.5.

(i) Practically all the transition metals possess oxidation states of +2 and +3. Both nickel(III) and copper(III) compounds are rather rare, while scandium(II) compounds do not exist.

(ii) There is a sequence of ions which is *formally* related to the noble gas argon. Since very small, highly charged ions exert such a powerful electrostatic field, however, V^{5+}, Cr^{6+} and Mn^{7+} are never observed free, and in aqueous solutions oxo-ions are formed.

Simple ions of the transition metals do not exist in aqueous solution: the ions which appear to be simple are in fact complexed with water, generally accepting six H_2O ligands at octahedral positions. Other ligands such as Cl^-, NH_3, CN^-, CNS^-, and $H_2N.CH_2.CH_2.NH_2$ readily form complexes with the transition metals (see later sections).

Complexes of the transition metals have properties which markedly distinguish them from those of B-metals:

(a) *They are generally coloured*, exceptions are those with completely filled *d*-orbitals, e.g. those of copper(I), and silver(I), or with empty *d*-orbitals, e.g. scandium(III).

(b) *They are frequently paramagnetic*, showing that there must be unpaired electrons present. Ions of *s*-block metals and B-metals do not show paramagnetic behaviour.

(c) *They often act as homogeneous catalysts in solution*. The metals and their solid compounds are also good heterogeneous catalysts for many gas phase reactions.

The transition metals occupy similar positions in the electrochemical series to the B-metals—in fact there is considerable overlap—indicating that the tendency of the metals to form aqueous ions is generally less strong than for the *s*-block metals.

Table 35.2 gives the colours of the +2 and +3 hydrated ions.

TABLE 35.2

	+2 ion	+3 ion
Ti	—	violet
V	violet	green
Cr	blue	dark green
Mn	pale pink	red
Fe	pale green	yellow-brown
Co	pink	—
Ni	green	—
Cu	blue	—
Zn	colourless	—

35.5 Redox potentials

It was noted in section 26.2 that the d^5 configuration (half-filled $3d$ shell) was stable relative to other configurations: the redox potentials for the formation of the +2 and +3 ions provide further evidence for this effect (Figs. 35.6 and 35.7).

When a transition metal forms an ion, it is the *s* electrons which are lost first, not the *d*. The electronic configurations of the +2 and +3 ions are shown in Table 35.3.

The redox data indicate that Mn^{2+}, Fe^{3+}, Cu^+ and Zn^{2+} compounds are more stable relative to their adjacent oxidation states than

TABLE 35.3

Metal	Electronic configuration (outer shell only; core = argon)	
	+2 state	+3 state
Sc	—	[argon]
Ti	$3d^2$	$3d^1$
V	$3d^3$	$3d^2$
Cr	$3d^4$	$3d^3$
Mn	$\mathbf{3d^5}$	$3d^4$
Fe	$3d^6$	$\mathbf{3d^5}$
Co	$3d^7$	$3d^6$
Ni	$3d^8$	$3d^7$
Cu	$3d^9$ [+1 state: $3d^{10}$]	—
Zn	$\mathbf{3d^{10}}$	—

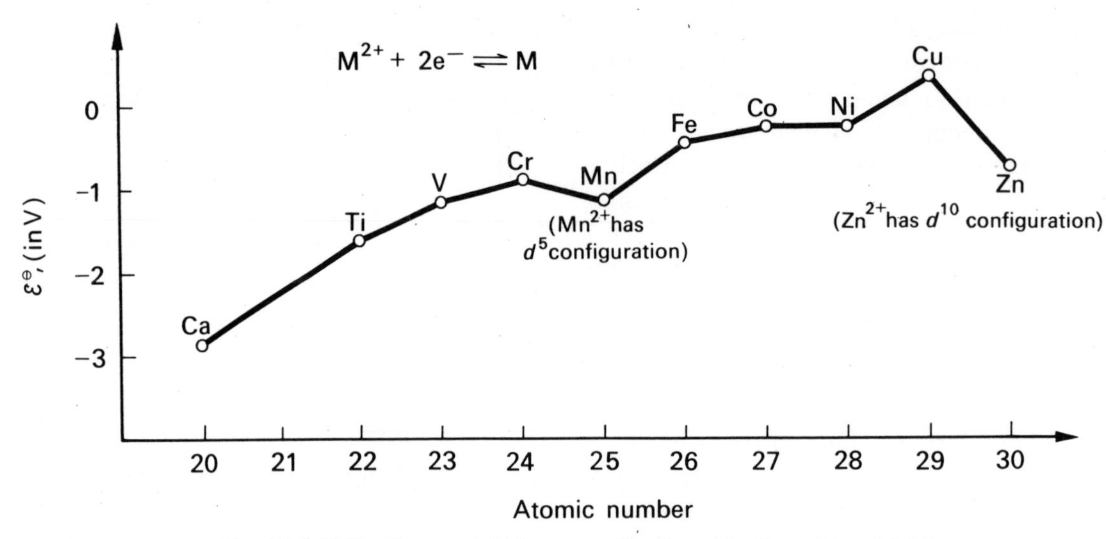

FIG. 35.6. Half-cell potentials (compare this plot with Figs. 35.1 and 35.2).

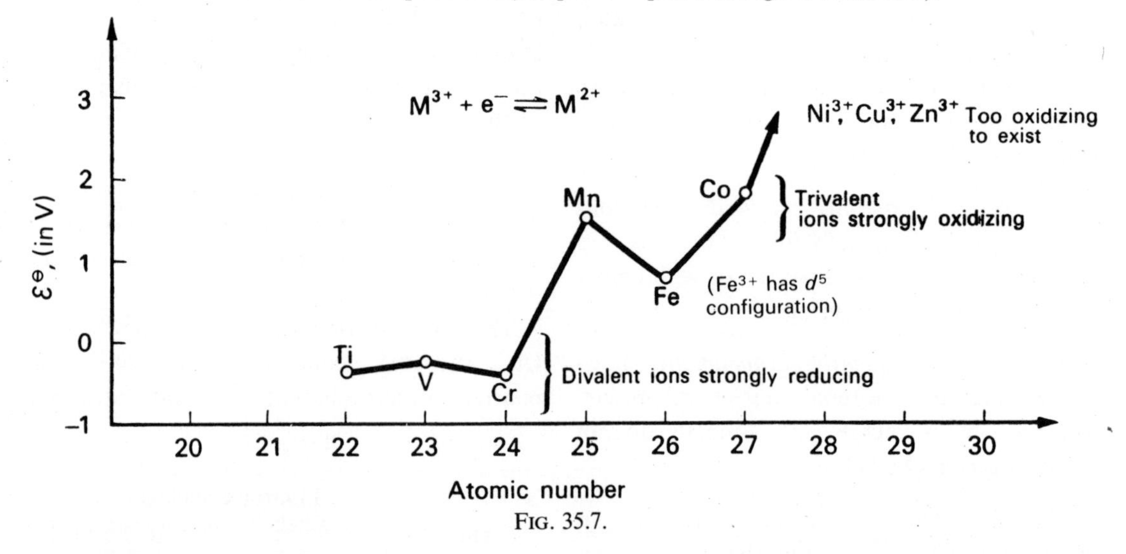

FIG. 35.7.

their position in the transition series would have suggested. Figure 35.6 should be compared with Figs. 35.1 and 35.2.

35.6 Oxides and hydroxides of the transition metals

All the transition metals form oxides. The affinity of transition metals for oxygen is in roughly the same order as the electrochemical series, and hence the metals like silver and gold have very low affinities for oxygen and do not combine easily with the gas.

Addition of $OH^-(aq)$ to the aqueous solution of a transition metal salt results in precipitation of the hydroxide in most cases. Amphoteric character is less common than among B-metals, though one or two hydroxides, for instance chromium(III), redissolve in excess alkali to form an anionic hydroxo-complex. Table 35.4 summarizes the action of alkali on aqueous

TABLE 35.4

	Precipitate on adding dilute alkali	Colour of precipitate	Comments
Cr^{3+}	$Cr(OH)_3.xH_2O$	Green	Highly insoluble in water but dissolves in conc. alkali (amphoteric)
Mn^{2+}	$Mn(OH)_2.xH_2O$	Buff	Not amphoteric
Fe^{2+}	$Fe(OH)_2.xH_2O$	Grey-green	Readily oxidized to $Fe(OH)_3$
Fe^{3+}	$Fe(OH)_3.xH_2O$	Brown	Not amphoteric
Co^{2+}	$Co(OH)_2.xH_2O$	Pink	Dissolves in conc. alkali
Ni^{2+}	$Ni(OH)_2.xH_2O$	Green	Not amphoteric
Cu^{2+}	$Cu(OH)_2.xH_2O$	Pale blue	Rapidly loses water forming black hydrated CuO
Ag^{+}	$Ag_2O.xH_2O$	Greyish-white	True hydroxide does not exist

solutions of some common transition metal cations.

The trivalent hydroxides are more insoluble in water than the divalent ones. Iron(III) and chromium(III) hydroxides are precipitated by a buffer solution of pH = 10, such as a mixture of ammonium chloride and ammonium hydroxide, whereas the divalent hydroxides do not precipitate under these conditions. This fact is made use of in analysis, for separating and identifying these two metals. The reason for the higher insolubility is probably the higher lattice energy of the hydroxides containing triply-charged ions.

The transition metal hydroxides form fairly gelatinous precipitates containing loosely bonded water in non-stoichiometric amounts, and if they are heated to remove this water decomposition to the oxide always occurs. Copper(II) hydroxide decomposes at room temperature on standing, and silver(I) hydroxide is not formed.

Transition metals form oxides which illustrate well the property of variable valency. Figure 35.8 gives the formulae of some of the common ones. In general these oxides are:

(i) basic for low oxidation number, acidic for high oxidation number;
(ii) insoluble in water;
(iii) highly coloured; or black;
(iv) macromolecular in structure, rather than truly ionic (section 22.12).

Basic or acidic character. Oxides lying below the line in Fig. 35.8 will dissolve in acids; the rate of dissolution is often slow unless the solid is finely divided. For instance, prolonged boiling with concentrated hydrochloric acid is necessary to dissolve chromium(III) oxide, Cr_2O_3, although this is a *rate* effect and the Cr^{3+}(aq) ions are perfectly stable once formed.

Oxides lying on the line are amphoteric in character. Manganese(IV) oxide, MnO_2, reacts slowly with fused alkali forming the

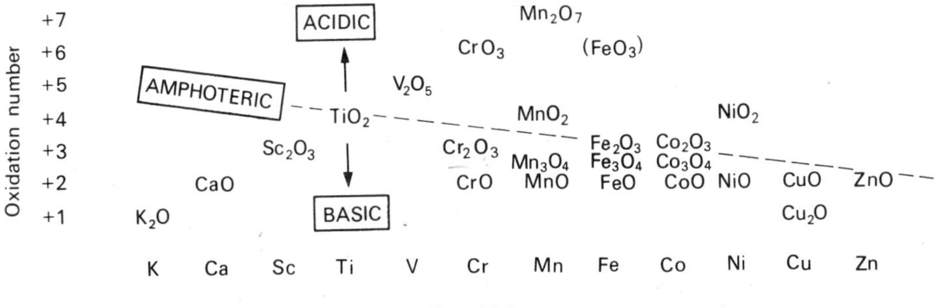

FIG. 35.8.

manganate(IV) ion:

$$MnO_2 + 2OH^- \rightarrow MnO_3^{2-} + H_2O$$

It reacts even more readily in the presence of oxygen, or an oxygen-donating substance such as potassium nitrate, for then it is oxidized up to the +6 state, forming manganate(VI):

$$MnO_2 + 2OH^- + [O] \rightarrow MnO_4^{2-} + H_2O.$$

Manganese(IV) oxide dissolves in ice-cold concentrated hydrochloric acid forming the chloride $MnCl_4$, but this is unstable and the more usual reaction which occurs at room temperature is the redox process:

$$MnO_2 + 4HCl(aq) \rightarrow Mn^{2+}(aq) + 2Cl^-(aq) + Cl_2(g) + H_2O.$$

The properties of manganese(IV) oxide are predominantly acidic and it will not react with acids unless it can oxidize them.

Chromium(VI) oxide, CrO_3, is a typical oxide of a transition metal in a high oxidation state. Unlike the lower oxides, it is soluble in water, in which it dissolves forming chromic(VI) acid:

$$H_2O + 2CrO_3 \rightarrow H_2Cr_2O_7.$$

The solution which results contains mainly the dichromate(VI) ion, $Cr_2O_7^{2-}$.

Structure of transition metal oxides. Many of the oxides have roughly cubic structures, but they are macromolecular rather than ionic. There is no evidence for ions being present, and the solids have very high melting points. Layer structures are frequently present and this accounts for the very high insolubility of these oxides; they are often non-stoichiometric.

Reactions of transition metal oxides. The chemical inertness of transition metal oxides may have an important bearing on the properties of the metals themselves, though the subject is far from being fully understood. Certain metals, notably iron and chromium, become "passive" in contact with nitric acid at certain concentrations, that is, they refuse to dissolve in it. The formation of a protective oxide coat may play a part in the mechanism of passivity.

All the transition metal oxides can be reduced by carbon though certain ones such as titanium(IV) oxide require an exceedingly high temperature (Chapter 13). Aluminium is also a reducing agent, which undergoes a Thermit reaction with transition metal oxides. Hydrogen reduces the oxides of metals low in the electrochemical series, but fails with metals higher up. For instance copper(II) oxide is reduced while chromium(III) oxide is not.

Oxides of mixed oxidation state occur among the transition metals, and these may be regarded as "salts" of two oxides (one basic and one acidic) of the same metal. For instance:

$Fe_3O_4 = FeO.Fe_2O_3$
iron(II) diiron(III) oxide

$Mn_3O_4 = 2MnO.MnO_2$
dimanganese(II) manganese(IV) oxide

35.7 Oxo-ions of transition metals

The higher oxidation states of transition metals exist only as anions, which means that the names of these compounds end in *-ate*. Some important oxo-salts are listed in Table 35.5 and these will be considered in this section.

The salts listed in Table 35.5 are all water-soluble, and are oxidizing agents. The oxidizing reactions of the anions are important in volumetric analysis.

TABLE 35.5

Name and formula of anion	Name and formula of a typical salt	Oxidation state of transition metal
Vanadate(V), VO_3^-	Ammonium vanadate(V) NH_4VO_3	+5
Chromate(VI), CrO_4^{2-}	Potassium chromate(VI) K_2CrO_4	+6
Dichromate(VI), $Cr_2O_7^{2-}$	Sodium dichromate(VI) $Na_2Cr_2O_7$	+6
Manganate(VI), MnO_4^{2-}	Potassium manganate(VI) K_2MnO_4	+6
Manganate(VII), MnO_4^-	Potassium manganate(VII) $KMnO_4$	+7

Vanadates. Vanadates(V) are salts of the acidic oxide V_2O_5, which is a yellow solid formed during the extraction of the metal. Dilute alkali dissolves V_2O_5, forming a series of vanadates corresponding in formula to the phosphates, section 23.9, of which the trioxovanadates are the most stable. Action of NH_4^+(aq) on a solution of a trioxovanadate, VO_3^-(aq), precipitates ammonium trioxovanadate, NH_4VO_3. Solutions of vanadates are used for volumetric analysis, and vanadium(V) oxide itself is used as a catalyst in the manufacture of sulphur trioxide and sulphuric acid.

Chromates and dichromates. Chromium(VI) forms a series of oxo-anions depending on the pH of the solution; in alkali, the chromate(VI) ion is formed when a Cr^{3+} salt is heated with sodium peroxide:

$$\underset{\text{green}}{2Cr^{3+}} + 4OH^- + \underset{\text{from peroxide}}{3O_2^{2-}} \rightarrow \underset{\text{yellow}}{2CrO_4^{2-}} + 2H_2O$$

On acidification of an aqueous solution of a chromate(VI), the ion "condenses" to form dichromate(VI) ion, and further to form trichromate(VI) and tetrachromate(VI) if the pH is low enough:

$$\underset{\text{yellow chromate(VI)}}{2CrO_4^{2-}} + 2H^+ \rightleftharpoons \underset{\text{orange dichromate(VI)}}{Cr_2O_7^{2-}} + H_2O; (\rightarrow \underset{\text{orange trichromate(VI)}}{Cr_3O_{10}^{2-}} \text{ etc.})$$

The structure of these poly-ions is similar to those formed by non-metals, such as silicon, phosphorus and sulphur, in that the element is alternately linked with oxygen atoms:

$$\left[\begin{array}{ccccc} & O & & O & \\ & \uparrow & & \uparrow & \\ O\leftarrow & Cr & —O— & Cr & \rightarrow O \\ & \downarrow & & \downarrow & \\ & O & & O & \end{array}\right]^{2-}$$

$$\left[\begin{array}{ccccccc} & O & & O & & O & \\ & \uparrow & & \uparrow & & \uparrow & \\ O\leftarrow & Cr & —O— & Cr & —O— & Cr & \rightarrow O \\ & \downarrow & & \downarrow & & \downarrow & \\ & O & & O & & O & \end{array}\right]^{2-}$$

Potassium chromate(VI) is a yellow crystalline solid which forms a useful analytical reagent. It forms precipitates with a number of metal ions, such as lead(II) chromate, $PbCrO_4$, which is the pigment *chrome yellow*, and silver(I) chromate, Ag_2CrO_4, which is a brick red solid formed at the end point when potassium chromate is used as indicator for silver nitrate titrations.

Sodium dichromate(VI) is an orange solid highly soluble in water and deliquescent. Its deliquescence makes it unsuitable as a standard for volumetric analysis and for this purpose the less soluble potassium dichromate(VI) is used. Potassium dichromate(VI) does not contain water of crystallization and when used as an oxidizing agent one mole of it will accept six moles of electrons from a reducing agent:

$$Cr_2O_7^{2-} + 14H^+ + 6e^- \rightleftharpoons 2Cr^{3+} + 7H_2O; \quad \mathcal{E}^\ominus = +1{\cdot}33\,V$$

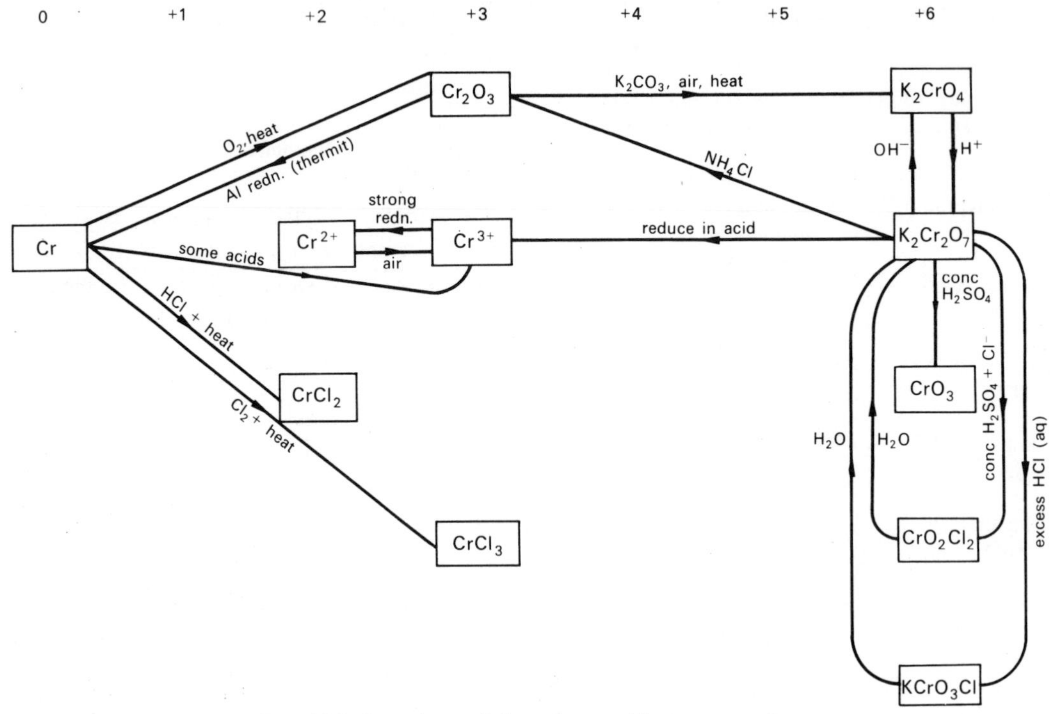

FIG. 35.9. Reactions of chromium and its compounds.

Although there is a colour change from orange to green at the endpoint this is not sufficiently distinct, and an indicator such as barium diphenylamine sulphonate is generally added.

Conversion of chromate(VI) to dichromate(VI) is *not* a redox process since the metal is in the same oxidation state in both ions.

Manganates. The manganate(VI) ion is dark green, and a dark green mass of K_2MnO_4 is formed when manganese(IV) oxide is fused with potassium nitrate (section 35.5). This ion is only stable in alkali, for in acid it disproportionates into manganate(VII) and manganese(IV):

$$\underset{\text{dark green}}{3MnO_4^{2-}(aq)} + 4H^+(aq) \rightarrow \underset{\text{black}}{MnO_2(s)} + \underset{\text{purple}}{2MnO_4^-(aq)} + 2H_2O$$

This is the usual way of preparing potassium manganate(VII): the salt crystallizes out on evaporation. An alternative way, which gives complete conversion instead of just two-thirds, is to oxidize the manganate(VI) ions with chlorine:

$$2MnO_4^{2-}(aq) + Cl_2(g) \rightarrow 2MnO_4^-(aq) + 2Cl^-(aq)$$

Figure 35.10 shows some of the reactions of the various oxidation states of manganese, showing how they are linked together. The +7 state of manganese is normally only accessible via the +4 and +6, though very powerful oxidizing agents are capable of oxidizing directly from manganese(II) to the manganate(VII) ion. A test for manganese is to add a solution of sodium bismuthate(V), "$NaBiO_3$", in nitric acid to the suspected solution; a pink colour of manganate(VII) ion shows the presence of manganese. It is the inert-pair effect which makes sodium bismuthate(V) such a powerful oxidizing agent:

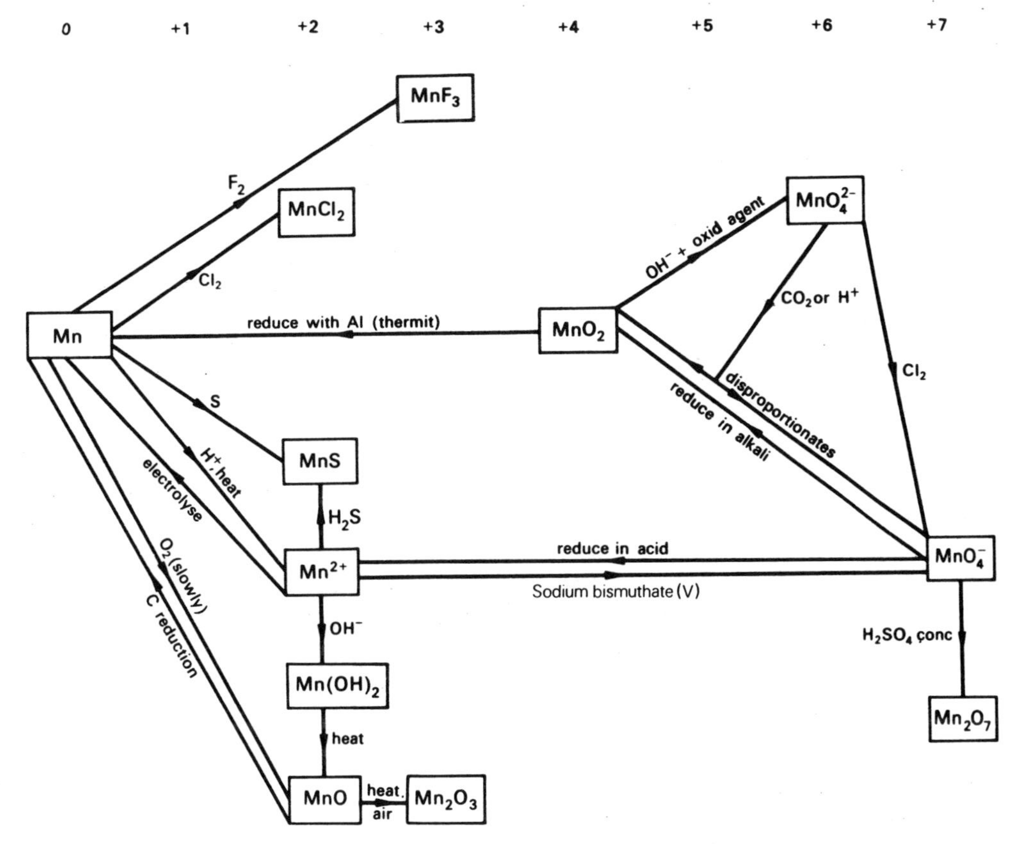

FIG. 35.10. Reactions of manganese and its compounds.

bismuth has a strong tendency to be reduced from the +5 state in bismuthate, to +3.

Potassium manganate(VII). Stable manganate(VII) salts of the alkali metals exist, but the one used for volumetric analysis is potassium manganate(VII), $KMnO_4$. It is less soluble than the sodium salt, and is easier to obtain and keep in the pure state. The extremely dark colour of manganate(VII) solutions is deceptive, for the potassium salt is not very soluble in water (a saturated solution is about 0·4 M at 20°C).

One mole of the manganate(VII) ion will take *five* moles of electrons from a reducing agent:

$$MnO_4^-(aq) + 8H^+ + 5e^- \rightleftharpoons Mn^{2+} + 4H_2O; \quad \mathcal{E}^\ominus = +1{\cdot}51\ V$$

The manganate(VII) ion is one of the most powerful oxidizing agents available for volumetric analysis, and its extremely high half-cell potential suggests that it ought to oxidize water to oxygen ($\mathcal{E}^\ominus$ of O_2/H_2O couple = +1·229 V); in fact this reaction does occur slowly in the presence of light. Potassium manganate(VII) solutions do not alter in concentration over a period of weeks if kept in dark bottles.

Among the oxidations for which potassium manganate(VII) solutions may be used are (a) iron(II) to iron(III), (b) ethanedioate ion, $C_2O_4^{2-}$, to carbon dioxide, (c) hydrogen peroxide to oxygen. All these reactions are used in volumetric analysis.

Along a transition series the highest oxidation

state becomes progressively more oxidizing. Hence the order of increasing oxidizing power is Sc(III) < Ti(IV) < V(V) < Cr(VI) < Mn(VII).

35.8 Transition metal halides

The transition metals readily form compounds with the halogens, the order of reactivity being roughly fluorine > chlorine > bromine > iodine. Fluorine reacts directly with all metals, even the noble metals, and forms a compound with the metal in a high oxidation state. Chlorine attacks all the elements of the first transition series; bromine and iodine also form compounds, though the reactions with iodine generally require the assistance of heat. Fluorine is more powerfully oxidizing than chlorine and sometimes produces a higher oxidation state, for instance MnF_3 and CoF_3, but $MnCl_2$ and $CoCl_2$.

Chlorides. Nearly all the anhydrous chlorides are soluble in water. Among later transition elements, copper(I) chloride, CuCl, silver(I) chloride, AgCl, and mercury(I) chloride, Hg_2Cl_2, are notable for being insoluble in water.

Hydrated chlorides are formed if aqueous solutions of transition metal chlorides are crystallized, but these are hydrolysed on evaporation. Iron(II) chloride is a typical example, for on evaporation it yields the basic chloride, Fe(OH)Cl:

$$FeCl_2(aq) + H_2O \rightleftharpoons Fe(OH)Cl(s) + HCl(g).$$

This property is not exclusive to the chlorides of transition metals: most metal ions are in fact hydrolysed in aqueous solution and hence give basic salts on evaporation, and indeed the salts of the alkali metals and alkaline earth metals are to some extent exceptional in *not* being hydrolysed.

Chlorine and iron filings react to give iron(III) chloride, a red solid which is volatile owing to its molecular structure. Molecules of formula Fe_2Cl_6 exist in the anhydrous solid, exactly analogous to Al_2Cl_6, but on adding water a solution containing $Fe^{3+}(aq)$ and $Cl^-(aq)$ ions is formed. The Fe^{3+} ions are hydrolysed.

The chlorides of the transition metals are coloured in the solid state. Ions are produced when they are dissolved in water; there is often a change of colour, for the aqueous solution takes its colour from the aqueous ion of the metal, while the solid chloride generally has a colour characteristic of a chloro-complex of the metal. Thus copper(II) chloride, $CuCl_2$, is green-blue in the solid state but dissolves to form a blue solution.

Chlorides of very low boiling point are molecular, e.g. $TiCl_4$ and Fe_2Cl_6, while the remainder have generally macromolecular structures. Ions may be produced on melting, as indicated by the fact that most of the fused chlorides can be electrolysed, though the structures of the solids suggest that the bonding there is more covalent in nature.

Figure 35.11 is a graph of heat of formation of the divalent chlorides, MCl_2, against atomic number. This graph shows a similar pattern to Fig. 35.2 because the heats of atomization of the metals are important in determining heat of formation of compounds.

Other halides. The properties of fluorides are such as to suggest that their bonding is generally predominantly ionic in nature.

Bromides and iodides of transition metals are often less soluble in water than the chlorides. Note the trend:

AgF	≫	AgCl	>	AgBr	>	AgI
soluble in water and in dilute $NH_3(aq)$		almost insoluble in water: soluble in dilute $NH_3(aq)$		soluble in conc. $NH_3(aq)$		insoluble in water; insoluble even in conc. $NH_3(aq)$

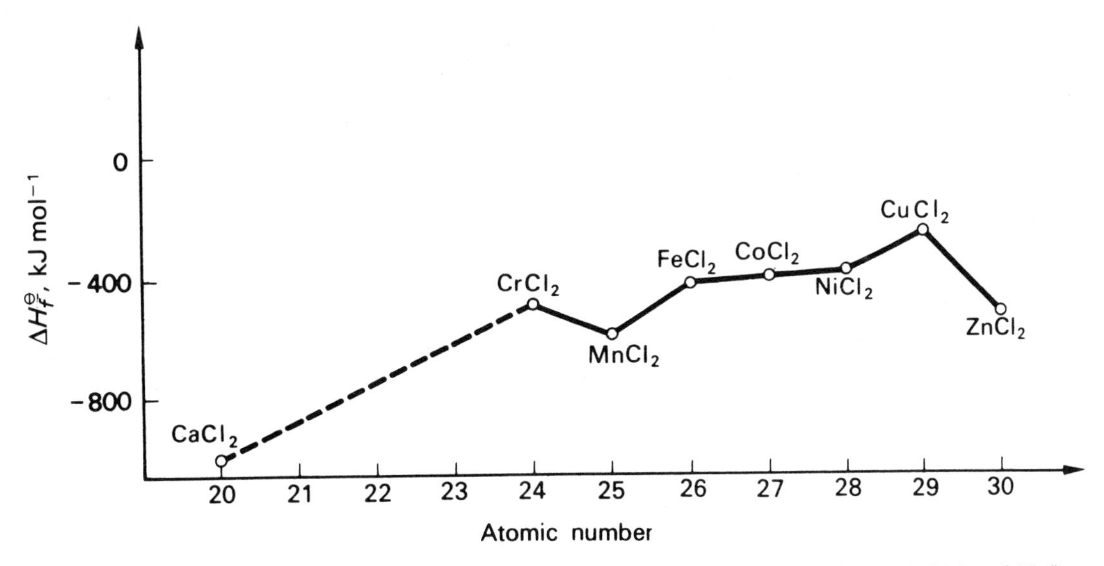

FIG. 35.11. Heat of formation of divalent chlorides (compare this plot with Figs. 35.1, 35.2 and 35.6).

TABLE 35.6

Aqueous ion	Halide complexes	Colour
Ti(IV)	$[TiCl_5.H_2O]^-$ and $TiCl_6^{2-}$	Colourless
Fe(III)	$[Fe(H_2O)_5Cl]^{2+}$, $[Fe(H_2O)_4Cl_2]^+$	Yellow
Co(II)	$CoCl_4^{2-}$	Deep blue
Cu(II)	$CuCl_4^{2-}$	Yellow
Zn(II)	$[ZnCl_3.H_2O]^-$ and $ZnCl_4^{2-}$	Colourless

Complex ions. The transition metals readily form halide complexes, and these usually have different colours from the aquo-complexes, for instance $CuCl_4^{2-}$ (yellow), produced by adding concentrated HCl(aq) to a solution of Cu^{2+}(aq).

35.9 Sulphides of the transition metals

It is a commonly observed phenomenon that elements which form polarizable ions tend to occur together in compounds in the Earth's crust. Transition metal ions are polarizable and so are sulphide ions, so it is not surprising that many of the ores of transition metals contain sulphide. Common ores are *iron pyrites*, FeS_2(s), and *copper pyrites*, $CuFeS_2$. Manganese, nickel, silver and mercury also occur as sulphides. These sulphides are highly coloured or black, macromolecular, and insoluble in water, the colour being due to charge transfer absorption.

All the metals of the first transition series react directly with sulphur to form a sulphide of fairly low oxidation state. Sulphur is not such a powerful oxidizing agent as oxygen or fluorine, so iron reacts to form iron(II) sulphide, FeS, whereas oxygen gives iron(III) oxide, Fe_2O_3.

The sulphide ores of the metals are often used for extraction. It is not practicable to reduce the sulphide ore directly with carbon (the theoretical reasons for this are outlined in Chapter 13) and so the sulphide ore is first converted to the oxide by **roasting** in air. This reaction forms the metal oxide and sulphur dioxide.

35.10 Ammine complexes

Complex ions in which the ligand is ammonia are known as ammines. Ammines are formed by all the transition metals, but many of them are not particularly stable in aqueous solution. Particularly stable ones are formed by the later members of the first transition series, and these are readily distinguished from the aquo-complexes by their different colours.

Cobalt ammines. Although $Co^{3+}(aq)$ is a very powerful oxidizing agent (Fig. 35.7), the addition of ammonia alters the redox potential of the cobalt(II)/cobalt(III) couple completely and stabilizes the +3 state.

$$Co^{3+}(aq)+e^{-} \rightleftharpoons Co^{2+}(aq); \quad \mathcal{E}^{\ominus} = +1{\cdot}84\,V$$

$$Co(NH_3)_6^{3+}+e^{-} \rightleftharpoons Co(NH_3)_6^{2+}; \quad \mathcal{E}^{\ominus} = +0{\cdot}10\,V$$

If ammonia is added to a solution of $Co^{2+}(aq)$, a blue precipitate of a basic cobalt(II) salt is first formed. On adding excess ammonia this precipitate dissolves. In the absence of air, a brownish-yellow solution of a cobalt(II) ammine is formed, but in the presence of air this is rapidly oxidized to a red solution of the cobalt(III) ammine.

This reaction affords a good illustration of the stabilization of an oxidation state by a ligand; cobalt(III) complexes are quite common even though the simple aquo-complex is unstable due to its ready reduction. If cobalt(II) chloride is reacted with ammonia and air a series of mixed complexes is formed, depending on the conditions:

$Co(NH_3)_6^{3+}3Cl^{-}$—all the Cl^{-} precipitated by silver nitrate.

$[Co(NH_3)_5Cl]^{2+}2Cl^{-}$—two-thirds of the chloride precipitated by silver nitrate.

$[Co(NH_3)_4Cl_2]^{+}Cl^{-}$—one-third of the chloride precipitated by silver nitrate.

This series of compounds illustrates the stepwise replacement of one ligand by another.

Copper ammines. On adding ammonia to a solution of copper(II) ions, a pale blue precipitate is first formed, consisting of a basic salt of copper(II). Excess ammonia dissolves this to form a very deep blue solution containing tetraammine copper(II) ion, $Cu(NH_3)_4^{2+}$. Like other complexes, this one is in fact formed stepwise by the addition of ligands in succession, though in equilibrium the complex with four ligands is the one which predominates.

The ion $Cu(NH_3)_4^{2+}$ is tetragonal (square planar) in shape. Although this seems to be a violation of the rules for molecular and ionic shapes (section 6.6) the reason is that there is a non-bonding electron left in the *d*-shell which repels the ligands away from tetrahedral positions.

35.11 Formulae and stability constants of complex ions

In order to find the formula of a complex ion, the equilibrium between the complex and its component species must be investigated quantitatively. Various methods are available, for instance:

(i) if the complex is highly coloured, a colorimetric method may be used;

(ii) if the ligand is soluble in an organic solvent, its concentration in the aqueous system may be determined by a partition method;

(iii) a solubility method may sometimes be used.

Method (iii) may be illustrated by the determination of the formula of the silver(I) ammine, which can be taken as $Ag(NH_3)_x^+$. This ammine is quite stable and soluble, and is formed when a precipitate of silver chloride is dissolved in ammonia. Silver chloride is very sparingly soluble in pure water (K_s = $[Ag^+][Cl^-]$ = about 10^{-10} mol^2 dm^{-6}) but dissolves readily in aqueous ammonia. The stability constant of the complex ion may be written as follows:

$$Ag^+(aq) + xNH_3(aq) \rightleftharpoons Ag(NH_3)_x^+;$$

$$K = \frac{[\text{complex}]}{[Ag^+][NH_3]^x}$$

The complex must be fairly stable, since addition of ammonia lowers the concentration $[Ag^+]$ sufficiently for the solubility product of silver chloride not to be exceeded.

The procedure is to measure the solubility of silver chloride in a series of ammonia solutions of varying concentrations; the more concentrated the ammonia, the more silver chloride is required to saturate the solution. The formula is then calculated as follows:

$$\text{Since } K_s = [Ag^+][Cl^-],$$

$$K = \frac{[\text{complex}][Cl^-]}{K_s.[NH_3]^x}$$

Since we know that the solubility of silver chloride in pure water is very low, there cannot be many Ag^+ ions present, therefore almost all the silver dissolved must be present as complex. For every mole of silver dissolved, one mole of chloride dissolves, therefore [complex] = $[Cl^-]$

$$\text{Therefore, } K = \frac{[Cl^-]^2}{K_s.[NH_3]^x}$$

$$\text{Therefore, } [Cl^-]^2 \propto [NH_3]^x$$

Cl^- may be determined readily by extracting ing a portion of the solution and acidifying it to reprecipitate the silver chloride which can then be weighed. $[NH_3]$ can be assumed to be the same as the concentration of the ammonia solution originally used, provided a large excess is employed. A graph of $\log_{10}[Cl^-]$ against $\log_{10}[NH_3]$ gives the value of x, which is found by experiment to equal 2. Substitution of actual values of $[Cl^-]$ and $[NH_3]$ leads to the numerical value of K, which turns out to be about 10^7 mol^{-2} dm^6 at 25°C.

35.12 Thermodynamic and kinetic stability of complexes

Complexes of cobalt(III) are stable in aqueous solution, usually because there are no suitable vacant electron orbitals available to accept an attacking species: they are said to be *kinetically* stable. Kinetic and thermodynamic stability can be distinguished as follows:

Kinetically stable: rate of decomposition slow, due to high energy of activation of reactions,

Thermodynamically stable: free energy of reaction under the given conditions (e.g. in presence of air, water, etc.) is positive.

There is an analogy between cobalt(III) complexes in the presence of water, and carbon compounds in the presence of air: both are often thermodynamically unstable but kinetically stable.

35.13 Bonding in transition metal complexes

The precise nature of the bonds between ligands and the central metal ion is not fully understood. Some ligands such as ammonia, water and chloride ion appear to be attached by what are essentially electrostatic forces, rather like the forces that hold together an ionic lattice. On the other hand there are ligands such as carbon monoxide and the cyanide ion where transfer of electrons appears to occur between the ligand and the metal. In such cases a pair of electrons is often donated by the ligand to form a weak bond, and this bond can be strengthened by the metal donating some of its *d*-electrons back to the ligand. This back-donation can only occur with multiple-bonded ligands, such as $C{\equiv}N^-$, $C{\equiv}O$, nitrogen oxide, NO, and alkenes.

Back-donation from the metal to the ligand occurs most readily from metals with a large number of *d*-electrons, namely those at the end of a given transition series. Examples of such compounds are ferrocene, $Fe(C_5H_5)_2$, dibenzenechromium(0), $Cr(C_6H_6)_2$, tetracarbonylnickel(0), $Ni(CO)_4$, and Zeise's salt, $K[PtCl_3(C_2H_4)]H_2O$, first made in 1829.

Ligands that are not unsaturated, but which have a large number of electrons available for donation, tend to form complexes with those metals with few *d*-electrons, namely those at the beginning of a given transition series. Thus titanium, vanadium and chromium all form strong complexes with the peroxide ion, $[O{-}O]^{2-}$, which can be used in the qualitative analysis of these elements.

The property of the later transition metals forming complexes with multiple bonded ligands makes them important catalysts. Nickel, palladium and platinum are widely used as hydrogenation catalysts in organic chemistry.

35.14 Cyano-complexes

All the transition metals form complex ions where the ligand is cyanide, CN^-, but the most stable ones are formed by the elements at the end of the 3*d*-series, since the cyanide ion contains a multiple bond $[C{\equiv}N]^-$, (section 35.13).

Iron forms two hexacyano complexes, with oxidation states of +2 and +3. The redox potential of the iron(II)/iron(III) couple is considerably altered by the addition of CN^- ligands:

$$Fe^{3+}(aq)+e^- \rightleftharpoons Fe^{2+}(aq); \quad \mathcal{E}^{\ominus} = +0{\cdot}77\ V$$

$$Fe(CN)_6^{3-}+e^- \rightleftharpoons Fe(CN)_6^{4-}; \quad \mathcal{E}^{\ominus} = +0{\cdot}36\ V$$

If a solution of iron(II) sulphate is boiled with potassium cyanide, $Fe(CN)_6^{4-}$, hexacyanoferrate(II), is the product. This is readily oxidized to the iron(III) complex by passing chlorine into the solution.

Most cyano-complexes are sufficiently stable for the reactions of the free ions to be masked entirely. For instance, solutions of $Fe(CN)_6^{3-}$ and $Fe(CN)_6^{4-}$ do not give precipitates with alkali or with hydrogen sulphide, and addition of acid does not liberate HCN(g).

The cyano-complexes of iron are octahedral in shape, the metal–carbon–nitrogen nuclei being collinear. The metal–carbon bonds have some multiple-bond character.

If iron(II) ions are added to $Fe(CN)_6^{3-}$, or alternatively iron(III) ions are added to $Fe(CN)_6^{4-}$ (that is, the oxidation states are mixed) a dark blue precipitate, *prussian blue* is formed. These reactions form the basis of sensitive tests for the two oxidation states of iron. The precipitate is non-stoichiometric in composition, and its constitution depends upon its method of preparation.

Most transition metals, and some non-transition metals, form precipitates with $Fe(CN)_6^{4-}$. A brown precipitate of copper(II) hexacyanoferrate(II), $Cu_2Fe(CN)_6(s)$, has been made use of as a semi-permeable membrane in osmosis.

35.15 Thiocyanate complexes

The ligand CNS^-, thiocyanate, is similar to cyanide in its ability to form complex ions. If a solution of potassium or ammonium thiocyanate is added to $Fe^{3+}(aq)$, a very deep blood red colour is obtained due to complexes formed. Iron(II) does not give this reaction, so the reagent may be used to test for iron(III).

The complexes form by stepwise replacement of H_2O ligands by CNS^-. If a trace of Fe^{3+}(aq) is added to a large excess of CNS^-(aq), the main species formed will be $Fe(CNS)_6^{3-}$, while if an excess of Fe^{3+}(aq) is used with a trace of CNS^-(aq), the main species will be $Fe(CNS)^{2+}$(aq). At an intermediate stage the zero-charged species $Fe(CNS)_3$ will be formed, and this, being molecular in nature can be extracted into an organic solvent such as ether.

The formation of thiocyanate complexes of iron(III) is used as the basis of a method of quantitative analysis based upon **colorimetry** (section 24.10). A dilute solution of the iron(III) salt is taken, and excess potassium or ammonium thiocyanate is then added to form the complex. A standard solution of iron(III) is similarly treated. The intensity of colour in the two solutions is then compared, either by using a photoelectric meter to measure the transmitted light, or by comparing the relative depths of the two solutions required to give the same degree of absorption of light.

35.16 Polydentate ligands

Transition metals form many stable complexes with ligands which have more than one point of attachment. Ethane-1,2-diamine, $H_2N.CH_2.CH_2.NH_2$, has already been mentioned as an example of a bidentate ligand. Another example is the ethanedioate ion, $C_2O_4^{2-}$, which forms chelate complexes with elements such as chromium. The salt $K_3[Cr(C_2O_4)_3]$ is an example.

Chelates (complexes in which the ligand forms a ring with the central metal atom) are generally more stable than simple complexes with monodentate ligands. The reason is that for the ligand to become detached it is necessary for all the bonds holding it to the metal atom to be broken simultaneously. If only one bond breaks, the complex as a whole does not decompose. Among the most stable of all complexes are those formed by the edta ion which has *six* points of attachment with the central atom (section 31.6).

35.17 Analytical use of complexes

Transition metal complexes are often used in analytical chemistry, both quantitatively and qualitatively.

(a) *Qualitative uses.* The colour of the complex formed can be used to show the presence of either a particular metal ion or a ligand. Table 35.7 summarizes some of the more common tests.

(b) *Quantitative uses.* All first-row transition metals form 1:1 complexes with edta, and this reaction can be used volumetrically with a suitable indicator, to estimate the concentration of a transition metal ion. Where the complex is insoluble, it can be precipitated and determined **gravimetrically**, i.e. by weighing. Many other metals can also be determined by gravimetric analysis, but a more convenient and extremely sensitive method is to measure the intensity of absorption of a suitable wavelength of light by

TABLE 35.7

Metal ion	Ligand	Colour
Fe^{3+}	CNS^-	deep red
Fe^{3+}	phenol	purple
Fe^{3+}	ethanoate and similar ions	red
Co^{2+}	water	pink
Ni^{2+}	butanedione dioxime (dimethyl glyoxime) HON:C(CH$_3$).C(CH$_3$):NOH	red solid (or colour when very dilute)
Cu^{2+}	water	blue
Cu^{2+}	ammonia	indigo
Cu^{2+}	amines, amino-acids	deep blue
Cu^{2+}	peptide links	violet

a soluble complex. Such colorimetric methods can however give unreliable results if other ions are also present.

35.18 Inner transition metals

The third and fourth transition series both start with a set of *f*-block elements, in which the 4*f* and 5*f* shells are filling with electrons. The 4*f*-series, lanthanum to ytterbium, is known as the **rare-earth series** or **lanthanide series**. The term "rare-earth" is better avoided, for many of these elements are far from rare; the term arose because for many years pure compounds of these elements were difficult to obtain. Since their outer electron configurations are the same throughout the series, their chemical properties are extremely similar and they are very difficult to separate. Ion exchange and countercurrent distribution are nowadays used when pure samples of the elements are required. Solutions of cerium(IV) can be used as an alternative to potassium manganate(VII) for volumetric oxidations.

The 5*f*-series is known as the **actinide series**, and it includes the artificially made transuranic elements (elements 93 onwards). All the actinides are radioactive and the series was completed with the synthesis of lawrencium, element 103, in 1966.

35.19 Summary of the properties of metals

TABLE 35.8

	s-block	B-metals	Transition metals
Physical properties	Soft, low m.p.	Harder and less easily melted than *s*-block	Considerably harder, with higher melting points, than B-metals
Oxidation states	Not variable—Group number only	Inert-pair effect	Variable, frequently differing by one unit
Properties of ions	Simple ions, noble-gas structures, not easily polarized; complexes not readily formed (except Be)	Simple ions have completed *d*-shell; easily polarized to form complex ions	Simple ions not formed; complexes readily formed
Complex ions	Not stable. Simple ions loosely hydrated, colourless	Formed in preference to simple ions; colourless	Readily formed, usually coloured
Bonding in compounds	Usually ionic	Usually covalent or complex ions	Usually covalent or complex ions
Reactivity with water	Often vigorous	Not with cold water except slowly	Not with cold water except slowly
Reactivity with non-metals	Vigorous	Less vigorous than *s*-block	About the same as B-metals
Reactivity with hydrogen	Saline hydrides formed	None as a rule	Some form interstitial hydrides

Study Questions

1. Give the outer electronic configuration of the following ions Ti^{4+}, Cr^{2+}, Mn^{2+}, Ni^{2+}, Ag^{+}.

2. (a) In older versions of the periodic table, chromium and sulphur were often placed in the same Group. In what ways are these elements and their compounds (i) similar, (ii) different?

(b) Can you discover any other reasonable analogies between transition metals and *s*- or *p*-block elements?

3. Of the transition elements (Ti–Cu) which would you expect to

(a) have the highest second ionization energy?
(b) form a colourless singly charged cation?
(c) form a carbonyl of formula $M(CO)_5$?
(d) form the smallest M^{2+} ion?
(e) form an ion $M_2O_7^{4-}$?
(f) form a volatile bromide of formula MBr_4?
(g) form a compound MO_3F?

4. (a) Which oxide of manganese is typically basic?

(b) Which oxide of manganese is molecular and acidic?

(c) Which ions are isoelectronic with the manganate(VII) ion?

(d) Are any ions isoelectronic with the manganate(VI) ion?

5. Give balanced equations for the reactions of potassium manganate(VII) in acid solution with (a) iron(II) giving iron(III); (b) ethanedioate ion giving carbon dioxide; (c) hydrogen peroxide giving oxygen; (d) iron(II) ethanedioate.

6. (a) How many isomers are there of the following? (i) $Co(NH_3)_6^{3+}$; (ii) $[Co(NH_3)_5Cl]^{2+}$; (iii) $[Co(NH_3)_4Cl_2]^{+}$; (iv) $Co(NH_3)_3Cl_3$. (Assume octahedral geometry about the cobalt atom.)

(b) There are two isomers of $Pt(NH_3)_2Cl_2$. What does this tell you about the geometry of this compound?

7. "The oxidation state +2 becomes more stable, while the oxidation state +3 becomes less stable with increasing atomic number across the first transition series."

(a) Which metals provide exceptions to this statement?

(b) Suggest reasons for the exceptions.

8. M $KMnO_4$(aq) was titrated against a given volume of reducing agent, first in acid, second in neutral, third in alkaline solution. In acid, 10 cm^3 of $KMnO_4$ were required, in neutral solution 16·7 cm^3 and in alkaline solution, 50 cm^3.

(a) What is the oxidation number of the reduction product in each case?

(b) Suggest a balanced equation for each half-reaction.

(c) How many cm^3 of M $K_2Cr_2O_7$ would have been required to react with the same volume of reducing agent in acid solution?

9. A dark brown solid, A, was fused with KOH and KNO_3 to give a melt from which a green solution of B was extracted. With carbon dioxide, B gave A and a purple solution of C. With reducing agents in acid, C gave solutions of D. With ammonium sulphide, D gave a buff precipitate of E. With sodium hydroxide solution, D gave a precipitate of F, which oxidized rapidly to G.

Identify the lettered compounds and suggest equations for the reactions where possible.

10. A green solution, H, of a transition metal chloride was heated with sodium hydroxide and hydrogen peroxide solutions to give a yellow solution of J. On acidification, J gave K, an orange solution. When crystals of K were treated with sodium chloride and concentrated sulphuric acid, a red volatile liquid L, was obtained. On adding ammonia to a strongly acid solution of K, orange crystals of M formed. M decomposed violently on heating to give steam, nitrogen and a green solid, N.

Identify the lettered compounds and give equations for the reactions where possible.

11. On heating, a pale green solid, P, gave carbon dioxide and a black solid, Q. With hydrochloric acid, P gave carbon dioxide and a green solution of R, which, with ammonia, eventually gave a blue solution of S. Electrolysis of R gave the grey metal, T, at the cathode.

(a) Identify the lettered compounds.
(b) How would you confirm your answers to (a)?

12. A transition metal A was heated in a stream of hydrogen chloride to give hydrogen and B, which dissolved in water to give a green solution. With potassium cyanide solution, B gave a brown precipitate of C, which dissolved in excess potassium cyanide solution to give a yellow solution of D. B reacted with chlorine to give E, which dissolved in water to give a reddish solution. When solutions of D and E were mixed, an intense blue precipitate of F formed. With potassium thiocyanate solution, E gave a deep red colour due to G, but B was unaffected. A solution of E can be converted to a solution of B by using tin(II) chloride.

Identify the lettered compounds and where possible give equations for the reactions.

13. A white solid A was dissolved in water and then reduced by zinc in hydrochloric acid to give, consecutively, a blue solution of B, a green solution of C and a violet solution of D. When A was heated, ammonia, water and a yellowish solid, E, formed. E is widely used as a catalyst. On heating with carbon and chlorine E gave a volatile liquid, F, of molecular weight 173. Using bromine instead of chlorine, G, a similar volatile liquid with a molecular weight of 307 was formed.

Suggest the nature of the lettered compounds.

14. When a blue solution, A, was treated with potassium iodide solution, iodine and a white compound, B, were formed. On dropping an iron nail into A, a red deposit of C formed on the nail. Addition of ammonia to A gave initially a pale blue precipitate of D, but, with excess ammonia, a deep blue solution of E. When solution A was evaporated to dryness and the resulting solid heated, oxygen, brown acidic fumes and a black solid F were obtained. F could be reduced in a stream of hydrogen to give C.

(a) Identify the lettered compounds.
(b) Give equations for the reactions that took place.

15. Some blue crystals of A dissolved in water to give a pink solution of B. With ammonium sulphide solution, B gave a black precipitate of C. C did not form when the solution of B was acid. When 2·6 g of A were dissolved in dilute nitric acid and then treated with silver nitrate solution, 5·73 g of a white precipitate were obtained. This precipitate dissolved in dilute ammonia.

Identify the lettered compounds.

16. (a) What factors are important in determining the heat of formation of the chlorides, MCl_2? (Draw an energy level cycle.)

(b) Can the abnormal heat of formation of $MnCl_2$ be accounted for by the abnormal heat of sublimation of Mn? (Figs. 26.2 and 26.11.)

(c) Why does manganese so often behave in an exceptional manner?

(d) Use a data book to list other abnormal properties of manganese and its compounds.

17. Explain why

(a) Copper is used in electrical wires.
(b) Tungsten is used in electric light bulbs.
(c) Nickel is used as a hydrogenation catalyst for unsaturated hydrocarbons.
(d) Gold is used in wedding rings.

18. Suggest explanations for the following:

(a) 1 mole of Cu^{2+} complexes with four moles of NH_3, but with only two moles of $H_2NCH_2.CH_2NH_2$ (ethane-1,2-diamine, "en").

(b) The overall stability constant for $Cu(en)_2^{2+}$ is 10^{20}, much higher than the value for $Cu(NH_3)_4^{2+}$, which is 10^{13}.

(c) Edta combines with first row transition metals in a 1:1 ratio only.

19. The following overall stability constants apply to red-brown Fe(III):

$Fe(CNS)_6^{3-}$	$Log_{10}K = 7$	Deep red.
FeF_6^{3-}	$Log_{10}K = 15$	Colourless.
$Fe(edta)^+$	$Log_{10}K = 25$	Green-brown.

(a) What would you observe if thiocyanate (CNS^-) ions were added to neutral iron(III) chloride?

(b) What would you expect to observe if excess fluoride ions were added to the solution left in (a)?

(c) What would you expect to observe if excess neutral edta was added to the solution left in (b)?

(d) The 2-hydroxybenzoate ion forms a deep purple complex with iron(III). Suggest simple experiments which would enable you to obtain a rough idea of the stability constant of this complex.

A-level examination questions

The Chapters which may be helpful in answering the questions are given in square brackets after each question.

Questions marked with an asterisk * are taken from the "structured questions" papers set by the University of London GCE Board for the Nuffield Advanced Science (Chemistry) project. Short answers only are required, and in the original papers space was provided for candidates to write in their answers on the paper itself.

Abbreviations:

Camb.: Cambridge Local Examinations.

Southern: Southern Universities' Joint Board for School Examinations.

Nuffield: University of London, Nuffield Science Teaching Project.

*1. This question principally concerns the rate of change of 4-hydroxybutanoic acid into water and a compound called a lactone, according to the equation:

$$\begin{array}{ll} OH & OH \\ | & | \\ CH_2 & C{=}O \\ | & | \\ CH_2 & -CH_2 \end{array} \longrightarrow \begin{array}{c} H_2C-O-C{=}O \\ | \qquad\quad | \\ H_2C-\!\!-CH_2 \end{array} + H_2O$$

The reaction is catalyzed by hydrogen ions. A mixture of 4-hydroxybutanoic acid and a known volume of standard hydrochloric acid was placed in a thermostat. At definite time intervals, a constant volume of reaction mixture was withdrawn and titrated with standard alkali. From the results the concentration of 4-hydroxybutanoic acid remaining at each time interval can be calculated.

If originally there were a moles of the 4-hydroxybutanoic acid per dm^3, then the time t taken for x moles of it to have reacted is given by the equation:

$$t = \frac{2{\cdot}3}{k} \lg \frac{a}{a-x}$$

where k is the rate constant.

The following table shows data obtained in an actual experiment:

Time/min	0	21	50	65	80	100
$\text{Lg} \dfrac{a}{a-x}$	0	0·061	0·139	0·176	0·216	0·270

(a) Plot a graph from which you would be able to find the value of the rate constant k.

(b) From your graph find the value of the rate constant k.

(c) Write the name and formula for the organic compound which would be formed when the lactone is refluxed with excess sodium hydroxide solution.

(d) Write down the structural formula of an isomer of 4-hydroxybutanoic acid which may be expected to be optically active.

(Nuffield 1973) [Ch. 17]

*2. In this question reference is made to the following data:

Electrode process	$\mathcal{E}^\ominus/V$
$\frac{1}{2}I_2 + e^- \rightarrow I^-$	$+0{\cdot}54$
$Fe^{3+} + e^- \rightarrow Fe^{2+}$	$+0{\cdot}77$
$\frac{1}{2}Cr_2O_7^{2-} + 7H^+ + 3e^- \rightarrow Cr^{3+} + 3\frac{1}{2}H_2O$	$+1{\cdot}33$
$BrO_3^- + 6H^+ + 5e^- \rightarrow \frac{1}{2}Br_2 + 3H_2O$	$+1{\cdot}52$

The scheme below represents a series of chromium compounds:

$$\underset{A}{CrCl_3(aq)} \rightarrow \underset{B}{Cr(OH)_3(s)} \rightarrow \underset{C}{K_2CrO_4(aq)} \rightarrow \underset{D}{K_2Cr_2O_7(aq)}$$

$$\downarrow$$

$$\underset{E}{Cr(NH_3)_6Cl_3(aq)}$$

(a) (i) What is the oxidation number of chromium in substances A and D?

(ii) Which step in the series involves the oxidation of chromium?

(b) (i) Name the complex ion present in substance *E*.

(ii) Draw a diagram to show the shape of the complex ion in *E*.

(c) (i) From the standard electrode potential data decide with which ONE of the following substances dichromate(VI) ions would react:

iodine solution; iron(II) ions; bromine water.

(ii) Under what conditions would the reaction take place?

(iii) Write a balanced equation for the reaction.

(d) (i) Draw a fully labelled diagram to show the construction of a voltaic cell in which the reaction in part (c) would occur.

(ii) What standard e.m.f. would you expect the cell to have?

(Nuffield 1973) [Ch. 11]

***3.** Water containing Ca^{2+} and Mg^{2+} ions is said to be "hard" and such water is very wasteful of soap as these ions react with soap to form an insoluble scum. One method for estimating the hardness of water is to titrate samples of the water with the disodium salt of ethylenediaminetetra-acetic acid (edta), and typical instructions for the estimation are as follows:

"To 50-cm^3 portions of hard water, add 2 cm^3 of buffer solution (pH 10) and 6 drops of Eriochrome Black T. Titrate with 0·01M edta solution. At the end-point all trace of red colour disappears and the solution becomes pure blue in colour."

(a) The ion of edta is known to be a hexadentate ligand, and its structure is:

```
      O                              O
      ||                             ||
O⁻—C.CH₂                     CH₂.C—O⁻
        \                   /
         N—CH₂—CH₂—N
        /                   \
O⁻—C.CH₂                     CH₂.C—O⁻
      ||                             ||
      O                              O
```

Mark with an asterisk all the atoms on the above structure which are responsible for the bonding with Ca^{2+} and Mg^{2+} ions.

(b) Fill in the two gaps in the following sentence: a suitable buffer solution for the reaction would be an aqueous solution containing   and

(c) The logarithms of the stability constants for the complexes formed between the two complexing agents and Ca^{2+} and Mg^{2+} ions are given in the following table:

	Ca^{2+}	Mg^{2+}
edta	10·7	8·7
Eriochrome Black T	5·4	7·0

Give a brief explanation of what occurs during the titration, stating the species responsible for the red and blue colours respectively. (The complexes formed between edta and the ions of the alkaline earth metals are colourless.)

(d) The method described gives a total value for the concentrations of Ca^{2+} and Mg^{2+} ions together. Suggest a general experimental method whereby one of these ions might be removed so that the concentration of the other alone could be determined by the above method.

(e) If, by another method, the total Ca^{2+} and Mg^{2+} concentration in a sample of water had been found to be 0·004M, and 20 cm^3 of 0·01M edta solution was required for the titration as described above, what would be the value of n in the formula of the complex ion:

$$[M^{2+}(edta^{4-})_n].$$

(Nuffield 1970) [Ch. 10 etc.]

***4.** The standard enthalpies of combustion, $\Delta H^{\ominus}_{298}$, of two isomeric liquid hydrocarbons *A* and *B* of molecular formula C_6H_{12} are -3825 and -4001 kJ mol^{-1} respectively. The standard enthalpies of combustion, $\Delta H^{\ominus}_{298}$, of hydrogen and carbon are -287 and -395 kJ mol^{-1} respectively. The bond energy terms in kJ mol^{-1} are

$\bar{E}(C—H) = 415$
$\bar{E}(C—C) = 347$
$\bar{E}(C{=}C) = 613.$

(a) Calculate the standard enthalpies of formation, $\Delta H^{\ominus}_{f,298}$, of the two compounds and write down this value for *A* and *B*.

(b) The most likely structures for *A* and *B* are:

I $CH_3.CH_2.CH_2.CH_2.CH{=}CH_2$

and II

```
        CH₂
      /     \
  H₂C        CH₂
    |         |
  H₂C        CH₂
      \     /
        CH₂
```

Use the bond energy terms given to calculate an approximate value for the energy required to atomize the compounds represented by I and II.

(c) From your calculations in (a) and (b), predict which of formulae I and II is likely to correspond with hydrocarbon *A*. Give your reasons.

(d) Name the most suitable physical method for the investigation of structure which could be used to confirm the formula of *A* and state very briefly why the method you suggest would distinguish between formulae I and II.

(Nuffield 1970) [Ch. 9]

***5.** Liquid ammonia, which boils at 240K, is an ionizing solvent. Salts are less ionized in liquid ammonia than they are in water but, owing to the lower viscosity, the movement of ions through liquid ammonia is much more rapid for a given potential gradient. The ionization of liquid ammonia

$$2NH_3 \rightleftharpoons NH_4^+ + NH_2^-$$

is very slight. The ionic product $[NH_4^+][NH_2^-] = 10^{-28}$ mol^2 dm^{-6} at the boiling point. Definitions of an acid and a base similar to those used for aqueous solutions can be used for solutes in liquid ammonia. This question is mainly about acid–base reactions in liquid ammonia as solvent.

(a) Write the formula of the solvated proton in the ammonia system.

(b) In the ammonia system state the bases corresponding to each of the following species in the water system: (i) H_2O, (ii) OH^-, (iii) O^{2-}.

(c) Write equations for the reactions in liquid ammonia of, (i) sodium to give a base and hydrogen; (ii) the neutralization reaction corresponding to

$$HCl(aq) + NaOH(aq) \rightarrow NaCl(aq) + H_2O(l)$$

(d) What would be the concentration of NH_2^- (in mol dm^{-3}) in a solution of liquid ammonia containing 0·01 mol dm^{-3} of ammonium ions?

(e) The dissociation constant of acetic acid in liquid ammonia is greater than it is in water. Suggest a reason for the difference.

(Nuffield 1972) [Ch. 19]

***6.** This set of questions concerns the following partially-complete synthetic route:

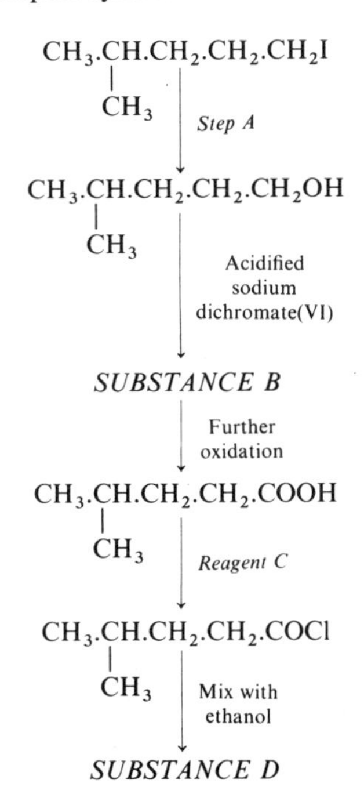

(a) Name the starting substance.

(b) Name the reagents and the conditions required for step *A*.

(c) Give the formula for substance *B*.

(d) Name the reagent *C*.

(e) Give the formula of the substance *D*. To what class of substance does it belong?

(Nuffield 1971) [Chs. 28, 29, 30]

7. "Wood-spirit" is the old name for methanol (b.p. 65°C) because it used to be manufactured from wood. The wood was heated out of contact with the air and the distillate allowed to settle, when the brown aqueous upper layer (referred to as pyroligeneous acid) was separated from the wood tar, neutralized to remove acetic acid, and then distilled. The distillate, crude wood alcohol, from which methanol of up to about 99% purity could be obtained, contained about 70% of methanol, 16% of acetone and other ketones, 8% of water and 6% of a mixture of aldehydes, methyl acetate, amines, and oily matter.

Imagine you have been asked to devise modern laboratory methods to demonstrate how this old manufacture was carried out, and describe the procedures you would adopt.

For each stage you should indicate the apparatus you would use, by means of diagrams or otherwise, and you should explain carefully what is being achieved by the procedure you adopt.

(Nuffield 1971) [Chs. 14, 30]

8. Information concerning the structure of a substance can be obtained when:

(a) Electromagnetic radiation is emitted or absorbed by the substance.

(b) Electromagnetic radiation or a stream of particles such as electrons interacts with the substance to give diffraction patterns.

(c) The substance interacts with an electric or magnetic field.

Show how *ONE* of these methods can yield information about the structure of a substance. You should give an account of the theoretical basis which relates observations to structure, the technique itself, and an illustration of the interpretation of results for *ONE* actual substance.

(Nuffield 1971) [Ch. 4]

9. None of the reactions listed below takes place. For any five examples, by explaining or defining the physical principles involved, explain why.

(a) $NaNO_3 + H_2O \rightarrow NaOH + HNO_3$
(b) $Cl_2 + 2HF \rightarrow F_2 + 2HCl$
(c) $K_4Fe(CN)_6 + 2KOH \rightarrow Fe(OH)_2 + 6KCN$

(d) $10MgSO_4 + 2KMnO_4 + 8H_2SO_4$
$\rightarrow 5Mg_2(SO_4)_3 + K_2SO_4 + 2MnSO_4 + 8H_2O$
(e) $CaCl_2 + H_2S \rightarrow CaS + 2HCl$
(f) $2Ag + H_2SO_4 \rightarrow Ag_2SO_4 + H_2$
(g) In electrolysis:

$$SO_4^{2-} - 2e^- \rightarrow SO_4$$

followed by

$$2SO_4 + 2H_2O \rightarrow 2H_2SO_4 + O_2$$

(Southern 1967) [Chs. 18, 19, 20, 21, 22, 26]

10. Discuss the meaning of the term *stability* as used by chemists. You should distinguish between kinetic and energetic stability, discuss the relationship between stability and temperature, and illustrate your answer with suitable examples supported by quantitative evidence where possible.

(Nuffield 1971) [Chs. 9, 12, 17]

11. Give an account of the formation, structure, and uses of the complexes formed between d-block elements and polydentate ligands. Illustrate your answer with suitable examples.

(Nuffield 1971) [Ch. 35]

12. A given solution is approximately 0·01 M with respect to each of the chlorides of magnesium, calcium and barium. Calculate approximate values for the solubility products of the carbonates, sulphates, chromates and oxalates of the three metals, and hence suggest how the three ions could be separately precipitated, making clear, with reasons, the molarities of any solutions you would use. Solubilities, in mol dm^{-3}, are as follows:

$MgCO_3$	$1·1 \times 10^{-3}$	$MgCrO_4$	8
$CaCO_3$	$6·7 \times 10^{-4}$	$CaCrO_4$	$8·5 \times 10^{-1}$
$BaCO_3$	$1·1 \times 10^{-4}$	$BaCrO_4$	1×10^{-4}
$MgSO_4$	3	MgC_2O_4	$4·7 \times 10^{-3}$
$CaSO_4$	$1·5 \times 10^{-2}$	CaC_2O_4	$5·3 \times 10^{-5}$
$BaSO_4$	1×10^{-3}	BaC_2O_4	$3·6 \times 10^{-4}$

(Nuffield 1972) [Ch. 10]

13. Equilibrium constants of gaseous reactions can be determined either directly or indirectly. Give an account of the determination of the equilibrium constant of one reaction by a direct method and of another reaction by an indirect method.

Discuss the way in which an equilibrium constant for a reaction varies with temperature.

(Nuffield 1972) [Ch. 10]

14. Explain the following observations.

(i) Hydrogen fluoride is both less volatile and a weaker acid than hydrogen chloride.

(ii) Silver chloride dissolves in ammonia solution; silver iodide does not.

(iii) Chlorine and methane react rapidly when exposed to strong sunlight.

(iv) Pure hydrogen bromide is not readily obtained by the action of concentrated sulphuric acid on sodium bromide.

(v) Sulphuric acid and hydrochloric acid are equally strong, yet hydrogen chloride is evolved by the action of concentrated sulphuric acid on sodium chloride.

(Camb. 1967) [Chs. 9, 12, 18, 21]

15. Comment critically on the following statements.

(a) If the standard free energy change ($\Delta G^{\ominus}$) for a process is negative, it will proceed spontaneously.

(b) Standard electrode potentials ($E^{\ominus}$) indicate the ease with which atoms lose or gain electrons.

(c) The rate of a chemical reaction is proportional to the collision rate of the reacting particles.

(Nuffield 1972) [Chs. 11, 12, 17]

16. Using your Book of Chemical Data where possible, but otherwise making your own estimate, give a value for the energy required (in kJ mol^{-1}) to remove one electron from each of the species shown below, and then arrange each group of three in order of increasing energy. (Estimated values should be shown with an asterisk.)

Attempt to explain the reason for the increasing energy in each case.

(a) Na, Mg, Al;
(b) K, Ca, Sc;
(c) K^+, Ca^+, Sc^+;
(d) K, Rb, Cs;
(e) Cu, Ag, Au;
(f) C, N, O;
(g) Na^+, Ne, F^-;
(h) Fe, Fe^{2+}, Fe^{3+}.

(Nuffield 1972) [Ch. 3]

17. Discuss the types of bonding that hold atoms and ions together in molecules and crystals. Include in your answer evidence for the existence of the bonds that you describe, and some indication of their relative strength.

(Nuffield 1973) [Chs. 5, 6, 9]

18. Give a full account of the general types of reaction which the —OH group undergoes in the many carbon compounds in which it occurs. You should illustrate each category of reaction with a particular example, show how the reactions of the group may be modified when attached to differing groups, and show how bonding theory provides explanations of the modifications.

(Nuffield 1972) [Ch. 30]

19. Each of the following pairs of equations represents one reaction which does proceed in the way indicated and one which does not. For *FIVE* of the following pairs of equations, say which represents a genuine reaction and discuss possible reasons why the other reaction does not in fact proceed in a similar way.

(a) (i) $Pb^{2+}(aq)+CrO_4^{2-}(aq) \rightarrow PbCrO_4(s)$
(ii) $Sn^{2+}(aq)+CrO_4^{2-}(aq) \rightarrow SnCrO_4(s)$

(b) (i) $2Al^{3+}(aq)+3CO_3^{2-}(aq) \rightarrow Al_2(CO_3)_3(s)$
(ii) $Ca^{2+}(aq)+CO_3^{2-}(aq) \rightarrow CaCO_3(s)$

(c) (i) $C_2H_5CO_2H(aq)+OH^-(aq)$
$\rightarrow C_2H_5CO_2^-(aq)+H_2O(l)$
(ii) $C_2H_5OH(aq)+OH^-(aq)$
$\rightarrow C_2H_5O^-(aq)+H_2O(l)$

(d)
(i) $(R)(H)C{=}O + HCN \rightarrow (R)(OH)C(H)(CN)$

(ii) $(R)(HO)C{=}O + HCN \rightarrow (R)(OH)C(HO)(CN)$

(e) (i) $Ag^+(aq)+F^-(aq) \rightarrow AgF(s)$
(ii) $Ag^+(aq)+Cl^-(aq) \rightarrow AgCl(s)$

(f) (i) $CH_3COCl(l)+H_2O(l)$
$\rightarrow CH_3CO_2H(aq)+HCl(aq)$
(ii) $C_6H_5Cl(l)+H_2O(l)$
$\rightarrow C_6H_5OH(aq)+HCl(aq)$

(Nuffield 'S' 1972) [Chs. 10, 11, 19, 28, 29, 30]

Appendix I *Half-cell potential data**

These values apply to solutions of pH = 0 (acid solutions), at 25°C. Although the data apply to standard conditions, they may be used as a rough guide as to whether reactions will proceed under normal experimental conditions, though of course they give no guide as to the *rate* of reaction.

It is important to bear in mind that the formation of a complex ion may have a considerable effect on the half-cell potential, e.g.

$$Fe^{3+}(aq)+e^- \rightleftharpoons Fe^{2+}(aq); \quad \mathcal{E}^\ominus = +0{\cdot}771 \text{ V}$$
$$Fe(CN)_6^{3-}+e^- \rightleftharpoons Fe(CN)_6^{4-}; \quad \mathcal{E}^\ominus = +0{\cdot}36 \text{ V}$$

The formation of stable, highly insoluble compounds is analogous to complex formation. For example, the redox data predict that $Cu^+(aq)$ is unstable. Nevertheless, *insoluble* compounds of copper(I) and complexes such as $Cu(CN)_4^{3-}$ and $CuCl_4^{3-}$, can be prepared.

$\mathcal{E}^\ominus$ (in V)	oxidized species + ne^- ⇌ reduced species
−3·05	$Li^+ + e^- \rightleftharpoons Li$
−2·925	$K^+ + e^- \rightleftharpoons K$
−2·925	$Rb^+ + e^- \rightleftharpoons Rb$
−2·923	$Cs^+ + e^- \rightleftharpoons Cs$
−2·90	$Ba^{2+} + 2e^- \rightleftharpoons Ba$
−2·89	$Sr^{2+} + 2e^- \rightleftharpoons Sr$
−2·87	$Ca^{2+} + 2e^- \rightleftharpoons Ca$
−2·714	$Na^+ + e^- \rightleftharpoons Na$
−2·37	$Mg^{2+} + 2e^- \rightleftharpoons Mg$
−2·25	$\frac{1}{2}H_2 + e^- \rightleftharpoons H^-$
−2·08	$Sc^{3+} + 3e^- \rightleftharpoons Sc$
−1·85	$Be^{2+} + 2e^- \rightleftharpoons Be$
−1·66	$Al^{3+} + 3e^- \rightleftharpoons Al$
−1·63	
−1·18	$Mn^{2+} + 2e^- \rightleftharpoons Mn$
−1·18	$V^{2+} + 2e^- \rightleftharpoons V$
−0·87	$H_3BO_3 + 3H^+ + 3e^- \rightleftharpoons B + 3H_2O$
−0·86	$SiO_2 + 4H^+ + 4e^- \rightleftharpoons Si + 2H_2O$
−0·763	$Zn^{2+} + 2e^- \rightleftharpoons Zn$
−0·74	$Cr^{3+} + 3e^- \rightleftharpoons Cr$
−0·60	$As + 3H^+ + 3e^- \rightleftharpoons AsH_3$
−0·53	$Ga^{3+} + 3e^- \rightleftharpoons Ga$
−0·51	$Sb + 3H^+ + 3e^- \rightleftharpoons SbH_3$
−0·49	$2CO_2 + 2H^+ + 2e^- \rightleftharpoons (COOH)_2$
−0·440	$Fe^{2+} + 2e^- \rightleftharpoons Fe$
−0·41	$Cr^{3+} + e^- \rightleftharpoons Cr^{2+}$
−0·403	$Cd^{2+} + 2e^- \rightleftharpoons Cd$
−0·37	$Ti^{3+} + e^- \rightleftharpoons Ti^{2+}$
−0·342	$In^{3+} + 3e^- \rightleftharpoons In$
−0·3363	$Tl^+ + e^- \rightleftharpoons Tl$
−0·277	$Co^{2+} + 2e^- \rightleftharpoons Co$
−0·276	$H_3PO_4 + 2H^+ + 2e^- \rightleftharpoons H_3PO_3 + H_2O$
−0·255	$V^{3+} + e^- \rightleftharpoons V^{2+}$
−0·250	$Ni^{2+} + 2e^- \rightleftharpoons Ni$
−0·136	$Sn^{2+} + 2e^- \rightleftharpoons Sn$
−0·126	$Pb^{2+} + 2e^- \rightleftharpoons Pb$
−0·15	$GeO_2 + 4H^+ + 4e^- \rightleftharpoons Ge + 2H_2O$
0·000	$\mathbf{2H^+ + 2e^- \rightleftharpoons H_2}$
+0·06	$P + 3H^+ + 3e^- \rightleftharpoons PH_3$
+0·10	$Co(NH_3)_6^{3+} + e^- \rightleftharpoons Co(NH_3)_6^{2+}$
+0·1	$TiO^{2+} + 2H^+ + e^- \rightleftharpoons Ti^{3+} + H_2O$
+0·141	$S + 2H^+ + 2e^- \rightleftharpoons H_2S$
+0·15	$Sn^{4+} + 2e^- \rightleftharpoons Sn^{2+}$
+0·152	$SbO^+ + 2H^+ + 2e^- \rightleftharpoons Sb + H_2O$
+0·153	$Cu^{2+} + e^- \rightleftharpoons Cu^+$
+0·17	$SO_4^{2-} + 4H^+ + 2e^- \rightleftharpoons H_2SO_3 + H_2O$
+0·247	$HAsO_2(aq) + 3H^+ + 3e^- \rightleftharpoons As + 2H_2O$
+0·32	$BiO^+ + 2H^+ + 3e^- \rightleftharpoons Bi + H_2O$
+0·337	$Cu^{2+} + 2e^- \rightleftharpoons Cu$
+0·36	$Fe(CN)_6^{3-} + e^- \rightleftharpoons Fe(CN)_6^{4-}$
+0·521	$Cu^+ + e^- \rightleftharpoons Cu$
+0·5355	$I_2 + 2e^- \rightleftharpoons 2I^-$
+0·559	$H_3AsO_4 + 2H^+ + 2e^- \rightleftharpoons HAsO_2 + 2H_2O$
+0·564	$MnO_4^- + e^- \rightleftharpoons MnO_4^{2-}$
+0·771	$Fe^{3+} + e^- \rightleftharpoons Fe^{2+}$
+0·789	$Hg_2^{2+} + 2e^- \rightleftharpoons 2Hg$
+0·7991	$Ag^+ + e^- \rightleftharpoons Ag$
+0·80	$2NO_3^- + 4H^+ + 4e^- \rightleftharpoons N_2O_4 + 2H_2O$

* Selected from W. M. Latimer, *Oxidation potentials*, 2nd ed., Prentice Hall, Englewood Cliffs, N.J., 1952.

$\mathcal{E}^\ominus$ (in V)	oxidized species + ne^- $\rightleftharpoons$ reduced species
+0·86	$Cu^{2+} + I^- + e^- \rightleftharpoons CuI$
+0·920	$2Hg^{2+} + 2e^- \rightleftharpoons Hg_2^{2+}$
+0·94	$NO_3^- + 3H^+ + 2e^- \rightleftharpoons HNO_2 + H_2O$
+0·96	$NO_3^- + 4H^+ + 3e^- \rightleftharpoons NO + 2H_2O$
+1·0652	$Br_2 + 2e^- \rightleftharpoons 2Br^-$
+1·03	$N_2O_4 + 4H^+ + 4e^- \rightleftharpoons NO + 2H_2O$
+1·07	$N_2O_4 + 2H^+ + 2e^- \rightleftharpoons 2HNO_2$
+1·19	$ClO_4^- + 2H^+ + 2e^- \rightleftharpoons ClO_3^- + H_2O$
+1·195	$IO_3^- + 6H^+ + 5e^- \rightleftharpoons \frac{1}{2}I_2 + 3H_2O$
+1·229	$O_2 + 4H^+ + 4e^- \rightleftharpoons 2H_2O$
+1·25	$Tl^{3+} + 2e^- \rightleftharpoons Tl^+$
+1·33	$Cr_2O_7^{2-} + 14H^+ + 6e^- \rightleftharpoons 2Cr^{3+} + 7H_2O$
+1·3595	$Cl_2 + 2e^- \rightleftharpoons 2Cl^-$
+1·455	$PbO_2 + 4H^+ + 2e^- \rightleftharpoons Pb^{2+} + 2H_2O$
+1·50	$Au^{3+} + 3e^- \rightleftharpoons Au$
+1·51	$Mn^{3+} + e^- \rightleftharpoons Mn^{2+}$
+1·51	$MnO_4^- + 8H^+ + 5e^- \rightleftharpoons Mn^{2+} + 4H_2O$
+1·61	$Ce^{4+} + e^- \rightleftharpoons Ce^{3+}$
+1·77	$H_2O_2 + 2H^+ + 2e^- \rightleftharpoons 2H_2O$
+1·82	$Co^{3+} + e^- \rightleftharpoons Co^{2+}$
+1·98	$Ag^{2+} + e^- \rightleftharpoons Ag^+$
+2·01	$S_2O_8^{2-} + 2e^- \rightleftharpoons 2SO_4^{2-}$
+2·07	$O_3 + 2H^+ + 2e^- \rightleftharpoons O_2 + H_2O$
+2·87	$F_2 + 2e^- \rightleftharpoons 2F^-$

Appendix II *International atomic weights (relative atomic masses) on the basis of* ^{12}C = 12·000 (1969)

The following values apply to elements as they exist in materials of terrestrial origin and to certain artificial elements. When used with due regard to the footnotes, they are considered reliable to ±1 in the last digit, or ±3 if that digit is in small type.

Element	Atomic weight	Element	Atomic weight	Element	Atomic weight
Actinium	—	Hafnium	178·49	Praseodymium	140·9077[a]
Aluminium	26·9815[a]	Helium	4·00260[b,c]	Promethium	—
Americium	—	Holmium	164·9303[a]	Protactinium	231·0359[a,f]
Antimony	121·7	Hydrogen	1·0080[b,d]	Radium	226·0254[a,f,g]
Argon	39·948[b,c,d,g]	Indium	114·82	Radon	—
Arsenic	74·9216[a]	Iodine	126·9045[a]	Rhenium	186·2
Astatine	—	Iridium	192·22	Rhodium	102·9055[a]
Barium	137·34	Iron	55·847	Rubidium	85·4678[c]
Berkelium	—	Krypton	83·80	Ruthenium	101·07
Beryllium	9·01218[a]	Lanthanum	138·9055[b]	Samarium	150·4
Bismuth	208·9806[a]	Lawrencium	—	Scandium	44·9559[a]
Boron	10·81[c,d,e]	Lead	207·2[d,g]	Selenium	78·96
Bromine	79·904[c]	Lithium	6·941[c,d,e]	Silicon	28·086[a]
Cadmium	112·40	Lutetium	174·97	Silver	107·868[c]
Caesium	132·9055[a]	Magnesium	24·305[c]	Sodium	22·9898[a]
Calcium	40·08	Manganese	54·9380[a]	Strontium	87·62[g]
Californium	—	Mendelevium	—	Sulphur	32·06[d]
Carbon	12·011[b,d]	Mercury	200·59	Tantalum	180·9479[b]
Cerium	140·12	Molybdenum	95·94	Technetium	98·9062[f]
Chlorine	35·453[c]	Neodymium	144·24	Tellurium	127·60
Chromium	51·996[c]	Neon	20·179[c]	Terbium	158·9254[a]
Cobalt	58·9332[a]	Neptunium	237·0482[b,f]	Thallium	204·37
Copper	63·546[c,d]	Nickel	58·71	Thorium	232·0381[a,f]
Curium	—	Niobium	92·9064[a]	Thulium	168·9342[a]
Dysprosium	162·50	Nitrogen	14·0067[b,c]	Tin	118·69
Einsteinium	—	Nobelium	—	Titanium	47·90
Erbium	167·26	Osmium	190·2	Tungsten	183·85
Europium	151·96	Oxygen	15·9994[b,c,d]	Uranium	238·029[b,c,e]
Fermium	—	Palladium	106·4	Vanadium	50·9414[b,c]

Fluorine	18·9984[a]	Phosphorus	30·9738[a]	Xenon	131·30
Francium	—	Platinum	195·09	Ytterbium	173·04
Gadolinium	157·25	Plutonium	—	Yttrium	88·9059[a]
Gallium	69·72	Polonium	—	Zinc	65·37
Germanium	72·59	Potassium	39·102	Zirconium	91·22
Gold	196·9665[a]				

Footnotes

[a] Mononuclidic element.

[b] Element with one predominant isotope (about 99–100 per cent abundance).

[c] Element for which the atomic weight is based on calibrated measurements.

[d] Element for which variation in isotopic abundance in terrestrial samples limits the precision of the atomic weight given.

[e] Element for which users are cautioned against the possibility of large variations in atomic weight due to inadvertent or undisclosed artificial isotopic separation in commercially available materials.

[f] Most commonly available long-lived isotope.

[g] In some geological specimens this element has a highly anomalous isotopic composition corresponding to an atomic weight significantly different from that given.

Appendix III *Successive ionization energies, in kJ, of the elements from hydrogen to sodium*

Element	1	2	3	4	5	6	7	8	9	10	11
H	1310										
He	2370	5250									
Li	520	7300	11 800								
Be	900	1760	14 900	21 100							
B	800	2430	3680	25 000	32 800						
C	1090	2350	4650	6180	37 800	47 600					
N	1400	2840	4520	7450	9450	53 000	64 200				
O	1310	3380	5310	7450	11 000	13 300	71 500	84 000			
F	1680	3370	6080	8400	11 000	15 200	17 800	92 000	106 000		
Ne	2080	3960	6200	9400	12 200	13 300	20 000	23 000	114 000	131 000	
Na	495	4560	6950	9550	13 400	16 600	20 100	25 500	28 900	141 000	158 000

Appendix IV *Physical constants, conversion factors and units*

There are seven basic units in the SI system (Système Internationale d'Unités):

Quantity	Name of unit	Symbol
length	metre	m
mass	kilogramme	kg
time	second	s
electric current	ampère*	A
thermodynamic temperature	kelvin*	K
luminous intensity	candela	cd
mole	mole	mol

In addition there are a number of derived units, including the following:

Quantity	Name of unit	Symbol	Dimensions
force	newton*	N	$kg\ m\ s^{-2}$
energy	joule*	J	$kg\ m^2\ s^{-2}$
power	watt*	W	$kg\ m^2\ s^{-3} = J\ s^{-1}$
electrical charge	coulomb*	C	A s
potential difference	volt*	V	$J\ A^{-1}\ s^{-1}$
electrical resistance	ohm*	Ω	$V\ A^{-1}$
frequency	hertz*	Hz	s^{-1}
customary temperature	degree Celsius*	°C	K − 273·15

* Note that units named after a person are abbreviated with a capital letter, though the unit itself must be written with a small letter if it is written out in full.

Special prefixes and symbols are used to indicate multiples and sub-multiples of the basic units, in powers of ten. The following are encountered in this book:

Multiple	Prefix	Symbol
10^6	mega	M
10^3	kilo	k
10^{-1}	deci	d
10^{-2}	centi	c
10^{-3}	milli	m
10^{-6}	micro	μ
10^{-9}	nano	n
10^{-12}	pico	p

Two common units which are frequently met in scientific literature, though they are avoided in this book, are the calorie (cal) which equals 4·184 J, and the angstrom unit (Å) which equals 10^{-10} m = 10^{-1} nm.

Some useful physical constants, expressed in SI units, are:

Quantity		Value
velocity of light, *c*		$2{\cdot}998 \times 10^8\ m\ s^{-1}$
Boltzmann constant, *k*		$1{\cdot}380 \times 10^{-23}\ J\ K^{-1}$
charge of electron, *e*		$1{\cdot}602 \times 10^{-19}\ C$
Planck's constant, *h*		$6{\cdot}626 \times 10^{-34}\ J\ s$
Avogadro constant, *L*		$6{\cdot}022 \times 10^{-23}\ mol^{-1}$
molar gas volume at s.t.p.		$22{\cdot}4\ dm^3\ mol^{-1}$
	or	$(2{\cdot}241 \times 10^{-2}\ m^3\ mol^{-1})$
gas constant, *R*		$8{\cdot}314\ J\ K^{-1}\ mol^{-1}$
Faraday constant, *F*		$9{\cdot}649 \times 10^4\ C\ mol^{-1}$

Appendix V *Chemical nomenclature*

The names of chemical substances given in this book are broadly in accordance with the rules laid down by the International Union of Pure and Applied Chemistry (I.U.P.A.C. Rules). Some details are given in the main body of the book, and a summary is given in this appendix. Alternative names of individual substances are given by means of cross-references in the index.

Naming inorganic compounds

Binary compounds

(1) The ending *-ide* is used.
(2) The *less* electronegative element is named *first*.
(3) Simple molecular compounds are named with prefixes to show molecular formula, the prefix *mono-* being generally omitted, e.g.

NO nitrogen oxide,
NO_2 nitrogen dioxide,
N_2O dinitrogen oxide.

(4) Macromolecular or complex molecular compounds are shown using an oxidation number (for rules, see section 11.13), e.g.

CuO copper(II) oxide,
P_4O_{10} phosphorus(V) oxide

Oxidation number is omitted for metals with only one common oxidation state, namely aluminium, zinc, and the elements of Groups IA and IIA.
(5) The ending *-ide* is also used for hydroxide, OH^- and cyanide, CN^-.

Complexes

A complex is loosely defined as a central atom with one or more ligands attached. The name is built up as follows:

(1) The names of the ligands are arranged alphabetically.
(2) Numerical prefixes are then added.
(3) The name of the central atom follows next.
(4) The ending *-ate* indicates that the whole species is an anion. (Older Latin names are retained in some cases, e.g. ferrate, cuprate, plumbate, stannate).
(5) The oxidation number of the central atom is placed last, e.g.

$Ni(CO)_4$	tetracarbonylnickel(0)
AlF_6^{3-}	hexafluoroaluminate
$Cu(NH_3)_4^{2+}$	tetraamminecopper(II)
$CuCl_4^{3-}$	tetrachlorocuprate(I)
$Fe(H_2O)_5NO^{2+}$	pentaaquanitrosyliron(II)

(6) The following is a list of some common ligands. Negatively charged ligands end in "o", hence *aqua-* (H_2O) and *ammine-* (NH_3) are preferred to aquo- and ammino-.

ammine	NH_3	cyano	CN^-	hydroxo	OH^-
aqua	H_2O	ethanoato	$CH_3.COO^-$	oxo	O^{2-}
bromo	Br^-	fluoro	F^-	thio	S^{2-}
chloro	Cl^-	hydrido	H^-		

Oxo-ions

An oxo-ion is named by regarding it as a complex in which the ligand is O^{2-}, oxo. The oxidation number of the central atom is calculated on this basis. Full names of oxo-ions are rarely used unless some special point of distinction has to be made. The I.U.P.A.C. system uses the ending -ate for *all* anions. However, one or two trivial names seem likely to persist for a while, and have been retained in this book, namely

NO_2^- (nitrite) HNO_2 (nitrous acid)
SO_3^{2-} (sulphite) H_2SO_3 (sulphurous acid)

Phosphorus oxo-acids have unusual structures and the ligand O^{2-} is not always present. Special names have therefore to be employed (see index, and section 23.9).

Formula	I.U.P.A.C. Name	Name used in this book
SO_4^{2-}	Tetraoxosulphate(VI)	Sulphate
SO_3^{2-}	Trioxosulphate(IV)	Sulphite
NO_3^-	Trioxonitrate(V)	Nitrate
NO_2^-	Dioxonitrate(III)	Nitrite
PO_4^{3-}	Tetraoxophosphate(V)	Phosphate(V)
$(PO_3^-)_n$	Polytrioxophosphate(V)	Trioxophosphate(V)
$P_2O_7^{4-}$	Heptaoxodiphosphate(V)	Heptaoxodiphosphate(V)
$HPO(OH)_2$	Phosphonic acid	Phosphonic acid
CrO_4^{2-}	Tetraoxochromate(VI)	Chromate(VI)
$Cr_2O_7^{2-}$	Heptaoxodichromate(VI)	Dichromate(VI)
MnO_4^-	Tetraoxomanganate(VII)	Manganate(VII)

The parent acid is named by substituting "*ic acid*" for "*ate*".

Peroxo-acids contain the link —O—O—, and are named e.g. H_2SO_5, peroxosulphuric(VI) acid; $S_2O_8^{2-}$, peroxodisulphate(VI) ion.

Acid salts are named by prefixing the anion name with "hydrogen-", "dihydrogen-" etc. without a space, e.g.

$NaHCO_3$ sodium hydrogencarbonate
NaH_2PO_4 sodium dihydrogenphosphate(V)
Na_2HPO_4 disodium hydrogenphosphate(V)

Salt hydrates

Named very simply e.g. $CuSO_4.5H_2O$ is copper(II) sulphate-5-water. The older names such as pentahydrate are thus discontinued.

Naming organic compounds

At present, the I.U.P.A.C. system allows alternative names for a compound, using prefixes to denote the *substitution* of hydrogen atoms by other groups (the term group including single atoms). "Additive" names e.g. ethylene dibromide, are discontinued. Individual examples of names appear in the main text, and the following is a highly condensed summary of the main rules.

(1) The name consists of **(prefixes)-stem-suffix**.

(2) The stem denotes the nature or number of carbon atoms in the skeleton, derived according to the rules set out in section 25.4.

(3) The stem ending is -an(e) for a saturated carbon chain,
-en(e) for a chain containing C=C,
-yn(e) for a chain containing C≡C.

A number can be added to show the position of a multiple bond (section 26.2).

(4) A suffix is added if a "principal" functional group (i.e. one in the following list) is present. If more than one group in the following list occurs, the one nearest the top counts as the "principal" group and is named in the suffix, any other groups being named by prefixes. If the "principal" functional group contains a carbon atom, there are two ways in which it can be denoted:

(a) its own carbon atom may be included as part of the backbone when naming the stem (this is always done if possible), or

(b) the stem is named, omitting the carbon atom of the principal functional group; this is needed for benzene derivatives for instance.

Examples of both systems are given in the list which follows.

Principal functional groups

	Example of suffix type (a)		*Example of suffix type (b)*	
$—COOH$	$CH_3.COOH$	ethan*oic acid*	$C_6H_5.COOH$	benzene*carboxylic acid*
$—COO(R)$	$CH_3.COO(R)$	(alkyl) ethan*oate*	$C_6H_5.COO(R)$	(alkyl) benzene*carboxylate*
$—COCl$	$CH_3.COCl$	ethan*oyl chloride*	$C_6H_5.COCl$	benzene*carbonyl chloride*
$—CONH_2$	$CH_3.CONH_2$	ethan*amide*	$C_6H_5.CONH_2$	benzene*carboxamide*
$—CHO$	$CH_3.CHO$	ethan*al*	$C_6H_5.CHO$	benzene*carbaldehyde*
$>C=O$	$CH_3.CO.CH_3$	propan*one*		
$—CN$	$CH_3.CN$	ethan*onitrile*	$C_6H_5.CN$	benzene*carbonitrile*
$—OH$	$CH_3.CH_2OH$	ethan*ol*		
$—NH_2$	$CH_3.CH_2NH_2$	ethan*amine** (ethylamine)		

(*The name ethanamine does not yet seem to be fully accepted, yet the system is used for difunctional compounds e.g. $H_2NCH_2.CH_2NH_2$ ethane-1,2-diamine. Conjugate acids of amines are named as alkylammonium ions e.g. $C_2H_5NH_3^+$ ethylammonium.)

Additional functional groups

These are named as prefixes, the following rules being applied:

(1) The prefixes are arranged alphabetically;

(2) Multiplying prefixes, di-, tri- etc. do not affect the alphabetical order;

(3) Numbers are assigned to give the position of substituent groups on the skeleton (section 25.5). The numbering is arranged to give:

(a) the lowest possible number to the "principal" functional group i.e. the suffix group, then

(b) the lowest individual numbers to prefixes.

The main prefixes, in alphabetical order, in use in this book are:

$—C_nH_{2n+1}$ alkyl (methyl, ethyl etc.), $—NH_2$ amino, —Br bromo, —COOH carboxy, —Cl chloro, —CN cyano, —F fluoro, —OH hydroxy, =O oxo, $—C_6H_5$ phenyl. In the above list, all the groups except oxo- substitute one hydrogen atom. "Oxo-" substitutes two, e.g. $CH_3.CO.CH_2.CH_2.COOH$ would be 4-oxopentan-1-oic acid. The name pentan-1-oic acid-4-one would be incorrect because a name cannot have two suffixes.

Appendix VI *Answers to kinetics questions not in the "Handbook with Answers"*

Chapter 17 Reaction Rates

The answers to questions 1 to 4, 6 and 7 are given in the handbook.

5. (a) The rate should slowly decrease with time.

(b) The rate would increase with time, until it fell quickly near the end of the reaction.

8. This was question 9 in the original edition. The answer is in the handbook.

Chapter 18 Hydrogen

11. The decomposition will be speeded up. A catalyst lowers the activation energy of a reaction and BOTH forward and backward reactions occur more quickly.

Chapter 19 Acids

18. (a) Concentration doubles; time is halved. First order.

(b) (i) Zero order (Use *B* and *D*).

(ii) First order (Use *A* and *C*).

(c) Rate = $k\,[CH_3COCH_3]\,[H^+]$.

(d) About 29 s.

(e) To avoid complications caused by conditions changing as the reaction proceeds, to avoid worrying about the reverse reaction and side reactions etc.

(f) Very little. $[H^+]$ only increases from 0.50 to 0.51 in the course of the reaction.

Chapter 30

13. (a) $CH_3COOCH_3 + OH^-(aq) \rightarrow CH_3COO^-$ and CH_3OH

The products are ethanoate ion and methanol.

(b) The alkali is in such great excess that it has effectively a constant concentration.

(c) The mechanism will be the same.

Chapter 33

5. (a) Because HCl is a product of the reaction.

(b) The rate increases and then levels off.

(c) (i) First. (ii) Zero.

(d) There are only a limited number of urease molecules available.

When these are all operating at maximum capacity, the concentration of urea becomes irrelevant.

6. (a) 2, 4, 8, 16, 32 units.

(b) 1.504 divided by (0.00296–0.00346)

$= 1.504/-0.0005 = -3.08 \times 10^3$

(c) $E = 3.08 \times 2.3 \times 8.31$ kJ $= 59$ kJ (approx.)

(d) The rate increase is not the same for each temperature difference.

Index

Page references in **bold figures** indicate items of some importance.
Note: This index may also be used as a reference guide for modern nomenclature. Entries are given page references under their modern names, but older names are cross-referenced to assist readers using older literature (*see also* Appendix V).

PERIODIC TABLE OF THE ELEMENTS

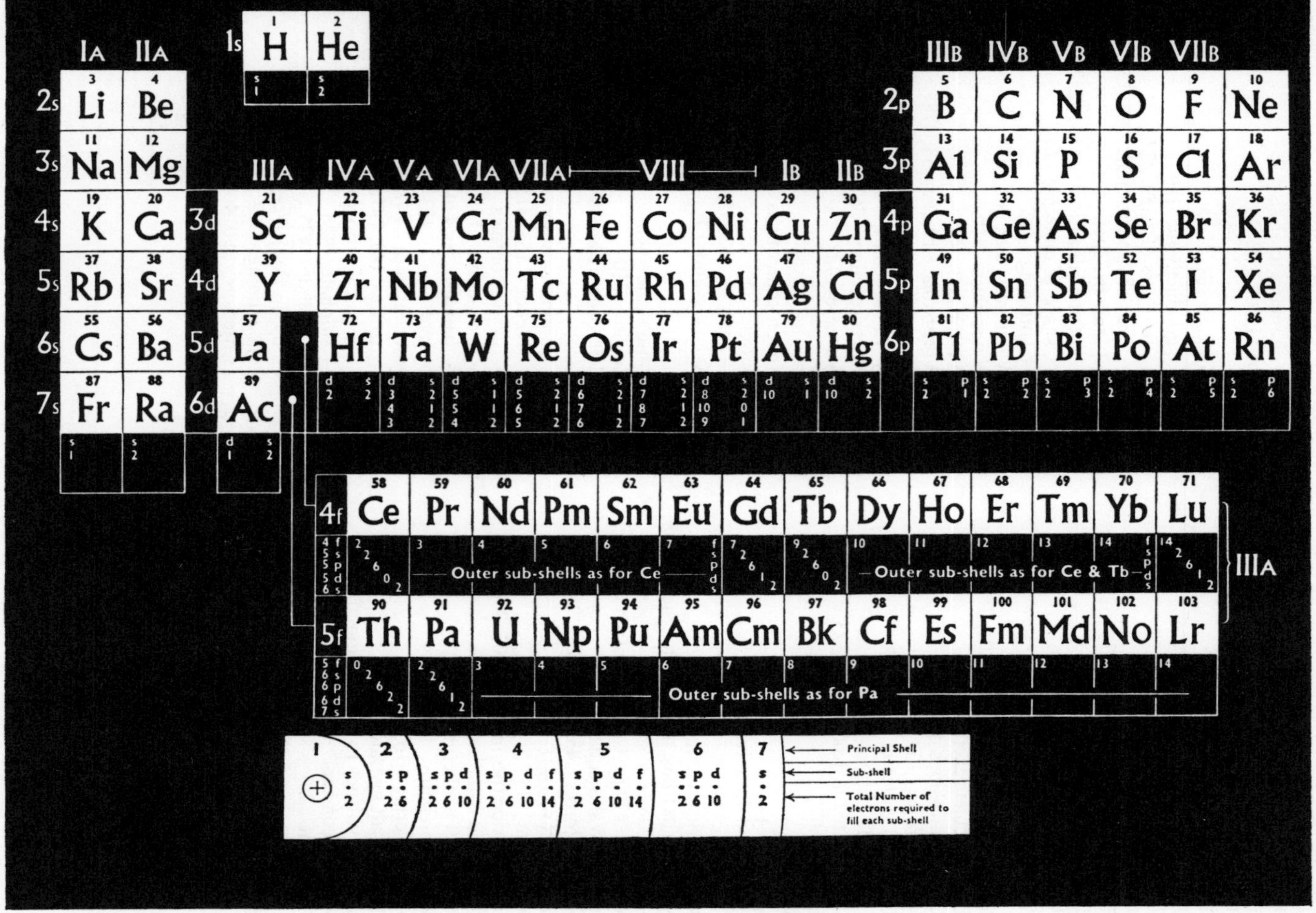

	IA	IIA		IIIA	IVA	VA	VIA	VIIA	VIII	VIII	VIII	IB	IIB		IIIB	IVB	VB	VIB	VIIB	
1s	1 H (s 1)	2 He (s 2)																		
2s	3 Li	4 Be												2p	5 B	6 C	7 N	8 O	9 F	10 Ne
3s	11 Na	12 Mg												3p	13 Al	14 Si	15 P	16 S	17 Cl	18 Ar
4s	19 K	20 Ca	3d	21 Sc	22 Ti	23 V	24 Cr	25 Mn	26 Fe	27 Co	28 Ni	29 Cu	30 Zn	4p	31 Ga	32 Ge	33 As	34 Se	35 Br	36 Kr
5s	37 Rb	38 Sr	4d	39 Y	40 Zr	41 Nb	42 Mo	43 Tc	44 Ru	45 Rh	46 Pd	47 Ag	48 Cd	5p	49 In	50 Sn	51 Sb	52 Te	53 I	54 Xe
6s	55 Cs	56 Ba	5d	57 La	72 Hf	73 Ta	74 W	75 Re	76 Os	77 Ir	78 Pt	79 Au	80 Hg	6p	81 Tl	82 Pb	83 Bi	84 Po	85 At	86 Rn
7s	87 Fr	88 Ra	6d	89 Ac																
	s 1	s 2		d 1, s 2	d 2, s 2	d 3 4 3, s 2 1 2	d 5 5 4, s 1 1 2	d 5 6 5, s 2 1 2	d 6 7 6, s 2 1 2	d 7 8 7, s 2 1 2	d 8 10 9, s 2 0 1	d 10, s 1	d 10, s 2		s 2, p 1	s 2, p 2	s 2, p 3	s 2, p 4	s 2, p 5	s 2, p 6

4f	58 Ce	59 Pr	60 Nd	61 Pm	62 Sm	63 Eu	64 Gd	65 Tb	66 Dy	67 Ho	68 Er	69 Tm	70 Yb	71 Lu	IIIA
4f 5s 5p 5d 6s	2, 2, 6, 0, 2	3	4	5	6	7	7, 2, 6, 1, 2	9, 2, 6, 0, 2	10	11	12	13	14	14, 2, 6, 1, 2	
5f	90 Th	91 Pa	92 U	93 Np	94 Pu	95 Am	96 Cm	97 Bk	98 Cf	99 Es	100 Fm	101 Md	102 No	103 Lr	
5f 6s 6p 6d 7s	0, 2, 6, 2, 2	2, 2, 6, 1, 2	3	4	5	6	7	8	9	10	11	12	13	14	

— Outer sub-shells as for Ce — (Pr to Eu); f s p d s

— Outer sub-shells as for Ce & Tb — (Dy to Yb); f s p d s

— Outer sub-shells as for Pa — (U to Lr)

Principal Shell	1	2		3			4				5				6			7
Sub-shell	s	s	p	s	p	d	s	p	d	f	s	p	d	f	s	p	d	s
Total Number of electrons required to fill each sub-shell	2	2	6	2	6	10	2	6	10	14	2	6	10	14	2	6	10	2